Lecture Notes in Computer Science 8327

Commenced Publication in 1973
Founding and Former Series Editors:
Gerhard Goos, Juris Hartmanis, and Jan van Leeuwen

Viliam Geffert   Bart Preneel
Branislav Rovan   Július Štuller
A Min Tjoa (Eds.)

SOFSEM 2014: Theory and Practice of Computer Science

40th International Conference on Current Trends
in Theory and Practice of Computer Science
Nový Smokovec, Slovakia, January 26-29, 2014
Proceedings

 Springer

Volume Editors

Viliam Geffert
P. J. Šafárik University, Košice, Slovakia
E-mail: viliam.geffert@upjs.sk

Bart Preneel
Katholieke Universiteit Leuven, Belgium
E-mail: bart.preneel@esat.kuleuven.be

Branislav Rovan
Comenius University, Bratislava, Slovakia
E-mail: rovan@dcs.fmph.uniba.sk

Július Štuller
Institute of Computer Science, Prague, Czech Republic
E-mail: stuller@cs.cas.cz

A Min Tjoa
Vienna University of Technology, Austria
E-mail: amin@ifs.tuwien.ac.at

ISSN 0302-9743   e-ISSN 1611-3349
ISBN 978-3-319-04297-8   e-ISBN 978-3-319-04298-5
DOI 10.1007/978-3-319-04298-5
Springer Cham Heidelberg New York Dordrecht London

Library of Congress Control Number: 2013956643

CR Subject Classification (1998): G.2.2, F.2.2, H.3.3-5, D.2.13, D.2.11, D.2.2,
I.2.3-4, H.5.3, H.4.3, E.1, F.1.1

LNCS Sublibrary: SL 1 – Theoretical Computer Science and General Issues

Preface

This volume contains the invited and contributed papers selected for presentation at SOFSEM 2014, the 40th Conference on Current Trends in Theory and Practice of Computer Science, which was held January 26–29, 2014, in Atrium Hotel, Nový Smokovec, High Tatras, in Slovakia. SOFSEM 2014 was jointly organized by the Institute of Computer Science of the P. J. Šafárik University, Košice, Slovakia, and by the Slovak Society for Computer Science. The conference is supported by the Czech Society for Cybernetics and Informatics.

SOFSEM (originally SOFtware SEMinar) is devoted to leading research and fosters cooperation among researchers and professionals from academia and industry in all areas of computer science. SOFSEM started in 1974 in former Czechoslovakia as a local conference and winter school combination. The renowned invited speakers and the growing interest of the authors from abroad gradually moved SOFSEM to an international conference with proceedings published in the Springer LNCS series. SOFSEM became a well-established and fully international conference maintaining the best of its original Winter School aspects, such as a number of invited talks (7 for the year 2014) and an in-depth coverage of novel research results in selected areas of computer science. SOFSEM 2014 was organized around the following four tracks:

- Foundations of Computer Science (chaired by Viliam Geffert)
- Software and Web Engineering (chaired by A Min Tjoa)
- Data, Information, and Knowledge Engineering (chaired by Július Štuller)
- Cryptography, Security, and Verification (chaired by Bart Preneel)

With its four tracks, SOFSEM 2014 covered the latest advances in research, both theoretical and applied, in leading areas of computer science. The SOFSEM 2014 Program Committee consisted of 86 international experts from 22 different countries, representing the track areas with outstanding expertise.

An integral part of SOFSEM 2014 was the traditional SOFSEM Student Research Forum (chaired by Roman Špánek, Institute of Computer Science of the Academy of Sciences of the Czech Republic in Prague), organized with the aim of presenting student projects on both the theory and practice of computer science, and to give the students feedback on the originality of their results. The papers presented at the Student Research Forum were published in separate local proceedings.

In response to the call for papers, SOFSEM 2014 received 104 submissions by 258 authors from 34 different countries of all continents (except for Antarctica). Two submissions were later withdrawn. The remaining submissions were distributed in the conference tracks as follows: 53 in the Foundations of Computer Science, 24 in the Software and Web Engineering, 18 in the Data, Information, and Knowledge Engineering, and 7 in the Cryptography, Security, and Verification. From these, 19 submissions fell in the student category.

After a detailed reviewing process, with 3.22 reviewers per paper on average, a careful selection procedure was carried out between September 27 and October 11, 2013, using the EasyChair Conference System for an electronic discussion. Following strict criteria of quality and originality, 40 papers were selected for presentation, namely: 23 in the Foundations of Computer Science, 8 in the Software and Web Engineering, 5 in the Data, Information, and Knowledge Engineering, and 4 in the Cryptography, Security, and Verification. From these, 6 were student papers.

Based on the recommendation of the Chair of the Student Research Forum and with the approval of the Track Chairs and Program Committee members, seven student papers were chosen for the SOFSEM 2014 Student Research Forum.

The editors of these proceedings are grateful to everyone who contributed to the scientific program of the conference, especially the invited speakers and all the authors of contributed papers. We also thank the authors for their prompt responses to our editorial requests. SOFSEM 2014 was the result of a considerable effort by many people. We would like to express our special thanks to the members of the SOFSEM 2014 Program Committee and all external reviewers for their precise and detailed reviewing of the submissions, Roman Špánek for his handling of the Student Research Forum, Springer's LNCS team for its continued support of the SOFSEM conferences, and the SOFSEM 2014 Organizing Committee for the support and preparation of the conference.

November 2013

Viliam Geffert
Bart Preneel
Branislav Rovan
Július Štuller
A. Min Tjoa

Organization

Program Committee

General Chair

Branislav Rovan

Foundations in Computer Science

Andris Ambainis	Kazuo Iwama	Rolf Niedermeier
Witold Charatonik	Klaus Jansen	Alexander Okhotin
Jürgen Dassow	Christos Kapoutsis	Jean-Éric Pin
Volker Diekert	Juhani Karhumäki	José Rolim
Martin Dietzfelbinger	Daniel Král	Kai T. Salomaa
Pavol Ďuriš	Andrzej Lingas	Igor Walukiewicz
Viliam Geffert (Chair)	Maciej Liśkiewicz	Thomas Zeugmann
Lane A. Hemaspaandra	Markus Lohrey	
Markus Holzer	Carlo Mereghetti	

Software and Web Engineering

Armin Beer	Volker Gruhn	Andreas Rausch
Miklós Biró	Michael Kläs	Barbara Russo
Matthias Book	Óscar Pastor López	A. Min Tjoa (Chair)
Ruth Breu	Rudolf Ramler	

Data, Information, and Knowledge Engineering

Bernd Amann	Siegfried Handschuh	Mark Roantree
Ioannis Anagnostopoulos	Theo Härder	Marie-Christine Rousset
Zohra Bellahsène	Hannu Jaakkola	Matthew Rowe
Mária Bieliková	Leszek Maciaszek	Harald Sack
Ivan Bratko	Yannis Manolopoulos	Umberto Straccia
Barbara Catania	Andreas Nürnberger	Július Štuller (Chair)
Richard Chbeir	Tadeusz Pankowski	Massimo Tisi
Johann Eder	Adrian Paschke	Remco Veltkamp
Uwe Egly	Tassilo Pellegrini	Manuel Wimmer
Johann Gamper	Dimitris Plexousakis	
Hele-Mai Haav	Jaroslav Pokorný	

Cryptography, Security, and Verification

Dieter Gollmann	Václav Matyáš	Jean-François Raskin
Thorsten Holz	Kaisa Nyberg	Tamara Rezk
Stefan Katzenbeisser	Kenny Paterson	Ahmad-Reza Sadeghi
Aggelos Kiayias	Krzysztof Pietrzak	Martin Stanek
Marek Klonowski	Mila Dalla Preda	Graham Steel
Markulf Kohlweiss	Bart Preneel (Chair)	Claire Vishik

External Reviewers

Amin Anjomshoaa	Ulrich Hertrampf	George Mertzios
Alla Anohina-Naumeca	Marc Hesenius	Tobias Mömke
Aleksandrs Belovs	Lucian Ilie	Irina-Mihaela Mustaţă
Amine Benelallam	Sebastian Jakobi	André Nichterlein
Marian Benner-Wickner	Philipp Kalb	Dirk Nowotka
Sebastian Berndt	Michael Kaufmann	Beatrice Palano
Maria Paola Bianchi	Jonathan Kausch	Alexandros Palioudakis
Matthias Book	Markus Kleffmann	Denis Pankratov
Vladimir Braverman	Kim-Manuel Klein	Dana Pardubská
Robert Bredereck	Dušan Knop	Theodore Patkos
Srečko Brlek	Magnus Knuth	Giovanni Pighizzini
Georg Buchgeher	Christian Komusiewicz	Lars Prädel
Boban Čelebić	Boris Köpf	Narad Rampersad
Kit Yan Chan	Jakub Kowalski	Daniel Reidenbach
Krishnendu Chatterjee	Mirosław Kowaluk	Christian Rosenke
Taolue Chen	Stefan Kraft	Peter Rossmanith
Jacek Cichoń	Rastislav Královič	Christian Salomon
Michael Domaratzki	Kyriakos Kritikos	Fernando Sánchez
Jürgen Etzlstorfer	Martin Kutrib	Villaamil
Marco Faella	Felix Land	Stefan Schirra
Martin Fleck	Kati Land	Thomas Schneider
Peter Floderus	Alexander Lauser	Martin R. Schuster
Hervé Fournier	Văn Bằng Lê	Juraj Šebej
Paweł Gawrychowski	Thierry Lecroq	Patrice Séébold
Annemarie Goldmann	Christos Levcopoulos	Dzmitry Sledneu
Petr Golovach	Jianyi Lin	Florian Stefan
Alexander Golovnev	Eva-Marta Lundell	Christoph Stockhusen
Simon Grapenthin	Marten Maack	Alexandre Termier
Torsten Hahmann	Salvador Martínez	Johannes Textor
Christian Haisjackl	Tanja Mayerhofer	Vojtěch Tůma
Matthew Hammer	Ian McQuillan	Torsten Ueckerdt
Florian Häser	Klaus Meer	Victor Vianu
Erik Hebisch	Stefan Mengel	Imrich Vrt'o

Jörg Waitelonis   Armin Weiss   Piotr Witkowski
Tobias Walter   Magdalena Widl   Jānis Zuters
Antonius Weinzierl

Steering Committee

Ivana Černá   Masaryk University, Brno, Czech Republic
Keith Jeffery   STFC Rutherford Appleton Laboratory, UK
Mirosław Kutyłowski   Wrocław University of Technology, Poland
Jan van Leeuwen   Utrecht University, The Netherlands
Branislav Rovan   Comenius University in Bratislava, Slovakia
Petr Šaloun   Observer, Technical University of Ostrava,
                                 Czech Republic
Július Štuller   Chair, Institute of Computer Science,
                                 Czech Republic

Organizing Committee

Gabriela Andrejková
Jozef Gajdoš
František Galčík
Peter Gurský
Tomáš Horváth
Ondrej Krídlo
Rastislav Krivoš-Belluš
Gabriel Semanišin (Chair)

Invited Talks

Episode-Centric Conceptual Modeling

Nicola Guarino

ISTC-CNR Laboratory for Applied Ontology, Trento, Italy
Nicola.Guarino@cnr.it

Most conceptual modeling and knowledge representation schemes focus on relations. The Entity-Relationship approach is a paradigmatic example in this respect. However, formal ontology distinguishes between two main kinds of relations: *formal relations*, which hold by necessity just in virtue of the existence of their relata, and *material relations*, which do not necessarily hold whenever their relata exist, but require in addition the occurrence of a specific condition. For example, *3 is greater than 2, the specific mass of lead is higher than that of iron, humans are different from tables* are all relations of the first kind, while *John works for Mary* is a relation of the second kind, since, in addition to the mere existence of John and Mary, it requires something to happen (at a specific time) in order for the relation hold. In other words, material relations presuppose a certain *temporal phenomenon* to occur whenever they hold.

In this talk I will argue in favor of the systematic, explicit representation of the temporal phenomena underlying material relations. The terminology used in the literature for such temporal phenomena is multifarious: events, processes, states, situations, occurrences, perdurants... I will focus here on a large relevant subclass of temporal phenomena, which I will call *episodes*. According to the Oxford Advanced Learning Dictionary, an episode is "an event, a situation, or a period of time that is *important or interesting in some way*". I will interpret the interestingness requirement as a requirement on the *maximal temporal connectedness* of the phenomenon at hand: an episode is a temporal phenomenon occurring in a maximally connected time interval. This means that, for instance, a sitting episode has a starting time which coincides with the transition between not being sitting and being sitting, and an end time which coincides with the opposite transition.

My main point will be that, whenever there is an episode which corresponds to a particular relationship, it is very useful to model such episode explicitly, putting it in the domain of discourse. Once we make this modeling choice, we can easily express relevant information which would otherwise be hidden: how long will the work relationship last? What is the different role of the participants? What are their mutual rights and obligations? Where does the work take place? In addition, we can easily describe the internal dynamics of the relationship, distinguishing for instance phases of hard work or vacation which will be modeled as sub-episodes of the main one.

Finally, I will investigate the technical implications of this approach in the light of current research on the ontology of temporal phenomena, and illustrate a number of open ontological issues concerning episodes, their participants, and the spatiotemporal context where the episode occur.

Open Services for Software Process Compliance Engineering

Miklos Biro

Software Competence Center Hagenberg GmbH,
Softwarepark 21,4232 Hagenberg, Austria
miklos.biro@scch.at
http://www.scch.at

Abstract. The paper presents an update of the change of expectations and most recent new approaches regarding software processes and their improvement following the Software Process Improvement Hype Cycle introduced earlier by the author as an extension of the Gartner Hype Cycle idea. Software process assessment and improvement can itself be considered on the more abstract level as a quest for compliance with best practices. Etics and regulatory regimes explicitly addressing safety-critical systems mean however stringent compliance requirements beyond the commitment to improve process capability. New approaches are necessary for software engineers to fulfill the considerably growing expectations regarding quality under much slower changing development budget and deadline constraints. Open Services for Lifecycle Collaboration (OSLC) is the emerging initiative inspired by the web which is currently at the technology trigger stage along its hype cycle with the potential to have a determining impact on the future of Software Process Compliance Engineering.

Towards a Higher-Dimensional String Theory for the Modeling of Computerized Systems

David Janin[*]

LaBRI, Université de Bordeaux,
351, cours de la libération
F-33405 Talence, France
`janin@labri.fr`

Abstract. Recent modeling experiments conducted in computational music give evidence that a number of concepts, methods and tools belonging to inverse semigroup theory can be attuned towards the concrete modeling of time-sensitive interactive systems. Further theoretical developments show that some related notions of higher-dimensional strings can be used as a unifying theme across word or tree automata theory. In this invited paper, we will provide a guided tour of this emerging theory both as an abstract theory and with a view to concrete applications.

[*] CNRS temporary researcher fellow (2013-2014).

Advice Complexity:
Quantitative Approach to A-Priori Information
(Extended Abstract)

Rastislav Královič

Comenius University, Bratislava, Slovakia
kralovic@dcs.fmph.uniba.sk

Abstract. We survey recent results from different areas, studying how introducing per-instance a-priori information affects the solvability and complexity of given tasks. We mainly focus on distributed, and online computation, where some sort of hidden information plays a crucial role: in the distributed computing, typically nodes have no or only limited information about the global state of the network; in online problems, the algorithm lacks the information about the future input. The traditional approach in both areas is to study how the properties of the problem change if some partial information is available (e.g., nodes of a distributed system have sense of direction, the online algorithm has the promise that the input requests come in some specified order etc.). Recently, attempts have been made to study this information from a quantitative point of view: there is an oracle that delivers (per-instance) best-case information of a limited size, and the relationship between the amount of the additional information, and the benefit it can provide to the algorithm, is investigated. We show cases where this relationship has a form of a trade-off, and others where one or more thresholds can be identified.

Matching of Images of Non-planar Objects with View Synthesis

Dmytro Mishkin and Jiří Matas*

Center for Machine Perception, Faculty of Electrical Engineering,
Czech Technical University in Prague
ducha.aiki@gmail.com, matas@cmp.felk.cvut.cz

Abstract. We explore the performance of the recently proposed two-view image matching algorithms using affine view synthesis – ASIFT (Morel and Yu, 2009) [14] and MODS (Mishkin, Perdoch and Matas, 2013) [10] on images of objects that do not have significant local texture and that are locally not well approximated by planes.

Experiments show that view synthesis improves matching results on images of such objects, but the number of "useful" synthetic views is lower than for planar objects matching. The best detector for matching images of 3D objects is the Hessian-Affine in the SPARSE configuration. The iterative MODS matcher performs comparably confirming it is a robust, generic method for two view matching that performs well for different types of scenes and a wide range of viewing conditions.

---

* Invited Speaker.

Agile Requirements Engineering: A Research Perspective

Jerzy Nawrocki*, Mirosław Ochodek, Jakub Jurkiewicz,
Sylwia Kopczyńska, and Bartosz Alchimowicz

Poznan University of Technology, Institute of Computing Science,
ul. Piotrowo 2, 60-965 Poznań, Poland
{Jerzy.Nawrocki,Miroslaw.Ochodek,Jakub.Jurkiewicz,Sylwia.Kopczynska,
Bartosz.Alchimowicz}@cs.put.poznan.pl

Abstract. Agile methodologies have impact not only on coding, but also on requirements engineering activities. In the paper agile requirements engineering is examined from the research point of view. It is claimed that use cases are a better tool for requirements description than user stories as they allow zooming through abstraction levels, can be reused for user manual generation, and when used properly can provide quite good effort estimates. Moreover, as it follows from recent research, parts of use cases (namely event descriptions) can be generated in an automatic way. Also the approach to non-functional requirements can be different. Our experience shows that they can be elicited very fast and can be quite stable.

---

* Invited Speaker.

Table of Contents

Open Services for Software Process Compliance Engineering

Miklos Biro

Software Competence Center Hagenberg GmbH,
Softwarepark 21,4232 Hagenberg, Austria
miklos.biro@scch.at
http://www.scch.at

Abstract. The paper presents an update of the change of expectations
and most recent new approaches regarding software processes and their
improvement following the Software Process Improvement Hype Cycle
introduced earlier by the author as an extension of the Gartner Hype
Cycle idea. Software process assessment and improvement can itself be
considered on the more abstract level as a quest for compliance with
best practices. Etics and regulatory regimes explicitly addressing safety-
critical systems mean however stringent compliance requirements beyond
the commitment to improve process capability. New approaches are necessary for software engineers to fulfill the considerably growing expectations regarding quality under much slower changing development budget and deadline constraints. Open Services for Lifecycle Collaboration
(OSLC) is the emerging initiative inspired by the web which is currently
at the technology trigger stage along its hype cycle with the potential to
have a determining impact on the future of Software Process Compliance
Engineering.

Keywords: software, process, improvement, safety, traceability, open,
lifecycle, hype, regulatory, compliance, productivity, www.

1 Introduction

The popular concept of Hype Cycle was coined in 1995 at the Gartner information technology research and advisory company based in the U.S. Referring
to the complete analysis of the subject to [2], the phases of this hype cycle,
originally applied to emerging technologies are:

1. "Technology Trigger"
2. "Peak of Inflated Expectations"
3. "Trough of Disillusionment"
4. "Slope of Enlightenment"
5. "Plateau of Productivity"

The software process improvement movement started with the CMM being a
significant innovation. The hype cycle was triggered at the end of the 1980s and

V. Geffert et al. (Eds.): SOFSEM 2014, LNCS 8327, pp. 1–6, 2014.

went through altered phases which do not purely follow the one promoted by Gartner. One of the reasons of the difference is the support and acceptance of the model, worked out and managed at the SEI, by the U.S. Department of Defence which helped the CMM avoiding the full trough of disillusionment by supporting continuous innovation in the form of the Capability Maturity Model Integration (CMMI) [4] for example. The other major parallel high impact initiative is the trial and adoption of the ISO/IEC 15504 (SPICE) international standard. SPICE and CMMI are both the results and the further catalysts of the spreading of process maturity models to other disciplines than software development. This plateau of spreading to other disciplines and models is followed by the trough of doubts and new triggers like the Agile Manifesto with a new hype cycle starting with new expectations and also the SPI Manifesto [5] . The altered hype cycle of software process improvement consists of the following phases analysed in more detail in [1] extended with the currently last hype promising new perspectives as discussed below:

1. Awareness of process capability weaknesses triggered by the software crisis and CMM.
2. SPI and ISO 9000 expectations.
3. Bridging the trough of disillusionment.
4. Enlightenment leading to further recognition of the importance of business goals
5. Plateau of spreading to other disciplines and models.
6. Trough of doubts and new triggers.
7. Plateau of reconciliation and industrial adoption.
8. Perspective: SPI and its industrial adoption to be fertilized by Web technology

2   Growing Expectations Regarding Safety-Critical Systems

The growing expectations regarding software components of safety-critical systems, to which the category of Programmable Electrical Medical Systems (PEMS) belongs, is a consequence of the changing impact of software on the consumer value of PEMS which was earlier fundamentally determined by hardware components with software primarily used for algorithmic tasks. Increasingly, embedded software is creating competitive differentiation for manufacturers [6].

Software, on the other hand, is perceived by business as more capable to be adapted to fluid requirements changes than hardware. In the software engineering discipline, it has for long been recognized however that the only way to achieve high reliability is to follow appropriately defined processes. The necessary processes are summarized in international standards (ISO/IEC 12207 for software, ISO/IEC 15288 for systems), while their assessment and improvement is facilitated by the ISO/IEC 15504 series of standards currently evolving into the ISO/IEC 330xx series.

Focusing on medical systems, the Association for the Advancement of Medical Instrumentation (AAMI) software committee reviewed ISO/IEC 12207:1995 and identified a number of shortcomings due to the fact that it was a generic standard. As a result a decision was taken to create a new standard which was domain specific to medical device software development, and in 2006, the new standard IEC 62304:2006 Medical device software - Software life cycle processes, was released. IEC 62304:2006 is approved today by the FDA (U.S. Food and Drug Administration) and is harmonized with the European MDD (Medical Device Directive). The quality management standard ISO 13485:2003, and the risk management standard ISO/IEC 14971:2007 are considered to be aligned with IEC 62304:2006 and their relationship is documented in IEC 62304:2006 itself.

An extensive revision of the ISO/IEC 12207 standard took place in its release in 2008. As a result, all derived standards, including IEC 62304:2006, are under review.

To facilitate the assessment and improvement of software development processes for medical devices, the MediSPICE model based upon ISO/IEC 15504-5 is being developed. The Process Reference Model (PRM) of MediSPICE will enable the processes in the new release of IEC 62304 to be comparable with those of ISO 12207:2008 [7]. The above points give just a glimpse of the changes heavily affecting software developers in the medical devices domain. Nevertheless, the key initial issues ultimately addressed by the standards, namely traceability in achieving and rigor in certifying the safety of PEMS must be kept in focus.

3   Traceability, Interoperability, OSLC

Traceability and even bilateral (ISO/IEC 15504) or bidirectional (CMMI) traceability are key notions of all process assessment and improvement models. Unfortunately, traceability as well as interoperability is difficult to achieve with the heterogeneous variety of application lifecycle management tools companies are faced with [8], [9].

Full traceability of a requirement throughout the development chain and even the entire supply chain is also a major focus point of the recently completed authoritative European CESAR project (Cost-Efficient Methods and Processes for Safety Relevant Embedded Systems) which adopted interoperability technologies proposed by the Open Services for Life-cycle Collaboration (OSLC) initiative [10].

OSLC is the recently formed cross-industry initiative aiming to define standards for compatibility of software lifecycle tools. Its aim is to make it easy and practical to integrate software used for development, deployment, and monitoring applications. This aim seems to be too obvious and overly ambitious at the same time. However, despite its relatively short history starting in 2008, OSLC is the only potential approach to achieve these aims at a universal level, and is already widely supported by industry. The OSLC aim is of course of utmost significance in the case of the Programmable Electrical Medical Systems.

The unprecedented potential of the OSLC approach is based on its foundation on the architecture of the World Wide Web unquestionably proven to be powerful and scalable and on the generally accepted software engineering principle to always focus first on the simplest possible things that will work.

The elementary concepts and rules are defined in the OSLC Core Specification [11] which sets out the common features that every OSLC Service is expected to support using the terminology and generally accepted approaches of the World Wide Web Consortium (W3C). One of the key approaches is Linked Data being the primary technology leading to the Semantic Web which is defined by W3C as providing a common framework that allows data to be shared and reused across application, enterprise, and community boundaries. And formulated at the most abstract level, this is the exact goal OSLC intends to achieve in the interest of full traceability and interoperability in the software lifecycle. Similarly to the example of the capability maturity models, the software focused OSLC is having a determining cross-fertilizing effect on the progress of the more general purpose Semantic Web itself.

The OSLC Core Specification is actually the core on which all lifecycle element (domain) specifications must be built upon. Examples of already defined OSLC Specifications include:

- Architecture Management
- Asset Management
- Automation
- Change Management
- Quality Management
- Requirements Management

Focusing on the example of the Requirements Management Specification whose version 2.0 was finalized in September 2012 [12], it builds of course on the Core, briefly introduced above, to define the resource types, properties and operations to be supported by any OSLC Requirements Definition and Management (OSLC-RM) provider.

Examples of possible OSLC Resources include requirements, change requests, defects, test cases and any application lifecycle management or product lifecycle management artifacts. Resource types are defined by the properties that are allowed and required in the resource.

In addition to the Core resource types (e.g. Service, Query Capability, Dialog, etc...), the minimal set of special Requirements Management resource types simply consists of:

- Requirements
- Requirements Collections.

The properties defined in the OSLC Requirements Management Specification describe these resource types and the relationships between them and all other resources. The relationship properties describe for example that

- the requirement is elaborated by a use case
- the requirement is specified by a model element
- the requirement is affected by another resource, such as a defect or issue

- another resource, such as a change request tracks the requirement
- another resource, such as a change request implements the requirement
- another resource, such as a test case validates the requirement
- the requirement is decomposed into a collection of requirements
- the requirement is constrained by another requirement

OSLC is currently at the technology trigger stage along its hype cycle, it is already clear however that it is the approach which has the potential to have a determining impact on the future of Software Process Compliance Engineering.

4   Conclusion

The original paradigm underlying software process improvement was based on the highest priority to be attributed to the organization of software development processes with the possession of the necessary methodological knowledge having high but lower significance, and technology considered to be indispensable of course, but having the lowest impact on the actual capability of software processes [3].

It turns out that considering the heterogeneity of application lifecycle management tools companies are using, it can be claimed that today, technology is becoming the bottleneck of improving traceability and interoperability being key elements of software process improvement and subject to intensively growing expectations regarding compliance to etical, regulatory, and business requirements.

The future will have to lead to the mutual appreciation of the special advantages of the different approaches to software process improvement positively cross-fertilized by evolving WWW technology.

References

1. Biro, M.: The Software Process Improvement Hype Cycle. Invited contribution to the Monograph: Experiences and Advances in Software Quality. In: Dalcher, D., Fernndez-Sanz, L. (Guest eds.) CEPIS UPGRADE, vol. X (5), pp. 14–20 (2009), http://www.cepis.org/files/cepisupgrade/issue%20V-2009-fullissue.pdf (accessed: October 21, 2013)
2. Fenn, J., Raskino, M.: Mastering the Hype Cycle. Harvard Business Press (2008)
3. Biro, M., Feuer, E., Haase, V., Koch, G.R., Kugler, H.J., Messnarz, R., Remzso, T.: BOOTSTRAP and ISCN a current look at the European Software Quality Network. In: Sima, D., Haring, G. (eds.) The Challenge of Networking: Connecting Equipment, Humans, Institutions, pp. 97–106. R.Oldenbourg, Wien (1993)
4. CMMI for Development, Version 1.3. SEI, (2010) http://www.sei.cmu.edu/reports/10tr033.pdf (accessed: October 21, 2013)
5. SPI Manifesto Version A.1.2.2010, EuroSPI (2010), http://www.eurospi.net/images/documents/spi_manifesto.pdf (accessed: October 21, 2013)
6. Bakal, M.: Challenges and opportunities for the medical device industry. IBM Software, Life Sciences, Thought Leadership White Paper (2011)

7. Casey, V., McCaffery, F.: The development and current status of mediSPICE. In: Woronowicz, T., Rout, T., O'Connor, R.V., Dorling, A. (eds.) SPICE 2013. CCIS, vol. 349, pp. 49–60. Springer, Heidelberg (2013)

8. Pirklbauer, G., Ramler, R., Zeilinger, R.: An integration-oriented model for application lifecycle management. In: Proceedings of the 11th International Conference con Enterprise Information Systems (ICEIS 2009), pp. 399–403. INSTICC (2009)

9. Murphy, T.E., Duggan, J.: Magic Quadrant for Application Life Cycle Management. Gartner (2012)

10. Open Services for Lifecycle Collaboration (2008), `http://open-services.net/`

11. Open Services for Lifecycle Collaboration Core Specification Version 2.0 (2013), `http://open-services.net/bin/view/Main/OslcCoreSpecification`

12. Open Services for Lifecycle Collaboration Requirements Management Specification Version 2.0 (2012), `http://open-services.net/bin/view/Main/RmSpecificationV2/`

Towards a Higher-Dimensional String Theory for the Modeling of Computerized Systems

David Janin*

LaBRI, Université de Bordeaux,
351, cours de la libération
F-33405 Talence, France
janin@labri.fr

Abstract. Recent modeling experiments conducted in computational music give evidence that a number of concepts, methods and tools belonging to inverse semigroup theory can be attuned towards the concrete modeling of time-sensitive interactive systems. Further theoretical developments show that some related notions of higher-dimensional strings can be used as a unifying theme across word or tree automata theory. In this invited paper, we will provide a guided tour of this emerging theory both as an abstract theory and with a view to concrete applications.

1 Introduction

As Mozart's character puts it in Milos Forman's masterpiece *Amadeus*, opera was for many years the only medium in which everyone could be talking at the same time and still manage to understand each other. Today, this has become common practice in communication networks.

Music theory has been developed empirically down the ages until it achieved status as a recognized discipline. It now provides us with the means for describing the underlying mechanisms and subtle interaction between musicians as they perform, based on complex combinatory rules relating to rhythm, harmony and melody. Computer science also aims to describe the subtle organization and treatment of data. It therefore follows that the study of modeling in the field of music might lead to the discovery of concepts, abstract tools and modeling principles which are applicable to modern computer engineering.

It is by developing language theoretical models of musical rhythmic structures that the author of this paper eventually re-discovered inverse semigroup theory: a part of semigroup theory that has been developed since the 50s. Further experiments, both in the field of computational music and in the field of formal language theory, provide evidence that such a theory can be further developed and attuned towards concrete computer engineering.

The purpose of this paper is to give an overview of these recent experiments and the underlying emerging ideas, methods and tools.

---

* CNRS temporary researcher fellow (2013-2014).

V. Geffert et al. (Eds.): SOFSEM 2014, LNCS 8327, pp. 7–20, 2014.

Mathematical Frameworks *for* Computer Science and Engineering.
In computer science and, more specifically, software or hardware engineering, *formal methods* are mathematically rigorous techniques for the specification, development and verification of computerized systems. Based on research fields in theoretical computer science as varied as type theory, automata and language theory, logic, these methods have already demonstrated their relevance for increasing the reliability of both software and hardware systems.

For instance, in functional programming, it is now common knowledge that proof and type theory does provide numerous metaphors and concepts that can be efficiently used by systems architects and developers. Typed functional languages such as *Haskell* or *OCAML* illustrate how deep mathematical theories can effectively be attuned towards relevant concepts which may be applied with ease. A similar observation can be made concerning data-base system design that now integrates highly usable methods and concepts that are deeply rooted in model theory.

Mathematical Frameworks *for* Interactive System Design.
One of the most demanding application areas of automata and formal language theories is the domain of interactive system design. These systems are commonly viewed as state/transition systems, that is automata, and the behavior of these systems is commonly represented by the sets of their possible execution traces; that is, formal languages. A formal method such as *event B* [1], whose applicability to industry is clearly demonstrated regards to its use in automated public transport, is based on state/transition formalisms. With many features inherited from *method B* [3], it offers a particularly good example of how topics as varied as logic, proof theory, automata theory and formal languages can be *combined* and *shaped* towards applications [36].

However, while the mathematical frameworks that are available for functional programming or database design have been considerably and successfully developed and attuned towards applications over the last decades, the mathematics for interactive system modeling does not yet seem to have reached the same level of maturity. Since the early 80s, developing what became the *theory of concurrency*, many authors promoted the idea of modeling interactive systems by means of two distinct operators: the *sequential composition* and the *parallel composition* of system behaviors (see e.g. [11,29]). However, examples of (distributed) system modeling show that such a distinction does not necessarily fit the abstraction/refinement methodology that system designers follow.

For instance, at the *abstract specification level* the well known distributed algorithm for leader election in a graph (see e.g. [4]) is based upon the *successive* execution of two global phases: the construction of a spanning tree followed by the pruning of that spaning tree. However, at the (distributed) *concrete execution level* these two phases overlap in time. Indeed, the pruning phase may start locally as soon as the spanning tree phase is locally completed. In other words, when composing two distributed algorithms one "after" the other, a composition of this type is neither purely parallel nor purely sequential: it is a sort of *mixed product* of some higher-dimensional models of spatiotemporal behaviors.

Towards a Theory of Higher Dimensional Strings.
This raises the question of whether there could be a mathematical theory of *higher-dimensional strings*. By describing quasi-crystals in physics [22], Kellendonk showed how such strings should be composed.

In one dimension, each string has two ends and, once a convention has been made on parity, a unique multiplication may be defined, called string concatenation, which enables pairs of strings to be multiplied, thereby giving rise to the structure of a free monoid. In higher dimensions, there is no uniquely defined way of composing patterns. Kellendonk hits upon the idea of labeling patterns by selecting two tiles of the pattern: one to play the role of the input and the other that of output. By using this labeling, a unique composition can be defined. The resulting semigroup is no longer free but is some kind of inverse semigroup [22,23].

In music system programming where the time vs space problematic also appears (see e.g. [14]), similar ideas, also rooted in inverse semigroup theory, have led to the definition of a higher abstraction layer into which both strings and streams may be embedded into a single object type: tiled streams [20]. As a result, this new programming layer allows for the combination of both (sequential) strings and (parallel) streams with a single mixed product: the tiled product.

By capitalizing on theoretical developments within the general theory of inverse semigroups (see [25] for instance) and the associated emerging notion of higher-dimensional strings [22,23], we thus aim at developing the idea of an inverse semigroup theory *for* computer science much in the same way that the development of logic *for* computer science is advocated in [36].

In this paper, we provide a guided tour of the first experiments conducted in this direction, both as an abstract theory and with a view to concrete applications.

2   From Partial Observations to Inverse Semigroups

On the footpath of Kellendonk for describing the local structures of quasi-crystals that leads to the discovery of (a notion of) tiling semigroups [22,23], we aim here at illustrating how inverse semigroup structures naturally arise by performing *partial observations* of (the local structure of) computerized *system behavior spaces*.

Partial Observations.
We assume that the behavior of a system can be viewed as the traversal of a complex space $\mathcal{S}$ of all potential behaviors that describe both the way data are structured in space but also the way they can evolve as time passes. Then, a *partial observation* is defined by the observation resulting of an out of time[1] traversal of such a space $\mathcal{S}$, that induces some *domain* of the traversal, together with the starting point of such a traversal, called the *input root*, and the end point of such a traversal, called the *output root*. Such a (meta) notion of partial observation may be depicted as in Figure 1. Of course, the exact nature of a

---

[1] Time flows does not matter at that stage.

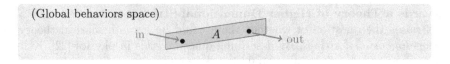

Fig. 1. A partial observation A

partial observation, that is, the nature of its domain and the associated input and input roots, depends on the mathematical nature of the space $\mathcal{S}$.

For instance, in the approach of Kellendonk [22], the space $\mathcal{S}$ is a quasi-crystal, that is a collection of distinct tiles that tiles the Euclidian space $\mathbb{R}^d$. Then, a partial observation is defined as a set T of tiles of $\mathcal{S}$ that cover a connected subset of $\mathbb{R}^d$ together with two distinguished tiles $I \in T$ and $O \in T$ acting as input and output roots (see [22] and [23] for more details).

In free inverse monoids [32,30], the space $\mathcal{S}$ is defined as the Cayley graph of the free group $FG(A)$ generated by a given alphabet A. Then, partial observations are defined as finite connected subgraphs of that Cayley graph with two distinguished vertices: one for the input root and the other for the output root.

In this case, partial observations are *birooted trees*, that is, finite directed tree-shaped graphs with deterministic and co-deterministic A-labeled edges. Extension with labeled vertices are considered in [18] in connection with tree walking automata and also in [16] in connection with non deterministic tree automata.

Restricting to linear birooted trees, we obtain what may be called *birooted words*, structures that already appear in the 70's as elements of the monoid of McAlister [26].

Observation Composition.
Two observations A and B can be composed in a two step procedure defined by, first, a *synchronization* that amounts to sewing the output root of A with the input root of B, followed secondly by a *fusion* that amounts to merging the sub-domains that may overlap. Such a (meta) notion of composition of observations can be depicted as in Figure 2.

Fig. 2. The product $A \cdot B$ of two partial observations A and B

In many settings, the fusion operation may fail hence, apriori, the product of two observations is a partial product. In such a case, it is completed by adding an undefined observation, denoted by 0, that simply acts as zero, i.e. $A \cdot 0 = 0 = 0 \cdot A$ for every observation A. It is an easy exercise to check that such a zero element is necessarily unique and idempotent, i.e. $0 \cdot 0 = 0$.

It may also be the case that the structure of observations allows for the definition of a special element, the unit, denoted by 1, that is neutral for the product, i.e. $A \cdot 1 = A = 1 \cdot A$ for every observation A. It is an easy exercise to check that, necessarily, such a neutral element is also unique and idempotent.

The Semigroup/Monoid of Partial Observations.
In general, nothing guarantees that such a product is associative so we simply assume this, that is, we assume that $(A \cdot B) \cdot C = A \cdot (B \cdot C)$ for every observation A, B and C. Though simple, let us insists on the fact that associativity is the first key property for a robust and usable implementation for it means that a complex product can be specified and computed in any order.

In algebra, the resulting structure is known as a *semigroup*. In the case there is also a unit, it is a *monoid*.

Without any further assumptions, two special observations can already be defined out of a given one. Indeed, given an observation A, we may also consider the observation A^R, called the *reset* or *right projection* of A, defined by moving the output root to the input root, and, symmetrically, the observation A^L, called the *co-reset* or *left projection* of A, defined by moving the input to the output root. The resulting observations are depicted in Figure 3. We observe that the

Fig. 3. The reset and co-reset operators

mapping $A \mapsto A^L$ and $A \mapsto A^R$ are indeed projection since, for every observation A, we have $(A^L)^L = A^L = (A^L)^R$ and $(A^R)^L = A^R = (A^R)^R$. The left and right projection are extended to zero by taking $0^L = 0 = 0^R$. It is then possible to define two derived operators, called *fork* and *join*, by $fork(A, B) = A^R \cdot B$ and $join(A, B) = A \cdot B^L$. These derived operators are depicted in Figure 4.

Fig. 4. The *fork* and *join* derived operators

Interpreting for a while the paths from input to output roots as time flows, in the *fork* operation, everything looks as if the two observations "start" at the same "time" in their input roots. Similarly, in the *join* operation, it looks as if the two observations "end" at the same "time" in their output roots.

More generally, at the *abstract syntactic level*, the composition $A \cdot B$ of two successive observations A and B, actually admits, at the *concrete model level*, some parallel flavored characteristics induced by the possible overlaps that may arise in the resulting structure. In other words, everything looks as if the observation product we are defining here at some meta level is a good candidate for the *mixed* sequential and parallel operator we are looking for as mentioned above.

Example. In computational music, such an approach is already used as follows. Every finite audio stream is enriched with input and output synchronization marks that, when mixing two streams, allows for automatically positioning them in time, one with respect to the other. Doing so, synchronization specifications have thus been internalized in some sense into the audio streams themselves, much like music bars in a music score. The resulting algebra [2], with product, left and right projections, and some other operators acting on tempos, provides a fairly robust and versatile language for music composition. Further implementation experiments have been conducted in [21].

Product by Synchronized Superposition.
In the concrete inverse semigroups mentioned above, the definition of the product of two partial observations can be refined as follows:

Definition 1. *The product $A \cdot B$ of two observations A and B is defined to be the observation C (assumed to be uniquely defined up to isomorphism when it exists) such that there exist two embeddings $\varphi_A : A \to C$ and $\varphi_B : B \to C$ so that:*

- *the input root of A (resp. the output root of B) is mapped to the input root (resp. the output root) of C,*
- *the output root of A and the input root of B are mapped to the same image,*
- *the domain of C is the union of the domains of $\varphi_A(A)$ and $\varphi_B(B)$,*

the product being completed by zero when no such observation C exists.

Remark. In other words, in that case, everything looks as if the product $A \cdot B$ is performed by translating the two observations A and B in such a way that their output and input root coincide. Then, the fusion just amounts to checking that the resulting overlapping subdomains are isomorphic.

More generally, Stephen's representation theorem of inverse semigroups [34] tells that every elements of every inverse semigroup have a *graphical representations*. In view of applications, this fact that is of high interest. However, it may be the case that the product induces a graph composition that does not fit the above definition.

In other words, the informal point of view that has been previously given is still worthy of being kept in mind when looking for new instances of models of partial observations and the related compositions.

Resulting Idempotents (System States).
When the observation product is defined by synchronized superposition, as well as with Stephen graphical representation of inverse semigroups, and probably in many other settings still to be defined and studied, elements for which the input and output root coincide plays an especially important rôle.

In a context even more general than inverse semigroups as shown in the next section, these elements are be called *subunits*. By definition, both the left and right projection A^L and A^R of an arbitrary observation A, are subunits.

The fact is that subunits are both *idempotent* and *commute*, that is, for every elements E and F with identical input and output roots, we have $E \cdot E = E$ and $E \cdot F = F \cdot E$. They can thus be partially ordered as follows. For every two subunit observation E and F, we say that E is smaller than F, which is written $E \leq F$, when $E = E \cdot F$.

One can easily check that this indeed defines a partial order relation, that is, a reflexive, transitive and antisymmetric relation. Moreover, the meet $E \wedge F$ of every two subunits E and F, that is the greatest subunit below both E and F, exists and can simply be computed by $E \wedge F = E \cdot F$. Such an order relation is called the *natural order* for it can be defined within the semigroup itself [31].

Example. Potential application oriented views of such an order relation are numerous. For instance, one may view every subunit as the partial description of the *state* of the modeled system. The underlying domain of a subunit E tels how far in the past, in the future and also in the present time the system has been observed in E.

Then, the natural order tells us how wide such an observation is: the wider the observation is, the lower it is in the natural order, until it becomes incoherent. Indeed, the lowest subunit is zero. Then, the product of two partial states/subunit observations simply describes the union of the descriptions associated with these two partial states.

Quite strikingly, as recently observed by Dominique Méry and the author, when observations are viewed as models of behaviors, realizing a system that goes from a state specification S_0 to a state specification S_1 amount to finding a system behavior X such that $0 < (S_0 \cdot X)^L \leq S_1$. Some modeling experiments of similar ideas have been conducted in [6].

The Inverse Semigroup of Partial Observations.
The inverse semigroup of observations arises when we assume, as is the case with the product by superposition, that, for every observation A, the observation A^{-1} obtained just by inverting the input and the output root of the observation A, as depicted in Figure 5, satisfies the following properties:

$$A \cdot A^{-1} \cdot A = A \text{ and } A^{-1} \cdot A \cdot A^{-1} = A^{-1} \qquad (1)$$

for every observation A, and,

$$A \cdot A^{-1} \cdot B \cdot B^{-1} = B \cdot B^{-1} \cdot A \cdot A^{-1} \qquad (2)$$

for every observation A and B.

In semigroup theoretical terms, Property (1) says that elements A and A^{-1} are *semigroup inverses* one with respect to the other. Property (2) says that the resulting idempotent elements of the form $A \cdot A^{-1}$ commute. Indeed, by associativity, we have $(A \cdot A^{-1}) \cdot (A \cdot A^{-1}) = (A \cdot A^{-1} \cdot A) \cdot A^{-1}$ hence, by applying Property (1), we have $(A \cdot A^{-1}) \cdot (A \cdot A^{-1}) = A \cdot A^{-1}$. Then we also

Fig. 5. An observation and its inverse

have $A^R = A \cdot A^{-1}$ and $A^L = A^{-1} \cdot A$, as depicted in Figure 6. Even more, we have $A^R \cdot A = A = A \cdot A^L$, i.e. these projections are (the least) local units of the observation A. Forming an inverse semigroup, the *natural order* on observations

Fig. 6. Inverses and left and right projections

is extended to all semigroup elements by letting $A \leq B$ when $A = A^R \cdot B$ or, equivalently, $A = B \cdot A^L$. This order relation is depicted in Figure 7. One can easily check that the natural order is stable under product, i.e. if $A \leq B$ then $A \cdot C \leq B \cdot C$ and $C \cdot A \leq C \cdot B$ for every observations A, B and C. In the concrete cases mentioned so far, we even have $A \leq B$ when there exists an embedding $\varphi : B \to A$ that preserves input and output roots. Let us mentioned that many

Fig. 7. The natural order

other properties of inverse semigroups can be found in [25] that provide a much deeper (and precise) description of such a peculiarly rich theory.

Remark. It may be the case that the partial observations described so far are defined on a structure space that forces input root and output root to be ordered, the input root always preceding or being equal to the output roots, as this can be the case with positive birooted words studied in [17] or positive birooted trees that are known to be the elements of the free ample monoids [10].

The resulting semigroups are no longer inverse semigroups for inverses themselves are not elements of these semigroups. However, these semigroups are *quasi-inverse* in some sense: the natural order and the left and right projections are still available.

It must mentioned that, since the late 70s in semigroups theory, Fountain et al. have develop a theory of semigroups with local units (see [9,24,12,5]), quasi-inverse in the above sense, which underlying concepts are of high interest for developing a language theory of higher-dimensional strings as in the next section.

3   On Languages of Higher-Dimensional Strings

In view of applications to computer science, there is a need for a language theory for observations. Indeed, when specifying the expected behavior of a system one generally describes some characteristic properties of its correct behavior. In general, especially in the context of an abstraction/refinement designed method, there is no reason for such specifications to characterize a single possible behavior. In this section, we thus aim at defining adequate algebraic tools for a language theory of observations.

Recognizability and Logic: The Classical Approaches.
In the absence of concrete structures, as it is the case in the general and abstract setting described here, algebraic recognizability is a major tool for defining languages, that is, subsets of peculiar semigroups or monoids.

More precisely, a mapping $\varphi : S \to T$ from a semigroup S to a semigroup T is a semigroup morphism when $\varphi(x \cdot y) = \varphi(x) \cdot \varphi(y)$ for every x and $y \in S$. When both S and T are monoid, the semigroup morphism φ is a monoid morphism when $\varphi(1) = 1$. A subset $L \subseteq S$ of a semigroup S (resp. monoid S) is said to be recognizable by a semigroup (resp. monoid) T when there exists a semigroup (resp. monoid) morphism such that $X = \varphi^{-1}(\varphi(X))$.

In the case of languages of words, that is subsets of the free monoid, the algebraic notion of recognizability does coincide with the notion of recognizability by finite state automata.

Indeed, given a morphism $\varphi : A^* \to S$ with finite monoid S, one can define the automaton $\mathcal{A}_S$ with set of state S, initial state 1, and deterministic transition $s \xrightarrow{a} t$ whenever $s \cdot \varphi(a) = t$. Then, for every $X \subseteq S$, the language $\varphi^{-1}(X)$ exactly correspond to the language of words recognized by the automaton $\mathcal{A}_S$ with accepting states X. Incidentally, this observation also gives a fairly efficient way to compute $\varphi(w)$ for every $w \in A^*$ given as input.

However, if we restrict target semigroups to be inverse as in [27], or if we consider languages of free inverse monoids as in [33], besides the inherent interest of the results established in these studies, the notion of recognizability by or even from inverse semigroups collapses. Indeed, when logical definability in monadic second order (MSO) logic [35] is available, as with birooted words [17] or birooted trees [16], the languages that are recognizable by means of finite monoids and morphisms are far simpler than the languages definable by means of an MSO formula (see [17,33,16]).

This comes from the fact that the direct image of an inverse semigroup by a semigroup morphism is an inverse semigroup hence the automaton $\mathcal{A}_S$ defined above is of a very special kind: it is reversible in some sense.

Relaxing Morphisms into Premorphisms.

A remedy to the above mentioned collapse is based on the fact that inverse monoids are partially ordered by means of their natural order. Then, we can relax the morphism condition into the following condition. A mapping $\varphi : S \to T$ from one partially ordered monoid S in a partially ordered monoid T is a *premorphism* (or ∨-premorphism in [13]) when $\varphi(x \cdot y) \leq \varphi(x) \cdot \varphi(y)$ for every x and $y \in S$ with $\varphi(1) = 1$.

Then, as with morphisms, it is tempting to define recognizability by partially ordered monoid via premorphism $\varphi : S \to T$. However, premorphism condition alone is too weak. It can be shown that there are premorphisms from finitely generated inverse monoids into finite monoids that are even not computable [15]. The difficulty thus lies in defining an adequate restriction of both premorphisms and partially ordered monoids that induces a notion of recognizability that is both sufficiently expressive and still computable.

Our proposal is based on the fact that, in a finitely generated inverse monoid such as monoids of birooted words [19] or monoids of birooted trees [16], every element can be defined out from the finite generators, a notion of disjoint product and left and right projections. In other words, neither arbitrary product nor inverses themselves are needed to generate these monoids.

Remark. Though premorphisms have been known and used for quite some time in inverse semigroup theory [28], it seems that their use for language recognizability has been first proposed by the author himself in [15].

Adequate Monoids and Premorphism.

The recognizers we use instead of semigroups (or monoids) are adequately ordered monoids. Following the research track initiated by Fountain, our proposal is based on ordered monoids that are quasi-inverse in the sense that, though without inverses themselves, these monoids are still equipped with left and right projection that behaves *like xx^{-1} and $x^{-1}x$.*

Definition 2. *A partially ordered* monoid is a monoid S *equipped with a stable partial order relation* $\leq$, *that is, for every x, y and $z \in S$, if $x \leq y$ then $xz \leq yz$ and $zx \leq zy$. Then, an* adequately ordered monoid *is a partially ordered monoid so that, for every element $x \in S$:*

- *if $x \leq 1$ then $xx = x$, i.e. subunits are idempotent,*
- *both $x^L = \min\{z \leq 1 : xz = x\}$ and $x^R = \min\{z \leq 1 : zx = x\}$ exist.*

It is an easy exercice to check that, by stability, subunits also commute and, ordered by the order relation, form a meet semi-lattice with product as meet. When S is finite, subunits even form a complete lattice which suffices to guarantee the existence of left and the right projections.

Examples. Every monoid S, trivially ordered with a new leaset element zero, is an adequately ordered monoid with projections $x^L = 1 = x^R$ for every $x \in S$. Also, as already observed, every inverse semigroup S is adequately ordered by the natural order with $x^L = x^{-1}x$ and $x^R = xx^{-1}$ for every $x \in S$. The monoid $\mathcal{P}(Q \times Q)$ of relations over a set Q and ordered by relation inclusion is also adequately ordered.

The notion of observation disjoint product, though defined in a quite adhoc way in each concrete settings, essentially amounts to saying that the product $A{\cdot}B$ of two observations A and B is disjoint when it is non zero and the intersection of (the embedding of) their domains in the resulting product is limited to the output root of A that has been sewn with the input root of B. An example of a disjoint product of this type is depicted in Figure 8.

Fig. 8. Disjoint product of two observations

Definition 3. *A adequate premorphism is a premorphism $\varphi : S \to T$ from a given concrete monoid S of observations into an adequately ordered monoid T such that:*

- *if $A \leq B$ then $\varphi(A) \leq \varphi(B)$,*
- *$\varphi(A^L) = (\varphi(A))^L$ and $\varphi(A^R) = (\varphi(A))^R$,*
- *if the product $A \cdot B$ is disjoint then $\varphi(A \cdot B) = \varphi(A) \cdot \varphi(B)$.*

for every observations A and $B \in S$.

Remark. An immediate consequence of such a definition is that in the case observations can be generated from a finite set of elementary observations, disjoint products and left and right projections, i.e. every observations admits a good representations, then the image $\varphi(A)$ by any adequate premorphism φ into a finite adequately ordered monoid is computable in linear time in the size of the good representations. This happens with birooted words or trees [15,16].

Quasi-recognizability.
We are now ready to define quasi-recognizability, that is, recognizability by adequate premorphisms and ordered monoids.

Definition 4. *A language $L \subseteq S - 0$ is quasi-recognizable when there is an adequate premorphism $\varphi : S \to T$ from S to a finite adequately ordered monoid T such that $L = \varphi^{-1}(\varphi(L))$.*

Such an extension of language recognizability has been proposed in [17] and studied further in [19] and [16]. It happens that it is quite robust in the sense that, in the studied case of birooted words and birooted trees where definability in MSO is available, we prove that:

Theorem 1 (see [19,16]). *When S is the concrete monoid of birooted words or (labeled) birooted trees, a language $L \subseteq S$ is quasi-recognizable if and only if it is a finite boolean combination of upward closed MSO definable languages.*

Proving these results is achieved via a fairly simple extension of the notion of finite state non deterministic automata to birooted structures. These finite state automata are shown to characterize MSO definable upward closed languages or, equivalently, upward closed quasi-recognizable languages.

Generalized to languages, finding a set X of possible behaviors from a set of initial states S_0 to a set of possible final states S_1 amounts still to solving an inequality of the form $0 < (S_0 \cdot X)^L \leq S_1$ with product and projections extended in a point wise manner.

Remark. It must be mentioned that walking automata on trees also induce premorphisms that must satisfy certain kinds of (greatest fixpoint) equations that, in turn, make the premorphism (simply) computable [18].

Quite interestingly, these premorphisms do not preserve disjoint products but, instead, some extension of the well known notion of restricted products in inverse semigroup theory. Since the theory of recognizability by premorphisms is still in its infancy, it may be the case that this fixpoint approach could well be much more fruitful than the one presented here.

4   Conclusion

It shall be clear that the higher-dimensional string theory proposed here is still in its early stages of development. Further studies are currently conducted towards deeper modeling experiments, extension to infinite structures [7] and towards developing the underlying algebraic tools [8].

For instance, our logical characterization stated above ensures that the quasi-recognizable languages of positive birooted words or trees are closed under product or iterated product. Yet, providing direct algebraic or automata theoretic proof of these facts has been surprisingly difficult [8]. This means that the algebraic setting that is proposed here needs to be understood in a much greater depth.

More generally, every mathematical framework that can be used in computer science can be seen as a spotlight that helps us understanding the nature of the objects computer science and engineering may handle. The various experiments and theoretical developments that have been sketched in this paper tend to prove that inverse semigroup theory may provide, in the long term, an especially bright light for such a purpose.

Acknowledgment. Developing a trans-disciplinary research program that also aims at achieving advances in each of the specific research fields that are covered would probably be simply impossible without the support, encouragement and knowledge of true experts in the covered fields. The author wishes to express his deepest gratitude to all those he has had the pleasure of meeting and discussing questions with during the last two years.

More specifically, to name but a few, grateful thanks to Myriam DeSainte-Catherine, Florent Berthaut, Jean-Louis Giavitto and Yann Orlarey in the field of Computational Music, Marc Zeitoun, Victoria Gould, Mark Lawson and Sylvain Lombardy in the field of Semigroup Theory, Anne Dicky and Dominique Méry in the field of Formal methods, Sylvain Salvati and Paul Hudak in the field of Typed Functional Programming, and, last, Lucy Edwards for her invaluable knowledge of the English language itself.

The pleasure and honor the author felt when being kindly invited by Professor Viliam Geffert to give a lecture at SOFSEM 2014 in the beautiful landscape of the High Tatras in Winter is to be shared with all of them.

References

1. Abrial, J.R.: Modeling in Event-B - System and Software Engineering. Cambridge University Press (2010)
2. Berthaut, F., Janin, D., Martin, B.: Advanced synchronization of audio or symbolic musical patterns: an algebraic approach. International Journal of Semantic Computing 6(4), 409–427 (2012)
3. Cansell, D., Méry, D.: Foundations of the B method. Computers and Informatics 22 (2003)
4. Chalopin, J., Métivier, Y.: An efficient message passing election algorithm based on mazurkiewicz's algorithm. Fundam. Inform. 80(1-3), 221–246 (2007)
5. Cornock, C., Gould, V.: Proper two-sided restriction semigroups and partial actions. Journal of Pure and Applied Algebra 216, 935–949 (2012)
6. Dicky, A., Janin, D.: Modélisation algébrique du diner des philosophes. Modélisation des Systèmes Réactifs (MSR). Journal Européen des Systèmes Automatisés (JESA) 47(1-2-3) (November 2013)
7. Dicky, A., Janin, D.: Embedding finite and infinite words into overlapping tiles. Technical report, LaBRI, Université de Bordeaux (October 2013)
8. Dubourg, E., Janin, D.: Algebraic tools for the overlapping tile product. Technical report, LaBRI, Université de Bordeaux (October 2013)
9. Fountain, J.: Right PP monoids with central idempotents. Semigroup Forum 13, 229–237 (1977)
10. Fountain, J., Gomes, G., Gould, V.: The free ample monoid. Int. Jour. of Algebra and Computation 19, 527–554 (2009)
11. Hoare, C.: Communicating Sequential Processing. International Series in Computer Science. Prentice-Hall International (1985)
12. Hollings, C.D.: From right PP monoids to restriction semigroups: a survey. European Journal of Pure and Applied Mathematics 2(1), 21–57 (2009)
13. Hollings, C.D.: The Ehresmann-Schein-Nambooripad Theorem and its successors. European Journal of Pure and Applied Mathematics 5(4), 414–450 (2012)
14. Hudak, P.: A sound and complete axiomatization of polymorphic temporal media. Technical Report RR-1259, Department of Computer Science, Yale University (2008)
15. Janin, D.: Quasi-recognizable vs MSO definable languages of one-dimensional overlapping tiles. In: Rovan, B., Sassone, V., Widmayer, P. (eds.) MFCS 2012. LNCS, vol. 7464, pp. 516–528. Springer, Heidelberg (2012)
16. Janin, D.: Algebras, automata and logic for languages of labeled birooted trees. In: Fomin, F.V., Freivalds, R., Kwiatkowska, M., Peleg, D. (eds.) ICALP 2013, Part II. LNCS, vol. 7966, pp. 312–323. Springer, Heidelberg (2013)

17. Janin, D.: On languages of one-dimensional overlapping tiles. In: van Emde Boas, P., Groen, F.C.A., Italiano, G.F., Nawrocki, J., Sack, H. (eds.) SOFSEM 2013. LNCS, vol. 7741, pp. 244–256. Springer, Heidelberg (2013)
18. Janin, D.: Overlapping tile automata. In: Bulatov, A.A., Shur, A.M. (eds.) CSR 2013. LNCS, vol. 7913, pp. 431–443. Springer, Heidelberg (2013)
19. Janin, D.: Walking automata in the free inverse monoid. Technical Report RR-1464-12, LaBRI, Université de Bordeaux (2013)
20. Janin, D., Berthaut, F., DeSainte-Catherine, M., Orlarey, Y., Salvati, S.: The T-calculus : towards a structured programming of (musical) time and space. In: Workshop on Functional Art, Music, Modeling and Design (FARM). ACM Press (2013)
21. Janin, D., Berthaut, F., DeSainteCatherine, M.: Multi-scale design of interactive music systems: The libTuiles experiment. In: Sound and Music Computing, SMC (2013)
22. Kellendonk, J.: The local structure of tilings and their integer group of coinvariants. Comm. Math. Phys. 187, 115–157 (1997)
23. Kellendonk, J., Lawson, M.V.: Tiling semigroups. Journal of Algebra 224(1), 140–150 (2000)
24. Lawson, M.V.: Semigroups and ordered categories. I. the reduced case. Journal of Algebra 141(2), 422–462 (1991)
25. Lawson, M.V.: Inverse Semigroups: The theory of partial symmetries. World Scientific (1998)
26. Lawson, M.V.: McAlister semigroups. Journal of Algebra 202(1), 276–294 (1998)
27. Margolis, S.W., Pin, J.E.: Languages and inverse semigroups. In: Paredaens, J. (ed.) ICALP 1984. LNCS, vol. 172, pp. 337–346. Springer, Heidelberg (1984)
28. McAlister, D., Reilly, N.R.: E-unitary covers for inverse semigroups. Pacific Journal of Mathematics 68, 178–206 (1977)
29. Milner, R.: Communication and concurrency. Prentice-Hall (1989)
30. Munn, W.D.: Free inverse semigroups. Proceeedings of the London Mathematical Society 29(3), 385–404 (1974)
31. Nambooripad, K.S.S.: The natural partial order on a regular semigroup. Proc. Edinburgh Math. Soc. 23, 249–260 (1980)
32. Scheiblich, H.E.: Free inverse semigroups. Semigroup Forum 4, 351–359 (1972)
33. Silva, P.V.: On free inverse monoid languages. ITA 30(4), 349–378 (1996)
34. Stephen, J.: Presentations of inverse monoids. Journal of Pure and Applied Algebra 63, 81–112 (1990)
35. Thomas, W.: Languages, automata, and logic. In: Handbook of Formal Languages, vol. III, ch. 7, pp. 389–455. Springer, Heidelberg (1997)
36. Thomas, W.: Logic for computer science: The engineering challenge. In: Wilhelm, R. (ed.) Informatics: 10 Years Back, 10 Years Ahead. LNCS, vol. 2000, pp. 257–267. Springer, Heidelberg (2001)

Advice Complexity:
Quantitative Approach to A-Priori Information

(Extended Abstract)

Rastislav Královič

Comenius University, Bratislava, Slovakia
kralovic@dcs.fmph.uniba.sk

Abstract. We survey recent results from different areas, studying how introducing per-instance a-priori information affects the solvability and complexity of given tasks. We mainly focus on distributed, and online computation, where some sort of hidden information plays a crucial role: in the distributed computing, typically nodes have no or only limited information about the global state of the network; in online problems, the algorithm lacks the information about the future input. The traditional approach in both areas is to study how the properties of the problem change if some partial information is available (e.g., nodes of a distributed system have sense of direction, the online algorithm has the promise that the input requests come in some specified order etc.). Recently, attempts have been made to study this information from a quantitative point of view: there is an oracle that delivers (per-instance) best-case information of a limited size, and the relationship between the amount of the additional information, and the benefit it can provide to the algorithm, is investigated. We show cases where this relationship has a form of a trade-off, and others where one or more thresholds can be identified.

1 Introduction

Computation is often thought of as information processing: The input instance contains some implicit, hidden information, and the role of the algorithm is to make this information explicit, which usually means to produce some specified form of output. While from the information-theoretic point of view, all the relevant information is contained in the input, some of this information may not be available to the algorithm due to its limited resources, or due the nature of the computational model. In the talk we mainly focus on two areas where the missing information plays a crucial role.

In online problems [13], the algorithm must make irreversible decisions based only on partial knowledge about the input. There has been an extensive research concerning the augmentation of the algorithm with some a-priori information about the input, an approach known as semi-online algorithms.

V. Geffert et al. (Eds.): SOFSEM 2014, LNCS 8327, pp. 21–29, 2014.
© Springer International Publishing Switzerland 2014

In distributed computing the global state of the system is usually not known to the computing entities, yet it often plays a crucial role in the efficiency (or even feasibility) of the solution. Many works have been studying the impact of knowledge of the network topology on the efficiency and feasibility of various distributed tasks. Other pieces of information that influence the distributed algorithm are the knowledge of (some) identifiers, and, possibly, the knowledge of failure patterns.

While in both areas the traditional approach is to consider various particular forms of a-priori information, there are recent attempts analyzing the impact of a-priori information in a quantitative way.

2 Online Computing

The notion of a competitive ratio was introduced by Sleator and Tarjan [80]: a (minimization) algorithm A is called c-competitive, if it always produces an output where $\mathrm{cost}(\mathtt{A}) \leq c \cdot \mathtt{Opt}$. Note that in online problems the main concern is not the computational complexity, but the inherent loss of performance due to the unknown future.

Online computation has received considerable attention over the past decades as a natural way of modeling real-time processing of data. Since the algorithm does not know the future input, and because it is compared to the offline optimum in the worst case, many problems have no good competitive algorithms. In order to make the situation less unfair for the algorithm, randomization is often employed. Here, the algorithm has additional access to a random string. In order to be c-competitive, it is sufficient that $\mathrm{E}\left[\mathrm{cost}(\mathtt{A})\right] \leq c \cdot \mathtt{Opt} + \alpha$ where the expectation is taken over all random strings.

Many results have been proven about enhancing the algorithm with a particular type of information about the input, see e. g. [1,14,58,74].

The first attempt to analyze the impact of added information quantitatively was due to Halldórsson et al [53]. The authors considered the problem of finding the maximum independent set online, and introduced a model where the algorithm can maintain a set of solutions. The final solution produced by the algorithm is the best one from the set at the time of the last input request. If the algorithm is allowed to maintain $r(n)$ solutions, this model can be interpreted as running the algorithm with $\log r(n)$ bits of advice describing the particular input.

In [33], the authors start a systematic quantitative treatment of the problem-specific information. In the model from [11], the algorithm has, from the beginning of the computation, access to a tape with the advice describing the particular input. The maximal number of bits the algorithm reads from the tape during the computation is the advice complexity of the algorithm.

A number of problems have been considered in this model, including paging [11], k-server [10], knapsack [12], set cover [61], metrical task systems [37] buffer management [35], job shop scheduling [11], independent sets in various classes of graphs (general graphs [53], interval graphs [11], bipartite graphs [32]), and various variants of online coloring (bipartite graphs [7], paths [43], 3-colorable graphs [78], $L(2, 1)$ coloring [8]).

In general, there are three questions that are usually asked about a problem:

- What advice is needed to get optimal solution?
- What advice is needed to get the competitive ratio of the best possible randomized algorithm?
- What is the relationship between the size of the advice and the competitive ratio?

The comparison of advice and randomization is an interesting point, since the two approaches use different properties of the solution space: to have a randomized algorithm with good expected performance, many good witnesses for each input instance are needed. On the other hand, for good performance of advice algorithms, only one witness is sufficient for a given instance, but the space of possible witnesses must be small. In general, it holds (see [10]) that a randomized algorithm can be turned into an almost equally good deterministic one using $O(\log \log |\mathcal{I}(n)|)$ bits of advice, where $\mathcal{I}(n)$ is the set of all inputs of length n. However, in many cases significantly smaller advice is sufficient to be on par with randomization.

The relationship between the size of the advice and the competitive ratio is a complex one. In some cases, a trade-off relation exists, where increasing the advice yields a better competitive ratio, as is, e.g., the case of constant competitive ratio of paging [11]. On the other hand, there are thresholds, where increasing the advice does not help (e.g.[12]).

Notable is also the approach from [9], where an artificial problem of string guessing is analyzed, and a reduction is used to prove lower bounds on the advice complexity of online set cover.

3 Distributed Computing

Distributed systems consist of independent entities connected in a network that can communicate by some form of exchange of messages. There are two basic views on such systems, which are in essence equivalent, but yield themselves to different types of questions – either the active components are the nodes of the network, and messages are pieces of data send among them, or the active components are the messages (agents) that traverse the network, and the nodes passively provide resources for computation and communication. Typical problems solved in the message-based systems include communication tasks such as broadcasting, wake-up, leader election, or computational problems where some graph-theoretic objects are to be constructed, like, e.g., various spanners, colorings, independent or dominating sets, etc. On the other hand, typical problems tackled in the agent-based view include many variants of graph exploration, map drawing, agent rendezvous, and similar.

The quantitative study of the topological information in message based systems was introduced in the work of Fraigniaud, Ilcinkas, and Pelc [48], where a distinction has been shown between two similar problems: broadcasting, and wakeup, for which results for specific forms of information (e. g.sense of direction) have been known (e. g., [27,34,41]). In [48] the authors model the topological information in the following way: a-priori, each node knows its identity, and the local labeling of incident edges. Before the algorithm starts, each node v is provided with a binary string $f(v)$. The function f assigning the strings is called an oracle, and the overall length of the strings is its size. The smallest number of messages, over all oracles of a given size, exchanged by the algorithm is considered as a complexity measure.

The same notion of oracle size has been addressed for a number of other problems in the synchronous setting. Fusco and Pelc [51] consider wakeup in a rooted tree in the one-port model, where each node may send in each step only one message, and the aim is to minimize the number of steps.

In the $\mathcal{LOCAL}$ model from [75],i. e.,, in a synchronous message-passing system with nodes that have unique identifiers proper 3-coloring of cycles and trees ([46]), and the construction of minimum spanning tree ([50]) have been studied.

Broadcasting in radio networks has been considered in [57], where a trade-off between the size of advice and broadcasting time has been devised.

Also, let us note that the above mentioned work is tightly connected with the study of informative labeling schemes (see, e. g., [18,20,44,62,63,64] and references therein): here, the aim is to label vertices of the graph in such a way that it is possible to extract, based solely on the labels of a subset of vertices $V' \subseteq V$ some parameter concerning V' (e. g., if V' is any two-element set, and the parameter is distance, the scheme is called distance labelling scheme).

In the agent-based systems, the main focus is given to various graph exploration problems by either a single agent or a team of co-operating agents [3,15,16,38,23,36,45,47,72,73,79]. Directed graphs have been treated, e. g., in [2,6,24,40]. Apart from the various variants of graph exploration, problems like rendezvous (e. g., [5,17,26,66]) or black hole search (e. g., [22,28,29]) have been investigated.

When considering the additional information, and how it affects the exploration, one should note that the local labelling of the incident links is a potential source of information. A series of papers [30,52,56,65,81] investigates how the properly chosen labeling may help the algorithm.

The oracle-based approach where additional advice strings can be placed in the nodes was applied in [19], where it is proven that 2 bits in every node are sufficient for a finite automaton to explore all graphs, a task that is not possible without any information.

In the same model, the problem of drawing a map of an unlabeled graph ([25]), and traversing an unknown tree ([49]) have been considered.

Contrary to the local port scenario, where the agent residing in a given node can locally distinguish the incident links, in the so-called fixed graph scenario introduced by Kalyanasundaram and Pruhs in [59], the nodes have identifiers, and when the agent arrives at a node $v \in G$, it learns all incident edges, their endpoints, and, if the graph is weighted, their weights. While learning the endpoints of the incident edges is stronger than the typical exploration scenario, it does have a justification (see [59] and [69]); it also corresponds to the previously studied neighbourhood sense of direction [42].

In [31], the following problem was addressed from the point of view of advice size: the agent starts at a node v of an undirected labeled graph with n nodes, where each edge has a non-negative cost. The agent has no knowledge about the graph, and has to visit every node of the graph and return to v. The agent can move only along the edges, each time paying the respective edge cost. The natural Nearest Neighbor heuristics has competitive ratio of $\Theta(\log n)$ ([77]), which is tight even on planar unit-weight graphs ([55]). Despite many partial results ([4,59,70,69]), the main question, whether there exists a constant-competitive algorithm is still open. The same concept of advice has been also applied to the graph searching problem in [71].

4   Conclusion

Recently, there have been several attempts to analyze the impact of the hidden information in a quantitative way. Although they are applied in different areas, they share a common framework: The algorithm is enhanced by some information about the unknown part of the input, which may be of any type, but of bounded size. This approach may deepen the understanding of the structure of the respective problems. Finally, we note that the term advice complexity has traditionally been used as a synonym for relativized complexity (i. e., a sequential computation where the Turing machine gets an advice that depends on the length of the input), which may cause some confusion. Also, we note similar approaches in the treatment of the problem of factorization ([21,54,68,76]) where the number of queries to a yes/no oracle needed to determine the factors of a number was studied.

References

1. Albers, S.: On the influence of lookahead in competitive paging algorithms. Algorithmica 18(3), 283–305 (1997)
2. Albers, S., Henzinger, M.R.: Exploring unknown environments. SIAM J. Comput. 29(4), 1164–1188 (2000)
3. Antelmann, H., Budach, L., Rollik, H.-A.: On universal traps. Elektronische Informationsverarbeitung und Kybernetik 15(3), 123–131 (1979)
4. Asahiro, Y., Miyano, E., Miyazaki, S., Yoshimuta, T.: Weighted nearest neighbor algorithms for the graph exploration problem on cycles. Information Processing Letters 110(3), 93–98 (2010)
5. Barrière, L., Flocchini, P., Fraigniaud, P., Santoro, N.: Rendezvous and election of mobile agents: Impact of sense of direction. Theory Comput. Syst. 40(2), 143–162 (2007)
6. Bender, M.A., Slonim, D.K.: The power of team exploration: Two robots can learn unlabeled directed graphs. In: FOCS, pp. 75–85. IEEE Computer Society (1994)

7. Bianchi, M.P., Böckenhauer, H.-J., Hromkovič, J., Keller, L.: Online coloring of bipartite graphs with and without advice. In: Gudmundsson, J., Mestre, J., Viglas, T. (eds.) COCOON 2012. LNCS, vol. 7434, pp. 519–530. Springer, Heidelberg (2012)
8. Bianchi, M.P., Böckenhauer, H.-J., Hromkovič, J., Krug, S., Steffen, B.: On the advice complexity of the online $L(2,1)$-coloring problem on paths and cycles. In: Du, D.-Z., Zhang, G. (eds.) COCOON 2013. LNCS, vol. 7936, pp. 53–64. Springer, Heidelberg (2013)
9. Böckenhauer, H.-J., Hromkovič, J., Komm, D., Krug, S., Smula, J., Sprock, A.: The string guessing problem as a method to prove lower bounds on the advice complexity. In: Du, D.-Z., Zhang, G. (eds.) COCOON 2013. LNCS, vol. 7936, pp. 493–505. Springer, Heidelberg (2013)
10. Böckenhauer, H.-J., Komm, D., Královič, R., Královič, R.: On the advice complexity of the k-server problem. In: Aceto, L., Henzinger, M., Sgall, J. (eds.) ICALP 2011, Part I. LNCS, vol. 6755, pp. 207–218. Springer, Heidelberg (2011)
11. Böckenhauer, H.-J., Komm, D., Královič, R., Královič, R., Mömke, T.: On the advice complexity of online problems. In: Dong, Y., Du, D.-Z., Ibarra, O. (eds.) ISAAC 2009. LNCS, vol. 5878, pp. 331–340. Springer, Heidelberg (2009)
12. Böckenhauer, H.-J., Komm, D., Královič, R., Rossmanith, P.: On the advice complexity of the knapsack problem. In: Fernández-Baca, D. (ed.) LATIN 2012. LNCS, vol. 7256, pp. 61–72. Springer, Heidelberg (2012)
13. Borodin, A., El-Yaniv, R.: Online Computation and Competitive Analysis. Cambridge University Press (1998)
14. Borodin, A., Irani, S., Raghavan, P., Schieber, B.: Competitive paging with locality of reference (preliminary version). In: Koutsougeras, C., Vitter, J.S. (eds.) STOC, pp. 249–259. ACM (1991)
15. Budach, L.: On the solution of the labyrinth problem for finite automata. Elektronische Informationsverarbeitung und Kybernetik 11(10-12), 661–672 (1975)
16. Budach, L.: Environments, labyrinths and automata. In: Karpinski, M. (ed.) FCT 1977. LNCS, vol. 56, pp. 54–64. Springer, Heidelberg (1977)
17. Chalopin, J., Das, S., Widmayer, P.: Rendezvous of mobile agents in directed graphs. In: Lynch, Shvartsman (eds.) [67], pp. 282–296
18. Chepoi, V., Dragan, F.F., Estellon, B., Habib, M., Vaxès, Y., Xiang, Y.: Additive spanners and distance and routing labeling schemes for hyperbolic graphs. Algorithmica 62(3-4), 713–732 (2012)
19. Cohen, R., Fraigniaud, P., Ilcinkas, D., Korman, A., Peleg, D.: Label-guided graph exploration by a finite automaton. ACM Transactions on Algorithms 4(4) (2008)
20. Cohen, R., Fraigniaud, P., Ilcinkas, D., Korman, A., Peleg, D.: Labeling schemes for tree representation. Algorithmica 53(1), 1–15 (2009)
21. Coppersmith, D.: Finding a small root of a bivariate integer equation; factoring with high bits known. In: Maurer, U.M. (ed.) EUROCRYPT 1996. LNCS, vol. 1070, pp. 178–189. Springer, Heidelberg (1996)
22. Czyzowicz, J., Dobrev, S., Královič, R., Miklík, S., Pardubská, D.: Black hole search in directed graphs. In: Kutten, S., Žerovnik, J. (eds.) SIROCCO 2009. LNCS, vol. 5869, pp. 182–194. Springer, Heidelberg (2010)
23. Das, S., Flocchini, P., Kutten, S., Nayak, A., Santoro, N.: Map construction of unknown graphs by multiple agents. Theor. Comput. Sci. 385(1-3), 34–48 (2007)
24. Deng, X., Papadimitriou, C.H.: Exploring an unknown graph. Journal of Graph Theory 32(3), 265–297 (1999)
25. Dereniowski, D., Pelc, A.: Drawing maps with advice. In: Lynch, Shvartsman (eds.) [67], pp. 328–342

26. Dessmark, A., Fraigniaud, P., Kowalski, D.R., Pelc, A.: Deterministic rendezvous in graphs. Algorithmica 46(1), 69–96 (2006)

27. Diks, K., Dobrev, S., Kranakis, E., Pelc, A., Ruzicka, P.: Broadcasting in unlabeled hypercubes with a linear number of messages. Inf. Process. Lett. 66(4), 181–186 (1998)

28. Dobrev, S., Flocchini, P., Kralovic, R., Ruzicka, P., Prencipe, G., Santoro, N.: Black hole search in common interconnection networks. Networks 47(2), 61–71 (2006)

29. Dobrev, S., Flocchini, P., Prencipe, G., Santoro, N.: Searching for a black hole in arbitrary networks: optimal mobile agents protocols. Distributed Computing 19(1) (2006)

30. Dobrev, S., Jansson, J., Sadakane, K., Sung, W.-K.: Finding short right-hand-on-the-wall walks in graphs. In: Pelc, A., Raynal, M. (eds.) SIROCCO 2005. LNCS, vol. 3499, pp. 127–139. Springer, Heidelberg (2005)

31. Dobrev, S., Královič, R., Markou, E.: Online graph exploration with advice. In: Even, G., Halldórsson, M.M. (eds.) SIROCCO 2012. LNCS, vol. 7355, pp. 267–278. Springer, Heidelberg (2012)

32. Dobrev, S., Královič, R., Královič, R.: Independent Set with Advice: The Impact of Graph Knowledge. In: Erlebach, T., Persiano, G. (eds.) WAOA 2012. LNCS, vol. 7846, pp. 2–15. Springer, Heidelberg (2013)

33. Dobrev, S., Královič, R., Pardubská, D.: Measuring the problem-relevant information in input. RAIRO Theoretical Informatics and Applications 43(3), 585–613 (2009)

34. Dobrev, S., Ružička, P.: Broadcasting on anonymous unoriented tori. In: Hromkovič, J., Sýkora, O. (eds.) WG 1998. LNCS, vol. 1517, pp. 50–62. Springer, Heidelberg (1998)

35. Dorrigiv, R., He, M., Zeh, N.: On the advice complexity of buffer management. In: Chao, K.-M., Hsu, T.-s., Lee, D.-T. (eds.) ISAAC 2012. LNCS, vol. 7676, pp. 136–145. Springer, Heidelberg (2012)

36. Dynia, M., Łopuszański, J., Schindelhauer, C.: Why robots need maps. In: Prencipe, G., Zaks, S. (eds.) SIROCCO 2007. LNCS, vol. 4474, pp. 41–50. Springer, Heidelberg (2007)

37. Emek, Y., Fraigniaud, P., Korman, A., Rosén, A.: Online computation with advice. Theoretical Computer Science 412(24), 2642–2656 (2011)

38. Euler, L.: Solutio problematis ad geometriam situs pertinentis. Novi Commentarii Academiae Scientarium Imperialis Petropolitanque 7, 9–28 (1758-1759)

39. Fiat, A., Karp, R.M., Luby, M., McGeoch, L.A., Sleator, D.D., Young, N.E.: Competitive paging algorithms. J. Algorithms 12(4), 685–699 (1991)

40. Fleischer, R., Trippen, G.: Exploring an unknown graph efficiently. In: Brodal, G.S., Leonardi, S. (eds.) ESA 2005. LNCS, vol. 3669, pp. 11–22. Springer, Heidelberg (2005)

41. Flocchini, P., Mans, B., Santoro, N.: On the impact of sense of direction on message complexity. Inf. Process. Lett. 63(1), 23–31 (1997)

42. Flocchini, P., Mans, B., Santoro, N.: Sense of direction in distributed computing. Theor. Comput. Sci. 291(1), 29–53 (2003)

43. Forišek, M., Keller, L., Steinová, M.: Advice complexity of online coloring for paths. In: Dediu, A.-H., Martín-Vide, C. (eds.) LATA 2012. LNCS, vol. 7183, pp. 228–239. Springer, Heidelberg (2012)

44. Fraigniaud, P.: Informative labeling schemes. In: Abramsky, S., Gavoille, C., Kirchner, C., Meyer auf der Heide, F., Spirakis, P.G. (eds.) ICALP 2010, Part II. LNCS, vol. 6199, pp. 1–1. Springer, Heidelberg (2010)

45. Fraigniaud, P., Gasieniec, L., Kowalski, D.R., Pelc, A.: Collective tree exploration. Networks 48(3), 166–177 (2006)
46. Fraigniaud, P., Gavoille, C., Ilcinkas, D., Pelc, A.: Distributed computing with advice: information sensitivity of graph coloring. Distributed Computing 21(6), 395–403 (2009)
47. Fraigniaud, P., Ilcinkas, D., Peer, G., Pelc, A., Peleg, D.: Graph exploration by a finite automaton. Theor. Comput. Sci. 345(2-3), 331–344 (2005)
48. Fraigniaud, P., Ilcinkas, D., Pelc, A.: Oracle size: a new measure of difficulty for communication tasks. In: Ruppert, E., Malkhi, D. (eds.) PODC, pp. 179–187. ACM (2006)
49. Fraigniaud, P., Ilcinkas, D., Pelc, A.: Tree exploration with advice. Inf. Comput. 206(11), 1276–1287 (2008)
50. Fraigniaud, P., Korman, A., Lebhar, E.: Local mst computation with short advice. Theory Comput. Syst. 47(4), 920–933 (2010)
51. Fusco, E.G., Pelc, A.: Trade-offs between the size of advice and broadcasting time in trees. In: auf der Heide, F.M., Shavit, N. (eds.) SPAA, pp. 77–84. ACM (2008)
52. Gasieniec, L., Klasing, R., Martin, R.A., Navarra, A., Zhang, X.: Fast periodic graph exploration with constant memory. J. Comput. Syst. Sci. 74(5), 808–822 (2008)
53. Halldórsson, M.M., Iwama, K., Miyazaki, S., Taketomi, S.: Online independent sets. Theor. Comput. Sci. 289(2), 953–962 (2002)
54. Herrmann, M., May, A.: On factoring arbitrary integers with known bits. In: Koschke, R., Herzog, O., Rödiger, K.-H., Ronthaler, M. (eds.) GI Jahrestagung, Part II. LNI, vol. 110, pp. 195–199 (2007)
55. Hurkens, C.A., Woeginger, G.J.: On the nearest neighbor rule for the traveling salesman problem. Operations Research Letters 32(1), 1–4 (2004)
56. Ilcinkas, D.: Setting port numbers for fast graph exploration. Theor. Comput. Sci. 401(1-3), 236–242 (2008)
57. Ilcinkas, D., Kowalski, D.R., Pelc, A.: Fast radio broadcasting with advice. Theor. Comput. Sci. 411(14-15), 1544–1557 (2010)
58. Irani, S., Karlin, A.R., Phillips, S.: Strongly competitive algorithms for paging with locality of reference. In: Frederickson, G.N. (ed.) SODA, pp. 228–236. ACM/SIAM (1992)
59. Kalyanasundaram, B., Pruhs, K.R.: Constructing competitive tours from local information. Theoretical Computer Science 130(1), 125–138 (1994)
60. Komm, D., Královič, R.: Advice complexity and barely random algorithms. In: Černá, I., Gyimóthy, T., Hromkovič, J., Jefferey, K., Královič, R., Vukolić, M., Wolf, S. (eds.) SOFSEM 2011. LNCS, vol. 6543, pp. 332–343. Springer, Heidelberg (2011)
61. Komm, D., Královič, R., Mömke, T.: On the advice complexity of the set cover problem. In: Hirsch, E.A., Karhumäki, J., Lepistö, A., Prilutskii, M. (eds.) CSR 2012. LNCS, vol. 7353, pp. 241–252. Springer, Heidelberg (2012)
62. Korman, A.: Labeling schemes for vertex connectivity. ACM Transactions on Algorithms 6(2) (2010)
63. Korman, A., Kutten, S., Peleg, D.: Proof labeling schemes. Distributed Computing 22(4), 215–233 (2010)
64. Korman, A., Peleg, D., Rodeh, Y.: Constructing labeling schemes through universal matrices. Algorithmica 57(4), 641–652 (2010)
65. Kosowski, A., Navarra, A.: Graph decomposition for memoryless periodic exploration. Algorithmica 63(1-2), 26–38 (2012)

66. Kowalski, D.R., Malinowski, A.: How to meet in anonymous network. Theor. Comput. Sci. 399(1-2), 141–156 (2008)
67. Lynch, N.A., Shvartsman, A.A. (eds.): DISC 2010. LNCS, vol. 6343. Springer, Heidelberg (2010)
68. Maurer, U.M.: On the oracle complexity of factoring integers. Computational Complexity 5(3/4), 237–247 (1995)
69. Megow, N., Mehlhorn, K., Schweitzer, P.: Online graph exploration: New results on old and new algorithms. In: Aceto, L., Henzinger, M., Sgall, J. (eds.) ICALP 2011, Part II. LNCS, vol. 6756, pp. 478–489. Springer, Heidelberg (2011)
70. Miyazaki, S., Morimoto, N., Okabe, Y.: The online graph exploration problem on restricted graphs. IEICE Transactions on Information and Systems 92(9), 1620–1627 (2009)
71. Nisse, N., Soguet, D.: Graph searching with advice. Theoretical Computer Science 410(14), 1307–1318 (2009)
72. Panaite, P., Pelc, A.: Exploring unknown undirected graphs. J. Algorithms 33(2), 281–295 (1999)
73. Panaite, P., Pelc, A.: Impact of topographic information on graph exploration efficiency. Networks 36(2), 96–103 (2000)
74. Pandurangan, G., Upfal, E.: Can entropy characterize performance of online algorithms? In: Kosaraju, S.R. (ed.) SODA, pp. 727–734. ACM/SIAM (2001)
75. Peleg, D.: Distributed Computing: A Locality-Sensitive Approach. Monographs on Discrete Mathematics and Applications. Society for Industrial and Applied Mathematics (2000)
76. Robertson, N., Seymour, P.D.: Graph minors. XIII. The disjoint paths problem. Journal of Combinatorial Theory, Series B 63(1), 65–110 (1995)
77. Rosenkrantz, D.J., Stearns, R.E., Lewis II., P.M.: An analysis of several heuristics for the traveling salesman problem. SIAM Journal on Computing 6(3), 563–581 (1977)
78. Seibert, S., Sprock, A., Unger, W.: Advice complexity of the online coloring problem. In: Spirakis, P.G., Serna, M. (eds.) CIAC 2013. LNCS, vol. 7878, pp. 345–357. Springer, Heidelberg (2013)
79. Shannon, C.E.: Presentation of a maze solving machine. In: von Foerster, H., Mead, M., Teuber, H.L. (eds.) Cybernetics: Circular, Causal and Feedback Mechanisms in Biological and Social Systems, Transactions Eighth Conference, March 15–16, pp. 169–181. Josiah Macy Jr. Foundation, New York (1951)
80. Sleator, D.D., Tarjan, R.E.: Amortized efficiency of list update and paging rules. Communications of the ACM 28(2), 202–208 (1985)
81. Steinová, M.: On the power of local orientations. In: Shvartsman, A.A., Felber, P. (eds.) SIROCCO 2008. LNCS, vol. 5058, pp. 156–169. Springer, Heidelberg (2008)

Matching of Images of Non-planar Objects with View Synthesis

Dmytro Mishkin and Jiří Matas*

Center for Machine Perception, Faculty of Electrical Engineering,
Czech Technical University in Prague
ducha.aiki@gmail.com, matas@cmp.felk.cvut.cz

Abstract. We explore the performance of the recently proposed two-view image matching algorithms using affine view synthesis – ASIFT (Morel and Yu, 2009) [14] and MODS (Mishkin, Perdoch and Matas, 2013) [10] on images of objects that do not have significant local texture and that are locally not well approximated by planes.

Experiments show that view synthesis improves matching results on images of such objects, but the number of "useful" synthetic views is lower than for planar objects matching. The best detector for matching images of 3D objects is the Hessian-Affine in the SPARSE configuration. The iterative MODS matcher performs comparably confirming it is a robust, generic method for two view matching that performs well for different types of scenes and a wide range of viewing conditions.

Keywords: feature detectors, view synthesis, image matching.

1 Introduction

The authors of the recently developed algorithms [14], [10] for wide baseline matching reported significant progress in the ability to deal with large viewpoint differences of matched images. The ASIFT method [14] generates synthetic affine transformed views of given images in order to improve the range of transformations handled the DoG detector. This idea has been further extended in [10] who incorporate multiple affine-covariant detectors and adopt an iterative approach that attempts to minimize the matching time.

However, the methodology and datasets [8], [14], [10] used in the evaluation are limited to images related by a homography. We study the performance of ASIFT and MODS on images of objects that do not have significant local texture and are not locally well approximated by planes. Such will be referred to as "3D objects" and image collections capturing such objects "3D datasets".

We have adopted MODS algorithm for the matching of images of such objects and present results of the matching performance evaluation obtained on 3D dataset [12]. We show, that affine view synthesis improves performance of 3D object matching, although the gain is significantly less pronounced than for planar objects.

---

* Invited Speaker.

V. Geffert et al. (Eds.): SOFSEM 2014, LNCS 8327, pp. 30–39, 2014.

Algorithm 1. ASIFT

Input: I_1, I_2 – two images.
Output: List of corresponding points.

---

for I_1 and I_2 separately **do**
    1 Generate synthetic views according to the tilt-rotation-detector setup.
    2 Detect and describe local features.
end for
3 Generate tentative correspondences for each pair of the synthesized views of
    I_1 and I_2 separately using the 2nd closest ratio.
4 Add correspondences to the general list.
    Reproject corresponding features to original images.
5 Filter duplicate, "one-to-many" and "many-to-one" matches.
6 Geometrically verify tentative correspondences using ORSA [11]
    while estimating F.

Table 1. ASIFT view synthesis configuration

Detector	View synthesis setup
DoG	$\{S\} = \{1\}$, $\{t\} = \{1; \sqrt{2}; 2; 2\sqrt{2}; 4; 4\sqrt{2}; 8\}$, $\Delta\phi = 72°/t$

Related Work. The performance of the detectors and descriptors for object on 3D scenes have been evaluated by Moreels and Perona [12],[13]. Authors tested distinctness of the detectors using large database of features contained both features from the related and unrelated images and did not evaluated performance of the whole matching system which involves geometric verification. The best performance was shown be Hessian-Affine [9] detector.

Dahl *et al.* [2] have tested matching performance of the descriptor-detector pair on synthetic dataset and reported the domination of the DoG [6] and MSER [7] as a detectors with SIFT [6] descriptor.

2 Tested Matchers

The MODS [10] and ASIFT [14] matchers and different view synthesis setups for Hessian-Affine [9], MSER [7] and DoG [6] interest point detectors proposed in [10] have been tested. In this section we shortly overview matching algorithms.

The ASIFT pipeline is presented in Alg. 1. The synthetic affine views are generated according to the affine transformation decomposition proposed in [14]:

$$A = H_\lambda R_1(\psi) T_t R_2(\phi) = \lambda \begin{pmatrix} \cos\psi & -\sin\psi \\ \sin\psi & \cos\psi \end{pmatrix} \begin{pmatrix} t & 0 \\ 0 & 1 \end{pmatrix} \begin{pmatrix} \cos\phi & -\sin\phi \\ \sin\phi & \cos\phi \end{pmatrix} \quad (1)$$

where $\lambda > 0$, R_1 and R_2 are rotations, and T_t is a diagonal matrix with $t > 1$. Parameter t is called the absolute tilt, $\phi \in \langle 0, \pi)$ is the optical axis longitude and $\psi \in \langle 0, 2\pi)$ is the rotation of the camera around the optical axis.

Algorithm 2. MODS-F

Input: I_1, I_2 – two images; θ_m – minimum required number of matches; $S_{\max}$ – maximum number of iterations.
Output: Fundamental matrix F; list of corresponding points.
Variables: N_{matches} – detected correspondences, Iter – current iteration.

---

while $(N_{\text{matches}} < \theta_m)$ **and** (Iter $< S_{\max}$) **do**
   for I_1 and I_2 separately **do**
      1 Generate synthetic views according to the
      scale-tilt-rotation-detector setup for the Iter (Tables 2, 3).
      2 Detect and describe local features.
      3 Reproject local features to original image.
      Add described features to general list.
   end for
   4 Generate tentative correspondences using the first geom. inconsistent rule.
   5 Filter duplicate matches.
   6 Geometrically verify tentative correspondences with DEGENSAC [1]
   while estimating F.
   7 Geometrically verify inliers with local affine frame shape.
end while

Table 2. MODS configuration [10]

Iter.	View synthesis setup
1	MSER,$\{S\} = \{1; 0.25; 0.125\}$, $\{t\} = \{1\}$, $\Delta\phi = 360°/t$
2	MSER,$\{S\} = \{1; 0.25; 0.125\}$, $\{t\} = \{1; 5; 9\}$, $\Delta\phi = 360°/t$
3	HessAff, $\{S\} = \{1\}$, $\{t\} = \{1; \sqrt{2}; 2; 2\sqrt{2}; 4; 4\sqrt{2}; 8\}$, $\Delta\phi = 360°/t$
4	HessAff , $\{S\} = \{1\}$, $\{t\} = \{1; 2; 4; 6; 8\}$, $\Delta\phi = 72°/t$

The parameters of synthesis are: set of scales $\{S\}$, $\Delta\phi_{\text{base}}$ step of longitude samples at tilt $t = 1$, and a set of simulated tilts $\{t\}$.

The feature detection and description are the same as in original SIFT [6]. The tentative correspondences are generated for each pair of generated views, i.e. matching stage consists of n^2 separated parts[1], which results are concatenated.

The *duplicate filtering* prunes correspondences with close spatial distance (2 pixels) of local features in both images – all this correspondences except one (random) are eliminated from the final correspondences list. "*One-to-many*" correspondence means situation when features which are close to each other (are situated in radius of $\sqrt{2}$ pixels) in one image correspond to the features situated in different locations in other images. All such correspondences are eliminated, although some of them can be correct. ORSA [11] is RANSAC-based method, which exploits an a-contrario statistic-based approach to detect incorrect epipolar geometry. Instead of having constant error threshold, ORSA looks for the matches that have the highest "diameter", i.e. matches which cover larger image

---

[1] n – number of synthesized views per image.

Table 3. Tested view synthesis configurations for single detectors [10]

Detector	View synthesis setup	
	SPARSE	DENSE
MSER	$\{S\} = \{1; 0.25; 0.125\}$, $\{t\} = \{1; 5; 9\}$, $\Delta\phi = 360°/t$	$\{S\} = \{1; 0.25; 0.125\}$, $\{t\} = \{1; 2; 4; 6; 8\}$, $\Delta\phi = 72°/t$
HessAff	$\{S\} = \{1\}$, $\{t\} = \{1; \sqrt{2}; 2; 2\sqrt{2}; 4; 4\sqrt{2}; 8\}$, $\Delta\phi = 360°/t$	$\{S\} = \{1\}$, $\{t\} = \{1; 2; 4; 6; 8\}$, $\Delta\phi = 72°/t$
DoG	$\{S\} = \{1\}$, $\{t\} = \{1; 2; 4; 6; 8\}$, $\Delta\phi = 120°/t$	$\{S\} = \{1\}$, $\{t\} = \{1; \sqrt{2}; 2; 2\sqrt{2}; 4; 4\sqrt{2}; 8\}$, $\Delta\phi = 72°/t$

area having minimum possible error and it estimates whether such inliers could be non-random.

Unlike the ASIFT, MODS (see Alg. 2) uses different detectors for the detection of local features – MSER and Hessian-Affine. The tentative correspondences are generated using kd-tree [15] which is build for all features detected in all views generated from the one image. The features detected in all views generated from the other image are used as a query. The first geometrically inconsistent distance ratio is used instead of 2nd closest distance ratio. Descriptors in one image are geometrically inconsistent, if the Euclidean distance between centers of the regions is ≥ 10 pixels [10]. The elimination of the duplicate matches is done the same way as in ASIFT. LO-RANSAC [4], used in the original MODS algorithm, was replaced with DegenSAC [1], which estimates fundamental matrix instead of homography.

Since that an epipolar geometry constraint is much less restrictive than a homography, wrong correspondences as well as correct ones can be consistent with some (random) fundamental matrix. The local affine frame consistency check (LAF-check) was applied for elimination of the incorrect correspondences. We use coordinates of the closest and furthest ellipse points from the ellipse center of both matched local affine frames to check whether the whole local feature is consistent with estimated geometry model. This check is performed for the selected geometry model and regions which do not pass the check are discarded from the list of inliers. If the number of correspondences after the LAF-check is fewer than the user defined minimum, matcher continues with the next step of view synthesis. Because LAF-check is performed after the RANSAC step, it helps to reject incorrect geometry models.

The single detector configurations specified in Table 3 were matched using MODS algorithm with single iteration.

3 Experiments

3.1 Evaluation Protocol

The evaluation dataset consists of 32 image sequences taken from the Turntable dataset [13] ("Bottom" camera) shown in Table 4. Eight image sets contain

Table 4. Reference views of the image sequences used in the evaluation (from [13])

Abroller	Bannanas	Camera2	Car
Car2	CementBase	Cloth	Conch
Desk	Dinosaur	Dog	DVD
FloppyBox	FlowerLamp	Gelsole	GrandfatherClock
Horse	Keyboard	Motorcycle	MouthGuard
PaperBin	PS2	Razor	RiceCooker
Rock	RollerBlade	Spoons	TeddyBear
Toothpaste	Tricycle	Tripod	VolleyBall

relatively large planar surfaces and the rest twenty-four are "full" 3D and low-textured.

The view marked as "0°" in the Turntable dataset was used as a reference view and $0 - 90°$ and 270-355 ° views with a 5° step were matched against it using the procedure described in Sec. 2, forming a $[-90°, 90°]$ sequence. Note that reference view is not usually the "frontal" or "side" view, but rather some intermediate view which caused asymmetry in results (see Fig. 1, Table 5).

The output of the matchers is a set of the correspondences and the estimated geometrical transformation. The accuracy of the matched correspondences was chosen as the performance criterion, similarly to in [2]. For all output correspondences, the symmetrical epipolar error [3] e_{SymEG} was computed according to the following expression:

$$e_{\text{SymEG}}(\mathbf{F}, \mathbf{u}, \mathbf{v}) = (\mathbf{v}^\top \mathbf{F}\mathbf{u})^2 \times \left(\frac{1}{(\mathbf{Fu})_1^2 + (\mathbf{Fu})_2^2} + \frac{1}{(\mathbf{F}^\top \mathbf{v})_1^2 + (\mathbf{F}^\top \mathbf{v})_2^2} \right), \quad (2)$$

where $\mathbf{F}$ – fundamental matrix, $\mathbf{u}$, $\mathbf{v}$ – corresponding points, $(\mathbf{Fu})_j^2$ – the square of the j-th entry of the vector $\mathbf{Fu}$.

The ground truth fundamental matrix was obtained from the difference in camera positions [3], assuming that turntable is fixed and the camera moved around object, according to the following equation:

$$F = K^{-\top} R K^\top [K R^\top t]_\times,$$

$$R = \begin{pmatrix} \cos\phi & 0 & -\sin\phi \\ 0 & 1 & 0 \\ \sin\phi & 0 & \cos\phi \end{pmatrix}, K = \begin{pmatrix} \frac{mf}{\text{FR}_X} & 0 & \frac{m}{2} \\ 0 & \frac{nf}{\text{FR}_Y} & \frac{n}{2} \\ 0 & 0 & 1 \end{pmatrix}, t = r \begin{pmatrix} \sin\phi \\ 0 \\ 1 - \cos\phi \end{pmatrix}, \quad (3)$$

where R is the orientation matrix of the second camera, K – the camera projection matrix, t – the virtual translation of the second camera, r – the distance from camera to the object, ϕ – the viewpoint angle difference, FR_X, FR_Y – the focal plane resolution, f – the focal length, m, n – the sensor matrix width and height in pixels. The last five are obtained from from EXIF data.

One of the evaluation problems is that background regions, i. e. regions that are not on the object placed on the turnable, are often detected and matched influencing the geometry transformation estimation. The matches are correct, but consistent with an identity transform of the (background of) the test images, not the fundamental matrix associated with the movement of the object on the turntable.

The median value of the correspondence errors was chosen as the measure of precision because of its tolerance to the low number of outliers (e.g. the above-mentioned background correspondences) and its sensitivity to the incorrect geometric model estimated by RANSAC.

An image pair is considered as *correctly matched if the median symmetrical epipolar error on the correspondences using ground truth fundamental matrix is ≤ 6 pixels.*

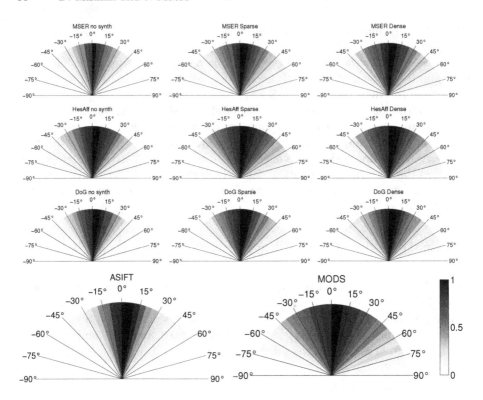

Fig. 1. A comparison of view synthesis configurations on the Turntable dataset [13]. The fraction of correctly matched images for a given viewpoint difference.

3.2   Results

Figure 1 and Table 5 show the percentage and the number of image sequences respectively for which the reference and tested views for the given viewing angle difference were matched correctly. On average, view synthesis adds $5 - 10°$ to the handled viewpoint difference for structured objects. The results show that the MODS matching algorithm used in DoG DENSE configuration outperforms the ASIFT algorithm with the same view synthesis setup.

The difference between DENSE and SPARSE configurations is small for structured scenes — unlike planar ones [10]. Difficulties in matching are caused not by the inability to detect distorted regions but by object self-occlusions. Therefore synthesis of the additional views does not bring more correspondences.

Experiments with view synthesis confirmed Moreels and Perona [13] results that the Hessian-Affine outperforms other detectors for matching of the structured scenes and can be used alone in such scenes.

All computations have been performed on Intel i7 3.9GHz (8 cores) desktop with 16Gb RAM. Examples of the matched images are shown in Fig. 2.

Fig. 2. Correspondences found by the MODS algorithm. Green – corresponding regions, cyan – epipolar lines.

4   Conclusion

We have shown that view synthesis improves matching performance of images of non-planar objects. The gain is less prominent than for matching of locally planar object, but yet significant.

The best detector for matching images of 3D objects is the Hessian-Affine in the SPARSE configuration, additional view synthesis beyond this configuration does not increase the number of correct correspondences.

The iterative MODS matcher performs comparably to the Hessian-Affine Sparse configurations but it is slower. The experiments thus confirm that MODS is a robust, generic method for two view matching that performs well for different types of scenes and a wide range of viewing conditions.

Table 5. The number of correctly matched image pairs and the runtime per pair

	Image sets solved (out of 32)										Time
	Viewpoint angular difference										
Matcher	0°	5°	10°	15°	20°	25°	30°	35°	40°	≥ 45°	[s]
MSER no synth	26	23	18	17	13	10	8	8	8	3	0.6
MSER Sparse	31	28	24	21	17	16	11	10	8	6	1.5
MSER Dense	**32**	28	25	24	19	17	12	11	10	7	9.5
HesAff no synth	**32**	28	27	24	22	18	13	11	10	6	0.8
HesAff Sparse	**32**	30	**29**	26	**25**	**21**	**19**	**16**	11	**11**	1.8
HesAff Dense	**32**	**31**	28	25	21	19	16	15	12	9	5.3
DoG no synth	**32**	29	24	23	16	13	10	7	4	4	0.9
DoG Sparse	**32**	**31**	25	22	17	17	12	11	7	5	4.0
DoG Dense	**32**	30	26	20	17	15	14	9	8	7	8.0
ASIFT	**32**	29	21	16	12	8	7	5	4	3	27.7
MODS	**32**	28	27	**27**	23	**21**	**19**	**16**	**16**	**11**	4.3

Acknowledgment. The authors were supported by The Czech Science Foundation Project GACR P103/12/G084.

References

1. Chum, O., Werner, T., Matas, J.: Two-view geometry estimation unaffected by a dominant plane. In: Proceedings of the 2005 IEEE Computer Society Conference on Computer Vision and Pattern Recognition (CVPR 2005), vol. 01, pp. 772–779. IEEE Computer Society, Washington, DC (2005)
2. Dahl, A.L., Aanaes, H., Pedersen, K.S.: Finding the best feature detector-descriptor combination. In: Proceedings of the 2011 International Conference on 3D Imaging, Modeling, Processing, Visualization and Transmission, 3DIMPVT 2011, pp. 318–325. IEEE Computer Society, Washington, DC (2011)
3. Hartley, R.I., Zisserman, A.: Multiple View Geometry in Computer Vision, 2nd edn. Cambridge University Press (2004)
4. Lebeda, K., Matas, J., Chum, O.: Fixing the Locally Optimized RANSAC. In: Proceedings of the British Machine Vision Conference, pp. 1013–1023 (2012)
5. Lepetit, V., Fua, P.: Keypoint recognition using randomized trees. IEEE Trans. Pattern Anal. Mach. Intell. 28(9), 1465–1479 (2006)
6. Lowe, D.: Distinctive image features from scale-invariant keypoints. Int. J. Comput. Vision 20, 91–110 (2004)
7. Matas, J., Chum, O., Urban, M., Pajdla, T.: Robust wide baseline stereo from maximally stable extrema regions. In: Proceedings of the British Machine Vision Conference, pp. 384–393 (2002)
8. Mikolajczyk, K., Tuytelaars, T., Schmid, C., Zisserman, A., Matas, J., Schaffalitzky, F., Kadir, T., Gool, L.V.: A comparison of affine region detectors. Int. J. Comput. Vision 65(1-2), 43–72 (2005)
9. Mikolajczyk, K., Schmid, C.: Scale & affine invariant interest point detectors. Int. J. Comput. Vision 60, 63–86 (2004)
10. Mishkin, D., Perdoch, M., Matas, J.: Two-view matching with view synthesis revisited. In: Proceedings of the 28th Conference on Image and Vision Computing, New Zealand (2013)

11. Moisan, L., Stival, B.: A probabilistic criterion to detect rigid point matches between two images and estimate the fundamental matrix. Int. J. Comput. Vision 57(3), 201–218 (2004)
12. Moreels, P., Perona, P.: Evaluation of features detectors and descriptors based on 3d objects. In: Proceedings of the Tenth IEEE International Conference on Computer Vision, pp. 800–807 (2005)
13. Moreels, P., Perona, P.: Evaluation of features detectors and descriptors based on 3d objects. Int. J. Comput. Vision 73(3), 263–284 (2007)
14. Morel, J.M., Yu, G.: ASIFT: A new framework for fully affine invariant image comparison. SIAM J. Img. Sci. 2(2), 438–469 (2009)
15. Muja, M., Lowe, D.: Fast approximate nearest neighbors with automatic algorithm configuration. In: VISAPP International Conference on Computer Vision Theory and Applications, pp. 331–340 (2009)

Agile Requirements Engineering: A Research Perspective

Jerzy Nawrocki*, Mirosław Ochodek, Jakub Jurkiewicz, Sylwia Kopczyńska,
and Bartosz Alchimowicz

Poznan University of Technology, Institute of Computing Science,
ul. Piotrowo 2, 60-965 Poznań, Poland
{Jerzy.Nawrocki,Miroslaw.Ochodek,Jakub.Jurkiewicz,Sylwia.Kopczynska,
Bartosz.Alchimowicz}@cs.put.poznan.pl

Abstract. Agile methodologies have impact not only on coding, but also
on requirements engineering activities. In the paper agile requirements
engineering is examined from the research point of view. It is claimed
that use cases are a better tool for requirements description than user
stories as they allow zooming through abstraction levels, can be reused
for user manual generation, and when used properly can provide quite
good effort estimates. Moreover, as it follows from recent research, parts
of use cases (namely event descriptions) can be generated in an automatic
way. Also the approach to non-functional requirements can be different.
Our experience shows that they can be elicited very fast and can be quite
stable.

Keywords: Requirements engineering, agility, use cases, non-functional
requirements, effort estimation, user manual.

1 Introduction

Agile methodologies, like XP [6] and Scrum [31], have changed our way of thinking about software development and are getting more and more popular. They
emphasize the importance of four factors: oral communication, orientation towards working software (main products are code and test cases), customer collaboration, and openness to changes.

Agility impacts not only design and coding but also concerns requirements
engineering [7,8]. In this approach classical requirements specification based on
IEEE Std. 830 [1] is replaced with user stories [6,12] and face-to-face communication. User stories can be used for effort estimation and planning. Effort estimation is based rather on personal judgement than on such methods as Function
Points [3] or Use-Case Points [17]. Moreover, in the agile approach, requirements
are not predefined – they emerge while software is being developed [8].

As pointed out by Colin Doyle[1], agile requirements engineering based on user
stories has some advantages and disadvantages . On the one hand, it encourages

---

* Invited Speaker.
[1] C. Doyle, Agile Requirements Management – When User Stories Are Not Enough,
http://www.youtube.com/watch?v=vHNr-amZDsU

V. Geffert et al. (Eds.): SOFSEM 2014, LNCS 8327, pp. 40–51, 2014.

effective communication and better adapts to change. On the other hand, it cannot be applied in a project where a request-for-tenders is used (this requires an up-front specified requirements), the customer is not always available for just-in-time requirements discussions, or the product/project is complex (many conflicting customers, lots of requirements). Another problem is human memory: the Product Owner cannot remember all the requirements and s/he is prone to forgetfulness. In such cases documented requirements would help a lot.

In this paper a research perspective concerning agile requirements engineering is discussed. We focus on the following questions:

- **Q1-UStories**: User stories have an old competitor: use cases invented by Ivar Jacobson in the 80s [15] and later elaborated upon by Alistair Cockburn et al. [2,11]. Are user stories really the best choice?
- **Q2-NFRs**: How can non-functional requirements be elicited to obtain a good balance between speed and quality?
- **Q3-Effort**: One of the key activities in release planning (e.g. performed by playing the Planning Game) is effort estimation. Can automatic effort estimation provide reliable enough estimates?
- **Q4-Manual**: Can the effort concerning auxiliary activities, like writing a user manual, be significantly reduced by using requirements specification?

Those issues are discussed in the subsequent sections of the paper.

2   Written vs. Oral Communication

Agile approaches emphasize oral communication when it comes to requirements elicitation. User stories, which are advocated by XP [6] and Scrum [30], are not supposed to provide complete requirements, they are rather "reminders to have a conversation" [12] with the customer. User Story answers three questions: *what action is possible?*, *who performs this action?* and *what is the business value of this action?* User Stories are short, they are usually expressed in one sentence. Example of a user story is presented in Figure 1A. Similar information is presented in UML Use Case Diagrams [28]. They convey information about actors and their goals (actions). Example of a UML Use Case diagram is presented in Figure 1B. Apparently, User Stories and UML Use Case Diagrams provide roughly the same amount of information about the requirements. Therefore, UML Use Case Diagrams can be used instead of user stories.

User Stories, as reminders for future conversation, are good when a customer is non-stop available for the software delivery team. This is supported by the XP *on-site customer* practice. Unfortunately, on-site customer is rarely available in real life projects, mostly due to high costs. Therefore, more information needs to be captured during the short times when customer is available. While, with user stories only so called *acceptance criteria* can be documented, use-cases allow for more precise capture of requirements. These requirements are documented in a textual main scenario - sequence of steps presenting interaction between actor

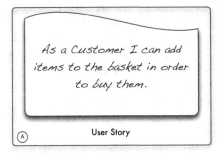
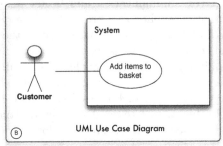

Fig. 1. A - Example of a user story; B - Example of a UML Use Case Diagram

and system. Main scenario should always demonstrate the interaction which leads to obtaining a goal by an actor.

Observation 1. Use cases are more general (flexible) than user stories as they provide an *abstraction ladder* (zooming).

The highest abstraction rung is the level of context diagram (just actors). Below is the level of use case diagrams (such a diagram can be zoomed-out to a context diagram). More details are available at the level of use case scenarios (they can be zoom-out to a context diagram), beneath which are events and alternative activities (they can be easily trunkated to main scenarios). Scenarios can be decorated with low-fidelity screen designs and that way one can generate mockups (see e.g. [22]). Those mockups can be used to elicit test cases from end users, what makes use cases testable.

2.1   HAZOP-Based Identification of Events in Use Cases

To have a complete use case one has to identify all events that can appear when use case steps are executed. A question arises: *how to identify events in use cases in effective and efficient way?*

Events in use cases resemble deviations in mission-critical systems. Therefore, a HAZOP-based method, called H4U [16], for events identification has beed proposed. HAZOP [29] is a method for hazard and deviations analysis in mission-critical systems. This method is based on sets of primary and secondary keywords which are used in brain-storming sessions in order to identify possible hazards. H4U is also built on the idea of keywords which help to identify possible events in use cases. The accuracy and speed of H4U were evaluated and compared to the *ad hoc* approach in two experiments: with 18 students and with 64 IT professionals. Based on the experiments it could be concluded that H4U method offers higher accuracy (maximum average accuracy was equal to 0.26) but lower speed (maximum average speed was equal to 1.04 steps per minute) comparing to the *ad hoc* approach (maximum average accuracy was equal to 0.19, maximum average speed was equal 2.23 steps per minute).

2.2   Automatic Identification of Events in Use Cases

Using H4U one can achieve higher accuracy of events identification, however, higher effort is required. What if events in use cases could be identified in an automatic way? This could reduce the effort and time required to identify events. Moreover, this would mean that Analyst can focus on actors' goals and positive scenarios which require a lot of creative work and the more tedious task of events identification could be done automatically. Results from our initial research show that around 80% of events of the benchmark requirements specification [4] can be identified in an automatic way. Morover, speed of automatic events identification was at the level of 10 steps per minute. This results were achieved with a prototype tool built with a knowledge base from real-life use-cases, inference engine, NLP and NLG tools. Comparing these results to the results from the ealier mentioned experiments with the *ad hoc* and H4U method, it can be concluded that events can be effectively idetified in an automatic way. This allows Analyst to focus on the core of the requirements (use-case names and main scenarios) and have descriptions of events generated automatically.

This research poses a more general research question:

Question 1. To what extend requirements specification can be supported by a computer?

3   Elicitation of Non-functional Requirements

Although user stories are considered by XP [6] and Scrum [30] as the main tool to record requirements, as we stated in Section 3, they may not be sufficient. At first, a user story is not supposed to express a complete requirement. Secondly, it focuses on what actions and activities are performed by a user in the interaction with a system. Such approach may lead to omission of the requirements regarding *how* the functions provided by the system should be executed. According to the results of the investigation of agile projects carried out by Cao and Ramesh [8], customers often "focus on core functionality and ignore NFRs". They also found that in many organizations non-functional requirements(NFRs) are frequently ill defined and ignored in early development stages [8]. It may have severe consequences, as in the cases of Therac-25[19], Arline 5 accident[23] etc., it may lead to excessive refactorings, or cease further system development as its architecture was ill-designed based on insufficient information. However, in agile methodologies, thorough analysis and completeness of requirements (e.g. required by IEEE 830[1]) are no longer a prerequisite to software design and coding - time to market is getting more and more important. Thus, in the context of NFRs, the challenge is how to achieve *"a proper balance between the cost of specifying them and the value of reducing the acceptance risk"* [9]. There exist a number of methods and frameworks which deal with non-functional requirements, e.g., NFR Framework [21], KAOS [32]. However, the existing methods are claimed to be too heavy-weight to be used in agile context [10], and there is little known about cost and value of using them.

To respond to the challenge and fill the gap we proposed a quick method, called SENoR (Structured Elicitation of Non-functional Requirements), dedicated to agile software development. It consists of 3 steps that are presented in Figure 2). The cornerstone of SENoR is *Workshop* which consists of: (1) a presentation of the business case of the project and of the already known functionality of the system, (2) a series of short brainstorming sessions driven by the quality subcharacteristics of ISO25010 [14] - this is the main part of the Workshop, (3) a voting regarding the importance of the elicited requirements.

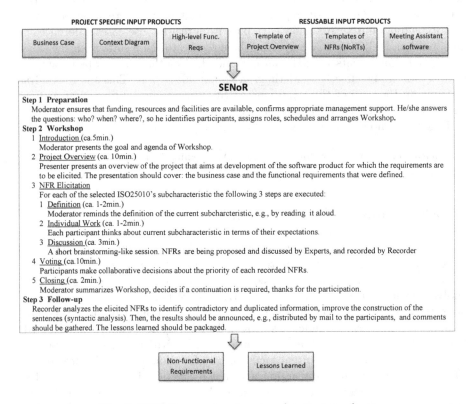

Fig. 2. SENoR – process, input and output products

SENoR workshops are supported with short definitions of quality subcharacteristics and templates of non-functional requirements. Each SENoR workshop lasts no longer than 2 hours.

The data about SENoR workshops have been collected since its first application in 2008 [18]. They have been used to improve the method. Recently,

7 agile projects run at the Poznan University of Technology have been observed. Those projects used SENoR workshops and their aim was to deliver internet applications to be used by the university for administrative purposes.

The average time of a SENoR workshop was ca. 1h 15min, (the shortest took 2854s and the longest — 7554s). 89% (34 out of 38) participants claimed that they are for organizing such workshops in their future projects. On average, 27 NFRs were defined in a workshop, and 92% participants regarded quality of the elicited NFRs good-enough (sufficiently correct and complete) to start architectural design. We also investigated stability of the elicited NFRs, i.e. how many of them 'survived' development changes. What is surprising, average stability of NFRs collected within SENoR workshops was at the level of 80%.

From the point of view of the Q2-NFRs question, the presented results are very promising:

Conjecture 1. A sequence of very short brainstorming sessions driven by quality characteristics can provide quite stable set of non-functional requirements and represents a good balance between cost (time) and quality (stability).

If the above conjecture was true, it would imply that software architecture, which strongly depends on NFRs, can also be quite stable.

4   Automatic Effort Estimation Based on Use Cases

Although project planning is not directly a part of the RE process, they both visibly relate to each other. In order to plan a project or development stage, one has to determine and analyze its scope. The connection between planning and RE is especially visible in the context of agile software development, where constant project planning is often placed next to the core RE practices [8].

Indisputably, in order to plan a project one has to first estimate the effort required to perform all the project's tasks. Depending on the chosen type of budget, an accurate effort estimation becomes more important at different stages of the project. For a fixed scope budget, obtaining an accurate estimate is already extremely important at the early stages of software development, when the crucial decisions about the overall budget are made. If an unrealistic assumption about the development cost is made, the project is in danger. Both underestimated and overestimated effort is harmful. Underestimation leads to a situation where a project's commitments cannot be fulfilled because of a shortage of time and/or funds. Overestimation can result in the rejection of a project proposal, which otherwise would be accepted and would create new opportunities for the organization. In the fixed budget approach the situation is different. The budget could be allocated in advance, but the project team is trying to incrementally understand the customer's needs and deliver as much business value as possible until the available resources are depleted.

When it comes to agile software development methods, they are naturally well suited to fixed budget projects. They assume that changes in requirements are an inherent property of software development projects, thus, it is not reasonable

to invest too much time in the preparation of comprehensive software requirements specification at the beginning of the projects, which can quickly become
obsolete. As a result, project planning in agile software development is more
oriented towards estimating and planning releases than the project as a whole
(e.g, using the planning game, story points, planning poker [6,12,13,20]). Therefore, an important question emerges about what to do in the case of fixed scope
projects being developed in an agile environment or when a customer agrees to
the fixed budget approach, but would also like to know if he/she can afford to
solve the problem.

An answer to the question would be to elicitate high-level requirements at
the project's beginning, and then to estimate its total effort assuming the requirements forming the scope of the project. According to Cao and Ramesh [8]
elicitating such requirements is not such an uncommon practice in agile projects.
Still providing an accurate estimate based on such requirements is a challenge.

One of the methods that could be used for effort estimation based on high-level
requirements is the TTPoints method [26,27]. It is designed to provide functional
size measurement based on use cases, but contrary to other use-case-based effort
estimation methods, such as Use Case Points [17], it is not strictly bounded to
the syntax of use cases. Instead it relies on the semantics of interaction presented
in use-case scenarios.

The main, considered unit of interaction in TTPoints is called semantic transaction. Empirical analysis of use cases has led us to define a catalogue of 12
semantic transactions-types presented in Figure 3. Each transaction type corresponds to the main intent of the interaction between the user and the system
under development. This enables their identification even if the details of a use-
case scenario are still to be determined. An example of a create-type transaction
identified in a use case and user story is presented in Figure 4. If fully-dressed
use cases are available transaction identification is more accurate and could even
be automated using NLP (natural language processing) tools [24,25]. It visibly
reduces the effort required to analyze a use-case-based requirements specification.

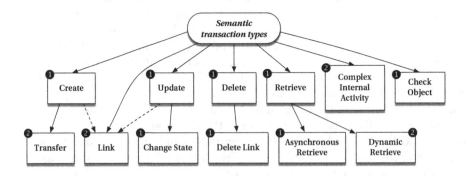

Fig. 3. Semantic transaction-types in use cases with the numbers of core actions

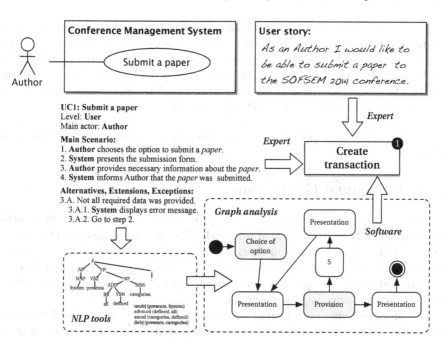

Fig. 4. An example of create semantic-transaction in a use case and user story

The next step is the analysis of each transaction. One has to determine the number of different actors that interact with the system under development and the number of domain objects being processed within a transaction.

Finally, one can calculate the functional size expressed in TTPoints according to the following formula:

$$TTPoints = \sum_{i=1}^{n} Core\_Actions_i \times Objects_i \times Actors_i \qquad (1)$$

where

- n is the number of semantic transactions within the scope of the count;
- $Core\_Actions_i$ is the number of *core actions* of the i-th transaction (see Figure 3);
- $Objects_i$ is the number of meaningful (independent) domain objects processed by the i-th transaction;
- $Actors_i$ is the number of actors in the i-th transaction which cooperate with the system under development.

The TTPoints size measure can be further used to estimate development effort. If historical data is available in organization one can construct a regression model or simply calculate average product delivery rate (PDR). According to our research the average PDR for TTPoints is around 25h/TTPoint.

The obvious drawback of early-effort estimation is its low accuracy due to the high level of uncertainty related to the final scope of the project. Still, the results we obtained for effort estimation based on TTPoints are promising [27]. We were able to estimate effort with on-average error (MMRE) at the level of ~0.26, which was on-average lower by 0.14 to 0.22 than the estimation error for different variants of the Use Case Points method.

Claim 1. Effort estimation based on use cases can provide reasonably good estimates.

5   User Manual Generation

Ongoing research shows that up-to-date project documentation can be beneficial, especially when it facilitates the work of a software development team in an automatic way. This section describes initial research concerning automatic generation of a user manual on the basis of software requirements and other information available in a project. An example of a tool that creates a description of fields used in forms is also presented.

The proposed approach focuses on web applications and static documents (documents that do not allow any interaction with users). It is also assumed that the generated user manual is aimed at IT-laymen, people whose computer knowledge is low.

Results of the conducted research show that on the basis of project documentation a number of descriptions can be generated and the produced elements can be used to create a user manual. Functional requirements in the form of use cases are especially helpful, since this form of representation contains an amount of information which is important from the point of view of end users. For example, it is possible to create a list of functionalities available to end users. Each function can be decorated with a description of actions required to get the desired results and a description of exceptions that may occur. To improve the usefulness of the user manual, the description of available functionality can be enriched by a set of screen shots taken from the running application. To get them in an automatic way one can use acceptance tests. After combining use cases and acceptance tests (such information can be got from tools used for traceability), one can easily obtain images.

The process of enhancement can go further, depending on the available information and the possibility of processing it. An example of additional information provided to end users is the field explanation presented in Figure 5. This figure is generated on the basis of a regular expression used in an application to check whether data provided by a user is a valid number of a credit card or not. Our research shows that regular expressions are sufficient to generate three types of description: an explanation in natural language, a diagram and a set of examples. To present regular expressions in an easy to understand way, a special version of syntax diagrams has been designed [5].

Conjucture 2. Use cases can significantly support generation of a user manual.

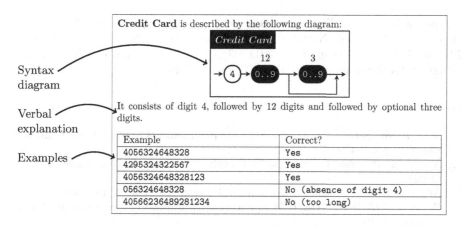

Fig. 5. Field explanation of a credit card number

6   Conclusions

In this paper we have discussed research issues concerning agile requirements engineering. Our findings are as follows:

- Use-cases provide a flexible way of describing functional requirement. They can be presented at several abstraction levels: from extremely concise context diagrams, through use-case diagrams, main scenarios, exceptions, down to the level of alternative steps. Some use-cases can be presented only at the level of use-case diagrams (i.e. lacking further description) while others can also have exceptions and alternative steps specified.
- People are not very good at identifying exceptions (events). Their effectiveness is below 30%. However, from our early experiments it follows that events identification performed by a computer can have effectiveness at the level of 80%. That is a good incentive for further research in this area.
- Non-functional requirements can be elicited in very short brain-storming sessions driven by ISO 25000 quality characteristics. The stability of non-functional requirements elicited in that way for Internet applications was at the level of 80%. It suggests that non-functional requirements are pretty stable and can be collected early.
- The TTPoints method of effort estimation fits use cases well and its average estimation error is below 30%. This is a pretty good result, but further research seems necessary.
- Taking care of requirements specification pays-off, and not only with effort estimates. When functional requirements (in the form of use cases) are complete, then a considerable part of a user manual can be generated. That option should be particularly interesting in Software Product Lines where there are many variants of the same systems and the same user manual.

Acknowledgments. This work has been partially supported by two projects financed by the Polish National Science Center based on the decisions DEC-2011/01/N/ST6/06794 and DEC-2011/03/N/ST6/03016, and by a Poznan University of Technology internal grant 91-518 / 91-549.

References

1. IEEE Recommended Practice for Software Requirements Specifications. IEEE Std 830-1998, pp. 1–40 (1998)
2. Adolph, S., Bramble, P.: Patterns for Effective Use Cases. Addison Wesley, Boston (2002)
3. Albrecht, A.J.: Measuring application development productivity. In: Proceedings of the Joint SHARE/GUIDE/IBM Application Development Symposium, pp. 83–92 (October 1979)
4. Alchimowicz, B., Jurkiewicz, J., Ochodek, M., Nawrocki, J.: Building benchmarks for use cases. Computing and Informatics 29(1), 27–44 (2010)
5. Alchimowicz, B., Nawrocki, J.: Generating syntax diagrams from regular expressions. Foundations of Computing and Decision Sciences 36(2), 81–97 (2011)
6. Beck, K., Andres, C.: Extreme Programming Explained: Embrace Change, 2nd edn. Addison-Wesley Professional (2004)
7. Bjarnason, E., Wnuk, K., Regnell, B.: A case study on benefits and siede-effects of agile practices in large-scale requirements engineering. In: Agile Requirements Engineering Workshop, Agile RE 2011, pp. 9–13. ACM (2011)
8. Cao, L., Ramesh, B.: Agile requirements engineering practices: An empirical study. IEEE Software 25(1), 60–67 (2008)
9. Cleland-Huang, J.: Quality Requirements and their Role in Successful Products. In: IEEE International Requirements Engineering Conf., pp. 361–361. IEEE (October 2007)
10. Cleland-Huang, J., Czauderna, A., Keenan, E.: A persona-based approach for exploring architecturally significant requirements in agile projects. In: Doerr, J., Opdahl, A.L. (eds.) REFSQ 2013. LNCS, vol. 7830, pp. 18–33. Springer, Heidelberg (2013)
11. Cockburn, A.: Writing Effective Use Cases. Addison Wesley, Boston (2000)
12. Cohn, M.: User Stories Applied: For Agile Software Development. Addison Wesley Longman Publishing Co., Inc., Redwood City (2004)
13. Haugen, N.C.: An empirical study of using planning poker for user story estimation. In: Agile Conference, pp. 25–34. IEEE (2006)
14. ISO. ISO/IEC 25010:2011 - Systems and software engineering – Systems and software Quality Requirements and Evaluation (SQuaRE) – System and software quality models. International Organization for Standardization, Geneva, Switzerland (2011)
15. Jacobson, I.: Object-Oriented Software Engineering. Addison-Wesley (1992)
16. Jurkiewicz, J., Nawrocki, J., Ochodek, M., Glowacki, T.: HAZOP-based identification of events in use cases. an empirical study. Empirical Software Engineering (2013) (accepted for publication), doi:10.1007/s10664-013-9277-5
17. Karner, G.: Metrics for objectory. No. LiTH-IDA-Ex-9344:21. Master's thesis, University of Linköping, Sweden (1993)
18. Kopczyńska, S., Maćkowiak, M., Nawrocki, J.: Structured meetings for non-functional requirements elicitation. Foundations of Computing and Decision Sciences 36(1), 35–56 (2011)

19. Leveson, N.G., Turner, C.S.: An investigation of the therac-25 accidents. Computer 26(7), 18–41 (1993)
20. Mahnič, V., Hovelja, T.: On using planning poker for estimating user stories. Journal of Systems and Software 85(9), 2086–2095 (2012)
21. Mylopoulos, J., Chung, L., Nixon, B.: Representing and using nonfunctional requirements: a process-oriented approach. IEEE Transactions on Software Engineering 18(6), 483–497 (1992)
22. Nawrocki, J.R., Olek, Ł.: UC workbench – A tool for writing use cases and generating mockups. In: Baumeister, H., Marchesi, M., Holcombe, M. (eds.) XP 2005. LNCS, vol. 3556, pp. 230–234. Springer, Heidelberg (2005)
23. Nuseibeh, B.: Ariane 5: Who dunnit? IEEE Software 14(3), 15–16 (1997)
24. Ochodek, M., Alchimowicz, B., Jurkiewicz, J., Nawrocki, J.: Improving the reliability of transaction identification in use cases. Information and Software Technology 53(8), 885–897 (2011)
25. Ochodek, M., Nawrocki, J.: Automatic transactions identification in use cases. In: Meyer, B., Nawrocki, J.R., Walter, B. (eds.) CEE-SET 2007. LNCS, vol. 5082, pp. 55–68. Springer, Heidelberg (2008)
26. Ochodek, M., Nawrocki, J.: Enhancing use-case-based effort estimation with transaction types. Foundations of Computing and Decision Sciences 35(2), 91–106 (2010)
27. Ochodek, M., Nawrocki, J., Kwarciak, K.: Simplifying effort estimation based on Use Case Points. Information and Software Technology 53(3), 200–213 (2011)
28. OMG. OMG Unified Modeling Language™(OMG UML), superstructure, version 2.3 (May 2010)
29. Redmill, F., Chudleigh, M., Catmur, J.: System safety: HAZOP and software HAZOP. Wiley (1999)
30. Schwaber, K.: Scrum development process. In: Proceedings of the 10th Annual ACM Conference on Object Oriented Programming Systems, Languages, and Applications, pp. 117–134 (1995)
31. Schwaber, K., Beedle, M.: Agile Software Development with Scrum. Prentice Hall (2001)
32. Van Lamsweerde, A.: Goal-oriented requirements engineering: A guided tour. In: Fifth IEEE International Symposium on Requirements Engineering, pp. 249–262. IEEE (2001)

Fitting Planar Graphs on Planar Maps

Md. Jawaherul Alam[1], Michael Kaufmann[2],
Stephen G. Kobourov[1], and Tamara Mchedlidze[3]

[1] Department of Computer Science, University of Arizona, USA
[2] Institute for Informatics, University of Tübingen, Germany
[3] Institute for Theoretical Informatics, Karlsruhe Institute of Technology, Germany

Abstract. Graph and cartographic visualization have the common objective to provide intuitive understanding of some underlying data. We consider a problem that combines aspects of both by studying the problem of fitting planar graphs on planar maps. After providing an NP-hardness result for the general decision problem, we identify sufficient conditions so that a fit is possible on a map with rectangular regions. We generalize our techniques to non-convex rectilinear polygons, where we also address the problem of efficient distribution of the vertices inside the map regions.

1 Introduction

Visualizing geographic maps may require showing relational information between entities within and between the map regions. We study the problem of fitting such relational data on a given map. In particular, we consider the problem of *fitting planar graphs on planar maps*, subject to natural requirements, such as avoiding edge crossings and ensuring that edges between points in the same region remain in that region.

Fitting planar graphs on planar maps is related to cluster planarity [2,3,12]. In cluster-planar drawing we are given the graph along with a clustering and the goal is to find disjoint regions in the plane for the clusters for a valid plane realization of the given graph. The realization is *valid* if all the vertices in a given cluster are placed in their corresponding region, and there are no edge-crossings or edge-region crossings (i.e., edges crossing a region in the map more than once).

In our setting (fitting graphs on maps), we are given both the graph and the regions embedded in the plane, and must draw the clusters in their corresponding regions. The regions form a proper partition of the plane, such that the adjacency between two clusters is represented by a common border between their corresponding regions.

1.1 Related Work

The concept of clustering involves the notion of grouping objects based on the similarity between pairs of objects. In graph theory, this notion is captured by a clustered graph. Clustering of graphs is used in information visualization [17], VLSI design [15], knowledge representation [18], and many other areas.

Feng *et al.* defined c-planarity as planarity for clustered graphs [13]; also see Section 2 for related definitions. For clustered graphs in which every cluster induces a connected subgraph, c-planarity can be tested in quadratic time. Without the connectivity

V. Geffert et al. (Eds.): SOFSEM 2014, LNCS 8327, pp. 52–64, 2014.

condition, the complexity of testing c-planarity is still an open problem. Algorithms for creating regions in the plane in which to draw c-planar graphs have also been studied. Eades *et al.* [10] presented an algorithm for constructing c-planar straight-line drawings of c-planar clustered graphs in which each cluster is drawn as a convex region, while Angelini *et al.* [1] show that every such c-planar clustered graph has a c-planar straight-line drawing where each cluster is drawn inside an axis-aligned rectangle.

Many visualizations take advantage of our familiarity with maps by producing map-like representations that show relations among abstract concepts. For example, treemaps [24], squarified treemaps [5] and news maps represent hierarchical information by means of space-filling tilings, allocating area in proportion to some metric. Concept maps [8] are diagrams showing relationships among concepts. Somewhat similar are cognitive maps and mind-maps that represent words or ideas linked to and arranged around a central keyword. GMap [17] uses the geographic map metaphor to visualize relational data by combining graph layout and graph clustering, together with the creation and coloring of regions/countries.

Also related is work on contact graphs, where vertices are represented by simple interior-disjoint polygons and adjacencies are represented by a shared boundary between the corresponding polygons. For example, every maximally planar graph has a contact representation with convex polygons with at most six sides, and six sides are also necessary [9]. Of particular interest are *rectilinear duals*, where the vertices are represented by simple (axis-aligned) rectilinear polygons. It is known that 8 sides are sometimes necessary and always sufficient [16,22,26]. If the rectilinear polygons are restricted to rectangles, the class of planar graphs that allows such *rectangular duals* is completely characterized [21,25] and can be obtained via bipolar orientation of the graph [14]; see Buchsbaum *et al.* [6] and Felsner [11] for excellent surveys.

1.2 Our Contributions

We first consider the question of testing whether a given planar clustered graph fits on a given planar map and show that the decision problem is NP-hard, even in the case where the map is made of only rectangular regions and each region contains only one vertex. Then we provide sufficient conditions that ensure such a fit on a rectangular map. Finally, we generalize the fitting techniques to rectilinear maps with rectangles, L-shaped and T-shaped polygons. In particular, we describe an efficient algorithm for distributing vertices appropriately in the case of maps with L-shaped polygons.

2 Preliminaries

In this section we introduce definitions used throughout the paper and then describe the properties of clustered graphs considered in the paper.

Let $G = (V, E)$ be a planar graph, with vertex set V partitioned into disjoint sets $\mathcal{V} = \{V_1, \ldots, V_k\}$. We call the pair $C = (G, \mathcal{V})$ a *planar clustered graph*. We consider the following partition of the edges of G that corresponds to the given partition of vertices $\mathcal{V} = \{V_1, \ldots, V_k\}$. Let E_i, for each i, $1 \leq i \leq k$ be the set of edges in E between two vertices of V_i and let E_{inter} be the set of all the remaining edges in E.

Note that $E = E_1 \cup E_2 \cup \ldots \cup E_k \cup E_{inter}$. We call $G_i = (V_i, E_i)$, $1 \leq i \leq k$, a *cluster* of G, the edges of E_i, $1 \leq i \leq k$, the *intra-cluster edges* and the edges of E_{inter} the *inter-cluster edges*.

The *cluster-graph* of a clustered graph $C = (G, \mathcal{V})$ is the graph $G_C = (\mathcal{V}, \mathcal{E})$, where the edge $(V_i, V_j) \in \mathcal{E}$, $1 \leq i, j \leq k$, $i \neq j$ if there exists an edge (u, w) in G so that $u \in V_i$ and $w \in V_j$. A clustered graph $C = (G, \mathcal{V})$ is said to be *connected* (resp. *biconnected*) if each of G_i, $1 \leq i \leq k$, is a connected (resp. *biconnected*) graph.

A *drawing of a planar clustered graph* $C = (G, \mathcal{V})$ is a planar straight-line drawing of G where each cluster G_i is represented by a simply-connected closed region R_i such that R_i contains only the vertices of G_i and the drawing of each edge e between two vertices of G_i is completely contained in R_i. An edge e and a region R have an *edge-region crossing* if the drawing of e crosses the boundary of R more than once. A drawing of a planar clustered graph C is *c-planar* if there is no edge crossing or edge-region crossing. If C has a c-planar drawing then we say that it is *c-planar*.

A *polygonal map* M is a set of interior-disjoint polygons on a plane. A *dual graph* G_M of M is a graph that contains one vertex for each polygon of M. Two vertices of G_M are connected by an edge if the corresponding polygons have a non-trivial common boundary. Given a planar graph G_M, a polygonal map M is called a *contact map* of G_M if G_M represents the dual graph of M. Let $C = (G, \mathcal{V})$ be a planar clustered graph. A polygonal map M which represents a contact map of the cluster-graph G_C is said to be *compatible* with C. Notice that this definition yields a correspondence between the clusters of C and polygons of M. In this paper we are interested in determining whether each cluster G_i of C can be drawn with straight-line edges inside its corresponding polygon in M, so that there is no edge crossing and no edge-region crossing. In case such a drawing exists we say that planar clustered graph C has a *straight-line planar fitting*, or just *planar fitting* on map M.

It is natural to consider all planar graphs, regardless of the clustering they come with. We preview the construction of a straight-line planar fitting and isolate the problem we are interested in. Recall that, by the definition of a planar fitting, each cluster has to be drawn inside a polygon, and there should be no edge crossings and no edge-region crossings. This implies that a clustered graph that has a planar fitting is also c-planar, so we consider only c-planar graphs. Unfortunately, the characterization of c-planar graphs is still an open problem. Thus we restrict ourselves to clustered graphs for which we know that c-planarity can be efficiently tested. We use the results of Feng et al. [13] who provide a polynomial-time algorithm to test whether a *connected* clustered graph is c-planar. Thus, in the rest of the paper we consider only *connected c-planar graphs*.

3   Fitting on a Rectangular Map

Here we consider the problem of deciding whether a connected c-planar graph G has a straight-line planar fitting on a given compatible rectangular map M. We first show that such a fitting does not always exist. To construct the counterexample we use a *wheel map*, which contains four "thin rectangles" that surrounds an inner rectangle; see Fig. 1.

Intuitively, the notion of a thin rectangle will be clear in the following constructions from the way it is used, but to be more precise, we formally define it. A *thin rectangle* is one whose larger side is at least 4 times its smaller side, i.e., it has aspect ratio at least 4. A thin rectangle is *horizontal* if its smaller side is its height; otherwise it is *vertical*. We assume all four thin rectangles in a wheel map have the same size (same length of larger sides, same length of smaller sides).

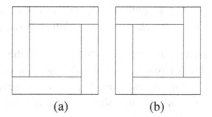

Fig. 1. Wheel maps cw (a), ccw (b)

Let $\{V_1, \ldots, V_k\}$ be the set of clusters of G and let (v_i, v_j) be an edge of G such that $v_i \in V_i$, $v_j \in V_j$, $1 \leq i, j \leq k$. Then there exists a common boundary between the polygons representing V_i and V_j in M. Call the common boundary the *door* for the edge (v_i, v_j). Consider a wheel map W and its dual G which has a simple clustering: each vertex constitutes a cluster. For the rest of the section we often assume that a wheel map is associated with this clustered graph. With this consideration in mind, each thin rectangle of W contains two doors, one for each incident thin rectangle. We define the *entry door* to be the one which contains a complete side of the rectangle, and the *exit door* to be the one that contains a complete side of a neighboring thin rectangle. We call a wheel map a *clockwise (cw) wheel* when going from the entry door to the exit door in each rectangle requires a clockwise walk through the wheel; see Fig. 1(a). A *counterclockwise (ccw) wheel* is defined analogously; see Fig. 1(b).

We now define the notion of "proximity region" for each door inside each thin rectangle of a wheel map. Given that the thin rectangles of a wheel map are of the same size, using basic geometry we can define inside each thin rectangle a triangular region of maximum area where these two conditions hold: (i) the triangular region inside each thin rectangle contains the entry door of this rectangle; and (ii) for each point inside the triangle of one of the four thin rectangles, there exists a point inside the triangle of each other rectangle such that the visibility line between the two points inside any pair of neighboring rectangles passes through the corresponding door. We call these triangular regions the *proximity regions* of the corresponding entry doors; see Fig. 2(a). For each exit door we can analogously define a quadrangular *proximity region*; see Fig 2(b). Next we state a simple observation that follows from these definitions:

Observation 1. *Let W be a wheel map and let G be its dual graph. In a straight-line planar fitting of G the vertices in the thin rectangles either all lie inside the proximity*

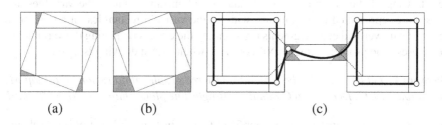

(a)   (b)   (c)

Fig. 2. (a)–(b) Proximity regions for the doors in a wheel map, (c) a clustered graph and a map with no fit (the visibility regions of the bridge are highlighted)

*region of the entry doors, or they all lie in the proximity region of the exit doors. There
exists a straight-line planar fitting in each case.*

Proof: The sufficiency follows from the definition of proximity regions. The necessity
follows from the fact that the proximity regions for the entry and the exit doors inside
each thin rectangle are disjoint since the aspect ratio of the thin rectangles is ≥ 4.   □

The next lemma shows that fitting a planar clustered graph on a compatible map is
not always possible.

Lemma 1. *There exist a planar clustered graph $C = (G, \mathcal{V})$ and a compatible rectangular map M, so that there is no straight-line planar fitting of C on M.*

Proof: Consider a rectangular map M made of two wheel maps (of the same size)
joined together by a thin horizontal rectangle, called a *bridge*; see Fig. 2(c). We choose
the height of the bridge to be at most the length of the smaller sides in the thin rectangles of the wheels and we attach the bridge in the middle of the neighboring two vertical
rectangles. For each of these two thin rectangles, the *visibility region* of the bridge is the
set of points in it from which the visibility line to at least one point in the proximity region of either of the entry or exit door of the rectangle passes through the door between
the bridge and the rectangle. We choose the length of the bridge to be long enough, such
that the two visibility regions for the two rectangles do not overlap.

Let G be the dual of M: two 4-cycles connected by a path of length two. Assume a
clustering of G where each vertex constitutes a separate cluster. Then G has no straight-
line planar fitting on M. If G had a straight-line planar fitting Γ on M, by Observation 1,
all the vertices inside the thin rectangles of both the wheels must be placed in the proximity regions of the doors in Γ. But then, there is no feasible position for the vertex that
represents the bridge since the two visibility regions of the bridge do not overlap.   □

3.1   Fitting Is NP-Hard

We show that deciding if a given planar clustered graph has a planar fitting inside a
given map is NP-hard, even for rectangular maps, with a reduction from Planar-3-SAT
which is known to be NP-complete [23]. *Planar-3-SAT* is defined analogously to 3-
SAT with the additional restriction that the *variable-clause bipartite graph* G_F for a
given formula F is planar. There is an edge (x_i, A_j) in G_F if and only if x_i or $\overline{x_i}$
appears in A_j. Knuth and Raghunathan [20] showed that one can always find a crossing-
free drawing of the graph G_F for a Planar-3-SAT instance, where the variables and
clauses are represented by rectangles, with all the variable-rectangles on a horizontal
line, and with vertical edge segments representing the edges connecting the variables to
the clauses. The problem remains NP-complete when such a drawing is given.

Theorem 1. *Let $C = (G, \mathcal{V})$ be a planar clustered graph with a rectangular map M,
compatible with C. Deciding if C admits a straight-line planar fitting on M is NP-hard.*

Proof: We reduce an instance of Planar-3-SAT to an instance (C, M) of our problem.
Let $F := A_1 \wedge \ldots \wedge A_m$ be an instance of a Planar-3-SAT, where each literal in each
clause A_i is a variable (possibly negated) from $U = \{x_1, \ldots, x_n\}$. Let Γ_F be the

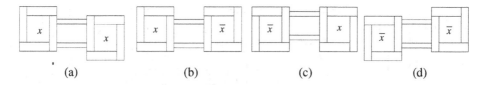

Fig. 3. Representation of variables

given planar rectilinear drawing for this instance, as defined in [20]. From Γ_F we first construct the rectangular map M, then take G as the dual of M, where each vertex constitutes a separate cluster. We represent each literal by a wheel map (of the same size) in M: a positive (negative) literal is a cw (ccw) wheel. From the two possible vertex configurations inside each wheel we take the one in which the corresponding literal assumes a true value when the vertices inside the thin rectangles of the wheel lie in the proximity region of the exit doors and the literal assumes a false value when they lie in the proximity region of the entry doors. Unlike in Γ_F, we use a distinct wheel for each literal in each clause. For each variable x, we draw the wheels for all the (positive and negative) literals for x appearing in different clauses in a left-to-right order, according to the ordering of the edges incident to the corresponding vertices in Γ_F. To maintain consistency, we ensure that a true (false) value for a literal x implies a true (false) value for every other instance of x and a false (true) value for each instance of $\overline{x}$. This is done by means of thin rectangular bridges between two consecutive literals; see Fig. 3. The size of the bridges is chosen equal to the thin rectangles in the wheels.

For each clause $A = (x \vee y \vee z)$ of F, with the corresponding vertex lying above the variables in Γ_F, we draw vertical rectangles l_x^A, l_y^A and l_z^A from the topmost rectangles T_x, T_y, T_z of the wheels for x, y and z, respectively, attached at the end that is not shared by other thin rectangles. (The case when the vertex lies below the variables in Γ_F is analogous.) We place l_x^A, l_y^A, l_z^A so that they are completely visible from all the points in the proximity of the exit doors of T_x, T_y, T_z, respectively. We choose the length of the rectangles l_x^A, l_y^A, l_z^A so that not all points inside them are visible from any point of the proximity region of the entry doors of T_x, T_y, T_z, respectively; see Fig. 5. We then draw a rectangle R for the clause and attach these three thin rectangles l_x^A, l_y^A and l_z^A to R. For z we attach the vertical rectangle l_z^A to the bottom of R, while for each of x and y, we attach horizontal rectangles h_x^A, h_y^A to R that also touch the vertical rectangles l_x^A and l_y^A coming from x or y, respectively. A point p in rectangle h_x^A (h_y^A) is *reachable* from a point q inside T_x (T_y) if there exists a point r inside l_x^A or l_y^A such that the two lines pr and rq pass through the doors between the corresponding rectangles. The *reachable region* of h_x^A (h_y^A) is the set of all points that are reachable from a point inside the proximity region of the entry

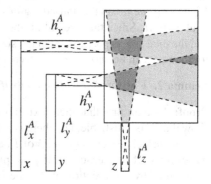

Fig. 4. Clause representation

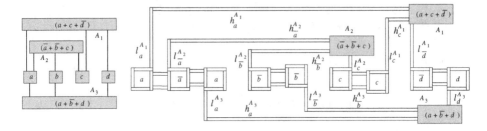

Fig. 5. Planar 3-SAT instance and corresponding map fitting instance

door of T_x (T_y). Similarly a point p inside l_z^A is
reachable from a point q inside rectangle T_z of
the wheel for z if the line pq passes through the door between the two rectangles. The
reachable region of l_z^A is the set of all points reachable from a point inside the proximity region of the entry door of T_z. Choose the lengths for the horizontal rectangles
h_x^A, h_y^A and the vertical rectangles l_x^A, l_y^A, l_z^A so that the reachable regions inside them
do not coincide with the entire inside of these rectangles. For this purpose it is sufficient
that the sizes of these rectangles are comparable to the sizes of the rectangles inside the
wheels. Thus the sizes of all the wheels and other rectangles are polynomial in the size
of the Planar-3-SAT instance.

Next we attach the thin rectangles h_x^A, h_y^A, l_z^A to R in such a way that the areas
visible from the reachable regions of h_x^A, h_y^A, l_z^A do not have a common intersection,
while every pair of them does have a common intersection; see Fig 4. We now observe
that the size of R does not have to be too big to ensure this. Specifically, we attach l_z^A
to the left half of the bottom line of R and choose the height of R small enough so that
the visible area from its reachable region is only in the left half of R. We also adjust
the vertical distance between the horizontal rectangles h_x^A and h_y^A and adjust the width
of R so that the areas visible from the reachable regions of h_x^A and h_y^A do not intersect
in the left half of R but they do intersect in the right half. This can be achieved if for
example we take the vertical distance between h_x^A and h_y^A to be a constant multiple of
their height, while we take the width of R to be a constant multiple of the length of h_x^A,
h_y^A. Finally we fill all the unused regions in the map with additional rectangles to get the
final map M. Since the sizes of all the rectangles are constant multiples of each other
and the total size is polynomial in the size of the Planar-3-SAT instance, the coordinates
for the map can be chosen to be polynomial in the size of the Planar-3-SAT instance.

Lemma 2. *F is satisfiable if and only if G has a straight-line planar fitting on M.*

Proof: Assume first that there exists a straight-line planar fitting Γ of G on M. We
show that F is satisfiable, i.e., there is a truth assignment for all the variables of F such
that for each clause $A = (x, y, z)$ of F, at least one of x, y and z is true. Let W_x be the
wheel for x. If the vertices in W_x are placed inside the proximity regions of the entry
doors, then by construction of M, the vertex in the horizontal rectangle h_x^A is placed
inside the reachable region. Thus this vertex can see only the highlighted visible area
in Fig. 4 inside the rectangle R for A. However if the vertices in W_x are placed in the

proximity regions of the exit doors, then the vertex in the horizontal rectangle h_x^A can be placed outside the reachable region so that it can see the entire interior of R. This is true for each of the three literals. Since the visible areas of the three literals have no common intersection, the vertices in the wheel for at least one of x, y and z must be placed in the proximity region of the exit door. We make each such literal true. This assignment has no conflict, because of the way the wheels for a particular variable are attached to each other. Furthermore, this assignment satisfies F.

Conversely if F is satisfiable, for each clause $A = (x, y, z)$ of F, at least one of x, y, z is true. Without loss of generality, assume that x is true. Place the vertices in the wheel of x in the proximity regions of the exit doors. Then the vertex in the h_x^A can be placed outside the reachable region and it can see the entire interior of R. Place the vertex for R in the intersection of the areas visible from reachable regions of h_y^A and l_z^A. This ensures that we can place the vertices in the wheel for y and z in the proximity regions of either the entry doors or the exit doors and we are still able to place the vertices in rectangles h_y^A, l_y^A, l_z^A so that all the straight-line edges create no area-region crossings. This yields the desired straight-line planar fitting of G on M.   □

The proof of Lemma 2 completes the NP-hardness proof. Fig. 5 illustrates a 3-SAT formula, its Planar-3-SAT realization with the conditions of [20], and the corresponding instance for the map fitting problem (rectangles filling up the holes are not shown).   □

Note that Bern and Gilbert [4] and recently Kerber [19] obtained NP-completeness results using similar techniques. In particular, Bern and Gilbert [4] consider the problem of drawing the dual on top of the drawing of a plane graph G, such that each dual vertex lies in the corresponding face of G, while each dual edge is drawn as a straight-line segment that crosses only its corresponding primal edge. They show that this problem is NP-complete and the techniques used are similar to ours, as this problem can be thought of as a special case of fitting a clustered graph on a map, where each cluster consists of a single vertex. However, we consider the more restricted class of rectangular maps instead of the generic drawing of a planar graph, and hence the NP-completeness of our problem is not implied by [4]. Kerber [19] considers the problem of embedding the dual on top of a primal partition of the d-dimensional cube into axis-aligned simplices and proved that this problem is NP-complete. In 2D, this problem is also a special case of our problem, with the exception that in Kerber's setting edge-region crossings are allowed. Thus the result in [19] also does not imply our results.

4   Sufficient Conditions for Fitting

We showed in Lemma 1 that not every c-planar connected graph admits a planar straight-line fitting on a compatible map even if each cluster is a single vertex. The counterexample relies on two facts: (1) there is a vertex in some cluster (the bridge) that is connected to vertices in two different clusters (the wheels); (2) its cluster-graph contains two cycles. By considering graphs that do not have at least one of the above characteristics we show planar straight-line fittings are always possible. In this sense the following two lemmas give tight sufficient conditions for graphs to admit planar straight-line fittings.

Lemma 3. *Let $C = (G, \mathcal{V})$ be a biconnected c-planar graph. Let M be a rectangular map compatible with C. If for each vertex v of G, all the vertices adjacent to v through an inter-cluster edge lie in the same cluster, C has a straight-line planar fitting on M.*

Proof: Let Γ be a c-planar drawing of C. Let $G_1, G_2, \ldots, G_k$ be the clusters of C and let $\mathcal{V} = \{V_1, V_2, \ldots, V_k\}$ be the corresponding vertex partition. For each rectangle R_i, $1 \le i \le k$, of M representing the cluster G_i, denote by O_i the ellipse inscribed in R_i. We first place the vertices on the outer boundary of G_i in Γ on O_i as follows. Consider two adjacent rectangles R_i and R_j in M. Let $v_{i_1}, \ldots, v_{i_r} \in V_i$ and $v_{j_1}, \ldots, v_{j_s} \in V_j$ be the vertices of V_i and V_j, incident to the inter-cluster edges between these two clusters, taken in the order they appear on the outer boundary of G_i and G_j, respectively. Define p_i, p_i' and p_j, p_j' to be points of O_i and O_j, respectively, such that the straight-line segments $\overline{p_i p_j}$ and $\overline{p_i' p_j'}$ cross the common border of R_i and R_j, without crossing each other. Place the vertices $v_{i_1}, \ldots, v_{i_r}$ of V_i and $v_{j_1}, \ldots, v_{j_s}$ of V_j on O_i and O_j, between points p_i, p_i' and p_j, p_j', respectively, so that all the inter-cluster edges between these vertices cross the common border of R_i and R_j. Repeat the above procedure for each pair of adjacent rectangles in M. Since each vertex thus placed is adjacent to a unique cluster, its position is uniquely defined. For each cluster G_i, $1 \le i \le k$, we have thus placed some vertices on the outer boundary of G_i in Γ on the ellipse O_i. Distribute the remaining vertices of the boundary of G_i on O_i, so that the order of the vertices is the same as in the boundary of G_i. Since the resulting drawing of the outer boundary of G_i is convex and G_i is biconnected, apply the algorithm for drawing a graph with a prescribed convex outer face [7] to complete the drawing of each cluster. □

Lemma 4. *Let $C = (G, \mathcal{V})$ be a biconnected c-planar graph. Let M be a rectangular map compatible with C. If each connected component of cluster-graph G_C contains at most one cycle, then C has a straight-line planar fitting on M.*

Proof Sketch: Assume that each connected component of G_C contains at most one cycle. Let $v_1, \ldots, v_k$ be the vertices of G_C that represent clusters $G_1, \ldots, G_k$ respectively. To complete the proof we go through the following steps:

(1) We show that G_C has a planar fitting on M.
(2) We blow up the drawing of G_C, so that the edges of G_C are represented by strips of width $\varepsilon > 0$, without creating edge-region crossings; see Fig. 6(a). For each vertex v_i of G_C, we draw a small circle $circ(G_i)$ centered at the intersection of the strip-edges that are adjacent to v_i.
(3) We draw the boundary of G_i on the circle $circ(G_i)$, $i = 1, \ldots, k$, so that the inter-cluster edges, when drawn with straight-line segments, intersect neither the boundaries of the clusters, nor each other; see Fig. 6(b).
(4) Since the boundary of each G_i is a convex polygon and G_i is biconnected, we can apply the algorithm for drawing a graph with a prescribed convex outer face [7] to complete the drawing of the clusters; see Fig. 6(c).

While steps (2) and (4) are straight-forward, steps (1) and (3) need to be proved. We provide a detailed proof of step (1) below. Step (3) is intuitively easy, however, its exact proof is quite technical; we omit the proof for this step here due to the space constraints.

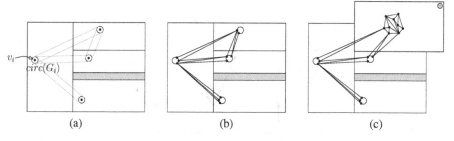

Fig. 6. (a) Drawing of G_C, each edge is represented by a strip of width $\varepsilon > 0$. (b) Placing the boundary vertices of the clusters on the corresponding circles. (c) Step 4 of the proof of Lemma 4.

Step (1). We show that $G_C = (V_C, E_C)$ has a straight-line planar fitting on M. Consider first the case when G_C is a tree and let $v_1 \in V_C$ be the root of G_C. We prove that even if the position of v_1 is fixed in its corresponding rectangle R_1, we can place the remaining vertices of G_C in their corresponding rectangles so that the resulting straight-line drawing is a planar fitting of G_C on M. Let $v_2, \ldots, v_f$ be the children of v_1 and let $R_2, \ldots, R_f$ be the corresponding rectangles of M. Place $v_2, \ldots, v_f$ inside $R_2, \ldots, R_f$, respectively so that the straight-line edges (v_1, v_i), $2 \le i \le f$ cross the common boundary of R_1 and R_i. Continue with the children of $v_2, \ldots, v_f$, recursively.

Assume now that each connected component of $G_C = (V_C, E_C)$ contains at most one cycle. We show how to draw a single connected component of G_C. Let $v_0, \ldots, v_m \in V_C$ induce the unique cycle of G_C and let $R_0, \ldots, R_m$ be the rectangles that correspond to them, so that R_i and $R_{(i+1) \bmod (m+1)}$, $0 \le i \le m$, are adjacent. Place v_i, $0 \le i \le m$ inside R_i such that for any point $p \in R_{(i+1) \bmod (m+1)}$, segment $\overline{pv_i}$ crosses the common boundary of R_i and $R_{(i+1) \bmod (m+1)}$. (For example placing v_i right on the door suffices.) The removal of the edges of this cycle results in several trees. Root the trees at the vertices $v_0, \ldots, v_m$, to which they are adjacent and apply the procedure described in the first part of the proof. This completes the construction. $\qquad\square$

Putting together the results in this section we obtain the following theorem:

Theorem 2. *Let $C = (G, \mathcal{V})$ be a biconnected c-planar graph. Let M be a rectangular map compatible with C. If (a) for each vertex v of G, all the vertices adjacent to v through an inter-cluster edge lie on the same cluster, or (b) each connected component of cluster-graph G_C contains at most one cycle, then C has a straight-line planar fitting on M. Moreover, there exist a c-planar graph C and a compatible map M which do not fulfill condition (a) and (b) and do not admit a planar straight-line fitting.*

5 Fitting Graphs on Rectilinear Maps

In this section we give a short overview of our results for more general rectilinear maps.

It is known that only a restricted class of planar graphs can be realized by rectangular maps. For general planar graphs, 8-sided polygons (T-shapes) are necessary and sufficient for contact maps [16]. In this section, we assume that the input is a rectilinear map, with rectangles, L- and T-shapes, together with a c-planar graph G with a cluster-graph

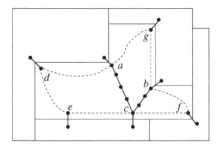

Fig. 7. Illustrating how vertices on the boundary of a T-shape are placed and how the cluster is partitioned by an $(a \rightsquigarrow c)$-path and a $(b \rightsquigarrow c)$-path, such that it fits into three adjacent convex polygons with the paths as common boundaries

G_C. The first condition that we require is that the subgraph induced by the inter-cluster edges is a matching. From Lemma 3 this condition is sufficient for rectangular maps. Now, we extend this to L-shaped and T-shaped polygons (maps). We impose several conditions under which we prove the existence of a fitting. Because of the presence of concave corners, we impose our second condition: none of the clusters contains a *boundary chord*, i.e., a non-boundary edge between two boundary vertices.

The idea is to apply the algorithm for drawing a graph with a prescribed convex outer face [7]. We partition the polygons into convex pieces. Since the polygons form a contact map, for each common boundary of adjacent polygons there is at least one edge between the corresponding clusters. Our last condition restricts this further: between any two adjacent clusters there exist at least two independent inter-cluster edges. We call a graph in which every pair of adjacent clusters has this property *doubly-interconnected*.

Note that the common boundary of two adjacent polygons contains at most two concave corners, one in each polygon. We place the vertices next to the common boundary on both sides of these concave corners. This ensures that the cycle spanned by the boundary vertices of the cluster is completely within the corresponding polygon and there are at most two concave corners along the cycle. Let a and b be the vertices at these corners; see Fig. 7. We choose a third boundary vertex c lying opposite a and b. Straight-line cuts between a, c and b, c define 3 convex regions. Now we compute an $(a \rightsquigarrow c)$-path and a $(b \rightsquigarrow c)$-path. These paths cannot have shortcuts, where a *shortcut* of a path P is an edge between vertices nonadjacent in P, so that we can place the vertices on these two paths on the two cuts between a, c and b, c. Note that such a path should not contain any other boundary vertex, already placed elsewhere.

The above strategy for T-shaped polygons can also be applied to L-shaped polygons, where the straight-line segment splitting this polygon into two convex parts is between the concave corner of the L-shaped polygon and its opposite convex corner of the neighboring polygon. Together, these yield the following theorem.

Theorem 3. *Let G be a doubly-interconnected and biconnected c-planar graph such that the inter-cluster edges of G form a matching and there is no boundary chord in any cluster. Then there exists a straight-line planar fitting of G on any compatible map with rectangular, L-shaped, or T-shaped polygons.*

Note that any algorithm that distributes vertices inside non-convex regions while preserving cluster-planarity must distribute the vertices among the convex components of the regions. It is only natural to try to make such distribution balanced. We define a measure called "imbalance" which captures the difference between the geometric partition of the non-convex regions into convex regions (e.g., region area) and the partition of the clusters into subclusters (e.g., subcluster size) corresponding to the convex regions.

First we consider the distribution inside one L-shaped polygon partitioned into two pieces by a straight-line cut; then we use this result to minimize the maximum imbalance in all L-shaped polygons in the map. We prove that the global imbalance minimization problem can be solved optimally, using dynamic programming and min-max shortest paths. These techniques are interesting by themselves.

Theorem 4. *Let G be a connected c-planar graph, G_C be the cluster-graph of G and let M be a rectilinear map of G_C with six-sided polygons such that M represents the contact map of G_C. Then one can split the regions of M in $O(n^4)$ time into convex shapes and distribute the vertices and faces of the clusters within the regions, such that the maximum imbalance is minimized.*

6 Conclusion and Future Work

We showed that fitting planar graphs on planar maps is NP-hard. The proof involves skinny regions; it is natural to ask whether the problem becomes easier if all regions are "fat". We presented necessary and sufficient conditions for the construction of planar straight-line fitting on rectangular map, for c-planar graphs with biconnected clusters. These conditions are tight, in the sense that violating them makes it possible to construct counterexamples. Relaxing the biconnectivity requirement is an open problem. Finally, we gave a rather restricted set of sufficient conditions for fitting planar graphs on maps with non-convex regions. It would be worthwhile to investigate whether these conditions can be relaxed. We gave an algorithm for finding a fitting with a "balanced distribution" of the vertices. Another interesting question is whether an exact bound on vertex resolution can be guaranteed.

References

1. Angelini, P., Frati, F., Kaufmann, M.: Straight-line rectangular drawings of clustered graphs. In: Symposium on Algorithms & Data Structures (WADS), pp. 25–36 (2009)
2. Battista, G.D., Drovandi, G., Frati, F.: How to draw a clustered tree. Journal of Discrete Algorithms 7(4), 479–499 (2009)
3. Battista, G.D., Frati, F.: Efficient c-planarity testing for embedded flat clustered graphs with small faces. Journal of Graph Algorithms and Applications 13(3), 349–378 (2009)
4. Bern, M.W., Gilbert, J.R.: Drawing the planar dual. Information Processing Letters 43(1), 7–13 (1992)
5. Bruls, M., Huizing, K., van Wijk, J.J.: Squarified treemaps. In: Joint Eurographics and IEEE TCVG Symposium on Visualization, pp. 33–42 (2000)
6. Buchsbaum, A., Gansner, E., Procopiuc, C., Venkatasubramanian, S.: Rectangular layouts and contact graphs. ACM Transactions on Algorithms 4(1) (2008)

7. Chambers, E., Eppstein, D., Goodrich, M., Löffler, M.: Drawing graphs in the plane with a prescribed outer face and polynomial area. Journal of Graph Algorithms and Applications 16(2), 243–259 (2012)

8. Coffey, J.W., Hoffman, R.R., Cañas, A.J.: Concept map-based knowledge modeling: perspectives from information and knowledge visualization. Information Visualization 5(3), 192–201 (2006)

9. Duncan, C., Gansner, E., Hu, Y., Kaufmann, M., Kobourov, S.G.: Optimal polygonal representation of planar graphs. Algorithmica 63(3), 672–691 (2012)

10. Eades, P., Feng, Q.-W., Lin, X., Nagamochi, H.: Straight-line drawing algorithms for hierarchical graphs and clustered graphs. Algorithmica 44(1), 1–32 (2006)

11. Felsner, S.: Rectangle and square representations of planar graphs. In: Pach, J. (ed.) Thirty Essays on Geometric Graph Theory. Springer (2012)

12. Feng, Q.-W., Cohen, R.F., Eades, P.: How to draw a planar clustered graph. In: Li, M., Du, D.-Z. (eds.) COCOON 1995. LNCS, vol. 959, pp. 21–30. Springer, Heidelberg (1995)

13. Feng, Q.-W., Cohen, R.F., Eades, P.: Planarity for clustered graphs. In: European Symposium on Algorithms (ESA), pp. 213–226 (1995)

14. Fusy, É.: Transversal structures on triangulations: A combinatorial study and straight-line drawings. Discrete Mathematics 309(7), 1870–1894 (2009)

15. Harel, D.: On visual formalisms. Communications of the ACM 31(5), 514–530 (1988)

16. He, X.: On floor-plan of plane graphs. SIAM Journal of Computing 28(6), 2150–2167 (1999)

17. Hu, Y., Gansner, E.R., Kobourov, S.G.: Visualizing graphs and clusters as maps. IEEE Computer Graphics and Applications 30(6), 54–66 (2010)

18. Kamada, T.: Visualizing Abstract Objects and Relations. World Scientific Series in Computer Science, vol. 5 (1989)

19. Kerber, M.: Embedding the dual complex of hyper-rectangular partitions. Journal of Computational Geometry 4(1), 13–37 (2013)

20. Knuth, D.E., Raghunathan, A.: The problem of compatible representatives. SIAM Journal on Discrete Mathematics 5(3), 422–427 (1992)

21. Koźmiński, K., Kinnen, E.: Rectangular duals of planar graphs. Networks 15, 145–157 (1985)

22. Liao, C.-C., Lu, H.-I., Yen, H.-C.: Compact floor-planning via orderly spanning trees. Journal of Algorithms 48, 441–451 (2003)

23. Lichtenstein, D.: Planar formulae and their uses. SIAM Journal on Computing 11(2), 329–343 (1982)

24. Shneiderman, B.: Tree visualization with tree-maps: 2-D space-filling approach. ACM Transactions on Graphics 11(1), 92–99 (1992)

25. Ungar, P.: On diagrams representing graphs. Journal of London Mathematical Sociery 28, 336–342 (1953)

26. Yeap, K.-H., Sarrafzadeh, M.: Floor-planning by graph dualization: 2-concave rectilinear modules. SIAM Journal on Computing 22, 500–526 (1993)

Minimum Activation Cost Node-Disjoint Paths in Graphs with Bounded Treewidth

Hasna Mohsen Alqahtani and Thomas Erlebach

Department of Computer Science, University of Leicester, Leicester, UK
{hmha1,t.erlebach}@leicester.ac.uk

Abstract. In *activation network* problems we are given a directed or undirected graph $G = (V, E)$ with a family $\{f_{uv} : (u, v) \in E\}$ of monotone non-decreasing activation functions from D^2 to $\{0, 1\}$, where D is a constant-size subset of the non-negative real numbers, and the goal is to find activation values x_v for all $v \in V$ of minimum total cost $\sum_{v \in V} x_v$ such that the activated set of edges satisfies some connectivity requirements. We propose algorithms that optimally solve the *minimum activation cost of k node-disjoint st-paths* (st-MANDP) problem in $O(tw((5 + tw)|D|)^{2tw+2}|V|^3)$ time and the *minimum activation cost of node-disjoint paths* (MANDP) problem for k disjoint terminal pairs $(s_1, t_1), \ldots, (s_k, t_k)$ in $O(tw((4 + 3tw)|D|)^{2tw+2}|V|)$ time for graphs with treewidth bounded by tw.

1 Introduction

In *activation network* problems, we are given an *activation network*, which is a directed or undirected graph $G = (V, E)$ together with a family $\{f_{uv} : (u, v) \in E\}$ of monotone non-decreasing activation functions from D^2 to $\{0, 1\}$, where D is a constant-size subset of the non-negative real numbers. The activation of an edge depends on the chosen values from the domain D at its endpoints. An edge $(u, v) \in E$ is *activated* for chosen values x_u and x_v if $f_{uv}(x_u, x_v) = 1$. An activation function f_{uv} for $(u, v) \in E$ is called *monotone non-decreasing* if $f_{uv}(x_u, x_v) = 1$ implies $f_{uv}(y_u, y_v) = 1$ for any $y_u \geq x_u$, $y_v \geq x_v$. The goal is to determine activation values $x_v \in D$ for all $v \in V$ so that the total activation cost $\sum_{v \in V} x_v$ is minimized and the activated set of edges satisfies some connectivity requirements. As activation network problems are computationally difficult in arbitrary graphs, it is meaningful to investigate whether restricted graph classes admit efficient algorithms. In this paper we consider the problems of finding *minimum activation cost k node-disjoint st-paths* (st-MANDP) and finding *minimum activation cost node-disjoint paths* (MANDP) between k disjoint terminals pairs, $(s_1, t_1), \ldots, (s_k, t_k)$, for graphs of bounded treewidth. Throughout this paper we consider undirected simple graphs.

The problem of finding k node/edge-disjoint paths between two nodes s and t in a given graph can be solved in polynomial time using network flow techniques [1], but no polynomial-time algorithm is known for the network activation setting. The problem of finding node/edge-disjoint paths (NDP/EDP) between

V. Geffert et al. (Eds.): SOFSEM 2014, LNCS 8327, pp. 65–76, 2014.

k terminals pairs, $(s_1, t_1), \ldots, (s_k, t_k)$, in a given graph is NP-complete if k is part of the input [5] (and already for $k = 2$ in directed graphs [4]). However, the problem is polynomial-time solvable for undirected graphs when k is fixed [11]. Scheffler [12] gave a linear-time algorithm that follows a classical bottom-up approach to solve the NDP problem for arbitrary k in graphs of bounded treewidth. In this paper we adapt Scheffler's algorithm [12] to the *activation network* version of node-disjoint path problems and devise polynomial-time optimal algorithms for solving the MANDP problem and the *st*-MANDP problem in graphs of bounded treewidth.

Related work. Activation network problems were introduced recently by Panigrahi [10]. The problem of finding the *minimum activation st-path* (MAP) can be solved optimally in polynomial-time [10]. However, the *minimum activation edge-disjoint st-paths* (*st*-MAEDP) problem is NP-hard [10]. The *minimum spanning activation tree* (MSpAT) problem is NP-hard to approximate within a factor of $o(\log n)$. The MSpAT problem is a special case of the problems of finding the *minimum Steiner activation network* (MSAN) and the *minimum edge/node-connected activation network* (MEAN/MNAN) (activating a network with k edge/node-disjoint paths between every pair of vertices). Therefore, these problems are also NP-hard to approximate within $o(\log n)$. As mentioned in [10], there exist $O(\log n)$-approximation algorithms for MSpAT, and for MEAN and MNAN in the case of $k = 2$. There is a 2-approximation algorithm for the *st*-MANDP problem and a $2k$-approximation algorithm for the *st*-MAEDP problem [8]. The *st*-MAEDP and *st*-MANDP problems when $k = 2$ have been studied recently by the authors [2]. They show that a ρ-approximation algorithm for the *minimum activation 2 node-disjoint st-paths* (*st*-MA2NDP) problem implies a ρ-approximation algorithm for the *minimum activation 2 edge-disjoint st-paths* (*st*-MA2EDP) problem. They also obtained a 1.5-approximation algorithm for the *st*-MA2NDP problem and hence for the *st*-MA2EDP problem. Furthermore, they showed that the *st*-MANDP problem for the restricted version of activation networks with $|D| = 2$ and a single activation function for all edges can be solved in polynomial time for arbitrary k (except for one case of the activation function, in which they require $k = 2$). However, the *st*-MAEDP problem remains NP-hard under this restriction [10,2]. It is not yet known whether the *st*-MANDP and *st*-MAEDP problems for an arbitrary constant-size D and fixed $k \geq 2$ are NP-hard.

Activation network problems can be viewed as a generalization of several known problems in wireless network design such as power optimization problems. In power optimization problems, we are given a graph $G = (V, E)$ and each edge $(u, v) \in E$ has a threshold power requirement θ_{uv}. In the undirected case we say the edge (u, v) is activated for chosen values x_u and x_v if each of these values is at least θ_{uv}, and in the directed case it is activated if $x_u \geq \theta_{uv}$. [7] shows that a simple reduction to the shortest *st*-path problem can solve the *minimum power st-path* (MPP) problem in polynomial time for directed/undirected networks. The problem of finding *minimum power k node-disjoint st-paths* (*st*-MPkNDP) in directed graphs can be solved in polynomial time [6,13]. However, the *minimum power k edge-disjoint st-paths* (*st*-MP*k*EDP) problem is unlikely to admit even a polylogarithmic approximation algorithm for both the directed and undirected variants [6]. In power optimization, the MSpAT problem is APX-hard and the MEAN and MNAN problems have 4-approximation and 11/3-approximation algorithms, respectively. See [2,8,9,10] for further motivation and applications of *activation network* problems.

Our results. We propose algorithms that optimally solve the *st*-MANDP problem in $O(tw((5 + tw)|D|)^{2tw+2}|V|^3)$ time and the MANDP in $O(tw\,(4 + 3tw)^{2tw+2}\,|D|^{2tw+2}\,|V|)$ time for graphs with treewidth bounded by tw.

This paper is organized as follows. In Section 2, we introduce some notations and definitions used throughout this paper. In Section 3, we propose an exact algorithm that solves the *st*-MANDP problem on graphs with bounded treewidth tw in polynomial-time. Section 4 presents an optimal algorithm that solves the MANDP problem on graphs with bounded treewidth in linear-time.

2   Preliminaries

The class of graphs of bounded treewidth [11] has attracted attention due to the fact that many NP-complete problems for arbitrary graphs are polynomial or even linear time solvable in graphs of bounded treewidth.

Definition 1. *A tree-decomposition for a given graph* $G = (V, E)$ *is a pair* $(\mathcal{X}, \mathcal{T})$ *of a tree* $\mathcal{T} = (I, F)$ *and a family* $\{X_i\}_{i \in I}$ *of subsets of* V *(called bags) satisfying the following two conditions: (1) For every edge* $(v, w) \in E$, *there exists an* $i \in I$ *with* $v \in X_i$ *and* $w \in X_i$. *(2) For every vertex* $v \in V$, *the nodes* $i \in I$ *with* $v \in X_i$ *form a subtree of* $\mathcal{T}$. *The width of a tree-decomposition is* $\max_{i \in I} |X_i| - 1$. *The treewidth* tw *of the graph* G *is the minimum width among all possible tree-decompositions of the graph.*

As shown in [3], there exists a linear-time algorithm that checks whether a given graph $G = (V, E)$ has treewidth at most tw, for fixed tw, and outputs the tree-decomposition $(\mathcal{X}, \mathcal{T})$ of G.

Definition 2. *A tree-decomposition* $(\mathcal{X}, \mathcal{T})$ *is called a* nice *tree-decomposition, if* $\mathcal{T}$ *is a binary tree rooted at some* $r \in I$ *that satisfies the following:*

- *Each node is either a leaf, or has exactly one or two children.*
- *Let* $i \in I$ *be a leaf. Then* $X_i \subseteq \{u : (u, v) \in E\} \cup \{v\} = N[v]$ *for some* $v \in V$.
- *For every edge* $(u, v) \in E$, *there is a leaf* $i \in I$ *such that* $\{u, v\} \subseteq X_i$.
- *Let* $j \in I$ *be the only child of* $i \in I$, *then either* $X_i = X_j \cup \{v\}$ *or* $X_i = X_j \setminus \{v\}$. *The node* i *is called an introduce node or forget node, respectively.*
- *Let* $j, j' \in I$ *be the two child nodes of a node* $i \in I$, *then* $X_j = X_{j'} = X_i$. *The node* i *is called a join node of* $\mathcal{T}$.

Scheffler [12] shows that any given tree-decomposition $(\mathcal{X}, \mathcal{T})$ can be easily converted into a nice tree-decomposition.

Theorem 1 ([12]). *A nice tree-decomposition of width* tw *and size* $O(|V(G)|)$ *can be constructed for a graph* $G = (V, E)$ *with treewidth at most* tw *in linear time, if* tw *is a fixed constant.*

To give a simpler description of our algorithms, we assume that every bag of a leaf of the nice tree-decomposition consists of two vertices that are connected by an edge in G, and that for every edge (u, v) of G there is exactly one leaf $i \in I$ such that $u, v \in X_i$ (and we say that the edge (u, v) is *associated* with that leaf $i \in I$). A nice tree-decomposition with this property can also be easily obtained from any given tree-decomposition in linear time. In the remainder of the paper, we assume that the input graph G is given as a simple undirected graph together with a nice tree-decomposition of width at most tw.

Let us define X_i^+ to be the set of all vertices in X_j for all nodes $j \in I$ such that $j = i$ or j is a descendant of i. Let G_i^+ describe the partial graphs of G. For a leaf node i, G_i^+ is the subgraph of G with vertex set X_i and the edge of G that is associated with i. For a non-leaf node i, G_i^+ is the graph that is the union of G_j^+ over all children j of i. Note that the graph G_r^+ for the root r of the tree-decomposition is equal to G.

3   Minimum Activation Cost k Node-Disjoint st-Paths

Let $G = (V, E)$ be an activation network and $s, t \in V$ be a pair of source and destination vertices. The goal of the st-MANDP problem is to find activation values $\{x_v : v \in V\}$ of minimum total cost $\sum_{v \in V} x_v$ such that the activated set of edges contains k node disjoint st-paths $\mathcal{P}^{st} = P_1, \ldots, P_k$. In this section we present a polynomial-time algorithm that solves the st-MANDP problem optimally in the case of graphs of bounded treewidth. The algorithm follows a bottom-up approach based on a nice tree-decomposition. In our algorithm, each node i of the tree-decomposition has a table tab_i to store its computed information. The algorithm computes a number of possible sub-solutions per tree node $i \in I$ based on the information computed previously for its children. Let $\mathcal{P} = P_1, \ldots, P_k$ be a solution for the st-MANDP problem. Let $\mathcal{P}^i = \mathcal{P}\left[G_i^+\right]$ be the induced solution in a partial graph G_i^+ (i.e., the set of vertices and edges that are both in $\mathcal{P}$ and in G_i^+). The interaction between $\mathcal{P}^i$ in G_i^+ and the rest of the graph happens only in vertices of X_i. The partial solution $\mathcal{P}^i$ can therefore be represented by an activation-value function $\Lambda_i : X_i \to D$ and a state function $\beta_i : X_i \to \{s, t, o, \infty\} \cup \{0, 1, \ldots, k\} \cup (X_i \times \{c\})$. As the algorithm needs to consider partial solutions in G_i^+ that cannot necessarily be completed to form a global solution, we call any subgraph of G_i^+ (with suitable activation values) a *partial solution* (or *sub-solution*) if it contains all vertices in $\{s, t\} \cap X_i^+$ and each connected component C satisfies one of the following conditions:

1. C is an isolated vertex $v \in X_i$.
2. C is a simple path having both end-vertices $v, w \in X_i$ and containing neither s nor t.
3. C contains s but not t and consists of several (at least one) paths that are node-disjoint (apart from meeting in s), each connecting s to a vertex in X_i (or the same condition with s and t exchanged).
4. C contains s and t and consists of some number of paths from s to t, some paths from s to a vertex in X_i, and some paths from t to a vertex in X_i.

All these paths are node-disjoint (apart from having s or t in common). If $s \notin X_i$, s has degree k, otherwise s has degree $\leq k$, and likewise for t.

Intuitively, partial solutions are subgraphs of G_i^+ that could potentially be completed to a global solution if the rest of the graph is of suitable form.

State Function. A partial solution $\mathcal{P}^i$ in G_i^+ can be represented by a state function $\beta_i : X_i \rightarrow \{s, t, o, \infty\} \cup \{0, 1, \ldots, k\} \cup (X_i \times \{c\})$ as follows:

- For any $v \in X_i$, we set $\beta_i(v) = 0$ iff $v \in \mathcal{P}^i$ and has degree zero in $\mathcal{P}^i$, i.e., v is a connected component of $\mathcal{P}^i$ that satisfies condition 1 above.
- For any $v \in X_i$, we set $\beta_i(v) = \infty$ iff $v \notin \mathcal{P}^i$.
- For any $v \in X_i \setminus \{s, t\}$, we set $\beta_i(v) = o$ iff v has degree 2 in $\mathcal{P}^i$.
- For any $v \in X_i \cap \{s, t\}$, we set $\beta_i(v) = k' \in \{1, \ldots, k\}$ iff a connected component C of $\mathcal{P}^i$ contains v together with k' incident edges, i.e., C satisfies condition 3 or 4 above. We call v an occupied vertex if $k' = k$.
- For any $v \in X_i \setminus \{s, t\}$, we set $\beta_i(v) = u \in \{s, t\}$ iff there is a non-empty path $u, \ldots, v$ not containing s or t as internal node that is a subgraph of a connected component of $\mathcal{P}^i$ that satisfies condition 3 or 4 above and v has degree 1 in that component.
- For any pair $u, v \in X_i \setminus \{s, t\}$, we set $\beta_i(u) = (v, c)$ and $\beta_i(v) = (u, c)$ iff u and v are connected with a path $u, \ldots, v$ that is a maximal connected component of $\mathcal{P}^i$ that does not contain s and t, i.e., the maximal connected component $u, \ldots, v$ satisfies condition 2 above.

3.1   Processing the Tree Decomposition

Let $val(\beta_i, \Lambda_i)$ denote the optimal cost of an assignment of activation values for G_i^+ which satisfies the restriction Λ_i and activates a partial solution satisfying β_i. In each row of the table tab_i of tree node $i \in I$, we store a solution of a unique combination of a state function β_i and a function of activation values Λ_i (each vertex of X_i has a state value and is assigned an activation value). Additionally, we store the activation cost value $val(\beta_i, \Lambda_i)$ for the solution. We can compute the sub-solution tables in a bottom-up approach.

Leaf. Let $i \in I$ be a leaf, $X_i = \{u, v\}$. Let (β_i, Λ_i) be any row of tab_i. We distinguish the following cases and define $val(\beta_i, \Lambda_i)$ for each case. Each case corresponds to a possible sub-solution in G_i^+. Recall that G_i^+ is a single edge. The sub-solution's cost $val(\beta_i, \Lambda_i)$ is set to $\sum_{v \in X_i} \Lambda_i(v)$ if one of the following cases applies:

- $\beta_i(w) \in \{0, \infty\}$ for all $w \in X_i$, and $\beta_i(w) = 0$ for $w \in X_i \cap \{s, t\}$. Intuitively, this means that the sub-solution has no edges.
- $\beta_i(u) = v$ and $\beta_i(v) = 1$ and $u \notin \{s, t\}$ and $v \in \{s, t\}$ and $f_{uv}(\Lambda_i(u), \Lambda_i(v)) = 1$. Intuitively, the sub-solution is a path with one edge and one endpoint equal to s or t. (The roles of u and v can be exchanged.)
- $\beta_i(u) = \beta_i(v) = 1$ and $u, v \in \{s, t\}$ and $f_{uv}(\Lambda_i(u), \Lambda_i(v)) = 1$. Intuitively, the sub-solution is a path with one edge containing s and t.

– $\beta_i(u) = (v, c)$ and $\beta_i(v) = (u, c)$ and $u, v \notin \{s, t\}$ and $f_{uv}(\Lambda_i(u), \Lambda_i(v)) = 1$.
  Intuitively, the sub-solution is a path with one edge not containing s or t.

If none of the above cases applies, $val(\beta_i, \Lambda_i) = +\infty$. In these cases, (β_i, Λ_i) does not represent a subgraph of G_i^+ that could be part of a global solution.

Introduce. Let $i \in I$ be an introduce node, and $j \in I$ its only child. We have $X_j \subset X_i$, $|X_i| = |X_j| + 1$ and let v be the additional isolated vertex in X_i. The vertex v always has state value $\beta_i(v) \in \{0, \infty\}$. For every row (β_j, Λ_j) in tab_j, there are $2|D|$ rows in tab_i such that $\beta_i(u) = \beta_j(u)$ and $\Lambda_i(u) = \Lambda_j(u)$ for all $u \in X_i \setminus \{v\}$. The sub-solution cost $val(\beta_i, \Lambda_i)$ for these rows is set to $val(\beta_j, \Lambda_j) + \Lambda_i(v)$ if $v \notin \{s, t\}$ or if $v \in \{s, t\}$ and $\beta_i(v) = 0$. Otherwise $val(\beta_i, \Lambda_i) = +\infty$.

Forget. Let $i \in I$ be a forget node, and $j \in I$ its only child. We have $X_i \subset X_j$, $|X_j| = |X_i| + 1$ and let v be the discarded vertex. For each (β_i, Λ_i), consider all (β_j, Λ_j) that agree with (β_i, Λ_i) for all $u \in X_i$ and satisfy that $\beta_j(v) = k$ if $v \in \{s, t\}$ and $\beta_j(v) \in \{o, \infty\}$ otherwise. The sub-solution's cost $val(\beta_i, \Lambda_i)$ is the minimum of $val(\beta_j, \Lambda_j)$ over all these (β_j, Λ_j).

Join. Let $i \in I$ be a join node, and $j, j' \in I$ its two children. We have $X_i = X_j = X_{j'}$. Let (β_j, Λ_j) and $(\beta_{j'}, \Lambda_{j'})$ be rows of tab_j and $tab_{j'}$, respectively. When we combine these solutions, their connected components may get merged into larger components, and we need to ensure that we do not create any cycles that do not contain s and t. For this purpose, we construct an auxiliary graph H_i with vertex set X_i and edge set $E_{H_i} = \{uv | \beta_j(v) = (u, c)\} \cup \{uv \mid \beta_{j'}(v) = (u, c)\}$ to help us detect cases where such a cycle would be created. The algorithm combines the sub-solutions (β_j, Λ_j) and $(\beta_{j'}, \Lambda_{j'})$ if both have the same activation-value function $(\Lambda_j(u) = \Lambda_{j'}(u)$ for all $u \in X_i)$ and the union does not satisfy any of the following conditions (where the roles of j and j' can be exchanged):

C_1. H_i contains a cycle (which may also consist of just one pair of parallel edges).
C_2. There is a vertex $v \in X_i$ with $\beta_j(v) = o$ and $\beta_{j'}(v) \neq 0$.
C_3. There is a vertex $v \in X_i$ with $\beta_j(v) = \infty$ and $\beta_{j'}(v) \neq \infty$.
C_4. There is a vertex $v \in X_i$ with $\beta_j(v) = \beta_{j'}(v) \in \{s, t\}$.
C_5. There is a vertex $v \in X_i \cap \{s, t\}$ with $\beta_j(v) = k'$ and $\beta_{j'}(v) = k''$ and $k' + k'' \notin \{0, 1, \ldots, k\}$.

We compute the state function β_i of the combination of β_j and $\beta_{j'}$. Consider any $v \in X_i$:

– If $\beta_j(v) = \beta_{j'}(v) = \sigma \in \{0, \infty\}$, then $\beta_i(v) = \sigma$.
– If $\beta_j(v) = \sigma \in \{o, s, t, k\}$ and $\beta_{j'}(v) = 0$, then $\beta_i(v) = \sigma$.
– If $\beta_j(v) = s$ and $\beta_{j'}(v) = t$, then $\beta_i(v) = o$.
– If $v \in \{s, t\}$ and $\beta_j(v) = k'$ and $\beta_{j'}(v) = k''$ for any $k' + k'' \in \{0, 1, \ldots, k\}$, then $\beta_i(v) = k' + k''$.
– If $\beta_j(v) = (x, c)$ and $\beta_{j'}(v) = (y, c)$ for any $x, y \in X_i$, then $\beta_i(v) = o$.
– If $\beta_j(v) = 0$ and $\beta_{j'}(v) = (x, c)$ and the maximal path in H_i that starts at v ends at u with $\beta_j(u) = 0$ and $\beta_{j'}(u) = (y, c)$ (or vice versa, i.e., the path ends at u with $\beta_j(u) = (y, c)$ and $\beta_{j'}(u) = 0$) for any $x, y \in X_i$, then $\beta_i(v) = (u, c)$ and $\beta_i(u) = (v, c)$.

- If $\beta_j(v) = s$ and $\beta_{j'}(v) = (x, c)$ and the maximal path in H_i that starts at v ends at u with $\beta_j(u) = (y, c)$ and $\beta_{j'}(u) = t$ (or vice versa) for any $x, y \in X_i$, then $\beta_i(v) = \beta_i(u) = o$.
- If $\beta_j(v) = \sigma \in \{s, t\}$ and $\beta_{j'}(v) = (x, c)$ and the maximal path in H_i that starts at v ends at u with $\beta_j(u) = (y, c)$ and $\beta_{j'}(u) = 0$ (or vice versa) for any $x, y \in X_i$, then $\beta_i(v) = o$ and $\beta_i(u) = \sigma$.

The value of the combined solution $val(\beta_i, \Lambda_i)$ is calculated as the minimum summation value over all pairs of sub-solutions that can be combined to produce (β_i, Λ_i) minus the activation cost of X_i.

Extracting the solution at the root. The algorithm checks all the solutions (β_r, Λ_r) of the root bag X_r such that vertices in $\{s, t\} \cap X_r$ are occupied and all other vertices have state value o or ∞. In this case (β_r, Λ_r) is a feasible solution. The output of the algorithm is the solution of minimum cost value among all the feasible solutions obtained at the root. For each row (β_i, Λ_i) of bag X_i, we store the rows of X_i's children that were used in the calculation of $val(\beta_i, \Lambda_i)$. Computing the optimum solution is possible by traversing top-down in the decomposition tree to the leaves (traceback).

3.2   Analysis

Let an instance of the problem be given by an activation network $G = (V, E)$ with bounded treewidth tw and terminals $s, t \in V$. Let $\mathcal{P}^{OPT}$ represent an optimal solution for this instance. We use $C(\mathcal{P}^i)$ to denote the activation cost of the induced solution $\mathcal{P}^i$ of $\mathcal{P}$ in a partial graph G_i^+.

Lemma 1. *The algorithm runs in $O(tw((5 + tw)|D|)^{2tw+2}|V|^3)$ time.*

Proof. The running-time of the algorithm depends on the size of the tables and the combination of tables during the bottom-up traversal. Each vertex $w \in X_i$ has $5 + (|X_i| - 1)$ possible state values from $\{0, s, t, o, \infty\} \cup ((X_i \setminus \{w\}) \times \{c\})$ if $w \notin \{s, t\}$ and $k + 1$ possible state values if $w \in X_i \cap \{s, t\}$. If $|X_i \cap \{s, t\}| \leq 1$, the table tab_i in a processed bag X_i contains no more than $O(k(5 + tw)^{tw+1}|D|^{tw+1})$ rows corresponding to the possible state functions and the $|D|$ possible activation values for each vertex of X_i. Assume that $\{s, t\} \subseteq X_i$ and $\beta_i(s) = k_s$, $\beta_i(t) = k_t$ and $c_q = \{v \in X_i : \beta_i(v) = q\}$ such that $k_s, k_t \in \{0, \ldots, k\}$ and $q \in \{s, t\}$. Since every path in the partial solution starting at s leads to t or to a vertex in X_i then $k_s - c_s = k_t - c_t$. Hence, there is at most one k_t when leaving the other values fixed. Therefore, the table tab_i in a processed bag X_i contains no more than $O(k(5 + tw)^{tw+1}|D|^{tw+1})$ rows. Considering all possible row combinations for two tables for a join node and noting that $k \leq |V|$, we see that the computation of the state functions needs $O(tw)$ time for each combination and $O(tw((5 + tw)|D|)^{2tw+2}|V|^3)$ time overall.   □

Lemma 2. *For any processed bag X_i, let $\mathcal{P}_i^{OPT}$ be the induced solution of $\mathcal{P}^{OPT}$ in G_i^+ and $(\beta_i^{OPT}, \Lambda_i^{OPT})$ be the corresponding state function and activation values, then $val(\beta_i^{OPT}, \Lambda_i^{OPT}) \leq C(\mathcal{P}_i^{OPT})$.*

Lemma 3. *For any processed bag X_i, any solution (β_i, Λ_i) where $val(\beta_i, \Lambda_i) = c_i < \infty$ corresponds to a partial solution $\mathcal{P}^i$ with the following properties:*

- *The state function of $\mathcal{P}^i$ in X_i is β_i.*
- *The activation values of $\mathcal{P}^i$ in X_i are Λ_i.*
- *The total activation cost in X_i^+ is c_i.*
- *$\mathcal{P}^i$ contains the terminal s if $s \in X_i^+$, and the terminal t if $t \in X_i^+$.*

Due to space restrictions, the proofs of Lemma 2 and Lemma 3 (which use induction over the tree-decomposition) are deferred to the full version.

We say that (β_i, Λ_i) is a feasible solution at the root node i iff (β_i, Λ_i) is a partial solution with $\beta_i(v) = k$ for $v \in \{s,t\} \cap X_i$ and $\beta_i(v) \in \{o, \infty\}$ for all $v \in X_i \setminus \{s,t\}$. That means the partial solution in G_i^+ consists of k node-disjoint paths between s and t. The algorithm outputs the solution of minimum activation cost among all feasible solutions obtained at the root. From the above lemmas and the extracting part in the algorithm, we get the following theorem.

Theorem 2. *The st-MANDP problem can be solved optimally in $O(tw\,(5 + tw)^{2tw+2}\,|D|^{2tw+2}\,|V|^3)$ time for graphs with treewidth bounded by tw.*

This algorithm can be used for the variant of the st-MANDP problem in graphs of bounded treewidth where s and t are assigned specified activation values $d \in D$ and $d' \in D$, respectively, by setting $\Lambda(s) = d$ and $\Lambda(t) = d'$. We recall the following theorem from [2]:

Theorem 3 ([2]). *If there is a ρ-approximation algorithm for the st-MA2NDP problem where s and t are assigned specified activation values, then there is a ρ-approximation algorithm for the st-MA2EDP problem.*

Corollary 1. *There exists a polynomial-time algorithm that optimally solves the st-MA2EDP problem for graphs of bounded treewidth.*

4 Minimum Activation Cost Node-Disjoint Paths between k Pairs of Terminals

Consider an instance of the problem given by an activation network $G = (V, E)$ and k disjoint pairs of terminals $(s_1, t_1), \ldots, (s_k, t_k)$. The goal of MANDP is to find activation values $\{x_v : v \in V\}$ of minimum total cost $\sum_{v \in V} x_v$ such that the activated set of edges contains k node disjoint paths $\mathcal{P} = P_1, \ldots, P_k$ in G where P_a is a path connecting s_a and t_a for all $a \in \{1, \ldots, k\}$. Define $\mathcal{S} = \{s_a | 1 \le a \le k\} \cup \{t_a | 1 \le a \le k\}$. Let $\mathcal{N}_a = (s_a, t_a)$, for $1 \le a \le k$. In this section we modify the linear-time algorithm proposed in [12] that solves the NDP problem on graphs of bounded treewidth to solve the problem in the activation network case. We assume that k is arbitrary. As for the st-MANDP algorithm in Section 3, the algorithm stores all computed sub-solutions for each node i of the nice tree-decomposition in a table tab_i. For every node i, we define $\mathcal{O}_i = \{\mathcal{N}_a | s_a \in X_i^+$ and $t_a \notin X_i^+$ or $s_a \notin X_i^+$ and $t_a \in X_i^+\}$ and $\mathcal{Q}_i = \{\mathcal{N}_a | s_a \in X_i^+$ and $t_a \in X_i^+\}$.

Let $\mathcal{P}^i$ be any subgraph of the partial graph G_i^+. If $\mathcal{P}^i$ is a partial solution, the algorithm characterizes $\mathcal{P}^i$ by an activation function $\Lambda_i : X_i \to D$ and a state function $\beta_i : X_i \to \{0, 1, \infty\} \cup \mathcal{O}_i \cup \mathcal{Q}_i \cup (X_i \times \{c, d\})$. Here, we say that $\mathcal{P}^i$ is a partial solution if it contains all $v \in X_i^+ \cap \mathcal{S}$ and any connected component C of $\mathcal{P}^i$ satisfies one of the following conditions:

1. C is a simple path P_a connecting s_a and t_a (and not containing any other vertex in $\mathcal{S}$).
2. C is an isolated vertex $v \in X_i$.
3. C is a simple path having both end-vertices $v, w \in X_i$ and not containing any vertices of $\mathcal{S}$.
4. C is a simple path that connects a source $s_a \in X_i^+$ with a vertex $s_a' \in X_i$. If $\mathcal{N}_a \in \mathcal{Q}_i$, there is another (possibly empty) path C' that connects the terminal $t_a \in X_i^+$ with a vertex $t_a' \in X_i$. (The roles of s_a and t_a can be exchanged.)

State Function. The algorithm characterizes $\mathcal{P}^i$ by an activation-value function $\Lambda_i : X_i \to D$ and a state function $\beta_i : X_i \to \{0, 1, \infty\} \cup \mathcal{O}_i \cup \mathcal{Q}_i \cup (X_i \times \{c, d\})$. Given a partial solution $\mathcal{P}^i$, the state function β_i is defined as follows:

- For any $v \in X_i$, we set $\beta_i(v) = 0$ iff $v \in \mathcal{P}^i$ and v has degree zero in $\mathcal{P}^i$.
- For any $v \in X_i$, we set $\beta_i(v) = \infty$ iff $v \notin \mathcal{P}^i$.
- For any $v \in X_i$, we set $\beta_i(v) = 1$ iff v is either an inner vertex of a path in $\mathcal{P}^i$ or $v \in \mathcal{S}$ and $\mathcal{P}^i$ contains v together with an incident edge. We call v an occupied vertex.
- For any $v \in X_i$, we set $\beta_i(v) = \mathcal{N}_a$ iff there is a path $s_a, \ldots, v$ (or a path $t_a, \ldots, v$) in $\mathcal{P}^i$, v has degree 1 in $\mathcal{P}_i$, and either $\mathcal{N}_a \in \mathcal{O}_i$ or we have that $\mathcal{N}_a \in \mathcal{Q}_i$ and t_a (or s_a) is in X_i and has degree 0 in $\mathcal{P}_i$.
- For any pair $u, v \in X_i$, we set $\beta_i(u) = (v, d)$ and $\beta_i(v) = (u, d)$ iff there are two paths $s_a, \ldots, u$ and $t_a, \ldots, v$ in $\mathcal{P}^i$ for any $\mathcal{N}_a \in \mathcal{Q}_i$ and u, v have degree 1. We say that s_a and t_a are a disconnected pair.
- For any pair $u, v \in X_i$, we set $\beta_i(u) = (v, c)$ and $\beta_i(v) = (u, c)$ iff u and v are connected with a path $u, \ldots, v$ that is a maximal connected component of $\mathcal{P}^i$ that does not contain any vertices from $\mathcal{S}$ and u, v have degree 1.

4.1   Processing the Tree Decomposition

As for the st-MANDP algorithm in Section 3, each tab_i of X_i stores multiple rows and each row represents a unique combination of a state function β_i and an activation function Λ_i (each vertex of X_i has a state and is assigned an activation value). The cost value $val(\beta_i, \Lambda_i)$ of the represented sub-solution is also stored in tab_i.

Leaf. Let $i \in I$ be a leaf, $X_i = \{u, v\}$ and $(u, v) \in E_i$, where E_i is the set of edges associated with i. We distinguish the following cases and define $val(\beta_i, \Lambda_i)$ for each case. Let (β_i, Λ_i) be any row of tab_i. If none of the following cases applies, $val(\beta_i, \Lambda_i) = +\infty$. The sub-solution's cost is $val(\beta_i, \Lambda_i) = \sum_{v \in X_i} \Lambda_i(v)$ if one of the following cases applies:

- $\beta_i(w) \in \{0, \infty\}$ for all $w \in X_i$ and $\beta_i(w) = 0$ for $w \in S$. Intuitively, the sub-solution has no edges.
- $\beta_i(u) = \beta_i(v) = 1$ and $u, v \in \mathcal{N}_a$ for any $\mathcal{N}_a \in \mathcal{Q}_i$ and $f_{uv}(\Lambda_i(u), \Lambda_i(v)) = 1$. Intuitively, the sub-solution is a path with one edge containing s_a and t_a for some $\mathcal{N}_a \in \mathcal{Q}_i$.
- $\beta_i(u) = \mathcal{N}_a$, $\beta_i(v) = 1$ and $u \notin S$ and $v \in \mathcal{N}_a$ for any $\mathcal{N}_a \in \mathcal{O}_i$ and $f_{uv}(\Lambda_i(u), \Lambda_i(v)) = 1$. Intuitively, the sub-solution is a path with one edge and one endpoint equal to s_a or t_a for any $\mathcal{N}_a \in \mathcal{O}_i$. (The roles of u and v can be exchanged.)
- $\beta_i(u) = (v, c)$, $\beta_i(v) = (u, c)$ and $u, v \notin S$ and $f_{uv}(\Lambda_i(u), \Lambda_i(v)) = 1$. Intuitively, the sub-solution is a path with one edge not containing any vertices of S.

Introduce and *Forget* nodes are processed in a similar way to Section 3.1.

Join. Let $i \in I$ be a join node, and $j, j' \in I$ its two children. We have $X_i = X_j = X_{j'}$. Let (β_j, Λ_j) and $(\beta_{j'}, \Lambda_{j'})$ be rows of tab_j and $tab_{j'}$, respectively. We consider an auxiliary graph H_i with vertex set X_i and edge set $E_{H_i} = \{uv | \beta_j(v) = (u, c)\} \cup \{uv \mid \beta_{j'}(v) = (u, c)\}$ to help computing the state function β_i of the combination of β_j and $\beta_{j'}$. The algorithm combines the sub-solutions (β_j, Λ_j) and $(\beta_{j'}, \Lambda_{j'})$ if both have the same activation-value function $(\Lambda_j(u) = \Lambda_{j'}(u)$ for all $u \in X_i$) and the union does not satisfy the following (the roles of j and j' could be exchanged):

C_1. H_i contains a cycle.
C_2. There is a vertex $v \in X_i$ with $\beta_j(v) = 1$ and $\beta_{j'}(v) \neq 0$.
C_3. There is a vertex $v \in X_i$ with $\beta_j(v) = \infty$ and $\beta_{j'}(v) \neq \infty$.
C_4. There is a vertex $v \in X_i$ with $\beta_j(v) = \mathcal{N}_a$ and $\beta_{j'}(v) = \mathcal{N}_b$ and $a \neq b$.
C_5. There is a vertex $v \in X_i$ such that $\beta_j(v) = \mathcal{N}_a$ and $\beta_{j'}(v) = (u, d)$ for any $u \in X_i$.
C_6. There is a vertex $v \in X_i$ with $\beta_j(v) = (u, d)$ and $\beta_{j'}(v) = (w, d)$ for any $u, w \in X_i$.

We compute the state function β_i of the combination of β_j and $\beta_{j'}$. Consider $v \in X_i$:

- If $\beta_j(v) = \beta_{j'}(v) = \sigma \in \{0, \infty\}$, then $\beta_i(v) = \sigma$.
- If $\beta_j(v) = 1$ and $\beta_{j'}(v) = 0$, then $\beta_i(v) = 1$.
- If $\beta_j(v) = \mathcal{N}_a$ and $\beta_{j'}(v) = 0$, we distinguish two cases: If there is another vertex $u \in X_i$ with $\beta_j(u) = 0$ and $\beta_{j'}(u) = \mathcal{N}_a$, then $\beta_i(u) = (v, d)$ and $\beta_i(v) = (u, d)$. Otherwise, $\beta_i(v) = \mathcal{N}_a$.
- If $\beta_j(v) = \mathcal{N}_a$ and $\beta_{j'}(v) = \mathcal{N}_a$, then $\beta_i(v) = 1$.
- If $\beta_j(v) = (x, c)$ and $\beta_{j'}(v) = (y, c)$ for any $x, y \in X_i$, then $\beta_i(v) = 1$.
- If $\beta_j(v) = (u, d)$ and $\beta_{j'}(v) = 0$ and $\beta_j(u) = (v, d)$ and $\beta_{j'}(u) = 0$ for any $u \in X_i$, then $\beta_i(u) = (v, d)$ and $\beta_i(v) = (u, d)$.
- If $\beta_j(v) = 0$ and $\beta_{j'}(v) = (x, c)$ and the maximal path in H_i that starts at v ends at u with $\beta_j(u) = 0$ and $\beta_{j'}(u) = (y, c)$ or $\beta_j(u) = (y, c)$ and $\beta_{j'}(u) = 0$ for any $x, y \in X_i$, then $\beta_i(v) = (u, c)$ and $\beta_i(u) = (v, c)$.

- If $\beta_j(v) = \mathcal{N}_a$ and $\beta_{j'}(v) = (x, c)$ and the maximal path in H_i that starts at v ends at u with $\beta_j(u) = (y, c)$ and $\beta_{j'}(u) = \mathcal{N}_a$ for any $x, y \in X_i$ and $a \in \{1, \ldots, k\}$, then $\beta_i(v) = \beta_i(u) = 1$.
- If $\beta_j(v) = (u, d)$ and $\beta_{j'}(v) = (x, c)$ and the maximal path in H_i that starts at v ends at u with $\beta_j(u) = (v, d)$ and $\beta_{j'}(u) = (y, c)$ for any $x, y \in X_i$, then $\beta_i(v) = \beta_i(u) = 1$.
- If $\beta_j(v) = \mathcal{N}_a$ and $\beta_{j'}(v) = (x, c)$ and the maximal path in H_i that starts at v ends at u with $\beta_j(u) = (y, c)$ and $\beta_{j'}(u) = 0$ for any $x, y \in X_i$ and $a \in \{1, \ldots, k\}$, then $\beta_i(v) = 1$ and $\beta_i(u) = \mathcal{N}_a$.
- If $\beta_j(v) = (u, d)$ and $\beta_{j'}(v) = (x, c)$ and the maximal path in H_i that starts at v ends at w with $\beta_j(w) = (y, c)$ and $\beta_{j'}(u) = 0$ for any $x, y \in X_i$, then $\beta_i(v) = 1$ and $\beta_i(u) = (w, d)$ and $\beta_i(w) = (u, d)$.
- If $\beta_j(v) = (u, d)$ and $\beta_{j'}(v) = (x, c)$ and $\beta_j(u) = (v, d)$ and $\beta_{j'}(u) = (y, c)$ the maximal path in H_i that starts at v ends at w with $\beta_j(w) = (z, c)$ and $\beta_{j'}(w) = 0$ and the maximal path in H_i that starts at u ends at q with $\beta_j(q) = (z', c)$ and $\beta_{j'}(q) = 0$ for any $x, y, z, z' \in X_i$, then $\beta_i(v) = \beta_i(u) = 1$, $\beta_i(q) = (w, d)$ and $\beta_i(w) = (q, d)$.

The value of the combined solution is the minimum summation value over all pairs of sub-solutions that can be combined to produce (β_i, Λ_i), minus the activation cost of X_i.

Extracting the solution at the root can be done similarly as in Section 3.1.

This completes the description of the MANDP algorithm. In this algorithm, each vertex $w \in X_i$ has $3 + 2(|X_i| - 1) + |X_i|$ possible state values from the set $\{0, 1, \infty\} \cup ((X_i \setminus \{w\}) \times \{c, d\}) \cup \{\mathcal{N}_a \mid \mathcal{N}_a \in \mathcal{O}_i\} \cup \{\mathcal{N}_a \mid \mathcal{N}_a \in \mathcal{Q}_i, \mathcal{N}_a \cap X_i \neq \emptyset\}$. (Note that the union of the latter two sets has cardinality at most $|X_i|$ if a feasible solution exists.) Then, the table tab_i contains no more than $O((4 + 3tw)^{tw+1}|D|^{tw+1})$ rows corresponding to the possible state functions and the $|D|$ possible activation values for each vertex of X_i. Each table has $O((4 + 3tw)^{tw+1}|D|^{tw+1})$ rows and the combination of a join node requires $O(tw(4 + 3tw)^{2tw+2}|D|^{2tw+2})$ time. The proofs of the running time and correctness of the MANDP algorithm are similar to the proofs of the running time and correctness of the st-MANDP algorithm. Due to space restrictions, the analysis is omitted from this paper, and we close this section with the following theorem.

Theorem 4. *The MANDP problem for graphs with bounded treewidth tw can be solved optimally in $O(tw(4 + 3tw)^{2tw+2}|D|^{2tw+2}|V|)$ time.*

5   Conclusion

We have presented a polynomial-time algorithm that optimally solves the st-MANDP problem for the case of graphs with bounded treewidth. We also showed that the linear-time algorithm for the NDP problem for graphs of bounded treewidth that has been presented in [12] can be modified to obtain a linear-time algorithm for the problem in activation networks. One open problem is to obtain a faster or even linear-time algorithm for the st-MANDP problem in graphs of bounded treewidth. It would also be interesting to investigate the st-MAEDP problem for graphs of bounded treewidth.

Acknowledgements. We would like to thank the anonymous reviewers for their thorough and helpful comments on an earlier version of this paper. Their comments helped making the paper clearer and improving the analysis of the running time of the algorithm in Section 3.

References

1. Ahuja, R.K., Magnanti, T.L., Orlin, J.B.: Network flows: Theory, Algorithms and Applications. Prentice Hall, New Jersey (1993)
2. Alqahtani, H.M., Erlebach, T.: Approximation Algorithms for Disjoint st-Paths with Minimum Activation Cost. In: Spirakis, P.G., Serna, M. (eds.) CIAC 2013. LNCS, vol. 7878, pp. 1–12. Springer, Heidelberg (2013)
3. Bodlaender, H.L.: A linear time algorithm for finding tree-decompositions of small treewidth. In: ACM STOC 1993, pp. 226–234 (1993)
4. Fortune, S., Hopcroft, J., Wyllie, J.: The directed subgraph homeomorphism problem. Theoretical Computer Science 10(2), 111–121 (1980)
5. Garey, M.R., Johnson, D.S.: Computers and Intractability. A Guide to the Theory of NP-Completeness. W. H. Freeman and Company, New York (1979)
6. Hajiaghayi, M.T., Kortsarz, G., Mirrokni, V.S., Nutov, Z.: Power optimization for connectivity problems. In: Jünger, M., Kaibel, V. (eds.) IPCO 2005. LNCS, vol. 3509, pp. 349–361. Springer, Heidelberg (2005)
7. Lando, Y., Nutov, Z.: On minimum power connectivity problems. In: Arge, L., Hoffmann, M., Welzl, E. (eds.) ESA 2007. LNCS, vol. 4698, pp. 87–98. Springer, Heidelberg (2007)
8. Nutov, Z.: Survivable network activation problems. In: Fernández-Baca, D. (ed.) LATIN 2012. LNCS, vol. 7256, pp. 594–605. Springer, Heidelberg (2012)
9. Nutov, Z.: Approximating Steiner networks with node-weights. SIAM J. Comput. 39(7), 3001–3022 (2010)
10. Panigrahi, D.: Survivable network design problems in wireless networks. In: 22nd Annual ACM-SIAM Symposium on Discrete Algorithms, pp. 1014–1027. SIAM (2011)
11. Robertson, N., Seymour, P.D.: Graph Minors XIII. The Disjoint Paths Problem. J. Comb. Theory, Ser. B 63, 65–110 (1995)
12. Scheffler, P.: A practical linear time algorithm for disjoint paths in graphs with bound tree-width. Technical Report 396, Dept. Mathematics, Technische Universität Berlin (1994)
13. Srinivas, A., Modiano, E.: Finding Minimum Energy Disjoint Paths in Wireless Ad-Hoc Networks. Wireless Networks 11(4), 401–417 (2005)

Tight Bounds for the Advice Complexity of the Online Minimum Steiner Tree Problem*

Kfir Barhum

Department of Computer Science, ETH Zurich, Switzerland
barhumk@inf.ethz.ch

Abstract. In this work, we study the advice complexity of the online minimum Steiner tree problem (ST). Given a (known) graph $G = (V, E)$ endowed with a weight function on the edges, a set of N terminals are revealed in a step-wise manner. The algorithm maintains a sub-graph of chosen edges, and at each stage, chooses more edges from G to its solution such that the terminals revealed so far are connected in it. In the standard online setting this problem was studied and a tight bound of $O(\log(N))$ on its competitive ratio is known. Here, we study the power of non-uniform advice and fully characterize it. As a first result we show that using $q \cdot \log(|V|)$ advice bits, where $0 \leq q \leq N - 1$, it is possible to obtain an algorithm with a competitive ratio of $O(\log(N/q))$. We then show a matching lower bound for all values of q, and thus settle the question.

Keywords: Online algorithms, advice complexity, minimum Steiner tree.

1 Introduction

Online algorithms are a realistic model for making decisions under uncertainty. As opposed to classical computational problems, in the online setting the full input to the problem is not known in advance, but is revealed in a step-wise manner, and after each step the algorithm has to commit to a part of its solution. The *competitive analysis* of an algorithm, as introduced first in [1], measures the worst-case performance of the algorithm compared with the optimal offline solution to the respective optimal offline (classical) computational problem. We refer to [2] for a detailed introduction and survey of many classical online problems.

In recent years, motivated, among others, by the fact that for some problems (e.g. Knapsack [3]) no deterministic algorithm can admit any competitive ratio, the natural question, "how much information about the future is needed in order to produce a competitive solution?", was posed by Dobrev et al. [4] and Böckenhauer et al. [5], and independently by Emek et al. [6]. In this work we use the framework of Hromkovic et al. [7], that unifies the models and allows posing the question in its full generality: "What is the exact power of advice bits for some specific online problem?".

---

* This work was partially supported by SNF grant 200021141089.

V. Geffert et al. (Eds.): SOFSEM 2014, LNCS 8327, pp. 77–88, 2014.

In online computation with advice, the algorithm's machine has access to a special infinite advice string ϕ, produced by an oracle that has access to the entire input. The general goal is to try to characterize the dependence of the achievable competitive ratio to the maximal number of advice bits read from the advice tape.

In this work, we focus on the advice complexity of the online version of the well-studied minimum Steiner tree problem. In the offline setting, an instance $\mathcal{I}$ to ST is a graph $G = (V, E)$ endowed with a weight function on the edges $w : E \to \mathbb{R}^+$ and a set $T \subseteq V$ of vertices called terminals. A subgraph σ of G is a solution to the instance if every pair of terminals is connected in it. The cost of a solution σ, denoted $\mathrm{cost}(\sigma)$, is the sum of the weights of the edges in it, and a solution is optimal if there exists no other solution with smaller cost.

Following previous work of Imase and Waxman [8], we consider the following natural online version of the minimum Steiner tree problem. Given a (known) weighted graph G, the terminals appear in a step-wise manner, and the algorithm maintains a subset of the edges as its solution. Upon receiving a new terminal, the algorithm extends the current solution so that the new terminal is connected to the old ones. The entire graph is known in advance, and only the specific subset of terminal vertices (and an ordering on it) is part of the instance.

More formally, given a ground graph G with a weight function w, an instance to $\mathsf{ST}(G, w)$ is an ordered list of vertices called terminals $[v_1, v_2, \ldots, v_N]$, where $v_i \in V$. At time step i, the algorithm receives terminal v_i and extends its current solution by choosing additional edges from G. The augmented solution computed by the algorithm by the end of step i is a solution to the offline problem on G with $\{v_1, \ldots, v_i\}$. As in the offline case, the cost of the solution is the total weight of edges chosen by the algorithm. An instance for $\mathsf{ST}(G, w)$ with N vertices is encoded canonically as a binary string of length $N \cdot \lceil \log(|V|) \rceil$.

An online algorithm with advice for $\mathsf{ST}(G, w)$ is strictly c-competitive using b advice bits if, for every instance, there exists an advice string ϕ such that the total weight of edges chosen by the algorithm during its computation with ϕ is at most c times the weight of the edges of an optimal solution to the offline problem, and at most b bits are read from ϕ. In general, $c = c(\cdot)$ and $b = b(\cdot)$ are function of some parameter of the input, typically the input length.

1.1   Our Contribution

We obtain a complete and exact characterization of the power of advice bits for the online Steiner tree problem.

In Section 2, we first give a variant to the greedy algorithm of [9] (without advice), which is $O(\log(N))$-competitive on an input with N terminals, and then show that our modified algorithm, which we call terminal-greedy algorithm, is $O(\log(\frac{N}{q}))$-competitive, utilizing an advice of size $q \cdot \log(|V|)$. Informally, the advice we employ is a description of the q most expensive terminals. Namely, the q terminals for which the terminal-greedy algorithm added the largest total weight of edges during its execution without advice.

In Section 3, we complement our algorithm with a matching lower bound, for the full range of advice bits.

We revisit the construction of [8], that shows a matching lower bound of $\Omega(\log(N))$ for the competitive ratio in the standard online setting (without advice). Inspired by their construction, we introduce Diamond graphs and study their properties. The construction they use can be viewed as a degenerated diamond graph. Our analysis takes a new approach using probabilistic arguments and requires a more general class of graphs in order to handle algorithms that use advice.

For every q s.t. $0 \leq q \leq N - 1$ and an online algorithm taking advice of size $q \cdot \log(|V|)$, we construct a *different* instance distribution on a suitable Diamond graph. We then employ the mechanism developed earlier in order to show that for this graph there exists an instance for which the algorithm is $\Omega(\log(\frac{N}{q}))$-competitive. Our lower bound here holds already for the unweighted case, where $w(e) = 1$ for every $e \in E$.

We observe (details are omitted in this extended abstract) that a partial result of a matching lower bound for some values of advice size $q \log(|V|)$ can be obtained using the original construction presented in [8], albeit using a different analysis. We emphasize that our new construction is essential for the proof of a matching lower bound for the full range $0 \leq q \leq N - 1$ of online algorithms using $q \cdot \log(|V|)$ advice bits.

1.2 Related Work

Imase and Waxman [8] were the first to study the Steiner Tree problem in the online setting and showed a tight bound of $\Theta(\log(N))$ for its competitive-ratio. Alon and Azar [9] show that almost the same lower bound holds also for the planar case, where the vertices are points in the Euclidean plane. Berman and Coulston [10] and Awerbuch et al. [11] study a generalized version of the problem and related problems. More recently, Garg et al. [12] considered a stochastic version of the online problem.

1.3 Some Notation

For two understood objects a and b, (i.e., instances, paths, etc.) we denote their concatenation by $a \circ b$. All logarithms are to base 2. For a non-empty set S we denote by $x \xleftarrow{\text{r}} S$ choosing an element x uniformly at random from S. For a positive natural number n, we denote $[n] \overset{\text{def}}{=} \{1, \ldots, n\}$. For a graph $G = (V, E)$ and two vertices $s, t \in V$, we denote by $s \rightsquigarrow t$ a simple path from s to t in G.

2   The Terminal-Greedy Algorithm

In this section, we present an $O(\log(\frac{N}{q}))$-competitive algorithm that utilizes $q' \overset{\text{def}}{=} q \cdot \log(|V|)$ advice bits, for any $q \in [N - 1]$.

Observe that an advice of size $(N-1)\log(|V|)$ is always sufficient in order to obtain an optimal solution, since the algorithm is required to make its first decision only upon receiving the second vertex. Therefore, one could canonically encode the rest of the input using $(N-1)\log(|V|)$ bits.

Recall that the online greedy algorithm that connects the current terminal v_i using the shortest weighted path to a vertex from the current solution is $O(\log(N))$-competitive. Our algorithm is obtained by a modification of the greedy algorithm. Whereas the greedy algorithm connects the next vertex to the current solution by using the shortest path to *any* vertex of the current solution, the terminal-greedy algorithm connects a new vertex using a shortest path to one of the terminals of the current input, ignoring possible shorter paths connecting to some non-terminal vertices already chosen by the solution.

The following lemma, whose proof is omitted in this extend abstract, was used in the proof of [9] for the standard greedy algorithm and still holds for our terminal-greedy algorithm.

Lemma 1. *Let* OptVal *denote the value of the optimal solution to an instance. Let $k \in \mathbb{N}$. The number of steps in which the terminal-greedy algorithm (without advice) adds edges of total weight more than $2 \cdot$ OptVal$/k$ is at most $k-1$.*

The terminal-greedy algorithm utilizes its advice as a list of q vertices from the instance. Intuitively, the vertices given as advice are the q most expensive ones for the input when given in an online fashion (without advice). The challenge is to show that no further costs are incurred by the algorithm using this approach.

Next, we describe the terminal-greedy algorithm with advice of size $q \cdot \log(|V|)$:[1] When the algorithm receives its first terminal vertex v_1 from the instance it computes the optimal (offline) Steiner Tree for the terminal set that consists of the q vertices given as advice along with v_1. Then, it sets the computed tree for this terminal set (which consists of $q+1$ vertices) as the current solution.

For $i \geq 2$, upon receiving a terminal v_i, the algorithm proceeds as follows: If v_i has already appeared in the advice, it does nothing. Otherwise, the algorithm computes the shortest path (in G) from v_i to all the terminals that have previously appeared as part of the instance $(v_1, \ldots, v_{i-1})$ and connects v_i using the shortest path among those $i-1$ paths (and to the lexicographically first in case that there is more than one).

Theorem 1. *Let $1 \leq q \leq N-1$. The terminal-greedy algorithm with $q \cdot \log(|V|)$ advice bits is $O(\log(\frac{N}{q}))$-competitive.*

Proof. Using induction one shows that, in every step, the chosen subgraph is a solution to the current instance. The rest of the proof is concerned with showing the bound on the cost of the algorithm.

---

[1] Fomally, following the model of Hromkovic et al. [7] the advice string is infinite, and therefore another $2\log(|V|)$ advice bits are given at the beginning of the input, encoding the value q. In our setting this can be ignored, incurring an additive constant imprecision of at most 1.

We show that, for every $q > 0$, there exists a set of size q such that, for every instance with N terminals with optimal (offline) solution of value OptVal, the solution computed by the algorithm has cost at most $O(\text{OptVal} \cdot \log(\frac{N}{q}))$.

For any $v_i \in \{v_1, \ldots, v_N\}$ we denote by $c(v_i)$ the cost incurred when adding vertex v_i according to the terminal-greedy algorithm (without advice). That is, $c(v_i)$ is the sum of the weights of all the edges chosen at step i in order to connect v_i to the solution. Let us sort the vertices of the instance according to their costs. That is, let $[v_1', v_2', \ldots, v_N']$ be the sorted permutation of $[v_1, \ldots, v_N]$, where $c(v_1') \geq c(v_2') \geq \cdots \geq c(v_N')$.

We claim that the terminal-greedy algorithm with advice $[v_1', \ldots, v_q']$ is $\log(\frac{N}{q})$-competitive. Indeed, the tree computed by the algorithm after v_1 is received has cost at most OptVal, as the optimal tree is one possible solution to it. Now, whenever a vertex v_i is received, it behaves exactly[2] as the greedy-terminal algorithm without advice would, and therefore its cost for this vertex is $c(v_i)$.

By Lemma 1, we know that for every $i \in [n]$ it holds that $c(v_i') \leq (2 \cdot \text{OptVal})/i$ as otherwise, since the v_i's are sorted, we get i vertices that each incurs a cost of more than $(2 \cdot \text{OptVal})/i$. Since $c(v_1) = c(v_n') = 0$, the total cost of the algorithm is bounded by

$$\text{OptVal} + \sum_{i=q+1}^{N-1} c(v_i') \leq \text{OptVal} + 2 \cdot \text{OptVal} \sum_{i=q+1}^{N-1} \frac{1}{i}$$

$$= \text{OptVal} \left(1 + 2 \left(\sum_{i=1}^{N-1} \frac{1}{i} - \sum_{i=1}^{q} \frac{1}{i} \right) \right)$$

$$< \text{OptVal} \left(1 + \frac{2}{\log(e)} \cdot \log(\frac{N-1}{q}) + \frac{1}{q} \right) ,$$

where in the last inequality we used the fact that $\sum_{i=1}^{k} \frac{1}{i} = \ln(k) + \gamma + \frac{1}{2k} \pm o(\frac{1}{k})$, where $\gamma \approx 0.5772$ is the Euler-Mascheroni constant. Finally, recall that a subset of the vertices of size q can be described using $q \cdot \log(|V|)$ bits. The bound follows. □

3   A Matching Lower Bound

In this section we show a lower bound matching the competitive ratio guarantee of the algorithm presented in Section 2. As mentioned, our construction holds already for the unweighted case where $w(e) = 1$, thus we omit w from our notation.

3.1   Edge-Efficient Algorithms

It will be useful for us to analyze the performance of algorithms that enjoy a canonical structure and have some guarantees on their behavior. We identify such

---

[2] Note that in general this does not hold for the 'standard' greedy-algorithm.

a class of algorithms next. An online algorithm A for ST is edge-efficient if, for every instance $\mathcal{I}$, when removing any edge from the solution $A(\mathcal{I})$, the resulting graph is not a solution. That is, removing any edge from $A(\mathcal{I})$ disconnects two terminals $v, v' \in \mathcal{I}$.

The next lemma shows that edge-efficient algorithms are as powerful as general algorithms and therefore we can focus our analysis on them. The proof of the following lemma is omitted in this extend abstract.

Lemma 2. *For every deterministic online algorithm A for ST there exists an edge-efficient algorithm A' such that, for every instance $\mathcal{I}$, we have $cost(A'(\mathcal{I})) \leq cost(A(\mathcal{I}))$.*

3.2   Diamond Graphs and Our Instance Distribution

For vertices s and t and a list of natural numbers $[\ell_1, \ell_2, \ldots, \ell_n]$, we define the **diamond graph** of level n on vertices s and t, denoted $D_n[\ell_1, \ldots, \ell_n](s, t)$, recursively as follows:

1. The graph $D_0[](s, t)$ (of level $n = 0$ with an empty list) consists of the vertices s and t and the single edge (s, t).
2. Given $G(s', t') \stackrel{\text{def}}{=} D_n[\ell_1, \ldots, \ell_n](s', t')$, a diamond graph of level n on vertices s', t', the graph $D_{n+1}[z, \ell_1, \ell_2, \ldots, \ell_n](s, t)$ is constructed as follows: We start with the vertices: s, t and $m_1, \ldots, m_z$. Next, we construct the following $2z$ copies of $G(s', t')$: $G(s, m_1), \ldots, G(s, m_z)$ and $G(m_1, t), \ldots, G(m_z, t)$, where $G(x, y)$ is a copy of the graph $G(s', t')$, where the vertices s' and t' are identified with x and y. Finally, the resulting graph is the union of the $2z$ diamond graphs $G(s, m_1), \ldots, G(s, m_z), G(m_1, t), \ldots, G(m_z, t)$.

We call the parameter ℓ_i the **width** of level i of the graph. and the vertices $m_1, \ldots, m_{\ell_1}$ the **middle vertices** of $D_n[\ell_1, \ldots, \ell_n](s, t)$. Note that the graphs in the union are almost disjoint, that is, any two of them share at most one vertex (and no edges).

For a fixed $n \in \mathbb{N}$ our instance distribution generates simultaneously an instance $\mathcal{I}$ that contains $N + 1 = 2^n + 1$ terminals, and a path P from s to t of length $N = 2^n$, which is an optimal solution to it.[3] The first two vertices are always s and t, and vertices along the path are chosen level by level, where choosing the vertices of level $i + 1$ can be thought of as a refinement of the path along the vertices of level i. The idea is that the algorithm has to connect all the level-i vertices before level-$(i + 1)$ vertices are revealed. Formally, the instance of $ST(D_n[\ell_1, \ldots, \ell_n](s, t))$ is generated according to Process 1.

The following propositions follow by simple induction and the definitions of $D_n[\ell_1, \ldots, \ell_n](s, t)$, GENERATEINSTANCE and GENERATEPATH.

Proposition 1. *The graph $D_n[\ell_1, \ldots, \ell_n](s, t)$ contains $2^n \cdot \prod_{i=1}^{n} \ell_i$ edges.*

---

[3] For simplicity of presentation, we use an instance of $N + 1$ instead of N terminals.

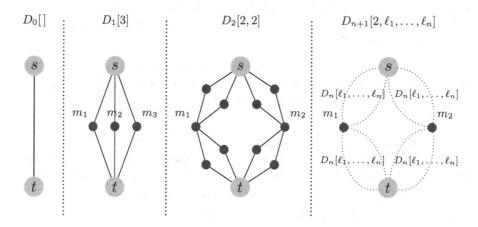

Fig. 1. Diamond Graphs

Process 1. GENERATEINSTANCE

Input: A graph $D_n[\ell_1, \ldots, \ell_n](s, t)$
Output: An instance $\mathcal{I}$ of $\mathbf{ST}(D_n[\ell_1, \ldots, \ell_n](s, t))$
1: $\mathcal{I} \leftarrow [s]$ ▷ Every instance starts with the vertex s
2: $\mathcal{I} \leftarrow \mathcal{I} \circ [t]$ ▷ followed by the vertex t.
3: $P \leftarrow$ GENERATEPATH$(D_n[\ell_1, \ldots, \ell_n](s, t))$
4:
5: **procedure** GENERATEPATH$(D_k[\ell'_1, \ldots, \ell'_k](u, v))$
6:     **if** $k = 0$ **then**
7:        **return** $e = (u, v)$
8:     **else**
9:        Choose $x \xleftarrow{\text{r}} \{m_1, \ldots, m_{\ell'_1}\}$ ▷ $m_1, \ldots, m_{\ell'_1}$ are the middle vertices of
10:       $\mathcal{I} \leftarrow \mathcal{I} \circ [x]$ ▷ $D_k[\ell'_1, \ldots, \ell'_k](u, v)$
11:       $P_1 \leftarrow$ GENERATEPATH$(D_{k-1}[\ell'_2, \ldots, \ell'_k](u, x))$
12:       $P_2 \leftarrow$ GENERATEPATH$(D_{k-1}[\ell'_2, \ldots, \ell'_k](x, v))$
13:       **return** $P_1 \circ P_2$
14:     **end if**
15: **end procedure**

Proposition 2. *Let $n \geq 1$. A simple path $s \rightsquigarrow t$ on $D_n[\ell_1, \ldots, \ell_n](s, t)$ is of the form $s \rightsquigarrow x \rightsquigarrow t$ for some $x \in \{m_1, \ldots, m_{\ell_1}\}$ and contains exactly 2^n edges.*

Proposition 3. *The path P computed during the execution of GENERATEINSTANCE$(D_n[\ell_1, \ldots, \ell_n](s, t))$ is a solution to the generated instance that contains exactly 2^n edges.*

Proposition 4. *During the run of GENERATEPATH$(D_k[\ell'_1, \ldots, \ell'_k](u, v))$, when the algorithm adds a vertex x as the next vertex of the instance $\mathcal{I}$ (Line 10), both u and v have already appeared in $\mathcal{I}$ and no other vertex from $D_k[\ell'_1, \ldots, \ell'_k](u, v)$ is contained in $\mathcal{I}$.*

Lemma 3. *Consider an execution of GENERATEINSTANCE$(D_n[\ell_1, \ldots, \ell_n](s, t))$, and let A be an edge-efficient algorithm. The number of edges added to the solution by A during every call to GENERATEPATH$(D_k[\ell'_1, \ldots, \ell'_k](u, v))$ is at least $W_k \stackrel{\text{def}}{=} \sum_{i=1}^{k} \left(\frac{2^k}{2^i} \sum_{j=1}^{2^{i-1}} X_{i,j} \right)$, where the $X_{i,j}$'s are independent Bernoulli random variables with $\Pr[X_{i,j} = 0] = 1/\ell'_i = 1 - \Pr[X_{i,j} = 1]$.*

Before proving the lemma, we prove the following proposition on the structure of the current solution restricted to the subgraph $D_k[\ell'_1, \ldots, \ell'_k](u, v)$ when GENERATEPATH(u, v) is called.

Proposition 5. *Let $k \in \{1, \ldots, n\}$ and let A be an edge-efficient algorithm. Consider an execution of GENERATEPATH$(D_n[\ell_1, \ldots, \ell_n](s, t))$. Whenever a call GENERATEPATH$(D_k[\ell'_1, \ldots, \ell'_k](u, v))$ is made, either (1) the current solution chosen by A restricted to the subgraph $D_k[\ell'_1, \ldots, \ell'_k](u, v)$ contains no edges, or, (2) $D_k[\ell'_1, \ldots, \ell'_k](u, v)$ contains a simple path of the form $u \rightsquigarrow y \rightsquigarrow v$ for some $y \in \{m_1, \ldots, m_{\ell'_1}\}$, where $m_1, \ldots, m_{\ell'_1}$ are the middle vertices $D_k[\ell'_1, \ldots, \ell'_k](u, v)$, and no other edges.*

Proof. By Proposition 4, we know that the vertices u and v have already appeared in the instance, and therefore they are connected in the current solution, and, in particular, by some simple path $u \rightsquigarrow v$. Consider the first edge (u, z) of this path. If z is contained in $D_k[\ell'_1, \ldots, \ell'_k](u, v)$, then, since the only way to reach a vertex outside of $D_k[\ell'_1, \ldots, \ell'_k](u, v)$ is through the vertices u and v, the entire path is contained in $D_k[\ell'_1, \ldots, \ell'_k](u, v)$. Conversely, if z is not contained in $D_k[\ell'_1, \ldots, \ell'_k](u, v)$, by the same argument, it follows that the path $u \rightsquigarrow v$ does not contain any inner vertex of $D_k[\ell'_1, \ldots, \ell'_k](u, v)$.

We argue that in both cases no other edges of the current solution are incident to $D_k[\ell'_1, \ldots, \ell'_k](u, v)$. Assume the contrary, and let $e \in D_k[\ell'_1, \ldots, \ell'_k](u, v)$ be such an edge (not in $u \rightsquigarrow v$ in the first case and any edge in $D_k[\ell'_1, \ldots, \ell'_k](u, v)$ in the second case).

Observe that, since e is an internal edge of $D_k[\ell'_1, \ldots, \ell'_k](u, v)$ not on the path $u \rightsquigarrow v$ and since by Proposition 4 at this point no other internal vertex is chosen to the current instance, the vertices of the current instance remain connected after removing e. This contradicts the edge-efficiency property of the current solution chosen by A. $\qquad\square$

Proof (of Lemma 3). We use induction on k, the parameter of the diamond subgraph. For $k = 0$, the claim holds trivially since $W_0 = 0$ and at least zero edges are added. Let $k > 0$, and assume that the claim holds for all $k' < k$. Let GENERATEPATH$(D_k[\ell'_1, \ldots, \ell'_k](u, v))$ be a call made during the execution of GENERATEINSTANCE$(D_n[\ell_1, \ldots, \ell_n](s, t))$. By Proposition 5, the solution chosen limited to $D_k[\ell'_1, \ldots, \ell'_k](u, v)$ either (1) has no edges, or, (2) has exactly one simple path between u and v, which by Proposition 2 has the form $u \rightsquigarrow y \rightsquigarrow v$, for $y \in \{m_1, \ldots, m_{\ell'_1}\}$. Without loss of generality, we assume that the path is of the form $u \rightsquigarrow m_1 \rightsquigarrow v$. In the first case, after lines 9 and 10, the algorithm connects the vertex x to the graph, which must be via the vertex u or v, in which case $\frac{2^k}{2}$ edges are added. In the second case, with probability $1 - 1/\ell'_1$ the vertex x is chosen from $m_2, \ldots, m_{\ell'}$, in which case as before, it must be connected using a path from u or v, which adds $\frac{2^k}{2}$ edges.

We conclude that the number of edges added to the solution due to the choice of the vertex x is at least $\frac{2^k}{2} X_{1,1}$, where $X_{1,1}$ is distributed according to $\Pr[X_{1,1} = 0] = 1/\ell'_1 = 1 - \Pr[X_{1,1} = 1]$.

Additionally, using the inductive hypothesis, the algorithm adds $W'_{k-1} = \sum_{i=1}^{k-1} \left(\frac{2^{k-1}}{2^i} \sum_{j=1}^{2^{i-1}} X'_{i,j} \right)$ and $W''_{k-1} = \sum_{i=1}^{k-1} \left(\frac{2^{k-1}}{2^i} \sum_{j=1}^{2^{i-1}} X''_{i,j} \right)$ edges during the executions of GENERATEPATH$(D_{k-1}[\ell'_2, \ldots, \ell'_k](s, x))$ and GENERATEPATH$(D_{k-1}[\ell'_2, \ldots, \ell'_k](x, t))$, respectively, where $X'_{i,j}$ and $X''_{i,j}$ are Bernoulli random variables distributed according to $\Pr[X'_{i,j} = 0] = \Pr[X''_{i,j} = 0] = 1/\ell'_{i+1} = 1 - \Pr[X''_{i,j} = 1] = 1 - \Pr[X'_{i,j} = 1]$.

Moreover, since the random choices of GENERATEPATH are independent, we have that the Bernoulli random variables are independent. Setting $X_{i+1,j} \overset{\text{def}}{=} X'_{i,j}$ and $X_{i+1,2^{i-1}+j} \overset{\text{def}}{=} X''_{i,j}$ for all $i \in \{1, \ldots, n\}$ and $j \in \{1, \ldots, 2^{i-1}\}$, we obtain that during the execution of GENERATEPATH the algorithm adds at least $\frac{2^k}{2} X_{1,1} + W'_{k-1} + W''_{k-1} = \sum_{i=1}^{k} \left(\frac{2^k}{2^i} \sum_{j=1}^{2^{i-1}} X_{i,j} \right)$ edges to the solution. The lemma is proved.   □

Corollary 1. *Any deterministic algorithm A for* ST, *when given an instance generated by* GENERATEPATH$(D_n[\ell_1, \ldots, \ell_n](s, t))$, *outputs a solution that contains at least*

$$\sum_{i=1}^{\log(N)} \left(\frac{N}{2^i} \sum_{j=1}^{2^{i-1}} X_{i,j} \right)$$

edges, where the $X_{i,j}$'s are as in Lemma 3.

Proof. By Proposition 2, we may assume that A is an edge-efficient algorithm. The corollary follows by Lemma 3.   □

We refer to an edge added due to some $X_{i,j} = 1$ as an edge of level i and say that in this case the algorithm made a wrong choice on $X_{i,j}$. Indeed, in this case it was possible to connect some vertices u and v through a middle vertex m such that the algorithm would not have had to add edges due to $X_{i,j}$.

3.3   Deriving the Lower Bound

In this section, we show that for every algorithm with advice size $q \cdot \log(|V|)$ the terminal-greedy algorithm is best possible.

The input distribution we use is a diamond graph with parameters that depend on the advice length of the specific algorithm it seeks to fail. Consider an algorithm taking $N \cdot 2^{-j(N)} \cdot \log(|V|)$ advice bits, where $\log(N) \geq j(N) \geq 0$.

We can assume that $\log(N) \geq j(N) > 10$ and, furthermore, that $j(N)$ is an even integer number. The first assumption is trivial to satisfy, since every algorithm is at most strictly 1-competitive and so for a constant $j(N)$ the asymptotic bound already holds. The second assumption incurs an additive term of 2 (recall that the bound we show is logarithmic). Therefore, both assumptions are made without loss of generality.

Set $j'(N) \stackrel{\text{def}}{=} \frac{j(N)}{2}$ and consider $D_n[\ell_1, \ldots, \ell_n](s, t)$, the diamond graph with $\log(N)$ levels, where the first $\log(N) - j'(N)$ levels are of width 2 and the last $j'(N)$ levels are of width N^2. That is, $\ell_1 = \cdots = \ell_{\log(N)-j'(N)} = 2$ and $\ell_{\log(N)-j'(N)+1} = \cdots = \ell_{\log(N)} = N^2$. For the rest of this section we refer to this graph as G.

We can show that for every online algorithm with $q \cdot \log(|V|)$ advice bits there exists an input on which it does not perform better than $\Omega(\log(\frac{N}{q}))$ compared to an optimal offline solution:

Theorem 2. *Let A be an online algorithm for* ST *taking* $q' \stackrel{\text{def}}{=} q \cdot \log(|V|)$ *advice bits, where* $q \stackrel{\text{def}}{=} N \cdot 2^{-j(N)}$ *and* $\log(N) \geq j(N) \geq 0$. *Then A has a competitive ratio of at least* $\Omega(\log(\frac{N}{q}))$.

We present an overview of the proof. Recall that for a fixed advice string $\phi \in \{0, 1\}^{q'}$, the algorithm A "hard-wired" with ϕ (denoted A_ϕ) is a deterministic online algorithm and therefore Corollary 1 establishes that on a random instance of GENERATEINSTANCE($D_n[\ell_1, \ldots, \ell_n](s, t)$) it chooses at least W_n edges.

We show that, a solution for an instance chosen by GENERATEINSTANCE contains, with very high probability, roughly $N \cdot \log(\frac{N}{q})$ edges, and then use the union bound to show that there exists an instance that makes all of the $2^{q'}$ algorithms choose this number of edges.

Using the machinery developed for general diamond graphs and the properties of GENERATEINSTANCE we show that, by our choice of $j'(N)$, it holds that $\log(|V|)$ is not too large, and for each of the last $j'(N)$ levels of the graph, a fixed deterministic algorithm chooses a linear (in N) number of edges with probability roughly $2^{-q'} = 2^{-q \cdot \log(|V|)}$.

Finally, we use the probabilistic method and show that there exists an instance on G, such that every A_ϕ chooses many edges on every level τ of the last $j'(N)$ levels in G.

Proof. Using Proposition 1 we have that the number of edges in G is $2^n \cdot 2^{(n-j'(N))} \cdot (N^2)^{j'(N)} < 4^n \cdot (N^2)^{j'(N)}$. Since the number of vertices in a graph is at most twice the number of its edges, we obtain that $|V|$, the number of vertices in G, is at most $2 \cdot 4^n \cdot (N^2)^{j'(N)}$, and therefore $\log(|V|) < 4 \cdot \log(N) \cdot j'(N)$. Therefore, we can bound the advice size by

$$q' = 2^{n-j(N)} \cdot \log(|V|) < 2^{n-j(N)} \cdot 4 \cdot \log(N) \cdot j'(N) = 4 \cdot 2^{n-2 \cdot j'(N)} \cdot n \cdot j'(N) \ . \quad (1)$$

By Lemma 3 and Corollary 1, for every level τ, where $n - j'(N) < \tau \le n$, the probability that an edge-efficient deterministic algorithm is correct on at least $\frac{2^{\tau-1}}{4}$ of its choices for level τ (i.e., at least this number of $X'_{\tau,j}$s are 0) can be computed as

$$\Pr\left[\exists S \subset [2^{\tau-1}] : |S| = \frac{2^{\tau-1}}{4} \wedge \forall p \in S : X_{\tau,p} = 0\right] < \binom{2^{\tau-1}}{\frac{2^{\tau-1}}{4}} \cdot \left(\frac{1}{N^2}\right)^{\frac{2^{\tau-1}}{4}}$$

$$\le (2^{\tau-1})^{\frac{2^{\tau-1}}{4}} \cdot \left(\frac{1}{N^2}\right)^{\frac{2^{\tau-1}}{4}}$$

$$= \left(\frac{2^{\tau-1}}{2^{2n}}\right)^{\frac{2^{\tau-1}}{4}}$$

$$\le \left(\frac{1}{2^n}\right)^{\frac{2^{\tau-1}}{4}}$$

$$= 2^{-\left(\frac{n \cdot 2^{\tau-1}}{4}\right)}$$

$$\le 2^{-\left(\frac{n \cdot 2^{n-j'(N)}}{4}\right)} \ .$$

Next we apply the union bound twice: The probability p that there exists a level $n - j'(N) < \tau \le n$ for which one of the $2^{q'}$ deterministic algorithms makes more than $\frac{2^{\tau-1}}{4}$ correct choices can be bounded as follows:

$$p < 2^{q'} \cdot j'(N) \cdot 2^{-\left(\frac{n \cdot 2^{n-j'(N)}}{4}\right)} \quad (2)$$

$$< 2^{\left(4 \cdot 2^{(n-2 \cdot j'(N))} \cdot n \cdot j'(N)\right)} \cdot 2^{\log(j'(N))} \cdot 2^{-\left(\frac{n \cdot 2^{n-j'(N)}}{4}\right)} \quad (3)$$

$$< 2^{\left(5 \cdot 2^{(n-2 \cdot j'(N))} \cdot n \cdot j'(N)\right)} \cdot 2^{-\left(\frac{n \cdot 2^{n-j'(N)}}{4}\right)} \quad (4)$$

$$= 2^{\left(\left(n \cdot 2^{n-j'(N)}\right)\left(5 \cdot 2^{-j'(N)} \cdot j'(N) - \frac{1}{4}\right)\right)} < 1 \quad (5)$$

In turn, observe that this implies that there exists a fixed instance $\mathcal{I}'$ such that the algorithm A, for every choice of advice of length q', and for every level τ in the range, makes at least $\frac{3 \cdot 2^{\tau-1}}{4}$ incorrect choices, each of which results in an addition of $\frac{N}{2^\tau}$ edges by the algorithm. Therefore, for this instance, the algorithm chooses a solution that contains at least

$$\sum_{\tau=\log(N)-j'(N)+1}^{\log(N)} \frac{N}{2^\tau} \cdot \frac{3 \cdot 2^{\tau-1}}{4} = \frac{3}{8} \cdot N \cdot j'(N) \in \Omega(N \cdot j(N)) = \Omega\left(N \cdot \log(\frac{N}{q})\right)$$

edges.

On the other hand, recall that, by Proposition 3, since $\mathcal{I}'$ is just one of the possible instances generated by GENERATEINSTANCE, there exists a solution that consists of N edges. The lower bound of $\Omega(\log(\frac{N}{q}))$ on the competitive ratio follows.   □

Acknowledgments. We would like to thank Juraj Hromkovič for suggesting this research direction and Hans-Joachim Böckenhauer for his very helpful comments and suggestions regarding the presentation of this work. We thank the anonymous reviewers for their helpful comments.

References

1. Sleator, D.D., Tarjan, R.E.: Amortized efficiency of list update and paging rules. Commun. ACM 28(2), 202–208 (1985)
2. Borodin, A., El-Yaniv, R.: Online computation and competitive analysis. Cambridge University Press (1998)
3. Böckenhauer, H.-J., Komm, D., Král1ovič, R., Rossmanith, P.: On the advice complexity of the knapsack problem. In: Fernández-Baca, D. (ed.) LATIN 2012. LNCS, vol. 7256, pp. 61–72. Springer, Heidelberg (2012)
4. Dobrev, S., Krállovič, R., Pardubská, D.: How much information about the future is needed? In: Geffert, V., Karhumäki, J., Bertoni, A., Preneel, B., Návrat, P., Bieliková, M. (eds.) SOFSEM 2008. LNCS, vol. 4910, pp. 247–258. Springer, Heidelberg (2008)
5. Böckenhauer, H.-J., Komm, D., Krállovič, R., Krállovič, R., Mömke, T.: On the advice complexity of online problems. In: Dong, Y., Du, D.-Z., Ibarra, O. (eds.) ISAAC 2009. LNCS, vol. 5878, pp. 331–340. Springer, Heidelberg (2009)
6. Emek, Y., Fraigniaud, P., Korman, A., Rosén, A.: Online computation with advice. Theor. Comput. Sci. 412(24), 2642–2656 (2011)
7. Hromkovič, J., Krállovič, R., Krállovič, R.: Information complexity of online problems. In: Hliněný, P., Kučera, A. (eds.) MFCS 2010. LNCS, vol. 6281, pp. 24–36. Springer, Heidelberg (2010)
8. Imase, M., Waxman, B.M.: Dynamic steiner tree problem. SIAM J. Discrete Math. 4(3), 369–384 (1991)
9. Alon, N., Azar, Y.: On-line steiner trees in the euclidean plane. In: Symposium on Computational Geometry, pp. 337–343 (1992)
10. Berman, P., Coulston, C.: On-line algorithms for steiner tree problems (extended abstract). In: Leighton, F.T., Shor, P.W. (eds.) STOC, pp. 344–353. ACM (1997)
11. Awerbuch, B., Azar, Y., Bartal, Y.: On-line generalized steiner problem. Theor. Comput. Sci. 324(2-3), 313–324 (2004)
12. Garg, N., Gupta, A., Leonardi, S., Sankowski, P.: Stochastic analyses for online combinatorial optimization problems. In: Teng, S.H. (ed.) SODA, pp. 942–951. SIAM (2008)

On the Power of Advice and Randomization for the Disjoint Path Allocation Problem*

Kfir Barhum[1], Hans-Joachim Böckenhauer[1], Michal Forišek[2], Heidi Gebauer[1],
Juraj Hromkovič[1], Sacha Krug[1], Jasmin Smula[1], and Björn Steffen[1]

[1] Department of Computer Science, ETH Zurich, Switzerland
{kfir.barhum,hjb,gebauerh,juraj.hromkovic,
sacha.krug,jasmin.smula,bjoern.steffen}@inf.ethz.ch
[2] Department of Computer Science, Comenius University, Bratislava, Slovakia
forisek@dcs.fmph.uniba.sk

Abstract. In the disjoint path allocation problem, we consider a path of $L+1$ vertices, representing the nodes in a communication network. Requests for an unbounded-time communication between pairs of vertices arrive in an online fashion and a central authority has to decide which of these calls to admit. The constraint is that each edge in the path can serve only one call and the goal is to admit as many calls as possible.

Advice complexity is a recently introduced method for a fine-grained analysis of the hardness of online problems. We consider the advice complexity of disjoint path allocation, measured in the length L of the path. We show that asking for a bit of advice for every edge is necessary to be optimal and give online algorithms with advice achieving a constant competitive ratio using much less advice. Furthermore, we consider the case of using less than $\log \log L$ advice bits, where we prove almost matching lower and upper bounds on the competitive ratio.

In the latter case, we moreover show that randomness is as powerful as advice by designing a barely random online algorithm achieving almost the same competitive ratio.

1 Introduction

Many important practical computational problems are best formulated in an online scenario, where the input arrives piecewise over time and an online algorithm has to irrevocably compute a part of the output for any given part of the input. One prominent example for such an online problem is call admission in communication networks, where a central authority has to admit or reject requests for communication between certain pairs of nodes in the network.

We consider a special case of the call admission problem in this paper, called the *disjoint path allocation problem*. Here, the communication network is simply modeled by a path of length L, where the $L+1$ vertices correspond to the nodes of the network which might want to communicate with each other using the

---

* This work was partially supported by SNF grant 200021-141089 and by VEGA grant V-12-031-00.

V. Geffert et al. (Eds.): SOFSEM 2014, LNCS 8327, pp. 89–101, 2014.

links modeled by the edges of the path. We assume that a call between two vertices is issued at some point in time, but is of unbounded duration. Moreover, we assume that any link of the path is only capable of serving one call. Thus, admitting a call between two nodes on the path prevents any node in between from participating in any communication. The goal is to admit as many calls as possible. This problem is well-studied, for an overview, see Section 13.5 in [4].

Classically, the hardness of online problems is measured using the competitive analysis introduced by Sleator and Tarjan [22] where the cost of the solution computed by an online algorithm is compared to the cost of an optimal (offline) algorithm that knows the complete input beforehand. Obviously, the offline algorithm has a big advantage by knowing the complete input. Thus, this way of measuring the hardness of online problems can be considered quite rough. Recently, advice complexity has been introduced and successfully used as a means for a more fine-grained complexity analysis of various online problems [8, 12, 14, 17]. The idea here is to measure how much information about the not yet revealed parts of the input is necessary and sufficient to be optimal or to reach a specific competitive ratio.

In the model of advice complexity, an online algorithm gets advice about the upcoming instance on an advice tape that has been prepared in advance by a clairvoyant and computationally unlimited oracle. The advice complexity is then the number of bits the algorithm reads from this advice tape. A number of online problems have already been analyzed within this model, such as paging [8], buffer management [13], job shop scheduling [8, 19], the k-server problem [7], online set cover [18], string guessing [5], metrical tasks systems [14], graph exploration [11], independent set [10], knapsack [6], bin packing [9], and graph coloring [2,3,15,21].

For a detailed introduction to the advice complexity of online problems, see [8, 17]. There is also an interesting relationship between advice complexity and randomized algorithms as discussed in [7, 14, 19].

The disjoint path allocation problem was among the first problems that were investigated using the model of advice complexity, but most upper and lower bounds were measured depending on the number of communication requests [8]. In contrast, most classical results in the competitive analysis of the disjoint path allocation problem were derived with respect to the size of the communication network, i. e., the length of the path. In this paper, we adopt this convention and analyze the advice complexity of the disjoint path allocation problem with respect to the path length. Our results can be summarized as follows. First, we prove that $L - 1$ advice bits are both necessary and sufficient to compute an optimal solution. While the upper bound is rather straightforward, we introduce a new technique to prove the lower bound which might be of independent interest. Second, we analyze the competitive ratio achievable by using a constant number of advice bits, or any advice of size $b \leq \log \log L$, respectively. Here, we are able to prove an upper bound of $\left((2^b + 1) \cdot \left(L^{\frac{1}{2^b+1}} + 2 \right) - 4 \right)$ and an almost matching lower bound. Then, we design several online algorithms with advice to achieve a constant competitive ratio. Some of our algorithms use the technique "Classify and Randomly Select" as introduced by Awerbuch et al. [1]. These upper bounds are complemented by a result of Gebauer et al. [16], who show that for every (not necessarily integer) constant c there is a $\delta = \delta(c)$ such that any c-competitive algorithm needs at least δL advice bits. We note that their result can also be generalized for slowly growing functions $c = c(L)$. In the last part, we prove that any number $b \le \log \log L$ of advice bits can be replaced by the same number of random bits while achieving (almost) the same competitive ratio in expectation. Thus, in some sense, a small number of random bits is as powerful as a small number of advice bits for this problem.

Due to space restrictions, some proofs have been omitted.

2   Preliminaries and Related Work

Let us first formally define the framework we are using. Consider an input sequence $I = (x_1, \ldots, x_n)$ of some maximization problem U with cost function $\text{cost}(\cdot)$. Let us denote by Opt an (offline) algorithm that outputs an optimal solution on every input I. We emphasize that Opt has access to the entire input sequence in advance and is computationally unbounded. Let $c \ge 1$.

Definition 1 (Online Algorithm). *An online algorithm* A *computes the output sequence* $A(I) = (y_1, \ldots, y_n)$, *where* $y_i = f(x_1, \ldots, x_i)$, *for some function* f *and* $1 \le i \le n$. *An algorithm* A *is* c-competitive *if there is a constant* α *such that* $\text{cost}(A(I)) \ge \text{cost}(\text{Opt}(I))/c - \alpha$. *We call* c *the* competitive ratio *of* A. *If* $\alpha = 0$, *then* A *is* strictly c-competitive. *We call* A optimal *if it is strictly 1-competitive.*

Definition 2 (Online Algorithm with Advice). *An online algorithm* A *with* advice *computes the output sequence* $A^\phi(I) = (y_1, \ldots, y_n)$, *where* ϕ *is an infinite bit string called the* advice *and* $y_i = f(\phi, x_1, \ldots, x_i)$, *for* $1 \le i \le n$. *The algorithm* A *is* c-competitive with advice complexity $s(n)$ *if there is a constant* α *such that, for every* n *and every input sequence* I *of length at most* n, *there is an advice* ϕ *such that* $\text{cost}(A^\phi(I)) \ge \text{cost}(\text{Opt}(I))/c - \alpha$ *holds and at most the first* $s(n)$ *bits of* ϕ *are accessed during the computation of* A^ϕ *on* I. *The definitions of* c-competitiveness and optimality are analogous to those in Definition 1.*

At times, when the input sequence I is clear from the context, we just write Opt and $\text{cost}(\text{Opt})$. For a given sequence σ, we denote by $[\sigma]_k$ the prefix of σ of length k and by $\log x$ the binary logarithm of x.

The *disjoint path allocation problem* is the following maximization problem. Given is a path and a sequence of *requests*, each of them being a subpath (given by two vertices of the path). The goal is to *admit* as many requests as possible, such that no two admitted requests share a common edge. We consider the *online* version of the problem, where the requests arrive sequentially, and the decision whether to admit a given request or not must be made before the next request arrives. Once the decision is made, it cannot be revoked later.

More formally, we define the disjoint path allocation problem as follows:

Definition 3. *The* disjoint path allocation problem (DPA), *is the following maximization problem on a path* $P = (v_0, \ldots, v_L)$. *In the first time step, the* *number L is revealed. In every successive time step t, for $2 \leq t \leq n+1$, a request is asked, represented by a subpath of P. The goal is to admit as many pairwise edge-disjoint requests as possible. An online algorithm has to decide immediately whether to admit a request.*

We write $[i, j]$ to denote the request for the subpath from v_i to v_j and we call i the start point and j the end point of the request. The *length* of $[i, j]$ is $j - i$.

We do not want to restrict ourselves to strict competitiveness for this problem. To see why, consider the following instance. An adversary Adv may first request the path P itself. If an algorithm A admits this request, then Adv requests all subpaths of P of length 1 and A achieves a competitive ratio of L. If A does not admit the first request, then Adv sends no more requests and A is not competitive at all. Thus, no deterministic algorithm can achieve a better strict competitive ratio than L. We avoid such pathological instances by setting $\alpha := 1$, as the algorithm is then free to reject the first request and still be competitive [20]. Note that, when computing the competitive ratio, this allows us to assume that any algorithm admits at least one request.

Also note that L is not a parameter of the problem, but is communicated as the first request instead. This avoids another pitfall: If an online algorithm A were designed for some specific constant L, it would always be 1-competitive when admitting at least one request, because for $\alpha := \binom{L+1}{2}$, we get

$$\text{cost}(A(I)) \geq \frac{\text{cost}(\text{Opt}(I))}{1} - \binom{L+1}{2},$$

as every subpath is, without loss of generality, requested at most once and therefore, there can be at most $\binom{L+1}{2}$ many requests in total.

The disjoint path allocation problem can be analyzed with respect to two different parameters: the number n of requested subpaths and the length L of the path. For the parameter n, the randomized setting was analyzed in [4] and the advice complexity in [8].

For L, only a lower bound on the advice complexity for optimality has previously been shown.

Theorem 1 (Komm [20]). *Every optimal online algorithm for DPA needs to read at least $L/2$ advice bits.*   □

We analyze DPA with respect to the parameter L. We establish upper and lower bounds for optimality and c-competitiveness.

3   A Technique to Prove Lower Bounds

We start by describing a generic technique to prove lower bounds on the advice complexity of online problems when solving them optimally. This technique has already been implicitly used by us and other authors for various lower bounds.

Our goal is however, to formalize this technique in an abstract way, such that it can be more generally used. The technique is based on a set of special instances such that no online algorithm can distinguish their prefixes of a certain length. Yet, an optimal online algorithm would need to behave differently on these instances, even before there is a distinction between them. Thus, the online algorithm is required to read advice in order to solve these instances optimally, as there is no other way to distinguish them.

The idea is to partition a set of instances into subsets, such that all input sequences of a given subset start with the same prefix of a given length for various prefix lengths. These partitions can be structured in a hierarchy of partitions, depending on the respective prefix length. Thus, we can describe the relationship between these instance sets with a *partition tree*.

Definition 4 (partition tree). *Consider some online problem U and a set $\mathcal{I}$ of input sequences of U. We define a partition tree of $\mathcal{I}$, denoted by $T(\mathcal{I})$, as a labeled rooted tree that satisfies the following properties:*

1. *Each vertex v of $T(\mathcal{I})$ is labeled by an instance set $\mathcal{I}_v \subseteq \mathcal{I}$ and by a natural number k_v, such that any two input sequences $I_1, I_2 \in \mathcal{I}_v$ have a common prefix of length at least k_v, i. e., $[I_1]_{k_v} = [I_2]_{k_v}$.*
2. *For every internal vertex v, the instance sets of its children form a partition of $\mathcal{I}_v$. Note that, for each child w, it holds that $k_w \geq k_v$.*
3. *For the root v of $T(\mathcal{I})$, we have $\mathcal{I}_v = \mathcal{I}$.*

Next we want to use a constructed partition tree to prove that an optimal online algorithm is required to read a certain number of advice bits. Here, the first property of a partition tree is crucial. This implies that no online algorithm can distinguish two given input sequences if it only sees their common prefix. Thus, an optimal online algorithm with advice has to use a different advice string for each set of the partition tree, as the following lemma shows.

Lemma 1. *Let $\mathcal{I}$ be a set of input sequences of some online problem U and let $T(\mathcal{I})$ be a partition tree of $\mathcal{I}$. Consider any two vertices v_1 and v_2 of $T(\mathcal{I})$, neither an ancestor of the other, with lowest common ancestor v and any two input sequences $I_1 \in \mathcal{I}_{v_1}$ and $I_2 \in \mathcal{I}_{v_2}$. Let $OPT(I_1)$ and $OPT(I_2)$ be the set of optimal output sequences for I_1 and I_2, respectively.*

If, for all $\pi_1 \in OPT(I_1)$ and all $\pi_2 \in OPT(I_2)$, $[\pi_1]_{k_v} \neq [\pi_2]_{k_v}$, then any optimal online algorithm needs a different advice for each of the two input sequences I_1 and I_2.

This leads to the following theorem that we can use to prove lower bounds.

Theorem 2. *Let $\mathcal{I}$ be a subset of the request sequences of some online problem U, and let $T(\mathcal{I})$ be a partition tree of $\mathcal{I}$ satisfying the prerequisite of Lemma 1.*

Then, any optimal online algorithm for U requires at least $\log m$ advice bits, where m is the number of leaves of $T(\mathcal{I})$. □

As there can be a lot of different optimal answer sequences associated with each leaf of the partition tree, it might be tedious to prove that the prerequisite of Lemma 1 is satisfied. Therefore, we usually construct a set $\mathcal{I}$ of input sequences and an associated partition tree with the property that (i) there is exactly one input sequence associated with each leaf of the partition tree, (ii) each of these input sequences has exactly one optimal output sequence, and (iii) the optimal output sequence for an input sequence is not optimal for all other input sequences.

4   Bounds for Achieving Optimality

Using the techniques from Section 3, we improve the known lower bound from Theorem 1 by a factor of 2. This new bound is tight.

Theorem 3. *To solve DPA optimally, $L - 1$ advice bits are necessary and sufficient.*

Proof. For the upper bound, consider an online algorithm A that reads $L - 1$ advice bits to learn whether a request that starts at the corresponding vertex should be admitted. Let $b_1 \ldots b_{L-1}$ denote the advice bits, and let $b_0 := 1$. Then, A admits a request $[i, j]$ if and only if $b_i = 1$ and $b_k = 0$, for $i < k < j$.

For the lower bound, we construct a set $\mathcal{I}$ of input sequences for which there is a partition tree $T(\mathcal{I})$ that satisfies the prerequisite of Lemma 1. Also we show that each input sequence has a unique optimal output sequence. Then, we only need to show that $T(\mathcal{I})$ has the desired number of leaves.

The requests of a particular input sequence are asked over a series of L phases, from phase L down to 1. In each phase p, all requests of length p are asked from the leftmost request to the rightmost one, except for some requests, such that each input instance is different. More specifically, for each input sequence, an associated bit string $b_0 \ldots b_L$ with $b_0 = 1$, $b_L = 1$ and $b_1 \ldots b_{L-1} \in \{0,1\}^{L-1}$ represents the optimal solution for this input sequence as described for the upper bound. If a request $[i, j]$ in phase $(j - i)$ is designated to be admitted in the optimal solution, that is, $b_i = 1$, $b_j = 1$ and $b_k = 0$, for $i < k < j$, then, in subsequent phases, no requests that overlap $[i, j]$ are requested. An example of such an input sequence is shown in Figure 1.

Now, we show that each input sequence has a unique optimal output sequence. Let $\mathsf{Opt}(I)$ be a optimal output sequence for a input sequence I. Assume that there is another output sequence $\mathsf{Opt}'(I)$ that differs at least by one answer. Two cases are possible. $\mathsf{Opt}'(I)$ refuses a request that is intended to be admitted. Then, by construction of the input sequences, no later requests overlap the subpath of the refused request. Thus $\mathsf{Opt}'(I)$ admits one request less and is therefore not optimal. In the other case, $\mathsf{Opt}'(I)$ admits a request $[i, j]$ that is not intended to be in the solution. Then $\mathsf{Opt}'(I)$ misses at least two requests, because in the bit string corresponding to I at least one bit is set to 1 between the bits b_i and b_j.

Lastly, we need to prove that the output sequences for two input sequences are already different before the corresponding input sequences differ. Let I and I' be two different input sequences and let p be the phase in which I and I' first differ. Because the requests in phase p of the two input sequences are different, there must be a request in the previous phase that should be admitted in one

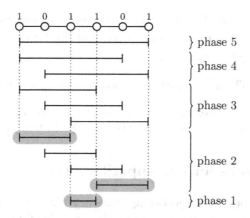

Fig. 1. Example of an input sequence represented by the bit string 101101. The unique optimal solution is highlighted in gray.

input sequence, but not in the other. Otherwise, the two input sequences would be the same, as the requests in phase p are dependent on whether some requests should be admitted in the previous phase. Hence, an optimal output sequence for I cannot also be optimal for I', and thus the prerequisite for Lemma 2 is given.

The $L-1$ bit strings define 2^{L-1} different input sequences, each belonging to a different leaf of $T(\mathcal{I})$. We have 2^{L-1} leaves and thus, from Lemma 1 and Theorem 2, it follows that an optimal online algorithm needs at least $\log(2^{L-1}) = L-1$ advice bits to be optimal for a given input sequence. □

5 Bounds for Small Advice

Our first approach for an upper bound is to divide incoming requests into classes according to their lengths and accept only requests from one class, based on the advice. For this, we need the following simple fact about the greedy algorithm that admits every possible request.

Lemma 2. *If the length of all requests is at least t and at most s, where $t \le s$, then the greedy algorithm is $(\lfloor (s-2)/t \rfloor + 2)$-competitive.*

Theorem 4. *There is a $\left((2^b+1) \cdot \left(L^{\frac{1}{2^b+1}} + 2 \right) - 4 \right)$-competitive algorithm that uses at most b advice bits.*

Proof Sketch. We divide the set of all possible requests into $2^b + 1$ classes. Class C_1 contains all requests of length at most $L^{\frac{1}{2^b+1}}$, and class C_i, for $2 \le i \le 2^b + 1$, contains all requests of length at least $L^{\frac{i-1}{2^b+1}} + 1$ and at most $L^{\frac{i}{2^b+1}}$.

The given advice is the index $j \in \{1, \ldots, 2^b\}$ of the class C_j that contains the most requests that can be accepted by a greedy algorithm, ignoring class C_{2^b+1}. The algorithm reads j and greedily admits the requests in C_j. □

Corollary 1. $2\sqrt{L}$-competitiveness can be achieved without advice, one advice bit is sufficient to be $\left(3\sqrt[3]{L} + 2\right)$-competitive, and $\lceil \log \log (L/2) \rceil$ advice bits are sufficient to be $(4 \log L - 4)$-competitive.   □

We complement Theorem 4 with an almost matching lower bound.

Theorem 5. Any algorithm that uses at most b advice bits cannot achieve a competitive ratio better than $\left(2^b + 1\right) L^{1/(2^b+1)} - 2^{b-1} \left(2^b + 1\right) - 3 \cdot 2^b$.

6   Upper Bounds for c-Competitiveness

In this section, we present several complementary upper bounds on the advice complexity to achieve c-competitiveness. The first algorithm that we present divides the path into segments and treats those segments separately.

Theorem 6. There is a c-competitive online algorithm A, for $c = 2\sqrt{k}$ with $k \in \mathbb{N}^+$, that uses at most $\left\lceil \lceil 4L/c^2 \rceil \log 3 \right\rceil$ advice bits.

Proof Sketch. We divide the path into $N := \lceil (L+1)/k \rceil$ consecutive vertex-disjoint subpaths (or *segments*) of length $k - 1$, i.e., every segment contains k vertices and $k - 1$ edges (except maybe the last one), and there is a separating edge between any two consecutive segments.

Let Opt be an arbitrary, but fixed optimal solution. For every segment S_i, $1 \le i \le N$, A reads a number $x_i \in \{0, 1, 2\}$ from the advice tape with the following meaning. If $x_i = 0$, then no request in Opt starts in S_i. (We then call S_i empty.) If $x_i = 1$, then at least one and at most $\lfloor \sqrt{k} \rfloor$ requests in Opt start in S_i. If $x_i = 2$, then more than $\lfloor \sqrt{k} \rfloor$ requests in Opt start in S_i. This needs at most $\lceil N \log 3 \rceil$ advice bits, and A reads the entire advice after the first request.[1]

Let r be an incoming request starting in segment S_i that can still be admitted, and let S_j be the first non-empty segment to the right of S_i, i.e., $x_{i+1} = x_{i+2} = \cdots = x_{j-1} = 0$ and $x_j \ne 0$. Furthermore, let S_l be the segment in which r ends. A inspects x_i and x_j and behaves as follows. (i) If $x_i = 1$ and $x_j = 1$, then A admits r if $l \le j$. (ii) If $x_i = 1$ and $x_j = 2$, then A admits r if $l < j$. (iii) If $x_i = 2$, then A admits r if $l < j$ and r contains at most $\lfloor \sqrt{k} \rfloor$ edges in segment S_i.

If there is no segment S_j, i.e., if $S_{i+1} = S_{i+2} = \cdots = S_N = 0$, then A admits r if $x_i = 1$, or if r has length of at most $\lceil \sqrt{k} \rceil$ and $x_i = 2$.   □

The following result provides a better upper bound for sufficiently large c.

Theorem 7. There is a c-competitive algorithm A, for $c \in \mathbb{N}^+$, that uses at most $\lceil 2(L-1)/c \rceil$ advice bits.

Proof Sketch. Consider the encoding of some fixed optimal solution Opt as used in the proof of Theorem 3, i.e., a bitstring B of length $L - 1$ with a 1 at every position where a request starts that is admitted in Opt. The oracle divides the

---

[1] A then knows L and k and thus also how many advice bits to read.

subpath $(v_1, \ldots, v_{L-1})$ into consecutive vertex-disjoint segments S_i. The edge starting in segment S_i and ending in S_{i+1} is numbered among the edges of S_i. Then, the oracle writes a shorter bitstring $d_1 \ldots d_b$ of length $b := \lceil (L-1)/(c/2) \rceil$ on the tape, where d_i is the bitwise OR of all bits of B corresponding to vertices in segment S_i.

A segment with corresponding advice bit 1 is called 1-segment. Then, A admits any satisfiable request in a 1-segment that has its end point in either this segment or the next 1-segment to the right.   □

Now we generalize the approach from Theorem 4 by applying it to edge-disjoint segments of the path.

Theorem 8. *There is a c-competitive algorithm, for $c = 4 \log k$ with $k \in \mathbb{N}^{\geq 2}$, that uses at most $\left\lceil \frac{L}{2^{c/4}} \cdot \left(\frac{c}{2} + \lceil \log c \rceil + 0.33 \right) \right\rceil$ advice bits.*

We already know how to construct a bit string B of length $L - 1$ that serves as advice to be optimal. Below, we show that approximate knowledge of B can still be used to guarantee a good competitive ratio. More precisely, we prove that we can get arbitrarily close to optimality using less than $L - 1$ advice bits.

Theorem 9. *For any $c = k/(k-1)$ with $k \in \mathbb{N}^{\geq 2}$, there is a c-competitive algorithm that uses $\lceil \log(c/(c-1)) \rceil + L - 1 - \lfloor \lfloor (c-1)(L-2)/c \rfloor \cdot (2 - \log 3) \rfloor$ advice bits.*

Proof Sketch. Let $B = b_1 \ldots b_{L-1}$ be the bit string corresponding to some fixed optimal solution computed by an optimal online algorithm Opt as described in the proof of Theorem 3. Our advice string is a shorter sequence created from B by taking some pairs of consecutive bits and adding them (producing a number between 0 and 2). For instance, instead of the sequence $(0, 1, 0, 0, 1, 0, 1, 1)$ we might use the sequence $(0, 1, 0+0, 1, 0, 1+1) = (0, 1, 0, 1, 0, 2)$. If there are $p > 0$ pairs of bits and the online algorithm knows their offsets, we need $L - 1 - 2p + \lceil p \log 3 \rceil \leq L - 1$ bits to store the resulting sequence.

Given such an advice sequence, an online algorithm A may behave as follows. First, it reconstructs the sequence B. This is only ambiguous if the sum of some two consecutive bits is 1. In that case, we assume that the bits are $0 + 1$ (and not $1 + 0$). Given the reconstructed sequence, we simulate Opt.

Let q be the number of pairs of bits we reconstructed incorrectly. We can easily see that A constructs a solution with at least $\text{cost}(\text{Opt}) - q$ accepted requests: The worst case is that we lose one accepted request per such pair.

We now consider k different strategies of creating the pairs of bits. For each i, $1 \leq i \leq k$, we consider the strategy where the added pairs of bits are $b_{i+ak} + b_{i+ak+1}$, for all $a \in \mathbb{N}^{\geq 0}$, $0 \leq a \leq \left\lceil \frac{L-2}{k} - 1 \right\rceil$. Hence, in each strategy, the number of such added pairs is $p \geq \left\lfloor \frac{L-2}{k} \right\rfloor$.

A detailed analysis of the strategies yields the theorem.   □

Corollary 2. *Fewer than $0.8L$ bits are sufficient to be 2-competitive. Fewer than $0.87L$ bits are sufficient to be 3/2-competitive.*   □

It is important to observe that indeed all four of the upper bounds established above are necessary, as they complement each other. In other words, each of them covers a particular range of c. More precisely, for $1 \leq c \leq 3$, Theorem 9 provides the best upper bound. (Note that the theorem actually only applies to competitive ratios in the range $1 \leq c \leq 2$, but the upper bound for 2 trivially propagates to all larger values.) In the range $3 \leq c \leq 2\sqrt{3}$, Theorem 7 is the best choice, and for $c \geq 2\sqrt{3}$, Theorem 6 is even better. For $c \geq 64$, however, they are all outperformed by the algorithm from Theorem 8.

7   On the Power of Random Bits

We continue the study of the exact power of randomness in online algorithms (focusing on DPA) and show a trade-off between the number of random bits available to the online algorithm and its competitive ratio.

Recall that an $O(\log(L))$-competitive algorithm for DPA can be implemented using $\lceil \log \log(L) \rceil$ random bits, and this matches the lower bound that can be obtained by any randomized online algorithm (Theorems 13.7 and 13.8 in [4]).

In this section, we obtain an online algorithm that uses only b random bits and enjoys a competitive ratio of $(L^{\frac{1}{2^b}} \cdot 2^{b+1})$ for any $b \in \{0, \ldots, \lceil \log \log(L) \rceil\}$. Indeed, for $b = 0$, we obtain the greedy algorithm, and for $b = \lceil \log \log(L) \rceil$, we obtain the randomized algorithm from [4].

In this section, we identify the edge (v_t, v_{t+1}) with its right vertex and call it edge $t + 1$. That is, a request $[i, j]$ contains the edges $\{i + 1, \ldots, j\}$. Requests *intersect* if they contain a mutual edge. A set of requests *covers* a request if every edge in the request is contained by some request from the set.

It will be useful to think of an edge t as the bit-string of length $\log(L + 1)$ that denotes its binary expansion. For ease of presentation, for the rest of this section, we assume that $L = 2^\ell - 1$ for some integer ℓ, but our results here hold for any L (just think of the natural embedding of the path to the first L vertices in a path of length $L' - 1$, where L' is the smallest power of 2 larger than L).

We partition the edges into ℓ levels, where an edge e belongs to level $\lambda(e)$, where $\lambda \colon E \to \mathbb{N}$ is given by $\lambda(e) := \max \{ t : 2^t \text{ divides } e \}$. Alternatively, $\lambda(e)$ is the largest t such that, in the binary representation of e, the t right-most bits of e are zero. That is, the edge $10^{\ell-1}$ is the only edge of level $\ell - 1$, the edges $10^{\ell-2}$ and $110^{\ell-2}$ are the only edges of level $\ell - 2$, and in general, there are exactly $2^{\ell-j-1}$ edges of level j.

It will be useful to consider a coarser partition to blocks of B levels. To this end, for every $B \in \mathbb{N}^+$, we define the B-*block* of an edge $\lambda_B \colon E \to \mathbb{N}$ by $\lambda_B(e) := \lfloor \frac{\lambda(e)}{B} \rfloor$. It is immediate that $\lambda_B(e) = i$ if and only if $\lambda(e) \in \{iB, iB + 1, \ldots, (i+1)B - 1\}$. We extend λ (resp., λ_B) to any request r by setting $\lambda(r) = \max_{e \in r} \lambda(e)$ (resp., $\lambda_B(r) = \max_{e \in r} \lambda_B(e)$). We call a request a *level-t* (resp., B-*block i*) request if $\lambda(r) = t$ (resp., $\lambda_B(r) = i$).

Let Opt be an optimal solution for a DPA instance with value cost(Opt). We denote by o_t the number of requests in Opt for which $\lambda(r) = t$. Similarly, we set

$o'_i := \sum_{j=0}^{B-1} o_{iB+j}$, the number of requests in Opt for which $\lambda_B(r) = i$. It holds that $\text{cost}(\text{Opt}) = \sum_{i=0}^{\lceil \ell/B \rceil - 1} o'_i = \sum_{t=0}^{\ell-1} o_t$. We make use of the following.

Proposition 1. *If a request contains two different edges of level t, then it contains an edge of level at least $t + 1$.*

Proposition 1 asserts that every request has exactly one edge of maximal level. We call this edge the *level-edge* of the request. Additionally, for any edge e, any solution to a DPA instance contains at most one request with e as its level edge (this is true for any edge, and in particular for the level edges).

Proposition 2. *Let r be a level-t request. Then, for any $t' \geq t$, any solution of a DPA instance contains at most one level-t' request that intersects with r.*

We now present and analyze the i-th B-block greedy algorithm. The algorithm $B\text{-Block-Greedy}_i$ takes all the requests offered from B-block i as long as they do not intersect with requests already chosen to the current solution.

Proposition 3. *For any instance of DPA, $B\text{-Block-Greedy}_i$ chooses at least $2^{-B} \cdot o'_i$ requests.*

Let $B \in \{1, \ldots, \ell\}$. The (randomized) $B\text{-Block-Greedy}$ algorithm chooses uniformly at random a block $i \leftarrow \{0, \ldots, \lceil \ell/B \rceil - 1\}$, and behaves according to $B\text{-Block-Greedy}_i$. The worst-case expected cost of this algorithm is

$$\underset{i \leftarrow \{0, \ldots, \lceil \ell/B \rceil - 1\}}{\mathbf{E}} \left[\text{cost}(B\text{-Block-Greedy}_i) \right] \geq \sum_{i=0}^{\lceil \ell/B \rceil - 1} \frac{2^{-B}}{\lceil \ell/B \rceil} \cdot o'_i$$

$$= \frac{2^{-B}}{\lceil \ell/B \rceil} \sum_{t=0}^{\ell-1} o_t = \frac{2^{-B}}{\lceil \ell/B \rceil} \cdot \text{cost}(\text{Opt}).$$

Put differently, we obtain a $(\lceil \ell/B \rceil \cdot 2^B)$-competitive algorithm. Note that choosing a random B-block out of the $\lceil \ell/B \rceil$ possible blocks is the only random choice of the algorithm and can be done using $\lceil \log(\lceil \ell/B \rceil) \rceil$ random bits.[2]

So far our analysis was made in terms of the block size B. Next, we present our main theorem for this section, which delineates explicitly the competitive ratio obtained as a function of the number of available random bits b.

Theorem 10. *The randomized $B\text{-Block-Greedy}$ algorithm that uses b random bits is $(L^{\frac{1}{2^b}} \cdot 2^{b+1})$-competitive.*

Acknowledgments. The authors would like to thank Maria Paola Bianchi, Daniel Graf, and Dennis Komm for valuable discussions.

---

[2] In fact, in general, an implementation that uses only $\lceil \log(\lceil \ell/B \rceil) \rceil$ bits obtains a $(\lceil \ell/B \rceil \cdot 2^{B+1})$-competitive ratio (that is, it incurs an additional factor of 2). However, whenever $\lceil \ell/B \rceil$ is a power of two, we can save this factor while still using $\lceil \log(\lceil \ell/B \rceil) \rceil$ random bits.

References

1. Awerbuch, B., Bartal, Y., Fiat, A., Rosén, A.: Competitive non-preemptive call control. In: Proc. of SODA 1994, pp. 312–320. ACM/SIAM (1994)
2. Bianchi, M.P., Böckenhauer, H.-J., Hromkovič, J., Keller, L.: Online coloring of bipartite graphs with and without advice. In: Gudmundsson, J., Mestre, J., Viglas, T. (eds.) COCOON 2012. LNCS, vol. 7434, pp. 519–530. Springer, Heidelberg (2012)
3. Bianchi, M.P., Böckenhauer, H.-J., Hromkovič, J., Krug, S., Steffen, B.: On the advice complexity of the online $L(2,1)$-coloring problem on paths and cycles. In: Du, D.-Z., Zhang, G. (eds.) COCOON 2013. LNCS, vol. 7936, pp. 53–64. Springer, Heidelberg (2013)
4. Borodin, A., El-Yaniv, R.: Online Computation and Competitive Analysis. Cambridge University Press (1998)
5. Böckenhauer, H.-J., Hromkovič, J., Komm, D., Krug, S., Smula, J., Sprock, A.: The string guessing problem as a method to prove lower bounds on the advice complexity. Electronic Colloquium on Computational Complexity (ECCC), TR12-162 (2012)
6. Böckenhauer, H.-J., Komm, D., Královič, R., Rossmanith, P.: On the advice complexity of the knapsack problem. In: Fernández-Baca, D. (ed.) LATIN 2012. LNCS, vol. 7256, pp. 61–72. Springer, Heidelberg (2012)
7. Böckenhauer, H.-J., Komm, D., Královič, R., Královič, R.: On the advice complexity of the k-server problem. In: Aceto, L., Henzinger, M., Sgall, J. (eds.) ICALP 2011, Part I. LNCS, vol. 6755, pp. 207–218. Springer, Heidelberg (2011)
8. Böckenhauer, H.-J., Komm, D., Královič, R., Královič, R., Mömke, T.: On the advice complexity of online problems. In: Dong, Y., Du, D.-Z., Ibarra, O. (eds.) ISAAC 2009. LNCS, vol. 5878, pp. 331–340. Springer, Heidelberg (2009)
9. Boyar, J., Kamali, S., Larsen, K.S., López-Ortiz, A.: Online bin packing with advice. Technical report, arXiv:1212.4016
10. Dobrev, S., Královič, R., Královič, R.: Independent set with advice: the impact of graph knowledge. In: Erlebach, T., Persiano, G. (eds.) WAOA 2012. LNCS, vol. 7846, pp. 2–15. Springer, Heidelberg (2013)
11. Dobrev, S., Královič, R., Markou, E.: Online graph exploration with advice. In: Even, G., Halldórsson, M.M. (eds.) SIROCCO 2012. LNCS, vol. 7355, pp. 267–278. Springer, Heidelberg (2012)
12. Dobrev, S., Královič, R., Pardubská, D.: Measuring the problem-relevant information in input. RAIRO Theoretical Informatics and Applications 43(3), 585–613 (2009)
13. Dorrigiv, R., He, M., Zeh, N.: On the advice complexity of buffer management. In: Chao, K.-M., Hsu, T.-s., Lee, D.-T. (eds.) ISAAC 2012. LNCS, vol. 7676, pp. 136–145. Springer, Heidelberg (2012)
14. Emek, Y., Fraigniaud, P., Korman, A., Rosén, A.: Online computation with advice. In: Albers, S., Marchetti-Spaccamela, A., Matias, Y., Nikoletseas, S., Thomas, W. (eds.) ICALP 2009, Part I. LNCS, vol. 5555, pp. 427–438. Springer, Heidelberg (2009)
15. Forišek, M., Keller, L., Steinová, M.: Advice complexity of online coloring for paths. In: Dediu, A.-H., Martín-Vide, C. (eds.) LATA 2012. LNCS, vol. 7183, pp. 228–239. Springer, Heidelberg (2012)
16. Gebauer, H., Královič, R., Královič, R.: On lower bounds for the advice complexity of the disjoint path allocation problem. Technical report (in preparation)

17. Hromkovič, J., Královič, R., Královič, R.: Information complexity of online problems. In: Hliněný, P., Kučera, A. (eds.) MFCS 2010. LNCS, vol. 6281, pp. 24–36. Springer, Heidelberg (2010)
18. Komm, D., Královič, R., Mömke, T.: On the advice complexity of the set cover problem. In: Hirsch, E.A., Karhumäki, J., Lepistö, A., Prilutskii, M. (eds.) CSR 2012. LNCS, vol. 7353, pp. 241–252. Springer, Heidelberg (2012)
19. Komm, D., Královič, R.: Advice complexity and barely random algorithms. Theoretical Informatics and Applications (RAIRO) 45(2), 249–267 (2011)
20. Komm, D.: Advice and Randomization in Online Computation. PhD Thesis, ETH Zurich (2012)
21. Seibert, S., Sprock, A., Unger, W.: Advice complexity of the online coloring problem. In: Spirakis, P.G., Serna, M. (eds.) CIAC 2013. LNCS, vol. 7878, pp. 345–357. Springer, Heidelberg (2013)
22. Sleator, D.D., Tarjan, R.E.: Amortized efficiency of list update and paging rules. Communications of the ACM 28(2), 202–208 (1985)

Goal-Based Establishment of an Information Security Management System Compliant to ISO 27001*

Kristian Beckers

Paluno, - The Ruhr Institute for Software Technology -, University of Duisburg-Essen, Germany
{firstname.lastname}@paluno.uni-due.de

Abstract. It is increasingly difficult for customers to understand complex systems like clouds and to trust them with regard to security. As a result, numerous companies achieved a security certification according to the ISO 27001 standard. However, assembling an Information Security Management System (ISMS) according to the ISO 27001 standard is difficult, because the standard provides only sparse support for system development and documentation.

Security requirements engineering methods have been used to elicit and analyse security requirements for building software. In this paper, we propose a goal-based security requirements engineering method for creating an ISMS compliant to ISO 27001. We illustrate our method via a smart grid example.

Keywords: security standards, requirements engineering, SI*.

1 Introduction

The increasing complexity of software systems and the surrounding environment is challenging to analyse with regard to security. Security standards, e.g. the ISO 27001 standards, offer a way to attain this goal. The ISO 27001 standard defines how to establish an information security management system (ISMS). This is a concern for the security needs of an organisation. Several relevant companies have taken this approach like Amazon[1]. However, the sparse descriptions in it makes the establishment of an ISO 27001 compliant ISMS difficult. For example, the standard contains a description of the scope and boundaries of the ISMS. The standard states only to consider "characteristics of the business, the organisation, its location, assets and technology" [1, p. 4].

Re-using well established methods security requirements engineering (SRE) methods, e.g., SI* [2] for establishing an ISMS according to the ISO 27001 is a possible solution. We provided a mapping from the ISO 27001 standards demands to the capabilities of SRE methods in a previous work [3].

This work is inspired by this mapping and shows how to use SI* for establishing an ISO 27001 ISMS. Our approach provides a structured refinement of the IT system's and stakeholders' information to assess the threats for a particular system. Our method

---

* This research was partially supported by the EU project Network of Excellence on Engineering Secure Future Internet Software Services and Systems (NESSoS, ICT-2009.1.4 Trustworthy ICT, Grant No. 256980). We thank Jorge Cuéller for his valuable feedback on our work.
[1] http://aws.amazon.com/security/

V. Geffert et al. (Eds.): SOFSEM 2014, LNCS 8327, pp. 102–113, 2014.

uses this information for risk assessment and security control selection according to the ISO 27001 standard. We also provide the required documentation of an ISMS for certification. We illustrate our approach by the example of a smart grid providing scalable energy infrastructure to consumers. We consider in particular the security of the smart metering gateway, the interface between the energy grid and the customer.

2 ISO 27001

The ISO 27001 standard is structured according to the "Plan-Do-Check-Act" (PDCA) model, the so-called *ISO 27001 process* [1]. In the *Plan* phase an ISMS is established, in the *Do* phase the ISMS is implemented and operated, in the *Check* phase the ISMS is monitored and reviewed, and in the *Act* phase the ISMS is maintained and improved. In the *Plan* phase, the *scope and boundaries* of the ISMS, its *interested parties, environment, assets*, and all the *technology* involved are defined. In this phase, also the ISMS *policies, risk assessments, evaluations*, and *controls* are defined. Controls in the ISO 27001 are measures to *modify risk*.

3 SI*

We use the SI* modeling language [2] for creating a refined ISMS scope definition, because SI* provides the means to model social dependencies between actors including security and trust relations. In SI* roles are abstractions of sets of actors, which are active entities that have goals. A goal is a state of affairs that the actor desires and that the system-to-be should possibly help to fulfill. Softgoals are similar, but have no clear criteria for stating if they are fulfilled or not. A resource is a physical or informational entity. Goals and resources can be refined using *AND decomposition*s, these have the word AND under a half circle. A *means-end* is an arrow that points towards a goal that provides the means to achieve a goal or the resources needed or produced by a goal.

Own relations denote that an actor owns a resource, or can decide if a goal is achieved. This relation is labeled with an **O**. Provide relations denote that an actor has the ability to achieve a goal or furnish a resource. This relation is labeled with a **P**. The own and provide relations are part of the so-called *Eco Model*.

SI* supports various trust relations, which are modelled as edges labeled with an abbreviation of the kind of trust relation it represents. Execution dependency **De** and permission delegation **Dp** allow the transfer of objectives and entitlements from an actor to another. Execution dependency **De** means that one actor appoints another actor to achieve a goal or furnish a resource. Permission delegation **Dp** indicates that an actor authorises another actor to achieve a goal or deliver a resource. Trust is a relation representing the expectations that an actor (the trustor) has in regards to the behavior of another actor (the trustee). A goal or a resource is part of a trust relation (the trustum). Trust of execution **Te** models the trustor's expectations regarding the ability and dependability of the trustee in fulfilling a goal or delivering a resource. Trusting in execution **Tp** means that the trustor is certain that the trustee accomplishes the trustum. Trust of permission models the trustor's expectations that the trustee does not abuse a goal or a resource. Trusting in permission means that the trustor is certain that the trustee does not misuse the (possible) received permission for accomplishing an aim different from the one for which the permission has been granted. Distrust execution **Se** models the explicit doubts about the behaviour of the trustor from the trustee about the abuse of a goal or a resource.

4   A Method for Goal-Based ISMS Establishment

We propose a method for creating an ISMS compliant to the ISO 27001 standard, which consists of the following steps:

Step 1: Get Management Commitment - The precondition for building an ISMS is that the management commits to it. Thus, we dedicated the first step of our method to elicit the management commitment of the project and the provision of adequate resources to do so. We create SI* diagrams that state the concerned roles and actors of an ISMS. The management commitment for an ISMS shall be granted for these roles and actors. The management commitment has to be gathered repeatedly when the ISMS is further described. However, starting from the initial definition of concerned stakeholders a management commitment should be given in written form. Without this commitment, insufficient resources will result in an insufficient ISMS.

Step 2: ISMS Scope Definition - We define the scope of the ISMS using the SI* diagrams created in the previous step. Although, we could have used other goal modeling notations, SI* provides the means to model trust into a goal model, which is essential for our asset identification and threat analysis. In addition, SI* is scalable, since it is possible to have multiple diagrams/views of the same model.

Step 3: Identify Assets - The entire ISMS scope description is the input for the asset identification. We identify all items of value of stakeholders by analyzing various relations in the SI* model. These range from resource, goal, stakeholder relations to the trust relations in the SI* model. These also help to clearly define the need for protection of the identified assets. In addition, a high level risk assessment of the assets is conducted. This step results in a list of assets, the stakeholders that own them, and initial risk levels for assets as an output.

Step 4: Analyze Threats - We conduct a threat analysis via modeling attackers to these threats in the SI* model. Attackers have to be of a specific type, which contains assumptions about the capabilities and motivations of the attacker. These attackers present threats to assets. The threats lead to the elicitation of security requirements. We use misuse cases [4] to map the threats to security requirements.

Step 5: Conduct Risk Assessment and Control Selection - The reasoning about controls starts with the risk assessment for each asset. For each asset the decision has to be made if the risk to that asset demands the inclusion of one or more controls of the ISO 270001 standard or if the risk levels are sufficient. For each asset we propose to compile a list that states why a list of each of the controls in the normative ANNEX A of the ISO 27001 should or should not be applied to the asset.

Step 6: Design ISMS Specification - The final step of our method concerns the ISO 27001 specification, an implementable description of the ISMS. We consider the ISO 27001 documentation demands and use the information elicited and documented in the previous steps of our method. This information is mapped to the required document types for certification.

5   Application of Our Method to a Smart Grid Scenario

We illustrate the benefits of our framework on a case study of a Smart Grid system. The case study was provided by the industrial partners of the EU project NESSoS[2]. A smart grid is a commodity network that intelligently manages the behavior and actions of its participants. The commodity consists of electricity, gas, water, or heat that is distributed via a grid (or network). The benefit of this network is envisioned to be a more economic, sustainable, and secure supply of commodities. Smart metering systems meter the consumption or production of energy and forward the data to external entities. This data can be used for billing and steering the energy production.

Step 1: Get Management Commitment - The ISO 27001 standard demands documentation of management commitment for the establishment of an ISMS. The demands are described in Sect. 5 of the standard. *Sect. 5.1 Management Commitment* concerns proof that the management shall provide for establishing an ISMS including objectives, plans, responsibilities and accepting risks. *Section 5.2 Resource Management* concerns the provision of resources for establishing the ISMS and the training of the members of the organization for security awareness and competence.

The management commitment for implementing an ISMS according to the ISO 27001 standard is of utmost importance, because without the commitment of sufficient staff and resources the ISMS implementation is doomed to fail. In addition, the publicly available sources of examples of ISMS implementations, e.g., the ISMS toolkit, define this also as the first step when implementing an ISMS[3].

The management commitment should be based upon a high level description of the part of an organization for which the management commits resources to build an ISMS. We propose to use SI* diagrams for this purpose, because these define stakeholders and operations for which an ISMS shall be established. A refined description using static and behavioral description is done during the ISMS scope refinement (see below). We propose to mark the ISMS scope in the diagram. and use a scenario-based elicitation of stakeholders. The management commitment for establishing an ISMS for the scope has to be presented in writing and in relation to a specific person, who is responsible for providing the required resources. The management commitment should relate to the use case diagram, e.g., let the management commitment state that the service provider can provide services in a secure environment.

Step 2: ISMS Scope Definition - After acquiring the management commitment, we have to provide a more detailed scope definition. Section 4 of the ISO 27001 standard describes the ISMS and in particular in Sect. 4.2 - Establishing and managing the ISMS - states the scope definition. Section 4.2.1 a demands to "Define the scope and boundaries of the ISMS in terms of the characteristics of the business, the organization, its location, assets and technology, and including details of and justification for any exclusions from the scope "[1, p.4]. In Sect. 4.2.1 d, which concerns risk identification, the scope definition is used to identify assets. Section 4.2.3 demands a management review of the ISMS that also includes to check for possible changes in the scope of the ISMS. Section 4.3 lists the documentation demands of the standard and Sect. 4.3.1 d requires a

---

[2] http://www.nessos-project.eu/
[3] http://www.iso27001security.com/html/iso27k_toolkit.html

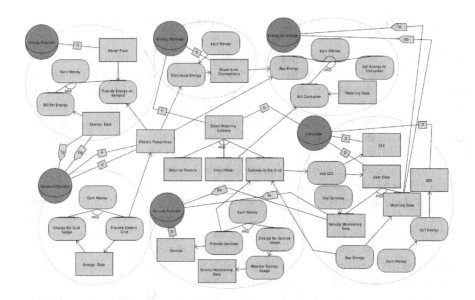

Fig. 1. Smart Grid SI* Diagram with Resources

documentation of the scope of the ISMS. The ISMS scope definition of the ISO 27001 standard is a vital step for its successful implementation, because all subsequent steps use it as an input.

We excluded the *Energy Market* from the scope of the ISMS and show the goals of the roles the ISMS is concerned with and their subgoals. These goals are shown in Fig. 1. Figure 1 presents a refined SI* diagram of our smart grid scenario. The *Energy Provider* owns *Power Plant*s and has the goal to *Earn Money*. The goal is decomposed into the subgoals to *Provide Energy on demand* and to *Bill for Energy* sales. In order to be able to bill for energy consumption, the *Energy Provider* has to acquire *Energy Data* that states the actual consumption of energy in the grid. Energy that is not consumed by a *Consumer* is an economic loss. Possible causes are loss of energy during transfer or just energy that fades in the grid, because of lack of energy storing capabilities. The *Energy Provider* has a delegation permission relation with the *Network Provider* for collecting the *Energy Data*. The *Energy Provider* trusts the *Network Provider*, because of long lasting partnership. Hence, both parties have a trust permission relation regarding the *Energy Data*.

The *Network Provider* also aims to *Earn Money* and this goal is decomposed into the goals *Charge for Grid Usage* and *Provide Electric Grid*. The *Provide Electric Grid* requires *Electric Powerlines*, which are owned and provided by the *Network Provider*. The *Energy Data* are an outcome of the realization of the goal *Provide Electric Grid*. The *Energy Data* are a means to achieve the goal *Charge for Grid Usage*.

The *Service Provider* also wants to *Earn Money* and does this via the subgoals *Provide Services* and *Charge for Service Usage*. The *Provide Services* goal leads to

Services, which are owned by the *Service Provider*. The *Charge for Service Usage* requires the subgoal *Monitor Service Usage*, which results in *Service Monitoring Data*. This data is collected from the *Consumer*. This is the reason why the *Service Provider* has an execution dependency relation with the *Consumer*. In addition, the *Consumer* is not familiar with the business practices of the *Service Provider*, which results in a distrust execution relation between these stakeholders.

The *Consumer* wants to *Buy Energy* and in addition *Use Services* and *Use CLS*, which are Controllable Local Systems (CLS). These are electronic components that use the *Smart Metering Gateway*. An example for a CLS is a controllable air conditioning. *CLS* devices are owned by the *Consumer*. The *Consumer* owns *User Data*. This data results from fulfilling the goal *Use CLS*. The *Use Services* goal produces *Service Monitoring Data* that the *Energy Distributer* generates for billing purposes. The *Buy Energy* goal also results in *Metering Data* about energy consumption. The *Metering Data* is shared with the *Energy Distributer* using a delegation permission. The *Consumer* trusts the *Energy Distributer* to use this data only for billing purposes. The *Consumer* has also the goal to *Earn Money* via the subgoal *Sell Energy*. The *Consumer* uses an *EPS* for this purpose, which is an Energy Producing System. The *Customer* owns the *EPS*, which can be for example solar panels.

The *Energy Distributer* aims to *Earn Money* by the goal *Sell Energy to Consumers*. The *Energy Distributer* wants to *Buy Energy*, which can be sold to *Consumers*. The *Energy Manager* provides the energy from the grid, so that the *Energy Distributer*'s customers can receive it. The goal *Bill Consumer* requires *Metering Data* that state the amount of energy used by *Consumers*. The *Energy Manager* owns *Power Line Connections* that are connected with the energy grid. These allow the *Energy Manager* to *Distribute Energy* throughout the grid.

The SI* model is an adequate description of an ISMS scope, because the SI* models describe the "characteristics of the business and the the organization" by analyzing and documenting goals, agents/roles including their relations. The standard also demands a documentation of the "technology" involved. These are included in the models via resources and its relations with goals and agents/roles.

The standard further demands the definition of "location". We propose to attach templates to the Si* model. The location template, shown in Tab. 1, lists the location of all Si* resources and agents/roles. Goals are not listed here, because these do not have physical locations. Moreover, the standard demands "details of and justification for any exclusions from the scope". We propose to use a scope exclusion template for that purpose that lists all resources and agents/roles that are excluded from the scope. We already excluded the *Service Provider* and *Energy Manager* from the scope of the ISMS. The template in Tab. 2 states the reasoning behind these decisions. We consider assets, which are also part of the scope of an ISMS, in the following part of our method.

Table 1. Instantiated Template for Locations of ISMS Elements

Si* Element	Location
power plant	Hannover, Germany
. . .	. . .

Table 2. Instantiated Template for Locations of ISMS Elements

Si* Element	Reason for Scope Exclusion
Energy Manager	The Energy Manager has already an ISO 27001 compliant ISMS in place.
Service Provider	The Service Provider offers software of different kinds to the Consumer. It is assumed that the Service Provider certifies all services compliant to the Common Criteria [5].
...	...

Step 3: Identify Assets - The design goal of the ISO 27001 ISMS is to protect assets with adequate security controls and this is stated already on page 1 of the standard. Section 4.2.1 a of the standard demands the definition of assets. Section 4.2.1 b concerns the definition of ISMS security policies and it demands that the policy shall consider assets. Section 4.2.1 d that concerns risk identification uses the scope definition to identify assets, to analyse threats to assets, and to analyse the impacts of losses to these assets. Section 4.2.1 e concerns risk analysis, which also clearly define to analyse assets and to conduct a vulnerability analysis regarding assets in light of the controls currently implemented. Thus, identification and analysis of assets is a vital part of establishing an ISO 27001 compliant ISMS. An asset is defined in the standard as "anything that has value to the organisation" [1, p. 2]. We propose the following steps for identifying assets, which concern resources in our SI* model. Thus, the following step aims to find resources and if the resources have a value for the asset owner, they are assets.

Investigate the Eco Model Relations. The relations of the *Eco Model*: *request, own,* and *provide* that consider a resource at one end reveal possible assets and in case of the *own* relation, also the asset owner.
Investigate Goal Relations. Means-end relations between a goal and a resource have to be investigated. In addition, for each goal we have to check if not a resource is missing that might be an asset.
Iterate over all Resources. In order not to miss any assets, an iteration of all resources in the model is done and a check is conducted if this is an asset.

For an accurate description of assets the following information has to be elicited for each asset.

State the Asset Owner. Check if the *own* relation of the Eco Model is set on an asset. If this is the case, the agent or role on that relation is the asset owner. If this relation is not set, it has to be included into the model.
Define the Need of Protection. We want to state the need for protection of an asset. This information can help to assess an initial risk level for an asset and serves as an input of the threat analysis. At this stage only the trust relations in the SI* model are considered. Any assets (resources) that have an *execution dependency* or *permission delegation* relation have an interaction with another agent or role. These can require a need of protection, which has to be described. The trust relations *trust of execution* or *trusting in execution* result in a limited need for protection, while a *distrust relation* requires a significant protection.
Assess Initial Risk. The description of assets and their need for protection entries shall be analysed by domain experts and initial risk values shall be assigned. These values are meant to categorise assets by risk level. We propose to limit the possible

labels to low (1), medium (2), and high (3) as proposed by the NIST 800-30 [6] standard for risk management. These values are later in the process refined in order to assess if an asset has an acceptable risk level in light of its threats or if additional controls are needed. We illustrate the resulting asset list in Tab. 3.

Table 3. Asset List

Asset	Asset Owner	Need for Protection	Risk Level
Power Plant	Energy Provider	The power plant produces the energy sold and consumed in the smart grid. Its availability is of utmost importance.	3
...	...	...	...

Step 4: Analyze Threats - The ISO 27001 standard concerns threat analysis in several sections for determining the risks to assets. Section 4.2.1 d demands a threat analysis for assets for the purpose of identifying risks and the vulnerabilities that might be exploited by those threats. Section 4.2.1 e concerns risk analysis and evaluation and demands to determine likelihoods and consequences for threats.

The ISO 27001 standard demands threat analysis in order to determine and analyse risks to assets. In particular, the standard mentions the importance of physical and network threat analysis. We consider four basic kinds of attackers for our threat analysis as proposed in [7]. These are *software attackers* that target software systems, *network attackers* that are reading or manipulating network traffic, *physical attackers* that are targeting hardware installations, and *social engineering attackers* that manipulate roles or agents. A study of the SANS Institute from 2006[4] revealed four fundamental motivations of social engineering attackers: Financial gain, self-interest, revenge, external pressure. We believe these motivations are generic enough to serve all types of IT attackers. We also added the motivation curiosity, which we identified in discussions with the industrial partners of the NESSoS project. We explain all of these motivations in the following: We model attacker motivations as soft goals of attackers, depicted in Fig. 2. The assumptions about each attacker are annotated using UML notes. The refined goals of attackers from their soft goals are threats. This refinement is modeled with *means-end* relationships, because the threats are a means to act upon the attacker's motivation. We use the *means-end* relationship to model relations between threats and resources, as well. The reason is that the exploit of a resource fulfills a threat. For simplicity's sake, we show only the elements of the SI* model necessary for the threat analysis in Fig. 2.

We consider two different *Network Attackers* in our analysis. One *Network Attacker* has the soft goal *External Pressure*. Hence, the *Network Attacker* has the capabilities to attack the network, but no motivation for doing so. We assume the attacker is pressured by a criminal organisation to *Access and Manipulate Network Traffic*. The resources this goal targets are the *Security Module*, the *Smart Meter*, and the *Gateway to the Grid*, because all of these are connected via a network and we assume the *Network Attacker*

---

[4] http://www.sans.org/reading_room/whitepapers/engineering/
social-engineering-means-violate-computer-system_529

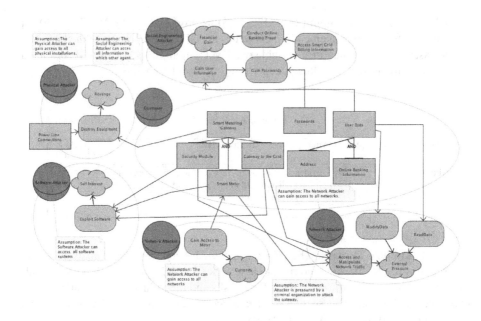

Fig. 2. SI* Diagram concerning goal-based Threat Analysis

can gain access to all networks. The *Network Attacker* is pressured to *Read Data* and *ModifyData*. *User Data* is *owned* by the customer and threatened by the *Network Attacker*, as well. These threats are a means to achieve the goals of the *Network Attacker*. A second *Network Attacker* acts out of *Curiosity* and gains access to the *Smart Meter*. The *Social Engineering Attacker* has the soft goal to get *Financial Gain* from attacking the *Smart Metering Gateway*. The attacker wants to *Conduct Online Banking Fraud* and for this purpose *Access the Smart Grid Billing Information*. The attacker aims to *Gain Passwords* of the *Consumer*. After the attacker has acquired the *Passwords* of the *Consumer*, the attacker can *Gain User Information*. The *Physical Attacker* is motivated by *Revenge* against the *Customer* and wants to *Destroy Equipment*. The attacker targets the *Power Line Connections* and the *Smart Metering Gateway*. The *Software Attacker* is motivated by *Self Interest* and wants to *Exploit Software* in order to hide data about his or her energy consumption. For simplicity's sake we do not show all possible attackers and their motivations. However, we show exemplary the exclusion of one attacker. The *Physical Attacker* with the motivation *Financial Gain* is not considered, because the effort and skill required to steal a *Smart Metering Gateway* is not worth the insignificant monetary value for it. In particular, because the *Energy Manager* has equipped the gateway with IDs and the *Energy Manager* can block the access of stolen gateways to the smart grid. We use the elicited threats as inputs for misuse cases [4]. These are textual representations of attacker's actions for threat identification. We use them to derive security requirements and check for missing threats. We propose a table as introduced by Deng et al. [8] that lists misuse cases and their corresponding security requirement.

Table 4. From Misuse Cases to Security Requirements

Misuse Case	Security Requirement
1. The confidentiality of the *Consumer's Passwords* might be compromised by a *Social Engineering Attacker*.	Ensure that the confidentiality of the *Passwords* is not compromised by a social engineering attack.
2. The availability and integrity of the *Smart Metering Gateway* can be compromised by a *Software Attacker*.	The *Smart Metering Gateway* has to be protected against *Software Attackers* that aim to execute exploits.
. . .	. . .

In contrast to the work of Deng et al., we do not consider solutions in this step. We discuss these during the selection of ISO 27001 security controls in the following. We illustrate several misuse cases in Tab. 4.

Step 5: Conduct Risk Assessment and Control Selection - Risk management is mentioned in numerous sections of the ISO 27001 standard. In the method risk is used to assess if an asset requires an additional control or not. We use the risk management technique proposed by Asnar et al. [9] for goal-based requirements engineering. For simplicity's sake, we do not explain it in detail in this work.

For each of the assets that has an unacceptable risk level controls have to be selected to reduce that risk. We use the resulting security requirements of the threat analysis as guidance for selecting controls. The numbering of the controls starts with A.5 and ends with A.15. The reason for not starting the numbering with A.1 is that the control numbering shall align with the controls listed in the ISO/IEC 17799:2005 standard. This standard provides guidelines on how to implement the controls, but it is not normative.

For the requirement 2 from Tab. 4 we choose adequate controls. The control *A.10 Communications and operations management* contains the sub control *A.10.4 Protection against malicious and mobile code*. Further sub controls are *A.10.4.1 Controls against malicious code*, which is described as "Detection, prevention, and recovery controls to protect against malicious code and appropriate user awareness procedures shall be implemented. "[1, p. 19]. In addition, another relevant sub control is *A.10.4.2 Controls against mobile code*: "Where the use of mobile code is authorized, the configuration shall ensure that the authorized mobile code operates according to a clearly defined security policy, and unauthorized mobile code shall be prevented from executing." [1, p. 19]. The selection of these controls is followed by selecting concrete measures. For example, we have to conduct penetration testing in order to find existing vulnerabilities in the software of the *Smart Metering Gateway* and fix these. For each asset, we have to iterate over all controls in the Appendix A of the ISO 27001 standard and state if a control is required or not for that asset. The resulting document is the so-called *Statement of Applicability*.

Step 6: Design ISMS Specification - The ISO 27001 standard demands a documentation of the ISMS. The standard demands several documents for each part of the ISMS, but the standard states no demands for the form or medium. Hence, we developed a mapping (see Tab. 5) of the generated artifacts from our method to the documentation demands.

Table 5. Support of our Method for ISO 27001 Documentation Demands

ISO 27001 Documentation Requirement	Artifacts of our Methods
ISMS policies and objectives	Misuse cases and Security Requirements
Scope and boundaries of the ISMS	SI* diagrams
Procedures and controls	Documentation of selected security controls and their implementation
The risk assessment methodology	Description of the method by Asnar et al. [9]
Risk assessment report	Results of asset identification and threat analysis including SI* models
Risk treatment plan	Risk Assessment and Control Selection
Information security procedures	Control Documentation of resulting security processes
Control and protection of records	Documentation of selected measures to control documents
Statement of Applicability	Reasoning about Controls

6   Discussion and Related Work

The procedure presented in this chapter was developed based on discussions with practitioners from security and especially ISO 27001 projects. Parts of our method was discussed with security consultants. The security consultants mentioned that this structured procedure

- Helps to describe the attackers' abilities in more detail,
- Supports the identification of all threats on the given assets,
- Supports the identification and classification of assets.
- Increases the use of models instead of texts in standards, which eases the effort of understanding the system documentation,
- Provides the means for abstraction of a complex system and structured reasoning for security based upon this abstraction.

One issue that needs further investigation is that of scalability, both in terms of the effort needed by the requirements engineer in order to enter all information about the organization and the threat analysis proposed. We will use the method for different scenarios to investigate if the method scales for complex goal models.

To the best of our knowledge no approach exist to use a goal-based security requirements engineering approach for ISO 27001 complaint ISMS establishment.

Mellado et al. [10] created the Security Requirements Engineering Process (SREP). SREP is an iterative and incremental security requirements engineering process. In addition, SREP is asset-based, risk driven, and follows the structure of the Common Criteria [11]. The work differs from ours, because the authors do not support the ISO 27001 standard.

7   Conclusion

We have presented a structured method to establish an Information Security Management System (ISMS) according to the ISO 27001 standard, which builds upon the security requirements engineering method SI*. Our method provides the means to elicit the context of an ISMS consider management commitment, threat and risk analysis, as well as security requirements-based control selection.

Our method offers the following main benefits:

- A structured method for describing the context, analyzing threats and risks, formulating security requirements, and selecting ISO 27001 controls,
- Re-using SRE methods to support the development of an ISO 27001 ISMS,
- Support for generating consistent ISMS documentation compliant to ISO 27001
- Re-using the structured techniques of SRE methods for analyzing complex systems and eliciting security requirements, to support the refinement of sparsely described sections of the ISO 27001 standard.

References

1. ISO/IEC: Information technology - Security techniques - Information security management systems - Requirements. ISO/IEC 27001, International Organization for Standardization (ISO) and International Electrotechnical Commission (IEC) (2005)
2. Massacci, F., Mylopoulos, J., Zannone, N.: Security requirements engineering: The SI* modeling language and the secure tropos methodology. In: Ras, Z.W., Tsay, L.-S. (eds.) Advances in Intelligent Information Systems. SCI, vol. 265, pp. 147–174. Springer, Heidelberg (2010)
3. Beckers, K., Faßbender, S., Heisel, M., Küster, J.-C., Schmidt, H.: Supporting the Development and Documentation of ISO 27001 Information Security Management Systems through Security Requirements Engineering Approaches. In: Barthe, G., Livshits, B., Scandariato, R. (eds.) ESSoS 2012. LNCS, vol. 7159, pp. 14–21. Springer, Heidelberg (2012)
4. Opdahl, A.L., Sindre, G.: Experimental comparison of attack trees and misuse cases for security threat identification. Inf. Softw. Technol. 51, 916–932 (2009)
5. ISO and IEC: Common Criteria for Information Technology Security Evaluation. ISO/IEC 15408, International Organization for Standardization (ISO) and International Electrotechnical Commission (IEC) (2009)
6. Stoneburner, G., Goguen, A., Feringa, A.: Risk management guide for information technology systems. NIST Special Publication 800-30, National Institute of Standards and Technology (NIST) (2002)
7. Beckers, K., Côté, I., Hatebur, D., Faßbender, S., Heisel, M.: Common Criteria CompliAnt Software Development (CC-CASD). In: Proceedings 28th Symposium on Applied Computing, pp. 937–943. ACM (2013)
8. Deng, M., Wuyts, K., Scandariato, R., Preneel, B., Joosen, W.: A privacy threat analysis framework: supporting the elicitation and fulfillment of privacy requirements. Requir. Eng. 16, 3–32 (2011)
9. Asnar, Y., Giorgini, P., Massacci, F., Zannone, N.: From trust to dependability through risk analysis. In: Proceedings of ARES, pp. 19–26 (2007)
10. Mellado, D., Fernandez-Medina, E., Piattini, M.: A comparison of the common criteria with proposals of information systems security requirements. In: ARES, pp. 654–661 (April 2006)
11. Mellado, D., Fernández-Medina, E., Piattini, M.: Applying a security requirements engineering process. In: Gollmann, D., Meier, J., Sabelfeld, A. (eds.) ESORICS 2006. LNCS, vol. 4189, pp. 192–206. Springer, Heidelberg (2006)

ProofBook: An Online Social Network
Based on Proof-of-Work and Friend-Propagation

Sebastian Biedermann[1], Nikolaos P. Karvelas[1], Stefan Katzenbeisser[1],
Thorsten Strufe[2], and Andreas Peter[3]

[1] Security Engineering Group
Technische Universität Darmstadt
{biedermann,karvelas,katzenbeisser}@seceng.informatik.tu-darmstadt.de
[2] P2P Networking Group
Technische Universtität Darmstadt
strufe@cs.tu-darmstadt.de
[3] Distributed and Embedded Security Group
University of Twente
a.peter@utwente.nl

Abstract. Online Social Networks (OSNs) enjoy high popularity, but
their centralized architectures lead to intransparency and mistrust in the
providers who can be the single point of failure. A solution is to adapt
the OSN functionality to an underlying and fully distributed peer-to-peer
(P2P) substrate. Several approaches in the field of OSNs based on P2P
architectures have been proposed, but they share substantial P2P weaknesses and they suffer from low availability and privacy problems. In this
work, we propose a distributed OSN which combines an underlying P2P
architecture with friend-based data propagation and a Proof-of-Work
(PoW) concept. ProofBook provides availability of user data, stability of
the underlying network architecture and privacy improvements while it
does not limit simple data sharing based on social relations.

Keywords: Online Social Network, Peer-to-Peer, Proof-of-Work.

1 Introduction

Popular Online Social Networks (OSNs) use a centralized design which sometimes leads to intransparent providers that can be the single point of failure. In
order to solve this problem, current research aims at either content confidentiality through encryption of posts, or at privacy through the removal of centralized
data storage control by introducing an underlying peer-to-peer (P2P) substrate.
However, P2P architectures lead to new problems which can have a strong impact on design principles of these OSNs. Availability and freshness of the users'
published content, is a very important property of any successful OSN. User
data has to be stored on the devices of other users alternating between an online
and offline status. In order to increase user data availability within the OSN,
the user data has to be stored on a great number of other users, which finally

V. Geffert et al. (Eds.): SOFSEM 2014, LNCS 8327, pp. 114–125, 2014.
© Springer International Publishing Switzerland 2014

results in the fact that each OSN participant has to provide a lot of storage. In general, OSNs which use other unknown P2P participants to store the users' sensitive data do not enjoy great popularity even if the stored data is encrypted. Furthermore, P2P substrates suffer from stability and security problems like Denial-of-Service (DoS) attacks (in which large amounts of requests are sent), Eclipse attacks [15], systematic content pollution (by introducing large amounts of fake data) or misuse for collusive piracy, many of which are exacerbated by the possibility to create a number of different zero-cost identities in the system (sybils).

In this paper, we assume an attacker model in which malicious users can perform DoS attacks, for example in the form of message flooding, and propose a fully distributed OSN architecture which nevertheless ensures privacy of each participant. We focus on the mitigation of these DoS attacks and propose a new OSN architecture, based on an underlying P2P substrate and an incentivised Proof-of-Work (PoW) concept which we call ProofBook. ProofBook has a decentralized architecture which nevertheless can ensure availability of published content, can increase anonymity and privacy of each user and can prevent manipulations and insider attacks of malicious participants as well as can mitigate DoS attacks. We can summarize the contributions of ProofBook as follows:

- *Availability:* User data is continuously available and as up-to-date as possible. This is realized with an update-on-request concept in which each user stores only the data of friends.
- *Stability:* Systematic content pollution, which is a major problem in P2P substrates, is mitigated in ProofBook by design, due to the implemented PoW concept which operates like a stamp used to pay for delivery of requests.
- *Privacy:* Friend relationships among ProofBook users are private and can neither be manipulated nor is any private user data revealed. This is enforced with the help of cryptographic techniques and hiding sources of delivered requests.

2   Related Work

2.1   Consolidations of P2P Substrates and OSNs

In general, data sharing performance in P2P networks can be increased based on social information about the participants. Chen Hua et al. [10] proposed "Maze", a hybrid P2P architecture which benefits from social information to help peers discover each other. Pouwelse et al. [13] proposed "Tribler" which is a consolidation of an underlying P2P architecture and social network data in order to increase usability. Graffi et al. [9] investigated security problems in P2P-based social networks. They proposed a P2P based social network with fine-grained user- and group-based access control to shared content.

In order to avoid the centralized architectures of OSNs, different OSN architectures based on an underlying P2P substrate have been proposed. The most popular architecture is Diaspora [5] which is an OSN based on a network of independent servers that are maintained by users who allow other users to store their data. Buchegger et al. [6] introduced PeerSoN, a distributed OSN based on a two-tiered P2P architecture which consists of peers communicating with each other and a separate look-up service. Cutillo et al. [7] proposed Safebook which exploits real-life trust relationships and maps these social links to a decentralized P2P network. Based on this, security mechanisms are implemented, while data integrity and availability are provided. In summary, these consolidations can achieve availability, but they often lack privacy.

2.2 Proof-of-Work Based Architectures

A lot of work has been done in the field of incentivised P2P substrates based on certain proofs for more accountability or e-cash ([3],[4]). Proof-of-Work (PoW) schemes are variants of cost-functions which are difficult to produce but trivial to verify. The degree of difficulty can vary depending on factors like the amount of participants. A PoW can be used to verify the existence of remote hardware resources that are controlled by a remote client. In particular, PoW approaches are used to combat spam mail, to mitigate DoS attacks or to control access to a shared resource [8]. Back [1] proposed "Hashcash" which is a PoW-based architecture that throttles systematic abuse of remote network resources. Bitcoin [12] is an electronic currency system based on a P2P substrate which prevents double spending of digital cash by adding a transaction with a PoW into a globally distributed chain of participants. The PoW is to find a hash for a given content which has a previously defined amount of initial zeros. This hash is calculated over data that includes the history of transactions. Since the currently most efficient way to find a valid hash are brute forcing techniques, the difficulty exponentially increases with the increasing amount of required initial zeros. Different hashes are created by changing an included nonce.

3   Overview of the Key Scheme

The architecture of ProofBook combines an underlying P2P substrate with a protocol that makes use of a PoW concept and incentivised cooperation to mitigate misuse. ProofBook provides standard OSN operations, integrates an incentive for participants without disrupting the utilization of the OSN and propagates new user data with the help of the user's friends.

3.1   User Registration and Friendships

ProofBook does not have a database which stores account information about users. Joining ProofBook and accordingly the underlying P2P substrate can be achieved by receiving the IP address of a participating peer from a secondary channel (for example via a web site). New friendships can be established by directly exchanging initial information. More precisely, each user U is associated with a public and private key pair (k_p, k_s) that will be used for signing certain information like U's data container. Additionally, U creates a symmetric "friend-key" k_f. The latter is distributed among U's group of friends and enables them to reply on requests which target U's data. Based on different "friend-keys", U can also maintain different group of friends (close friends, colleagues, etc). A friend relationship to U is established by retrieving U's public key k_p and subsequently U's data container which also includes the symmetric friend-key k_f. Obtaining U's public key enables a friend to identify oneself as member of U's group of friends. We do not treat the exchange of this data further, as this can be done with standard cryptographic techniques using a secondary channel.

3.2 User Data Propagation and Availability

The data of each ProofBook user U is saved in a container structure which can include up-to-date status information, personal information and pictures. U's data container is identified by U's ID and a signature on the content under U's secret key k_s to verify U as the owner. The data container's structure is illustrated in Figure 1. U's data container also includes the IDs of U's current friends. This way, a friend of U can contact other friends of U even if they have no established friendship themselves. A ProofBook container is separated into a redundant array of 8 data blocks (block-level striping with double distributed parity). This offers the opportunity to restore the whole data of U's container even if 2 sub-containers are not available while the storage efficiency is still 75%. Furthermore, each sub-container includes a timestamp (last-modified). The key scheme of ProofBook is based on a simple fact which can guarantee availability of U's data even if U is offline: *If U has retrieved and viewed the data container from an arbitrary friend once, U can locally save this data forever and this event cannot be undone.* Based on this, users directly save all data containers of their friends locally and ProofBook benefits from this approach, because users can not only request the friend's data from the target friend itself, but also from other friends of this friend who have a locally stored the data containers of their friends as well. In addition, there is no need to encrypt a container's data since it is only stored on users who are allowed to access this data anyway.

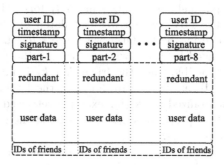

Fig. 1. A container which is published by U and can be requested by friends of U

ProofBook users can perform update requests when viewing the content of a friend's data container. If a newer data container of this friend is currently available, it can be requested and used to replace the locally saved one. The updated data container is again available for update-requests of other friends of this friend. The functionality of ProofBook is based on the simple update requests which are delivered by the underlying P2P substrate and which can be answered by friends of the target friend or the target friend himself. The content of a ProofBook update request can be seen in Figure 2.

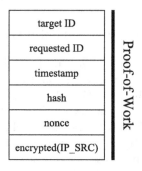

Fig. 2. A ProofBook update request

A ProofBook request includes a target user ID which can be the ID of the target friend or the ID of a friend of this friend, retrieved from the target friend's data container. Furthermore, there is the ID of the friend whose data container is requested, a timestamp and a hash (SHA256) which actually represents a PoW that is calculated over the content of the request and that needs to have a previously defined amount of initial zeros. A valid hash can be found by changing the nonce which is included in the content. Additionally, the request includes the encrypted source IP address of the requesting user which can only be decrypted by users who have the target friend's friend key k_f. The request is anonymous since there is no plain information about its source, only about the target. To mitigate DoS attacks, systematic content pollution and to limit users obtaining multiple identities in parallel, the requests are only forwarded by the underlying P2P substrate if the PoW, which was calculated by the source user, can be verified by the delivering peers. Otherwise, the requests are dropped. The request cannot be changed by peers without redoing the PoW and successful modifications are not possible since the friend key k_f is not available to arbitrary delivering peers. To ensure the freshness of the PoW, the peers use the timestamp and a synchronized clock (for example retrieved from a fixed clock server). New data of U is propagated stepwise among the group of U's friends based on these update requests.

Figure 3 shows the initial steps of a request sent by a user Alice to retrieve a new data container of her friend Bob. In an update procedure, Alice sends multiple requests targeting Bob as well as Bob's friends. If the PoW of the request

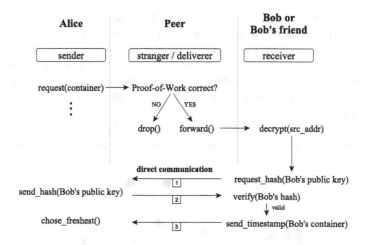

Fig. 3. Alice performs an update request which is forwarded to retrieve timestamps of Bob's currently available data containers

can be verified by the peers which are on the route, they forward the request to the target user. Receivers decrypt Alice's source IP address using Bob's friend key k_f and directly reply with a request for a hash over Bob's public key to verify that Alice is a friend of Bob. If Alice replies with a valid hash, the receiver sends the timestamp of Bob's stored data container. Alice chooses the most up-to-date data container and directly downloads all sub-containers from the chosen target. This target can be Bob himself (if Bob is currently online). Alice verifies the signature of each sub-container of Bob's container with the help of Bob's public key k_p. The reply procedure uses directly established connections and the P2P substrate is only used in the update request delivering procedure (one-way).

3.3 The ProofBook Payment Scheme

In Bitcoin [12], the amount of initial zeros of the required hashes which is used as degree of difficulty for the PoW is dynamically arranged. In ProofBook, there are only three different static levels $L_i, i \in \{1, 2, 3\}$ implemented. They enforce three degrees of difficulty on two different points.

First, since the update requests from a specific source are usually delivered on equal paths in the underlying P2P substrate within same periods of time, L_i depends on the amount of requests a peer has already delivered to the same target. A peer will only deliver a specific amount of requests to a target peer with the initial PoW difficulty L_1 within a fixed implemented time slot. If more requests have to be delivered within the same time slot, the difficulty to send a request to the same target becomes harder (L_2) and finally very hard (L_3). The increasing difficulty levels do not depend on the sources of the requests, rather on the target. Sending update requests to a specific target gets more expensive, the more requests are sent to the same target within the same time slot. This way, several attacks based on huge amount of requests can be mitigated.

Second, the last peer on the path to an arbitrary target obtains information about the target's hardware device, retrieved from the client once enrolled in the P2P substrate. If the target device is a desktop computer, the peer delivers every request, if it is a notebook only with L_2 or higher and if it is smartphone only with L_3.

This way, there is another load balance, which enforces more load on devices having better performance than others because they are cheaper to request.

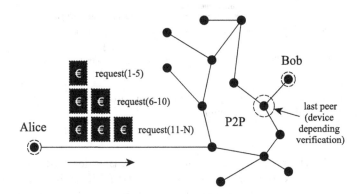

Fig. 4. Utilization of different example PoW levels L_i comparable to amounts of required stamps in the real world

Figure 4 illustrates the utilization of L_i at the two different points in Proof-Book. Like a stamp which is used to send a letter in the physical world, these values are used to pay different prices for the update requests to a specific target. The PoW acts as a buffer and reduces the amount of requests which can be delivered to the same target within a time slot depending on the amount of requests and the target's device.

It shall be noted that a botnet owner could compute the PoWs which he could use for continuous high amounts of requests to a single target in order to make it also costly for others to reach this user. However, the underlying P2P substrate is dynamic and uses different paths for the requests, depending on the entry point of the sender. That way, an attacker would require a high degree of knowledge about the current internal topology as well as many user IDs of users who are not necessarily connected with each other to mount such an attack successfully.

In order to implement an incentive for the replier of an update request, who finally sends the data container to the requesting user, we use a stepwise strategy in a commitment scheme. For each sub-container Bob sends to Alice, Bob receives a PoW-token from Alice (L_1) which he can use to decrease PoWs of own requests at a later point in time (Figure 5). If it so happens that Alice or Bob skips during the container transfer, Bob has at least retrieved some usable tokens and Alice can restore the data container even if up to two of the eight sub-containers are

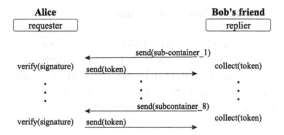

Fig. 5. For each verified received sub-container, Alice sends a PoW token to Bob

missing using the redundant structure of the containers. Bob can use the PoW-tokens in later requests to decrease his required PoW. This can be reached by extending an update request (Figure 6) with these tokens and calculating the PoW over the request and the extending tokens (max 8). The tokens can be used within a period of time in the future, defined by their timestamps. Furthermore, they can only be used in requests targeting the data container of the same user who Alice has requested and with whom Bob is a friend as well.

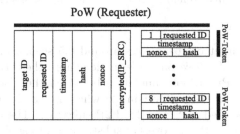

Fig. 6. An update request extended with previously earned PoW-tokens

Delivering peers only accept two different extended update requests, either with at least 6 or 8 additional PoW tokens targeting the same requested user ID. 6 tokens can decrease one level of PoW (e.g. from L_2 to L_1) and 8 tokens can decrease two levels (from L_3 to L_1). With this strategy, ProofBook actually gives an incentive for Bob to send that much sub-containers which Alice needs since Bob can perform these extended requests faster.

4   Evaluation

In the evaluation, we investigated two questions: First, how much time is required for an update request, since delays are caused by the PoW calculations and verifications. Second, in which periods of time data containers are propagated among a group of friends based on the update-on-request approach with the help of friends.

4.1   ProofBook Update Request Timings

First, we evaluated how much time the PoW computations require in our setup (Intel(R) Core(TM) i7 M620 with 2.67Ghz). We calculated PoWs for random update requests depending on different levels (amount of required initial zeros). Since the calculations become exponentially more difficult, the required time becomes quickly very large (Figure 7).

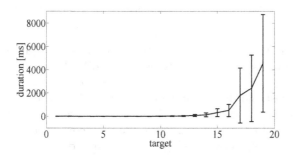

Fig. 7. Timings required to find a PoW with different targets (1024 test-runs)

A PoW with at target of 16 requires around 0.5 second on average having a standard deviation of 0.5 second. We implemented these timings as well as their deviations in a ProofBook simulation based on the discrete event-based P2P network simulation engine Oversim [2] which is based on Omnet++ [16]. For the test-runs, we used a Pastry [14] P2P substrate. The expected amount of forwarding peers in a Pastry overlay substrate is $O(\log N)$ where N is the number of participating peers. We investigated how much time a ProofBook update request requires including the PoW calculations and the "on-the-way" verifications.

Finally, we compared the results to other test-runs with normal request in the same Pastry P2P substrate. We used a P2P overlay with 1000 peers each of them running the ProofBook application on top of the Pastry P2P substrate and a ProofBook update request message size of 256 Bytes. A random ProofBook user is continuously chosen and sends an update request to another random ProofBook user. We have chosen two levels L_1 with a target of 17 and L_2 with a target of 19. Results of these test-runs can be seen in Table 1.

Table 1. Mean of delivered requests/s between random users (1000 participants)

PoW L_i (target)	L_0 (0)	L_1 (17)	L_2 (19)
Delivered requests/s	0.99301	0.35602	0.24580

The amount of involved peers lead to an average hop count of 2.51 ± 0.03 peers for a one-way delivered request. Table 1 shows the requests in comparison to requests without any PoWs in the same P2P Pastry substrate. With L_1, an update request can reach the target only every three seconds in average and with L_2 an update request can reach the target only every four seconds in average. The PoW scheme successfully limits the amount of delivered requests.

4.2   User Data Container Propagation

We executed other test-runs to investigate how fast a randomly chosen updated data container of an arbitrary user U propagates within U's group of friends. Since the network protocols are negligible in this case, we developed a ProofBook data container propagation simulation in Java. For a realistic scenario, we first needed to determine several values:

First, based on [11], we assume a power law probability distribution (Zipf's law) for the amount of friends each user has. We use a maximum of 500 and an α of 0.5 which leads to a numerical mean of 341 friends which is consistent with a recent Facebook study[1], where the median of the users' friends was found to be 342. Second, in reality, OSN users are continuously interested in up-to-date data of just a few of their "best" friends and in the data of most of their other friends only sporadically. We subdivide each user's friends into three equal subgroups. A random friend from the first subgroup (best-friends) is requested with a probability of $p = 0.5$, from the second subgroup with $p = 0.35$ and from the third subgroup with $p = 0.15$. Third, a user is not continuously online. Because of the increasing number of involved mobile online devices, we assumed a probability P_{on} of 33% for U to be online in any time slot of 6 minutes which leads to an average online time for a user of 7 hours a day.

In the test-runs, we monitored a randomly chosen user U_m who updated his data and we monitored all of U_m's friends. In a time slot (6 minutes), each friend F_i of U_m can perform an amount of R_n update request. Requests of F_i are only performed if F_i is online as well as replies are only sent if the target is online. F_i's chosen targets depend on F_i's "best-friends" subgroups. After each slot, all friends of U_m have adapted their online status. In multiple test-runs, we monitored the time which is required to distribute U_m's new data container among at least 33% of U_m's friends. This is feasible in our scenario since we also assume that not more than 33% of U_m's friends belong to U_m's best friends. Table 2 shows results. In average, U_m's new data container was distributed in 11.91 hours if U_m's friends request their friends for updates every minute depending on their best-friend subgroups. In order to improve the container propagation, we enforced additional requests of U_m's friends in each time slot. Multiple update requests are feasible since the most requests do not finally lead to the transfer of a data container. If we assume that the clients of U_m's friends additionally perform a request to one of their friends (randomly chosen) every minute ($+ \ random_6$), the container of U_m is distributed within 5.47 hours. If we

---

[1] http://blog.stephenwolfram.com/2013/04/

Table 2. Time required to propagate an arbitrary user's data container within 33% of this user's friends (100 test-runs)

R_n within time slot	mean [h]	std [h]
6	11.91	5.86
$6 + random_6$	5.47	2.67
$6 + random_{24}$	2.45	2.15

increase the number of enforced requests to one request every $15s$ ($+ random_{24}$), U_m's container is distributed in 2.45 hours.

5   Conclusion

To the best of our knowledge, ProofBook is the first OSN architecture which is based on a combination of two approaches: First, a Proof-of-Work (PoW) architecture is implemented which mitigates certain network attacks like Denial-of-Service as well as allows preferring users with better network performance in data transfers. In order to combine OSN mentality and the PoW, incentives are introduced and up-to-date data containers can be only retrieved unhindered if the users follow the underlying protocol. Second, ProofBook benefits from a friend-based data propagation approach which is based on the ability of friends of a user replying to update requests targeting the data container of this user. The privacy level is enhanced since the users' data is only stored on the users' friends rather than on unknown peers. The sources of requests are anonymous and can only be decrypted by the target group of friends. In an evaluation, we showed that the PoW scheme helps to limit the amount of requests that can be successfully performed and accordingly mitigates certain attacks. We showed that the friend-based data propagation scheme is feasible under realistic conditions. An implementation and a user-interface are crucial for the success of any OSN and required to perform further evaluations of the proposed architecture. This as well as a more large-scaled evaluation belong to our future work.

Acknowledgments. Thorsten Strufe is partially supported by MSIP (Ministry of Science, ICT & Future Planning), Korea in the ICT R&D Program 2013. Andreas Peter is supported by the THeCS project as part of the Dutch national program COMMIT.

References

1. Back, A.: Hashcash: A denial of service counter-measure (2002)
2. Baumgart, I., Heep, B., Krause, S.: OverSim: A flexible overlay network simulation framework. In: Proceedings of 10th IEEE Global Internet Symposium (GI 2007) in conjunction with IEEE INFOCOM 2007, Anchorage, AK, USA, pp. 79–84 (2007)

3. Belenkiy, M., Chase, M., Erway, C.C., Jannotti, J., Küpçü, A., Lysyanskaya, A.:
Incentivizing outsourced computation. In: Proceedings of the 3rd International
Workshop on Economics of Networked Systems, NetEcon 2008, pp. 85–90. ACM,
New York (2008)

4. Belenkiy, M., Chase, M., Erway, C.C., Jannotti, J., Küpçü, A., Lysyanskaya, A.,
Rachlin, E.: Making p2p accountable without losing privacy. In: Proceedings of
the 2007 ACM Workshop on Privacy in Electronic Society, WPES 2007, pp.
31–40. ACM, New York (2007)

5. Bielenberg, A., Helm, L., Gentilucci, A., Stefanescu, D., Zhang, H.: The growth
of diaspora - a decentralized online social network in the wild. In: 2012 IEEE
Conference on Computer Communications Workshops (INFOCOM WKSHPS), pp.
13–18 (2012)

6. Buchegger, S., Schiöberg, D., Vu, L.-H., Datta, A.: Peerson: P2p social networking:
early experiences and insights. In: Proceedings of the Second ACM EuroSys Workshop on Social Network Systems, SNS 2009, pp. 46–52. ACM, New York (2009)

7. Cutillo, L.A., Molva, R., Strufe, T.: Safebook: a privacy preserving online social
network leveraging on real-life trust. IEEE Communications Magazine 47(12) (December 2009), Consumer Communications and Networking Series

8. Dwork, C., Naor, M.: Pricing via processing or combatting junk mail. In: Brickell,
E.F. (ed.) CRYPTO 1992. LNCS, vol. 740, pp. 139–147. Springer, Heidelberg (1993)

9. Graffi, K., Mukherjee, P., Menges, B., Hartung, D., Kovacevic, A., Steinmetz, R.:
Practical security in p2p-based social networks. In: IEEE 34th Conference on Local
Computer Networks, LCN 2009, pp. 269–272 (October 2009)

10. Hua, C., Mao, Y., Jinqiang, H., Haiqing, D., Xiaoming, L.: Maze: a social peer-to-
peer network. In: IEEE International Conference on E-Commerce Technology for
Dynamic E-Business, pp. 290–293 (September 2004)

11. Mislove, A., Marcon, M., Gummadi, K.P., Druschel, P., Bhattacharjee, B.: Mea-
surement and analysis of online social networks. In: Proceedings of the 7th ACM
SIGCOMM Conference on Internet Measurement, IMC 2007, pp. 29–42. ACM,
New York (2007)

12. Nakamoto, S.: Bitcoin: A peer-to-peer electronic cash system (2008)

13. Pouwelse, J.A., Garbacki, P., Wang, J., Bakker, A., Yang, J., Iosup, A., Epema,
D.H.J., Reinders, M.J.T., Steen, M.R.V., Sips, H.J.: TRIBLER: a social-based
peer-to-peer system. Concurrency and Computation: Practice and Experience 20,
127–138 (2008)

14. Rowstron, A.I.T., Druschel, P.: Pastry: Scalable, Decentralized Object Location,
and Routing for Large-Scale Peer-to-Peer Systems (2001)

15. Singh, A., Castro, M., Druschel, P., Rowstron, A.I.T.: Defending against eclipse
attacks on overlay networks. In: SIGOPS European Workshop (2004)

16. Varga, A.: Using the omnet++ discrete event simulation system in education. IEEE
Transactions on Education 42(4), 11 (1999)

Platform Independent Software Development Monitoring: Design of an Architecture

Mária Bieliková, Ivan Polášek, Michal Barla, Eduard Kuric,
Karol Rástočný, Jozef Tvarožek, and Peter Lacko

Institute of Informatics and Software Engineering, Faculty of Informatics
and Information Technologies, Slovak University of Technology
Ilkovičova 2, 842 16 Bratislava, Slovakia
{name.surname}@stuba.sk
http://perconik.fiit.stuba.sk

Abstract. Many of software engineering tools and systems are focused
to monitoring source code quality and optimizing software development.
Many of them use similar source code metrics to solve different kinds
of problems. This inspired us to propose an environment for platform
independent code monitoring, which supports employment of multiple
software development monitoring tools and sharing of information among
them to reduce redundant calculations. In this paper we present design of
an architecture of the environment, whose main contribution is employing (acquiring, generating and processing) information tags - descriptive
metadata that indirectly refer source code artifacts, project documentations and developers activity via document models and user models.
Information tags represent novel concept unifying traditional content
based software metrics with recently developed activity-based metrics.
We also describe prototype realization of the environment within project
PerConIK (Personalized Conveying Information and Knowledge), which
proves feasibility and usability of the proposed environment.

Keywords: Information tag, Source code, Software development, Developer's expertise, Interaction data, Software metrics.

1 Introduction

Source code quality and optimization of software development process fall within
long-term problems of software engineering. Many of proposed methods that
aim to solve these problems utilize source code metrics (e.g., LLOC, CLOC) [7],
watch activities of developers and process of the development [6] and visualize results in different views that simplifies identification of problematic source
code and communication with stakeholders [5]. These methods have different
strengths and weaknesses related to particular problems. Thus by combining
several methods in one environment we can achieve a more robust solution.
Even more, we can move from a separate execution of distinct methods into an
"orchestration", where particular methods take advantage of and reuse shared
"knowledge" base about source codes, project documentations and developers.

V. Geffert et al. (Eds.): SOFSEM 2014, LNCS 8327, pp. 126–137, 2014.

Current trends of information and knowledge modelling utilize ontologies for sharing and storing information [20]. Ontologies (either lightweight or heavyweight) are also often used in web information services that describe webpage artifacts, which are in some aspects similar to software source code artifacts. This description of webpage artifacts is provided via semantic annotations that can refer to concepts of external ontologies [1] or they can directly contain fragments of an ontology in form of bags of triples [18]. These semantic annotations can be put directly into web pages as HTML tag attributes, practically visible only to machines and not distracting user in any way. Such approach however cannot be used for source codes, as they would quickly become unclear, hard to read and understand for a programmer.

Another approach is to store metadata in external files with proper references to described resources. This approach suffers by granularity issues – metadata can often refer only the whole file [15], what is insufficient for source codes, or they straightly refer line numbers of annotated source code artifacts [10] what significantly decreases maintainability of metadata (after each source code file modification, all references have to be updated to current line numbers).

In this paper we introduce design of an architecture of a novel environment for code monitoring which employs *information tags* as descriptive metadata over document model and user model. This way we contribute to solving a problem of computational redundancy and increase cooperation among services and tools for software development support.

2 Architecture Overview

The proposed environment has to be able process data created by developers directly as well as indirectly (by observing and logging their actions) and to provide added value in real-time. For this reason we divide processing data to multiple partial processes that are sub-processes of two main processes – *data acquisition* and *added value provision*.

2.1 Data Acquisition

Decomposition of data processing into multiple processes gives us a possibility to break complex problems of processing big data to several smaller and less complex problems that can be distributed over multiple machines and processed in real time. This can be especially notable in the process of data acquisition, which needs to cope with stream of events about modifications in source code, projects documentations and learning materials and also activities of developers and team leaders. Thus we decompose data acquisition to four horizontal layers – *data source*, *data*, *metadata* and *metadata processing* based on granularity of the data and to three vertical layers - *tags*, *documents* and *logs* about activities based on the character of the data (see Fig. 1).

Data acquisition starts in data source – users' *working environments*. Data from working environments are collected via a set of pre-installed tools that

128   M. Bieliková et al.

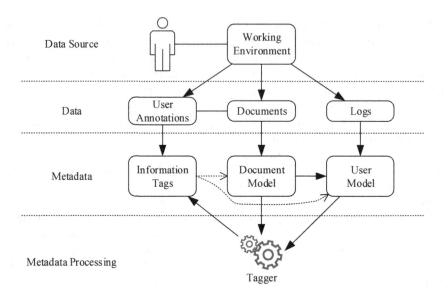

Fig. 1. Horizontal layers of data acquisition view. All elements are independent software systems with SOAP a REST-based interfaces that make stored objects accessible via URI identifiers and queries.

monitor activities of developers in their integrated development environments, web browsers and office tools and collect contextual (possibly including also biometrics) information. The collected data are stored into repositories at the data layer, from which they are processed into *document models* and *user models*. All modifications in models are streamed as events to the *tagger*, a component which processes them and creates information tags - descriptive metadata assigned to the objects from document models and user models (we discuss "information tags" and motivation for creating them in the section 3.3).

2.2   Added Value Provision

Processing the acquired data to metadata in formats of models and information tags unburdens individual methods/services from recurrent preprocessing of raw data and redundant calculations. This does not only save computation resources, but it also saves data traffic as they do not have to access whole raw data but they only query for necessary metadata. Sources of queried data depend on roles of end users – consumers of methods and services. In our environment we differentiate two main roles of users - *developers* and *team leaders* (see Fig. 2).

Developers work directly with software project documents (e.g., source code, documentations). The most of documents used by developers are stored in developer' machines in needed versions or they are synchronized with documents repositories by specialized tools (e.g., IDE, office tools). Therefore services of the environment will do redundant processing if they load and work directly

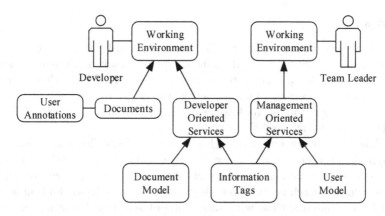

Fig. 2. Enriching users' working environment with added value provided by the code monitoring environment

with these documents. It is more efficient to perform calculations and processing over document models and information tags and then send results to developers' machines, where they can be merged back into the documents.

Management oriented services work similarly to *developer oriented services.* Both work only with metadata without necessity of an access to the whole raw data. Difference is in metadata stored in models that are processed by management oriented services. These services do not need to work with document models. Managers do not need to read documentations or source code, they need information about their teams (e.g., skills of developers). Therefore management oriented services query the user model and information tags and present results through specialized tools or web applications.

3   Vertical Layers

3.1   Documents

During software development, developers write and analyze source code, create and study different kinds of texts (e.g., specifications, API, tutorials), or use Q&A sites and community forums for finding/providing solutions (e.g., code examples). It is rich information space which includes resources in a natural language as well as a "spiderweb" of software artifacts. Whether it is a file/text in a natural language or a programming language or it is a resource located on the Web or in a local repository, everything is a document.

We classify documents into three classes: *source code, web pages* and *project documentation.* Although, this classification seems to be straightforward, note that there are also different relationships among the documents. For example, a code snippet can be copied from a web page into local code or documentation. The relationship information between documents (source, target) is captured in user activity logs and stored in information tags.

3.2   Logs

We have developed several agents (tools) that collect and process (existing) documents and user activities. They allow us to capture, track, analyze and evaluate different events. We focus on monitoring developers' work in IDE, their activities in a web browser and events of an operating system. To capture coding/working activities we use supporting tools/plug-ins (e.g., plug-ins for Microsoft Visual Studio IDE, Eclipse IDE and Firefox).

In an IDE we capture activities such as open/add/edit file (associated with a solution/project), check-in, debug, copy/paste, code focus and selection, built-in find, code refactoring, and stream of edits. In a web browser we capture activities such as search on the Web (keywords, target URLs), find on a page, entering URLs, manipulation with tabs, content selection, creating and using bookmarks. For monitoring other activities we use OS Monitor. It allows us to capture and monitor running applications, utilization of hardware resources, biometric information (keyboard, mouse) and performed activities in office tools (e.g., Microsoft Office Word, Microsoft Office OneNote).

Each agent collects activities, transforms them to logs and sends the logs to the Local Logging Service (LLS). The task of LLS is to gather the logs. A delivered log from an agent to LLS can be set so-called "flag milestone". It means that LLS sets up a package of the gathered logs and sends the package to the Server Logging Service (SLS) that stores the logs into a database (see Fig. 3).

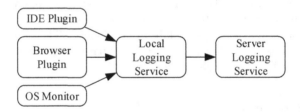

Fig. 3. Dataflow of collected logs

3.3   Information Tags

Information tag is a descriptive metadata with a semantic relation to a tagged content. It is an extension to the common concept of a tag as a simple keyword or phrase assigned to an artifact. Information tag adds additional value (semantics) to the software artifact itself. For example, information tags can be product of monitoring signals of explicit and implicit feedback generated by developers working on source code. Information tag is defined by a triple of [11]:

- *Type* - defines a type and a meaning of the information tag;
- *Anchoring* - identifies a tagged information artifact;
- *Body* - represents a structured information, a structure of which corresponds to the type of the information tag.

We distinguish information tags according their source: user created and machine generated information tags. *User created information tags* are created by users via specialized tools integrated into their working environment (e.g., in a form of a plug-in for IDE). They are directly readable for users. This makes user created information tags easily understandable and usable, what gives the environment advantage of naturally collecting users' knowledge about tagged objects (e.g., ratings of classes). In this manner user created information tags generally have a clear meaning for our users (e.g., provide review feedback attached to a source code). They can also be utilized to train or evaluate methods for enhancement of software development (e.g., automatic identification of unreliable or risky source code).

Machine generated information tags are an analogy of user created information tags for programme services. It is a tag which contains structured machine-readable information which has a meaning with its interconnection with a tagged content and/or its context (e.g., environment, history). For example, if the information tag "Edited 23 times" is defined, its information has meaning for us only if we look up at tagged method in a source code file, which has been edited. This way information tags represent form of lightweight semantics, in which any service can store and share its information related to objects (or any part of an object) of an information space (e.g., source code fragments). As a result, information tags decrease redundancy of data processing when some services need common partial results and also allow employing data mining techniques.

Not all source code annotations created by a user (*user annotations* in Fig. 1) can be considered as information tags, i.e. a descriptive metadata expressed in defined structure. The diversity of machine generated and user created information tags is in the logical level, in which user created information tags are always metadata straightly understandable to users. In the realization of the environment (see following section) we implement common model, repository and maintaining services for user created and machine generated information tags.

4 Case Study: Code Monitoring in Software House Environment

To evaluate feasibility of the proposed environment we have developed its prototype realization within the research project PerConIK (**Per**sonalized **Con**veying of **I**nformation and **K**nowledge). The project is focused on support of enterprise applications development in a software company by considering a software system as a web of information artifacts [3]. We experiment also with the development of students' team projects in master study programmes in Information Systems and Software Engineering at the slovak University of Technology in Bratislava. Project leader is a medium size software company so that all monitoring is realized in accordance with defined company policies.

In this section we describe core parts of the realized environment (e.g., information tags management) and several methods/services with partial results.

4.1   Infrastructure and Metadata Management

Information Tags Repository. The main innovation in architecture of the presented environment lies in employing information tags. To utilize advantages of information tags we proposed information tags repository which respects following requirements [4]:

– The repository has to be scalable - it has to have good read and modify performance despite of nontrivial number of stored information tags.
– The repository has to be able to store information tags in a freeform model which can be easily extended with new information tag types.
– Due to semantic meaning of information tags, inference has to be supported.

To fulfill these requirements we combine advantages of RDF and document stores. RDF has advantage in possibility of freeform data modeling and inference possibilities but it is at the expense of time complexity in the case when whole information has to be loaded (multiple SPARQL queries have to be processed while each query can take several seconds [14]). On the other side, document stores have good access to whole objects but they do not support inference [19].

Our repository is based on MongoDB[1] database which stores information tags in the object model based on Open Annotation Data Model[2]. We utilized standardized Open Annotation Data Model for its prevalence in annotation systems and straight analogy between information tags and annotations (both have a type, a body and an anchoring). We solved problem of missing support for inference in MongoDB by proposition of MapReduce-based SPARQL query processing algorithm [4]. We also performed several usability evaluations with our prototype realization. Executed test cases proved that proposed information tag repository provide enough performance for use cases of the environment and also that proposed SPARQL query processing algorithm reached almost same time as processing SPARQL queries as native horizontally scalable RDF storage BigData[3]. Our results are in detail described in [4].

Information Tags Maintenance. Information tags are linked with problem of their maintenance. This is especially visible in the case of information tags anchored to source code. Source code files are continuously modified, deleted and created. It leads to several problems of information tags maintenance:

– Generating missing information tags - newly written source code files or their parts have to be tagged with information tags that describe new source code.
– Repairing affected information tags - each modification in a source code file can affect validity of information tags at two levels - validity of body and anchoring of information tags. In addition, information tags' bodies can be affected by time - information that are stored in them can become obsolete.

---

[1] http://www.mongodb.org
[2] http://openannotation.org/spec/core/
[3] http://www.systap.com/bigdata.htm

– Removing invalid information tags - unrepairable information tags or information tags those targets are missing (have been deleted) have to be deleted from the information tags repository.

The first step of information tags maintenance is repairing invalidated information tag anchoring. This has to be done because we have to be able locate right source code artifacts where information tags have to be anchored before providing necessary maintenance of affected information tags. To solve this problem we do not employ any special process or service. We made a decision to design a robust location descriptor suitable for source code with algorithms for its building and interpreting [12]. It give us possibility to recalculate positions in real time (during editing code) without necessity to load previous versions of code.

Remaining maintenance tasks are provided by *tagger*. Tagger is a rule-based service which collects streams of events about modifications in user and document models and in case of fulfillment of a rule's condition it performs actions described in the rule. The tagger's core is based on linked stream data processing [9]. We employ C-SPARQL engine [2], which processes RDF graphs of events from models updating services. Employment of linked stream data [16] increases inference possibilities over events and decreases memory complexity of rule execution (incremental events processing). In addition tagger uses inferred results of fulfilled C-SPARQL query-based conditions in simple rules' actions.

Presentation of Tags. Information tags are primarily designed for services, but some of them can have direct added value for developers too. E.g., a service, which watches developers' activity, can automatically assign information tags with authorship to source code artifacts. Such information tags can be important for a developer that has to refactor older source code.

In addition some information tags and especially user generated information tags could not be maintained automatically. In these cases developers have to manually maintain invalidated information tags. For these reasons we implemented the plug-in for Microsoft Visual Studio 2012 for visualizing information tags (see Fig 4). Small graphic symbols (on the left side) pointed to concrete tags in the source code: green triangles for single tags and two *halftriangles* with connector for the pair tags in relation. On the right-hand side of the editor we can display the labels with the content of the tag: authors of the particular source code (after the keyword *by*), users of the source code (after the keyword *used by*), ranking (green stars are positive ranking and red stars are negative), topics, patterns or antipatterns. On the bottom of the editor we built filters and also we can activate infotip on the information tag label which completes whole information (here in the figure for example only the number of commits, the *author* of the tag (generated tag by the *tagger*), creation time, etc.).

4.2   Infrastructure Usage Possibilities

The information tags allow us to design and develop wide range of methods/services focused, e.g., to automatically evaluate developers' expertise, to

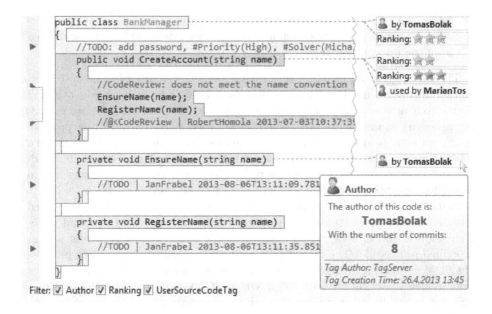

Fig. 4. Visualization of information tags in Microsoft Visual Studio 2012

discover a degree of developers' productivity and effectiveness, to reveal developers' practices and habits or to establish quality of created code.

Modeling Developer's Karma. One possible employment of proposed infrastructure for software development monitoring is to model developer's expertise (karma). It is valuable in real (internal) environment of a software company, but also in academic environment. Determining a developer's expertise [13] in a software company allows for example managers and team leaders to look for specialists with desired abilities, form working teams or compare candidates for certain positions. In academic environment, automatic establishment of students' expertise allows a teacher to evaluate students' knowledge and know-how. Based on it, e.g., the teacher can adapt and modify his teaching practices. On the contrary of a software company, where software is created by professionals, in academic environment, students learn how to design and develop software. Therefore, the establishment of developer's expertise requires different approaches.

One possible scenario we work on is to establish automatically developer's karma based on monitoring his working activities during coding in IDE, analyzing and evaluating the (resultant) code he creates and commits to a local repository. To establish the overall developer's karma for a software project we investigate information tags on software artifacts (components), which the developer creates. We take into account developer's "karma elements" as:

– *degree of authorship* – the developer's code contributions and the way how the contributions were created to a component;

- *authorship duration and stability* – the developer's know-how persistency about a component;
- *technological know-how* – the level of how the developer knows the used technologies (libraries), i.e., broadly/effectively, this includes also estimating quality of developed source code;
- *degree of productiveness* - a degree of difference between the real generated and finally used code lines in a component;
- *component importance* - a degree of importance of a component in the software project.

We established these particular developer's karma elements (metrics) based on our observation of developers' activities (logs). The overall developer's karma is calculated as a linear combination of these karma elements. Each karma element is calculated based on information tags generated while the developer works on source code or by off-line analyses of source code (those indicating quality of code developed by the developer). By applying the metrics we are able to observe and evaluate different indicators. For example, we can sight the developer who often copies and pastes source code from an external source (Web). Contributions of such developer can be relative to the software project, moreover, it can reveal a reason of his frequent mistakes or low productivity in comparison with other developers.

Search in Source Code Based on Reputation Ranking. Code search engines help developers to find and reuse software components. To support search-driven development it is not sufficient to implement a "mere" full text search over a base of code, human factors have to be taken into account as well. Reputation ranking can be a plausible way to rank code results using social factors. Trustability of code (developer's/author's reputability) is a big issue for reusing code (software components). When a developer reuses code from an external source he has to trust the work of external developers that are unknown to him. It can be supported by using an externalized model of each developer's expertise of a particular code (software component).

In search-driven development we apply our model and approach for automatic establishing developer's karma. It allows developers to rank code results not only based on relevance but also authors' reputation of the results. We exploited our know-how in implementing a search engine which in addition to relevance of code (software component) establishes its importance [8].

5   Conclusions and Future Work

We have introduced an approach to code monitoring in software projects based on information tags as descriptive metadata that provide a unifying element for reasoning on source code and developer activities represented by document and user models.

Information tags provide a basis for reasoning useful information to developers and managers similarly as metadata do for the applications on the Web. Examples are identification of bad practices, evaluation of source code quality based on an estimation of the current user state followed his activity, recommendation of good programming practices and tricks/ snippets used by colleagues. They also serve as an input for reasoning on properties of software artifacts such as similarity with code smells, estimation of developer skill and proficiency [3]. Information tags are stored in the information tags repository which is designed with great emphasis on scalability and the ability to store information tags in a freeform model which can be easily extended with new information tag types. Information tags are linked with a problem of their maintenance. We have introduced solutions for repairing and removing invalid information tags.

Our approach allows to design and develop wide range of methods/services, e.g. an automatic evaluation of developers' expertise, an establishment of quality of created code or an identification of duplicated code [17]. Although, in this paper we present an approach for modeling developer's karma, in our research we also use and apply the proposed approach in modeling developer's emotion and investigation of the influence of the detected emotion on the quality of created code and recommendation of software artifacts to a developer during working in IDE.

In future work, our primary goal is to finish the implementation of our core services and to perform their final evaluation. Next we plan to deploy the implemented prototype in a software company and at the University in the subject called "Team project" where students develop relatively large software systems. We also plan to propose and realize additional supporting services, e.g., a service for establishment of code quality based on context - i.e. developer's position (work, home) or weather. Our final aim is to improve development efficiency and software quality during its evolution.

Acknowledgments. This contribution is the partial result of the Research & Development Operational Programme for the project Research of methods for acquisition, analysis and personalized conveying of information and knowledge, ITMS 26240220039, co-funded by the ERDF and the project No. APVV-0233-10.

References

1. Araujo, S., Houben, G.J., Schwabe, D.: Linkator: Enriching web pages by automatically adding dereferenceable semantic annotations. In: Benatallah, B., Casati, F., Kappel, G., Rossi, G. (eds.) ICWE 2010. LNCS, vol. 6189, pp. 355–369. Springer, Heidelberg (2010)
2. Barbieri, D.F., Braga, D., Ceri, S., Grossniklaus, M.: An execution environment for c-sparql queries. In: Proc. of the 13th Int. Conf. on Extending Database Tech., pp. 441–452. ACM, New York (2010)
3. Bieliková, M., Návrat, P., Chudá, D., Polášek, I., Barla, M., Tvarožek, J., Tvarožek, M.: Webification of software development: General outline and the case of enterprise application development. In: Proc. of 3rd World Conf. on Inf. Tech (WCIT 2012), pp. 1157–1162. University of Barcelon, Barcelona (2013)

4. Bieliková, M., Rástočný, K.: Lightweight semantics over web information systems content employing knowledge tags. In: Castano, S., Vassiliadis, P., Lakshmanan, L.V.S., Lee, M.L. (eds.) ER 2012 Workshops 2012. LNCS, vol. 7518, pp. 327–336. Springer, Heidelberg (2012)

5. Bohnet, J., Döllner, J.: Monitoring code quality and development activity by software maps. In: Proc. of the 2nd Workshop on Managing Technical Debt, pp. 9–16. ACM, New York (2011)

6. Fritz, T., Murphy, G.C., Hill, E.: Does a programmer's activity indicate knowledge of code? In: Proc. of the the 6th Joint Meeting of the European Soft. Eng. Conf. and the ACM SIGSOFT Symposium on The Foundations of Soft. Eng., pp. 341–350. ACM, New York (2007)

7. Kothapalli, C., Ganesh, S.G., Singh, H.K., Radhika, D.V., Rajaram, T., Ravikanth, K., Gupta, S., Rao, K.: Continual monitoring of code quality. In: Proc. of the 4th India Software Eng. Conf., pp. 175–184. ACM, New York (2011)

8. Kuric, E., Bieliková, M.: Search in source code based on identifying popular fragments. In: van Emde Boas, P., Groen, F.C.A., Italiano, G.F., Nawrocki, J., Sack, H. (eds.) SOFSEM 2013. LNCS, vol. 7741, pp. 408–419. Springer, Heidelberg (2013)

9. Le-Phuoc, D., Xavier Parreira, J., Hauswirth, M.: Linked stream data processing. In: Eiter, T., Krennwallner, T. (eds.) Reasoning Web 2012. LNCS, vol. 7487, pp. 245–289. Springer, Heidelberg (2012)

10. Priest, R., Plimmer, B.: Rca: experiences with an ide annotation tool. In: Proc. of the 7th ACM SIGCHI New Zealand Chapter's Int. Conf. on HCI: Design Centered HCI, pp. 53–60. ACM, New York (2006)

11. Rástočný, K., Bieliková, M.: Maintenance of human and machine metadata over the web content. In: Grossniklaus, M., Wimmer, M. (eds.) ICWE Workshops 2012. LNCS, vol. 7703, pp. 216–220. Springer, Heidelberg (2012)

12. Rástočný, K., Bieliková, M.: Metadata anchoring for source code: Robust location descriptor definition, building and interpreting. In: Decker, H., Lhotská, L., Link, S., Basl, J., Tjoa, A.M. (eds.) DEXA 2013, Part II. LNCS, vol. 8056, pp. 372–379. Springer, Heidelberg (2013)

13. Robbes, R., Röthlisberger, D.: Using developer interaction data to compare expertise metrics. In: Proc. of the 10th Working Conf. on Mining Soft, pp. 297–300. IEEE Press, Piscataway (2013)

14. Rohloff, K., Dean, M., Emmons, I., Ryder, D., Sumner, J.: An evaluation of triplestore technologies for large data stores. In: Meersman, R., Tari, Z. (eds.) OTM-WS 2007, Part II. LNCS, vol. 4806, pp. 1105–1114. Springer, Heidelberg (2007)

15. Schandl, B., King, R.: The semdav project: metadata management for unstructured content. In: Proc. of the 1st Int. Workshop on Context. Attention Metadata: Coll., Managing and Exploiting of Rich Usage Inf., pp. 27–32. ACM, New York (2006)

16. Sequeda, J.F., Corcho, O.: Linked stream data: A position paper. In: Proc. of the 2nd Int. Workshop on Sem. Sensor Net., SSN 2009. CEUR-WS, Washington (2009)

17. Súkeník, J., Lacko, P.: Search in code duplicates. In: Proc. of the WIKT 2012, STU, Bratislava, pp. 189–192 (2012) (in Slovak)

18. Tallis, M.: Semantic word processing for content authors. In: Proc. of the 2nd Int. Conf. on Knowledge Capture, Sanibel (2003)

19. Tiwari, S.: Professional NoSQL. John Wiley & Sons, Inc., Indianapolis (2011)

20. Woitsch, R., Hrgovcic, V.: Modeling knowledge: an open models approach. In: Proc. of the 11th Int. Conf. on Knowledge Management and Knowledge Tech., pp. 20:1–20:8. ACM, New York (2011)

Towards Unlocking the Full Potential
of Multileaf Collimators*

Guillaume Blin[1], Paul Morel[1], Romeo Rizzi[2], and Stéphane Vialette[1]

[1] Université Paris-Est, LIGM - UMR CNRS 8049, France
{gblin,paul.morel,vialette}@univ-mlv.fr
[2] Department of Computer Science - University of Verona, Italy
romeo.rizzi@univr.it

Abstract. A central problem in the delivery of intensity-modulated radiation therapy (IMRT) using a multileaf collimator (MLC) relies on finding a series of leaves configurations that can be shaped with the MLC to properly deliver a given treatment. In this paper, we analyse, from an algorithmic point of view, the impact of using dual-layer MLCs and Rotating Collimators for this purpose.

1 Radiation Therapy Planning

Radiation therapy is one of the most commonly used cancer treatments and has been shown to be effective. The radiation treatment poses a tuning problem: the radiation needs to be effective enough to kill the tumor while sparing healthy tissues and organs close to the tumor – so-called *organs at risk*. Towards this goal, the design of a radiation treatment has to be specifically customized for each patient. Once both tumor and organs at risk have been delineated, the radiation oncologist will prescribe minimal, maximal and mean irradiation quantity for each of them. The amount of radiation is measured in gray (Gy). For example, typical dose for a tumor ranges from 60 Gy to 80 Gy (a minimal dose that the treatment should achieve), whereas healthy organs should not receive more than a given threshold of radiation – for example, 20 Gy for lungs, 50 Gy for bones or 12 Gy for eye lenses. Usually, the overall treatment dose is fractionated – e.g. 1.8 to 2 Gy per day, five days a week for an adult.

Each fraction is delivered by a linear accelerator (linac) using a cone beam that rotates around the patient; achieving a concentric irradiation converging in the tumor site. In the so-called "Step-And-Shoot" technique, the treatment design specifies some specific angles where the linac successively stops to irradiate the patient. For each of these angles, a specific intensity distribution across the radiation beam (later on referred to as intensity matrix) is computed (for instance, with the multicriteria approach to radiation therapy planning of Hamacher and Küfer [7]) in order to achieve the desired overall dosage of the

---

* Work partially supported by ANR project BIRDS JCJC SIMI 2-2010. Some supplementary materials are available on a companion website (http://igm.univ-mlv.fr/~gblin/MOD).

V. Geffert et al. (Eds.): SOFSEM 2014, LNCS 8327, pp. 138–149, 2014.

fraction. An illustration is provided in Figure 1a. The radiation generated by the accelerator is uniform. Therefore, in order to achieve the varying intensity, this radiation needs to be modulated. For this purpose, each intensity matrix is delivered through a multileaf collimator (MLC). An MLC is a device composed of parallel pairs (referred to as rows) of facing tungsten strips (referred to as leaves) that can block the radiation by moving towards each other from left and right (see Figure 1c). However, radiation can pass through the open gap between the leaves endpoint. Each intensity matrix is realized by a sequence of MLC configurations (*i.e.* specific leaves positions for each row of the MLC) each of which is maintained for a certain amount of time (corresponding to the intensity). In the static case, the radiation is switched off while the collimator leaves are moving. The so-called *gantry* denotes the whole device including the linac and the MLC.

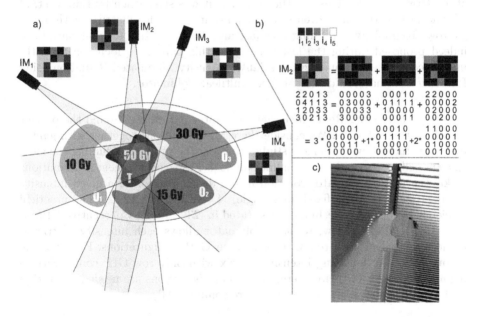

Fig. 1. a) IMRT with some intensity matrices – shown in grayscale coded grids with 5 intensities (the lighter the color the higher the radiation intensity). b) A realization of IM_2 with $i_1 = 0$, $i_2 = 1$, $i_3 = 2$, $i_4 = 3$, $i_5 = 4$. c) MLC illustration from Varian.

From an algorithmic point of view, the corresponding problem is a matrix decomposition problem where each intensity matrix is given as a positive integer matrix that has to be decomposed into a weighted sum of binary matrices (each binary matrix denotes an MLC configuration and the weight represents the associated intensity). These binary matrices are consecutive ones matrices (the 1s occur consecutively as a single block in each row) since MLC leaves are moving from left and right sides of the device at each row. For example, the intensity matrix IM_2 of Figure 1b can be decomposed into three configurations.

Of course, there are many ways of decomposing a given intensity matrix. It is desirable to select the decomposition that can be delivered the most efficiently. The two main efficiency criteria that play a role are the *total beam-on time, i.e.,* the total amount of time that the patient is being irradiated, and the *total setup time, i.e.,* the total amount of time that is spent shaping the apertures. The former metric is proportional to the sum of intensities used in the decomposition, while the latter is (approximately) proportional to the number of matrices used in the decomposition. Although closely related, these two efficiency criteria are not equivalent. The intensity matrix IM_2 decomposition shows a decomposition using only 3 apertures with a beam-on time of 6. However, the minimum beam-on time for this intensity matrix is 5, which can be realized by 5 apertures. Actually, it turns out that, while minimizing the total beam-on time is solvable in linear time [1] , minimizing the total setup time is NP-hard for matrices with at least two rows [5]. This NP-Hardness result was strengthen by Baatar et. al [2] who proved that it is even strongly NP-hard, even for matrices with a single row. Technology is running very fast and different MLC settings have been indeed proposed during the last decade. In this paper, we focus on algorithmic aspects of two technological variants: Intensity Modulated Radiation using Rotating Collimator and Multi-Layer Multileaf Collimator.

Rotating Collimator. Remember that the gantry is rotating around the patient to deliver the radiation. We consider here MLC rotations together with gantry rotations. Indeed, whereas most studies consider the problem of finding the most efficient collimator angle for each linac angle [3,15,17], some recent contributions tackle the use of collimator rotation in the decomposition of a given intensity matrix (*i.e.* for a given fixed linac angle) [4,6,12,13,16,18,19,22]. The practical efficiency of this latter technique was stated in [23]. From a Consecutive 1 Property (C1P) point of view, rotating collimator allows each intensity matrix to be decomposed in both row C1P or column C1P configurations. For example, decomposing the following intensity matrix with only row C1P configurations requires at least 8 of them whereas only 6 configurations are needed if rotation is allowed (the last two configurations are column C1P).

$$
\begin{bmatrix} 1\,4\,2\,5 \\ 1\,3\,3\,2 \\ 1\,3\,5\,5 \\ 6\,4\,6\,0 \end{bmatrix} = \begin{bmatrix} 0\,0\,0\,1 \\ 0\,0\,1\,1 \\ 0\,0\,1\,1 \\ 1\,1\,1\,0 \end{bmatrix} + \begin{bmatrix} 0\,0\,0\,1 \\ 0\,1\,1\,1 \\ 0\,0\,1\,1 \\ 1\,1\,1\,0 \end{bmatrix} + \begin{bmatrix} 0\,1\,1\,1 \\ 1\,1\,1\,0 \\ 0\,1\,1\,1 \\ 1\,1\,1\,0 \end{bmatrix} + \begin{bmatrix} 1\,1\,1\,1 \\ 0\,0\,0\,0 \\ 1\,1\,1\,1 \\ 1\,1\,1\,0 \end{bmatrix} + \circlearrowleft + \begin{bmatrix} 0\,1\,0\,1 \\ 0\,0\,0\,0 \\ 0\,0\,0\,0 \\ 1\,0\,1\,0 \end{bmatrix} + \begin{bmatrix} 0\,1\,0\,0 \\ 0\,1\,0\,0 \\ 0\,1\,0\,1 \\ 1\,0\,1\,0 \end{bmatrix}
$$

Multi-Layer Multileaf Collimator. Using multiple layers of leaves has been originally patented by [25] in 1997 and has been intensively studied since 2003. Most studies consider two orthogonal layers, referred in the literature as the dual-MLC [8,9,10,11,14]. Topolnjak et al. investigated the use of three layers placed every 60 degrees [20,21]. The state-of-the-art is presented in [24] which claims the efficiency of the gear. For example, decomposing the following intensity matrix with only C1P matrices (even allowing MLC rotations) requires a linear number of configurations whereas only 2 are needed when using a dual-MLC ($\uparrow$ and $\rightarrow$ represent resp. vertical and horizontal blocking leaves).

$$
\begin{bmatrix}
1\,0\,1\,0\,1\ldots 0 \\
0\,1\,0\,1\,0\ldots 1 \\
1\,0\,1\,0\,1\ldots 0 \\
0\,1\,0\,1\,0\ldots 1 \\
1\,0\,1\,0\,1\ldots 0 \\
\vdots\ \vdots\ \vdots\ \vdots\ \vdots\ \ddots\ \vdots \\
0\,1\,0\,1\,0\ldots 1
\end{bmatrix}
=
\begin{bmatrix}
1\uparrow 1\uparrow 1\ldots \uparrow \\
\leftarrow + \leftarrow + \leftarrow \ldots + \\
1\uparrow 1\uparrow 1\ldots \uparrow \\
\leftarrow + \leftarrow + \leftarrow \ldots + \\
1\uparrow 1\uparrow 1\ldots \uparrow \\
\vdots\ \vdots\ \vdots\ \vdots\ \vdots\ \ddots\ \vdots \\
\leftarrow + \leftarrow + \leftarrow \ldots +
\end{bmatrix}
+
\begin{bmatrix}
+ \rightarrow + \rightarrow + \ldots \rightarrow \\
\uparrow 1\uparrow 1\uparrow \ldots 1 \\
+ \rightarrow + \rightarrow + \ldots \rightarrow \\
\uparrow 1\uparrow 1\uparrow \ldots 1 \\
+ \rightarrow + \rightarrow + \ldots \rightarrow \\
\vdots\ \vdots\ \vdots\ \vdots\ \vdots\ \ddots\ \vdots \\
\uparrow 1\uparrow 1\uparrow \ldots 1
\end{bmatrix}
$$

To the best of our knowledge, all these contributions mentioned the complexity increase of the problem when considering rotation or multiple layers without being able to state it formally. In this contribution, we prove formally the algorithmic hardness of the corresponding problems. More precisely, we study the Matrix Orthogonal Decomposition (MOD) problem introduced in [6] which consider the decomposition problem of an intensity matrix using a unique 90° rotation and the Dual-MLC Decomposition (DMD) problem which consider two orthogonal layers of MLCs. We prove that both problems are still NP-hard when minimizing total-setup time. Finally we prove that MOD becomes NP-hard even when minimizing the total beam-on time but is approximable.

2   Dual-MLC Decomposition Minimizing Total Setup-Time

In order to prove the hardness of the problem, we will use the construction of Baatar et. al [2] as a gadget. Therefore, let us first present briefly a slightly modified version of their proof. As a reminder, they originally proved the hardness of total setup-time decomposition even for matrices with a single row by a reduction from the NP-complete 3-Partition problem where one has to partition $3Q$ positive numbers – say $S = (b_1, b_2, \ldots, b_{3Q})$ – (allowing duplicates) into Q triples – say $\{T_1, T_2, \ldots, T_Q\}$, such that each triple has the same sum. Considering that all $3Q$ numbers sum to N, then every triple should have a sum of $B = \frac{N}{Q}$ (we may assume that $\frac{B}{4} < b_i < \frac{B}{2}$ for every $b_i \in S$).

From any instance S of the 3-Partition problem, one can construct in polynomial time an integer vector $A = x_1\,x_2\,\ldots x_{3Q}\,y_Q\,y_{Q-1}\ldots y_1\,z_0$ such that $x_i = \sum_{j=1}^{i} b_j$, $y_i = i \cdot B$ and $z_0 = b_1$[1] and asks for a decomposition with at most $3Q$ MLC configurations. As a reminder, any solution of the problem is a set of C1P vectors (i.e. configurations) provided with corresponding intensity. Therefore, for each configuration, we will denote by left (resp. right) endpoint the first (resp. last) position of a 1 in the corresponding vector. First, notice that since $\{x_i | 1 \leq i \leq 3Q\}$ is a set of $3Q$ different values, any decomposition of A will need at least $3Q$ configurations and thus $3Q$ corresponding intensities

---

[1] We added z_0 to the original construction for ease of proof demonstration.

(among which one is b_1 due to x_1) each having their left endpoint disjointly in one of $\{x_i | 1 \leq i \leq 3Q\}$ positions. Moreover, the configuration with the intensity b_1 is defined as a totally open configuration (*i.e.* a vector of $4Q + 1$ $1's$). Indeed, among the at most $Q \cdot B$ radiation doses that can go through x_{3Q}, exactly B of them are needed for irradiating y_1. To respect the consecutiveness of the $1's$, whatever left endpoints of the configurations used, the corresponding configurations will contribute exactly B to each of $\{y_i | 2 \leq i \leq Q\}$. Repeating this last argument over $\{y_i | 2 \leq i \leq Q\}$, one can prove that each y_i is the right endpoint of some configurations, whose overall contribution sums to B. Moreover, by construction, z_0 needs b_1 irradiation doses that has to be included into the B needed by y_1. Consequently, since one of the configurations contributed b_1 to all positions of A, x_2 now only needs an extra b_2 contribution which should be delivered at once. Repeating this argument over $\{x_i | 3 \leq i \leq 3Q\}$, one can prove that the set of configuration intensities is indeed $\{b_i | 1 \leq i \leq 3Q\}$. Since, for any $1 \leq i \leq 3Q$, $\frac{B}{4} < b_i < \frac{B}{2}$, any y_i will need to be the right endpoint of exactly 3 configurations to get an overall irradiation summing to B (recall that there are at most $3Q$ configurations). Provided with these properties, one can easily prove that ($\Leftarrow$) given a solution to the 3-Partition problem such that, w.l.o.g., $b_1 \in T_1$, for all $2 \leq i \leq 3Q$, irradiating the interval $[x_i, y_q]$ with intensity b_i if $b_i \in T_q$ and irradiating the full vector with b_1 leads to a valid decomposition of exactly $3Q$ configurations. Moreover, ($\Rightarrow$) considering any solution of the decomposition problem, defining the triples $\{T_i | 1 \leq i \leq Q\}$ such that $b_j \in T_q \Leftrightarrow$ there exists a configuration of intensity b_i with resp. left and right endpoints x_i and y_q leads to a solution to 3-Partition.

Back to our original problem, we will use a slightly similar reduction using A as a gadget. We, thus now consider the decomposition of a matrix. For ease, in the rest of the paper, $[x]^k$ will denote a sequence of k copies of element x. The reduction is again from the 3-Partition problem. From any instance S, one can construct in polynomial time a matrix $M = (R_1, R_2, \ldots R_{6Q+3})$ (illustrated in Figure 2 in the companion website) composed of $6Q + 3$ rows, where for all $1 \leq i \leq 3Q$, $R_i = R_{6Q+4-i} = [x_{3Q+1-i}]^{4Q+1}$, $R_{3Q+1} = R_{3Q+3} = [0]^{4Q+1}$ and R_{3Q+2} is the vector A designed in the previous proof and ask again for a decomposition with at most $3Q$ Dual-MLC configurations. Roughly, the vector A is vertically surrounded by two opposed sorted heaps of vectors (increasing, when going away from A) filled with the $\{x_i | 1 \leq i \leq 3Q\}$ values defined in the previous proof and two null rows. The correctness of the proof relies on proving that, whereas one may use the vertical leaves to make a different set of configurations for realizing the peculiar row A, this would not lead to a valid solution. Indeed, the rows R_{3Q+1} and R_{3Q+3} ensures that if a vertical leaf was used to tune the irradiation configuration used for A – say the one in column j – then the corresponding intensity could not be used for any element of column j except in A. Since, by construction, yet again exactly $3Q$ configurations are required, there will exist at least one row in the end with a non null value on the column j. This property, ensures that if the vertical leaves are used, this is not to disturb the configurations plan of row R_{3Q+2}. Provided with these properties, one can easily prove that ($\Leftarrow$) given a solution to the 3-Partition problem such that, w.l.o.g., $b_1 \in T_1$, for all $2 \leq j \leq 3Q$, irradiating the interval $R_{3Q+2}[x_j, y_q]$ and fully the rows R_{3Q-j} and R_{6Q+3-j} with intensity b_j if $b_j \in T_q$ and irradiating the full rows R_{3Q}, R_{3Q+2} and R_{3Q+4} with b_1 leads to a valid decomposition of exactly $3Q$ configurations. Moreover, ($\Rightarrow$) considering any solution of the decomposition problem, defining the triples $\{T_i | 1 \leq i \leq Q\}$ such that $b_j \in T_q \Leftrightarrow$ there exists a configuration of intensity b_j with resp. left and right endpoints in x_j and y_q in row R_{3Q+2} leads to a solution to 3-Partitioning.

Theorem 1. *The Dual-MLC Decomposition problem is NP-hard when minimizing the total setup time.*

3   Matrix Orthogonal Decomposition

This section is devoted to proving a stronger result for the MOD problem: minimizing the total setup time is NP-hard even if the intensity matrix is binary. This result shows that the problem is also NP-hard when one wants to minimize the beam-on time (whereas it is polynomial when rotation is not changed during decomposition). Fortunately, we will also prove that the problem is however approximable in this later case. For ease of presentation, we will first give a construction using a positive integer matrix and show afterwards how to make it binary. In order to prove the hardness of the problem, we define a reduction from the NP-complete 3-Hitting Set problem: given a collection $C = \{C_1, \ldots, C_m\}$ of m subsets of size at most three of a finite set $S = \{x_1, \ldots x_n\}$ of n elements and a positive integer k, the problem asks for a subset $S' \subseteq S$ with $|S'| \leq k$ such that S' contains at least one element from each C_i's.

From any instance (C, k) of the 3-Hitting Set problem, one can construct in polynomial time a square matrix M composed of two rows and columns independent submatrices – a submatrix of $2n + 9$ columns and $2n + 8$ rows referred to as M_{DHV}, defined below, is placed top-right of M whereas another submatrix of $3n + 2$ columns and $m + 4$ rows named M_{3HS}, and defined later on, is placed bottom-left; the rest of the matrix M is filled with 0's in order to obtain a square matrix – and one can ask for a decomposition with at most $n + 3$ MLC configurations. The submatrix M_{DHV} is designed in such a way that it will ensure that any solution to the decomposition problem will use only one vertical configuration and $(n + 2)$ horizontal ones. Indeed, since, by construction, a) there are $(n + 4)$ horizontal blocks of single $1's$ in the first row, and b) all the $2(n + 2)$ last columns are each composed of $(n + 4)$ vertical blocks of single $1's$, any solution (*i.e.* not inducing more than $(n + 3)$ configurations) has to have at least one horizontal and one vertical configurations. Moreover, any solution has to use exactly one vertical configuration. Indeed, suppose, aiming at a contradiction, that a given solution uses more – say k' vertical configurations, then at most k' $1's$ from each column can be irradiated by those k' configurations. Unfortunately, since $k' \leq (n + 2)$, at least two $1's$ per column (except the five leftmost ones) will subsist. In order for the solution to be feasible, one would then have to irradiate the remaining $1's$ with a unique horizontal configuration.

To do so, the remaining $1's$ should be placed in order not to have more than one 1 per row; a contradiction since we have at least $2 \times (2n + 4)$ $1's$ and at most $2n + 8$ rows. We just proved that the submatrix M_{DHV} will force any solution to use exactly one vertical configuration and $(n + 2)$ horizontal ones, each with an intensity of 1.

$$M = \begin{bmatrix} 0 & M_{DHV} \\ M_{3HS} & 0 \end{bmatrix} \text{ with } M_{DHV} = \begin{bmatrix} 0\,1\,0\,1\,0\,1\,0\,1\,0\,\ldots\,1\,0 \\ \begin{array}{|c|} \hline 0\,0\,1\,0\,1\,\ldots\,0\,1 \\ 0\,1\,0\,1\,0\,\ldots\,1\,0 \\ \vdots\;\vdots\;\vdots\;\vdots\;\vdots\;\ldots\;\vdots\;\vdots \\ \mathbf{0} \quad 0\,1\,0\,1\,0\,\ldots\,1\,0 \\ 0\,0\,1\,0\,1\,\ldots\,0\,1 \\ 0\,1\,0\,1\,0\,\ldots\,1\,0 \\ \hline \end{array} \end{bmatrix} \text{ and }$$

$$M_{3HS} = \begin{bmatrix} CTRL_U^V\{ & (n+3) & 0 & 1 & 1 & 0 & 1 & 1 & 0 & \ldots & 1 & 1 & 0 \\ GATE\{ & 0 & 0 & 0 & 0 & 0 & 0 & 0 & 0 & \ldots & 0 & 0 & 0 \\ CTRL_D^V\{ & (n+2) & 0 & 0 & 0 & 1 & 0 & 0 & 1 & \ldots & 0 & 0 & 1 \\ CTRL^{max}\{ & (n+2-k) & 0 & 0 & 1 & 1 & 1 & 1 & 1 & \ldots & 1 & 1 & 1 \\ C_1\{ & c_1 & 0 & 0 & 1 & x_1^1 & 0 & 1 & x_1^2 & \ldots & 0 & 1 & x_1^n \\ C_2\{ & c_2 & 0 & 0 & 1 & x_2^1 & 0 & 1 & x_2^2 & \ldots & 0 & 1 & x_2^n \\ & \vdots & & & & \vdots & & & & & & & \\ C_m\{ & c_m & 0 & 0 & 1 & x_m^1 & 0 & 1 & x_m^2 & \ldots & 0 & 1 & x_m^n \\ & & & & & \underbrace{\quad}_{x_1} & & & \underbrace{\quad}_{x_2} & \ldots & & & \underbrace{\quad}_{x_n} \end{bmatrix} \Big\} C$$

Let us now describe the submatrix M_{3HS} which is totally independent of M_{DHV} but which will inherit the repartition of the vertical and horizontal configurations that we just showed. M_{3HS} is defined as pictured above and where $x_i^j = 2$ if $x_j \in C_i$; $x_i^j = 1$ otherwise and $c_i = 0$ if $|C_i| = 3$; $c_i = 1$ otherwise. The submatrix M_{3HS} is designed to encode the 3-Hitting Set instance. Roughly, each subset C_i of C is encoded by a row whereas each element x_i of S is encoded by a column[2]. Let us now prove some interesting properties of this construction.

Let us have a look at the constraints of the unique vertical configuration. Since $CTRL_U^V[0]$ is set to $(n+3)$ – that is both the maximal number of configurations and intensity, the first column of the vertical configuration will have to irradiate this last and so does the corresponding row (i.e. $(2n + 9)^{\text{th}}$ of M) of all the horizontal configurations. It has many consequences: a) all the $1's$ of $CTRL_U^V$ will have to be irradiated during the vertical configuration (inducing that no other $1's$ in the corresponding columns can be irradiated during the vertical configuration – namely columns in $\{3j - 1, 3j | 1 \le j \le n\}$) and b) $CTRL_D^V[0]$, $CTRL^{max}[0]$ and any $C_i[0]$ will all only be irradiated by horizontal configurations[3]. Since

---

[2] Due to space consideration, we will not include a full illustration of a construction but one may build one from our companion website.

[3] For ease, in the above description of M_{3HS} matrix, all the cells which will not have a contribution from the vertical configuration have been put in gray.

$CTRL_D^V[0]$ is set to $(n+2)$, all the horizontal configurations for this specific row $CTRL_D^V$ will need to be dedicated to $CTRL_D^V[0]$. This implies, in turn, that all remaining $1's$ of $CTRL_D^V$ – that is $\{CTRL_D^V[i]|i = 2 + 3j, 1 \leq j \leq n\}$ – would have to be vertically irradiated. On the whole, except for the set of bottom leaves for the columns $\{i = 2 + 3j|1 \leq j \leq n\}$, we know exactly what is the endpoint position of each leave (top and bottom) of the vertical configuration in M_{3HS}: a) for all $i \in \{0, 3j, 3j - 1|1 \leq j \leq n\}$ the ith top leave (resp. bottom one) precisely blocks all the rows preceding (resp. succeeding) $CTRL_U^V$, b) column 1 is totally blocked and c) for all $i \in \{3j + 1|1 \leq j \leq n\}$ the ith top leave precisely blocks all the rows preceding $CTRL_D^V$.

Now, notice that any C_i needs at least $(n+3)$ configurations to be realized. This implies that any C_i should be irradiated at least in one of its column by the vertical configuration. By construction, this irradiation can only occur in $(x_i^j)'s$ positions (i.e. $\{3j + 1|1 \leq j \leq n\}$) moreover set to **2** (otherwise it will not help decreasing the total irradiation needed to realize C_i) – later referred to as target positions. The $CTRL^{max}$ row is designed to ensure that at least $n-k$ cells among $\{CTRL^{max}[3j+1]|1 \leq j \leq n\}$ will be blocked by bottom leaves. In other words, at least $n-k$ bottom leaves will block all the succeeding rows of $CTRL^{max}$. Thus, at most k bottom leaves would be able to allow vertical irradiation contribution for the target positions. As we just prove, the only differences between both solutions to the decomposition problem is the position of bottom leaves endpoints for positions in $\{3j + 1|1 \leq j \leq n\}$; we will thus characterize any such solution as a set of n positions in $[2n + 13, 2n + m + 14]$ (corresponding to all possible solutions – $2n + m + 14$ being a leave not used at all).

Provided with these properties, one can easily prove that ($\Leftarrow$) given a solution $(S' \subseteq S)$ to 3-Hitting Set problem, for each $1 \leq i \leq n$, if $x_i \in S'$, $P[i] = 2n + m + 14$; $P[i] = 2n + 13$ otherwise. We claim that P corresponds to valid positions for the bottom leaves in $\{3j + 1|1 \leq j \leq n\}$. Since S' is a hitting set of size at most k, we can ensure that at least one element of each subset in C belongs to S'. This guarantees that all $C_i's$ rows and $CTRL^{max}$ are realized. Moreover, ($\Rightarrow$) considering any solution to the decomposition problem, one can define the hitting set S' such that $x_j \in S' \Leftrightarrow$ the position of the $(3j + 1)^{th}$ bottom leaf is strictly greater than $2n + 13$.

Let us now try to transform the construction in order to obtain a binary matrix with the same property. First, one can encode the $x_i's$ using two columns rather than one as follows. Insert a column just before each actual column representing an x_i, fill it with $0's$ except on the $CTRL_U^V$ row which has to be set to **1** and the $C_j's$ rows where the corresponding columns have to be set to 0 **1** if $x_i \in C_j$; 1 **1** otherwise[4]. This update is clearly not changing the original proof. The tricky part stands in the replacement of $CTRL_U^V[0]$. Indeed, one wants that it still requires all the horizontal configurations and the vertical one. To do so, one can design a submatrix of $2(n + 4)$ rows defined as follows: a) each odd row is filled with $0's$ and b) the ith even row is defined as $[0]^{i-1}$ **1** $0[1\ 0]^{n+2}[0]^{n+3-i}$.

---

[4] Due to space consideration, please consider checking the construction on our companion website.

Roughly, the block representing $(n + 3)$ (*i.e.* $[1\ 0]^{n+3}$) is shifted right of one position every new even row. This ensures that no one under the position of this gadget will be able to have a vertical contribution and that in any of the corresponding rows, all the remaining $1's$ will need to be irradiated vertically. We just showed that the properties of the original gadget are preserved. Both $CTRL_D^V[0]$ and $CTRL^{max}[0]$ can be easily replaced by resp. $[0\ 1]^{n+2}[0]^{3n+7}$ and $[0\ 1]^{n+2-k}[0]^{3n+7}$. Again, the properties of the original gadgets are preserved. This concludes the proof of the following theorem.

Theorem 2. *The Matrix Orthogonal Decomposition problem is NP-hard when minimizing either the total setup or the total beam-on time.*

Now, let us prove that there exists an algorithm based on linear programming and rounding techniques that produces an approximate solution for minimizing the total beam-on time. First, recall that for horizontal configurations, rows can be dealt with separately. It is also the case for vertical configurations and columns. Indeed, an intensity matrix is realized by a sequence of MLC configurations each of which is maintained for a certain amount of time (corresponding to the intensity). Since the problem is to minimize the sum of intensities and not the number of configurations, one can always consider that configurations can be changed every unit of time. This implies that any row (column) can be processed independently of the others and that the overall beam-on time will be deduced by the (most) expensive row (column). The problem can be phrased, as an Integer Linear Programming, as defined in Figure 2.

minimize $H + V$

subject to $\forall 1 \leq k \leq m,$ $\qquad\qquad\qquad\qquad \sum_{i \leq j} H_{ij}^k \leq H \quad (1)$

$\forall 1 \leq k \leq m,$ $\qquad\qquad\qquad\qquad \sum_{i \leq j} V_{ij}^k \leq V \quad (2)$

$\forall k, k' \in \{1, \ldots m\}^2,$ $\quad \sum_{i \leq k' \leq j} H_{ij}^k + \sum_{i' \leq k \leq j'} V_{i'j'}^{k'} = M[k][k'] \quad (3)$

$\forall i, j, k,$ $\qquad\qquad\qquad\qquad H_{ij}^k \geq 0, V_{ij}^k \geq 0$

$1 \leq i \leq j \leq m, 1 \leq i' \leq j' \leq m, H \geq 0, V \geq 0.$

Fig. 2. Integer Linear Program minimizing the total beam-on time for MOD

For any row of the intensity matrix M, let H_{ij}^k be a variable indicating the amount of time the following horizontal configuration is maintained: considering the kth pair of leaves, the left one's endpoint is at position $i-1$ and the right one's endpoint is at position $j+1$ (therefore irradiating any position between i and j in row k). Similarly, for any column of the intensity matrix M, let V_{ij}^k be a variable indicating the amount of time the following vertical configuration is maintained: considering the kth pair of leaves, the left one's endpoint is at position $i-1$ and the right one's endpoint is at position $j+1$ (therefore irradiating any position between i and j in column k). Finally, variables H and V are respectively horizontal and vertical costs of a solution computed respectively as $\max_k \sum_{i \leq j} H_{ij}^k$ and $\max_k \sum_{i \leq j} V_{ij}^k$ (which is encoded by constraints (1) and (2)). Constraint (3) ensures that the desired intensity matrix is realized. Indeed, $\sum_{i \leq k' \leq j} H_{ij}^k$ (resp. $\sum_{i' \leq k \leq j'} V_{i'j'}^{k'}$) represents the overall contribution of all the horizontal (resp. vertical) configurations contributing to the entry $M[k][k']$. There are about $2m^3 + 2$ variables, $2n$ inequalities and n^2 equalities. Our linear programming problem can be rewritten with only inequalities. Indeed, each equality constraint may be removed, by solving it for variable $H_{0k'}^k$ and substituting this solution into the corresponding form of constraint (1) (i.e. for the corresponding k).

Of course Integer Linear Programming is NP-hard. Therefore, we relax the integrality constraint, that is, allowing all variables to take a non-integral but still positive value. We end-up with a fractional linear program that can be solved in polynomial time. Notice that the solution provided by this linear program cannot be greater than the optimal integer one, since we only allow more solutions to become feasible. We apply a rounding of the fractional solution to obtain an integral feasible solution not too far from optimal.

Assume that $f_L : \{V, H, H_{ij}^k, V_{ij}^k | 1 \leq i \leq j \leq m, 1 \leq k \leq m\} \rightarrow \mathbb{R}$ is an optimal fractional solution of the relaxed version of our problem. If one slightly modifies the values of H_{ij}^k's then due to constraint (3) the values of $V_{i'j'}^{k'}$'s will need to be modified accordingly and with a comparable amount. The basic idea is to provide an integral rounding of the horizontal configurations and compute polynomially the corresponding vertical configurations while guaranteeing that the corresponding solution is a good approximation of the optimal one.

Let us present the rounding technique for a single row – say the kth. Considering all the corresponding variables $\{H_{ij}^k | 1 \leq i \leq j \leq m\}$, one can represent each non-null variable H_{ij}^k by an interval $[i, j]$ over the real line on $[1, m]$ weighted by H_{ij}^k (illustrated in Figure 3 in companion website). Let us transform this set of intervals $\mathcal{I}$ into a set $\mathcal{I}'$, where given any pair of intervals either one is included into the other or they are disjoint. To do so, we process $\mathcal{I}$ with the following algorithm. While there exists two intervals $[i, j]$ and $[k, l]$ with respective weights w_1 and w_2 such that $i < k < j < l$ (i.e. crossing) remove $[i, j]$ and $[k, l]$ from $\mathcal{I}$ and add $[i, k-1]$, $[k, j]$ and $[j+1, l]$ with respective weights w_1, $w_1 + w_2$ and w_2. Now that all intervals are nested or independent, while there exists three intervals $[x, y]$, $[i, j]$ and $[k, l]$ with respective weights w_1, w_2 and w_3 such that $x \leq i < j \leq k < l \leq y$, if $j < k$ then remove $[x, y]$ from $\mathcal{I}$ and add $[x, j]$ and $[j, y]$ both weighted by w_1; otherwise $(j = k)$ remove $[i, j]$ and $[k, l]$ from $\mathcal{I}$ and if $w_2 < w_3$ then add $[i, l]$ and $[k, l]$ with respective weights w_2 and $w_3 - w_2$; otherwise add $[i, l]$ and $[i, j]$ with respective weights w_3 and $w_2 - w_3$. Moreover given two copies of an interval $[i, j]$ with respective weights w_1 and w_2, remove them and add an interval $[i, j]$ with weight $w_1 + w_2$.

We end up with a set of independent subsets of nested intervals (later referred to as a *stack*). Note that there are at most m such stacks. We will proceed to the rounding of each stack separately. We will do so while ensuring that the sum of the original weights is smaller than the sum of the rounded ones with a gap of at most 1. This will induce that for a given row of the horizontal configuration, we manage to get an integral solution with an at most m extra cost. For ease, considering the stack as increasingly sorted by interval size and let w_i and w_i' denote respectively the original and rounded weights of the ith interval of the stack. The rounding algorithm proceeds as follows: start from the wider interval and round up w_1[5]. Then consider iteratively each remaining interval – say the jth, round it down if possible – that is if $\lfloor w_j \rfloor + \sum_{i=1}^{j-1} w_i' \geq \sum_{i=1}^{j} w_i$ – round it up otherwise.

Applying the rounding to each row of the horizontal configurations leads to an integral solution for the horizontal configurations that we can subtract for the original intensity matrix. Then, we compute in polynomial time the vertical configurations on the resulting matrix. We claim that the overall solution is at most $2m$ from the optimal solution. Indeed, in the resulting matrix, each cell is at most greater by one than the fractional matrix. This means that the sum of the elements in any column is at most greater by m than the fractional matrix. Thus we loose at most m with the rounding of the horizontal configurations plus at most m for adjusting the vertical configurations, for a total of a $2m$ additional cost.

Acknowledgement. We wish to thank members of the Radiotherapy Group of Institut Bergonié (Bordeaux, France) and of the Dpt of Radiation Oncology (University of Iowa, USA) – especially Guy Kantor, Christina Zacharatou, Xiaodong Wu, Dongxu Wang and Ryan T. Flynn – for their valuable technical feedback on Radiotherapy.

References

1. Ahuja, R., Hamacher, H.: A network flow algorithm to minimize beam-on time for unconstrained multileaf collimator problems in cancer radiation therapy. Networks 45(1), 36–41 (2005)
2. Baatar, D., Hamacher, H.W., Ehrgott, M., Woeginger, G.J.: Decomposition of integer matrices and multileaf collimator sequencing. Discrete Applied Mathematics 152(1-3), 6–34 (2005)
3. Beavis, A., Ganney, P., Whitton, V., Xing, L.: Optimisation of MLC orientation to improve accuracy in the static field delivery of IMRT. In: Proceedings of 22nd Annual International Conference of the IEEE Engineering in Medicine and Biology Society, vol. 4, pp. 3086–3089 (2000)
4. Broderick, M., Leech, M., Coffey, M.: Direct aperture optimization as a means of reducing the complexity of Intensity Modulated Radiation Therapy plans. Radiation Oncology 4, 8 (2009)

---

[5] One cannot do differently since there exists at least one position not covered by another interval.

5. Burkard, R.: Open problem session, Oberwolfach Conference on Combinatorial Optimization, November 24-29 (2002)
6. Dou, X., Wu, X., Bayouth, J.E., Buatti, J.M.: The Matrix Orthogonal Decomposition Problem in Intensity-Modulated Radiation Therapy. In: Chen, D.Z., Lee, D.T. (eds.) COCOON 2006. LNCS, vol. 4112, pp. 156–165. Springer, Heidelberg (2006)
7. Hamacher, H.W., Küfer, K.-H.: Inverse radiation therapy planning - a multiple objective optimization approach. Discrete Applied Mathematics 118(1-2), 145–161 (2002)
8. Hughes, J.: US Patent 6,600,810: Multiple layer multileaf collimator design to improve resolution and reduce leakage (2003)
9. Jarray, F., Picouleau, C.: Minimum decomposition into convex binary matrices. Discrete Applied Mathematics 160(7-8), 1164–1175 (2012)
10. Liu, Y., Shi, C., Lin, B., Ha, C.: Delivery of four-dimensional radiotherapy with TrackBeam for moving target using a dual-layer MLC: dynamic phantoms study. Journal of Applied Clinical Medical Physics 10(2), 1–21 (2009)
11. Liu, Y., Shi, C., Tynan, P., Papanikolaou, N.: Dosimetric characteristics of dual-layer multileaf collimation for small-field and intensity-modulated radiation therapy applications. Journal of Applied Clinical Medical Physics 9(2), 2709 (2008)
12. Milette, M.: Direct optimization of 3D dose distributions using collimator rotation. PhD thesis, University of British Columbia (2008)
13. Milette, M., Rolles, M., Otto, K.: TU-C-224A-06: Exploiting the Full Potential of MLC Based Aperture Optimization Through Collimator Rotation. Medical Physics 33(6), 2191 (2006)
14. Oh, S., Jung, W., Suh, T.: SU-FF-T-28: A New Concept of Multileaf Collimator (dual-Layer MLC). Medical Physics 34(6), 2407 (2007)
15. Otto, K.: Intensity modulation of therapeutic photon beams using a rotating multileaf collimator. PhD thesis, University of British Columbia (2003)
16. Otto, K.: US Patent 6,907,105: Methods and apparatus for planning and delivering intensity modulated radiation fields with a rotating multileaf collimator (2005)
17. Otto, K., Clark, B.: Enhancement of IMRT delivery through MLC rotation. Physics in Medicine and Biology 47(22), 3997–4017 (2002)
18. Otto, K., Milette, M., Schmuland, M.: SU-FF-T-104: Rotating Aperture Optimization - Planning and Delivery Characteristics. Medical Physics 32(6), 1973 (2005)
19. Schmuland, M.: Dose verification of rotating collimator intensity modulated radiation thearpy. PhD thesis, University of British Columbia (2006)
20. Topolnjak, R., van der Heide, U., Lagendijk, J.: IMRT sequencing for a six-bank multi-leaf system. Physics in Medicine and Biology 50(9), 2015–2031 (2005)
21. Topolnjak, R., van der Heide, U., Raaymakers, B., Kotte, A., Welleweerd, J., Lagendijk, J.: A six-bank multi-leaf system for high precision shaping of large fields. Physics in Medicine and Biology 49(12), 2645–2656 (2004)
22. Wang, D., Hill, R., Lam, S.: US Patent 7,015,490: Method and apparatus for optimization of collimator angles in intensity modulated radiation therapy treatment (2006)
23. Webb, S.: Does the option to rotate the Elekta Beam Modulator MLC during VMAT IMRT delivery confer advantage? - a study of 'parked gaps'. Physics in Medicine and Biology 55(11), N303–N319 (2010)
24. Webb, S.: A 4-bank multileaf collimator provides a decomposition advantage for delivering intensity-modulated beams by step-and-shoot. Physica Medica 28(1), 1–6 (2012)
25. Yao, J.: US Patent 5,591,983: Multiple layer multileaf collimator (1997)

Parameterized Complexity of the Sparsest k-Subgraph Problem in Chordal Graphs*

Marin Bougeret, Nicolas Bousquet, Rodolphe Giroudeau, and Rémi Watrigant

LIRMM, Université Montpellier 2, France

Abstract. In this paper we study the SPARSEST k-SUBGRAPH problem which consists in finding a subset of k vertices in a graph which induces the minimum number of edges. The SPARSEST k-SUBGRAPH problem is a natural generalization of the INDEPENDENT SET problem, and thus is $\mathcal{NP}$-hard (and even $W[1]$-hard) in general graphs. In this paper we investigate the parameterized complexity of both SPARSEST k-SUBGRAPH and DENSEST k-SUBGRAPH in chordal graphs. We first provide simple proofs that DENSEST k-SUBGRAPH in chordal graphs is FPT and does not admit a polynomial kernel unless $\mathcal{NP} \subseteq co\mathcal{NP}/poly$ (both parameterized by k). More involved proofs will ensure the same behavior for SPARSEST k-SUBGRAPH in the same graph class.

1 Introduction

Presentation of the Problem. Given a simple undirected graph $G = (V, E)$ and an integer k, the SPARSEST k-SUBGRAPH problem asks to find k vertices in G inducing the minimum number of edges. The decision version asks if there exists a k-subgraph inducing at most C edges. As a generalization of the classical INDEPENDENT SET problem (for $C = 0$), SPARSEST k-SUBGRAPH is $\mathcal{NP}$-hard in general graphs, as well as $W[1]$-hard when parameterized by k (as INDEPENDENT SET is $W[1]$-hard [11]). In addition, there is an obvious XP algorithm for SPARSEST k-SUBGRAPH when parameterized by k, as all subsets of size k can be enumerated in $\mathcal{O}(n^k)$ time, where n is the number of vertices in the graph.

Related Problems. Several problems closely related to SPARSEST k-SUBGRAPH have been extensively studied in the past decades. Among them, one can mention the maximization version of SPARSEST k-SUBGRAPH, namely the DENSEST k-SUBGRAPH, for which several results have been obtained in general or restricted graphs. In [9], the authors showed that DENSEST k-SUBGRAPH remains $\mathcal{NP}$-hard in bipartite, comparability and chordal graphs, and is polynomial-time solvable in trees, cographs, and split graphs. The complexity status of DENSEST k-SUBGRAPH in interval graphs, proper interval graphs and planar graphs is left as an open problem, and is still not answered yet. More recently, [6] improved some of these results by showing that both DENSEST k-SUBGRAPH and SPARSEST k-SUBGRAPH are polynomial-time solvable in bounded clique-width

---

* This work has been funded by grant ANR **2010** BLAN **021902**.

graphs, and [3] developed exact algorithms for SPARSEST k-SUBGRAPH, DENS-EST k-SUBGRAPH and other similar problems in general graphs parameterized by k and the maximum degree Δ of the graph. During the past two decades, a large amount of work has been dedicated to the approximability of DENSEST k-SUBGRAPH in general graphs. So far, the best approximation ratio is $O(n^\delta)$ for some $\delta < 1/3$ [10], while the only negative result is due to Khot [15] ruling out a PTAS under some complexity assumptions. Still concerning DENSEST k-SUBGRAPH but in restricted graph classes, [17] developed a PTAS in interval graphs, and [8,16] developed constant approximation algorithms in chordal graphs. In [18], we recently proved that SPARSEST k-SUBGRAPH remains $\mathcal{NP}$-hard in chordal graphs and admits a 2-approximation algorithm. We can also mention the dual version of SPARSEST k-SUBGRAPH, namely the MAXIMUM PARTIAL VERTEX COVER problem, for which we are looking for k vertices in the input graph which *cover* the maximum number of edges. Very recently [1] and [14] independently proved the $\mathcal{NP}$-hardness of MAXIMUM PARTIAL VERTEX COVER in bipartite graphs, which directly transfers to SPARSEST k-SUBGRAPH (since finding k vertices covering the maximum number of edges is equivalent to find $(n - k)$ vertices inducing the minimum number of edges).

More generally, SPARSEST k-SUBGRAPH, DENSEST k-SUBGRAPH and MAXIMUM PARTIAL VERTEX COVER fall into the family of *cardinality constrained optimization problems* introduced by Cai [7]. In its survey, the author proved that these three problems are $W[1]$-hard in regular graphs, and gives an XP algorithm for general graphs with a better running time than the trivial algorithm.

As mentioned previously, SPARSEST k-SUBGRAPH and DENSEST k-SUBGRAPH are natural generalizations of k-INDEPENDENT SET and k-CLIQUE, and are thus important both from a theoretical and practical point of view. Our motivation is to study their computational (parameterized) complexity in graph classes where they remains $\mathcal{NP}$-hard whereas k-INDEPENDENT SET and k-CLIQUE are polynomial-time solvable, such as the well-known class of perfect graphs and some of its subclasses. To that end, we study their parameterized complexity in the class of chordal graphs, an important subclass of perfect graphs which arises in many practical situations [13]. More precisely, we prove that both SPARSEST k-SUBGRAPH and DENSEST k-SUBGRAPH in chordal graphs are fixed-parameter tractable and do not admit a polynomial kernel under some classical complexity assumptions. As we will see, the results are quite easy to obtain for DENSEST k-SUBGRAPH, but require some efforts for SPARSEST k-SUBGRAPH.

Organization of the Paper. The paper is organized as follows: in the following section (Section 2), we recall the classical definitions of parameterized complexity and chordal graphs. Our two main results, namely the FPT algorithm and kernel lower bound for SPARSEST k-SUBGRAPH in chordal graphs, are presented respectively in Sections 4 and 5. Before all these, we study as an appetizer the parameterized complexity of DENSEST k-SUBGRAPH in chordal graphs in Section 3. Due to space restrictions, some proofs and figures were omitted. They can be found in the long version of the paper available in [5].

2   Parameterized Algorithms, Chordal Graphs

Parameterized Algorithms. An interesting way to tackle $\mathcal{NP}$-hard problems is parameterized complexity. A parameterized problem Q is a subset of $\Sigma^* \times \mathbb{N}$, where the second component is called the *parameter* of the instance. A *fixed-parameter tractable* (*FPT* for short) problem is a problem for which there exists an algorithm which, given $(x, k) \in \Sigma^* \times \mathbb{N}$, decides whether $(x, k) \in Q$ in time $f(k)|x|^{O(1)}$ for some computable function f. Such an algorithm becomes efficient with an hopefully small parameter. A *kernel* is a polynomial algorithm which, given $(x, k) \in \Sigma^* \times \mathbb{N}$, outputs an instance (x', k') such that $(x, k) \in Q \Leftrightarrow (x', k') \in Q$ and $|x'| + k' \leq f(k)$ for some computable function f. The existence of a kernel is equivalent to the existence of an FPT-algorithm. Nevertheless one can ask the function f to be a polynomial. If so, then the kernel is called a *polynomial kernel*. If a problem admits a polynomial kernel, then it roughly means that we can, in polynomial time, compress the initial instance into an instance of size $poly(k)$ which contains all the hardness of the instance. In order to rule out polynomial kernels, we will use the recent technique of cross-composition [2].

Roughly speaking, a cross-composition is a polynomial reduction from t instances of a (non-parameterized) problem A to a single instance of a parameterized problem B such that the constructed instance is positive iff one of the input instances is positive. In addition, the parameter of the constructed instance must be of size polynomial in the maximum size of the input instances and the logarithm of t. It is known that if A is $\mathcal{NP}$-hard and A cross-composes into B, then B cannot admit a polynomial kernel under some complexity assumptions. For a stronger background concerning the parameterized complexity in general and to cross-compositions in particular, we refer the reader respectively to [11,2].

Chordal Graphs. A graph $G = (V, E)$ is a chordal graph if it does not contain an induced cycle of length at least four. As said previously, chordal graphs form an important subclass of perfect graphs. One can also equivalently define chordal graphs in terms of a special tree decomposition. Indeed, it is known [12] that a graph $G = (V, E)$ is a chordal graph if and only if one can find a tree $T = (\mathcal{X}, A)$ with $\mathcal{X} \subseteq 2^V$ such that for all $v \in V$, the set of nodes of T containing v, that is $\mathcal{X}_v = \{X \in \mathcal{X} : v \in X\}$, induces a (connected) tree, and such that for all $u, v \in V$ we have $\{u, v\} \in E$ if and only if $\mathcal{X}_u \cap \mathcal{X}_v \neq \emptyset$. Moreover, given a chordal graph, this corresponding tree can be found in polynomial time. From this definition, it is clear that each $X \in \mathcal{X}$ induces a clique in G.

3   Appetizer: Parameterized Complexity of Densest k-Subgraph in Chordal Graphs

Let us discuss the parameterized status of DENSEST k-SUBGRAPH in chordal graphs. First, it is well-known that computing a maximal clique in chordal graphs can be done in polynomial time. Hence, if the input graph contains a clique of size k or more, we can immediately output it. Otherwise, we use another classical property of chordal graphs stating that their tree-width equals the size of their maximal clique minus one. Thus, the tree-width of the input graph is bounded by $(k-1)$. Finally, using a classical dynamic programming on a tree decomposition, such as the one described in [4], we can compute an optimal solution in *FPT* time.

On the negative side, we can easily cross-compose DENSEST k-SUBGRAPH in chordal graphs into itself, by taking the disjoint union of t chordal graphs, and adding sufficiently enough universal vertices to each connected component. Due to space restrictions, this construction cannot be formally defined and proved here. To summarize, we have the following result:

Theorem 1. DENSEST k-SUBGRAPH *in chordal graphs is FPT and does not admit a polynomial kernel unless* $\mathcal{NP} \subseteq co\mathcal{NP}/poly$, *both parameterized by* k.

4   FPT Algorithm for Sparsest k-Subgraph in Chordal Graphs

Definitions and Notations. Let $G = (V, E)$ be a chordal graph and $T = (\mathcal{X}, A)$ be its corresponding tree decomposition as defined in Section 2. Recall that for each $X \in \mathcal{X}$, X induces a clique in G.

We denote respectively by $\mathcal{L}$ and $\mathcal{I}$ the set of leaves and internal nodes of T (we have $\mathcal{X} = \mathcal{L} \cup \mathcal{I}$). In the following we suppose that T is rooted at an arbitrary node X_r. Let $X \in \mathcal{X}$, we denote by $pred(X)$ the unique predecessor of X in T (by convention $pred(X_r) = \emptyset$), and by $succ(X)$ the set of successors of X in T. For a vertex $v \in V$ (resp. a node $X \in \mathcal{X}$), we denote by $d(v)$ (resp. $d(X)$) its degree in G (resp. in T). For a set of vertices $U \subseteq V$ (resp. set of nodes $A \subseteq \mathcal{X}$), we denote by $G[U]$ (resp. $T[A]$) the subgraph of G induced by U (resp. the subforest of T induced by A). We say that a vertex $v \in V$ is a *lonely*[1] vertex (*resp. almost lonely* vertex) if $|\mathcal{X}_v| = 1$ (resp. $|\mathcal{X}_v| = 2$), *i.e.* if it appears in only one (resp. two) nodes of T. For $x \in V$, $N(x)$ denotes the neighborhood of x.

First Observations. A maximum independent set can be computed in polynomial time in chordal graphs (since chordal graphs are perfect). Hence, we first determine if there exists an independent set of size k. In this case, we return this set which is naturally an optimal solution. Thus, we assume in the following that the graph G does not contain an independent set of size k.

Notice that we can assume that for every leaf L of the tree we do not have $L \subseteq pred(L)$ (otherwise we can contract the two nodes). Therefore, for each leaf L of the tree, there is a vertex $x \in L$ such that $x \notin pred(L)$, i.e. x is a lonely vertex. Since there is no independent set of size k in G, and since lonely vertices of leaves are pairwise non adjacent, we have the following:

---

[1] Notice that every lonely vertex is a so-called "simplicial vertex" (a vertex whose neighborhood is a clique). However, if a node of the tree is contained in another node, a simplicial vertex may not be a lonely one. Since we do not make any supposition on the tree T (we will in particular duplicate nodes during the algorithm), we will prefer the term "lonely".

Observation 1. *We can assume that $|\mathcal{L}| < k$.*

Let us now state a simple property verified by optimal solutions. Let S be a set of k vertices. Assume that there are vertices $x \in S$ and $y \in V \setminus S$ such that $\mathcal{X}_y \subsetneq \mathcal{X}_x$. Then it means that $N(y) \subseteq N(x)$, Thus, if we replace x by y in the solution, the number of edges in the solution cannot increase. A set S is *closed under inclusion* if there is no vertex x in S such that there exists $y \in V \setminus S$ such that $\mathcal{X}_y \subsetneq \mathcal{X}_x$. So there always exists an optimal solution closed under inclusion.

Idea of the Algorithm. Our goal is to find an optimal solution closed under inclusion. First note that any optimal solution closed under inclusion must contain a lonely vertex per leaf of T. Indeed, as each leaf L is not included in $pred(L)$, there exists a lonely vertex x in L. Thus, either the solution intersects L, and since it is closed by inclusion it contains a lonely vertex, or we can take a vertex of the solution and replace it by x, which does not create any additional edge (since no vertex of $N(x) = L \setminus \{x\}$ was in the solution).

Our method can be summarized as follows. First, we take a lonely vertex in each leaf and guess a binary flag $w(L) \in \{0, 1\}$ for each leaf L which indicates whether another vertex of L has to (with value 1) or does not have to (with value 0) be taken in the solution. The width of such a branching is bounded according to Observation 1. Then, given a leaf L with $w(L) = 1$, we first try to add to the solution the "most interesting" vertex of the leaf (for example a lonely vertex). When this is not possible (the neighborhood of the vertices of L can be incomparable if these vertices appear on incomparable subtrees), we apply some branching rules that re-structure the tree and create new "interesting vertices".

Terminology for the Algorithm. The algorithm is a branching algorithm composed of pre-processing rules (which do not require branching) and branching rules. When a rule is applied, we assume that previous rules cannot be applied.

During the algorithm, a partial solution S (initialized to $\emptyset$) will be constructed, and the input graph $G = (V, E)$ together with k, T (and thus $\mathcal{X}$, $\mathcal{L}$ and $\mathcal{I}$) will be modified. To avoid heavy notation we will keep these variables to denote the current input, and denote by G_0 the original graph, and by N_0 the neighborhood function of G_0.

In the following, *taking* a vertex $v \in V$ in the solution means that v is added to S, and v is removed both from the graph G and the tree T (removing each of its occurrences). *Deleting* a vertex means removing the vertex from G and from T. If a leaf of T becomes empty after taking or deleting a vertex, then simply remove the leaf.

Let $F \in \mathcal{I}$ be a leaf of $T[\mathcal{I}]$ (*i.e.* a node of T of which all successors are leaves). The node F is a *bad father* if there exists a vertex u which appears in at least two leaves of $succ(F)$. So a node is a bad father if the leaves attached to it are not vertex disjoint. We denote by $\#BF$ the number of bad fathers of the tree. Finally, we denote by $\#AL$ (for "almost leaf") the number of internal nodes of T such that at least one successor is a leaf. Notice that $\#AL, \#BF \leq |\mathcal{L}|$.

In addition, as said previously, we will put "flags" on some leaves $\mathcal{L}^* \subseteq \mathcal{L}$ by introducing a boolean function $w : \mathcal{L}^* \to \{0,1\}$, which indicates whether it intersects the solution (value 1) or not (value 0). At the beginning of the algorithm we have $\mathcal{L}^* = \emptyset$. For a solution $S \subseteq V_0$, we say that S *respects* the flags w if for all $L \in \mathcal{L}^*$, $w(L) = 0$ iff $S \cap L = \emptyset$. During the algorithm we will use the term "guessing" the value $w(L)$ of $L \in \mathcal{L}$. By this, we mean that we try the two possible choices (consistent with the previous ones), creating at most two distinct executions of the algorithm. Notice that $\mathcal{L}^*$ will be implicitly updated (*i.e.* L belongs to $\mathcal{L}^*$ in the next executions if we have guessed $w(L)$).

We also add a function $g : \mathcal{L}^* \to 2^S$. Roughly speaking, we will modify g during the algorithm such that g remembers the neighbors of the remaining vertices V in the partial solution S already constructed. Notice that we introduced g only for the analysis, and more precisely for maintaining our invariants (see bellow).

Correctness and Time Complexity. As usually in a branching algorithm, we bound the time complexity by bounding both the depth and the maximum degree of the search tree. More precisely, we will show that:

- Each rule can be applied in FPT time.
- The branching degree of each branching rule is a function of k.
- Any branching rule strictly decreases $(k, \#AL, \#BF)$ using the lexicographic order, whose initial value only depends on the initial value of k (by Observation 1).
- Any pre-processing rule does not increase $(k, \#AL, \#BF)$ and decreases $|V| + |\mathcal{I}|$.

Thus, the number of branching steps of the search tree is a function of k only, whereas the number of steps between two branchings is polynomial in n (recall that $|\mathcal{X}|$ is polynomial in n), which leads to an *FPT* running time.

Recall that S denotes the partial current solution. Concerning the correctness of the algorithm, we will say that a rule is *safe* if it preserves all the following invariants:

1. The tree T is still a tree decomposition (as defined in 2) of G, which is an induced subgraph of G_0.
2. If a vertex of the partial solution is adjacent to a "surviving" vertex $v \in V$, then v must appear in a leaf where a flag is defined, *i.e.*:
   $N_0(S) \cap V \subseteq \bigcup_{L \in \mathcal{L}^*} L$.
3. The neighborhood of a surviving vertex u in the partial solution is defined by the union of $g(L)$ for each L in which u appears, *i.e.* $g : \mathcal{L}^* \to 2^S$ is such that $\forall u \in V$ we have $N_0(u) \cap S = \bigcup_{L \in \mathcal{X}_u \cap \mathcal{L}^*} g(L)$.
   In particular, this invariant implies that if there are $u, v \in V$ such that $\mathcal{X}_u \cap \mathcal{L}^* \subseteq \mathcal{X}_v \cap \mathcal{L}^*$ (*i.e.* v appears in at least as many labelled leaves as u), then we must have $N_0(u) \cap S \subseteq N_0(v) \cap S$ (*i.e.* v is adjacent to at least as many vertices of the solution as u).
4. If there is an optimal solution (closed under inclusion) $S^* \subseteq V$ such that $S \subseteq S^*$, and S^* respects the flags w, then one of the branching will output an optimal solution.

Reduction Rules. Notice that each of the following rules defines a new value for variables k, T, S, w and g. However, for the sake of readability we will not mention variables that are not modified. Due to space restrictions, all safeness proofs were omitted.

Pre-Processing Rule 1: useless duplicated node.

If there exists $X \in \mathcal{X}$ such that $X \notin \mathcal{L}^*$ and $X \subseteq pred(X)$, then contract X and $pred(X)$ (i.e. delete X, and connect every $Y \in succ(X)$ to $pred(X)$).

Pre-Processing Rule 2: removing a (almost) lonely vertex.

If there exists $L \in \mathcal{L}^*$ such that $w(L) = 0$, then if L contains a lonely vertex v, delete v. Otherwise, if L contains an almost lonely vertex v, then delete v.

Branching Rule 1: taking a lonely vertex.

If there exists $L \in \mathcal{L}^*$ such that $w(L) = 1$ and L contains a lonely vertex v, then take v in the solution and decrease k by one. In addition, add v into $g(L)$, and if L does not become empty, then guess a new value $w(L)$.

Remark 1. At this point, since *Pre-Processing Rule 1* does not apply, it is clear that every leaf $L \in \mathcal{L} \setminus \mathcal{L}^*$ contains a lonely vertex. The following branching rule aims to process these leaves.

Branching Rule 2: processing leaves with no flag.

If there exists $L \in \mathcal{L} \setminus \mathcal{L}^*$, then take a lonely vertex $v \in L$ in the solution and decrease k by one. In addition, add v into $g(L)$, and if L does not become empty, guess a value $w(L)$.

Remark 2. At this point, notice that $\mathcal{L}^* = \mathcal{L}$, i.e. a flag has been assigned to each leaf. Indeed, suppose that there exists $L \in \mathcal{L} \setminus \mathcal{L}^*$. If L contains a lonely vertex, then *Branching Rule 1* must apply. Otherwise, *Pre-Processing Rule 1* must apply. In addition, there is no lonely vertex in the leaves, as otherwise *Branching Rule 1* or *Pre-Processing Rule 2* would apply.

Branching Rule 3: partitioning leaves of a bad father.

If there exists a bad father $F \in \mathcal{X}$, let $\mathcal{L}'$ be the set of leaves in $succ(F)$ and $C = \bigcup_{L \in \mathcal{L}'} L$ be the set of vertices contained these leaves. Partition C into the equivalence classes $C_1, ..., C_t$ of the following equivalence relation: two vertices $u, v \in C$ are equivalent if $\mathcal{X}_u \cap \mathcal{L}' = \mathcal{X}_v \cap \mathcal{L}'$ (i.e. u and v appear in the same subset of leaves of F). For all $i \in \{1, ..., t\}$, let $\mathcal{L}^i \subseteq \mathcal{L}'$ denote the subset of leaves in which vertices of C_i were before the partitioning. Then, replace the leaves of F by $C_1, ..., C_t$, and for all $i \in \{1, ..., t\}$, guess $w(C_i)$ and set $g(C_i) = \bigcup_{L \in \mathcal{L}^i} g(L)$.

Let us give the intuition of *Branching Rule 3*. This rule ensures that the set of leaves attached to a same node are vertex disjoint and that the partition was made in such a way that two vertices in the same leaf after the application of the rule were in the same subset of leaves before it. Notice that the remaining Branching Rules can create bad fathers, but decrease k.

Branching Rule 4: taking a lonely vertex in a father.
If there exists $L \in \mathcal{L}$ such that $pred(L)$ contains a lonely vertex v, then take v in the solution, delete k by one, and create a new leaf N adjacent to $pred(L)$ and containing vertices of $L \setminus \{v\}$. Finally, guess a value for $w(N)$ and set $g(N) = \{v\}$.

Branching Rule 5: taking an almost lonely vertex in a leaf.
If there exists $L \in \mathcal{L}$ such that $w(L) = 1$ and L contains an almost lonely vertex v (thus contained in L and $pred(L)$), then take v in the solution, decrease k by one, and create a new leaf N adjacent to $pred(L)$ and containing vertices of $pred(L) \setminus \{v\}$. If L does not become empty, then guess a new value $w(L)$. Finally, guess a value $w(N)$, add v into $g(L)$, and set $g(N) = \{v\}$.

End of the Algorithm.

Lemma 1. *If no rule can be applied then either G is empty or $k = 0$.*

According to the introduction and all the safeness lemmas, the size of the search tree is a function of k. Then, let us remark that all rules can be applied in *FPT* time. This is clear for *Pre-Processing Rules 1* and *2*, as well as for *Branching Rules 1, 2, 4* and *5*. Concerning *Branching Rule 3*, which consists in partitioning a subset of leaves, it runs in *FPT* time as long as $|\mathcal{L}|$ is a function of k. This is obviously the case at the beginning of the algorithm (since $|\mathcal{L}| < k$), and the number of leaves only increase by one in *Branching Rule 4* and *5*, and by a function of the previous number of leaves in *Branching Rule 3*. Since the branching rules are applied at most $f(k)$ times, we get the desired result.

Theorem 2. *There is an FPT algorithm for* SPARSEST k-SUBGRAPH *in chordal graphs, parameterized by k.*

However, the running time of the algorithm may be a tower of 2 of height k, since *Branching Rule 3* may create 2^t new leaves, where t is the number of previous leaves of the node F. Nevertheless, we can slightly modify the algorithm in order to obtain a $O^*(2^{k^2})$ running time[2]. Indeed, after the application of this rule, all leaves L such that $w(L) = 0$ can be gathered into one leaf, since all these vertices are not in the solution. And since all leaves are vertex disjoint, the number of leaves L such that $w(L) = 1$ is at most k (since one vertex of each leaf is in the solution). Hence, the number of leaves of F after the application of *Branching Rule 3* can actually be bounded by $k+1$. Then, as said previously, the only other branching rules which increase the number of leaves are *Branching Rules 4* and *5*, which both add at most one leaf when they are applied. However, since these branching rules are decreasing k, the maximum number of leaves of a node F before the application of *Branching Rule 3* is $2k$. Hence, this rule (which upper bounds the running time of the algorithm) runs in time $O^*(2^{O(2k^2)})$ (we have at most 2^{2k} leaves and we choose at most k leaves such that $w(L) = 1$). For sake of readability, the presented algorithm does not contain this slight modification.

---

[2] The $O^*(.)$ notation avoids polynomial terms.

5   Kernel Lower Bound of Sparsest k-Subgraph in Chordal Graphs

Intuition of the Proof. The following kernel lower bound is obtained using a cross-composition. It is an extension of our previous work [18], showing the $\mathcal{NP}$-hardness of SPARSEST k-SUBGRAPH in chordal graphs. Let us first give the intuition of this result, and then explain the modification we apply which leads to the kernel lower bound. We then explicit the whole construction of the cross-composition and give a formal proof of the result.

The $\mathcal{NP}$-hardness proof is a reduction from the classical k-CLIQUE problem in general graphs and roughly works as follows. Given an input instance $G = (V, E)$, $k \in \mathbb{N}$ of k-CLIQUE, we first build a clique A representing the vertices of G. We also represent each edge $e_j = \{u, v\} \in E$ by a gadget F_j, and connect the representative vertices of u and v in A to some vertices of F_j (see the left of Figure 1). The reduction will force the solution to take in A the representatives of $(n - k)$ vertices of G (corresponding to the complement of a solution S of size k in G), and also to take the same number of vertices among each gadget. The key idea is that the cost of a gadget F_j increases by one if it is adjacent to one of the selected vertices of A. Thus, since the goal is to minimize the cost, we will try to maximize the number of gadgets adjacent to the representatives of S (*i.e.* vertices we did not pick in A), the maximum being reached when S is a clique in G.

To adapt this reduction into a cross-composition, we add an *instance selector* composed of $2 \log t$ gadgets adjacent to A (where t is the number of input instances of the cross-composition) which encodes the binary representation of each instance index. These gadgets have the same structure as the F_j. For technical reasons, this instance selector has to be duplicated many times, as well as the clique A which we must duplicate t times in order to encode the vertex set of each instance. The right of Figure 1 represents the construction in a simplified way. Let us now define formally the gadgets and state their properties.

Definition of a Gadget. Let $T \in \mathbb{N}$ (we will set the value of T later). The vertex set of each gadget is composed of three sets of T vertices X, Y and Z, with $X = \{x_1, ..., x_T\}, Y = \{y_1, ..., y_T\}$ and $Z = \{z_1, ..., z_T\}$. The set X induces an independent set, the set Z induces a clique, and there is a $(T - 1)$-clique on $\{y_2, ..., y_T\}$. In addition, for all $i \in \{1, ..., T\}$, we connect y_i to all vertices of Z and to x_i. The left of Figure 1 summarizes the construction.

In the following cross-composition, we will force the solution to take $2T$ vertices among each gadget F. It is easy to see that the sparsest $2T$-subgraph of F is composed of the sets X and Z, which induces $\binom{T}{2}$ edges. In addition, if we forbid the set Z to be in the solution (if the gadget is adjacent to some picked vertices of A), then the remaining $2T$ vertices (namely X and Y) induce $\left(\binom{T}{2} + 1\right)$ edges.

Theorem 3. SPARSEST k-SUBGRAPH *does not admit a polynomial kernel in chordal graphs unless* $\mathcal{NP} \subseteq co\mathcal{NP}/poly$ *(parameterized by k).*

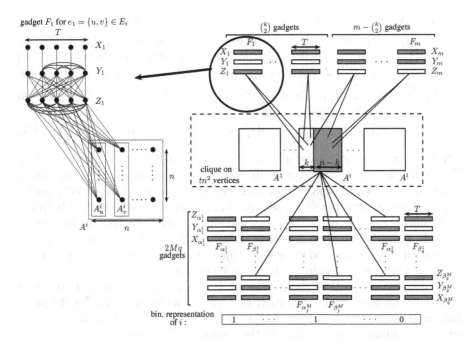

Fig. 1. Schema of the cross-composition (right) and a detailed gadget (left). Grey rectangles represent vertices of the solution, supposing that G_i contains a clique of size k. Notice that gadgets of the bottom have been drawn in the reverse direction (*e.g.* $X_{\beta_1^1}$ is below $Y_{\beta_1^1}$). Edges of the clique A have not been drawn for sake of clarity.

Proof. Let $(G_1, k_1), ..., (G_t, k_t)$ be a sequence of t instances of k-CLIQUE, with $G_i = (V_i, E_i)$ for all $i \in \{1, ..., t\}$. W.l.o.g. we suppose that $t = 2^q$ for some $q \in \mathbb{N}$, and define $T = n(n-k)$ and $M = n^6$.

Our polynomial equivalence relation is the following: for $1 \leq i, j \leq t$, (G_i, k_i) is equivalent to (G_j, k_j) if $|V_i| = |V_j| = n$, $|E_i| = |E_j| = m$ and $k_i = k_j = k$. One can verify that this relation is a polynomial equivalence relation. In what follows we suppose that all instances of the sequence are in the same equivalence class. The output instance $G' = (V', E'), k', C'$ is defined as follows (see Figure 1):

- For each $i \in \{1, ..., t\}$ we construct a clique A^i on n^2 vertices, where A^i is composed of n subcliques $A_1^i, ..., A_n^i$. We also add all possible edges between all cliques $(A^i)_{i=1..n}$. Hence, $A = \bigcup_{i=1}^{t} A^i$ is a clique of size tn^2.
- Since all instances have the same number of edges, we construct m gadgets $(F_j)_{j=1..m}$, where each F_j is composed of X_j, Y_j and Z_j as described previously. For all $i \in \{1, ..., t\}$, if there is an edge $e_j = \{u, v\} \in E_i$, then we connect all vertices of Z_j to all vertices of A_u^i and A_v^i. Let us define $\mathcal{F} = \bigcup_{j=1}^{m} F_j$ the subgraph of all gadgets of the "edge selector".
- We add $2qM$ gadgets $(F_{\alpha_j^h})_{j=1..q}^{h=1..M}$ and $(F_{\beta_j^h})_{j=1..q}^{h=1..M}$, where all gadgets are isomorphic to the edge gadgets, and thus composed of $X_{\alpha_j^h}, Y_{\alpha_j^h}$ and $Z_{\alpha_j^h}$

(resp. $X_{\beta_j^h}, Y_{\beta_j^h}$ and $Z_{\beta_j^h}$) for all $h \in \{1, ..., M\}$ and all $j \in \{1, ..., q\}$. Let $i \in \{1, ..., t\}$, and consider its binary representation $b \in \{0, 1\}^q$. For all $j \in \{1, ..., q\}$, if the jth bit of b equals 0, then connect all vertices of A^i to all vertices of $\bigcup_{h=1}^M Z_{\alpha_j^h}$. Otherwise, connect all vertices of A^i to all vertices of $\bigcup_{h=1}^M Z_{\beta_j^h}$. Let us define $\mathcal{B} = \bigcup_{h=1}^M \bigcup_{j=1}^q (F_{\alpha_j^h} \cup F_{\beta_j^h})$ the subgraph of all gadgets of the "instance selector".

- We set $k' = T + 2Tm + 4TqM$ and $C' = \binom{T}{2} + \binom{T}{2}(m + 2Mq) + (m - \binom{k}{2}) + Mq$.

It is clear that G', k' and C' can be constructed in time polynomial in $\sum_{i=1}^t |G_i| + k_i$. Then, one can verify that G' is a chordal graph. Indeed, it is known [13] that a graph is chordal if and only if one can repeatedly find a simplicial vertex (a vertex whose neighborhood is a clique) and delete it from the graph until it becomes empty. Such an ordering is called a simplicial elimination order. It is easily seen that for each gadget, X, Y and then Z is a simplicial elimination order (each gadget is only adjacent to the clique A *via* its set Z). Finally it remains the clique A which can be eliminated.

In addition, notice that the parameter k' is a polynomial in n, k and $\log t$ only and thus respect the definition of a cross-composition. We finally prove that there exists $i \in \{1, ..., t\}$ such that G_i contains a clique K of size k if and only if G' contains a set K' of k' vertices inducing C' edges or less.

Lemma 2. *If there exists $i \in \{1, ..., t\}$ such that G_i contains a k-clique, then G' contains k' vertices inducing at most C' edges.*

Proof. Suppose that $K \subseteq V_i$ is a clique of size k in G_i. W.l.o.g. suppose that $K = \{v_1, ..., v_k\}$, and that $\{\{u, v\}, u, v \in K\} = \{e_1, ..., e_{\binom{k}{2}}\}$. Let $b \in \{0, 1\}^q$ be the binary representation of i. We build K' as follows (see Figure 1).

- For all $j \in \{1, ..., \binom{k}{2}\}$, K' contains X_j and Z_j ($2T$ vertices inducing $\binom{T}{2}$ edges for each gadget F_j).
- For all $j \in \{\binom{k}{2} + 1, ..., m\}$, K' contains X_j and Y_j. ($2T$ vertices inducing $(\binom{T}{2} + 1)$ edges for each gadget F_j).
- For all $u \notin \{1, ..., k\}$, K' contains A_u^i (T vertices inducing $\binom{T}{2}$ edges).
- For all $h \in \{1, ..., M\}$, and all $j \in \{1, ..., q\}$, K' contains $X_{\alpha_j^h}$ and $X_{\beta_j^h}$. Moreover, if the jth bit of b equals 1, then K' contains $Y_{\beta_j^h}$ and $Z_{\alpha_j^h}$, otherwise K' contains $Z_{\beta_j^j}$ and $Y_{\alpha_j^h}$ ($4T$ vertices inducing $(2\binom{T}{2} + 1)$ edges for each pair of gadgets $F_{\alpha_j^h}$ and $F_{\beta_j^h}$).

One can easily verify that K' is a set of k' vertices inducing C' edges.   □

The following lemma, which proof has been omitted, terminates the proof.

Lemma 3. *If G' contains k' vertices inducing at most C' edges, then $\exists i \in \{1, ..., t\}$ such that G_i contains a k-clique.*

References

1. Apollonio, N., Simeone, B.: The maximum vertex coverage problem on bipartite graphs. Discrete Applied Mathematics (in press, 2013)
2. Bodlaender, H.L., Jansen, B.M.P., Kratsch, S.: Cross-composition: A new technique for kernelization lower bounds. In: STACS, pp. 423–434 (2011)
3. Bonnet, É., Escoffier, B., Paschos, V.T., Tourniaire, É.: Multi-parameter complexity analysis for constrained size graph problems: Using greediness for parameterization. In: Gutin, G., Szeider, S. (eds.) IPEC 2013. LNCS, vol. 8246, pp. 66–77. Springer, Heidelberg (2013)
4. Bourgeois, N., Giannakos, A., Lucarelli, G., Milis, I., Paschos, V.T.: Exact and approximation algorithms for densest k-subgraph. In: Ghosh, S.K., Tokuyama, T. (eds.) WALCOM 2013. LNCS, vol. 7748, pp. 114–125. Springer, Heidelberg (2013)
5. Bousquet, N., Bougeret, M., Giroudeau, R., Watrigant, R.: Parameterized complexity of the sparsest k-subgraph problem in chordal graphs. Technical Report RR-13033, LIRMM (2013)
6. Broersma, H., Golovach, P.A., Patel, V.: Tight complexity bounds for FPT subgraph problems parameterized by clique-width. In: Marx, D., Rossmanith, P. (eds.) IPEC 2011. LNCS, vol. 7112, pp. 207–218. Springer, Heidelberg (2012)
7. Cai, L.: Parameterized complexity of cardinality constrained optimization problems. Computer Journal 51(1), 102–121 (2008)
8. Chen, D.Z., Fleischer, R., Li, J.: Densest k-subgraph approximation on intersection graphs. In: Jansen, K., Solis-Oba, R. (eds.) WAOA 2010. LNCS, vol. 6534, pp. 83–93. Springer, Heidelberg (2011)
9. Corneil, D.G., Perl, Y.: Clustering and domination in perfect graphs. Discrete Applied Mathematics 9(1), 27–39 (1984)
10. Feige, U., Kortsarz, G., Peleg, D.: The dense k-subgraph problem. Algorithmica 29(3), 410–421 (2001)
11. Flum, J., Grohe, M.: Parameterized Complexity Theory. Springer (2006)
12. Gavril, F.: The intersection graphs of subtrees in trees are exactly the chordal graphs. Journal of Combinatorial Theory, Series B 16(1), 47–56 (1974)
13. Golumbic, M.C.: Algorithmic graph theory and perfect graphs. Academic Press, New York (1980)
14. Joret, G., Vetta, A.: Reducing the rank of a matroid. CoRR, abs/1211.4853 (2012)
15. Khot, S.: Ruling out ptas for graph min-bisection, dense k-subgraph, and bipartite clique. SIAM Journal of Computing 36, 1025–1071 (2004)
16. Liazi, M., Milis, I., Zissimopoulos, V.: A constant approximation algorithm for the densest k-subgraph problem on chordal graphs. Information Processing Letters 108(1), 29–32 (2008)
17. Nonner, T.: PTAS for densest k-subgraph in interval graphs. In: Dehne, F., Iacono, J., Sack, J.-R. (eds.) WADS 2011. LNCS, vol. 6844, pp. 631–641. Springer, Heidelberg (2011)
18. Watrigant, R., Bougeret, M., Giroudeau, R.: Approximating the sparsest k-subgraph in chordal graphs. To Appear in WAOA 2013 (2013)

Error-Pruning in Interface Automata[*]

Ferenc Bujtor and Walter Vogler

Inst. f. Informatik, Universität Augsburg

Abstract. De Alfaro and Henzinger introduced interface automata to model and study behavioural types. These come with alternating simulation as refinement and with a specific parallel composition: if one component receives an unexpected input, this is regarded as an error and the resp. error states are removed with a special pruning operation. In this paper, we return to the foundations of interface automata and study how refinement and parallel composition should be defined best. We take as basic requirement that an implementation must be error-free, if the specification is. For three variants of error-free, we consider the coarsest precongruence for parallel composition respecting the basic requirement. We find that pruning proves to be relevant in all cases; we also point out an error in an early paper by de Alfaro and Henzinger.

1 Introduction

Interface automata as introduced by de Alfaro and Henzinger e.g. in [5] are an abstract description of the communication behaviour of a system or component in terms of input and output actions. Based on this behavioural type, one can study whether two systems are compatible if put in parallel, and one can define a refinement for specifications. Essential for such a setting is that the refinement relation is a precongruence for parallel composition.

A basic intuition here is that outputs are under the control of the respective system: if one component in a composition provides an output for another, the latter must synchronize by performing the same action as input; if this is not possible, the whole system might malfunction – such a catastrophic error state has to be avoided. In contrast to the I/O-automata of [10], interface automata are not input enabled. Instead, a missing input in a state corresponds to the requirement that an environment must not send this input to this state.

There are two essential design decisions in the approach of [5] that we will scrutinize in this paper. First, the approach is optimistic: an error state is not a problem, if it cannot be reached in a helpful environment. This is reflected in the details of parallel composition, where from a standard product automaton all states are removed that can reach an error state just by local, i.e. output and internal actions (often called *pruning*). Although this definition has some intuitive justification, its details appear somewhat arbitrary. This is also the case for the second decision to take some alternating simulation as refinement

---

[*] This research was supported by DFG-project VO 615/12-1.

V. Geffert et al. (Eds.): SOFSEM 2014, LNCS 8327, pp. 162–173, 2014.

relation. Actually, the same authors used a slightly different relation for a slightly larger class of automata in the earlier [4]; no argument is given for the change.

Here, we will work out to what degree these design decisions can be *justified* from some more basic and, hopefully, more agreeable ideas. We model components as labelled transition systems (LTSs) with disjoint input and output actions and an internal action τ, quite like the interface automata of [5]. So as not to exclude any possibilities prematurely, our LTS have explicit error states. For these Error-IO-Transition Systems (EIO), we consider a standard parallel composition where additionally error states occur as described above; error states are also inherited from the components.

An undisputable requirement for a refinement relation $\sqsubseteq$ is that an error-free specification should only be refined by an error-free system. This can be understood as a basic refinement relation, which is parametric in the exact meaning of error-free: in the optimistic view, error-free means that no error state can be reached by locally controlled actions only; in the pessimistic view (cf. e.g. [1]), a system is error-free only if no error state is reachable at all.

For modular reasoning, which is at the heart of the approach under study, $\sqsubseteq$ must be a precongruence: if a component of a parallel composition is replaced by a refinement, the composition itself gets refined, i.e. $S_1 \sqsubseteq S_2$ implies $S_1 \mid S \sqsubseteq S_2 \mid S$. Since the basic relations fail to be precongruences in each case, we will characterize (or at least approximate) the resp. coarsest precongruence that is contained in the basic relation. Such a *fully abstract* precongruence is optimal for preserving error-freeness, since it does not distinguish components for no semantic reason.

In the optimistic case, the precongruence can be characterized as (componentwise) inclusion for a pair of trace sets; the definition of one of these uses pruning on traces. With this chacterisation we can prove that, essentially, each EIO is equivalent w.r.t the precongruence to one without error states, where the latter can be obtained by pruning the former as in [5]. Thus, we can work with EIO without error states, i.e. with interface automata and with the parallel composition of [5], but our pruning is *proven* to be correct.

While this justifies the first design decision in [5], our precongruence shows that alternating simulation is unnecessarily strict. This is not really new. A setting with input and outputs where unexpected inputs lead to errors has been studied long before [5] for speed-independent (thus asynchronous) circuits by Dill in [6]. The difference is that Dill does not start from an operational model as we do (in particular, there is no parallel composition for LTS), but on a semantic level with pairs of trace sets; he requires these pairs to be input enabled. On this semantic level, he also uses pruning; a normalized form of his pairs coincides with our pairs. Essentially, the full abstraction result can also be found in [3], though for a slightly different parallel composition and only for a congruence. Since that paper starts from a declarative approach, our presentation and proofs are more direct, and they prepare the reader for the other sections.

In [3], EIO (called Logic IOLTS there) are seen as an alternative framework to interface automata, and an error state is actually added to normalize an EIO.

We see error states only as a tool to study interface automata and would prefer to remove them in the end; with this view, we discovered a subtle point about pruning. Interface automata in [5] are deterministic w.r.t. input actions. Since we do not require this here, our pruning is a bit different from the one in [5]. In fact, the interface automata in [4] are also not input deterministic, but pruning used there is the same as the one in [5]. As a consequence, Theorem 1 of [4] claiming associativity for parallel composition is wrong; in our setting, it is easily proven.

It might seem that we have actually prescribed pruning in our optimistic approach since we consider only locally reachable errors as relevant and pruning removes exactly those states that can reach an error locally. To fortify the justification of pruning, we turn to a 'hyper-optimistic' approach next, where only internally reachable errors are relevant. With this more generous notion of error-free we obtain a slightly stricter precongruence and characterize it. The characterization is again based on pruning; the new idea is to extend traces with a set of outputs removed during pruning. This is an interesting precongruence but, compared to our first one, it looks unnecessarily complicated.

Finally, in the pessimistic approach every reachable error is relevant as advocated e.g. in [1]. Here, we describe a precongruence contained in the resp. basic relation, which is based on three trace sets and again employs pruning. We sketch how one might get the fully abstract precongruence, but this will be technically so involved as to make it unattractive. Even without characterising this precongruence, we can show that it is stricter than the optimistic one.

The next section will explain some basic notions. Sections 3 to 5 describe in turn our results for the optimistic, hyper-optimistic and pessimistic approach. Finally, in Section 6, we conclude with a comparison and give arguments as to why we will prefer the optimistic variant in the future. Due to lack of space, proofs and most examples had to be omitted.

2   Definitions and Notation

First we define our scenario. In interface automata, internal actions have different names. These sets of names must be disjoint for two automata to be composable; hence, standard α-conversion is not fully supported. To improve our EIOs have just one internal, unobservable actions τ, and they have a separate set of error states; such states can be created in a parallel composition.

Definition 1. An *Error-IO-Transition-System (EIO)* is defined as a tuple $S = (Q, I, O, \delta, q_0, E)$, with Q: a set of states, I, O: disjoint sets of (visible) input and output actions, $\delta \subseteq Q \times (I \cup O \cup \{\tau\}) \times Q$: a transition relation, $q_0 \in Q$: an initial state, $E \subseteq Q$: a set of *error* states. The *actions* of S are $\Sigma := I \cup O$, and its signature is $Sig(S) = (I, O)$. We call S *closed*, if $\Sigma = \emptyset$.

We write Q_1, I_1, δ_1 etc. for components of the EIO S_1 and Q_2, I_2, δ_2 etc. for S_2 and so on. We extend this for semantics and e.g. write ET_1 for $ET(S_1)$ as defined later on. We also write $q \xrightarrow{a} p$ for $(q, a, p) \in \delta$ and $q \xrightarrow{a}$ for $\exists p : (q, a, p) \in \delta$. Extending this to sequences $w \in (\Sigma \cup \{\tau\})^*$, we write $q \xrightarrow{w} p$, $(q \xrightarrow{w})$ whenever

$q \xrightarrow{a_1} \xrightarrow{a_2} \cdots \xrightarrow{a_n} p$ (called a *run*), ($q \xrightarrow{a_1} \xrightarrow{a_2} \cdots \xrightarrow{a_n}$) with $w = a_1 \cdots a_n$. Furthermore, $w|_B$ denotes the action sequence obtained from w by deleting all actions not in $B \subseteq \Sigma$. We write $q \xRightarrow{w} p$ for $w \in \Sigma^*$ if $\exists w' \in (\Sigma \cup \{\tau\})^* : w'|_\Sigma = w \wedge q \xrightarrow{w'} p$, and $q \xRightarrow{w}$ if some p with $q \xRightarrow{w} p$ exists. The *basic language* of S is $L(S) = \{w \in \Sigma^* \mid q_0 \xRightarrow{w}\}$, and it consists of the *traces* of S.

In a parallel composition, all common actions are synchronized and then hidden. Two EIOs can only be composed, if their input and output actions fit together, i.e. the EIOs have neither common inputs nor common outputs. A state of the composition is an error state if one component is in an error state (*inherited* error) or if one component sends an output to the other one, which is not ready to receive it (*new* error).

Definition 2. Two EIOs S_1, S_2 are *composable* if $I_1 \cap I_2 = \emptyset = O_1 \cap O_2$. The *parallel composition* is defined for two composable EIOs as $S_1 \mid S_2 = (Q, I, O, \delta, q_0, E)$, where $Q = Q_1 \times Q_2$, $I = (I_1 \backslash O_2) \cup (I_2 \backslash O_1)$, $O = ((O_1 \backslash I_2) \cup (O_2 \backslash I_1))$, and $q_0 = (q_{01}, q_{02})$. Furthermore, with $Synch(S_1, S_2) = (I_1 \cap O_2) \cup (I_2 \cap O_1)$ being the set of synchronized actions, we define

$$\delta = \{((q_1, q_2), \alpha, (p_1, q_2)) \mid (q_1, \alpha, p_1) \in \delta_1, \alpha \in (\Sigma_1 \cup \{\tau\}) \backslash Synch(S_1, S_2)\} \cup$$
$$\{((q_1, q_2), \alpha, (q_1, p_2)) \mid (q_2, \alpha, p_2) \in \delta_2, \alpha \in (\Sigma_2 \cup \{\tau\}) \backslash Synch(S_1, S_2)\} \cup$$
$$\{((q_1, q_2), \tau, (p_1, p_2)) \mid (q_1, \alpha, p_1) \in \delta_1, (q_2, \alpha, p_2) \in \delta_2, \alpha \in Synch(S_1, S_2)\}$$

$$E = (Q_1 \times E_2) \cup (E_1 \times Q_2) \qquad \text{'inherited errors'}$$
$$\cup \{(q_1, q_2) \mid \exists a \in O_1 \cap I_2 : q_1 \xrightarrow{a} \wedge q_2 \xslashedrightarrow{a}\}$$
$$\cup \{(q_1, q_2) \mid \exists a \in I_1 \cap O_2 : q_1 \xslashedrightarrow{a} \wedge q_2 \xrightarrow{a}\} \qquad \text{'new errors'}$$

We introduce S_{12} as shorthand for $S_1 \mid S_2$ and similarly for its components and semantics. We call an EIO S a *partner* of an EIO S' if their parallel composition is closed, i.e. if they have dual signatures $Sig(S') = (I, O)$ and $Sig(S) = (O, I)$. For our results we also define $\mid$ on traces.

Definition 3. Given two composable EIOs S_1, S_2, $w_1 \in \Sigma_1^*, w_2 \in \Sigma_2^*, W_1 \subseteq \Sigma_1^*$ and $W_2 \subseteq \Sigma_2^*$, we define $w_1 \mid w_2 = \{w|_{\Sigma_{12}} \mid w|_{\Sigma_1} = w_1 \wedge w|_{\Sigma_2} = w_2\}$ and $W_1 \mid W_2 = \bigcup \{w_1 \mid w_2 \mid w_1 \in W_1 \wedge w_2 \in W_2\}$.

We will base our semantics on traces that can lead to error states. In this context, we will use a pruning function, which removes all output actions from the end of a trace. We also define a function for arbitrary continuation of traces; for trace sets, this generalizes to describing the continuation or suffix closure.

Definition 4. For an EIO S, we define:
- $prune : \Sigma^* \to \Sigma^*$, $w \mapsto u$, where $w = uv$, ($u = \varepsilon \vee u \in \Sigma^* \cdot I$) and $v \in O^*$
- $cont : \Sigma^* \to \mathfrak{P}(\Sigma^*)$, $w \mapsto \{wu \mid u \in \Sigma^*\}$
- $cont : \mathfrak{P}(\Sigma^*) \to \mathfrak{P}(\Sigma^*)$, $L \mapsto \bigcup_{w \in L} cont(w)$

3 Optimistic Approach: Local Errors

We are now ready to consider some basic refinement relations. We will use variations of the notation '$Impl \sqsubseteq^B Spec$' to denote that $Impl$ in some basic sense is an implementation of, i.e. refines, the specification $Spec$.

3.1   Precongruence for Local Errors

In this section, we start with a variant based on *local* (i.e. internal and output) actions. We consider the following requirement: An implementation can only have an error state reachable by local actions if the specification has one as well. This is an optimistic view: It only considers processes to be dangerous, if they can run into an error on their own, i.e. using only local actions. Formally:

Definition 5. An *error is locally reachable* in an EIO S, if $\exists w \in O^* : q_0 \overset{w}{\Rightarrow} q \in E$. For EIOs *Impl* and *Spec* with the same signature, we write $Impl \sqsubseteq_{loc}^B Spec$, whenever an error is locally reachable in *Impl* only if an error is locally reachable in *Spec*.

By $\sqsubseteq_{loc}^c$ we denote the *fully abstract* precongruence with respect to $\sqsubseteq_{loc}^B$ and $|$, i.e. the coarsest precongruence with respect to $|$ that is contained in $\sqsubseteq_{loc}^B$.

In order to characterize this coarsest precongruence, we will need several trace sets. Naturally, we are interested in those traces that can reach an error state, the so-called strict error traces. If an EIO can perform a trace w such that input a is not possible in the state reached, and the environment allows this state to be reached by providing the necessary inputs and then performs a as an output, a new error state arises in the composition. Thus, we are also interested in the sequence wa and call it a missing-input trace.

Definition 6. We define the following trace sets for an EIO S:
- *strict error traces:* $StT(S) = \{w \in \Sigma^* \mid q_0 \overset{w}{\Rightarrow} q \in E\}$
- *pruned error traces:* $PrT(S) = \{prune(w) \mid w \in StT(S)\}$
- *missing-input traces:* $MIT(S) = \{wa \in \Sigma^* \mid q_0 \overset{w}{\Rightarrow} q \wedge a \in I \wedge q \overset{a}{\not\rightarrow}\}$

The characterization we are looking for will be provided by the following local error semantics; the intuitions are as follows. Errors arise in a composition because a component cannot accept some input after a trace or because it performs a strict error trace; in the latter case, the error is already unavoidable if the error state can be reached by local actions only. Hence, we consider the trace sets PrT and MIT in the definition of ET below. But as already explained above, the other component must take part in such problematic behaviour, hence we are also interested in the basic language of a component.

If *along* an action sequence an error *can* occur, it does not matter whether the sequence can be performed at all, and if so, whether it leads to an error state. Thus, we want to obliterate this information about such a trace; for this purpose, we close the set of problematic traces under continuation, and we also include this extended set in the language; this technique of flooding is well known e.g. in the context of failures semantics [2].

It will turn out that we can characterize $\sqsubseteq_{loc}^c$ as componentwise set inclusion for pairs $(ET(S), EL(S))$, denoted by $\sqsubseteq_{loc}$.

Definition 7. Let S be an EIO.
- The set of *error traces* of S is $ET(S) = cont(PrT(S)) \cup cont(MIT(S))$;
- the *flooded language* of S is $EL(S) = L(S) \cup ET(S)$.

For two EIOs *Impl* and *Spec* with the same signature, we write $Impl \sqsubseteq_{loc}$ *Spec* if $ET(Impl) \subseteq ET(Spec)$ and $EL(Impl) \subseteq EL(Spec)$ *Impl* and *Spec* are *local-error equivalent*, $Impl =_{loc} Spec$, if $Impl \sqsubseteq_{loc} Spec$ and $Spec \sqsubseteq_{loc} Impl$.

Our first result shows that the local error semantics is compositional.

Theorem 8. *For two composable EIOs S_1, S_2 and $S_{12} = S_1 \mid S_2$ we have:*

- $ET_{12} = cont\Big(prune\big((ET_1 \mid EL_2) \cup (EL_1 \mid ET_2)\big)\Big)$
- $EL_{12} = (EL_1 \mid EL_2) \cup ET_{12}$

Hence, $\sqsubseteq_{loc}$ is a precongruence; to show that it is the coarsest one, we have constructed an environment for each relevant trace w of *Impl* that reveals that w is also an appropriate trace of *Spec*.

Theorem 9. *For EIOs Impl and Spec with the same signature, $Impl \sqsubseteq_{loc}^c Spec \Leftrightarrow Impl \sqsubseteq_{loc} Spec$.*

3.2   Comparison to Interface Automata

We will show now that, up to local-error equivalence, we can essentially work with EIOs without error states. Such EIOs are the same as the *interface automata* of [5], if they additionally are *input-deterministic*: if $q \xrightarrow{a} q'$ and $q \xrightarrow{a} q''$ for some $a \in I$, then $q' = q''$. The only difference is that, in a setting with EIOs without error states, we do not have EIOs anymore that are an error initially.

Theorem 10. *Let S be an EIO, and obtain $prune(S)$ from S by removing the illegal states in $illegal(S) = \{q \in Q \mid$ an error state is reachable from q by local actions$\}$, their in- and out-going transitions and all transitions $q \xrightarrow{a} q'$ where $q \xrightarrow{a} q''$ with $q'' \in illegal(S)$ for some $a \in I$. If $q_0 \notin illegal(S)$, $prune(S)$ is an EIO and local-error equivalent to S.*

The resp. pruning in the definition of parallel composition in [5] only removes transitions from legal to illegal states. (Since then the illegal states are unreachable, they can be removed as well.) The additional removal of transitions $q \xrightarrow{a} q'$ as described in the theorem is obviously redundant in case of input determinism.

According to Theorem 10, we could work with EIOs without error states; whenever we put such EIOs in parallel, we have to normalize the result taking $prune(S_1 \mid S_2)$ as parallel composition. To ensure well-definedness, we call EIOs S_1 and S_2 *compatible*, if the initial state of $S_1 \mid S_2$ is not illegal, and we only apply the new parallel composition to compatible S_1 and S_2. For this, we have:

Proposition 11. *If Spec and Spec' are compatible EIOs and $Impl \sqsubseteq_{loc} Spec$, then also Impl and Spec' are compatible.*

In our setting, we have proved pruning as introduced in [5] correct also on the level of transition systems. But the refinement relation in [5] is somewhat arbitrarily too strict, as we will show below. To the best of our knowledge, alternating simulation for refinement has first been considered for modal transition systems [8]; see [9] for a comparison to the setting of interface automata.

Since the refinement relation of [5] is a precongruence, one might believe that it should imply $\sqsubseteq_{loc}$ due to our coarsest precongruence result. This is not really so obvious: we have considered parallel components that are not interface automata (since they violate input determinism), and this could have forced us to be too strict w.r.t alternating simulation. But actually, this is not the case:

Definition 12. For EIOs S_1 and S_2 with the same signature, an *alternating simulation relation* from S_1 to S_2 is some $\mathcal{R} \subseteq Q_1 \times Q_2$ with $(q_{01}, q_{02}) \in \mathcal{R}$ such that for all $(q_1, q_2) \in \mathcal{R}$ we have:
1. If $q_2 \xrightarrow{a} q_2'$ and $a \in I_1$, then $q_1 \xrightarrow{a} q_1'$ and $(q_1', q_2') \in \mathcal{R}$.
2. If $q_1 \xrightarrow{a} q_1'$ and $a \in O_1$, then $q_2 \xrightarrow{\varepsilon} \xrightarrow{a} q_2'$ and $(q_1', q_2') \in \mathcal{R}$.
3. If $q_1 \xrightarrow{\tau} q_1'$, then $q_2 \xrightarrow{\varepsilon} q_2'$ and $(q_1', q_2') \in \mathcal{R}$.

Thus, implementation S_1 must match a prescribed input immediately, while an output or τ is allowed for S_1 if S_2 can match it using internal steps.

Proposition 13. *If there exists some alternating simulation relation for interface automata S_1 and S_2, then $S_1 \sqsubseteq_{loc} S_2$. This implication is strict.*

Next is associativity for parallel composition. As mentioned in the introduction, Theorem 1 of [4] claiming this associativity is wrong there. We have an example, where S_1 and $S_2 \mid S_3$ are compatible, but $S_1 \mid S_2$ and S_3 are not, i.e. their composition is undefined. The reason is that pruning in [4] does not remove the transitions $q \xrightarrow{a} q'$ mentioned in Theorem 10; with this removal, associativity would be correct. This problem disappears in later work on interface automata, where the automata are required to be input deterministic.

This demonstrates the danger when one develops an unorthodox definition, like pruning in $\mid$, justified with informal intuitive arguments only. In the present paper, pruning on EIOs is *proven correct* in Theorem 10, and this proof would fail with some incorrect definition of pruning. In our setting with error states associativity is easy, because the two systems are easily seen to be isomorphic. Hence, associativity holds for any sensible equivalence on EIOs.

Theorem 14. *For pairwise composable EIOs S_1, S_2 and S_3, $S_1 \mid (S_2 \mid S_3)$ and $(S_1 \mid S_2) \mid S_3$ are isomorphic and in particular local-error equivalent.*

According to this theorem, also other equivalences in this paper make $\mid$ associative, and commutativity is obvious. In this context, it is useful to mention the following general result; see e.g. [11, Sec. 3.2] for similar considerations.

Theorem 15. *Let $\sqsubseteq^B$ be a preorder on some set E such that operation $\mid$ is commutative and associative for the related equivalence $=^B$, and there exists $Nil \in E$ with $S \mid Nil =^B S$ for all $S \in E$. Let preorder $\sqsubseteq$ satisfy for all Impl and Spec in E: $Impl \sqsubseteq Spec \Leftrightarrow U \mid Impl \sqsubseteq^B U \mid Spec$ for all composable U. Then $\sqsubseteq$ is $\sqsubseteq^c$, the fully abstract precongruence for $\sqsubseteq^B$ and $\mid$.*

This theorem tells us that, for proving that $\sqsubseteq_{loc}$ is fully abstract, it suffices to consider parallel compositions with two components instead of arbitrary ones.

For the present setting, we have adapted this to a partial composition to prove Theorem 26. From the observation that only partners as defined after Definition 2 are used in the proof of Theorem 9, we get an even simpler characterization.

Corollary 16. *For EIOs Impl and Spec, we have Impl $\sqsubseteq_{loc}$ Spec if and only if $U \mid Impl \sqsubseteq_{loc}^{B} U \mid Spec$ for all partners U.*

This result is relevant for the following reason: when a (possibly composed) system is finally put to use, it is composed in parallel with a user, resulting in a closed system. In other words, a user is a partner U and $U \mid Impl \sqsubseteq_{loc}^{B} U \mid Spec$ means that the user is as happy with *Impl* as she was with *Spec*. For some people, a relation with such a characterization is of foremost interest (see e.g.[7] where a "happy" partner is called a strategy), even though it is not necessarily a precongruence. The corollary will be essential for proving below that the third precongruence $\sqsubseteq_{act}^{c}$ is strictly finer than $\sqsubseteq_{loc}$.

4 Hyper-Optimistic Approach: Internal Errors

To obtain an even *better justification* for pruning, we now will focus on errors reached by internal actions only. The view that only such errors count is even more optimistic than our first one, since errors reachable by output actions are no longer considered dangerous. The new idea for the resulting semantics is that each error trace is annotated with a set of output actions; traces are pruned again and, then, the set contains the output actions that are needed to reach an error state; the intuition is: if the system performs the trace while synchronizing with another one on the given output actions, then an error state can be reached internally afterwards. If some action o of these actions is not synchronized, the error state is only reached by performing the still visible o.

Our base relation is now defined as:

Definition 17. An *error is internally reachable* in S, if $\varepsilon \in StT(S)$. For EIOs *Impl* and *Spec* with the same signature, we write $Impl \sqsubseteq_{int}^{B} Spec$ whenever an error is internally reachable in *Impl* only if an error is internally reachable in *Spec*; $\sqsubseteq_{int}^{c}$ denotes the fully abstract precongruence with respect to $\sqsubseteq_{int}^{B}$ and $\mid$.

Definition 18. Given an EIO S, we define $out : \Sigma^* \to \mathfrak{P}(O)$ such that $out(w)$ consists of all output actions in w. An *error pair over a signature* (I, O) is a pair $(w, X) \in (I \cup O)^* \times \mathfrak{P}(O)$ with $out(w) \subseteq X$.

Given two composable EIOs S_1, S_2, we define for an error pair (w, X) over (I_1, O_1) and $v \in \Sigma_2^*$: $(w, X) \mid v = \{(z, Y) \mid z \in w \mid v, Y = (X \cup out(v)) \cap \Sigma_{12}\}$

It is easy to see that this set consist of error pairs over the signatures of $S_{12} = S_1 \mid S_2$. On error pairs over some (I, O), we define:
- $prune(w, X) := (prune(w), X)$ (an error pair again)
- $cont(w, X) := \{(v, Y) \mid v \in cont(w), X \subseteq Y\}$ (consisting of error pairs)

Definition 19. We define the following sets of error pairs for an EIO S:
- *strict error pairs*: $StP(S) = \{(w, X) \mid w \in StT(S), out(w) = X\}$

– *pruned error pairs*: $PrP(S) = \{prune(w, X) \mid (w, X) \in StP(S)\}$
– *missing-input pairs*: $MIP(S) = \{(w, X) \mid w \in MIT(S), out(w) = X\}$

It is easy to see that these sets indeed consist of error pairs over (I, O), and that they are an enhanced version of similar sets defined in the previous section. It will turn out that we can characterize $\sqsubseteq_{int}^c$ as componentwise set inclusion for pairs $(EP(S), EPL(S))$, where the latter is the basic language of S flooded with a set of traces derived from $EP(S)$.

Definition 20. Let S be an EIO.
– the set of *error pairs* of S is $EP(S) = cont(PrP(S)) \cup cont(MIP(S))$;
– the set of *error pair traces* of S is $EPT(S) = \{w \mid (w, out(w)) \in EP(S)\}$;
– the flooded language, or *error pair language*, is $EPL(S) = L(S) \cup EPT(S)$.
For two EIOs *Impl* and *Spec* with the same signature, we write:
$Impl \sqsubseteq_{int} Spec$ if $EP(Impl) \subseteq EP(Spec)$ and $EPL(Impl) \subseteq EPL(Spec)$.

For the characterization result, it is again crucial that the internal error semantics is compositional. This implies that $\sqsubseteq_{int}$ is a precongruence, and it is indeed the coarsest one:

Theorem 21. *a) For two composable EIOs S_1, S_2 and $S_{12} = S_1 \mid S_2$ we have:*
– $EP_{12} = cont\left(prune\left((EP_1 \mid EPL_2) \cup (EPL_1 \mid EP_2)\right)\right)$
– $EPL_{12} = (EPL_1 \mid EPL_2) \cup EPT_{12}$
b) For two systems Impl and Spec with the same signature, we have $Impl \sqsubseteq_{int}^c$ Spec $\Leftrightarrow$ Impl $\sqsubseteq_{int}$ Spec.

Thus, we have characterized the fully abstract precongruence for $\sqsubseteq_{int}^B$ and $\mid$. From a more general perspective, two points are notable: Although outputs do not play a special role for $\sqsubseteq_{int}^B$, pruning of outputs on traces is again essential for our characterization. Since the concept of error-freeness underlying $\sqsubseteq_{int}^B$ is more liberal than the one for $\sqsubseteq_{loc}^B$ it is maybe surprising that the resulting precongruence is strictly finer.

Proposition 22. *The internal precongruence $\sqsubseteq_{int}$ is strictly finer than the local precongruence $\sqsubseteq_{loc}$ i.e. for all EIOs Impl and Spec with the same signature, Impl $\sqsubseteq_{int}$ Spec implies Impl $\sqsubseteq_{loc}$ Spec, but not the other way round.*

5   Pessimistic Approach: Reachable Errors

Now we turn to the pessimistic approach, which has already been discussed in the literature e.g. in [1], and consider only those systems error-free that do not have any reachable error states.

Definition 23. An *error is reachable* in an EIO S, if $\exists w \in \Sigma^* : w \in StT(S)$. For EIOs *Impl* and *Spec* with the same signature, we write $Impl \sqsubseteq_{act}^B Spec$, whenever an error is reachable in *Impl* only if an error is reachable in *Spec*. We denote the fully abstract precongruence with respect to $\sqsubseteq_{act}^B$ and $\mid$ by $\sqsubseteq_{act}^c$.

A pessimistic person might argue that systems with an error should just not be used at all (such a view is presumably taken in [1]), and that it does not make sense to distinguish between

P: x?   Q: i?;x!   R: i!

$\rightarrow \cdot$ $\rightarrow \cdot \underset{i?}{\rightarrow} \cdot \underset{x!}{\rightarrow} \cdot$ $\rightarrow \cdot$

erroneous systems, as we will do with $\sqsubseteq^c_{act}$. This has the severe disadvantage that parallel composition is not associative. Consider P, Q and R and their inputs and outputs as shown above; in figures, $i?$ ($i!$) indicates that i is an input (output), and a box denotes an error state. $P \mid Q$ has a reachable error, so $(P \mid Q) \mid R$ is not considered – in contrast to $P \mid (Q \mid R)$, since $Q \mid R$ and $P \mid (Q \mid R)$ are error-free and consist just of the initial state.

Our local error semantics does not suffice for $\sqsubseteq^c_{act}$, since it does not differentiate between a missing input and an input leading to an error state. But we can adapt it to get a precongruence for the pessimistic approach, albeit not the coarsest one. CPT deals with the real errors and is based on pruning of outputs again; MIC additionally considers the missing-input traces, this time without closing under continuation; another subtle point is that both, MIC and L, are flooded with CPT.

Definition 24. Let S be an EIO.
– The set of *continued pruned traces* of S is $CPT(S) = cont(prune(StT(S)))$;
– the set of *flooded missing-input traces* of S is $MIC(S) = MIT(S) \cup CPT(S)$;
– the *CPT-flooded language* of S is $LCP(S) = L(S) \cup CPT(S)$.
For two EIOs *Impl* and *Spec* with the same signature, we write *Impl* $\sqsubseteq_{act}$ *Spec* if and only if $CPT(Impl) \subseteq CPT(Spec)$, $MIC(Impl) \subseteq MIC(Spec)$ and $LCP(Impl) \subseteq LCP(Spec)$.

Theorem 25. *For two composable EIOs* S_1, S_2 *and* $S_{12} = S_1 \mid S_2$, *we define* $\Gamma_1 := \Sigma^*_1(I_1 \cap O_2)$ *and* $\Gamma'_1 = \Sigma^*_1(I_1 \setminus O_2)$ *for the sets of traces of* S_1 *ending on synchronizing inputs, on non-synchronizing resp.* Γ_2 *and* Γ'_2 *are defined analogously. Then we have:*
– $CPT_{12} = cont\big(prune(((MIC_1 \cap \Gamma_1) \mid LCP_2) \cup (LCP_1 \mid (MIC_2 \cap \Gamma_2)) \cup (CPT_1 \mid LCP_2) \cup (LCP_1 \mid CPT_2)))$
– $MIC_{12} = \bigcup\{(w_1 \mid w_2)a \mid w_1a \in MIC_1 \cap \Gamma'_1, w_2 \in LCP_2\} \cup$
$\qquad\qquad \bigcup\{(w_1 \mid w_2)a \mid w_1 \in LCP_1, w_2a \in MIC_2 \cap \Gamma'_2\} \cup CPT_{12}$
– $LCP_{12} = (LCP_1 \mid LCP_2) \cup CPT_{12}$
Thus, $\sqsubseteq_{act}$ *is a precongruence, and it refines* $\sqsubseteq^B_{act}$.

But $\sqsubseteq_{act}$ is not fully abstract regarding $\mid$ and $\sqsubseteq^B_{act}$. We have $P \not\sqsubseteq_{act} Q$ for the two EIOs P and Q with $I = \{a, b\}$ and $O = \emptyset$ shown below, because $ba \in CPT(P) \setminus CPT(Q)$. To show that $P \sqsubseteq^c_{act} Q$, it suffices to prove that there is no U with $P \mid U \not\sqsubseteq^B_{act} Q \mid U$, cf. Theorem 15. If U is not error free, then both $P \mid U$ and $Q \mid U$ have an error: both, P and Q, have no output actions, hence cannot prevent U to perform a run ending with an error state. So assume that U is error-free.

If $a, b \notin O_U$, then P and Q can reach an error by performing ab autonomously. If $a \in O_U$ and $b \notin O_U$, U might never perform a and neither $P \mid U$ nor $Q \mid U$ can reach

an error. If a is performed, then b can be performed autonomously by P or Q resp., leading to an error. The case that only $b \in O_U$ is analogous. Let $a, b \in O_U$; if U can perform neither one a followed by b nor vice versa, all errors are prevented; otherwise, an error occurs in both $P \mid U$ and $Q \mid U$ – where $P \mid U$ has an inherited and $Q \mid U$ a new error if U performs b before a.

Thus, to characterize $\sqsubseteq_{act}^c$, it seems that some missing-input traces (like ba for Q) have to be added to CPT. This appears to be the case if the missing-input trace and some error trace are the same when projecting the missing action away. An extended example (omitted) shows that adding one trace can lead to the addition of another, additions need to be done iteratively.

Hopefully, these considerations have convinced the reader that a characterization of the coarsest refinement will be overly complicated and not really worth the effort to work it out in detail. Even though we do not have a characterization of $\sqsubseteq_{act}^c$, we can compare it it to the local precongruence $\sqsubseteq_{loc}$ using Corollary 16; in contrast to the previous section, we have made the notion of error-freeness much *stricter*, but it turns out that again this leads to a *finer* precongruence.

Theorem 26. *The coarsest pessimistic precongruence $\sqsubseteq_{act}^c$ is strictly finer than the local precongruence $\sqsubseteq_{loc}$, i.e. $\sqsubseteq_{act}^c \subsetneq \sqsubseteq_{loc}$. Hence, this also applies for $\sqsubseteq_{act}$.*

6   Conclusion

To study the foundations of interface automata, we have chosen a variant with explicit error states and a standard parallel composition extended according to the characteristic idea: an output that is not expected by the recipient creates an error. To determine an appropriate refinement relation, we started from the basic idea that an error-free specification can only have error-free implementations and then considered the coarsest precongruence respecting this basic requirement. We have done this for three variants of error-freeness and characterized or at least approximated the precongruence with an essentially trace-style semantics.

For the optimistic view, where errors only count if they are reached locally, the simulation-style refinement of [5] is unnecessarily strict semantically, but the pruning integrated into the parallel composition of [5] is justified. Then, we looked at a hyper-optimistic version (only internally reachable errors count) and a pessimistic version (all reachable errors count). Surprisingly, both variants lead to a stricter precongruence, and both the semantics are also based on the same idea of pruning outputs (justifying it further). Since they are more complicated, one might prefer the local variant for its simplicity.

More intuitively, we believe that it also is based on the right concept. At the heart of interface automata is the idea that each system controls its outputs and internal actions, so a locally reachable error can indeed not be prevented by the environment. The hyper-optimistic view is less intuitive, but at least it served to show that output pruning does not rely so much on the idea that only locally reachable errors count. Also the characterising semantics is based on an idea that might be useful elsewhere. The pessimistic view has the plausibility of controlling the worst case; but comparing a state where input i is missing with a state where it leads to an error state, we see that both just formulate the same requirement for the environment: the environment must take this state into account and avoid producing i – there is no difference at all. Put another way, input transitions are only taken if the input is provided; for the two states mentioned, nothing bad will happen without this.

A final argument concerns the approach we described at the end of Section 3.2; assume we call $Impl$ better than $Spec$ if each $user$ (partner) U encounters an error with $Impl$ only if the same can happen with $Spec$. The three variants for what it means to encounter an error agree for closed systems like $Impl \mid U$ and $Spec \mid U$. Hence, there is only one meaning for "better-than", and this is the precongruence $\sqsubseteq_{loc}$ of the first variant due to Corollary 16. It is conceptually easy to decide $\sqsubseteq_{loc}$ for finite-state EIOs with automata-theoretic methods.

Interface automata have been extended and studied in various ways. In the future, we will consider how these extensions work mainly on the basis of the trace-based view we developed for the first variant.

References

1. Bauer, S.S., Mayer, P., Schroeder, A., Hennicker, R.: On weak modal compatibility, refinement, and the MIO workbench. In: TACAS 2010. LNCS, vol. 6015, pp. 175–189. Springer, Heidelberg (2010)
2. Brookes, S.D., Hoare, C.A.R., Roscoe, A.W.: A theory of communicating sequential processes. J. ACM 31(3), 560–599 (1984)
3. Chen, T., Chilton, C., Jonsson, B., Kwiatkowska, M.: A compositional specification theory for component behaviours. In: Seidl, H. (ed.) ESOP 2012. LNCS, vol. 7211, pp. 148–168. Springer, Heidelberg (2012)
4. de Alfaro, L., Henzinger, T.A.: Interface automata. In: ESEC/FSE 2001, pp. 109–120. ACM (2001)
5. de Alfaro, L., Henzinger, T.A.: Interface-based design. In: Engineering Theories of Software Intensive Systems. NATO Science Series, vol. 195, pp. 1–148. Springer (2005)
6. Dill, D.: Trace Theory for Automatic Hierarchical Verification of Speed-Independent circuits. MIT Press, Cambridge (1989)
7. Stahl, C., Massuthe, P., Bretschneider, J.: Deciding substitutability of services with operating guidelines. In: Jensen, K., van der Aalst, W.M.P. (eds.) ToPNoC II. LNCS, vol. 5460, pp. 172–191. Springer, Heidelberg (2009)
8. Larsen, K.G.: Modal specifications. In: Sifakis, J. (ed.) CAV 1989. LNCS, vol. 407, pp. 232–246. Springer, Heidelberg (1990)
9. Larsen, K.G., Nyman, U., Wąsowski, A.: Modal I/O automata for interface and product line theories. In: De Nicola, R. (ed.) ESOP 2007. LNCS, vol. 4421, pp. 64–79. Springer, Heidelberg (2007)
10. Lynch, N.: Distributed Algorithms. Morgan Kaufmann Publishers (1996)
11. Vogler, W.: Modular Construction and Partial Order Semantics of Petri Nets. LNCS, vol. 625. Springer, Heidelberg (1992)

Aspect-Driven Design of Information Systems

Karel Cemus and Tomas Cerny

Department of Computer Science and Engineering,
Czech Technical University, Charles square 13, 121 35 Prague 2, CZ
{cemuskar,tomas.cerny}@fel.cvut.cz

Abstract. Contemporary enterprise web applications deal with a large stack of different kinds of concerns involving business rules, security policies, cross-cutting configuration, etc. At the same time, increasing demands on user interface complexity make designers to consider the above concerns in the presentation. To locate a concern knowledge, we try to identify an appropriate system component with the concern definition. Unfortunately, this is not always possible, since there exist concerns cross-cutting multiple components. Thus to capture the entire knowledge we need to locate multiple components. In addition to it, often, we must restate the knowledge in the user interface because of technological incompatibility between the knowledge source and the user interface language. Such design suffers from tangled and hard to read code, due to the cross-cutting concerns and also from restated information and duplicated knowledge. This leads to a product that is hard to maintain, a small change becomes expensive, error-prone and tedious due to the necessity of manual changes in multiple locations.

This paper introduces a novel approach based on independent, description of all orthogonal concerns in information systems and their dynamic automated weaving according to the current user's context. Such approach avoids information restatement, speeds up development and simplifies maintenance efforts due to application of automated programming and runtime weaving of all concerns, and thus distributes the knowledge through the entire system.

Keywords: Aspect-oriented design, Business logic, Model-driven development, Reduced maintenance and development efforts.

1 Introduction

Contemporary design of information systems face multiple challenges caused by inability of present-day technologies to encapsulate business logic and implement self-maintainable user interfaces. It is a common practice [1] to capture business rules and security policy tangled through all layers of the application. This is mostly notable in the presentation layer, which results with its source code barely maintainable. Such design then exhibits issues with consistency. For example, a minor change to the application data model, e.g. a field type, or its constraint, requires manual change propagation into multiple other locations. This becomes

V. Geffert et al. (Eds.): SOFSEM 2014, LNCS 8327, pp. 174–186, 2014.

very tedious and error-prone especially for presentations built with languages without type safety. A model change then requires the developer to issue a text search for its references in the presentation and to adjust the code. In [10] authors discuss that information restatement in the presentation is the source of increased maintenance. Although, information restatement is not the only problem involved in extensive maintenance. Application design becomes hard to maintain when it exhibits cross-cutting concerns [2], such as business rules or security concerns.

In this paper we introduce a novel aspect-driven approach to deal with the above mentioned issues. The fundamental idea of the concept lies in the separation of application business rules from the application code base. Instead an aspect-language is used to capture these rules, which allows us to encapsulate all application rules to a central location. Next, it allows us to automatically enforce application business rules or to generate context-aware user interfaces. As a consequence, a data model change directly propagates to the rest of the system which enforces its integrity. Such approach leads to efficient maintenance of the entire system business logic and application user interface (UI).

The main benefit of the presented approach is prevented information restatement achieved by having a single location for definition of each system knowledge. Such knowledge can be automatically distributed to other parts of the system at runtime. This reduces system element coupling, supports reuse and rule enforcement at the same time. Furthermore, such separation of concerns allows us to reuse system elements and transform them to automated UIs presentation of data elements.

This paper is organized as follows: In section 2 we provide the background of the contemporary approaches. The related work is discussed in section 3. Description of our approach is provided in section 4. Next, a case study and its evaluation in section 5. We end the paper with conclusion and future work.

2 Background

Existing development approaches often use the three-layered architecture model [5], which divides the functionality into three encapsulated components. The lowest layer is responsible for data persistence and accessing third-party services. The middle one encapsulates business logic, which exposes through public interfaces to the top, presentation layer. The top layer deals with UI, provides access to the functionality and different views on data, communicates to user and processes his responses.

Despite clear definition of responsibilities for each layer, there exists a certain group of problems that cannot be simply encapsulated in any of these layers. Furthermore, it must be noted that contemporary development approaches such as Object-oriented design does not provide effective mechanisms to encapsulate these problems [11,14]. This group of problems is in literature called cross-cutting concerns [13] due to their characteristics - they cut across the whole application and have to be considered at multiple places in all these layers, we can look at them as on layers orthogonal to this linear architecture. For example, consider logging, exception management, security, etc.

Business logic, also a cross-cutting concern [1], must be considered at multiple places such as in the UI (rendering and input validation phases), or as an input validation policy of business operations or as a data model integrity check [2]. Based on our research that involved inspection of a large information system ACM-ICPC, we found out that there exists three categories of high-level rules: *contextless*, *contextual* and *cross-cutting* rules. The contextless rules are understood as validation rules, which have to be satisfied always and everywhere otherwise it would break data consistency or integrity. For example: *'Username must be at least 6 characters long'*. The second group consists of contextual rules. These rules are considered within given context, such as: *Student's GPA must be higher than 3.8*, when he applies for a scholarship. The last group are rules that apply under given conditions nearly everywhere. These rules are often parametrized by application state or user's profile. For example, consider a rule: *'System is not locked for editing'*. Such rule has a vast area of impact, because there may exist a deadline date after which no changes are allowed.

Listing 1. Scenario: User registration

```
preconditions:
 - username is not empty
 - password has at least 6 characters
 - email address is valid
successful path: ...
```

Next, consider a business operation described as a single use-case or a user scenario. For example, consider the implementation in Listing 2, which is a common example of the contemporary approach in development. This operation defines multiple pre-conditions that determine position of the operation in an application domain. It means, the operation can only be performed when all the pre-conditions are satisfied. From another perspective this is nothing else than a few business rules. Thus we can look at it from the perspective of a business context of a given operation. The context then compounds multiple business rules. With clearly defined application contexts every operation can address one.

Listing 2. Standard scenario implementation

```
void register(User user) {
    // Check for violation of any rule. The validation is vastly duplicated
    // across whole application code. It can be reduced but not avoided.
    if ( isEmpty( user.username ) ) { ... }
    if ( ! isStrongPassword( user.password ) ) { ... }
    if ( ! isEmail( user.email ) ) { ... }
    // user is valid, register him
}
```

The role of business rules in an application is complex, because there are many locations where they have to be considered. For example, every form in the UI has to include a proper validation that is determined by the field data type and business policy. During a page rendering phase, every UI element has to determine whether it should apply and if so then how, which bases on the user context, the application state, security policy and the business rules that are usually parametrized by user's profile. Unfortunately, contemporary technolo-

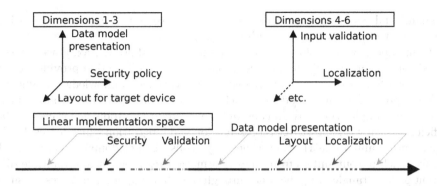

Fig. 1. Facets considered in user interface

gies does not think of business rules as about a crucial part of the application, which requires a special support, so they do not implement mechanisms providing efficient business policy handling. In consequence, the only business policy implementation in Java EE available by now [4] builds on high information restatement and source code duplication, as is shown in Listing 2.

User interface is the most difficult and complex part of every software application. Its quality decides about products attractiveness, usability and marketability, because even the best functionality without user friendly interface is nearly worthless. In conclusion, UI development is crucial and important part of software life-cycle. Its source code comprises up to 48% of total application code [10], thus any inefficiency in UI development can cause significant issues.

Unfortunately, UI implementation is highly difficult task because there have to be considered multiple concerns. Final UI appearance is affected by the data model presentation, application state, current user's profile, target device layout, security policy, input validation, localization, etc [2]. In deeper analysis we find out that all these concerns can be seen as mutually orthogonal domains, which have to be serialized into renderable linear source code, as is illustrated on Fig. 1. Such code involves for each field more than just a style of presentation and visibility status, but for example also the whole page must be structured into some layout derived from target device. Unfortunately, all these parameters depend on multiple variables from orthogonal domains.

When we think about this background carefully, we realise that each of this mentioned concerns can be described completely in its own, nearly independent, way. For each facet can be defined e.g. set of layout templates, set of security rules, localization files, component templates, etc. so we can look at it as on building multidimensional space where each request gain multiple values from distinct domains. It is just like placing the single point into multidimensional space. The only thing connecting all these distinct planes is the current context, which parametrizes all of them and is brought by a request.

In the light of this domain decomposition we can see that we face to the new problem: linearisation. We need to figure out how to seamlessly, automatically and efficiently integrate all these given heterogeneous descriptions of all distinct domains and construct the renderable linear source code parameterizable with the current user's and application's context.

Contemporary technologies focus on performance, short syntax, powerful isolated components (e.g.: graphical components, caching, injection providers etc.) but only a few of them actually realise the importance of knowledge caught inside of application source code and such tools bring only partial solutions which cover only simple scenarios. In consequence, there are no tools nor ways how to efficiently deal with knowledge representation and integration which forces developers into manual linearisation of heterogeneous distinct domains as described above. Such output suffers from high information restatement and multiple different policies tangled together and through the whole application. In result, such code is difficult to maintain and any change means localization and modification of multiple places, which is tedious, expensive and error-prone.

3 Related Work

This paper addresses multiple issues related to business policy management, application design and source code generation to provide efficient way how to deal with knowledge representation without unnecessary information restatement.

Importance and complexity of business rules discuss authors in [1]. Together with analysis of three-tiered information systems they introduce couple solutions how to reduce maintenance effort of business rules in the middle layer. The proposed solutions proceed from design patterns [5][7], but still their solution suffers from some code duplication.

Standardized Java EE development [4] brings up usage of *Chain of Responsibility* [7] design pattern inserting external processors before business operation invocation. Such approach would allow e.g. input validation invocation before the operation calls, but due to detached connection to the method, it is not invoked during intra-class operation calls. Besides that, there is no good design support for target method inspection and invocation flow modification. In conclusion, this method is not suitable for application to the whole system, rather only to the business layer.

Inconvenience of manual code duplication and information restatement is discussed in [12][3]. Model driven architecture (MDA) brings up the idea of system modelling in a graphical language, such as UML. Such model represents the knowledge and main source of information as well as a convenient input for tools generating source code. Although MDA generates source code, the whole process is run just once, because its output must be extended with another code snippets. That makes the model updates complicated as the modified code gets lost upon subsequent model to code transformations. Furthermore, the whole concept is derived form object oriented paradigm, which suffers from inability to reflect cross-cutting concerns, thus also this approach preserves code duplication and information restatement.

A technique of automatic programming is also based on reality modelling and following source code generation, but with two significant differences. The first one is a language. While the MDA usually uses UML and OCL for model and knowledge representation, automatic programming uses domain specific languages [6], which are efficient in target domain description. For example, they involve templates and special mark-up derivatives. Another difference is in the way of code generation. While the MDA uses static generation, that runs usually once or only a few times, the automatic programming uses dynamic source code generation executed on each project compilation or even at runtime. Such improvement brings more dynamics into models, because on any change in the domain description the target source code can be automatically updated. With a smart and powerful generator and suitable domain description we are able to completely avoid manual code duplication and achieve simply and neatly modifiable application implementation. The whole concept is introduced in [9].

Object oriented paradigm (OOP) dominates to systems design and brings multiple efficient principles. But there is no silver bullet and it has limitations, such as dealing with the cross-cutting concerns. The OOP is unable to encapsulate it into single component [14], which results in vast code duplication and tangling, which is tedious, inefficient, error-prone and complicates the maintenance. The aspect oriented programming (AOP) introduced in [11] presents an alternative concept focused on realisation of cross-cutting concerns. This paradigm comes with a new terminology. It considers two types of components, besides the OOP (or general) components, there are also components that are understood as a single cross-cutting concern, called *aspects*. They comprises of *pointcuts* and *advices*. A pointcut is an expression addressing all locations in the OOP components where and when the aspect should apply. Such spots addressed by pointcuts are called *join-points*. In [14] authors differentiate between static join-points, characterised as locations in the program's source code, and dynamic join-points, defined as points in the execution of a program that consider runtime context. When the aspect's pointcut applies, it uses an *advice*, the extension of the functionality invoked when the execution-flow reaches the addressed join-points, or a composition rules [14]. In general, the goal of advising is modification, integration or an extension of the wrapped functionality. Aspects are weaved to the OOP components through a simple compiler or an integration tool called *aspect weaver*. Aspects are often described by domain specific languages and typical instances of them are security managers, cache providers and exception handlers. Both paradigms are usable together as a multiparadigmatic design.

When we are looking for existing and verified solution efficiently dealing with business rules we have to look at rule-driven production systems [8], which represents target domain by huge set of business rules comprised of preconditions and actions. Each rule represents description of some situation or event and also contains a properly defined reaction to do when it occurs. Strong advantage of these systems is knowledge stored outside the compiled source code and usage of domain specific languages, which allows delegation of rules design and management to domain experts and frees developers. The comparison of knowledge represented inside of production systems to business policy of information systems must conclude that the business policy is light version, subset, of production systems. This awareness opens many possibilities like usage of production system engine and kind of knowledge representation in new context.

4   Aspect Approach

Each software system consists of a static and a dynamic part. As the static part we perceive the data model, its fields and their types, which is immutable and stable concern. From the data model we derive the UI elements for data visualization. Set of business rules such as validation, integrity constrains, context verification policy etc. is also stable and usually immutable, so we involve it into the static part. Processing of this concerns can be done once, for example at the compile time. On the other hand, the output of each request differs in the exact composition of UI, in validation rules, presented data and available actions. All these dynamic aspects depend on a position of user's context in a multidimensional space. Efficient and proper processing needs runtime evaluation and integration of all those concerns, including preprocessed the static ones.

Considering the example from section 2 and conclusion from previous paragraph we are looking for a way allowing us to statically integrate business rules into the whole system and generate a proper UI structure based on the data model. In addition, all these parts must consider concerns such as target device, security, etc. and weave them together at runtime on each request. The AOP, which is designed to handle cross-cutting concerns and weave heterogeneous sources, together with a technique of automatic programming seems to be a convenient way to process all descriptions of distinct domains and weave them into a resulting source code.

4.1   Transformation into Terms of AOP

Application of AOP requires transformation of the domain into a new terminology. The AOP comes with multiple terms which we must define in examined domains: *join-point, pointcut, advice, aspect language* and finally *aspect weaver.*

UI concerns involve both static and dynamic parts of the system. For example, we can think of a UI form/table as of data visualization. Such form must reflect the data structure and provide presentation for each its field, although we must consider also dynamic information such as security, selected layout, conditional rendering, etc., which may influence the resulting presentation.

In the visualization we use a data structure as the base component used for AOP composition. To visualize the data in the UI we use its static structure as a source for static join-points. Among them we consider *entity* and *field names*, their *data types*, and *field annotations* with their *parameters*. Considered join-points also include dynamic join-points to integrate dynamic aspects. In order to keep our approach general it is possible to pass to the weaving context a *runtime information* such as user access rights, geo-location, local context for presentation, device screen size, etc. All dynamic join-points are used in the visualization. The goal of this process is composition similar to Hyper/J aspect oriented system. The first level aspect is field presentation, for this we use *presentation rules*. Its *pointcuts* can query the visualized data and context for given properties such as whether a given visualized field is a *string* with given *length* restriction or whether it is *'Friday'* and the end user is *'administrator'* with *small screen* size. Pointcuts can address all join-points, logical combination and use arithmetical operations. The *advice* provides an integration template that uses the target presentation language and composition rules to integrate second-level aspects, such as validation, business rules, internationalization, etc. These aspects use *pointcuts* identical with the first-level and their *advice* define a composition rules. After all data fields are processed by first and second level aspects the result is decorated by a layout template providing a third-level aspect the layout. The aspect weaver is attached to the presentation language as a special component. Designer then uses this component to visualize data at the page and defines the aspect integration through the weaver configuration and the integration templates. Dynamic join-points can be passed to the weaver through the special component, or by a system context. The life-cycle then looks as follows. A page renderer processes the view code, once it reaches a special component it calls its custom handler that integrates the weaving. It takes the data instance passed to the component and gets its static join-points, it may pass the dynamic join-points to the weaver at this point. Next, the weaver walks through the data structure and for each field applies presentation rules. Based on field properties and the context it selects a composition template that defines the presentation and extends it with other aspects. The result of this transformation is decorated by layout. The generated target code is compiled and embedded to the page.

Business rules as a unit must be considered on multiple types of locations, e.g. input validation in business operations, input validation in UI or access restriction to data and operations based on context and security policy. In all these locations it is necessary to apply these rules. In terminology of AOP, such places are *join-points*. Although business rules can be defined as a standalone unit describing single domain, in each context they apply a bit differently. For example, in business layer they are evaluated against given context and input values, but in the UI they must be decomposed and transformed into a client-side validation. In consequence, rule transformation must be handled in multiple ways based on target context. Each way then applies on a subset of join-points which is selected through the *pointcut* and the concrete rule transformation is performed by the *advice* extending target locations. Kind of transformation differs per target context, because for input validation in business layer we demand input and context evaluation against business policy, while for client-side validation in the UI we require rules decomposition and transformation into e.g. a scripting language. As mentioned earlier, verified solution for business rules representation are domain specific languages used in a core of rule-driven systems, which stand here for the *aspect language*. Finally, as an *aspect weaver* can be considered any processor enhancing target join-points with new functionality. In a single system can be recognized multiple weavers, because the UI is usually described by different language than the business layer, thus weavers must support composition into these output languages. While for executable code would be enough to automatically prepend invocation of input validation mechanism, for the UI described using a mark-up language must be the source code completely recreated and completed with new checkers, validators and converters.

Appearance of the UI differs per request based on the current context, evaluation of business rules and many other disjunct concerns. Basically, the resulting UI becomes variable and dynamic, and the final product must be composed from components on each request. Applying component preparation and final integration using automated programming and complex aspect weavers for all the domains, we can achieve the desired appearance and behaviour because the runtime weaving relieves us from the duty of manual information restatement.

4.2   Implementation

Based on the informal specification we implemented an aspect oriented framework for automatic integration of business rules and dynamic constructing of context-aware user interfaces. Each independent concern is described in the most convenient, domain specific language to give us the best possible performance and effortless maintenance. Usage of a core of rule-driven system brings us an efficient way of business policy representation as is shown in Listing 3 and Listing 4. Furthermore, its engine is designed on high evaluation performance.

Listing 3. Simple entity constraint

```
rule "[Team] Standard validation"
   when Team(
        !isEmpty( name ),
        coach != null,
        members.size() == 3,
        members not contains coach
   ) then
end
```

Listing 4. Security check rule

```
global Set<String> security
rule "[Security] Admin"
   when eval( security.contains(
        SecurityRole.ADMIN
   ) )
        Person() || ... || Contest()
   then
end
```

One of significant issues we faced was operational context addressing. As is noted in section 2, each operation and action stands in its specific business context in an application domain. Three-layered system architecture recognizes multiple types of entry points, e.g. actions on controllers in the UI, web service actions and operations on services in the business layer. Each of these actions may stand in different context, so it is crucial to match them to contexts with this granularity. We decided for application of meta-instructions such as Java annotations. This construct opens efficient way to matching an operation to its context externally defined as a set of rules and groups of rules. Afterwards, inspection of method signatures allows extraction of their context address as is illustrated by Listing 5 and Listing 6. Efficient definition of business context opens the way to application of AOP to enhance chosen spots by application of additional instructions, such as input validation.

The data visualization becomes trivial. Listing 7 presents a component that takes a data instance (an data object) as a parameter, in our case *user*, and visualizes it on the page. This produces an UI form or a table depending on the component properties. The weaving process considers each data field and

Listing 5. Descriptive business context

```
@Validate @StandardValidation
@RequiredRules("Password is strong")
void register(User user) {
   // user is valid, register him
}
```

Listing 6. External bussines context

```
@Validate
@BusinessContext("user/register")
void register(User user) {
   // user is valid, register him
}
```

applies all three levels of aspects (presentation, field extension and layout). A presentation rule example with a *pointcut* and an *advise* is shown in Listing 9, it shows presentation of *String*-typed field and advices to composition templates. An example composition template with shorten version of the *pointcut/advise* is shown in Listing 8. This template shows the representation an extension of its properties by considered aspects and supplying the data context to the template. The layout integration through a layout template is shown in Listing 10.

Listing 7. Aspect-driven UI

```
<af:ui instance="#{user}"
       renderPassword="false"
       layout="#{device.layout}"/>
<!-- context-aware action -->
<h:button value="Ban user"
rendered="#{g:ctx(bean,'ban',user)}"
/>
```

Listing 8. Composition template

```
<x:inputText id="#{prefix}$field$"
    label="#{text['$entity$.$field$']}"
       edit="#{empty edit$field$
          ? edit : edit$field$}"
    validate="$businessRules.toJS()$"
       size="$size$" req="$required$"
       value="#{instance.$field$}" ../>
```

Listing 9. Presentation rules

```
<type>String</type>
<default tag="textTag.xhtml"/>
<cond expr="${email==true}"
       tag="emailTag.xhtml"/>
<cond expr="${link==true}"
       tag="linkTag.xhtml"/>
<cond expr="${maxLength>255}"
       tag="textAreaTg.xhtml"/>
```

Listing 10. Two column layout template

```
<table class="two-column-layout">
   <af:iteration-part>
      <tr>
         <td>$af:next$</td>
         <td>$af:next$</td>
      </tr>
   </af:iteration-part>
</table>
```

5 Case Study

Frameworks designed with respect to the informal specification described in previous section was applied in an experimental application. The goal is to receive preliminary statistics on efficiency in comparison to the common approach. In the measurement we focused on SLOC (Source Lines Of Code) and maintenance effort. Besides that we wanted to evaluate the amount of remaining restated information to lay down our future work.

For the application domain we choose inspiration in the ACM-ICPC programming contest, for which we designed a light registration system. As an implementation platform was chosen Java EE with the support of Spring framework and JavaServer Faces for the presentation layer. The main goal of this designed system was handling of state-flow of team applications to local contests. To design a real application we considered four security roles: member, coach, manager and administrator; the system's domain model is captured on Fig. 2. To illustrate the power of approach in automatic transformation we designed 32 business rules including all kinds of conditions, from simple constraints to complex time-conditioned cross-cutting rules, such as *'Applications are opened'*.

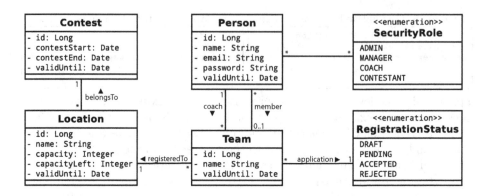

Fig. 2. Data model of registration system to programming contest

To gain preliminary evaluation of this approach we manually developed the same system with the same coding standards and technologies, but without the application of proposed solution. The final comparison of SLOC is covered in Table 1. The results clearly point out reduction of the amount of source code in business and presentation layers, but in return of increased amount of meta-instructions. Such results were expected, because due to the automated programming there is dynamically generated significant amount of code at run-time, which comes out of the weaving of multiple disjunct domains, where some of them are connected via business rules addresses in meta-instructions. We must consider that our application is rather lite, thus the code statistics would become considerably better with the growing size of the project as the reuse of various aspects is supported.

Table 1. Efficiency of compared approaches to system design

SLOC \Approach	Common	Novel	Layer
Java model	201	217	persistent
Java service	203	147	business
Java annotation	13	36	business
Java UI	490	414	presentation
Java annotation	48	116	presentation
UI XML	912	574	presentation

As our implementations are in alpha versions, there is still some kind of information restatement. But, instead of knowledge restatement it is duplication of knowledge addresses e.g. due to weak support of addressing of external context. Nevertheless, there is no knowledge restatement. Each piece of information such as layout, condition and security is captured only once in the source code and referenced from multiple places. That significantly reduces maintenance efforts, because every change of any aspect requires only a single place modification.

6   Conclusion

In this paper we introduced a novel aspect-oriented approach to deal with information system development, while avoiding information restatement, code duplication and problems related to cross-cutting concerns. The fundamental idea is to divide system concerns, describe them separately and let them compose together through an aspect weaver. Our focus in this approach considered application business logic domain and user interface data visualization, which both represent cross-cutting concerns. For both approaches we designed and implemented an aspect weaver and evaluated our approach in a case study. Our approach reduces development and maintenance efforts, it supports component reuse and improves system readability, since each system concern gathers its knowledge at a single location. Furthermore, our approach reduces manual mistakes and inconsistencies among system layers, it enforces change propagation to all coupled components and eliminated tedious work related change propagation.

In future work we aim to improve our framework and minimize remaining code duplication then conduct a study on a production level system. Next, we plan to evaluate performance and apply this approach to the domain of adaptive UIs. Besides that we plan to integrate our solution into web services and to persistence layer to prevent knowledge duplication in database queries.

Acknowledgments. This research was supported by the Grant Agency of the Czech Technical University in Prague, grant No. SGS12/147/OHK3/2T/13.

References

1. Cerny, T., Donahoo, M.J.: How to reduce costs of business logic maintenance. In: 2011 IEEE International Conference on Computer Science and Automation Engineering (CSAE), vol. 1, pp. 77–82. IEEE (2011)
2. Cerny, T., Donahoo, M.J., Song, E.: Towards effective adaptive user interfaces design. In: Proceedings of the 2013 Research in Applied Computation Symposium (RACS 2013) (October 2013)
3. Cerny, T., Song, E.: Model-driven rich form generation. International Information Institute(Tokyo). Information 15(7), 2695–2714 (2012)
4. Chinnici, R., Shannon, B.: JSR 316: Java^{TM} Platform, Enterprise Edition (Java EE) Specification, v6 (2009)
5. Fowler, M.: Patterns of enterprise application architecture. Addison-Wesley Longman Publishing Co., Inc. (2002)
6. Fowler, M.: Domain-specific languages. Pearson Education (2010)
7. Gamma, E., Helm, R., Johnson, R., Vlissides, J.: Design patterns: Abstraction and reuse of object-oriented design. Springer (2001)
8. Hayes-Roth, F.: Rule-based systems. Communications of the ACM 28(9), 921–932 (1985)
9. Kelly, S., Tolvanen, J.P.: Domain-specific modeling: enabling full code generation. Wiley. com (2008)
10. Kennard, R., Steele, R.: Application of software mining to automatic user interface generation (2008)
11. Kiczales, G., Lamping, J., Mendhekar, A., Maeda, C., Lopes, C., Loingtier, J.M., Irwin, J.: Aspect-oriented programming. Springer (1997)

12. Kleppe, A.G., Warmer, J.B., Bast, W.: MDA explained, the model driven architecture: Practice and promise. Addison-Wesley Professional (2003)
13. Laddad, R.: Aspectj in action: enterprise AOP with spring applications. Manning Publications Co (2009)
14. Stoerzer, M., Hanenberg, S.: A classification of pointcut language constructs. In: Workshop on Software-engineering Properties of Languages and Aspect Technologies (SPLAT) Held in Conjunction with AOSD (2005)

Exact Algorithms to Clique-Colour Graphs

Manfred Cochefert and Dieter Kratsch

Laboratoire d'Informatique Théorique et Appliquée, Université de Lorraine, 57045
Metz Cedex 01, France
{manfred.cochefert,dieter.kratsch}@univ-lorraine.fr

Abstract. The clique-chromatic number of a graph $G = (V, E)$ denoted by $\chi_c(G)$ is the smallest integer k such that there exists a partition of the vertex set of G into k subsets with the property that no maximal clique of G is contained in any of the subsets. Such a partition is called a k-clique-colouring of G. Recently Marx proved that deciding whether a graph admits a k-clique-colouring is Σ_2^p-complete for every fixed $k \geq 2$. Our main results are an $O^*(2^n)$ time inclusion-exclusion algorithm to compute $\chi_c(G)$ exactly, and a branching algorithm to decide whether a graph of bounded clique-size admits a 2-clique-colouring which runs in time $O^*(\lambda^n)$ for some $\lambda < 2$.

1 Introduction

A k-clique-colouring of an undirected graph $G = (V, E)$ is a partition of the vertex set V of size k, i.e. k colour classes, such that no maximal clique is monochromatic, in other words, every maximal clique of G contains two vertices of different colour with respect to the partition. The clique-chromatic number $\chi_c(G)$ of an undirected graph $G = (V, E)$ is the smallest integer k such that the graph has a k-clique-colouring. The clique-colouring problem was introduced by Duffus, Sands, Sauer and Woodrow in 1991, and can be seen as a special case of the hypergraph colouring problem since it is equivalent to colouring the clique hypergraph of G in which the hyperedges are the maximal cliques of G [6]. It is easy to see that every graph without isolated vertices admits an n-clique-colouring, and thus every such graph admits an optimal clique-colouring and a well-defined clique-chromatic number.

Previous Work. The complexity of the clique-colouring problem is well-studied. In 2011 Marx proved that deciding whether a graph admits a k-clique-colouring is Σ_2^p-complete for every fixed $k \geq 2$, and if the input is the clique hypergraph, i.e. the graph is given by a list of all of its maximal cliques, then the problem is NP-complete [13]. In 2004 Bacsó et al. showed that it is coNP-complete to decide whether a given vertex colouring is indeed a 2-clique-colouring [1]. The clique-colouring problem has been studied on various graph classes. In 2002 Kratochvíl and Tuza showed that it is NP-complete to decide whether a perfect graph has a 2-clique-colouring [10]. The clique chromatic number of perfect graphs and its subclasses has also been studied from a graph-theoretic point of view. The main motivation of this research is the long standing open conjecture that the clique-chromatic number of any perfect graph is at most three.

V. Geffert et al. (Eds.): SOFSEM 2014, LNCS 8327, pp. 187–198, 2014.

Currently, it is not even known whether there is a constant upper bound for the clique-chromatic number of perfect graphs. Defossez provides a collection of 2-clique-colourable graph classes [5], among them in particular strongly perfect graphs. Consequently all graphs in the following graph classes are 2-clique-colourable: bipartite graphs, comparability graphs, chordal graphs, cobipartite graphs, cocomparability graphs, see e.g. [2].

Kratochvíl and Tuza showed that the clique-chromatic number can be found in polynomial time for planar graphs [10]. They also proved that it can be decided in polynomial time whether a planar graph admits a 2-clique-colouring, and combined this with a result of Mohar and Škrekovski stating that every planar graph admits a 3-clique-colouring [14]. Recently, Klein and Morgana showed that there exists a polynomial time algorithm to compute the clique-chromatic number of graphs which contain few P_4's [9].

Our Results. Algorithms solving clique-colouring problems exactly are the main results presented in this paper. The first one is an $O^*(2^n)$ time inclusion-exclusion algorithm solving the Σ_2^p-complete problem asking to compute $\chi_c(G)$ (Section 3). The second one is a branching algorithm solving the NP-complete problem, asking to decide whether a graph of bounded clique-size admits a 2-clique-colouring (Section 5), in time faster than 2^n. The main structural tools of our algorithms are (minimal) transversals and (maximal) obliques of hypergraphs. Up to our knowledge, no non-trivial exact algorithm to compute the clique-chromatic number or to decide 2-clique-colourability was known prior to our work.

Due to space restrictions various proofs had to be omitted.

2 Preliminaries

Throughout this paper we denote by $G = (V, E)$ an undirected graph without isolated vertices, where V is the set of vertices, and E is the set of edges of G. We adopt the notation $n = |V|$ and $m = |E|$. For every $X \subseteq V$, we denote by $G[X] = (X, E_X)$ the subgraph of G induced by X. A subset $X \subseteq V$ is a clique of G iff for all $x, y \in X$ $\{x, y\} \in E$. X is a maximal clique of G iff X is a clique and for all $v \in V \backslash X$ the set $X \cup \{v\}$ is not a clique. We denote by $\omega(G)$ the maximum size of a clique in G.

Using the notation of Berge's book [3], a hypergraph $\mathcal{H} = \{X_1, \ldots, X_q\}$ is a set of subsets of a finite ground set that we usually denoted by V. Each $X_i \subseteq V$ is a hyperedge of $\mathcal{H}$. The rank $r(\mathcal{H})$ of $\mathcal{H}$ is the maximum size of its hyperedges. We denote by $\mathcal{H}_c(G)$ the clique hypergraph of a graph $G = (V, E)$, that is for all $X \subseteq V$ it holds $X \in \mathcal{H}_c(G)$ iff X is a maximal clique of G.

Let k be a positive integer. A k-partition of the vertex set of G, also called a partition of size k, can be viewed as a function $\varphi : V \longrightarrow \{1, \ldots, k\}$. For all $i \in \{1, \ldots, k\}$, we define $V_i = \{x \in V \ / \ \varphi(x) = i\}$. Such a function φ is called a k-colouring of G, and the sets V_i are its colour classes. We also denote a k-partition of V by $(V_1, \ldots, V_k)$.

A k-clique-colouring of G is a k-colouring of G such that there is no monochromatic maximal clique in G. Hence $(V_1, \ldots, V_k)$ is a k-clique-colouring of G iff for all $X \in \mathcal{H}_c(G)$ there exist $x, y \in X$ such that $\varphi(x) \neq \varphi(y)$, which is equivalent to the property that no hyperedge of $\mathcal{H}_c(G)$ is a subset of any colour class V_i of the k-colouring.

A subset of vertices $X \subseteq V$ hits a hyperedge C of a hypergraph iff $X \cap C \neq \emptyset$. A subset $T \subseteq V$ of vertices is a transversal of a hypergraph $\mathcal{H}$ iff T hits every hyperedge, that is for all $X \in \mathcal{H}$ it holds that $X \cap T \neq \emptyset$. T is a minimal transversal of a hypergraph $\mathcal{H}$ iff it is a transversal of $\mathcal{H}$ and every proper subset of T is not a transversal of $\mathcal{H}$. The set of all minimal transversals of a hypergraph $\mathcal{H}$ is denoted by $Tr(\mathcal{H})$ and forms a hypergraph on the same vertex set as $\mathcal{H}$.

Let $\mathcal{F} \subseteq 2^V$. We denote by $\mathcal{F}_\downarrow$ the down-closure of $\mathcal{F}$, and by $\mathcal{F}_\uparrow$ the up-closure of $\mathcal{F}$, which are defined as follows: $\mathcal{F}_\downarrow = \{X \subseteq V \mid \exists\, Y \in \mathcal{F}, X \subseteq Y\}$ and $\mathcal{F}_\uparrow = \{X \subseteq V \mid \exists\, Y \in \mathcal{F}, Y \subseteq X\}$.

3   Computing the Clique-Chromatic Number Exactly

Inclusion-exclusion is a powerful tool for designing exact exponential algorithms, and one chapter of the book "Exact Exponential Algorithms" [7] is dedicated to inclusion-exclusion algorithms. Björklund, Husfeldt and Koivisto [4] used inclusion-exclusion to design an $O^*(2^n)$ time [1] algorithm computing the number of (unordered) k-partitions of a family $\mathcal{F}$ of subsets of an n-vertex ground set S. In their algorithm the family $\mathcal{F}$ of subsets of S is explicitly given and part of the input. We use their algorithm in the following setting where the family $\mathcal{F}$ is not explicitly given and not part of the input. Hence computing $\mathcal{F}$ in a preprocessing step is an essential part of the algorithm.

Theorem 1. *Given a graph $G = (V, E)$ and a positive integer k as an input, there is an inclusion-exclusion algorithm to decide whether there is a k-partition $(V_1, V_2, \ldots, V_k)$ of V such that all colour classes V_i belong to a (implicitly given) family $\mathcal{F}$ of subsets of V. The running time of the algorithm is $t(n) + O^*(2^n)$, where $t(n)$ is the time to enumerate all elements of $\mathcal{F}$.*

Consider the well-known chromatic number problem. A graph $G = (V, E)$ has chromatic number at most k iff there is a k-partition of the set $S = V$ where the family $\mathcal{F} \subseteq 2^S$ of all possible colour classes in such a k-partition of V is the family of all independent sets of G. Note that this particular family $\mathcal{F}$ can easily be computed in time $O^*(2^n)$. In fact, in many applications one has $t(n) = O^*(2^n)$. The following lemma characterizes the family $\mathcal{F}$ needed in an inclusion-exclusion algorithm to compute a k-clique-colouring of a graph G.

Lemma 1. *$(V_1, \ldots, V_k)$ is a k-clique-colouring of G if and only if $(V_1, \ldots, V_k)$ is a k-partition of G and, for all $1 \leq i \leq k$, $\overline{V_i} = V \setminus V_i$ is a transversal of $\mathcal{H}_c(G)$.*

Proof. Let $(V_1, \ldots, V_k)$ be a k-clique-colouring. Assume that the complement of some color class $\overline{V_i}$ is not a transversal of $\mathcal{H}_c(G)$, $1 \leq i \leq k$. This implies

---

[1] We use the O^* notation: $f(n) = O^*(g(n))$ if $f(n) = O(g(n)p(n))$ for some polynomial $p(n)$.

that its complement $V \setminus \overline{V_i} = V_i$ contains a maximal clique, and thus $(V_1, \ldots, V_k)$ is not a k-clique-colouring, a contradiction. Conversely, let $(V_1, \ldots, V_k)$ be a k-partition of G such that each $\overline{V_i}$ is a transversal of $\mathcal{H}_c(G)$. Suppose there exists a monochromatic maximal clique $C \in \mathcal{H}_c(G)$, that is $C \subseteq V_i$ for some $1 \leq i \leq k$. Then $C \cap \overline{V_i} = \emptyset$, and thus $\overline{V_i}$ is not a transversal, a contradiction.   □

By Theorem 1 and Lemma 1, it is sufficient to apply the inclusion-exclusion algorithm to input G and k, where the family $\mathcal{F}$ consists of all complements of tranversals of the clique hypergraph $\mathcal{H}_c(G)$. It turns out that the bottleneck is the enumeration of all complements of transversals of the clique hypergraph. Our early efforts to design an algorithm to enumerate all transversals of a clique hypergraph established an algorithm of running time $O(2.4423^n)$.

In order to enumerate the family of transversals faster, the key idea was to use obliques, which are subsets of the vertices of a hypergraph that are not a transversal. This allows us to establish an $O^*(2^n)$ algorithm to enumerate all transversals of a clique hypergraph; one of the main results of this section.

Definition 1. *A subset of vertices $O \subseteq V$ is an oblique of a hypergraph $\mathcal{H}$ if it is not a transversal of $\mathcal{H}$. Hence $O \subseteq V$ is an oblique of $\mathcal{H}$ if there is an $X \in \mathcal{H}$ such that $O \cap X = \emptyset$. An oblique $O \subseteq V$ is a maximal oblique of $\mathcal{H}$ if it is not properly contained in any other oblique of $\mathcal{H}$.*

For the rest of the paper, we will denote by $\mathcal{T}_c$ and $\mathcal{O}_c$, respectively, the family of all transversals and the family of all obliques, respectively, of the clique hypergraph $\mathcal{H}_c(G)$. We will sometimes use the notation $\mathcal{T}$ and $\mathcal{O}$ when speaking of general hypergraphs $\mathcal{H}$. The definition directly implies various properties. Clearly each subset of vertices $X \subseteq V$ in a hypergraph is either an oblique or a transversal. Thus $(\mathcal{T}, \mathcal{O})$ is a partition of the family of all subsets of the vertices $\mathcal{P}(V)$ of a hypergraph. We list a few easy-to-prove properties of obliques and transversals which are fundamental for our work:

Property 1. $O \subseteq V$ is an oblique of a hypergraph $\mathcal{H}$ iff O is a subset of the complement of a hyperedge X of $\mathcal{H}$.

Property 2. $O \subseteq V$ is an oblique of an hypergraph $\mathcal{H}$ iff for all $X \subseteq O$ the set X is an oblique.

Property 3. $T \subseteq V$ is a transversal of an hypergraph $\mathcal{H}$ iff for all X such that $T \subseteq X$ the set X is a transversal.

In the following we study the recognition and enumeration of obliques and maximal obliques, respectively, of a clique hypergraph $\mathcal{H}_c(G)$. They are needed to establish our inclusion-exclusion algorithm to compute the clique-chromatic number of a graph G in time $O^*(2^n)$.

Lemma 2. *There exists an algorithm taking as an input a graph $G = (V, E)$ and a subset $X \subseteq V$ which decides in time $O(n|X|)$ whether X is properly contained in a clique of G.*

It is known that all maximal cliques of a graph can be enumerated in time $O^*(3^{n/3})$. To see this one may combine the combinatorial bound of Moon and Moser [15], saying that any graph on n vertices has at most $3^{n/3}$ maximal cliques, with a polynomial delay algorithm to enumerate all maximal cliques by Tsukiyama et al. [16].

Lemma 3. *There exists an $O^*(3^{(n-|X|)/3})$ algorithm which decides for a given graph $G = (V, E)$ and a subset $X \subseteq V$ whether X is an oblique of the clique hypergraph $\mathcal{H}_c(G)$.*

When used in a direct fashion, Lemma 3 gives an $O^*(2.4423^n)$ time algorithm to enumerate all elements of $\mathcal{O}_c$ for a given graph $G = (V, E)$. Indeed, this enumeration algorithm uses Lemma 3 to verify for every subset $X \subseteq V$ whether X is an oblique. Doing this for every subset of vertices of size i, from $i = 1$ up to n, the running time of the algorithm is bounded by $\sum_{i=0}^{n} \binom{n}{i} 3^{(n-i)/3} = (1 + 3^{1/3})^n = O^*(2.4423^n)$.

Now we show how to improve the running time of the enumeration of all obliques of $\mathcal{H}_c(G)$ by using the family of all maximal obliques of $\mathcal{H}_c(G)$. It is worth mentioning that, while the maximal obliques of a hypergraph are relatively easy to enumerate, the same is not true for its minimal transversals.

Theorem 2. *The maximal obliques of the clique hypergraph $\mathcal{H}_c(G)$ are exactly the complements of the maximal cliques of G.*

Proof. "$\Leftarrow$" Let C be a maximal clique of G and thus $C \in \mathcal{H}_c(G)$. Let $X \subseteq V$ be the complement of the maximal clique C. Hence $X \cap C = \emptyset$. By property 1, X is an oblique of $\mathcal{H}_c(G)$. We claim that X hits every maximal clique of $\mathcal{H}_c(G)$ but C. To show this by contradiction, assume that there is a $C" \in \mathcal{H}_c(G)$ such that $X \cap C" = \emptyset$. Hence $C" \subseteq \overline{X}$. Since $\overline{X} = C$, and $C" \neq C$, we have that $C" \subseteq C$, and thus $C"$ is not a maximal clique of G, a contradiction.

"$\Rightarrow$" Let $O \subseteq V$ be a maximal oblique of $\mathcal{H}_c(G)$. By definition, there exists a $C \in \mathcal{H}_c(G)$ such that $O \subseteq \overline{C}$. We claim that $O = \overline{C}$ and prove this by contradiction. Thus, assume that O is properly contained in $\overline{C}$, thus there is a $Y \subseteq \overline{C}$ such that $O \cup Y = \overline{C}$ and $Y \neq \emptyset$. By our construction and the definition of an oblique, $O \cup Y \subseteq \overline{C}$ implies that $O \cup Y$ is an oblique. Thus O is not a maximal oblique, a contradiction. $\qquad\square$

Corollary 1. *There exists an $O(n^2 - n|X|)$ time algorithm which decides for a given graph $G = (V, E)$ and a set $X \subseteq V$ whether X is a maximal oblique of the clique hypergraph $\mathcal{H}_c(G)$.*

Corollary 1 provides an easy algorithm of running time $O^*(2^n)$ to compute the family of all maximal obliques of a clique hypergraph. This can be improved by using another immediate consequence of Theorem 2.

Corollary 2. *The family of maximal obliques of $\mathcal{H}_c(G)$ can be enumerated in time $O^*(3^{n/3})$.*

The following lemma is crucial for our algorithms and of independent interest.

Lemma 4. *There are algorithms taking as input a family $\mathcal{F} \subseteq 2^V$ and enumerate the down-closure $\mathcal{F}_\downarrow$ in time $O^*(|\mathcal{F}_\downarrow|)$, and the up-closure $\mathcal{F}_\uparrow$ in time $O^*(|\mathcal{F}_\uparrow|)$, respectively.*

Proof. Both algorithms are based on the same idea which is to generate the corresponding closure level by level: in decreasing size of the sets when computing the elements of the down-closure $\mathcal{F}_\downarrow$, and in increasing size of the sets when computing the elements of the up-closure $\mathcal{F}_\uparrow$. There are various ways to implement this approach depending on the choice of the data structures. We give one that simplifies the description and runs within the claimed time.

Any family $\mathcal{G} \subseteq 2^V$ of vertices can be encoded into a subset of the integer set $\{0, \ldots, 2^n - 1\}$. We use a data structure D to store integer subsets $F \in 2^V$ in this way. Also we make sure that no key already stored in D is again inserted, i.e. D contains only unique keys. We use a red-black tree as data structure D, see [12]. Red-black trees support operations searching a key and inserting a key in time $O(\log p)$, where p is the number of stored keys. Whenever a key is supposed to be inserted into D, we first do a search operation, and only if search fails the key will be inserted. Hence such a modified insert can be done in time $O(p \cdot \log p)$, and guarantees that there are no multiple keys in D. Furthermore we use a queue Q to store subsets of D of size i, $1 \leq i \leq n$. We initialize D by passing through the list of $\mathcal{F}$ such that D contains exactly the keys of elements of $\mathcal{F}$. Initially Q is empty.

Now both algorithms are easy to present. To compute the down-closure $\mathcal{F}_\downarrow$, we loop for i from n to 1. In round i we pass through D and store in Q all elements of size i. Then for all $X \in Q$ and all $x \in X$, we insert each $X \backslash \{x\}$ into D. At the end of round i, Q is emptied. To compute the up-closure $\mathcal{F}_\uparrow$, we loop for i from 0 to $n - 1$. We pass through D and store in Q all elements of size i. Then for all $X \in Q$, we insert each $X \cup \{x\}$ with $x \in V \backslash X$ into D. At the end of round i we empty Q.

Consequently, at termination of both algorithms, the keys of the data structure D encode the elements of the corresponding closure $\mathcal{F}_\downarrow$ and $\mathcal{F}_\uparrow$. Any traversal on D can now be used to enumerate these closures in time $O(|\mathcal{F}_\downarrow|)$ or $O(|\mathcal{F}_\uparrow|)$. The correctness of the algorithms can easily be shown by induction on the size of the subsets of vertices (in the corresponding order).

Due to similarity of the algorithms, we show how to analyze the running time of the algorithm to compute $\mathcal{F}_\downarrow$ only. Clearly D can be initialized in time $O(|\mathcal{F}| \cdot \log|\mathcal{F}|)$. Since $\mathcal{F} \subseteq \mathcal{F}_\downarrow$, this first step takes time $O(|\mathcal{F}_\downarrow| \cdot \log|\mathcal{F}_\downarrow|)$. Then at each round i of the loop, the construction of Q takes time $O(|D|)$. Finally note that for any $1 \leq i \leq n$, we try to insert a subset $X \in \mathcal{F}_\downarrow$ of size $i - 1$ into D no more than $n - (i - 1) \leq n$ times. Since for any $1 \leq i \leq n$, we have $|D| \leq |\mathcal{F}_\downarrow|$, all operations on Q take in total time $O(n \cdot |\mathcal{F}_\downarrow|)$, and the (modified) inserting operations into D take in total time $O(n \cdot |\mathcal{F}_\downarrow| \cdot \log|\mathcal{F}_\downarrow|)$. Summarizing, the algorithm has an overall running time of $O(n \cdot |\mathcal{F}_\downarrow| \cdot \log|\mathcal{F}_\downarrow|)$. Finally, $|\mathcal{F}_\downarrow| \leq 2^n$ implies $\log|\mathcal{F}_\downarrow| \leq n$. Consequently the algorithm runs in time $O^*(|\mathcal{F}_\downarrow|)$.   □

Trivially Lemma 4 implies that both algorithms run in time $O^*(2^n)$.

Lemma 5. *There exists an algorithm which, given a hypergraph $\mathcal{H}$, computes the family $\mathcal{O}$ of all obliques of $\mathcal{H}$ in time $O^*(2^n + t(n))$, where $t(n)$ is the running time of an algorithm to enumerate all maximal obliques of $\mathcal{H}$. Furthermore, the family of all obliques of a clique hypergraph $\mathcal{H}_c(G)$ can be enumerated in time $O^*(2^n)$.*

Proof. Using property 2, we have that the family of obliques of any hypergraph $\mathcal{H}$ is the down-closure of the family of all maximal obliques of $\mathcal{H}$. By our assumption there is an algorithm to enumerate all maximal obliques of $\mathcal{H}$ in time $t(n)$. Then by Lemma 4, the family of all obliques of $\mathcal{H}$ can be computed in time $O^*(2^n)$, which implies overall running time $O^*(2^n + t(n))$.

In the special case of a clique hypergraph $\mathcal{H}_c(G)$, we have that $t(n) = O^*(3^{n/3})$ due to Corollary 2, and thus we can conclude that there is a $O^*(2^n)$ time algorithm to enumerate the family of obliques of the clique hypergraph. □

Now we are ready to present our inclusion-exclusion algorithm to compute the clique-chromatic number of a graph.

Theorem 3. *There is an algorithm to compute the clique-chromatic number of a given graph $G = (V, E)$ in time $O^*(2^n)$.*

Proof. The algorithm starts by building the family $\mathcal{F}$ of possible colour classes for any clique-partition of V. By Lemma 1, $\mathcal{F}$ is the family of complements of transversals of the clique-hypergraph $\mathcal{H}_c(G)$. Since the complement of the family of all obliques $\mathcal{O}_c$ of $\mathcal{H}_c(G)$ is the family of all tranversals $\mathcal{T}_c$ of $\mathcal{H}_c(G)$, i.e. $2^V \setminus \mathcal{O}_c = \mathcal{T}_c$, we obtain $\mathcal{F} = \{\overline{X}/X \in \mathcal{T}_c\} = \{X/X \in \mathcal{O}_c\}$. By Lemma 5, the family of all obliques of a clique hypergraph $\mathcal{H}_c(G)$ can be enumerated in time $O^*(2^n)$, and thus $\mathcal{F}$ can be enumerated in time $O^*(2^n)$.

Finally, once $\mathcal{F}$ has been enumerated, by Theorem 1, an inclusion-exclusion algorithm can count, for fixed k, the number the partitions of size k in which every colour class belongs to $\mathcal{F}$ in time $O^*(2^n)$. Running this algorithm for $k = 1$ up to n, we find the smallest k for which the number of k-clique colourings is not zero, and thus G has a k-clique-colouring and its clique-chromatic number is k. To analyse the running time, observe that the family $\mathcal{F}$ of possible colour classes can be constructed in time $O^*(2^n)$, and the inclusion-exclusion algorithm for given $\mathcal{F}$ also runs in time $O^*(2^n)$. □

Our algorithm to compute the clique-chromatic number of a graph has the same running time as the best known algorithm to compute the chromatic number of a graph. To improve upon this one, completely new ideas seem to be needed.

4   Minimal Transversals

The notion of transversals has been studied extensively in Berge's book on hypergraphs [3]. It is worth mentioning that in his book Berge even presents an algorithm to enumerate all minimal transversals of a given hypergraph (though without time analysis). In this section, we are interested in the recognition of transversals and minimal transversals, and the enumeration of all transversals of a hypergraph, when its family of minimal transversals is given. Those auxiliary results will be needed to build up our branching algorithms in Section 5. The three algorithms are quite similar to algorithms described in the previous section, and thus we only state the corresponding lemmas.

Lemma 6. *There is an $O^*(3^{(n-|X|)/3})$ time algorithm to decide for input $X \subseteq V$ and $G = (V, E)$ whether X is a transversal of the clique hypergraph $\mathcal{H}_c(G)$.*

Lemma 7. *There is an $O^*(3^{(n-|X|)/3})$ time algorithm to decide for given $X \subseteq V$ and $G = (V, E)$ whether X is a minimal transversal of $\mathcal{H}_c(G)$.*

Lemma 8. *There is an algorithm to enumerate the family $\mathcal{F}$ of all transversals of a clique hypergraph $\mathcal{H}_c(G)$ in time $O^*(2^n + t(n))$, where $t(n)$ is the time to enumerate all minimal transversals of $\mathcal{H}_c(G)$.*

5   2-Clique Colourability of Graphs of Bounded Clique Size

In 2006 Défossez proved that it is NP-hard to decide whether a K_4-free perfect graph admits a 2-clique-colouring [5]. In this section, we present an exact algorithm to compute a 2-clique-colouring, if there is one, assuming that the maximal cliques of the input graph G have bounded size, i.e., $\omega(G) \leq c$ for some positive constant c. We show how this problem on clique hypergraphs can be transformed into the problem to enumerate all minimal transversals of a hypergraph of bounded rank c. This leads to a running time better than those of the previously established $O^*(2^n)$ time inclusion-exclusion algorithm of Section 3. More precisely, for every $c \geq 2$, there exists a constant λ_c such that all minimal transversals of a hypergraph whose hyperedges have size bounded by c can be enumerated in time $O^*((\lambda_c)^n)$, where $\lambda_c < 2$.

While the algorithm of Section 3 used inclusion-exclusion, the main algorithm design technique of this section is branching. Branching algorithms are one of the main tools to design fast exact exponential algorithms [7]. A problem is solved by recursively decomposing it into subproblems of smaller sizes. This is done by applying certain branching and reduction rules. The execution of a branching algorithm can be illustrated by a search tree. Analysing the running time of a branching algorithm can be done by determining the maximum number of nodes in a search tree.

To do this, let $t(n)$ be an upper bound for the running time of the algorithm when applied to an instance of size n. Consider any branching rule. Let $n-c_1, \ldots, n-c_b$ be the sizes of the instances in the subproblems of the branching rule, where for all $1 \leq i \leq b$, $c_i \in \mathbb{R}_+^*$. Then the running time satisfyies $t(n) \leq \sum_{i=1}^{b} t(n - c_i)$.

All basic solutions of the corresponding homogeneous linear recurrence are of the form λ^n, where λ is a complex number. The value of λ we are interested in is the unique positive real root of the polynomial $\lambda^n - \sum_{i=1}^{b} \lambda^{n-c_i}$. One can use Newton's method to compute λ. It is common to denote this value by $\tau(c_1, \ldots, c_b)$, while $(c_1, \ldots, c_b)$ is the corresponding branching vector.

Several properties of branching vectors and branching numbers are mentioned in [7,11]. Let us mention the following rules that will be useful in the time analysis of our branching algorithms. Branching vectors satisfy the permutation rule, that is for any permutation π of the integer set $\{1, \ldots, k\}$, we have $\tau(c_1, \ldots, c_k) = \tau(c_{\pi(1)}, \ldots, c_{\pi(k)})$. Branching vectors satisfy the extension rule, that is for every $c_{k+1} > 0$, $\tau(c_1, \ldots, c_k) < \tau(c_1, \ldots, c_k, c_{k+1})$. Finally they satisfy the substitution rule, that is if $\alpha = \tau(c_1, \ldots, c_k)$ and $\alpha = \tau(c'_1, \ldots, c'_{k'})$, then for every k satisfying $1 \le i \le k$ it holds $\alpha = \tau(c_1, \ldots, c_{i-1}, c_i + c'_1, \ldots, c_i + c'_{k'}, c_{i+1}, \ldots, c_k)$.

The following lemma of independent interest is crucial for our work.

Lemma 9. *Every graph $G = (V, E)$ admits an optimal clique-colouring such that at least one colour class is the complement of a minimal transversal of the clique hypergraph $\mathcal{H}_c(G)$.*

Proof. Let $G = (V, E)$ be a graph and $k = \chi_c(G)$ its clique-chromatic number. Let $(V_1, \ldots, V_k)$ be an optimal clique-colouring of G. By Lemma 1, the complement of each colour class V_i is a transversal of $\mathcal{H}_c(G)$. Let us assume that none of the colour classes V_i is the complement of a minimal transversal. The idea is to transform the first colour class into the complement of a minimal transversal. Since we assumed that $\overline{V_1}$ is a transversal which is not minimal, there exists an $X_1 \subset \overline{V_1}$ such that X_1 is a minimal transversal of $\mathcal{H}_c(G)$. We set $V'_1 = \overline{X_1}$, and for all $i \ge 2$, we set $V'_i = V_i \backslash V'_1$. During this operation, for all $i \ge 2$, we have not increased the size of V_i, and thus we have not decreased the size of their complement which were already transversals. Thus for all $i \ge 2$, the complement of V'_i is a transversal. By the construction of V'_1, and by Lemma 1, we obtain that $(V'_1, V'_2, \ldots, V'_k)$ is a clique-colouring of G.   □

The previous lemma is optimal in the sense that there are graphs not having any clique-colouring of minimal size in which two colour classes are the complement of a minimal transversal of the clique hypergraph.

Since a hypergraph of rank 2 is equivalent to a graph, enumerating all minimal transversals of a given hypergraph of rank 2 is equivalent to enumerating all minimal vertex covers of a graph, and all minimal vertex covers can be enumerated in time $O^*(3^{n/3})$; which indeed is optimal.

In his master thesis [8], Gaspers showed that for every fixed integer $c \ge 3$ there is a branching algorithm to compute a minimum transversal of a hypergraph of rank c in time $O^*((\alpha_c)^n)$, where $\alpha_c = \tau(1, 2, \ldots, c) < 2$. See also Table 1 for some values of α_c. His approach can be extended to the enumeration of all minimal transversals.

Theorem 4. *Let $c \ge 3$. There is an algorithm to enumerate all minimal transversals of a hypergraph of rank c in time $O^*((\alpha_c)^n)$, where $\alpha_c = \tau(1, 2, 3, \ldots, c) < 2$.*

Table 1. The constants α_c of Theorem 4 and α'_c of Theorem 5 for $c = 3, \ldots, 10$

c	3	4	5	6	7	8	9	10
α_c	1.8393	1.9276	1.9660	1.9836	1.9920	1.9961	1.9981	1.9991
α'_c	1.7693	1.8994	1.9536	1.9779	1.9893	1.9947	1.9974	1.9987

Theorem 5. *Let $c \geq 3$. There is a $O^*((\alpha'_c)^n)$ time branching algorithm to enumerate all partitions of V into one minimal transversal and one transversal of a given hypergraph of rank c, where $\alpha'_c = \tau(2, 2, 3, 3, 4, 4, \ldots, c-1, c-1, c, c)$.*

Proof. Let $\mathcal{H} = \{X_1, \ldots, X_q\}$ be the input hypergraph, where $|X_i| \leq c$ for every hyperedge X_i of $\mathcal{H}$. The algorithm enumerates all partitions of V into two colour classes (V_1, V_2) such that the first colour class is a minimal transversal of $\mathcal{H}$ and the second colour class is a transversal of $\mathcal{H}$. An instance of our branching algorithm, which describes a subproblem and is associated to a node of the search tree, consists of a triplet $(T, W, \mathcal{R})$. $T \subseteq V$ is a partial transversal of $\mathcal{H}$, $W \subseteq V$ is the set of those vertices that might still be added to T to built the colour class V_1, and thus $T \cap W = \emptyset$. $\mathcal{R} = \{Y_1, \ldots, Y_{q'}\}$ is the hypergraph consisting of those hyperedges of $\mathcal{H}$ which are not hit by T. Since hyperedges of $\mathcal{R}$ might contain vertices not belonging to W, it holds that $W \subseteq \cup_{i=1}^{q'} Y_i$.

During the execution of the branching algorithm and the construction of the corresponding search tree, each leaf of the search tree will be marked either GOOD or BAD. A solution of the problem (to be enumerated) will be a couple $(T, V \backslash T)$ associated to a leaf which is not marked BAD. The algorithm uses two basic procedures to create a subproblem, either ADD (*add*) a subset $Z \subseteq W$ to the partial transversal T, or RMV (*remove*) Z from the current W-set which is equivalent to adding $Z \subseteq W$ to $\overline{T}$. Note that each procedure reduces the size of the W-set of the instance.

ADD(Z) : $W \leftarrow W \backslash Z$, $\mathcal{R} \leftarrow \mathcal{R} \backslash \cup_{i=1}^{q'} \{Y_i \mid Z \cap Y_i \neq \emptyset\}$ and $T \leftarrow T \cup Z$.

RMV(Z) : $W \leftarrow W \backslash Z$.

Now we describe our branching algorithm by listing its reduction and branching rules in the order they are to be applied. This means that a rule can only be applied if all previous ones fail to be applicable. Note that the first three rules are reduction rules and only rules R5.4 and R5.5 are branching rules. If the algorithm stops there will be a leaf in the search tree (to be marked either GOOD or BAD). Otherwise the algorithm branches into subproblems according to the rule applied; in the search tree this creates children of the current node.

R5.1: if $\mathcal{R} = \emptyset$, then STOP (GOOD).

R5.2: if $\exists\, Y_i \in \mathcal{R}$ such that $Y_i \cap W = \emptyset$, then STOP (BAD).

R5.3: if $\exists\, Y_i \in \mathcal{R}$ such that $Y_i \cap W = \{x\}$ then ADD($\{x\}$).

R5.4: if $\exists\, Y_i \in \mathcal{R}$ such that $Y_i \backslash W \neq \emptyset$ then let $W_i = \{c_1, \ldots, c_p\} = Y_i \cap W$. The idea is that in all the obtained subproblems the new partial transversal T needs to hit W_i. To obtain the first subproblem c_1 is added to T. To obtain the second subproblem c_1 is removed from W (and thus added to $\overline{T}$) and c_2 is added to T. In the third subproblem c_1 and c_2 are removed from W and c_3 is added to T. In this manner p subproblems are constructed, the p-th one is obtained by removing $c_1, c_2, \ldots, c_{p-1}$ from W and adding c_p to T. Furthermore for each subproblem $\mathcal{R}$ is updated by removing all hyperedges hit by the new set T.

R5.5: Let $C = \{c_1, \ldots, c_p\} \in \mathcal{R}$. Note that rule R5.4 did not apply, and thus $C \subseteq W$ which implies $C \cap T = \emptyset$ and $C \cap \overline{T} = \emptyset$ in the current instance. The branching rule guarantees that in every generated subproblem C is hit by both, partial transversal T and its complement $\overline{T}$. This means to add at least one W-vertex to T, and to remove at least one vertex from W; the latter is equivalent to adding a vertex to $\overline{T}$. First we keep all subproblems from rule R5.4 except the very first one. Finally we replace this subproblem by the following one. The first one adds c_1 to T and removes c_2 from W. The second adds c_1, c_2 to T and removes c_3 from W. The p-th one adds $c_1, c_2 \ldots, c_p$ to T and removes c_{p+1} from W. In this manner we have replaced the first subproblem of rule R5.4 by $p - 1$ new subproblems.   □

Lemma 10. *The algorithm of Theorem 5 is correct and has a branching number* $\alpha'_c = \tau(2, 2, 3, 3, 4, 4, \ldots, c-1, c-1, c, c)$ *satisfying* $\alpha'_c < \alpha_c < 2$.

One may show that $\alpha_c = \tau(1, 2, \ldots, c) = \tau(2, 3, \ldots, c+1, 2, 3, \ldots, c) = \tau(2, 2, 3, 3, \ldots, c-1, c-1, c, c, c+1) > \tau(2, 2, 3, 3, \ldots, c-1, c-1, c, c) = \alpha'_c$ using the abovementioned substitution, permutation, and extension rules for branching numbers. To evaluate the improvement we refer to some values of the branching numbers α_c and α'_c given in Table 1.

Theorem 6. *There is an algorithm to decide 2-clique-colourability of graphs of maximum clique size* $\omega(G) \leq c$ *in time* $O^*((\lambda_c)^n)$, *where* $\lambda_c \leq \alpha'_c < 2$.

Proof. Let $G = (V, E)$ be a graph such that $\omega(G) \leq c$. Consequently, every hyperedge of $\mathcal{H}_c(G)$ has size at most c, and the number of maximal cliques of G is at most $\binom{n}{c} = O(n^c)$. The polynomial delay algorithm to enumerate all maximal cliques in [16] when applied to the graph G is indeed an algorithm to enumerate all maximal cliques of G of running time polynomial in the number of vertices of G. By Lemma 1, G admits a 2-clique-colouring (V_1, V_2) iff $\overline{V_1} = V_2$ and $\overline{V_2} = V_1$ are transversals of the clique hypergraph $\mathcal{H}_c(G)$, that is both V_1 and V_2 must be transversals of $\mathcal{H}_c(G)$. Moreover by Lemma 9, it is sufficient to enumerate all minimal transversals X of $\mathcal{H}_c(G)$ as class V_1. Using the observations above, one can use either the algorithm from Theorem 4, running in time $O^*(\alpha_c^n)$, $\alpha_c < 2$, or the algorithm from Theorem 5, running in time $O^*((\alpha'_c)^n)$, where $\alpha'_c < \alpha_c < 2$, to establish an algorithm deciding 2-clique-colourability.   □

6   Conclusions

The main structural tools of our algorithms are the (minimal) transversals and (maximal) obliques of the clique hypergraph $\mathcal{H}_c(G)$. Enumerating all (minimal) transversals and all (maximal) obliques of a clique hypergraph are important procedures of our main algorithms. Our main results are algorithms solving clique-colouring problems exactly. The first one is an $O^*(2^n)$ time inclusion-exclusion algorithm to compute the clique-chromatic number of a graph. The second one is a branching algorithm deciding for an input graph G of clique size at most c whether G is 2-clique-colourable in time $O^*((\lambda_c)^n)$ where $\lambda_c < 2$, and computing a 2-clique-colouring of G, if there is one.

It is natural to ask whether there is a $O^*(\alpha^n)$ time algorithm with $\alpha < 2$ to compute the clique chromatic number of a graph or to decide its 2-clique colorability. Let us mention a question strongly related to our work. Is there an algorithm to compute the family of all minimal transversals of a clique hypergraph $\mathcal{H}_c(G)$ in time $O^*(\alpha^n)$ for some $\alpha < 2$? Note that if G is the disjoint union of K_3's; then its clique hypergraph has $3^{n/3}$ minimal transversals. Finally, what is the maximum number of minimal transversals in a clique hypergraph on n vertices?

References

1. Bacsó, G., Gravier, S., Gyárfás, A., Preissmann, M., Sebö, A.: Coloring the maximal cliques of graphs. SIAM Journal on Discrete Mathematics 17(3), 361–376 (2004)
2. Berge, C., Duchet, P.: Strongly perfect graphs. Annals of Discrete Mathematics 21, 57–61 (1984)
3. Berge, C.: Hypergraphs: combinatorics of finite sets. North holland (1984)
4. Björklund, A., Husfeldt, T., Koivisto, M.: Set Partitioning via Inclusion-Exclusion. SIAM J. Comput. 39(2), 546–563 (2009)
5. Défossez, D.: Clique-coloring some classes of odd-hole-free graphs. Journal of Graph Theory 53(3), 233–249 (2006)
6. Duffus, D., Sands, B., Sauer, N., Woodrow, R.: Two-colouring all two-element maximal antichains. Journal of Combinatorial Theory, Series A 57(1), 109–116 (1991)
7. Fomin, F.V., Kratsch, D.: Exact exponential algorithms. Springer (2011)
8. Gaspers, S.: Algorithmes exponentiels. Master's thesis, Université de Metz (June 2005)
9. Klein, S., Morgana, A.: On clique-colouring of graphs with few P4's. Journal of the Brazilian Computer Society 18(2), 113–119 (2012)
10. Kratochvíl, J., Tuza, Z.: On the complexity of bicoloring clique hypergraphs of graphs. Journal of Algorithms 45(1), 40–54 (2002)
11. Kullmann, O.: New methods for 3-SAT decision and worst-case analysis. Theor. Comput. Sci. 223(1-2), 1–72 (1999)
12. Leiserson, C.E., Rivest, R.L., Stein, C., Cormen, T.H.: Introduction to algorithms. The MIT press (2001)
13. Marx, D.: Complexity of clique coloring and related problems. Theoretical Computer Science 412(29), 3487–3500 (2011)
14. Mohar, B., Skrekovski, R.: The Grötzsch theorem for the hypergraph of maximal cliques. The Electronic Journal of Combinatorics 6(R26), 2 (1999)
15. Moon, J., Moser, L.: On cliques in graphs. Israel Journal of Mathematics 3(1), 23–28 (1965)
16. Tsukiyama, S., Ide, M., Ariyoshi, H., Shirakawa, I.: A new algorithm for generating all the maximal independent sets. SIAM Journal on Computing 6(3), 505–517 (1977)

Supporting Non-functional Requirements in Services Software Development Process: An MDD Approach

Valeria de Castro[1], Martin A. Musicante[2], Umberto Souza da Costa[2], Plácido A. de Souza Neto[3], and Genoveva Vargas-Solar[4]

[1] Universidad Rey Juan Carlos – Móstoles, Spain
Valeria.deCastro@urjc.es
[2] Federal University of Rio Grande do Norte (UFRN) – Natal-RN, Brazil
{mam,umberto}@dimap.ufrn.br
[3] Federal Technological Institute of Rio Grande do Norte (IFRN) – Natal-RN, Brazil
placido.neto@ifrn.edu.br
[4] French Council of Scientific Research (CNRS) – Grenoble, France
Genoveva.Vargas-Solar@imag.fr

Abstract. This paper presents the πSOD-M method, an extension to the Service-Oriented Development Method (SOD-M) to support the development of services software by considering their functional and non-functional requirements. Specifically, πSOD-M proposes: *(i)* meta-models for representing non-functional requirements at different abstraction levels; *(ii)* model-to-model transformation rules, useful to semi-automatically refine Platform Independent Models into Platform Specific Models; and *(iii)* rules to transform Platform Specific Models into concrete implementations. In order to illustrate our proposal, the paper also describes how to apply the methodology to develop a proof of concept.

Keywords: MDD, Service Oriented Applications, Non-functional Properties.

1 Introduction

Model Driven Development (MDD) [12] is a top-down approach proposed by the Object Management Group (OMG)[1] for designing and developing software systems. MDD provides a set of guidelines for structuring specifications by using *models* to specify software systems at different levels of abstraction or *viewpoints*:

- *Computation Independent Models (CIM)*: this viewpoint represents the software system at its highest level of abstraction. It focusses on the system environment and on business and requirement specifications. At this level, the structure of the system and its processing details are still unknown or undetermined.
- *Platform Independent Models (PIM)*: this viewpoint focusses on the system functionality, hiding the details of any particular platform.

---

[1] http://www.omg.org/mda

V. Geffert et al. (Eds.): SOFSEM 2014, LNCS 8327, pp. 199–210, 2014.

- *Platform Specific Models (PSM)*: this viewpoint focusses on the functionality, in the context of a particular implementation platform. Models at this level combine the platform-independent view with specific aspects of the target platform in order to implement the system.

Besides the notion of models at each level of abstraction, MDD requires *model transformations* between levels. These transformations may be automatic or semi-automatic and implement the refinement process between levels.

MDD has been applied for developing service-oriented applications. In Service-Oriented Computing [20], pre-existing services are combined to produce applications and provide the business logic. The selection of services is usually guided by the functional requirements of the application being developed. Some methodologies and techniques have been proposed to help the software developer in the specification of functional requirements of the business logic, such as the Service-Oriented Development Method (SOD-M) [9].

Ideally, non-functional requirements such as security, reliability, and efficiency would be considered along with all the stages of the software development. The adoption of non-functional specifications from the early states of development can help the developer to produce applications that are capable of dealing with the application context. Non-functional properties of service-oriented applications have been addressed in academic works and standards. Dealing with these kind of properties involves the use of specific technologies at different layers of the SOC architecture, for instance during the description of service APIs (such as WSDL[8] or REST [13]) or to express service coordinations (like WS-BPEL [1]).

Protocols and models implementing non-functional properties assume the existence of a global control of the artefacts implementing the application. They also assume that each service exports its interface. So, the challenge of supporting non-functional properties is related to *(i)* the specification of the business rules of the application; and *(ii)* dealing with the technical characteristics of the infrastructure where the application is going to be executed.

This paper presents πSOD-M, a methodology for supporting the development of service-oriented applications by taking into account both functional and non-functional requirements. This methodology aims to: *(i)* improve the development process by providing an abstract view of the application and ensuring its conformance to the business requirements; *(ii)* reduce the programming effort through the semi-automatic generation of models for the application, to produce concrete implementations from high-abstraction models. Accordingly, the remainder of the paper is organized as follows: Section 2 gives an overview of the SOD-M approach; Section 3 presents πSOD-M, the methodology that we propose to extend SOD-M; Section 4 shows the development of a proof of concept; Section 5 describes related works; and Section 6 concludes the paper.

2   SOD-M

The Service-Oriented Development Method (SOD-M) [9] adopts the MDD approach to build service-based applications. SOD-M considers two points of view:

(i) business, focusing on the characteristics and requirements of the organization; and *(ii) system requirements*, focusing on features and processes to be implemented in order to build service-based applications in accordance to the business requirements. In this way, SOD-M simplifies the design of service-oriented applications, as well as their implementation using current technologies.

SOD-M provides a framework with models and standards to express functionalities of applications at a high-level of abstraction. The SOD-M meta-models are organized into three levels: CIM (*Computational Independent Models*), PIM (*Platform Independent Models*) and PSM (*Platform Specific Models*). Two models are defined at the CIM level: *value model* and *BPMN model*. The PIM level models the entire structure of the application, as the PSM level provides transformations towards more specific platforms. The PIM-level models are: *use case, extended use case, service process* and *service composition*. The PSM-level models are: *web service interface, extended composition service* and *business logic*.

At the CIM level, the *value model* describes a business scenario as a set of values and activities shared by business actors. The *BPMN model* describes a business process and the corresponding environment. At the PIM level, the *use case model* represents a business service, as the *extended use case model* contains behavioral descriptions of features to be implemented. The *service process model* describes a set of activities to be performed in order to implement a business service. Finally, the *service composition model* represents the complete flow of a business system. This model is an extension of the service process model.

The SOD-M approach includes transformations between models: *CIM-to-PIM, PIM-to-PIM* and *PIM-to-PSM* transformations. Given an abstract model at the CIM level, it is possible to apply transformations for generating a model of the PSM level. In this context, it is necessary to follow the process activities described by the methodology. These three SOD-M levels have no support for describing non-functional requirements. The following section introduces πSOD-M, the extension that we propose for supporting these requirements.

3   πSOD-M

πSOD-M provides an environment for building service compositions by considering their non-functional requirements. πSOD-M proposes the generation of a set of models at different abstraction levels, as well as transformations between these models. πSOD-M includes non-functional specifications through four meta-models that extend the SOD-M meta-models at the PIM and PSM levels (see Figure 1): *π-UseCase, π-ServiceProcess, π-ServiceComposition* and *π-PEWS*.

The *π-UseCase* meta-model describes functional and non-functional requirements. Non-functional requirements are defined as *constraints* over processing steps and data. The *π-ServiceProcess* meta-model defines the concept of *service contract* to represent restrictions over data and actions that must be performed upon certain conditions. The *π-ServiceProcess* meta-model gathers the constraints described in the *π-UseCase* model into contracts that are associated with services. The *π-ServiceComposition* meta-model provides the concept

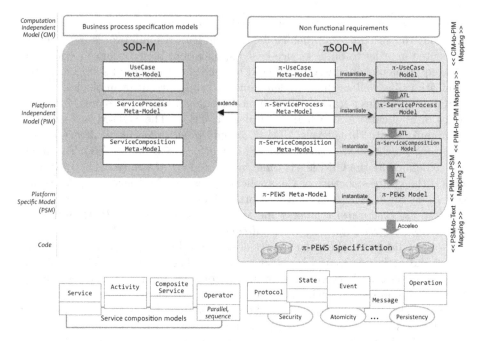

Fig. 1. πSOD-M

of *Policy* [11] which put together contracts with similar non-functional requirements. For instance, security and privacy restrictions may be grouped into a security policy. π-*ServiceComposition* models can be refined into PSMs. Policies are associated to service operations and combine *constraints* and *reactive recovery actions*. Constraints are restrictions that must be verified during the execution of the application. Failure to verify some constraint will trigger an exceptional behavior to execute the corresponding recovery action. An example of policy is the requirement of authentication for executing some of the system functions. The action associated to this policy may perform the authentication of the user. At the PSM level we have lower-level models that can be automatically translated into actual computer programs. The π-*PEWS* meta-model is the PSM adopted in this work (see Figure 1). π-*PEWS* models are textual descriptions of service compositions that can be translated into PEWS [3] code.

Thus, πSOD-M proposes a development process based on the definition of models (instances of the meta-models) and transformations between models. There are two kinds of transformations: model-to-model transformations, used to refine the specification during the software process; and model-to-text transformations, used to generate code in the last step of the software process. Notice that other composition languages, such as BPEL [1], can be used as target of the proposed software process. Another target language can be supported by defining the corresponding PIM-to-PSM and PSM-to-text transformations.

πSOD-M environment is built on the top of Eclipse. We also used the Eclipse Modelling Framework (EMF) to define, edit and handle (meta)-models. To automate the transformation models we use ATL [15] and Acceleo.

In the next section we develop an example, in order to illustrate our proposal. The example will show the actual notation used for models.

4   Proof of Concept: *Tracking Crimes*

Consider a tracking crime application where civilians and police share information about criminality in given zones of a city. Civilian users signal crimes using Twitter. Police officers can notify crimes, as well as update information about solving cases. Some of these information are confidential while other can be shared to the community of users using this application. Users can track crimes in given zones. Crime information stored by the system may be visualized on a map. Some users have different access rights than others. For example, police officers have more access rights than civilians.

In order to provide these functionalities, the application uses pre-existing services to provide, store and visualize information. The business process defines the logic of the application and is specified in terms of tasks. Tasks can be performed by people or computers.

The business process and requirements specifications presented in Figure 2 are instances of the Computation-Independent models of Figure 1. The business process is represented as a graph while requirements are given as text boxes.

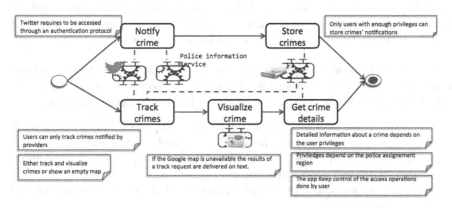

Fig. 2. Business process for the tracking crime example

In our example, crime processing can start with one of two tasks: *(i) notify a crime*, or *(ii) track a crime*. Notified crimes are stored in a database. Tracked crimes are visualized on a map. The used can ask for detailed information. The application is built upon four services: Twitter and an ad-hoc police service, for notifying crimes; Amazon, used as persistence service; and Google Maps, for locating and displaying crimes on a map.

Non-functional requirements are specified by rules and conditions to be verified during the execution of tasks. In our example we have the following non-functional requirements:

- Twitter requires authentication and allows three login failures before blocking.
- Crime notification needs privileged access.
- Civilian users can only track crimes for which they have clearance: civilian population cannot track all the crimes notified by the police.
- If Google Maps is unavailable, the results are delivered as text.
- Querying about crimes without having proper clearance yields an empty map.
- Access rights to detailed information depends on user clearance and zone assignment for police officers.
- The application maintains a detailed log.

The idea about these requirements is to leave the application logic expressed by functional requirements as independent as possible from exceptional situations (like the unavailability of a service) and from conditions under which services are called (for instance, through an authentication protocol). These requirements can be weaved as activities and branches of the composition or implemented apart. The second option is better because it makes the maintenance and evolution of applications easier. For instance, the services called by the application are not hard coded (Twitter and Google Maps in the example), neither the actions to deal with exceptions (replacing another map service or doing nothing).

All the system restrictions are modelled as constraints in this example. πSOD-M provides three types of constraints: *value*, *business* and *exceptions behaviour* constraints. Each use case (model) can be associated to one or more constraints[2].

π-UseCase model: our example has five use cases which represent the functions (tasks) and constraints of the system (Figure 3). We do not detail the functional part of the specification, due to lack of space. The constraints defined are:

- The Notify crime task requires that the user is logged in. This is an example of a *value constraint*, where the value associated to the condition depends on the semantics of the application. In this case, it represents the maximum number of allowed login attempts;
- The Store crimes task requires the verification of the user's clearance (also a value constraint).
- In order to perform the Track crimes task, the contact list of the requesting user must include the user that notified the crime. This is an example of *business constraint*. Additionally the requesting user must be logged in.
- For the View crimes' map task, the specification defines that if the service Google Maps is not available, the result is presented as text. This is an example of *exceptional behaviour constraint*. The availability of the Google Maps service is verified by a *business constraint*.

---

[2] For a more comprehensive account of πSOD-M the reader can refer to [23].

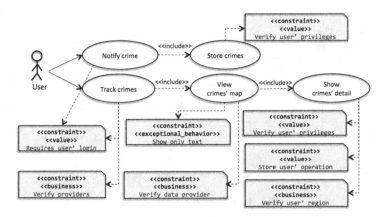

Fig. 3. π-UseCase Model

– The **Show crimes'** details task is specified to have three constraints: A *value constraint* is defined to verify the user's clearance level; a *business constraint* is used to ensure that the user's clearance is valid for the geographic zone of the crime; another *value constraint* defines that the log is to be maintained.

A model-to-model transformation is defined to refine the π-*UseCase* model into the corresponding π-*ServiceProcess* model, a more detailed representation. This (semi-automatic) transformation is supported by a tool (described in [23]).

π-*ServiceProcess model:* in this model task nodes of the π-UseCase model are better detailed, by refining the control and data flows, and its constraints are transformed into *contracts* (pre- and post-conditions). The π-ServiceProcess model describes the activities of the application and defines contracts for each activity or parts of the application (Figure 4). The main transformations are:

– Tasks of the previous model are transformed into *actions*;
– Actions are grouped into *activities* (in accordance to the business logic).
– Constraints of the π-*UseCase* model are transformed into assertions.

π-*ServiceComposition model:* this model refines the previous model by using the activities to produce the workflow of the application. The model serves to identify those entities that collaborate with the service process by providing services to execute actions. This model identifies the services and functions that correspond to each action in the business process.

In the case of our crime tracking example, the model produced from the π-*ServiceProcess* model of Figure 4 is given in Figures 5a and 5b. Figure 5a shows how the crime application interacts with its *business collaborators* (external services and entities). The interaction occurs by means of function calls (denoted by dotted lines in the figure). Figure 5b shows the definition of three *policies*, which define rules for service execution. In our case we have policies for *Security*, *Performance* and *Persistence*.

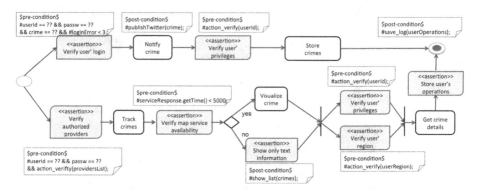

Fig. 4. π-*ServiceProcess* Model

π-*PEWS Model:* this model is produced by a model-to-text transformation that takes a π-*ServiceComposition* model and generates π-*PEWS* specification code. This code is a service composition program that can be compiled into an executable composition. π-*PEWS* models are expressed in a variant of the PEWS composition language. The π-PEWS program generated from the model in Figure 5 is partially presented in Figure 6. The figure shows a simplified program code, produced in accordance to the following guidelines:

- Namespaces, identifying the addresses of external services are produced from the Business Collaborators of the higher-level model. We define four of them, corresponding to the Police, Twitter, Google Map and Amazon partners.
- Specific operations exported by each business collaborator are identified as an *operation* of the program (to each operation is given an `alias`).
- The workflow in Figure 5a is translated into the text at line 11.
- *Contracts* are defined in π-PEWS as having pre-conditions (`requires`), post-conditions (`ensures`) and actions (`OnFailureDo`) to be executed whenever a condition is not verified. Contracts are generated from Policies (Figure 5b).

5   Related Work

Over the last years, a number of approaches have been proposed for the development of web services. These approaches range from the proposal of new languages for web service descriptions [1,21] to techniques to support phases of the development cycle of this kind of software [6]. In general, these approaches concentrate on specific problems, like supporting transactions or QoS, in order to improve the security and reliability of service-based applications. Some proposals address service composition: workflow definition [25,18] or semantic equivalence between services [3].

Works dealing with non-functional properties in service-oriented development can be organized in two main groups: those working on the *modelling of particular non-functional properties or QoS attributes* and those proposing *architectures or frameworks to manage and validate QoS attributes in web service*

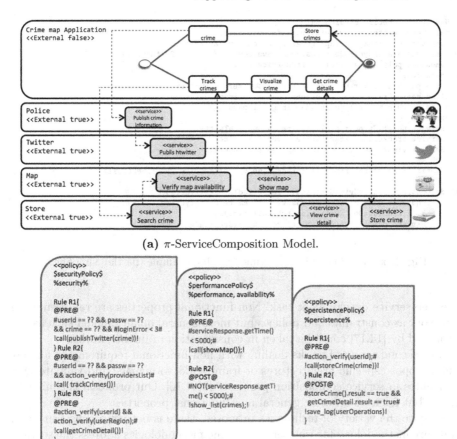

(a) π-ServiceComposition Model.

(b) π-ServiceComposition Policies.

Fig. 5. Service Composition and Policies

composition processes. The first group considers specific non-functional concerns (e.g., security) which is modelled and then associated to functional models of the application. The work of Chollet et al. [7] defines a proposal to associate non-functional quality properties (security properties) to functional activities in a web service composition model. Schmeling et al. [22] present an approach and also a toolset for specifying and implementing non-functional concerns in web service compositions. Non-functional concerns are modelled and then related to a service composition represented in a BPMN diagram. Ovaska et al. [19] present an approach to support quality management at design time. Quality requirements are modelled in a first phase and then represented in an architectural model where quality requirements are associated to some components of the model. We propose to model functional and non-functional properties at the same time during the software process. We claim what this approach simplifies

```
//Namespaces specify service URI
1 namespace twitter = www.twitter.com/service.wsdl
2 namespace googlemaps = maps.googleapis.com/maps/api/service
3 namespace amazondynamodb = rds.amazonaws.com/doc/2010-07-28/AmazonRDSv4.wsdl
4 namespace police = www.police.fr/service.wsdl
//Operations
5 alias publishTwitter = portType/publishTwitter in twitter
6 alias searchCrime = portType/searchCrime in amazondynamodb
7 alias showMap = portType/showMap in googlemaps
//Services
8  service notifyCrime = publishCrime  . publishTwitter
9  service trackCrime= searchCrime . verifyService
10 Service visualizeCrime =   showMap . getCrimeDetail
//Path
11 (notifyCrime.storeiCrime) || (trackCrime.visualizeCrime.getCrimeDetail)
//Contracts
12 defContract notifyCrimeContract{  isAppliedTo: notifyCrime
13    requires: userId == ?? && passw == ?? && req(notifyCrime)  < 3
14        (OnFailureDo: NOT(action_publish(crime));
15    ensures: publishTwitter(crime) == true  (OnFailureDo: skip); }
```

Fig. 6. π-PEWS code for the crime tracking example (partial, simplified)

the web service development task. Non-functional properties are represented in our work as constraints and policies but more general quality model such as the proposed by [14,17] could be taken in consideration in further works.

The second group of works dealing with non-functional requirements for services propose specific architectures or frameworks to manage and validate QoS attributes in service composition processes [26,4,16]. Our proposal is similar to these, but focusses on more general non-functional properties.

Despite the variety of techniques proposed, there is not yet a consensus on a software methodology for web services. Some methodologies address the service-based development towards a standard or a new way to develop reliable applications. SOD-M and SOMF [5] are MDD approaches for web services; S-Cube [20] is focused on the representation of business processes and service-based development; SOMA [2] is a methodology for SOA solutions; DEVISE [10] is a methodology for building service-based infrastructure for collaborative enterprises. Other proposals include the WIED model [24], that acts as a bridge between business modelling and design models on one hand, and traditional approaches for software engineering applied to SOC on the other hand.

6 Conclusions

This paper presented the πSOD-M software method for specifying and designing service-based applications in the presence of non-functional constraints. Our proposal enhances the SOD-M method with constraints, policies and contracts to consider non-functional constraints of applications. We implemented the proposed meta-models on the Eclipse platform and we illustrated the approach by developing a simple application.

πSOD-M is being used in an academic environment. So far, the preliminary results indicate that πSOD-M approach is useful for the development of complex web service applications. We are now working on the definition of a PSM-level meta-model to generate BPEL programs (instead of π-PEWS) in order to be able to generate code for the de-facto service coordination standard.

Acknowledgements. This research is partly supported by the National Institute of Science and Technology for Software Engineering (INES[3]), funded by CNPq (Brazil), grants 573964/2008-4 and 305619/2012-8; CAPES/UdelaR (Brazil/Uruguay) grant 021/ 2010; CAPES/STIC-AmSud (Brazil) grant 020/2010); MASAI project (TIN-2011-22617) financed by the Spanish Ministry of Science and Innovation and the Spanish Network on Service Science (TIN2011-15497-E) financed by the Spanish Ministry of Competitiveness and Economy.

References

1. Andrews, T., Curbera, F., Dholakia, H., Goland, Y., Klein, J., Leymann, F., Liu, K., Roller, D., Smith, D., Thatte, S., Trickovic, I., Weeranwarana, S.: Business Process Execution Language for Web Services (2003), http://www-128.ibm.com/developerworks/library/specification/ws-bpel/
2. Arsanjani, A.: SOMA: Service-Oriented Modeling and Architecture. Technical report, IBM (2004), http://www.ibm.com/developerworks/library/ws-soa-design1
3. Ba, C., Halfeld-Ferrari, M., Musicante, M.A.: Composing Web Services with PEWS: A Trace-Theoretical Approach. In: ECOWS 2006, pp. 65–74 (2006)
4. Babamir, S.M., Karimi, S., Shishechi, M.R.: A Broker-Based Architecture for Quality-Driven Web Services Composition. In: Proc. CiSE 2010 (2010)
5. Bell, M.: Service-Oriented Modeling (SOA): Service Analysis, Design, and Architecture. John Wiley (2008)
6. Börger, E., Cisternino, A. (eds.): Advances in Software Engineering. LNCS, vol. 5316. Springer, Heidelberg (2008)
7. Chollet, S., Lalanda, P.: An Extensible Abstract Service Orchestration Framework. In: Proc. ICWS 2009, pp. 831–838. IEEE (2009)
8. Christensen, E., Curbera, F., Meredith, G., Weerawarana, S.: Web Services Description Language (WSDL) 1.1. Technical report, World Wide Web Consortium (2001), http://www.w3.org/TR/wsdl
9. de Castro, V., Marcos, E., Wieringa, R.: Towards a service-oriented MDA-based approach to the alignment of business processes with IT systems: From the business model to a web service composition model. IJCIS 18(2) (2009)
10. Dhyanesh, N., Vineel, G.C., Raghavan, S.V.: DEVISE: A Methodology for Building Web Services Based Infrastructure for Collaborative Enterprises. In: Proc. WETIC 2003. IEEE Computer Society, USA (2003)
11. Espinosa-Oviedo, J.A., Vargas-Solar, G., Zechinelli-Martini, J.L., Collet, C.: Policy driven services coordination for building social networks based applications. In: Proc. of SCC 2011, Work-in-Progress Track. IEEE, USA (2011)
12. Favre, L.: A Rigorous Framework for Model Driven Development. In: Advanced Topics in Database Research, vol. 5, ch. I, USA, pp. 1–27 (2006)

---

[3] www.ines.org.br

13. Fielding, R.T.: REST: Architectural Styles and the Design of Network-based Software Architectures. Doctoral dissertation, University of California, Irvine (2000)
14. Goeb, A., Lochmann, K.: A software quality model for soa. In: Proc. WoSQ 2011, pp. 18–25. ACM (2011)
15. Group, A.: ATL: Atlas Transformation Language. Technical report, ATLAS Group, LINA & INRIA (February 2006)
16. Karunamurthy, R., Khendek, F., Glitho, R.H.: A novel architecture for Web service composition. J. of Network and Computer Applications 35(2), 787–802 (2012)
17. Klas, M., Heidrich, J., Munch, J., Trendowicz, A.: Cqml scheme: A classification scheme for comprehensive quality model landscapes. In: SEAA 2009, pp. 243–250 (2009)
18. Musicante, M.A., Potrich, E.: Expressing workflow patterns for web services: The case of pews. J.UCS 12(7), 903–921 (2006)
19. Ovaska, E., Evesti, A., Henttonen, K., Palviainen, M., Aho, P.: Knowledge based quality-driven architecture design and evaluation. Information & Software Technology 52(6), 577–601 (2010)
20. Papazoglou, M.P., Pohl, K., Parkin, M., Metzger, A. (eds.): Service Research Challenges and Solutions for the Future Internet. LNCS, vol. 6500. Springer, Heidelberg (2010)
21. Salaün, G., Bordeaux, L., Schaerf, M.: Describing and Reasoning on Web Services using Process Algebra. In: Proc. IEEE International Conference on Web Services, ICWS 2004. IEEE Computer Society, Washington, DC (2004)
22. Schmeling, B., Charfi, A., Mezini, M.: Composing Non-functional Concerns in Composite Web Services. In: Proc. ICWS 2011, pp. 331–338 (July 2011)
23. Souza Neto, P.A.: A methodology for building service-oriented applications in the presence of non-functional properties. PhD thesis, Federal University of Rio Grande do Norte (2012), http://www3.ifrn.edu.br/~placidoneto/thesisPlacidoASNeto.pdf
24. Tongrungrojana, R., Lowe, D.: WIED: A Web Modelling Language for Modelling Architectural-Level Information Flows. J. Digit. Inf. 5(2) (2004)
25. Van Der Aalst, W.M.P., Ter Hofstede, A.H.M., Kiepuszewski, B., Barros, A.P.: Workflow Patterns. Distrib. Parallel Databases 14(1), 5–51 (2003)
26. Xiao, H., Chan, B., Zou, Y., Benayon, J.W., O'Farrell, B., Litani, E., Hawkins, J.: A Framework for Verifying SLA Compliance in Composed Services. In: ICWS (2008)

Safety Contracts for Timed Reactive Components in SysML

Iulia Dragomir, Iulian Ober, and Christian Percebois

Université de Toulouse - IRIT
118 Route de Narbonne, 31062 Toulouse, France
{iulia.dragomir,iulian.ober,christian.percebois}@irit.fr

Abstract. A variety of system design and architecture description languages, such as SysML, UML or AADL, allows the decomposition of complex system designs into communicating timed components. In this paper we consider the contract-based specification of such components. A contract is a pair formed of an assumption, which is an abstraction of the component's environment, and a guarantee, which is an abstraction of the component's behavior given that the environment behaves according to the assumption. Thus, a contract concentrates on a specific aspect of the component's functionality and on a subset of its interface, which makes it relatively simpler to specify. Contracts may be used as an aid for hierarchical decomposition during design or for verification of properties of composites. This paper defines contracts for components formalized as a variant of timed input/output automata, introduces compositional results allowing to reason with contracts and shows how contracts can be used in a high-level modeling language (SysML) for specification and verification, based on an example extracted from a real-life system.

1 Motivation and Approach

The development of safety critical real-time embedded systems is a complex and costly process, and the early validation of design models is of paramount importance for satisfying qualification requirements, reducing overall costs and increasing quality. Design models are validated using a variety of techniques, including design reviews [24], simulation and model-checking [19, 25]. In all these activities system requirements play a central role; for this reason processes-oriented standards such as the DO-178C [22] emphasize the necessity to model requirements at various levels of abstraction and ensure their traceability from high-level down to detailed design and coding.

Since the vast majority of systems are designed with a component-based approach, the mapping of requirements is often difficult: a requirement is in general satisfied by the collaboration of a set of components and each component is involved in satisfying several requirements. A way to tackle this problem is to have partial and abstract component specifications which concentrate on specifying how a particular component collaborates in realizing a particular requirement; such a specification is called a *contract*. A contract is defined as a pair formed of an *assumption*, which is an abstraction of the component's environment, and a *guarantee*, which is an abstraction of the component's behavior given that the environment behaves according to the assumption.

The justification for using contracts is therefore manyfold: support for requirement specification and decomposition, mapping and tracing requirements

V. Geffert et al. (Eds.): SOFSEM 2014, LNCS 8327, pp. 211–222, 2014.

to components and even for model reviews. Last but not least, contracts can support formal verification of properties through model-checking. Given the right composability properties, they can be used to restructure the verification of a property by splitting it in two steps: (1) verify that each component satisfies its contract and (2) verify that the network of contracts correctly assembles and satisfies the property. Thus, one only needs to reason on abstractions when verifying a property, which potentially induces an important reduction of the combinatorial explosion problem.

Our interest in contracts is driven by potential applications in system engineering using SysML [23], in particular in the verification of complex industrial-scale designs for which we have reached the limits of our tools [16]. In SysML one can describe various types of communicating timed reactive components; for most of these, their semantics can be given in a variant of Timed Input/Output Automata (TIOA) [21]. For this reason, in this paper we concentrate on defining a contract theory for TIOA. This contract theory is applied on a SysML case study extracted from real-life.

2   A Meta-theory for Contract-Based Reasoning

Our contract theory is an instance of the *meta-theory* proposed in [27] and later detailed in [26], which formalizes the relations that come into play in a contract theory and the properties that these relations have to satisfy in order to support reasoning with contracts. The term *meta-theory* refers to the fact that the formalism used for component specification is not fixed, nor the exact nature of certain relations defined on specifications (conformance, refinement under context). In order to obtain a concrete contract theory for a particular specification formalism one has to define these relations such that certain properties, pre-required by the meta-theory, are satisfied. In return, this meta-theory provides a ready-to-use contract-based methodology.

The purpose of this methodology is to support reasoning with contracts in a system obtained by hierarchical composition of components. At any level of the hierarchy, n components $K_1, ..., K_n$ are combined to form a composite component $K_1 \parallel ... \parallel K_n$, where $\parallel$ denotes the parallel composition operator. Then verifying that the composite satisfies a global property φ runs down to checking that the contracts implemented by $K_1, ..., K_n$ combine together correctly to ensure φ. This avoids the need to directly model-check the composite to establish φ and, so, alleviates the combinatorial explosion of the state space. The contracts being specified by more abstract automata, one can assume that their composition will be reduced.

The reasoning proceeds as follows: for each component K_i, a contract C_i is given which consists of an abstraction A_i of the behavior of K_i's environment, and an abstraction G_i that describes the expected behavior of K_i given that the environment acts as A_i. Fig. 1 presents three components K_1, K_2 and K_3 and a corresponding set of contracts C_1, respectively C_2 and C_3. Step 1 of the reasoning consists in verifying that each component is a correct implementation of the contract, i.e. the component *satisfies* its contract. The *satisfaction* relation is directly derived from a more general one named *refinement under context*. The purpose of the latter is to model that a component K_i is a correct refinement of the specification K_j in the given environment E_k. Thus, a component implements

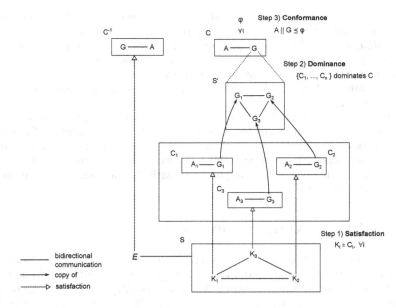

Fig. 1. Contract-based reasoning for a three-component subsystem ([26])

a contract if and only if *refinement under context* holds between the component K_i and the guarantee G_i in the environment A_i.

Step 2 of the reasoning consists in defining a contract $C = (A, G)$ for the composition $K_1 \parallel ... \parallel K_n$ and proving that the set of contracts $\{C_1, C_2, ..., C_n\}$ *implies* C. To do so, the meta-theory introduces a hierarchy relation between contracts, called *dominance*: a set of contracts $\{C_1, C_2, ..., C_n\}$ *dominates* a contract C if and only if the composition of any valid implementations of $C_1, C_2, ..., C_n$ is an implementation of C. In a multi-level hierarchy, the second step can be applied recursively up to a top-level contract, i.e. a contract for the whole (sub)system.

Finally, in the third step one has to prove that the top contract $C = (A, G)$ implies the specification φ. This is done by verifying that $A \parallel G \preceq \varphi$, where $\preceq$ is a *conformance* relation. This step is sufficient for proving that the whole system satisfies φ if and only if either the assumption A is empty as it is the case when the property is defined for the entire closed modeled system or A is a correct abstraction of the environment E with which S communicates given that S behaves like G, i.e. E *satisfies* the "mirror" contract $C^{-1} = (G, A)$.

The reasoning strategy presented here assumes that the system designer defines all the contracts necessary for verifying a particular requirement φ. How these contracts are produced is an interesting question but it is outside the scope of this paper.

The theoretical contribution of this paper is the instantiation of this meta-theory for a variant of Timed Input/Output Automata [21], by choosing the appropriate refinement relations and proving that they satisfy certain properties needed for the meta-theory to be applied. Concretely, the difficulties of the theoretical work consisted in defining the *conformance* and *refinement under context* relations such that one can reason in a contract framework only by handling assumptions and guarantees once satisfaction has been proven. The refinement relation has to satisfy the *compositionality* property and the soundness of *circular reasoning*, detailed in §4. *Compositionality* allows for *incremental design* by incorporating parts of the environment into the component under study while refinement under context holds at every step. *Circular reasoning* allows for *independent implementability* by breaking the dependency between the component and the environment.

The practical contribution of this paper is the application of the contract framework to a case study modeled in SysML, which can be found in §5. Due to space limitations we skip the syntactic details of how contracts are expressed in SysML and we concentrate on describing the example, the property of interest and the contracts involved in proving it. The relatively complex SysML language aspects are detailed in [17].

3   Timed Input/Output Automata

Many mathematical formalisms have been proposed in the literature for modeling communicating timed reactive components. We choose to build our framework on a variant of Timed Input/Output Automata of [21] since it is one of the most general formalisms, thoroughly defined and for which several interesting compositional results are already available.

A *timed input/output automaton* (or *component*) defines a set of internal variables of arbitrary types and clocks and a set of *input* actions I, *output* actions O, *visible* actions V and *internal* actions H. We denote by $E = I \cup O \cup V$ the set of external actions that also gives the *interface* of the component and by $A = E \cup H$ the set of actions. The state space of an automaton is described by the set of possible valuations of the set of variables. The state evolves either by discrete transitions or by *trajectories*. A discrete transition instantly changes the state and is labeled with an action from the mentioned sets. Trajectories change the state continuously during a time interval.

The behavior of a TIOA is described by an *execution fragment* which is a finite or infinite sequence alternating trajectories and discrete transitions. The *visible* behavior of a TIOA is given by a *trace*, which is a projection of an execution fragment on external actions and in which, from trajectories, only the information about the elapsed time is kept, while information about the variable valuations is abstracted away. For full definitions of all these notions, the reader is referred to the long version of this paper [18] and to [21].

The difference between the TIOA of [21] and our variant is that in addition to *inputs* and *outputs*, we allow for another type of *visible* actions; this is because, in [21], when composing two automata, an *output* of one matched by an *input* of the other becomes an *output* of the composite, which does not correspond to our needs when using TIOA for defining the semantics of usual modeling languages like SysML. Indeed, the matching between an input and an output results in a visible action in SysML that is not further involved in any synchronizations.

Moreover, in the following we will limit our attention to trajectories that are constant functions for discrete variables, and linear functions with derivative equal to 1 for clocks, while [21] allows more general functions to be used as trajectories. This restriction makes the model expressiveness equivalent to that of Alur-Dill timed automata [1], and will be important later on as it opens the possibility to automatically verify simulation relations between automata (simulation is undecidable for the TIOA of [21]). However, this hypothesis is not needed for proving the compositional results in §4.

The parallel composition operator for TIOA is based on binary synchronization of corresponding inputs and outputs and on the interleaved execution of other actions, like in [21]. The only difference is related to the interface of the composite timed input/output automata: the input and output sets of the composite are given by those actions not matched between components, while all matched input-output pairs become visible actions. Two automata can be composed if and only if they do not share variables and internal and visible actions. Two automata that can be composed are called *compatible components*. As in [21], we use timed trace inclusion as the refinement relation between components.

Definition 1 (Comparable components). *Two components K_1 and K_2 are comparable if they have the same external interface, i.e. $E_{K_1} = E_{K_2}$.*

Definition 2 (Conformance). *Let K_1 and K_2 be two comparable components. K_1 conforms to K_2, denoted $K_1 \preceq K_2$, if $traces_{K_1} \subseteq traces_{K_2}$.*

The conformance relation is used in the third step for verifying the satisfaction of the system's properties by the top contract: $A \parallel G \preceq \varphi$, where $A \parallel G$ and φ have the same interface. It can be easily shown that conformance is a preorder. The following useful compositional result (*theorem 8.5* of [21]) can be easily extended to our variant of TIOA:

Theorem 1. *Let K_1 and K_2 be two comparable components with $K_1 \preceq K_2$ and E a component compatible with both K_1 and K_2. Then $K_1 \parallel E \preceq K_2 \parallel E$.*

4   Contracts for Timed Input/Output Automata

In this section we formalize contracts for TIOA and the relations described in §2 and we list the properties that we proved upon these and that make contract-based reasoning possible.

Definition 3 (Environment). *Let K be a component. An environment E for the component K is a timed input/output automaton compatible with K for which the following hold: $I_E \subseteq O_K$ and $O_E \subseteq I_K$.*

Definition 4 (Closed/open component). *A component K is closed if $I_K = O_K = \emptyset$. A component is open if it not closed.*

Closed components result from the composition of components with complementary interfaces.

Definition 5 (Contract). *A contract for a component K is a pair (A, G) of TIOA such that $I_A = O_G$ and $I_G = O_A$ (i.e. the composition pair $A \parallel G$ defines a closed component) and $I_G \subseteq I_K$ and $O_G \subseteq O_K$ (i.e. the interface of G is a subset of that of K). A is called the* assumption *over the environment of the component and G is called the* guarantee. *The interface of a contract is that of its guarantee.*

The first step of the verification, as presented in §2, is to prove that the modeled components are an implementation of the given contracts. For this, we firstly define the *refinement under context* preorder relation which, in our framework, is further based on conformance. The complete formal definition of *refinement under context* can be found in [18].

Definition 6 (Refinement under context). *Let K_1 and K_2 be two components such that $I_{K_2} \subseteq I_{K_1} \cup V_{K_1}$, $O_{K_2} \subseteq O_{K_1} \cup V_{K_1}$ and $V_{K_2} \subseteq V_{K_1}$. Let E be an environment for K_1 compatible with both K_1 and K_2. We say that K_1 refines K_2 in the context of E, denoted $K_1 \sqsubseteq_E K_2$, if*

$$K_1 \parallel E \parallel E' \preceq K_2 \parallel E \parallel K' \parallel E'$$

where

- *E' is a TIOA defined such that the composition $K_1 \parallel E \parallel E'$ is closed. E' consumes all outputs of K_1 not matched by E and may emit all inputs of K_1 not appearing as outputs of E.*
- *K' is a TIOA defined similarly to E' such that the composition of $K_2 \parallel E \parallel K' \parallel E'$ is closed and comparable to $K_1 \parallel E \parallel E'$.*

Since we want to take into account interface refinement between components and conformance imposes comparability, we have to compose each member of the conformance relation obtained from refinement under context with an additional timed input/output automaton E', respectively K', such that they both define closed comparable systems. Both automata are uniquely defined by their interfaces and can be automatically computed.

Furthermore, the particular inclusion relations between the interfaces of K_1 and K_2 in the previous definition are due to the fact that both K_1 and K_2 can be components obtained from compositions, e.g., $K_1 = K_1' \parallel K_3$ and $K_2 = K_2' \parallel K_3$, where $I_{K_2'} \subseteq I_{K_1'}$, $O_{K_2'} \subseteq O_{K_1'}$ and $V_{K_2'} \subseteq V_{K_1'}$. This happens in particular when K_2' is a contract guarantee for K_1'. Then, by composition, actions of K_3 may be matched by actions of K_1' but have no input/output correspondent in K_2'.

Theorem 2. *Given a set $\mathcal{K}$ of comparable components and a fixed environment E for their interface, refinement under context $\sqsubseteq_E$ is a preorder over $\mathcal{K}$.*

We derive the *satisfaction* relation from refinement under context.

Definition 7 (Contract satisfaction). *A component K satisfies (implements) a contract $C = (A, G)$, denoted $K \models C$, if and only if $K \sqsubseteq_A G$.*

We have introduced the notions and relations in order to verify that a component is a correct implementation of a contract. The second step of the contract-based verification methodology relies on the notion of dominance, introduced informally in §2, which is formally defined in [26] as follows:

Definition 8 (Contract dominance). *Let C be a contract with the interface $\mathcal{P}$ and $\{C_i\}_{i=1}^n$ a set of contracts with the interface $\{\mathcal{P}_i\}_{i=1}^n$ and $\mathcal{P} \subseteq \bigcup_{i=1}^n \mathcal{P}_i$. Then $\{C_i\}_{i=1}^n$ dominates C if and only if for any set of components $\{K_i\}_{i=1}^n$ such that $\forall i$, $K_i \models C_i$, we have $(K_1 \parallel K_2 \parallel \cdots \parallel K_n) \models C$.*

In order to ease dominance verification by discarding components from now on and to be able to apply the meta-theory, we prove that the following compositional results hold in our framework.

Theorem 3 (Compositionality). *Let K_1 and K_2 be two components and E an environment compatible with both K_1 and K_2 such that $E = E_1 \parallel E_2$. Then $K_1 \sqsubseteq_{E_1 \parallel E_2} K_2 \Leftrightarrow K_1 \parallel E_1 \sqsubseteq_{E_2} K_2 \parallel E_1$.*

Theorem 4 (Soundness of circular reasoning). *Let K be a component, E its environment and $C = (A, G)$ the contract for K such that K and G are compatible with each of E and A. If (1) traces$_G$ is closed under limits, (2) traces$_G$ is closed under time-extension, (3) $K \sqsubseteq_A G$ and (4) $E \sqsubseteq_G A$ then $K \sqsubseteq_E G$.*

The definitions of closure under limits and closure under time-extension for a set of traces are those given in [21]. Closure under limits informally means that any infinite sequence whose finite prefixes are traces of G is also a trace of G, while closure under time-extension denotes that any finite trace can be extended with time passage to infinity. By making these hypotheses on G, G can only express safety properties on K and cannot impose stronger constraints on time passage than K. The proofs of all theorems presented here can be found in [18].

Then based on theorems 2, 3 and 4, the following theorem also proved in [18] which is a variant of *theorem 2.3.5* from [26] holds:

Theorem 5 (Sufficient condition for dominance). $\{C_i\}_{i=1}^n$ *dominates* C *if, $\forall i$, traces$_{A_i}$, traces$_{G_i}$, traces$_A$ and trace$_G$ are closed under limits and under time-extension and*

$$\begin{cases} G_1 \parallel ... \parallel G_n \sqsubseteq_A G \\ A \parallel G_1 \parallel ... \parallel G_{i-1} \parallel G_{i+1} \parallel ... \parallel G_n \sqsubseteq_{G_i} A_i, \ \forall i \end{cases}$$

The above theorem specifies the proof obligations that have to be satisfied by a system of contracts in order to be able to infer dominance in the second step of the verification methodology presented in §2.

5 Application to a SysML Model: the ATV Solar Wing Generation System Case Study

The contract-based reasoning method previously described is partially supported by the OMEGA-IFx Toolset [7] for SysML models. The details of the SysML language extended with contracts are left aside for space reasons and can be found in [17]. In the following we present a case study extracted from the industrial-grade system model of the Automated Transfer Vehicle (ATV) and we show how contracts can be used for property verification. This case study consists of the Solar Wing Generation System (SGS) [16] responsible for the deployment and management of the solar wings of the vehicle. The SysML model used in the following, provided by Astrium Space Transportation, was obtained by reverse engineering the actual SGS system for the purpose of this study.

The ATV system model illustrated in Fig. 2 summarizes the three main components involved in the case study and the bidirectional communications between them: the mission and vehicle management (*MVM*) part that initiates the two functionalities of the SGS (wing deployment and rotation), the *SOFTWARE* part of the *SGS* that based on commands received from the *MVM* executes the corresponding procedures and the *HARDWARE* part that consists of the four wings. We focus here on the wing deployment mode on which we want to verify the following property φ: after 10 minutes from system start-up, all four wings are deployed.

The system explicitly models the redundancy of the hardware equipments which aims to ensure fault tolerance. There are 56 possible failures (14 per wing)

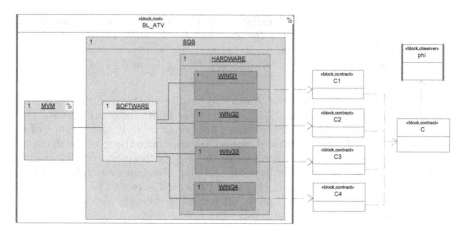

Fig. 2. Architecture of the SGS system including contracts (simplified view)

grouped in 3 classes depending on their target (thermal knives, hold-down and release system and solar array driving group). The following hypothesis is made: throughout any execution of the system, at most one hardware fault may occur (1-fault tolerance). We are interested in verifying φ by taking into consideration this hypothesis. But applying model-checking directly on the system leads to combinatorial explosion and the verification does not finish. To give an idea about the complexity of the model at runtime, the system contains 96 objects that communicate through 661 port instances and 504 connectors. In [16] we have shown the motivation for using contracts and we have sketched a proof that remained to be formalized and verified. In the following we complete this case study by proving the property φ with formalized contract-based reasoning.

Since the property φ is expressed with respect to the behavior of the four wings that are contained in the $HARDWARE$ block, with regard to the methodology of Fig. 1, the subsystem S can be identified in our case study with $HARDWARE$ and the components K_i are represented by $WINGi$, $i = \overline{1,4}$. The environment of the subsystem is given by the parts with which it communicates: bidirectional communication is directly established between $SOFTWARE$ and $HARDWARE$, while $SOFTWARE$ depends on the behavior of MVM. So, the environment E of Fig. 1 is represented here by the composition of MVM and $SOFTWARE$.

The first step of the methodology consists in defining a contract $C_i = (A_i, G_i)$ for each $WINGi$, and next proving that $WINGi$ satisfies C_i, $i = \overline{1,4}$. This step checks the validity of the dependency relations between the wings and their corresponding contracts. In order to model a contract, first we need to identify the environment of the component to which the contract is associated and to build an abstraction from the point of view of the component. Thus, for $WINGi$ the environment is given by the environment of the subsystem $HARDWARE$ and all $WINGj$ with $j \neq i$. We propose the following abstraction WAj for $WINGj$: the wing has a not deployed status for at most 400 seconds and a deployed status after 130 seconds, while all other received requests are consumed. The assumption A_i is then given by the parallel composition of MVM, $SOFTWARE$ and WAj with $j \neq i$. This abstraction of the environment is sufficient to drastically reduce the state space of the verification model, since the exponential explosion in the original model is mainly due to the parallelism of the hardware pieces which are abstracted to the three leaf parts WAj. We want to guarantee that even if $WINGi$ exhibits a failure it ends up being deployed after 400 seconds.

Contract $C_i = (A_i, G_i)$ where

- $A_i = MVM \parallel SOFTWARE \parallel (\parallel_{j \neq i} WAj)$.
- G_i: the wing answers to requests about its status with *not deployed* from startup up to 400 seconds or with *deployed* after 130 seconds and ignores all other requests. Between 130 and 400 seconds it can answer either, non-deterministically.

Since A_i is partially given by the concrete environment ($MVM \parallel SOFTWARE$) and C_i has to define a closed system, we have to manually model the behavior of G_i for all received requests. This constraint imposes to add as consuming transitions in every state all requests corresponding to wing deployment process. Furthermore, one can remark that this guarantee is stronger than the projection of the property φ on $WINGi$. The abstraction WAj can also be subject to one failure since this case was not excluded from its behavior. Then the fault tolerance property that we verify via contracts is stronger than the initial hypothesis: we guarantee that the system is 4-fault tolerant if faults occur in separate wings.

The second step consists in defining a global contract $C = (A, G)$ for *HARDWARE* and to prove that the contract is dominated by $\{C_1, C_2, C_3, C_4\}$ represented in Fig. 2 by the dependency relation between contracts. We use as assumption A the concrete environment of *HARDWARE*. The guarantee G is the composition of the four G_i. All A_i, G_i, A and G as defined satisfy the conditions for applying theorem 5.

Contract $C = (A, G)$ where

- $A = MVM \parallel SOFTWARE$
- G : for each wing status interrogation it answers as *not deployed* for at most 400 seconds and as *deployed* after at least 130 seconds, while all other requests are ignored.

The last step consists in verifying that the composition $A \parallel G$ conforms to φ, illustrated by the dependency between the contract and the property. Verifying that the environment satisfies the "mirror" contract is trivial since the assumption A is the environment itself.

The proofs of all three steps have been automatically verified within the OMEGA-IFx Toolset which translates SysML models in communicating timed automata [7]. Since trace inclusion is undecidable, we use a stronger *simulation* relation whose satisfaction implies trace inclusion. So, verifying simulation is sufficient (albeit not necessary) in order to prove the satisfaction of the conformance relation. A variant of this verification algorithm is implemented in the IFx Toolset.

For each step of the verification methodology we have manually modeled the contracts: assumptions as blocks that we had to connect via ports with the other components and guarantees as independent components. The first step gave 4 possible configurations with one concrete wing and 3 abstract ones that were each verified with respect to all 14 possible failures. The average time in seconds needed for the verification of the satisfaction relation for each contract with respect to each class of failures is presented in Table 1. Even though the system model looks symmetrical, some hardware pieces not represented here do not have a symmetrical behavior and due to their interconnections with the wings

Table 1. Average verification time for each contract C_i per induced failure group

Type of induced failure	Average verification time (s)			
	Wing 1	Wing 2	Wing 3	Wing 4
Thermal knife	13993	6869	18842	11412
Hold-down and release system	12672	6516	16578	9980
Solar array driving group	11527	5432	13548	6807

the state space of system's abstraction for $WING1$ and $WING3$ is larger than the one of $WING2$ and $WING4$. For the second step, only one model is created on which we verified all 5 proof obligations given by theorem 5: the automatic validation of the global guarantee G and the automatic validation of assumptions A_i. Modeling the assumptions A_i that play the role of guarantees for dominance verification shows the symmetry of the MVM and $SOFTWARE$ behavior. This means that only one verification is in fact sufficient for proving all 4 relations, verification that was realized in 9 seconds. The verification of the guarantee G needed 1 second. Finally, the same model was used for verifying φ that took 1 second.

6   Related Work

Contract-based specification and reasoning for communicating components has been subject to intensive research recently. As mentioned in the beginning, our contract theory for TIOA is an instance of the meta-theory of [26], which has previously been applied for a number of other components formalisms: Labeled Transition Systems (with priorities) [26], Modal Transition Systems [27], BIP framework [2, 26] and Heterogeneous Rich Components [5]. To the best of our knowledge, this is the first documented application of this meta-theory to Timed Input/Output Automata.

Contract meta-theories have also been built on specification theories. The aim of a specification theory is to provide a complete algebra with powerful operators for logical compositions of specifications and for synthesis of missing components (quotient), the main goal being to provide substitutability results allowing for compositional verification. The meta-theory of [3] falls under this category. The main differences with respect to [26] concern (1) the definition of contracts that do not support signature refinement (a partial solution for this problem has been proposed in [4]) and (2) the method for reasoning with contracts which relies on a *contract composition* operator that is partially defined. However, the meta-theory of [4] does not alleviate this problem. None of these two meta-theories supports circular arguments nor defines a methodology for reasoning with contracts. Moreover, the meta-theories do not provide means to formalize requirements and to verify their satisfaction. Other approaches for reasoning with contracts have been developed for interface theories [11, 12] and for logical specifications [10]. A complete comparison between (meta-)theories is available in [18].

Only the meta-theory of [3] has been instantiated for a variant of TIOA in [14, 15, 12] which is implemented in the ECDAR tool [13]. However, several aspects of this specification theory make it unsuitable for representing the semantics of timed components described in SysML or UML. The synchronization between an input of one component and an output of another component becomes an output of the composite, which equates to considering outputs as broadcasts and which is not consistent with the UML/SysML semantics. Moreover, the formalism forbids non-determinism due to the timed game semantics [6] and does not handle *silent* transitions, which is problematic for representing the semantics of complex components performing internal computation steps.

In addition, contracts in UML/SysML have until now been explored for the specification of composition compatibility of components via interfaces [28] and for the verification of pre/post conditions of operations as presented by [20]. Recent work covers the use of pre/post condition contracts for modeling transformation of models [9] and for modeling the execution semantics of UML elements [8]. To the best of our knowledge, our work is the first on using assume/guarantee behavioral contracts for the verification of UML/SysML model requirements.

7 Conclusions

We have presented a contract framework for Timed Input/Output Automata and results which allow contract-based reasoning for verifying timed safety properties of systems of TIOA components. We have illustrated the method on a case study extracted from an industrial-scale system model and we have showed how contract-based reasoning can alleviate the problem of combinatorial explosion for the verification of large systems.

The present work is a step further towards introducing contracts in SysML and providing a full solution to that problem. In [17] we defined a suitable syntax for contracts in SysML and a set of well-formedness rules that system models must satisfy for reasoning with contracts. For the moment, some steps of the method applied on SysML remain manual like modeling individual systems for each contract satisfaction relation or for each dominance proof obligation. Future work includes: (1) formalizing the semantic mapping between SysML components and contracts and their TIOA counterparts and (2) providing means for automatic verification by automated generation of proof obligations.

References

[1] Alur, R., Dill, D.L.: A Theory of Timed Automata. Theor. Comput. Sci. 126(2), 183–235 (1994)
[2] Basu, A., Bozga, M., Sifakis, J.: Modeling Heterogeneous Real-time Components in BIP. In: SEFM 2006, pp. 3–12 (2006)
[3] Bauer, S.S., David, A., Hennicker, R., Guldstrand Larsen, K., Legay, A., Nyman, U., Wąsowski, A.: Moving from Specifications to Contracts in Component-Based Design. In: de Lara, J., Zisman, A. (eds.) FASE 2012. LNCS, vol. 7212, pp. 43–58. Springer, Heidelberg (2012)
[4] Bauer, S., Hennicker, R., Legay, A.: Component Interfaces with Contracts on Ports. In: Păsăreanu, C.S., Salaün, G. (eds.) FACS 2012. LNCS, vol. 7684, pp. 19–35. Springer, Heidelberg (2013)
[5] Benvenuti, L., Ferrari, A., Mangeruca, L., Mazzi, E., Passerone, R., Sofronis, C.: A contract-based formalism for the specification of heterogeneous systems. In: FDL 2008. Forum on, pp. 142–147. IEEE (2008)
[6] Bourke, T., David, A., Larsen, K.G., Legay, A., Lime, D., Nyman, U., Wąsowski, A.: New Results on Timed Specifications. In: Mossakowski, T., Kreowski, H.-J. (eds.) WADT 2010. LNCS, vol. 7137, pp. 175–192. Springer, Heidelberg (2012)
[7] Bozga, M., Graf, S., Ober, I., Ober, I., Sifakis, J.: The IF toolset. In: Bernardo, M., Corradini, F. (eds.) SFM-RT 2004. LNCS, vol. 3185, pp. 237–267. Springer, Heidelberg (2004)

[8] Cariou, E., Ballagny, C., Feugas, A., Barbier, F.: Contracts for model execution verification. In: France, R.B., Kuester, J.M., Bordbar, B., Paige, R.F. (eds.) ECMFA 2011. LNCS, vol. 6698, pp. 3–18. Springer, Heidelberg (2011)

[9] Cariou, E., Belloir, N., Barbier, F., Djemam, N.: OCL contracts for the verification of model transformations. ECEASST 24 (2009)

[10] Chilton, C., Jonsson, B., Kwiatkowska, M.: Assume-Guarantee Reasoning for Safe Component Behaviours. In: Păsăreanu, C.S., Salaün, G. (eds.) FACS 2012. LNCS, vol. 7684, pp. 92–109. Springer, Heidelberg (2013)

[11] Chilton, C., Kwiatkowska, M., Wang, X.: Revisiting Timed Specification Theories: A Linear-Time Perspective. In: Jurdziński, M., Ničković, D. (eds.) FORMATS 2012. LNCS, vol. 7595, pp. 75–90. Springer, Heidelberg (2012)

[12] David, A., Guldstrand Larsen, K.G., Legay, A., Møller, M.H., Nyman, U., Ravn, A.P., Skou, A., Wasowski, A.: Compositional verification of real-time systems using ECDAR. STTT 14(6), 703–720 (2012)

[13] David, A., Larsen, K.G., Legay, A., Nyman, U., Wąsowski, A.: ECDAR: An Environment for Compositional Design and Analysis of Real Time Systems. In: Bouajjani, A., Chin, W.-N. (eds.) ATVA 2010. LNCS, vol. 6252, pp. 365–370. Springer, Heidelberg (2010)

[14] David, A., Larsen, K.G., Legay, A., Nyman, U., Wąsowski, A.: Methodologies for Specification of Real-Time Systems Using Timed I/O Automata. In: de Boer, F.S., Bonsangue, M.M., Hallerstede, S., Leuschel, M. (eds.) FMCO 2009. LNCS, vol. 6286, pp. 290–310. Springer, Heidelberg (2010)

[15] David, A., Larsen, K.G., Legay, A., Nyman, U., Wasowski, A.: Timed I/O automata: a complete specification theory for real-time systems. In: HSCC 2010, pp. 91–100. ACM (2010)

[16] Dragomir, I., Ober, I., Lesens, D.: A case study in formal system engineering with SysML. In: ICECCS 2012, pp. 189–198. IEEE Computer Society (2012)

[17] Dragomir, I., Ober, I., Percebois, C.: Integrating verifiable Assume/Guarantee contracts in UML/SysML. In: ACES-MB 2013. CEUR Workshop Proceedings (2013)

[18] Dragomir, I., Ober, I., Percebois, C.: Safety Contracts for Timed Reactive Components in SysML. Technical report, IRIT (2013), http://www.irit.fr/~Iulian.Ober/docs/TR-Contracts.pdf

[19] Emerson, E.A., Clarke, E.M.: Characterizing Correctness Properties of Parallel Programs Using Fixpoints. In: de Bakker, J.W., van Leeuwen, J. (eds.) ICALP 1980. LNCS, vol. 85, pp. 169–181. Springer, Heidelberg (1980)

[20] Hoare, C.A.R.: An Axiomatic Basis for Computer Programming. Commun. ACM 12(10), 576–580 (1969)

[21] Kaynar, D.K., Lynch, N., Segala, R., Vaandrager, F.: The Theory of Timed I/O Automata, 2nd edn. Morgan & Claypool Publishers (2010)

[22] RTCA Inc. Software Considerations in Airborne Systems and Equipment Certification. Document RTCA/DO-178C (2011)

[23] OMG. Object Management Group – Systems Modeling Language (SysML), v1.1 (2008), http://www.omg.org/spec/SysML/1.1

[24] Parnas, D., Weiss, D.: Active Design Reviews: Principles and Practices. In: ICSE 1985. IEEE Computer Society (1985)

[25] Queille, J.-P., Sifakis, J.: Specification and verification of concurrent systems in CESAR. In: Dezani-Ciancaglini, M., Montanari, U. (eds.) Programming 1982. LNCS, vol. 137, pp. 337–351. Springer, Heidelberg (1982)

[26] Quinton, S.: Design, vérification et implémentation de systèmes à composants. PhD thesis, Université de Grenoble (2011)

[27] Quinton, S., Graf, S.: Contract-Based Verification of Hierarchical Systems of Components. In: SEFM 2008, pp. 377–381 (2008)

[28] Weis, T., Becker, C., Geihs, K., Plouzeau, N.: A UML Meta-model for Contract Aware Components. In: Gogolla, M., Kobryn, C. (eds.) UML 2001. LNCS, vol. 2185, pp. 442–456. Springer, Heidelberg (2001)

Graph Clustering with Surprise: Complexity and Exact Solutions*

Tobias Fleck, Andrea Kappes, and Dorothea Wagner

Institute of Theoretical Informatics, Karlsruhe Institute of Technology, Germany

Abstract. Clustering graphs based on a comparison of the number of links within clusters and the expected value of this quantity in a random graph has gained a lot of attention and popularity in the last decade. Recently, Aldecoa and Marín proposed a related, but slightly different approach leading to the quality measure *surprise*, and reported good behavior in the context of synthetic and real world benchmarks. We show that the problem of finding a clustering with optimum surprise is $\mathcal{NP}$-hard. Moreover, a bicriterial view on the problem permits to compute optimum solutions for small instances by solving a small number of integer linear programs, and leads to a polynomial time algorithm on trees.

1 Introduction

Graph clustering, i.e., the partitioning of the entities of a network into densely connected groups, has received growing attention in the literature of the last decade, with applications ranging from the analysis of social networks to recommendation systems and bioinformatics [8]. Mathematical formulations thereof abound; for an extensive overview on different approaches see for example the reviews of Fortunato [8] and Schaeffer [16].

One line of research that recently gained a lot of popularity is based on *null models*, the most prominent objective function in this context being the *modularity* of a clustering [13]. Roughly speaking, the idea behind this approach is to compare the number of edges within the same cluster to its expected value in a random graph that inherits some properties of the graph given as input.

In a wider sense, the measure called *surprise* that has recently been suggested as an alternative to modularity is also based on a null model, although, compared to modularity and its modifications [8], it uses a different tradeoff between the observed and expected number of edges within clusters. Surprise is used as a quality function in the tools UVCLUSTER and Jerarca to analyze protein interaction data [5,1]. The authors' main arguments for using surprise instead of modularity is that it exhibits better behavior with respect to synthetic benchmarks and, empirically, it does not suffer to the same extent from the *resolution limit* of modularity [9], i.e. the tendency to merge small natural communities into larger ones [2,3,4]. However, these results are hard to assess, since a metaheuristic is used instead of directly optimizing the measure. It chooses among a

---

* This work was partially supported by the DFG under grant WA 654/19-1.

V. Geffert et al. (Eds.): SOFSEM 2014, LNCS 8327, pp. 223–234, 2014.

set of clusterings produced by general clustering algorithms the one that is best with respect to surprise.

In this work, we take first steps towards a theoretical analysis of surprise. We show that the problem of finding a clustering with optimal surprise is $\mathcal{NP}$-hard in general and polynomially solvable on trees. Moreover, we formulate surprise as a bicriterial problem, which allows to find provably optimal solutions for small instances by solving a small number of integer linear programs.

Notation. All graphs considered are unweighted, undirected and simple, i.e. they do not contain loops or parallel edges. A clustering ζ of a graph $G = (V, E)$ is a partitioning of V. Let $n := |V|$ and $m := |E|$ denote the number of vertices and edges of G, respectively. If C is a cluster in ζ, $i_e(C)$ denotes the number of *intracluster edges* in C, i.e., the number of edges having both endpoints in C. Similarly, $i_p(C) := \binom{|C|}{2}$ is the number of vertex pairs in C. Furthermore, let $p := \binom{n}{2}$ be the number of vertex pairs in G, $i_p(\zeta) := \sum_{C \in \zeta} i_p(C)$ be the total number of intracluster vertex pairs and $i_e(\zeta) := \sum_{C \in \zeta} i_e(C)$ the total number of intracluster edges. If the clustering is clear from the context, we will sometimes omit ζ and just write i_p and i_e. To ease notation, we will allow binomial coefficients $\binom{n}{k}$ for all n and $k \in \mathbb{N}$. If $k > n$, $\binom{n}{k} = 0$ by definition.

2   Definition and Basic Properties

Let ζ be a clustering of a graph $G = (V, E)$ with i_e intracluster edges. Among all graphs labeled with vertex set V and exactly m edges, we draw a graph $\mathcal{G}$ uniformly at random. The surprise $S(\zeta)$ of this clustering is then the probability that $\mathcal{G}$ has at least i_e intracluster edges with respect to ζ. The lower this probability, the more *surprising* it is to observe that many intracluster edges within G, and hence, the better the clustering. The above process corresponds to an urn model with $i_p(\zeta)$ white and $p - i_p(\zeta)$ black balls from which we draw m balls without replacement. The probability to draw at least i_e white balls then follows a hypergeometric distribution, which leads to the following definition[1]; the lower $S(\zeta)$, the better the clustering:

$$S(\zeta) := \sum_{i=i_e}^{m} \frac{\binom{i_p}{i} \cdot \binom{p-i_p}{m-i}}{\binom{p}{m}}$$

Basic Properties. For a fixed graph, the value of S only depends on two variables, i_p and i_e. To ease notation, we will use the term $S(i_p, i_e)$ for the value of a clustering with i_p intracluster pairs and i_e intracluster edges. The urn model view yields some simple properties that lead to a better understanding of how surprise behaves, and that are heavily used in the $\mathcal{NP}$-hardness proof.

Lemma 1. *Let i_e, i_p, p and m be given by a clustering, i.e. $0 \leq i_e \leq i_p \leq p$, $i_e \leq m$ and $m - i_e \leq p - i_p$. Then, the following statements hold:*

---

[1] This is the definition used in the original version [5]; later on, it was replaced by maximizing $-\log_{10} S(\zeta)$, which is equivalent with respect to optimum solutions.

(i) $S(i_p, i_e + 1) < S(i_p, i_e)$.
(ii) If $i_e > 0$, then $S(i_p - 1, i_e) < S(i_p, i_e)$
(iii) If $p - i_p > m - i_e$, then $S(i_p + 1, i_e + 1) < S(i_p, i_e)$.

Proof. Statement (i) is obvious. Similarly, statement (ii) is not hard to see if we recall that $S(i_p - 1, i_e)$ corresponds to the probability to draw at least i_e white balls after replacing one white ball with a black one.

For statement (iii), we show that the number k_1 of m-element subsets of the set of all balls containing at least i_e white balls is larger than the number k_2 of m-element subsets containing at least $i_e + 1$ white balls after painting one black ball b white. Any subset A that contributes to k_2 also contributes to k_1, as at most one ball in A got painted white. On the other hand, every m-element subset not containing b that contains exactly i_e white balls contributes to k_1, but not to k_2. As there are at least i_e white balls, and $p - i_p > m - i_e$ implies that there are at least $m - i_e + 1$ black balls, there is at least one subset with these properties. Hence $k_1 > k_2$, which is equivalent to $S(i_p + 1, i_e + 1) < S(i_p, i_e)$. $\square$

In other words, the value of surprise improves the more edges and the less vertex pairs within clusters exist. Moreover, part (iii) shows that if we increase the number of intracluster edges such that the number of *intracluster non-edges*, i.e., vertex pairs within clusters that are not linked by an edge, does not increase, this leads to a clustering with strictly smaller surprise. This immediately yields some basic properties of optimal clusterings with respect to surprise. Part (i) of the following proposition is interesting as it shows that optimal clusterings always fulfill the assumptions of Lemma 1(ii)-(iii).

Proposition 2. *Let $G = (V, E)$ be a graph that has at least one edge and that is not a clique and ζ be an optimal clustering of G with respect to surprise. Then,*

(i) $i_e(\zeta) > 0$ and $p - i_p(\zeta) > m - i_e(\zeta)$
(ii) $1 < |\zeta| < |V|$
(iii) ζ contains at least as many intracluster edges as any clustering ζ' of G into cliques.
(iv) Any cluster in ζ induces a connected subgraph.

Proof. (i): If $i_e(\zeta) = 0$ or $p - i_p(\zeta) = m - i_e(\zeta)$, it can be easily seen that $S(\zeta) = 1$. On the other hand, let us consider a clustering ζ' where each cluster contains one vertex, except for one cluster that contains two vertices linked by an edge e. As $m < p$, there is at least one labeled graph on V with m edges that does not contain e.

(ii): If $|\zeta| = 1$, $p - i_p(\zeta) = 0 = m - i_e(\zeta)$ and if $|\zeta| = |V|$, $i_e(\zeta) = 0$. The statement now follows from (i).

(iii): Let us assume that $i_e(\zeta) < i_e(\zeta')$. Lemma 1(ii) can be used to show that $S(\zeta) = S(i_p(\zeta), i_e(\zeta)) \geq S(i_e(\zeta), i_e(\zeta))$ and from Lemma 1(iii), it follows that $S(i_e(\zeta), i_e(\zeta)) > S(i_e(\zeta'), i_e(\zeta')) = S(\zeta')$.

(iv): Follows from Lemma 1(ii) and the fact that splitting a disconnected cluster into its connected components decreases the number of intracluster pairs and does not affect the number of intracluster edges. $\square$

Bicriterial View. From Lemma 1, it follows that an optimal solution with respect to surprise is *pareto optimal* with respect to (maximizing) i_e and (minimizing) i_p. Interestingly, this also holds for a simplification of modularity whose null model does not take vertex degrees into account and that was briefly considered by Reichardt and Bornholdt [15,14], although the tradeoff between the two objectives is different. Hence, an optimal clustering can be found by solving the following optimization problem for all $0 \leq k \leq m$ and choosing the solution that optimizes surprise.

Problem 3 (minIP). Given a graph G and an integer $k > 0$, find a clustering ζ with $i_e(\zeta) = k$, if there exists one, such that $i_p(\zeta)$ is minimal.

Unfortunately, the decision variant of minIP is $\mathcal{NP}$-complete even on bipartite graphs, as it is equivalent to the unweighted Minimum Average Contamination problem [12]. However, the formulation of minIP does not involve binomial coefficients and is thus in some aspects easier to handle. For example, in contrast to surprise, it can be easily cast into an integer linear program. We will use this in Sect. 4 to compute optimal solutions for small instances.

One might guess from the $\mathcal{NP}$-completeness of minIP that surprise minimization is also $\mathcal{NP}$-complete. However, there is no immediate reduction from minIP to the decision variant of surprise optimization, as the number of intra-cluster edges in an optimal clustering with respect to surprise is not fixed. In the following section, we will therefore give a proof for the hardness of finding a clustering with optimal surprise.

3   Complexity

We show $\mathcal{NP}$-completeness of the corresponding decision problem:

Problem 4 (Surprise Decision (SD)). Given a graph G and a parameter $k > 0$, decide whether there exists a clustering ζ of G with $S(\zeta) \leq k$.

As S can be clearly evaluated in polynomial time, SD is in $\mathcal{NP}$. To show $\mathcal{NP}$-completeness, we use a reduction from Exact Cover by 3-Sets [10]:

Problem 5 (Exact Cover by 3Sets (X3S)). Given a set $\mathcal{X}$ of elements and a collection $\mathcal{M}$ of 3-element subsets of $\mathcal{X}$, decide whether there is a subcollection $\mathcal{R}$ of $\mathcal{M}$ such that each element in $\mathcal{X}$ is contained in exactly one member of $\mathcal{R}$.

Let $I = (\mathcal{X}, \mathcal{M})$ be an instance of X3S. The reduction is based on the idea of implanting large disjoint cliques in the transformed instance that correspond to the subsets in $\mathcal{M}$. The size of these cliques is polynomial in $|\mathcal{M}|$, but large enough to ensure that they can neither be split nor merged in a clustering with low surprise. Hence, each of these cliques induces a cluster. The transformed instance further contains a vertex for each element in $\mathcal{X}$

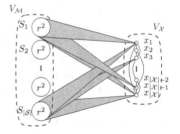

Fig. 1. Illustration for reduction

that is linked with the cliques corresponding to the subsets it is contained in. The idea is to show that in a clustering ζ with low surprise, each of these vertices is contained in a cluster induced by exactly one subset, and each cluster contains either three "element vertices" or none, which induces an exact cover of $\mathcal{X}$.

In the following, we will assume without loss of generality[2] that each element of $\mathcal{X}$ belongs to at least one set in $\mathcal{M}$, hence $|\mathcal{X}| \leq 3|\mathcal{M}|$. We construct an instance $I' = (G, k)$ of SD in the following way. Let $r := 3|\mathcal{M}|$. First, we map each set M in $\mathcal{M}$ to an r^2-clique $C(M)$ in G. Furthermore, we introduce an $|\mathcal{X}|$-clique to G, where each of the vertices $v(x)$ in it is associated with an element x in $\mathcal{X}$. We link $v(x)$ with each vertex in $C(M)$, if and only if x is contained in M. Let $V_{\mathcal{X}}$ be the set containing all vertices corresponding to elements in $\mathcal{X}$, and $V_{\mathcal{M}}$ the set of vertices corresponding to subsets. Fig. 1 illustrates the reduction, clearly, it is polynomial. In the proof, we will frequently use the notion *for large r, statement A(r) holds*. Formally, this is an abbreviation for the statement that there exists a constant $c > 0$ such that for all $r \geq c$, $A(r)$ is true. Consequently, the reduction only works for instances that are larger than the maximum of all these constants, which suffices to show that SD is $\mathcal{NP}$-complete[3].

Lemma 6. *Let ζ be an optimal clustering of G with respect to S. Then, $i_e(\zeta) \geq |\mathcal{M}| \cdot \binom{r^2}{2}$.*

Proof. Follows from Proposition 2(iii) and the fact that the clustering whose clusters are the cliques in $V_{\mathcal{M}}$ and the singletons in $V_{\mathcal{X}}$ is a clustering into cliques with $|\mathcal{M}| \cdot \binom{r^2}{2}$ intracluster edges.   □

Next, we give an upper bound on the number of *intracluster non edges*, i.e., vertex pairs within clusters that are not linked by an edge, in an optimal clustering of G. Its (rather technical) proof makes use of the asymptotic behavior of binomial coefficients and can be found in our technical report [7].

Lemma 7. *Let ζ be an optimal clustering of G with respect to surprise. Then, for large r, $i_p(\zeta) - i_e(\zeta) \leq \frac{r^4}{2}$.*

This can now be used to show that an optimal clustering of G is a clustering into cliques. We start by showing that the cliques in $V_{\mathcal{M}}$ cannot be split by an optimal clustering.

Lemma 8. *Let r be large and ζ be an optimal clustering of G with respect to S. Then, the cliques $C(M)$ in $V_{\mathcal{M}}$ are not split by ζ.*

Proof. Assume that there is at least one clique that is split by ζ. ζ induces a partition of each clique that it splits. We call the subsets of this partition the *parts* of the clique.
   Claim 1: Every clique $C(M)$ contains a part with at least $r^2 - 6$ vertices.

---

[2] Otherwise, the instance is trivially non-solvable.
[3] Smaller instances have constant size and can therefore be trivially solved by a brute-force algorithm.

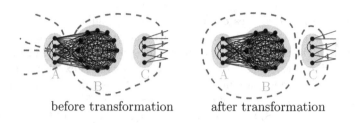

before transformation   after transformation

Fig. 2. Illustration for proof of Lemma 8

Proof of Claim 1: Assume that there is a clique K where each part has at most $r^2 - 7$ vertices. We can now greedily group the parts in two roughly equal sized regions, such that the smaller region contains at least 7 vertices and the larger region at least $r^2/2$ vertices. Let us look at the clustering we get by removing the vertices in K from their clusters and cluster them together. The vertices in K have in total $3r^2$ edges to vertices outside K and we gain at least $7/2 \cdot r^2$ new intracluster edges between the regions. Hence, the number of intracluster edges increases and the number of intracluster non-edges can only decrease. By Lemma 1(iii) and Lemma 1(i), it can be seen that this operation leads to a clustering with better surprise, which contradicts the optimality of ζ.

Let us now call the parts with size at least $r^2 - 6$ *large parts* and the other parts *small parts.*

Claim 2: No two large parts are clustered together.

Proof of Claim 2: Assume that there is a cluster that contains more than one large part. This cluster induces at least $(r^2 - 6)^2$ intracluster non-edges. For large r, this is larger than $r^4/2$ and Lemma 7 tells us that ζ was not optimal.

A simple counting argument now yields the following corollary.

Corollary: There must exist a large part B contained in a split clique whose cluster contains at most $|B| + 6$ vertices in $V_{\mathcal{M}}$.

Let B as in the corollary and A be the set of the vertices that are in the same clique as B but not in B and C be the set of vertices that are in the same cluster as B but not in B. Fig. 2 illustrates this case. We consider the clustering that we get by removing the vertices in A and B from their cluster and cluster them together. The number of vertices in A and C, respectively, is at most 6, and each of these vertices has at most 3 neighbors in $V_{\mathcal{X}}$. Hence, we lose at most 36 intracluster edges by this operation. On the other hand, we gain at least $r^2 - 6$ intracluster edges between A and B, thus, for large r, the number of intracluster edges increases. Again, the number of intracluster non-edges can only decrease and by Lemma 1(iii) and Lemma 1(i), we get that this operation leads to a clustering with better surprise, which contradicts the optimality of ζ.   □

Lemma 9. *Let r be large and ζ be an optimal clustering of G with respect to S. Then, no two of the cliques in $V_{\mathcal{M}}$ are contained in the same cluster.*

Proof. A cluster that contains two cliques in $V_{\mathcal{M}}$ induces at least r^4 intracluster non-edges. The statement now follows from Lemma 7.   □

Lemma 10. *Let r be large and ζ an optimal clustering of G with respect to S. Then, each $v(x)$ in $V_{\mathcal{X}}$ shares a cluster with a clique $C(M)$ such that $x \in M$.*

Proof. From Lemma 8 and Lemma 9 we know that ζ clusters the vertices in $V_{\mathcal{M}}$ according to the cliques we constructed. Assume that there is a vertex $v(x)$ in $V_{\mathcal{X}}$ that is not contained in any of the clusters induced by the sets containing x. Since each element in $\mathcal{X}$ is contained in at least one set in $\mathcal{M}$, there exists a clique K in $V_{\mathcal{M}}$ that contains r^2 neighbors of $v(x)$. As $v(x)$ has at most $|\mathcal{X}| - 1$ neighbors in its own cluster, removing it from its cluster and moving it to the cluster of K increases the number of intracluster edges. On the other hand, x is linked with all vertices in its new cluster and thus, the number of intracluster non-edges cannot increase. Hence, this operation leads to a clustering with better surprise, which contradicts the optimality of ζ. $\qquad\square$

Theorem 11. *For large r, $I = (\mathcal{X}, \mathcal{M})$ has a solution if and only if there exists a clustering ζ of G with $S(\zeta) \leq k := \binom{p}{m}^{-1} \cdot \left(\dfrac{\binom{|\mathcal{M}| \cdot r^2 + |\mathcal{X}|}{2} - |\mathcal{M}| \cdot \binom{r^2}{2} - |\mathcal{X}| \cdot r^2 - |\mathcal{X}|}{(3|\mathcal{M}| - |\mathcal{X}|) \cdot r^2 + \binom{|\mathcal{X}|}{2} - |\mathcal{X}|} \right)$.*

Proof. $\Rightarrow$: Let R be a solution of I. R induces a clustering of G in the following way: For each $M \in \mathcal{M} \setminus R$ we introduce a cluster $C_M = C(M)$ and for each $M' \in R$ a cluster $C_{M'} = C(M') \cup \{v(x) \mid x \in M'\}$. As R is an exact cover, this is a partition ζ of the vertex set. It is $p = \binom{|\mathcal{M}| \cdot r^2 + |\mathcal{X}|}{2}$, $m = |\mathcal{M}| \cdot \binom{r^2}{2} + 3 \cdot |\mathcal{M}| \cdot r^2 + \binom{|\mathcal{X}|}{2}$ and $i_p(\zeta) = i_e(\zeta) = |\mathcal{M}| \cdot \binom{r^2}{2} + |\mathcal{X}| \cdot r^2 + |\mathcal{X}|$. It can be easily verified that $S(\zeta) = k$.

$\Leftarrow$: Let ζ be an optimal clustering of G with respect to surprise and assume that $S(\zeta) \leq k$. From Lemma 8, Lemma 9 and Lemma 10, we know that, for large r, we have one cluster for each set M in $\mathcal{M}$ that contains $C(M)$ and each vertex $v(x)$ in $V_{\mathcal{X}}$ shares a cluster with a clique $C(M)$ such that $x \in M$. In particular, all clusters in ζ are cliques and hence $\binom{i_p(\zeta)}{i_e(\zeta)} = 1$. It follows that $\binom{p}{m} \cdot k \geq \binom{p}{m} \cdot S(\zeta) = \binom{p - i_e(\zeta)}{m - i_e(\zeta)}$. This term is strictly decreasing with $i_e(\zeta)$ and the above bound is tight for $i_e(\zeta) = |\mathcal{M}| \cdot \binom{r^2}{2} + |\mathcal{X}| \cdot r^2 + |\mathcal{X}| := t$. Hence, ζ contains at least t intracluster edges. The number of intracluster edges within $V_{\mathcal{M}}$ is exactly $|\mathcal{M}| \cdot \binom{r^2}{2}$ and the number of intracluster edges linking $V_{\mathcal{M}}$ with $V_{\mathcal{X}}$ is exactly $|\mathcal{X}| \cdot r^2$. The only quantity we do not know is the number of intracluster edges within $V_{\mathcal{X}}$, which we denote by $i_e(V_{\mathcal{X}})$. As $i_e(\zeta) \geq t$, it follows that $i_e(V_{\mathcal{X}}) \geq |\mathcal{X}|$. Thus, every vertex in $V_{\mathcal{X}}$ has in average two neighbors in $V_{\mathcal{X}}$ that are in the same cluster. On the other hand, vertices in $V_{\mathcal{X}}$ can only share a cluster if they are "assigned" to the same clique $C(M)$. As the sets in $\mathcal{M}$ only contain three elements, vertices in $V_{\mathcal{X}}$ can only have at most two neighbors in $V_{\mathcal{X}}$ in their cluster. It follows that ζ partitions $V_{\mathcal{X}}$ into triangles. Hence, the set of subsets R corresponding to cliques $C(M)$ whose clusters contain vertices in $V_{\mathcal{X}}$ form an exact cover of $\mathcal{X}$. $\qquad\square$

We now have a reduction from X3S to SD that works for all instances that are larger than a constant $c > 0$. Hence, we get the following corollary.

Corollary 12. SURPRISE DECISION *is $\mathcal{NP}$-complete.*

To show that an optimal clustering with respect to surprise can be found in polynomial time if G is a tree, we consider the following problem MACP [12]:

Problem 13 (MACP). Given a graph $G = (V, E)$ together with a weight function $w : V \to \mathbb{Q}_{\geq 0}$ on V and a parameter k. Find a clustering ζ of G such that $m - i_e(\zeta) = k$ and $\sum_{C \in \zeta} \left(\sum_{v \in C} w(v) \right)^2$ is minimal.

For the special case that $w(v)$ equals the degree of v and G is a tree, Dinh and Thai give a dynamic program that solves MACP for all $0 \leq k \leq m$ simultaneously [6]. This yields an $O(n^5)$ algorithm for modularity maximization in (unweighted) trees. In the context of surprise, we are interested in the special case that $w(v) = 1$ for all $v \in V$. The following conversion shows that this is equivalent to minIP with respect to optimal solutions:

$$i_p(\mathcal{C}) = \sum_{C \in \mathcal{C}} \frac{|C|\,(|C| - 1)}{2} = \frac{1}{2} \sum_{C \in \mathcal{C}} |C|^2 - \underbrace{\frac{1}{2}|V|}_{=\text{const.}} \tag{1}$$

The dynamic program of Dinh and Thai has a straightforward generalization to general vertex weights, which is polynomial in the case that each vertex has weight 1. For completeness, our technical report [7] contains a description of the dynamic program in this special case, together with a runtime analysis.

Theorem 14. *Let $T = (V, E)$ with $n := |V|$ be an unweighted tree. Then, a surprise optimal clustering of T can be calculated in $O(n^5)$ time.*

4   Exact Solutions

In this section, we give an integer linear program for minIP and discuss some variants of how to use this to get optimal clusterings with respect to surprise.

Linear Program for minIP. The following ILP is very similar to a number of linear programs used for other objectives in the context of graph clustering and partitioning, in particular, to one used for modularity maximization [6]. It uses a set of $\binom{n}{2}$ binary variables $\mathcal{X}_{uv}$ corresponding to vertex pairs, with the interpretation that $\mathcal{X}_{uv} = 1$ iff u and v are in the same cluster. Let $\text{Sep}(u, v)$ be a minimum u-v vertex separator in G if $\{u, v\} \notin E$ or in $G' = (V, E \setminus \{u, v\})$, otherwise. The objective is to

$$\text{minimize} \sum_{\{u,v\} \in \binom{V}{2}} \mathcal{X}_{uv} \tag{2}$$

such that

$$\mathcal{X}_{uv} \in \{0, 1\}, \quad \{u, v\} \in \binom{V}{2} \tag{3}$$

$$\mathcal{X}_{uw} + \mathcal{X}_{wv} - \mathcal{X}_{uv} \leq 1, \quad \{u, v\} \in \binom{V}{2}, w \in \text{Sep}(u, v) \tag{4}$$

$$\sum_{\{u,v\} \in E} \mathcal{X}_{uv} = k \tag{5}$$

Dinh and Thai consider the symmetric and reflexive relation induced by $\mathcal{X}$ and show that Constraint (4) suffices to enforce transitivity in the context of modularity maximization [6]. Their proof solely relies on the following argument. For an assignment of the variables $\mathcal{X}_{uv}$ that does not violate any constraints, let us consider the graph G' induced by the vertex pairs $\{u,v\}$ with $\mathcal{X}_{uv} = 1$. Now assume that there exists a connected component in G' that can be partitioned into two subsets A and B such that there are no edges in the original graph G between them. Setting $\mathcal{X}_{ab} := 0$ for all $a \in A$, $b \in B$ never violates any constraints and strictly improves the objective function. It can be verified that this argument also works in our scenario. Hence, a solution of the above ILP induces an equivalence relation and therefore a partition of the vertex set. As $\mathrm{Sep}(u,v)$ is not larger than the minimum of the degrees of u and v, we have $O(nm)$ constraints over $O(n^2)$ variables.

Variants. We tested several variants of the approach described in Sect. 1 to decrease the number of ILPs we have to solve.

- *Exact(E)*: Solve m times the above ILP and choose among the resulting clusterings the one optimizing surprise.
- *Relaxed(R)*: We relax Constraint (5), more specifically we replace it by

$$\sum_{\{u,v\} \in E} \mathcal{X}_{uv} \geq k \tag{6}$$

Lemma 1(i) tells us that the surprise of the resulting clustering is at least as good as the surprise of any clustering with exactly k intracluster edges. Moreover, by Lemma 1(ii), if i_p is the value of a solution to the modified ILP, $S(i_p, k')$ is a valid lower bound for the surprise of any clustering with $k' \geq k$ intracluster edges. In order to profit from this, we consider all possible values for the number of intracluster edges in increasing order and only solve an ILP if the lower bound is better than the best solution found so far.
- *Gap(G)*: Similarly to the relaxed variant, we replace Constraint (5) by (6) and modify (2) to

$$\text{minimize} \sum_{\{u,v\} \in \binom{V}{2}} \mathcal{X}_{uv} - \sum_{\{u,v\} \in E} \mathcal{X}_{uv} \tag{7}$$

By Lemma 1(ii), if g is the objective value and i_e the number of intracluster edges in a solution to the modified ILP, $S(k' + g, k')$ is a valid lower bound for the surprise of any clustering with $k' \geq k$ intracluster edges. Moreover, by Lemma 1(iii), we know that $S(i_e + g, i_e)$ is not larger than the surprise of any clustering with exactly k intracluster edges. Again, we consider all k in increasing order and try to prune ILP computations with the lower bound.

Case Study. Table 1 shows an overview of running times and the number of solved ILPs of the different strategies on some small instances. karate($n =$

Table 1. Number of linear programs solved and running times in seconds of successive ILP approach, different strategies.

variant	karate ILP	t(s)	lesmis ILP	t(s)	grid6 ILP	t(s)	dolphins ILP	t(s)
Exact	79	51	255	1192	61	470	160	494
Relaxed	49	21	176	282	42	449	107	163
Gap	39	15	112	205	37	401	91	147

$34, m = 78$), dolphins($n = 62, m = 159$) and lesmis($n = 77, m = 254$) are real world networks from the website of the 10th DIMACS implementation Challenge[4] that have been previously used to evaluate and compare clusterings, whereas grid6($n = 36, m = 60$) is a 2 dimensional grid graph. We used the C++-interface of gurobi5.1 [11] and computed the surprise of the resulting clusterings with the help of the GNU Multiple Precision Arithmetic Library, in order to guarantee optimality. The tests were executed on one core of an AMD Opteron Processor 2218. The machine is clocked at 2.1 GHz and has 16 GB of RAM. Running times are averaged over 5 runs.

It can be seen that the gap variant, and, to a smaller extent, the relaxed variant, are able to prune a large percentage of ILP computations and thus lead to less overall running time. These running times can be slightly improved by using some heuristic modifications described and evaluated in [7].

Properties of Optimal Clusterings. Fig. 3 illustrates optimal clusterings with respect to surprise and modularity on the test instances, Table 2 summarizes some of their properties. We also included one slightly larger graph, football($n = 115, m = 613$), as it has a known, well-motivated *ground truth clustering* and has been evaluated in [2]. The surprise based clusterings contain significantly more and smaller clusters than the modularity based ones, being *refinements* of the latter in the case of karate and lesmis. Another striking observation is that the surprise based clusterings contain far more *singletons*, i.e. clusters containing only one vertex with usually low degree; this can be explained by the fact that surprise does not take vertex degrees into account and hence, merging low degree vertices into larger clusters causes larger penalties. It reconstructs the ground-truth clustering of the football graph quite well. This confirms the observations of Aldecoa and Marín based on heuristically found clusterings [2]; in fact, we can show that for karate, this clustering was already optimal.

5   Conclusion

We showed that the problem of finding a clustering of a graph that is optimal with respect to the measure surprise is $\mathcal{NP}$-hard. The observation that surprise

[4] http://www.cc.gatech.edu/dimacs10/

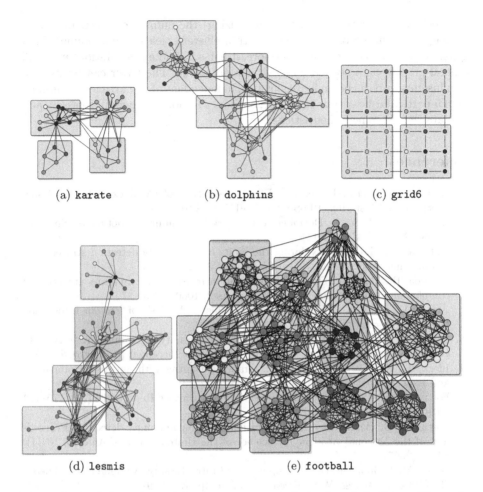

(a) `karate`   (b) `dolphins`   (c) `grid6`

(d) `lesmis`   (e) `football`

Fig. 3. Optimal clusterings with respect to surprise(colors) and, for (a) to (d), modularity(grouping). The grouping in (e) represents the *ground-truth clustering*, i.e. the mapping of teams to conferences.

Table 2. Properties of optimal clusterings with respect to surprise. S' denotes the surprise as defined by Aldecoa and Marín [2], i.e. $S'(\zeta) = -\log_{10} S(\zeta)$. S_o denotes the clustering with optimum surprise, S_h the heuristically found clusterings from [2], if this information was available, and M_o the modularity optimal clustering.

instance	i_e	i_p	$S(S_o)$	$S'(S_o)$	$S'(S_h)$	$|S_o|$	$|S_h|$	$|M_o|$
`karate`	29	30	$2{,}02 \cdot 10^{-26}$	25.69	25.69	19	19	4
`grid6`	36	54	$2{,}90 \cdot 10^{-29}$	28.54	-	9	-	4
`dolphins`	87	121	$9{,}93 \cdot 10^{-77}$	76.00	-	22	-	5
`lesmis`	165	179	$1{,}54 \cdot 10^{-184}$	183.81	-	33	-	6
`football`	399	458	$5{,}65 \cdot 10^{-407}$	406{,}25	-	15	15	10

is pareto optimal with respect to (maximizing) the number of edges and (minimizing) the number of vertex pairs within clusters yields a (polynomial time) dynamic program on trees. Furthermore, it helps to find exact solutions in small, general graphs via a sequence of ILP computations. The latter can be used to gain insights into the behavior of surprise, independent of any artifacts stemming from a particular heuristic. Moreover, optimal solutions are helpful to assess and validate the outcome of heuristics.

References

1. Aldecoa, R., Marín, I.: Jerarca: Efficient Analysis of Complex Networks Using Hierarchical Clustering. PLoS ONE 5, e11585 (2010)
2. Aldecoa, R., Marín, I.: Deciphering Network Community Structure by Surprise. PLoS ONE 6, e24195 (2011)
3. Aldecoa, R., Marín, I.: Exploring the limits of community detection strategies in complex networks. Nature Scientific Reports 3, 2216 (2013)
4. Aldecoa, R., Marín, I.: Surprise maximization reveals the community structure of complex networks. Nature Scientific Reports 3, 1060 (2013)
5. Arnau, V., Mars, S., Marín, I.: Iterative Cluster Analysis of Protein Interaction Data. Bioinformatics 21(3), 364–378 (2005)
6. Dinh, T.N., Thai, M.T.: Towards Optimal Community Detection: From Trees to General Weighted Networks. Internet Mathematics (accepted pending revision)
7. Fleck, T., Kappes, A., Wagner, D.: Graph Clustering with Surprise: Complexity and Exact Solutions. ArXiv e-prints (October 2013)
8. Fortunato, S.: Community detection in graphs. Physics Reports 486(3-5), 75–174 (2010)
9. Fortunato, S., Barthélemy, M.: Resolution limit in community detection. Proceedings of the National Academy of Science of the United States of America 104(1), 36–41 (2007)
10. Garey, M.R., Johnson, D.S.: Computers and Intractability. A Guide to the Theory of $\mathcal{NP}$-Completeness. W.H. Freeman and Company (1979)
11. I. Gurobi Optimization. Gurobi optimizer reference manual (2013)
12. Li, A., Tang, L.: The Complexity and Approximability of Minimum Contamination Problems. In: Ogihara, M., Tarui, J. (eds.) TAMC 2011. LNCS, vol. 6648, pp. 298–307. Springer, Heidelberg (2011)
13. Newman, M.E.J., Girvan, M.: Finding and evaluating community structure in networks. Physical Review E 69(026113), 1–16 (2004)
14. Reichardt, J., Bornholdt, S.: Detecting Fuzzy Community Structures in Complex Networks with a Potts Model. Physical Review Letters 93(21), 218701 (2004)
15. Reichardt, J., Bornholdt, S.: Statistical Mechanics of Community Detection. Physical Review E 74(016110), 1–16 (2006)
16. Schaeffer, S.E.: Graph Clustering. Computer Science Review 1(1), 27–64 (2007)

On Lower Bounds for the Time and the Bit Complexity of Some Probabilistic Distributed Graph Algorithms
(Extended Abstract)

Allyx Fontaine, Yves Métivier, John Michael Robson, and Akka Zemmari

Université de Bordeaux - LaBRI UMR CNRS 5800
351 cours de la Libération, 33405 Talence, France
{fontaine,metivier,robson,zemmari}@labri.fr

Abstract. This paper concerns probabilistic distributed graph algorithms to solve classical graph problems such as colouring, maximal matching or maximal independent set. We consider anonymous networks (no unique identifiers are available) where vertices communicate by single bit messages. We present a general framework, based on coverings, for proving lower bounds for the bit complexity and thus the execution time to solve these problems. In this way we obtain new proofs of some well known results and some new ones.

1 Introduction

The Problems. For many problems on graphs, lower bounds on the bit complexity and on the execution time of a probabilistic distributed algorithm can be obtained in a simple way by considering disconnected graphs. These results may be considered unsatisfactory since we are normally interested in connected graphs and networks. In this paper we present a general framework, based on coverings, for proving such results for (dis)connected graphs and apply it to problems such as colouring, maximal matching, maximal independent set (MIS for short), or some generalisations such as the following. The MIS problem or the colouring problem may be generalised to a distance k for any positive integer k. More precisely, let $G = (V, E)$ be a graph and let k be a positive integer; a k-independent set is a subset M of V such that the distance between any two vertices of M is at least $k + 1$. If M is maximal for this property M is said to be a maximal k-independent set (k-MIS for short). A distance-k colouring of G is a colouring of vertices of G such that any two different vertices connected by a path of at most k edges have different colours. The distributed complexity of problems given above is of fundamental interest for the study and analysis of distributed algorithms. Usually, the topology of a distributed system is modelled by a graph and paradigms of distributed systems are represented by classical problems in graph theory cited above. Each solution to one of these problems is a building block for many distributed algorithms: symmetry breaking, topology control, routing, resource allocation or network synchronisation.

V. Geffert et al. (Eds.): SOFSEM 2014, LNCS 8327, pp. 235–245, 2014.

Network, Time Complexity, Bit Complexity, Network Knowledge. (A general presentation may be found in [Tel00]). We consider the standard message passing model for distributed computing. The communication model consists of a point-to-point communication network described by a connected graph $G = (V, E)$, where the vertices V represent network processes and the edges E represent bidirectional communication channels. Processes communicate by message passing: a process sends a message to another by depositing the message in the corresponding channel. The state of each process, with respect to a distributed algorithm, is represented by a label $\lambda(v)$ associated to the corresponding vertex $v \in V$. We denote by $\mathbf{G} = (G, \lambda)$ such a labelled graph. We assume the system is fully synchronous, namely, all processes start at the same time and time proceeds in synchronised rounds.

A round (cycle) of each process is composed of the following three steps. Firstly, it sends messages to (some) neighbours ; secondly, it receives messages from (some) neighbours ; thirdly, it performs some local computation. As usual the time complexity is the number of rounds needed until every node has completed its computation. By definition, in a bit round each vertex can send/receive at most 1 bit from each of its neighbours. The bit complexity of algorithm $\mathcal{A}$ is the number of bit rounds to complete algorithm $\mathcal{A}$ (see [KOSS06]).

The network $G = (V, E)$ is anonymous: unique identities are not available to distinguish the processes. We do not assume any global knowledge of the network, not even its size or an upper bound on its size. The processes do not require any position or distance information. Each process knows from which channel it receives or to which it sends a message, thus one supposes that the network is represented by a connected graph with a port numbering function defined as follows (where $I_G(u)$ denotes the set of vertices of G adjacent to u): given a graph $G = (V, E)$, a **port numbering** function δ is a set of local functions $\{\delta_u \mid u \in V\}$ such that for each vertex $u \in V$, δ_u is a bijection between $I_G(u)$ and the set of natural numbers between 1 and $\deg_G(u)$.

The network is anonymous; thus two processes with the same degree are identical. Note that we consider only reliable systems: no fault can occur on processes or communication links.

A probabilistic algorithm is an algorithm which makes some random choices based on some given probability distributions. A distributed probabilistic algorithm is a collection of local probabilistic algorithms. Since the network is anonymous, if two processes have the same degree then their local probabilistic algorithms are identical and have the same probability distribution. We assume that choices of vertices are independent. A Las Vegas algorithm is a probabilistic algorithm which terminates with a positive probability (in general 1) and always produces a correct result. In this paper, results on graphs having n vertices are expressed with high probability (w.h.p. for short), meaning with probability $1 - o(n^{-1})$.

Our Contribution. The main contributions of this work are general constructions, based on coverings, which present in a unified way proofs of lower bounds for the time complexity and the bit complexity of some graph problems. Some of these lower bounds are already known, others are new. More precisely, thanks to coverings we build infinite families of disconnected or connected graphs with some port numberings, such that for each graph K having n vertices in those families, any Las Vegas distributed algorithm which uses messages of one bit cannot break some symmetries inside K after $c\sqrt{\log n}$ or after $c\log n$ rounds (for a certain constant c) with high probability. (i.e., some vertices remain in the same state for at least $c\log n$ rounds w.h.p.). From these constructions and results we deduce that:

- solving problems such as MIS[1], colouring[1], maximal matching[1], 2-MIS or distance-2 colouring takes $\Omega(\log n)$ rounds w.h.p. for an infinite family of disconnected graphs;
- solving the MIS problem or the maximal matching problem takes $\Omega(\sqrt{\log n})$ rounds w.h.p. for an infinite family of rings;
- solving problems like MIS[1], colouring[1], maximal matching, 2-MIS or distance-2 colouring takes $\Omega(\log n)$ rounds w.h.p. for an infinite family of connected graphs.

We deduce also that the maximal matching (resp. colouring, resp. MIS) Las Vegas distributed algorithm presented in [II86] (resp. [MRSDZ10], resp. [MRSDZ11]) is optimal (time and bit) modulo multiplicative constants. More precisely, the bit complexity of solutions presented in these papers is $O(\log n)$ w.h.p. for anonymous graphs having n vertices. These results can be summarised by:

Theorem 1.1. *The bit complexity of the MIS problem, the colouring problem, the maximal matching problem and the distance-2 colouring problem is $\Theta(\log n)$ w.h.p. for anonymous graphs with n vertices.*

If we consider the particular case of rings, we prove [FMRZar] that: $O(\sqrt{\log n})$ rounds are sufficient w.h.p. to compute a MIS or a maximal matching in a ring with n vertices. Thus, the bit complexity of the MIS problem or the maximal matching problem is $\Theta(\sqrt{\log n})$ w.h.p. for anonymous rings with n vertices.

Related Work. Bit complexity is considered as a finer measure of communication complexity and it has been studied for breaking and achieving symmetry or for colouring in [BMW94, KOSS06, DMR08]. Dinitz et al. explain in [DMR08] that it may be viewed as a natural extension of communication complexity (introduced by Yao [Yao79]) to the analysis of tasks in a distributed setting. An introduction to this area can be found in Kushilevitz and Nisan [KN99].

Kothapalli et al. consider the family of anonymous rings and show in [KOSS06] that if only one bit can be sent along each edge in a round, then every Las Vegas distributed vertex colouring algorithm (in which every node has the same initial state and initially only knows its own edges) needs $\Omega(\log n)$ rounds with high probability to colour the ring of size n with any finite number of colours. With the same assumptions, as is explained in [MRSDZ10], from this result we also deduce that every distributed algorithm that computes a MIS needs, in general, $\Omega(\log n)$ rounds with high probability.

---

[1] Already known result ([KOSS06],[MRSDZ10]).

2   Preliminaries

We will consider digraphs with multiple arcs and self-loops. A **digraph** $D = (V(D), A(D), s_D, t_D)$ is defined by a set $V(D)$ of vertices, a set $A(D)$ of arcs and by two maps s_D and t_D that assign to each arc two elements of $V(D)$: a source and a target.

A **symmetric** digraph D is a digraph endowed with a symmetry, that is, an involution $Sym : A(D) \to A(D)$ such that for every $a \in A(D)$, $s(a) = t(Sym(a))$. In a symmetric digraph D, the degree of a vertex v is $\deg_D(v) = |\{a \mid s(a) = v\}| = |\{a \mid t(a) = v\}|$.

A **homomorphism** between two digraphs maps vertices to vertices, arcs to arcs while preserving the incidence relation. More precisely, a homomorphism γ between the digraph D and the digraph D' is a mapping $\gamma : V(D) \cup A(D) \to V(D') \cup A(D')$ such that for each arc $a \in A(D)$, $\gamma(s(a)) = s(\gamma(a))$ and $\gamma(t(a)) = t(\gamma(a))$. A homomorphism $\gamma : D \to D'$ is an **isomorphism** if γ is bijective.

Throughout the paper, we will consider digraphs where the vertices and the arcs are labelled with labels from a recursive label set L. A digraph D labelled over L will be denoted by (D, λ), where $\lambda : V(D) \cup A(D) \to L$ is the **labelling function**. A mapping $\gamma : V(D) \cup A(D) \to V(D') \cup A(D')$ is a homomorphism from (D, λ) to (D', λ') if γ is a digraph homomorphism from D to D' which preserves the labelling, i.e., such that $\lambda'(\gamma(x)) = \lambda(x)$ for every $x \in V(D) \cup A(D)$. Labelled graphs or digraphs will be designated by bold letters such as $\mathbf{D}, \mathbf{D}', \ldots$

Let (G, λ) be a labelled graph with the port numbering δ. We will denote by $(\text{Dir}(\mathbf{G}), \delta')$ the symmetric labelled digraph $(\text{Dir}(G), (\lambda, \delta'))$ constructed in the following way. The vertices of $\text{Dir}(G)$ are the vertices of G and they have the same labels in $\mathbf{G}$ and in $\text{Dir}(\mathbf{G})$. Each edge $\{u, v\}$ of G is replaced in $(\text{Dir}(\mathbf{G}), \delta')$ by two arcs $a_{(u,v)}, a_{(v,u)} \in A(\text{Dir}(G))$ such that $s(a_{(u,v)}) = t(a_{(v,u)}) = u$, $t(a_{(u,v)}) = s(a_{(v,u)}) = v$, $\delta'(a_{(u,v)}) = (\delta_u(v), \delta_v(u))$ and $\delta'(a_{(v,u)}) = (\delta_v(u), \delta_u(v))$. These arcs correspond for each vertex to input ports and output ports. By extension, the labelling δ' of arcs is called a port numbering. (see Figure 1). Note that this digraph does not contain loops or multiple arcs. The object we use for our study is $(\text{Dir}(G), (\lambda, \delta'))$ and some results are stated with symmetric labelled digraphs.

Given a labelled graph $\mathbf{G} = (G, \lambda)$ with a port numbering δ, let $\mathbf{D} = (\text{Dir}(\mathbf{G}), \delta')$ be the corresponding labelled digraph $(\text{Dir}(G), (\lambda, \delta'))$. Let $\mathcal{A}$ be a synchronous distributed algorithm. We speak indifferently of an execution of $\mathcal{A}$ on $(\mathbf{G}, \delta)$ or on $\mathbf{D}$. The state of each process is represented by the label $\lambda(v)$ of the corresponding vertex v. Let $\mathbf{D}' = (Dir(\mathbf{G}'), \delta')$ be the labelled digraph obtained by the application of a step of $\mathcal{A}$ to $\mathbf{D}$. We recall that the system is synchronous, thus each vertex of D has a new state computed by a transition function which depends on the state of the vertex and the messages it has received. This transition is denoted by: $(Dir(\mathbf{G}), \delta') \underset{\mathcal{A}}{\Longrightarrow} (Dir(\mathbf{G}'), \delta')$ or $(\mathbf{G}, \delta) \underset{\mathcal{A}}{\Longrightarrow} (\mathbf{G}', \delta)$.

Let v be a vertex of $\mathbf{G}$. We denote by $(v, \lambda(v)) \underset{\mathcal{A}}{\Longrightarrow} (v, \lambda'(v))$ the transition associated to the vertex v of G.

Let r be a non-negative integer. A sequence $(\mathbf{D}_i)_{0 \le i \le r}$ of labelled digraphs is called an *A-execution* (or an *execution* when $\mathcal{A}$ is clear from the context) of length r if $\mathbf{D}_{i+1}$ is obtained from $\mathbf{D}_i$ in one step of a run of $\mathcal{A}$; this is denoted by $\mathbf{D}_i \mathcal{A} \mathbf{D}_{i+1}$ for every $0 \le i < r$. An execution of length 1 is a *step*. Furthermore if $\mathcal{A}$ is a probabilistic synchronous distributed algorithm and if the execution of this step has probability p, then it will be denoted by: $\mathbf{G} \underset{p}{\Longrightarrow} \mathbf{G}'$.

Let $\mathcal{A}$ be a probabilistic synchronous distributed algorithm. Let G be a graph. Let δ be a port numbering of G. Let λ be a labelling of G. We have: $(\mathbf{G}, \delta) \underset{p}{\Longrightarrow} (\mathbf{G}', \delta)$ (with $\mathbf{G}' = (G, \lambda')$) if and only if p equals the product over $V(G)$ of the probabilities of the transitions $(v, \lambda(v)) \underset{\mathcal{A}}{\Longrightarrow} (v, \lambda'(v))$.

2.1 Coverings and Synchronous Distributed Algorithms

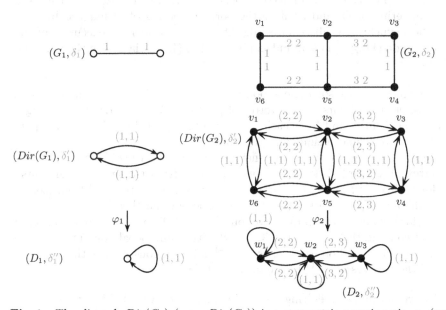

Fig. 1. The digraph $Dir(G_1)$ (resp. $Dir(G_2)$) is a symmetric covering via φ_1 (resp. φ_2) of D_1 (resp. D_2) where: φ_1 maps the vertices of $Dir(G_1)$ on the unique vertex of D_1 and $\varphi_2(v_1) = \varphi_2(v_6) = w_1$, $\varphi_2(v_2) = \varphi_2(v_5) = w_2$, and $\varphi_2(v_3) = \varphi_2(v_4) = w_3$. The number of sheets of these two coverings is 2. In grey, the port numbering δ_1'' (resp. δ_2'') of D_1 (resp. D_2) induces via φ_1^{-1} (resp. φ_2^{-2}) the port numbering δ_1' (resp. δ_2') of $Dir(G_1)$ (resp. $Dir(G_2)$) and thus the port numbering δ_1 (resp. δ_2) of G_1 (resp. G_2).

Definitions and principal properties of coverings are presented in [BV02]. A labelled digraph $\mathbf{D}$ is a *covering* of a labelled digraph $\mathbf{D}'$ via φ if φ is a homomorphism from $\mathbf{D}$ to $\mathbf{D}'$ such that for each arc $a' \in A(D')$ and for each vertex $v \in \varphi^{-1}(t(a'))$ (resp. $v \in \varphi^{-1}(s(a'))$), there exists a unique arc $a \in A(D)$ such that $t(a) = v$ (resp. $s(a) = v$) and $\varphi(a) = a'$.

The **fibre** over a vertex v' (resp. an arc a') of D' is defined as the set $\varphi^{-1}(v')$ of vertices of D (resp. the set $\varphi^{-1}(a')$ of arcs of D). An interesting property satisfied by coverings is that all the fibres of a given digraph have the same cardinality, that is called the **number of sheets** of the covering.

A symmetric labelled digraph $\mathbf{D}$ is a **symmetric covering** of a symmetric labelled digraph $\mathbf{D}'$ via φ if $\mathbf{D}$ is a covering of $\mathbf{D}'$ via φ and if for each arc $a \in A(D)$, $\varphi(Sym(a)) = Sym(\varphi(a))$. The homomorphism φ is a **symmetric covering projection** from $\mathbf{D}$ to $\mathbf{D}'$. Two examples are given in the parts in black of Figure 1.

Let D be a symmetric covering of D' via the homomorphism φ. Any port numbering on D' induces naturally via φ^{-1} a port numbering on D. Conversely, a port numbering on D induces a port numbering on D' via φ. (It is illustrated in grey in Figure 1).

Remark 2.1. Let $\mathbf{D}$ be a symmetric labelled covering of $\mathbf{D}'$ via φ. Let δ' be a port numbering of $\mathbf{D}'$ and let δ be the port numbering of $\mathbf{D}$ induced by φ^{-1}. Let $\mathcal{A}$ be a synchronous distributed algorithm. We consider the execution of one round of $\mathcal{A}$ on $(\mathbf{D}', \delta')$. This round can be lifted to $(\mathbf{D}, \delta)$ via φ^{-1} in the following way:

1. "send messages to (some) neighbours of $v' \in V(D')$" becomes "send messages to (some) neighbours of each vertex of $\varphi^{-1}(v') \subseteq V(D)$" (the same messages are sent from v' and from each vertex w of $\varphi^{-1}(v')$ through the same port numbers to (some) neighbours of v' and w);
2. "receive messages from (some) neighbours of $v' \in V(D')$" becomes "each vertex of $\varphi^{-1}(v') \subseteq V(D)$ receives messages from (some) its neighbours" (for each vertex w of $\varphi^{-1}(v')$ the same messages are received from (some) neighbours of v' and of w through the same port numbers);
3. "perform some local computation on $v' \in V(D')$" becomes "perform some local computation on $\varphi^{-1}(v') \subseteq V(D)$" (the same local computations are perform on v' and on each vertex of $\varphi^{-1}(v')$, as a consequence the vertex v' and the vertices of $\varphi^{-1}(v')$ are in same state).

Finally, we obtain the following classical lemma (see [Ang80, CM07]):

Lemma 2.2. *Let $\mathbf{D}$ be a symmetric labelled covering of $\mathbf{D}'$ via φ. Let δ' be a port numbering of $\mathbf{D}'$ and let δ be the port numbering of $\mathbf{D}$ induced by φ^{-1}. Then any execution of a distributed algorithm $\mathcal{A}$ on $(\mathbf{D}', \delta')$ can be lifted to an execution on $(\mathbf{D}, \delta)$, such that at the end of the execution, for any $v \in V(D)$, v is in the same state as $\varphi(v)$.*

As a direct consequence we have. An execution of a Las Vegas distributed algorithm $\mathcal{A}$ on a triangle induces, via the natural morphism, an execution on an hexagon. Hence, the following impossibility result holds:

Proposition 2.3. *Let k be a positive integer such that $k \geq 3$. There is no Las Vegas distributed algorithm for solving the distance-k colouring problem or the k-MIS problem.*

We are interested, in particular, in some problems like the computation of an MIS, colouring vertices or the computation of a maximal matching. In each case we have to break some symmetries inside a labelled graph. To break symmetries inside a labelled graph it suffices to distinguish a vertex. This is precisely the aim of an election algorithm. A distributed algorithm solves the election problem if it always terminates and in the final configuration, exactly one process is marked as *elected* and all the other processes are *non-elected*. Moreover, it is required that once a process becomes *elected* or *non-elected* then it remains in such a state until the end of the execution of the algorithm. The election problem is closely related to coverings. Indeed, a symmetric labelled digraph **D** is **symmetric covering prime** if there does not exist any symmetric labelled digraph **D'** not isomorphic to **D** such that **D** is a symmetric covering of **D'**. We have [CM07]:

Theorem 2.4. *Given a connected graph G, there exists an election algorithm for G if and only if $Dir(G)$ is symmetric covering prime.*

We study some graph problems which need to break some initial symmetries; these symmetries are precisely encoded by some non-covering-prime digraphs and correspond to vertices inside a fibre. Thus in the sequel we consider non-covering-prime digraphs.

3 Obtaining Lower Bounds by Considering Disconnected Graphs

Proposition 3.1. *Let G be a graph having n_G vertices. Assume $Dir(G)$ is not symmetric covering prime. Let δ_1 be a port numbering of G. Let (K, δ) be the graph K with the port numbering δ formed by α copies of (G, δ_1) and let $n = \alpha n_G$. Then, there exists a constant $c > 0$ such that, for any Las Vegas distributed algorithm $\mathcal{A}$ that uses messages of 1 bit, there is at least one copy of (G, δ_1) such that for each fibre all its vertices are in the same state for at least $c \log n$ rounds w.h.p.*

The idea of the proof of Proposition 3.1 is based on the assumption that $Dir(G)$ is not covering prime. Then it is a covering of a non-isomorphic graph, say D'. Any execution of the Las Vegas algorithm $\mathcal{A}$ on D' induces an execution of $\mathcal{A}$ on $Dir(G)$. We consider the execution having the highest probability. We show that, *w.h.p.*, there exists a constant c such that there is a fibre in one of the copies of $Dir(G)$ in which all vertices are in the same state for at least $c \log n$ rounds.

Corollary 3.2. *For every Las Vegas distributed algorithm $\mathcal{A}$ there is an infinite family $\mathcal{F}$ of disconnected graphs such that $\mathcal{A}$ has a bit complexity $\Omega(\log n)$ w.h.p. on $\mathcal{F}$ to solve either: the colouring problem, the MIS problem, the maximal matching problem, the 2-MIS problem, the distance-2 colouring problem.*

The idea of the proof is based on the choice of G. Take G, for each problem respectively as a: 1. single edge, 2. single edge, 3. triangle, 4. single edge, 5. single edge.

4   Obtaining Lower Bounds of the Form $\Omega(\sqrt{\log n})$ for Connected Graphs

The previous definition of coverings for digraphs becomes in the case of graphs: a graph G is a covering of a graph H via the homomorphism φ if φ is a homomorphism from G onto H such that for every vertex v of V the restriction of φ to neighbours of v is a bijection between neighbours of v and neighbours of $\varphi(v)$. The construction we give now has been presented by Reidemeister [Rei32] to describe all coverings of a given graph. We present briefly a precise description given by Bodlaender in [Bod89]. Let G be a graph. Consider a spanning tree (and more generally a spanning graph) S_T of G. Make α copies of S_T. For every edge in G that is not an edge of S_T, there are α copies of both its endpoints. We connect on a one-to-one basis every copy of the first endpoint to a unique copy of the second endpoint (see Figures 2 and 3). This construction builds a covering of G. Furthermore, given a spanning tree S_T of G, each covering of G can be obtained in this way from S_T. The number of copies is the number of sheets of the covering.

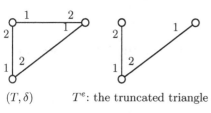
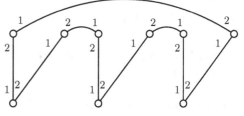

(T, δ) T^e: the truncated triangle

Fig. 2. A triangle T with a port numbering δ and the associated truncated triangle

Fig. 3. Reidemeister's construction with 3 copies of the truncated triangle T^e. We obtain $R(T, e, 3)$ and the port numbering induced by δ: a covering of the triangle (T, δ) with 3 sheets.

In the sequel, we use a particular case of this construction. Let G be a connected graph. Let e be an edge of G. Let G^e be the subgraph of G obtained by deleting the edge e. We denote by $R(G, e, \alpha)$ the graph constructed as: make α copies of G^e denoted G_i^e, $1 \le i \le \alpha$; for each i such that $1 \le i \le \alpha$ and $i + 1$ computed modulo α, connect the first endpoint of e in G_i^e to the second endpoint of e in G_{i+1}^e (see Figures 2 and 3). This construction is extended in a natural way to graphs with a port numbering. From this construction, by considering a sequence of consecutive copies and analysing what is produced in the middle in such a sequence, we can state the following lower bound:

Proposition 4.1. *Let G be a non-covering-prime graph having n_G vertices and at least one cycle. Let δ be a port numbering on G. Let e be an edge of G such that G^e, the graph obtained by deleting the edge e, is connected. Let α be a nonnegative integer. Let $n = \alpha n_G$. Let δ' be the port numbering induced by δ on $R(G, e, \alpha)$. Then there exists a constant $c > 0$ such that, for any Las Vegas distributed algorithm $\mathcal{A}$ that uses messages of 1 bit, there is at least one copy of*

G^e of $R(G, e, \alpha)$ such that for each fibre all its vertices are in the same state for at least $c\sqrt{\log n}$ rounds w.h.p.

By applying Proposition 4.1 to the triangle $(G = T)$ as is explained by Figures 2 and 3, we obtain:

Corollary 4.2. *For every Las Vegas distributed algorithm $\mathcal{A}$ there is an infinite family of rings $\mathcal{R}$ such that $\mathcal{A}$ has a bit complexity $\Omega(\sqrt{\log n})$ w.h.p. on $\mathcal{R}$ to compute a MIS or a maximal matching.*

5   Obtaining Lower Bounds of the Form $\Omega(\log n)$ for Connected Graphs

Let $G = (V, E)$ be a graph such that $D = Dir(G)$ is not covering prime. Let D' be a symmetric digraph such that D is a covering of D' via the homomorphism φ and D is not isomorphic to D'. Let F be a fibre of D, and let w be the corresponding vertex of D', i.e., $F = \varphi^{-1}(w)$. Let b be the size of the fibre $F = \{f_1, \ldots, f_b\}$. As D is not isomorphic to D' the size b of F is greater than or equal to 2. We can note also that F is a subset of the set of vertices of G. Let α be a positive integer. We denote by H the graph formed by α copies of G. Let u be a vertex which does not belong to H. Now we define the graph K obtained from H by adding a new vertex, the vertex u, and by adding an edge between u and all vertices of all copies of the fibre F. Let n be the number of vertices of K, we have: $n = \alpha n_G + 1$. We consider a port numbering of D'; it induces via φ^{-1} a port numbering on D thus on G and on H. Let d be the degree of any vertex of the fibre F. We assign to any port corresponding to u incident to any vertex of any copy of the fibre F the number $d + 1$. The numbers of the ports incident to u are chosen randomly, uniformly among the permutations of $[1, \alpha \cdot b]$. Let $\mathcal{A}$ be a Las Vegas distributed algorithm. The aim of this section is to prove that if $\mathcal{A}$ halts on K w.h.p. in a time less than $c \log n$, it has a high probability of giving incorrect output. It follows from this that there exists some port numbering for which the algorithm does not compute a correct result in time less than or equal to $c \log n$ w.h.p. Let $D'_{w'}$ be the graph obtained by adding a new vertex, denoted w', to the graph D' and by adding an edge between w' and w (the image of the fibre F by φ). Let D_F be the graph we obtain by adding b new vertices f'_i, $1 \leq i \leq b$, to D and by adding an edge between f_i and f'_i ($1 \leq i \leq b$). In the same manner we define G_F. It is easy to verify that D_F is a covering of $D'_{w'}$ via the morphism γ defined by: the restriction of γ to D is equal to φ and $\gamma(f'_i) = w'$ (for $1 \leq i \leq b$). An execution $(\mathbf{D}'_{w'i})_{0 \leq i \leq \ell}$ of $\mathcal{A}$ on $\mathbf{D}'_{w'}$ (with $\mathbf{D}'_{w'0} = \mathbf{D}'_{w'}$) induces an execution $(\mathbf{D}_{Fi})_{0 \leq i \leq \ell}$ of $\mathcal{A}$ on $\mathbf{D}_F$ (with $\mathbf{D}_{F0} = \mathbf{D}_F$) via γ. It induces also an execution on G_F. For this kind of execution, vertices of D_F, thus of G_F, which belong to the same fibre are in the same state at each step. Let D_f and G_f be the graphs obtained from D_F and G_F by fusing vertices $f'_1, \ldots f'_b$ into the same vertex, denoted f. Thus in D_f and G_f there is an edge between vertices $f_1, \ldots, f_b$ and f. Now we consider an execution of $\mathcal{A}$ on D_f (thus on G_f); if we assume that this execution is induced by an execution of $\mathcal{A}$ on $D'_{w'}$

for vertices different from f and the vertex f sends the same bit to every vertex of F then vertices of a given fibre of D_f (thus of G_f) are in the same state after each step of the execution. Finally, steps occur as if the execution is induced by a covering relation. By definition, an execution on D_f thus on G_f which satisfies this property is called uniform and, by extension, D_f is uniform. By choosing a constant $c_1 > 0$ depending on G, we show that after r rounds of $\mathcal{A}$, there is a uniform set of copies of G in K, denoted U_r of size nc_1^r. Choosing a suitable c such that the random variable $|U_{c \log n}| = \Omega(n^{1/2})$ gives the following result:

Proposition 5.1. *Let $\mathcal{A}$ be a Las Vegas distributed algorithm that produces at each round on each port the bit 1 or the bit 0. Let G be a connected graph having n_G vertices. Assume $Dir(G)$ is not symmetric covering prime. Let α be a non-negative integer. Let K be the connected graph defined above (a vertex added to α copies of G) and with the associated port numbering. Let $n = \alpha n_G$. There exists a constant $c > 0$ such that for at least $c \log n$ rounds of execution of Algorithm $\mathcal{A}$ on K, with high probability, there is a copy of G in K for which the execution is uniform.*

Corollary 5.2. *For every Las Vegas distributed algorithm $\mathcal{A}$ there is an infinite family $\mathcal{C}$ of connected graphs such that $\mathcal{A}$ has a bit complexity $\Omega(\log n)$ w.h.p. on $\mathcal{C}$ to solve either: the colouring problem, the MIS problem, the maximal matching problem, the 2-MIS problem or the distance-2 colouring problem.*

References

[Ang80]   Angluin, D.: Local and global properties in networks of processors. In: Proceedings of the 12th Symposium on Theory of Computing, pp. 82–93 (1980)

[BMW94]   Bodlaender, H.L., Moran, S., Warmuth, M.K.: The distributed bit complexity of the ring: from the anonymous case to the non-anonymous case. Information and Computation 114(2), 34–50 (1994)

[Bod89]   Bodlaender, H.-L.: The classification of coverings of processor networks. J. Parallel Distrib. Comput. 6, 166–182 (1989)

[BV02]   Boldi, P., Vigna, S.: Fibrations of graphs. Discrete Math. 243, 21–66 (2002)

[CM07]   Chalopin, J., Métivier, Y.: An efficient message passing election algorithm based on Mazurkiewicz's algorithm. Fundam. Inform. 80(1-3), 221–246 (2007)

[DMR08]   Dinitz, Y., Moran, S., Rajsbaum, S.: Bit complexity of breaking and achieving symmetry in chains and rings. Journal of the ACM 55(1) (2008)

[FMRZar]   Fontaine, A., Métivier, Y., Robson, J.M., Zemmari, A.: The bit complexity of the MIS problem and of the maximal matching problem in anonymous rings. Information and Computation (to appear)

[II86]   Israeli, A., Itai, A.: A fast and simple randomized parallel algorithm for maximal matching. Information Processing Letters 22, 77–80 (1986)

[KN99]   Kushilevitz, E., Nisan, N.: Communication complexity. Cambridge University Press (1999)

[KOSS06]   Kothapalli, K., Onus, M., Scheideler, C., Schindelhauer, C.: Distributed coloring in $O(\sqrt{\log n})$ bit rounds. In: Proceedings of the 20th International Parallel and Distributed Processing Symposium (IPDPS 2006), Rhodes Island, Greece, April 25-29. IEEE (2006)

[MRSDZ10] Métivier, Y., Robson, J.M., Saheb-Djahromi, N., Zemmari, A.: About randomised distributed graph colouring and graph partition algorithms. Information and Computation 208(11), 1296–1304 (2010)

[MRSDZ11] Métivier, Y., Robson, J.M., Saheb-Djahromi, N., Zemmari, A.: An optimal bit complexity randomized distributed MIS algorithm. Distributed Computing 23(5-6), 331–340 (2011)

[Rei32] Reidemeister, K.: Einführung in die Kombinatorische Topologie. Vieweg, Brunswick (1932)

[Tel00] Tel, G.: Introduction to distributed algorithms. Cambridge University Press (2000)

[Yao79] Yao, A.C.: Some complexity questions related to distributed computing. In: Proceedings of the 11th ACM Symposium on Theory of Computing (STOC), pp. 209–213. ACM Press (1979)

Active Learning of Recursive Functions
by Ultrametric Algorithms*

Rūsiņš Freivalds[1] and Thomas Zeugmann[2]

[1] Institute of Mathematics and Computer Science, University of Latvia
Raiņa bulvāris 29, Riga, LV-1459, Latvia
`Rusins.Freivalds@mii.lu.lv`
[2] Division of Computer Science, Hokkaido University
N-14, W-9, Sapporo 060-0814, Japan
`thomas@ist.hokudai.ac.jp`

Abstract. We study active learning of classes of recursive functions by
asking value queries about the target function f, where f is from the
target class. That is, the query is a natural number x, and the answer
to the query is $f(x)$. The complexity measure in this paper is the worst-
case number of queries asked. We prove that for some classes of recursive
functions *ultrametric active learning* algorithms can achieve the learning
goal by asking *significantly fewer* queries than deterministic, probabilistic, and even nondeterministic active learning algorithms. This is the first
ever example of a problem where ultrametric algorithms have advantages
over nondeterministic algorithms.

1   Introduction

Inductive inference has been studied intensively. Gold [12] defined *learning in the
limit*. The learner is a deterministic algorithm called *inductive inference machine*
(abbr. IIM), and the objects to be learned are recursive functions. The information source are growing initial segments $(x_0, f(x_0)), \ldots, (x_n, f(x_n))$ of ordered
pairs of the graph of the target function f. It is assumed that every pair $(x, f(x))$
appears eventually. As a hypothesis space one can choose any *Gödel numbering*
$\varphi_0, \varphi_1, \varphi_2, \ldots$ of the set of all partial recursive functions over the natural numbers $\mathbb{N} = \{0, 1, 2, \ldots\}$ (cf. [27]). If an $i \in \mathbb{N}$ is such that $\varphi_i = f$ then we call i a
φ-*program* of f. An IIM, on input an initial segment $(x_0, f(x_0)), \ldots, (x_n, f(x_n))$,
has to output a natural number i_n which is interpreted as φ-program. An IIM
learns f if the sequence $(i_n)_{n \in \mathbb{N}}$ of all computed φ-programs converges to a
program i such that $\varphi_i = f$.

Every IIM M learns some set of recursive functions which is denoted by
$EX(M)$. The family of all such sets, over the universe of effective algorithms
viewed as IIMs, serves as a characterization of the learning power inherent in

---

* The research was supported by Grant No. 09.1570 from the Latvian Council of
Science and the Invitation Fellowship for Research in Japan S12052 by Japan Society
for the Promotion of Science.

V. Geffert et al. (Eds.): SOFSEM 2014, LNCS 8327, pp. 246–257, 2014.

the Gold model. This family is denoted by EX (short for explanatory) and it is defined by $EX = \{\mathcal{U} \mid \exists M (\mathcal{U} \subseteq EX(M))\}$. Many studies of inductive inference set-theoretically compare the family EX with families that arise from considering other models (cf., e.g., [30]). One such model is *finite learning*, where the IIM either requests a new input and outputs nothing, or it outputs a program i, and stops. Again we require that program i is correct for f, i.e., $\varphi_i = f$.

The models described so far are models of *passive* learning, since the IIM has no influence on the order in which examples are presented. In contrast, the learning model considered in the present paper is an *active* one. This model goes back to Angluin [3] and is called *query learning*. In the query learning model the learner has access to a teacher that truthfully answers queries of a prespecified type. In this paper we only consider *value queries*. That is, the query is a natural number x, and the answer to the query is $f(x)$. A query learner is an algorithmic device that, depending on the answers already received, either computes a new value query or it returns a hypothesis i and stops. As above, the hypothesis is interpreted with respect to a fixed Gödel numbering φ and it is required that the hypothesis returned satisfies $\varphi_i = f$. So active learning is *finite learning*.

As in the Gold [12] model, we are interested in active learners that can infer whole classes of recursive functions. The complexity measure is then the *worst-case number of queries asked* to identify all the functions from the target class $\mathcal{U}$. We refer to any query learner as *query inference machine* (abbr. QIM).

Automata theory and complexity theory have considered several natural generalizations of deterministic algorithms, namely, nondeterministic and probabilistic algorithms. In many cases these generalized algorithms allow for computations having a complexity that is strictly less than their deterministic counterpart. Such generalized algorithms attracted considerable attention in learning theory, too. Many papers studied learnability by nondeterministic algorithms [1, 5, 11, 29] and probabilistic algorithms [14, 17, 21, 22, 25, 26].

Definition 1. *We say that a nondeterministic QIM learns a function f if*

(1) *there is at least one computation path such that the QIM produces a correct result on f, i.e., program j such that $\varphi_j = f$;*
(2) *at no computation path the QIM produces an incorrect result on f.*

Definition 2. *We say that a probabilistic QIM learns a function f with a probability p if*

(1) *the sum of all probabilities of all leaves which produce a correct result on f, i.e., a number j such that $\varphi_j = f$, is no less than p;*
(2) *at no computation path the QIM produces an incorrect result on f.*

Recently, Freivalds [7] introduced a new type of indeterministic algorithms called *ultrametric* algorithms. An extensive research on ultrametric algorithms of various kinds is performed by him and his co-authors (cf. [4, 15]). So, ultrametric algorithms are a very new concept and their potential still has to be explored. This is the first paper showing a problem where ultrametric algorithms *have advantages* over nondeterministic algorithms. Ultrametric algorithms are very similar to probabilistic algorithms but while probabilistic algorithms use *real* numbers r with $0 \leq r \leq 1$ as parameters, ultrametric algorithms use *p-adic* numbers as parameters. The usage of p-adic numbers as *amplitudes* and the ability to perform *measurements* to transform amplitudes into real numbers are inspired by quantum computations and allow for algorithms not possible in classical computations. Slightly simplifying the description of the definitions, one can say that ultrametric algorithms are the same as probabilistic algorithms, only the *interpretation* of the probabilities is *different*.

The choice of p-adic numbers instead of real numbers is not quite arbitrary. Ostrowski [24] proved that any non-trivial absolute value on the rational numbers $\mathbb{Q}$ is equivalent to either the usual real absolute value or a p-adic absolute value. This result shows that using p-adic numbers was not merely one of many possibilities to generalize the definition of deterministic algorithms but rather the only remaining possibility not yet explored.

The notion of p-adic numbers is widely used in science. String theory [28], chemistry [19] and molecular biology [6, 16] have introduced p-adic numbers to describe measures of indeterminism. Indeed, research on indeterminism in nature has a long history. Pascal and Fermat believed that every event of indeterminism can be described by a real number between 0 and 1 called *probability*. Quantum physics introduced a description in terms of complex numbers called *amplitude of probabilities* and later in terms of probabilistic combinations of amplitudes most conveniently described by *density matrices*. Using p-adic numbers to describe indeterminism allows to explore some aspects of indeterminism but, of course, does not exhaust all the aspects of it.

There are many distinct p-adic absolute values corresponding to the many prime numbers p. These absolute values are traditionally called *ultrametric*. Absolute values are needed to consider *distances* among objects. We are used to rational and irrational numbers as measures for distances, and there is a psychological difficulty to imagine that something else can be used instead of rational and irrational numbers, respectively. However, there is an important feature that distinguishes p-adic numbers from real numbers. Real numbers (both rational and irrational) are linearly ordered, while p-adic numbers *cannot* be linearly ordered. This is why *valuations* and *norms* of p-adic numbers are considered.

The situation is similar in Quantum Computation (see [23]). Quantum amplitudes are complex numbers which also cannot be linearly ordered. The counterpart of valuation for quantum algorithms is *measurement* translating a complex number $a + bi$ into a real number $a^2 + b^2$. Norms of p-adic numbers are rational numbers. We continue with a short description of p-adic numbers.

2 p-adic Numbers and Ultrametric Algorithms

Let p be an arbitrary prime number. A number $a \in \mathbb{N}$ with $0 \leq a \leq p - 1$ is called a *p-adic digit*. A *p-adic integer* is by definition a sequence $(a_i)_{i \in \mathbb{N}}$ of p-adic digits. We write this conventionally as $\cdots a_i \cdots a_2 a_1 a_0$, i.e., the a_i are written from left to right.

If n is a natural number, and $n = \overline{a_{k-1}a_{k-2}\cdots a_1 a_0}$ is its p-adic representation, i.e., $n = \sum_{i=0}^{k-1} a_i p^i$, where each a_i is a p-adic digit, then we identify n with the p-adic integer (a_i), where $a_i = 0$ for all $i \geq k$. This means that the natural numbers can be identified with the p-adic integers $(a_i)_{i \in \mathbb{N}}$ for which all but finitely many digits are 0. In particular, the number 0 is the p-adic integer all of whose digits are 0, and 1 is the p-adic integer all of whose digits are 0 except the right-most digit a_0 which is 1.

To obtain p-adic representations of all rational numbers, $\frac{1}{p}$ is represented as $\cdots 00.1$, the number $\frac{1}{p^2}$ as $\cdots 00.01$, and so on. For any p-adic number it is allowed to have infinitely many (!) digits to the left of the "p-adic" point but only a finite number of digits to the right of it.

However, p-adic numbers are not merely a generalization of rational numbers. They are related to the notion of *absolute value* of numbers. If X is a nonempty set, a distance, or metric, on X is a function d from $X \times X$ to the nonnegative real numbers such that for all $(x, y) \in X \times X$ the following conditions are satisfied.

(1) $d(x, y) \geq 0$, and $d(x, y) = 0$ if and only if $x = y$,
(2) $d(x, y) = d(y, x)$,
(3) $d(x, y) \leq d(x, z) + d(z, y)$ for all $z \in X$.

A set X together with a metric d is called a *metric space*. The same set X can give rise to many different metric spaces. If X is a linear space over the real numbers then the *norm* of an element $x \in X$ is its distance from 0, i.e., for all $x, y \in X$ and α any real number we have:

(1) $\|x\| \geq 0$, and $\|x\| = 0$ if and only if $x = 0$,
(2) $\|\alpha \cdot y\| = |\alpha| \cdot \|y\|$,
(3) $\|x + y\| \leq \|x\| + \|y\|$.

Note that every norm induces a metric d, i.e., $d(x, y) = \|x - y\|$. A well-known example is the metric over $\mathbb{Q}$ induced by the ordinary absolute value. However, there are other norms as well. A norm is called *ultrametric* if Requirement (3) can be replaced by the stronger statement: $\|x+y\| \leq \max\{\|x\|, \|y\|\}$. Otherwise, the norm is called *Archimedean*.

Definition 3. *Let $p \in \{2, 3, 5, 7, 11, 13, \ldots\}$ be any prime number. For any nonzero integer a, let the p-adic ordinal (or valuation) of a, denoted $\mathrm{ord}_p a$, be the highest power of p which divides a, i.e., the greatest number $m \in \mathbb{N}$ such that $a \equiv 0 \pmod{p^m}$. For any rational number $x = a/b$ we define $\mathrm{ord}_p x =_{df} \mathrm{ord}_p a - \mathrm{ord}_p b$. Additionally, $\mathrm{ord}_p x =_{df} \infty$ if and only if $x = 0$.*

For example, let $x = 63/550 = 2^{-1} \cdot 3^2 \cdot 5^{-2} \cdot 7^1 \cdot 11^{-1}$. Thus, we have

$$\mathrm{ord}_2 x = -1 \qquad \mathrm{ord}_7 x = +1$$
$$\mathrm{ord}_3 x = +2 \qquad \mathrm{ord}_{11} x = -1$$
$$\mathrm{ord}_5 x = -2 \qquad \mathrm{ord}_p x = 0 \quad \text{for every prime } p \notin \{2, 3, 5, 7, 11\} .$$

Definition 4. *Let $p \in \{2, 3, 5, 7, 11, 13, \ldots\}$ be any prime number. For any rational number x, we define its p-norm as $p^{-\mathrm{ord}_p x}$, and we set $\|0\|_p =_{df} 0$.*

For example, with $x = 63/550 = 2^{-1} \cdot 3^2 \cdot 5^{-2} \cdot 7^1 \cdot 11^{-1}$ we obtain:

$$\|x\|_2 = 2 \qquad\qquad \|x\|_7 = 1/7$$
$$\|x\|_3 = 1/9 \qquad\qquad \|x\|_{11} = 11$$
$$\|x\|_5 = 25 \qquad\qquad \|x\|_p = 1 \quad \text{for every prime } p \notin \{2,3,5,7,11\}\,.$$

Rational numbers are p-adic integers for all prime numbers p. Since the definitions given above are all we need, we finish our exposition of p-adic numbers here. For a more detailed description of p-adic numbers we refer to [13, 18].

We continue with *ultrametric algorithms*. In the following, p always denotes a prime number. Ultrametric algorithms are described by finite directed acyclic graphs (abbr. DAG), where exactly one node is marked as root. As usual, the root does not have any incoming edge. Furthermore, every node having outdegree zero is said to be a *leaf*. The leaves are the output nodes of the DAG.

Let v be a node in such a graph. Then each outgoing edge is labeled by a p-adic number which we call *amplitude*. We require that the sum of all amplitudes that correspond to v is 1. In order to determine the *total amplitude* along a computation path, we need the following definition.

Definition 5. *The total amplitude of the root is defined to be 1. Furthermore, let v be a node at depth d in the DAG, let α be its total amplitude, and let $\beta_1, \beta_2, \cdots, \beta_k$ be the amplitudes corresponding to the outgoing edges $e_1, \ldots, e_k$ of v. Let $v_1, \ldots, v_k$ be the nodes where the edges $e_1, \ldots, e_k$ point to. Then the total amplitude of v_ℓ, $\ell \in \{1, \ldots, k\}$, is defined as follows.*

(1) *If the indegree of v_ℓ is one, then its total amplitude is $\alpha\beta_\ell$.*
(2) *If the indegree of v_ℓ is bigger than one, i.e., if two or more computation paths are joined, say m paths, then let $\alpha, \gamma_2, \ldots, \gamma_m$ be the corresponding total amplitudes of the predecessors of v_ℓ and let $\beta_\ell, \delta_2, \ldots, \delta_m$ be the amplitudes of the incoming edges The total amplitude of the node v_ℓ is then defined to be $\alpha\beta_\ell + \gamma_2\delta_2 + \cdots + \delta_m\gamma_m$.*

Note that the total amplitude is a p-adic integer.

We refer the reader to the proof of Theorem 7 for an example.

It remains to define what is meant by saying that a p-ultrametric algorithm produces a result with a certain probability. This is specified by performing a so-called *measurement* at the leaves of the corresponding DAG. Here by measurement we mean that we transform the total amplitude β of each leaf to $\|\beta\|_p$. We refer to $\|\beta\|_p$ as the *p-probability* of the corresponding computation path.

Definition 6. *We say that a p-ultrametric algorithm produces a result m with a p-probability q if the sum of the p-probabilities of all leaves which correctly produce the result m is no less than q.*

Definition 7. *We say that a p-ultrametric QIM learns a function f with a p-probability q if*

(1) *the sum of the p-probabilities of all leaves which produce a correct result on f, i.e., a number j such that $\varphi_j = f$, is no less than q;*
(2) *at no computation path the QIM produces an incorrect result on f.*

3   Results

As explained in the Introduction we are interested in the number of queries a QIM has to ask in the worst-case in order to infer all recursive functions from a prespecified class $\mathcal{U}$. The hypothesis space will always be a Gödel numbering φ (cf. [27]). This is no restriction of generality since all natural programming languages provide Gödel numberings of recursive functions.

The complexity of learning recursive functions has been an important topic for several decades [2, 8, 10, 30]. In this paper we compare the query complexity of deterministic, nondeterministic, probabilistic, and ultrametric QIMs.

Our results are somewhat unexpected. Usually, for various classes of problems, nondeterministic algorithms provide the smallest complexity, deterministic algorithms provide the largest complexity and probabilistic algorithms provide some medium complexity. In [4, 7, 15] ultrametric algorithms also gave medium complexity sometimes better and sometimes worse than probabilistic algorithms. Our results in this paper show that, for learning recursive functions from value queries, there are classes $\mathcal{U}$ of recursive functions such that ultrametric QIMs have a much smaller complexity than even nondeterministic QIMs.

To show these results we use a combinatorial structure called the *Fano plane*. It is one of *finite geometries* (see [20]). The Fano plane consists of seven *points* $0, 1, 2, 3, 4, 5, 6$ and seven *lines* $(0,1,3)$, $(1,2,4)$, $(2,3,5)$, $(3,4,6)$, $(4,5,0)$, $(5,6,1)$, $(6,0,2)$. For any two points i, j with $i \neq j$, in this geometry there is exactly one line that contains these points (cf. Figure 1). For any two different lines in this geometry there is exactly one point contained in these two lines. In our construc-

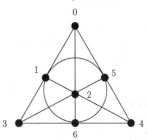

Fig. 1. The Fano Plane

tion the points $0, 1, 2, 3, 4, 5, 6$ are interpreted as colored in two colors RED and BLUE, respectively.

Lemma 1 ([20]). *For an arbitrary coloring of the Fano plane there is at least one line the 3 points of which are colored by the same color.*

Lemma 2 ([20]). *For any coloring of the Fano plane there cannot exist two lines colored in opposite colors.*

Proof. Any two lines intersect at some point.   □

To simplify notation, in the following we use $\mathcal{P}$ and $\mathcal{R}$ to denote the set of all partial recursive functions and of all recursive functions of one variable over $\mathbb{N}$, respectively. Let φ be a Gödel numbering of $\mathcal{P}$. We consider the following class $\mathcal{U}_7$ of recursive functions. Each function $f \in \mathcal{U}_7$ is such that $f \in \mathcal{R}$ and:

(1) every $f(x)$ where $0 \leq x \leq 6$ equals either 2^s or 3^t, where $s, t \in \mathbb{N}$, $s, t \geq 1$,
(2) if $0 \leq x_1 < x_2 \leq 6$, $f(x_1) = 2^s$ and $f(x_2) = 2^t$, then $f(x_1) = f(x_2)$,

(3) if $0 \le x_1 < x_2 \le 6$, $f(x_1) = 3^s$ and $f(x_2) = 3^t$, then $f(x_1) = f(x_2)$,
(4) there is a line (i, j, k) in the Fano plane such that $f(i) = f(j) = f(k) = 2^s$
    and $\varphi_s = f$ or there exists a line (i, j, k) in the Fano plane such that $f(i) = f(j) = f(k) = 3^t$ and $\varphi_t = f$.

Comment. In our construction of the class $\mathcal{U}_7$ the points $0, 1, 2, 3, 4, 5, 6$ can be interpreted as colored in two colors. Some points $f(i)$ are such that $f(i) = 2^s$ (these points are described below as RED) while some other points j are such that $f(j) = 3^t$ (these points are described below as BLUE). The properties of the Fano plane ensure that for every such coloring in two colors there exists a line such that the three points on this line are colored in the same color, and there cannot exist two lines colored in opposite colors.

Definition 8. *A partial coloring C of a Fano plane is an assignment of colors RED, BLUE, NONE to the points of the Fano plane.*
   A partial coloring C_2 is an extension *of a partial coloring C_1 if every point colored RED or BLUE in C_1 is colored in the same color in C_2.*
   A partial coloring C of a Fano plane is called complete *if every point is colored RED or BLUE.*

Lemma 3. *Given any partial coloring C of the points in the Fano plane assigning colors RED and BLUE to some but not all points such that no line contains three points in the same color, there exists*

(1) *a complete extension of the given coloring C such that it contains a line with three RED points, and*
(2) *a complete extension of the given coloring C such that it contains a line with three BLUE points.*

Proof. Color all the not colored points RED for the first function, and BLUE for the second function.   □

Lemma 4. *Given any partial coloring C of points in the Fano plane assigning colors RED and BLUE to some but not all points such that no line contains 3 points in the same color, there exist numbers $k_*, \ell_* \in \mathbb{N}$ and*

(1) *a function $f_{RED} \in \mathcal{U}_7$ defined as $f(x) = 2^{\ell_*}$ for all x colored RED in C such that f_{RED} contains a line with three RED points, and all the points $0, 1, 2, 3, 4, 5, 6$ are colored RED or BLUE,*
(2) *a function $f_{BLUE} \in \mathcal{U}_7$ defined as $f(x) = 3^{k_*}$ for all x colored BLUE in C such that f_{BLUE} contains a line with three BLUE points, and all the points $0, 1, 2, 3, 4, 5, 6$ are colored RED or BLUE.*

Proof. The assertions of the lemma can be shown by using the fixed point theorem [27] and by using Lemma 3.   □

Theorem 1. *There is a deterministic QIM M that learns the class $\mathcal{U}_7$ with 7 queries.*

Proof. The desired QIM M queries the points $0, 1, \ldots, 6$. After having received $f(0), f(1), f(2), \ldots, f(6)$, it checks at which line all points have the same color, and outputs the φ-program corresponding to this line. Note that by Lemmata 1 and 2 there is precisely one such line. By the definition of the class $\mathcal{U}_7$ one can directly output a correct φ-program for the target function f.   □

Theorem 2. *There exists no deterministic QIM learning $\mathcal{U}_7$ with 6 queries.*

Proof. The proof is by contradiction. Using Smullyan's double fixed point theorem [27] one can construct two functions f and $\tilde{f}$ such that both are in $\mathcal{U}_7$ but at least one of them is not correctly learned by the QIM M.   □

Theorem 3. *There is a nondeterministic QIM M learning $\mathcal{U}_7$ with 3 queries.*

Proof. The QIM M starts with a nondeterministic branching of the computation into 7 possibilities corresponding to the 7 lines in the Fano plane. In each case, all 3 points i, j, k are queried. If $f(i), f(j), f(k)$ are not of the same color then the computation path is aborted. If they are of the same color, e.g., $f(i) = 2^{s_i}, f(j) = 2^{s_j}, f(k) = 2^{s_k}$ then the definition of the class $\mathcal{U}_7$ ensures that $s_i = s_j = s_k$ and the QIM M outputs s_i which is a correct program computing the function f.   □

Theorem 4. *There is no nondeterministic QIM learning $\mathcal{U}_7$ with 2 queries.*

Proof. By Lemma 4, there are two distinct functions in the class $\mathcal{U}_7$ with the same values queried by the nondeterministic algorithm. The output is not correct for at least one of them.   □

Theorem 5. *There is a probabilistic QIM M learning $\mathcal{U}_7$ with probability $\frac{1}{7}$ with 3 queries.*

Proof. The algorithm starts with branching its computation into 7 possibilities corresponding to the 7 lines in Fano plane. Each branch is reached with probability $1/7$. In each branch, all 3 points i, j, k are queried. If $f(i), f(j), f(k)$ are not of the same color then the computation path is aborted. If they are of the same color, e.g., $f(i) = f(j) = f(k) = 2^s$, then s is output. By definition of the class $\mathcal{U}_7$ the result is a correct program computing the function f.   □

Theorem 6. *There is a probabilistic QIM M learning $\mathcal{U}_7$ with probability $\frac{4}{7}$ with 6 queries.*

Theorem 7. *For every prime number p, there is a p-ultrametric QIM M learning $\mathcal{U}_7$ with p-probability 1 with 2 queries.*

Proof. The desired QIM M branches its computation path into 7 branches at the root, where each branch corresponds to exactly one line of the Fano plane. We assign to each edge the amplitude $1/7$. At the second level, each of these branches is branched into 3 subbranches each of which is assigned the amplitude $1/3$. So far we have at level three 21 nodes denoted by $v_1, \ldots, v_{21}$ (cf. Figure 2). For each of these nodes we formulate two queries. Let v be such that its father

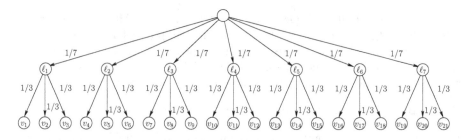

Fig. 2. The first three levels of the DAG representing the computation of the QIM M

node corresponds to the line containing the point i, j, k of the Fano plane, where we order these points such that $i < j < k$. If v is the leftmost node then we query (i, j), if v is the middle node then we query (j, k) and if v is the rightmost node then we query (i, k). Every triple of nodes having the same father share a register, say r_{ijk}. Initially, the register contains the value $\uparrow$ which stands for "no output." The node activated when reached in the computation path sends the following value to r_{ijk}. After having received the answer to its queries, e.g., $f(i) = 2^s$ and $f(j) = 3^t$ then it writes 0 in r_{ijk}, and if the values coincide, e.g., $f(i) = 3^t$ and $f(j) = 3^t$, then it writes t in r_{ijk}.

Looking at any triple of nodes having a common father at the third level, we note that the following 8 cases may occur as answer. We use again the corresponding colors, where R and B are shortcuts for RED and BLUE, respectively.

(i,j)	(j,k)	(i,k)	(i,j)	(j,k)	(i,k)
(B,B)	(B,R)	(B,R)	(R,R)	(R,B)	(R,B)
(B,B)	(B,B)	(B,B)	(R,R)	(R,R)	(R,R)
(B,R)	(R,R)	(B,R)	(R,B)	(B,B)	(R,B)
(B,R)	(R,B)	(B,B)	(R,B)	(B,R)	(R,R)

Thus, we need for each node at the third level 8 outgoing edges as the table above shows. If the edge corresponds to a pair (R, R) or (B, B) then we assign the amplitude $1/2$ and otherwise the amplitude $-1/4$. Note that sum of these amplitudes is again 1.

Finally, we join each triple as shown in table above into one node, e.g., the edges corresponding to (B, B), (B, R), and (B, R) are joined. If the total amplitude of such a node at the third level is different from zero, then the node produces as output the value stored in register r_{ijk}. Figure 3 shows the part of the DAG for the queries performed for the first line of the Fano plane, i.e., for the line $(0, 1, 3)$. So this part starts at the nodes v_1, v_2 and v_3 shown in Figure 2. For the sake of readability, we show the queries asked at each node, i.e., $(0, 1)$ at node v_1, $(1, 3)$ at node v_2, and $(0, 3)$ at node v_3. A dashed (blue) edge denotes the case that both answers to the queries asked at the corresponding vertex returned a value of f indicating that the related nodes of the first line of the Fano plane are blue. This result is then propagated along the dashed (blue) edges. Analogously, a dotted (red) edge indicates that both answers corresponded to a red node of the first line of the Fano plane. If the answers returned function values indicating that the colors of the queried nodes of the first line of the Fano plane have different colors then the edge is drawn in black. Dashed (blue) and dotted (red) edges have the amplitude $1/2$ and the black edges have the the amplitude $-1/4$.

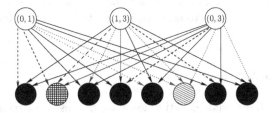

Fig. 3. The part of the DAG representing the computation of the QIM M for the line $(0, 1, 3)$ starting at the nodes of the third level

It remains to show that the QIM M has the desired properties. By construction, at every computation path exactly two queries are asked.

Next, by Definition 5 it is obvious that the total amplitude of each node at the second level is $1/21$. Next, we consider any node at the third level. If a triple (B, B), (B, B), and (B, B) is joined then the total amplitude is

$$\frac{1}{21} \cdot \frac{1}{2} + \frac{1}{21} \cdot \frac{1}{2} + \frac{1}{21} \cdot \frac{1}{2} = \frac{1}{2 \cdot 7} \, .$$

The same holds for (R, R), (R, R), and (R, R) (cf. Definition 5). Figure 3 shows the corresponding leaves in a squared pattern and lined pattern, respectively.

If a triple has a different form than considered above, e.g., (B, B), (B, R), and (B, R) then, again by Definition 5, we have for the total amplitude

$$\frac{1}{21} \cdot \frac{1}{2} - \frac{1}{21} \cdot \frac{1}{4} - \frac{1}{21} \cdot \frac{1}{4} = 0 \, .$$

One easily verifies that all remaining total amplitudes are also 0. Finally, we perform the measurement. Clearly, for each leaf which has a total amplitude 0 the measurement results in $\|0\|_p = 0$. For the remaining leaves we obtain $\|\frac{1}{2 \cdot 7}\|_p$ which is 1 for every prime p such that $p \notin \{2, 7\}$. If $p = 2$ then we have $\|\frac{1}{2 \cdot 7}\|_2 = 2$ and for $p = 7$ we get $\|\frac{1}{2 \cdot 7}\|_7 = 7$.

By Lemma 1 there must be at least one line such that all nodes have the same color, and by Lemma 2 it is not possible to have a line colored in RED and a line colored in BLUE simultaneously. So at least one node has p-probability at least 1, and the result output is correct by the definition of the class $\mathcal{U}_7$.

If there are several lines colored in the same color then distinct but correct results may be produced, since any two lines share exactly one point. Thus, the resulting p-probability is always no less than 1. $\qquad\qquad\square$

The idea of this paper can be extended to obtain even more spectacular advantages of ultrametric algorithms over nondeterministic ones. It is proved that there exist finite projective geometries with $n^2 + n + 1$ points and $n^2 + n + 1$ lines such that any two lines have exactly one common point and any two points lie on a common line. This allows us to construct a class $\mathcal{U}_m$ of recursive functions similar to the class $\mathcal{U}_7$ above, where $m =_{df} q^2 + q + 1$ for any prime power q. The counterpart of Lemma 1 does not hold but this demands only an additional requirement for the function in the class to have a line colored in one color. Due to the lack of space, we have to omit these results here, but refer the interested reader to [9].

4   Conclusions

In this paper we have studied active learning of classes of recursive functions from value queries. We compared the query complexity of deterministic, nondeterministic, probabilistic, and ultrametric QIM and showed the somehow unexpected result that p-ultrametric QIM can learn classes of recursive function with significantly fewer queries than nondeterministic, probabilistic QIM can do.

References

[1] Ambainis, A., Apsītis, K., Freivalds, R., Smith, C.H.: Hierarchies of probabilistic and team *FIN*-learning. Theoret. Comput. Sci. 261(1), 91–117 (2001)

[2] Ambainis, A., Smotrovs, J.: Enumerable classes of total recursive functions: Complexity of inductive inference. In: Arikawa, S., Jantke, K.P. (eds.) AII and ALT 1994. LNCS (LNAI), vol. 872, pp. 10–25. Springer, Heidelberg (1994)

[3] Angluin, D.: Queries and concept learning. Machine Learning 2(4), 319–342 (1988)

[4] Balodis, K., Beriņa, A., Cīpola, K., Dimitrijevs, M., Iraids, J., Jēriņš, K., Kacs, V., Kalējs, J., Krilšauks, R., Lukstiņš, K., Raumanis, R., Scegulnaja, I., Somova, N., Vanaga, A., Freivalds, R.: On the state complexity of ultrametric finite automata (unpublished manuscript 2013)

[5] Denis, F., Lemay, A., Terlutte, A.: Learning regular languages using non deterministic finite automata. In: Oliveira, A.L. (ed.) ICGI 2000. LNCS (LNAI), vol. 1891, pp. 39–50. Springer, Heidelberg (2000)

[6] Dragovich, B., Dragovich, A.Y.: A p-adic model of DNA sequence and genetic code. p-Adic Numbers, Ultrametric Analysis, and Applications 1(1), 34–41 (2009)

[7] Freivalds, R.: Ultrametric automata and Turing machines. In: Voronkov, A. (ed.) Turing-100. EPiC Series, vol. 10, pp. 98–112. EasyChair (2012)

[8] Freivalds, R., Bārzdiņš, J., Podnieks, K.: Inductive inference of recursive functions: Complexity bounds. In: Barzdins, J., Bjorner, D. (eds.) Baltic Computer Science. LNCS, vol. 502, pp. 111–155. Springer, Heidelberg (1991)

[9] Freivalds, R., Zeugmann, T.: Active learning of classes of recursive functions by ultrametric algorithms. Tech. Rep. TCS-TR-A-13-68, Division of Computer Science, Hokkaido University (2013)

[10] Freivalds, R., Kinber, E., Smith, C.H.: On the impact of forgetting on learning machines. J. ACM 42(6), 1146–1168 (1995)

[11] García, P., de Parga, M.V., Álvarez, G.I., Ruiz, J.: Learning regular languages using nondeterministic finite automata. In: Ibarra, O.H., Ravikumar, B. (eds.) CIAA 2008. LNCS, vol. 5148, pp. 92–101. Springer, Heidelberg (2008)

[12] Gold, E.M.: Language identification in the limit. Inform. Control 10(5), 447–474 (1967)

[13] Gouvea, F.Q.: p-adic Numbers: An Introduction (Universitext), 2nd edn. Springer, Berlin (1983)

[14] Greitāne, I.: Probabilistic inductive inference of indices in enumerable classes of total recursive functions. In: Jantke, K.P. (ed.) AII 1989. LNCS (LNAI), vol. 397, pp. 277–287. Springer, Heidelberg (1989)

[15] Jēriņš, K., Balodis, K., Krišlauks, R., Cīpola, K., Freivalds, R.: Ultrametric query algorithms (unpublished manuscript 2012)

[16] Khrennikov, A.: Non-Archimedean Analysis: Quantum Paradoxes, Dynamical Systems and Biological Models. Kluwer Academic Publishers (1997)

[17] Kinber, E., Zeugmann, T.: One-sided error probabilistic inductive inference and reliable frequency identification. Inform. Comput. 92(2), 253–284 (1991)

[18] Koblitz, N.: p-adic Numbers, p-adic Analysis, and Zeta-Functions, 2nd edn. Springer, Berlin (1984)

[19] Kozyrev, S.V.: Ultrametric analysis and interbasin kinetics. In: American Institute of Physics Conference Proceedings of the 2nd International Conference on p-Adic Mathematical Physics, Belgrade, Serbia and Montenegro, September 15-21, vol. 826, pp. 121–128 (2006)

[20] Meserve, B.E.: Fundamental Concepts of Geometry. Dover Publications, New York (1983)

[21] Meyer, L.: Probabilistic language learning under monotonicity constraint. Theoret. Comput. Sci. 185(1), 81–128 (1997)

[22] Meyer, L.: Aspects of complexity of probabilistic learning under monotonicity constraints. Theoret. Comput. Sci. 268(2), 275–322 (2001)

[23] Nielsen, M.A., Chuang, I.L.: Quantum Computation and Quantum Information. Cambridge University Press (2000)

[24] Ostrowski, A.: Über einige Lösungen der Funktionalgleichung $\varphi(x) \cdot \varphi(y) = \varphi(xy)$. Acta Mathematica 41(1), 271–284 (1916)

[25] Pitt, L.: Probabilistic inductive inference. J. ACM 36(2), 383–433 (1989)

[26] Pitt, L., Smith, C.H.: Probability and plurality for aggregations of learning machines. Inform. Comput. 77(1), 77–92 (1988)

[27] Rogers, Jr., H.: Theory of Recursive Functions and Effective Computability. McGraw-Hill (1967); reprinted. MIT Press (1987)

[28] Vladimirov, V.S., Volovich, I.V., Zelenov, E.I.: P-Adic Analysis and Mathematical Physics. World Scientific, Singapore (1994)

[29] Wu, Y.-C., Lee, Y.-S., Yang, J.-C., Yen, S.-J.: An integrated deterministic and nondeterministic inference algorithm for sequential labeling. In: Cheng, P.-J., Kan, M.-Y., Lam, W., Nakov, P. (eds.) AIRS 2010. LNCS, vol. 6458, pp. 221–230. Springer, Heidelberg (2010)

[30] Zeugmann, T., Zilles, S.: Learning recursive functions: A survey. Theoret. Comput. Sci. 397(1-3), 4–56 (2008)

Efficient Error-Correcting Codes
for Sliding Windows

Ran Gelles[1], Rafail Ostrovsky[1,2,⋆], and Alan Roytman[1]

[1] Department of Computer Science, University of California, Los Angeles
[2] Department of Mathematics, University of California, Los Angeles
{gelles,rafail,alanr}@cs.ucla.edu

Abstract. We consider the task of transmitting a data stream in the sliding window model, where communication takes place over an *adversarial* noisy channel with noise rate up to 1. For any noise level $c < 1$ we design an efficient encoding scheme, such that for any window in which the noise level does not exceed c, the receiving end decodes at least a $(1-c-\varepsilon)$-prefix of the window, for any small $\varepsilon > 0$. Decoding more than a $(1-c)$-prefix of the window is shown to be impossible in the worst case, which makes our scheme optimal in this sense. Our scheme runs in polylogarithmic time per element in the size of the window, causes constant communication overhead, and succeeds with overwhelming probability.

1 Introduction

As data continues to grow in size for many real world applications, streaming algorithms play an increasingly important role. Big data applications, ranging from sensor networks [9] to analyzing DNA sequences [5], demand streaming algorithms in order to deal with the tremendous amount of information being generated in a very short period of time. For certain applications, it is useful to only maintain statistics about recent data (rather than the entire stream). For instance, we may be interested in analyzing stock market transactions within the last hour, or monitoring and analyzing packets entering a network to detect suspicious activity, or identifying patterns in genomic sequences. This model is known as the *sliding windows model*, in which we care about a fixed-length window of time (say, 1 hour), which "slides" forward as time moves on (e.g., the last one hour, etc.).

While in the standard sliding window model, the aim is to maintain some statistics of an input stream (usually, using only polylogarithmic memory), our

---

⋆ Research supported in part by NSF grants CNS-0830803; CCF-0916574; IIS-1065276; CCF-1016540; CNS-1118126; CNS-1136174; US-Israel BSF grant 2008411, OKAWA Foundation Research Award, IBM Faculty Research Award, Xerox Faculty Research Award, B. John Garrick Foundation Award, Teradata Research Award, and Lockheed-Martin Corporation Research Award. This material is also based upon work supported by the Defense Advanced Research Projects Agency through the U.S. Office of Naval Research under Contract N00014-11-1-0392. The views expressed are those of the authors and do not reflect the official policy or position of the Department of Defense or the U.S. Government.

V. Geffert et al. (Eds.): SOFSEM 2014, LNCS 8327, pp. 258–268, 2014.
© Springer International Publishing Switzerland 2014

focus is applications in which the stream is generated in one place and then communicated to another place, as is the case for sensor networks, for instance. Inevitably, such applications are particularly vulnerable to data corruption while transmitting symbols: sensors are often placed in harsh environments connected to a base unit by a wireless channel which is susceptible to various kinds of errors (a weak signal, noise from other nearby transmitters, physical blockage of the transmitting medium, or even physical damage to the sensor). A natural question arises, how can we handle this constant stream of corrupted data while still being able to make sense of the information?

Our paper precisely aims to answer this question. We study encoding schemes for transmitting data streams in the sliding windows setting. We aim to design encoding schemes that can tolerate a high amount of errors while keeping the added redundancy low. Informally, we mainly consider noise rates above $1/2$, while adding only a constant overhead.[1] Another requirement of our scheme is to be efficient, that is, the encoding and decoding time (per element) must be at most polylogarithmic in the window size N.[2]

Before we describe our result in more details, let us explain why simple solutions for this task fail to work. Assume that an encoding scheme is meant to protect against a burst of noise of length n. A straightforward solution would be to cut the stream into chunks of size $N = (2 + \varepsilon)n$ and encode each chunk using some standard error-correcting code that can deal with a fraction of errors of almost $1/2$ (e.g., a Shannon error-correcting code [12] with good distance). This solution has three downsides. First, if the noise within one block is even slightly more than n, there are some messages for which the entire block will be incorrectly decoded. Second, the receiving side will be able to decode a block only after receiving the block entirely, which implies a delay of $2n$. Last, and maybe the most important downside, is that the rate of errors in one window using this solution cannot exceed $1/2$. We can replace the Shannon code with a code that resists a fraction of noise arbitrarily close to 1, such as the code of Micali, Peikert, Sudan, and Wilson [8], however this code uses an alphabet size that grows with the block size, and the obtained scheme would have a super-constant overhead. No codes over a constant-size alphabet are known to resist more than $1/2$ fraction of errors.

Contributions and Techniques. We provide a family of constant-rate encoding schemes in the sliding window model such that for any $c < 1$, if the rate of noise in the last window is less than c, the decoding of a $(1 - c)$-prefix of the window succeeds with overwhelming probability. Our scheme has polylogarithmic time complexity per element and linear space complexity. Formally,

---

[1] Loosely speaking, the *overhead* of a scheme is the amount of data communicated in one window divided by the amount of input data in that window, as a function of the windows size N.

[2] A similar polylogarithmic bound on the memory consumption would also be desired. However, at least for the sender, we show that the memory consumption must be at least linear in the window size, $\Omega(N)$.

Theorem 1. *For any $c < 1$ and any $\varepsilon > 0$, there exists an efficient encoding/decoding scheme for streams in the sliding window model with the following properties.*

For any window W of size N, if the fraction of errors within W is less than c, then except with negligible probability $N^{-\omega(1)}$, the decoder correctly decodes a prefix of W of size at least $(1 - c - \varepsilon)N$. The decoding takes $poly(\log N)$ time per element and the encoding takes $O(1)$ amortized time.

Clearly, for a noise rate c, there is no hope of decoding more than a $(1 - c)$-prefix of the window (e.g., consider decoding at time N when the channel was "blocked," i.e. fully corrupted, starting at time $(1 - c)N$), and our protocol is optimal in this aspect. Although a suffix of the current window may not be decoded, the data is not lost and will be decoded eventually as the window slides (if the noise rate is below c). Moreover, the delay of our scheme is $(c + \varepsilon)N$, since we only guarantee that the receiver correctly decodes the first $(1 - c - \varepsilon)N$ symbols in each window, even though N symbols have already been received. Given that our goal is to decode as large a prefix as possible, and that we cannot hope to decode more than a $(1 - c)$-prefix of the window, any scheme must incur a delay of at least cN, and hence the delay of our scheme is near-optimal.

To provide some intuition and insight into the techniques we use, we begin with the work of Franklin, Gelles, Ostrovsky, and Schulman for encoding streams in the unbounded model [1]. Roughly speaking, their scheme works as follows. At any particular time n (i.e., the length of the stream seen so far), the stream is split into blocks of size $\log n$, each of which is encoded using an online coding scheme called a *tree-code* [10,11]. At each time step, a constant number of blocks are chosen uniformly at random and the next symbol is sent (in each chosen block) according to the encoding determined by the tree code (it is assumed that the sender and receiver have access to a shared random string, so the receiver knows which random blocks are chosen by the sender). This encoding is concatenated with a weak message-authentication code, that detects a large fraction of errors and effectively makes the scheme resistant to higher noise rates. We observe that tree codes have several shortcomings. First, decoding a codeword using a tree code takes exponential time in the block size. Second, tree codes (with constant rate) are known to exist, but take exponential time to construct (though efficient relaxations exist, see [2]). These shortcomings imply that any efficient scheme is restricted to using tree codes on words of at most logarithmic length, and thus obtain at least a polynomially small failure probability.

Our scheme takes inspiration from the ideas in [1]. We also divide our window into blocks of equal size, however we consider only blocks that are in the window. The key observation is that we no longer need the "online" property of tree-codes. This is because, in the sliding window model, we care only about a fixed window size N, and the blocks within this window are well defined in advance. Hence, the blocks do not increase in size when more elements arrive, as in the unbounded case [1]. To completely avoid the need for an "online" coding, our scheme waits until an entire block arrives and then encodes it using an error-correcting block code.[3] This allows us to replace the tree-code with a very efficient block code, e.g., the almost-optimal linear-time code of Guruswami and Indyk [3], and improve upon the efficiency in [1] to constant amortized time for the sender and polylogarithmic time for the receiver per time step. Moreover, we are not bound to logarithmic block size anymore, and by using polylogarithmic block size we reduce the failure probability from polynomially small as in [1] to negligible[4], while keeping the scheme time-efficient.

The resulting scheme is very similar to, and can be seen as an extension of the "code scrambling" method suggested by Lipton [7]. Lipton's scheme encodes a single message by chopping the message into blocks, with encoding and decoding being done per block. Then, the scheme permutes the entire message and adds a random mask to the permuted string, using some shared randomness. The scheme has a negligible failure probability for any random, computationally bounded channel with error probability $p < 1/2$. In contrast, our scheme works in the sliding window model and potentially encodes an infinite message. This requires a more clever "scrambling" technique than simply permuting the entire message. The random mask performed in [7] is replaced with an error-detection code, which increases the resilience of our scheme. We analyze our scheme against an unbounded adversary and show that it can handle a corruption rate that is arbitrarily close to 1 while guaranteeing a constant rate and a negligible failure probability.

While our scheme resists fully adversarial channels, it requires a large amount of shared-randomness. We can reduce the amount of shared randomness via standard techniques (same as in [7]), under the assumption of a computationally bounded channel. In this case a short random key is assumed to be shared between the parties from which randomness is expanded as needed via a pseudo-random generator.

Other Related Work. Coding schemes that assume the parties pre-share some randomness first appeared in [13], and were greatly analyzed since. The main advantage of such codes is that they can deal with adversarial noise, rather than random noise. Langberg [6] showed codes that approach Shannon's bound and require only $O(\log n)$ randomness for a block size of n, as well as an $\Omega(\log n)$ lower bound on the amount of necessary randomness. The construction of Langberg also implies an efficient code with $O(n \log n)$ randomness. This result was improved to $n + o(n)$ randomness by Smith [14]. Explicit constructions with $o(n)$ randomness are not yet known (see [14]).

2   Preliminaries

For a number n we denote by $[n]$ the set $\{1, 2, \ldots, \lfloor n \rfloor\}$, and for a finite set Σ we denote by $\Sigma^{\leq n}$ the set $\cup_{k=1}^{n} \Sigma^k$. Throughout the paper, $\log()$ denotes the

---

[3] For a block size B, this incurs an additional delay of $O(B)$, but since our goal is to decode a $(1 - c - \varepsilon)$-prefix of the window, we already have an inherent delay of cN in any case (clearly, B will be sub-linear in N).

[4] See Section 2 for a formal definition of a negligible function.

binary logarithm (base 2). A *data stream* S is a (potentially infinite) sequence of elements $(a_0, a_1, a_2 \ldots)$ where element $a_t \in \{1, \ldots, u\}$ arrives at time t. In the *sliding window model* we consider at each time $t \geq N$ the last N elements of the stream, i.e. the window $W = \{a_{t-(N-1)}, \ldots, a_t\}$. These elements are called *active*, whereas elements that arrived prior to the current window $\{a_i \mid 0 \leq i < t - (N - 1)\}$ are *expired*. For $t < N$, the window consists of all the elements received so far, $\{a_0, \ldots, a_t\}$. Finally, we say that a function $f(N)$ is *negligible* if for any constant $c > 0$, $f(N) < \frac{1}{N^c}$ for sufficiently large N.

Shared Randomness Model. We assume the following model known as the *shared-randomness model*. The legitimate users (the sender and the receiver) have access to a random string Rand of unbounded length, which is unknown to the adversary. Protocols in this model are thus *probabilistic*, and are required to succeed with high probability over the choice of Rand. We assume that all the randomness comes from Rand and that for a fixed Rand the protocols are deterministic.

Error-detection Codes: The Blueberry Code. The Blueberry-Code (BC) is a randomized error-detection code introduced by Franklin, Gelles, Ostrovsky, and Schulman [1], in which each symbol is independently embedded into a larger symbol space via a random mapping. Since the mapping is random and unknown to the adversary, each corruption is detected with some constant probability.

Definition 2. *For $i \geq 1$ let* $\mathsf{BC}_i : \Sigma_I \to \Sigma_O$ *be a random and independent mapping. The* Blueberry code *maps a string* $x \in \Sigma_I^*$ *into a string* $\mathsf{BC}(x) \in \Sigma_O^*$ *of the same length where* $\mathsf{BC}(x) = \mathsf{BC}_1(x_1)\mathsf{BC}_2(x_2) \cdots \mathsf{BC}_n(x_n)$.

Conditioned on the fact that the legitimate users use a message space $\Sigma_I \subset \Sigma_O$, each corruption can independently be detected with probability $1 - q$, where $q = \frac{|\Sigma_I| - 1}{|\Sigma_O| - 1}$ (hence, q is the probability that it remains undetected). The users are assumed to be sharing the random mappings BC_is at the start of the protocol.

Linear-time Error-correcting Codes. We will use error-correcting codes which are very efficient (i.e., linear-time in the block size for encoding and decoding). Such codes were initially (explicitly) constructed by Spielman [16] (see also [15]). Specifically, we use a linear-time error-correcting code with almost optimal rate given by Guruswami and Indyk [3]:

Lemma 3 ([3]). *For every rate $0 < r < 1$, and all sufficiently small $\delta > 0$, there exists an explicit family of error-correcting codes* $\mathsf{ECC} : \Sigma^{rn} \to \Sigma^n$, $\mathsf{ECC}^{-1} : \Sigma^n \to \Sigma^{rn}$ *over an alphabet of size* $|\Sigma| = O_{\varepsilon,r}(1)$ *that can be encoded and (uniquely) decoded in time linear in n from a fraction e of errors and d of erasures provided that $2e + d \leq (1 - r - \delta)$.*

Communication and Noise Model. Our communication model consists of a channel $ch : \Sigma \to \Sigma$ subject to corruptions made by an adversary (or by the channel itself). For all of our applications we assume that, at any given time slot (i.e., for any arriving element of the stream), a constant number R of channel instantiations are allowed. We say that R is the blowup or overhead of our scheme.

The noise model is such that any symbol σ sent through the channel can turn into another symbol $\tilde{\sigma} \in \Sigma$. It is not allowed to insert or delete symbols. We assume the noise is *adversarial*, that is, the noise is made by an all-powerful entity that is bounded only in the amount of noise it is allowed to introduce (the adversarial noise model subsumes the common noise models of random-error and burst-error). We say that the *corruption rate* in the window is c, if the fraction of corrupted transmissions over the last N time steps is at most c.

3   A Polylogarithmic Sliding Window Coding Scheme

We consider the problem of streaming authentication in the sliding windows model. In this setting, we have a fixed window size N (assumed to be known in advance). At each time step, one element expires from this window and one element arrives.

The idea is the following. The sender maintains blocks of size s of elements from the current window, which means there are $\frac{N}{s}$ blocks in the window. All the blocks have the same amount of active elements, except the last block which may only be partially full, and the first one which may have several elements that have already expired. When all the elements of the first block have expired we remove it and re-number the indices; additionally, when the last block becomes full we introduce a new (empty) block to hold the arriving elements.

When the sender has received an entire block[5] from the stream, the block is encoded using a linear-time error-correcting code [3]. At each time step, one of the $\frac{N}{s}$ blocks is chosen uniformly at random and the next (unsent) symbol of the encoded string is communicated over the channel after being encoded via an error-detection code (this gives us better rate, and maximizes the amount of decoded information). The protocol is described in Algorithm 1.

We now continue to analyze the properties of Algorithm 1, and show how to fix its parameters so it will satisfy the conditions of Theorem 1.

Proposition 4. *Suppose that the rate of corruptions in the window, for a given time t, is at most $c < 1$. Fix a small enough $\varepsilon > 0$. Denote by TOTAL_k the number of total transmissions for B_k up to time t; by $\mathsf{CORRUPT}_k$ the number of transmissions in B_k up to time t which are corrupted; and by ERR_k the number of errors in B_k up to time t (i.e., corrupted transmissions that were not identified by the error-detection code BC). Then, for any $k \in [(1 - c - \varepsilon)\frac{N}{s}]$ the following holds.*

1. *$(c + \varepsilon)Rs \leq \mathbb{E}[\mathsf{TOTAL}_k] \leq Rs$.*
2. *$\mathbb{E}[\mathsf{CORRUPT}_k] \leq cRs$.*
3. *$\mathbb{E}[\mathsf{ERR}_k] \leq cqRs$.*

---

[5] We assume only complete blocks, that is, the scheme skips the last block until it contains exactly s elements. This causes an additional delay of s time slots.

Let the parameters of the protocol $c, \varepsilon < 1$ be fixed. Let $s, R \in \mathbb{N}$ be such that $s < N$. Assume an online (symbol-wise) error-detection code BC with failure probability q per corrupted symbol. Assume a linear-time error-correction code $\mathsf{ECC}() : \Sigma^s \to \Sigma^{s/r}$ with a rate $r < 1$ to be fixed later.

<u>Sender:</u> Maintain blocks B_k of size at most s for $1 \le k \le \lfloor \frac{N}{s} \rfloor + 1$; any arriving element is appended to the last non-empty block.[6] If all elements in the first block expire, add a new (empty) block and remove B_1. Reindex the blocks so that the first block in the window is B_1 and the last is $B_{\lfloor \frac{N}{s} \rfloor + 1}$.

Maintain a counter $count_k$ for each block (initialized to 0 when the block is added).

foreach *time step t* **do**
    for $j = 1, \ldots, R$ **do**
        Choose $k \in [\frac{N}{s}]$ uniformly at random.
        $count_k \leftarrow count_k + 1$.
        Send the next symbol of $\mathsf{BC} \circ \mathsf{ECC}(B_k)$ which has not yet been sent
        according to $count_k$ (send $\bot$, if all the symbols were communicated).
    end
end

<u>Receiver:</u> The receiver maintains the same partition of the stream into blocks (with consistent indexing).

foreach *time step t* **do**
    for $j = 1, \ldots, R$ **do**
        Assume the sender, at iteration (t, j), sent a symbol from B_k. Let x^k
        denote the string obtained by concatenating all the symbols received so
        far that belong to the same B_k (preserving their order of arrival).
        Let $B'_k = \mathsf{ECC}^{-1} \circ \mathsf{BC}^{-1}(x^k)$.
    end
end
Output B'_k for any $k \in [\frac{N}{s}]$.

Algorithm 1. Sliding window error-correcting scheme

Proof.

1. Define TOTAL_k to be the number of times bucket B_k is chosen up to time t (here, we mean the same bucket of data, regardless of the fact that its index k is changed over time). Since the number of elements which appear in the window after B_k is at least $N - ks$ and at most $N - (k-1)s$, and because B_k is chosen with probability exactly $\frac{s}{N}$, we have that $\mathbb{E}[\mathsf{TOTAL}_k] = \Theta(R(N - ks) \cdot \frac{s}{N})$. Specifically, for any k, $R(N - ks)\frac{s}{N} \le \mathbb{E}[\mathsf{TOTAL}_k] \le R(N - ks + s)\frac{s}{N}$. The result follows since $k \le (1 - c - \varepsilon)\frac{N}{s}$.

---

[6] For the very first elements of the stream, we artificially create a window of size N with, say, all 0's (as if the scheme had already been running for N time steps) and similarly divide it up into blocks. This is done to keep notation consistent.

2. $\mathsf{CORRUPT}_k$ is the amount of corrupted transmissions in B_k up to time t. The number of corrupted transmissions in the window is at most cRN, and since each transmission has probability s/N of belonging to B_k, we get $\mathbb{E}[\mathsf{CORRUPT}_k] \leq RcN \cdot \frac{s}{N} = cRs$.
3. We use an error-detection code in which any change is caught with probability $1 - q$ but makes an error with probability q. Assuming that the rate of corruptions in the window is at most c, we have $\mathbb{E}[\mathsf{ERR}_k] \leq cRN \cdot q \cdot \frac{s}{N} = cqRs$. $\square$

Proposition 5. *Suppose the current window's corruption rate is at most c for some constant $c > 0$, and let $\varepsilon > 0$ be any sufficiently small constant. Suppose we divide up the stream into blocks of size s. Then there exist constants $R, q, r, \delta = O_{c,\varepsilon}(1)$ such that, except with probability $N2^{-\Omega(s)}$, B_k is correctly decoded by the receiver for every $k \in [(1 - c - \varepsilon)\frac{N}{s}]$.*

Proof. We consider the worst-case scenario in which the adversary corrupts a c-fraction of the transmissions, and moreover all corruptions occur in the last cRN transmissions (as this simultaneously maximizes the expected number of corruptions in each block B_k with $k \in [(1 - c - \varepsilon)\frac{N}{s}]$, since the expected number of corruptions of such blocks grows with time as long as the block remains in the window).

For a specific k, the probability of incorrectly decoding block B_k is bounded by $\Pr[2\mathsf{ERR}_k + \mathsf{DEL}_k > (1 - r - \delta)\frac{s}{r}]$. Here, r and δ are the two parameters of the error-correction scheme specified in Lemma 3. Namely, $\frac{1}{r}$ is the overhead incurred due to encoding each block of size s and δ is a parameter which trades off error tolerance against alphabet size.

By Proposition 4, we know $\mathbb{E}[\mathsf{ERR}_k] = cqRs$. Hence, using Chernoff bounds we know that for any $\xi > 0$, we have $\Pr[\mathsf{ERR}_k \geq (1 + \xi)\mathbb{E}[\mathsf{ERR}_k]] \leq e^{-\frac{\xi^2}{3}cqRs}$. Deletions in block B_k come from two sources. The first source, denoted by D_k^1, stems from choosing block B_k less than s/r times (i.e., TOTAL_k is small). The second source, denoted by D_k^2, comes from the BC code detecting corruptions (note that $\mathsf{DEL}_k = \mathsf{D}_k^1 + \mathsf{D}_k^2$).

Note that $\mathsf{D}_k^1 = \max(s/r - \mathsf{TOTAL}_k, 0)$, and in order to make it small with high probability, we can require $\mathbb{E}[\mathsf{TOTAL}_k] > s/r$. Using Proposition 4, $\mathbb{E}[\mathsf{TOTAL}_k] \geq (c + \varepsilon)Rs$, thus by choosing $R = \frac{(1+\xi)}{r(c+\varepsilon)}$ for some $\xi > 0$, and applying Chernoff bounds, we get that

$$\Pr\left[\mathsf{TOTAL}_k \leq \frac{s}{r}\right] \leq \Pr\left[\mathsf{TOTAL}_k < \left(1 - \frac{\xi}{1+\xi}\right)\mathbb{E}[\mathsf{TOTAL}_k]\right] < e^{-\frac{\xi^2 s}{2(1+\xi)r}}.$$

Hence, except with exponentially small probability, we know $\mathsf{TOTAL}_k \geq \frac{s}{r}$, which implies that $\mathsf{D}_k^1 = 0$. Next, observe that $\mathsf{ERR}_k + \mathsf{D}_k^2 = \mathsf{CORRUPT}_k$ and that by Proposition 4 and by applying the Chernoff bound we know that for any $\xi > 0$, $\Pr[\mathsf{CORRUPT}_k > (1 + \xi)cRs] < e^{-\frac{\xi^2}{3}cRs}$.

Putting these bounds together, we know that for any constant $\xi > 0$,

$$2\mathsf{ERR}_k + \mathsf{DEL}_k = \mathsf{ERR}_k + \mathsf{CORRUPT}_k \leq (1 + \xi)cqRs + (1 + \xi)cRs \quad (1)$$

(except with probability exponentially small in s). As long as this term is smaller than $(1-r-\delta)\frac{s}{r}$, then by Lemma 3 block B_k will be decoded correctly. Recalling that we set $R = \frac{(1+\xi)}{r(c+\varepsilon)}$ and substituting into the right-hand side of Eq. (1), we get the following constraint on r:

$$r \leq 1 - \delta - \frac{c}{c+\varepsilon}(1+\xi)^2(1+q). \tag{2}$$

Hence, for any constant rate $0 < r < 1 - \frac{c}{c+\varepsilon}$, we can always choose sufficiently small constants $\delta, q, \xi = O_{c,\varepsilon}(1)$ so that the constraint in Eq. (2) is satisfied.

So far we have only argued that we can decode *one* block B_k correctly except with probability exponentially small in s. We simply apply the union bound to get that, except with probability $N2^{-\Omega(s)}$, *every* block B_k for $k \in [(1-c-\varepsilon)\frac{N}{s}]$ can be decoded correctly. $\square$

Hence, with very high probability, we are able to guarantee that the entire prefix of the window can be decoded correctly at each time step. We now seek to analyze the efficiency of our scheme.

Proposition 6. *For any time t, the time complexity of Algorithm 1 is $O(1)$ (amortized) for the sender, and $O(s)$ for the receiver.*

Proof. Omitted. $\square$

Finally, we set $s = \omega(\log N)$ and obtain Theorem 1 immediately from Proposition 5 and Proposition 6.

Memory Consumption. We now consider the memory requirements of the sender and receiver. We focus on the space required for the "work" memory, which we consider to be any memory required to perform computation that is separate from the space for input, output, and randomness bits.

It is easy to see that, in our scheme, both the sender and receiver take linear space $O(N)$. For the sender, we obtain a matching $\Omega(N)$ lower bound: consider the case that the channel is completely "jammed" between time 0 and cN (each symbol is replaced with a random symbol), and then no more errors happen until time N. Since the noise rate in the first window is c, we expect the decoder to correctly output elements $a_1, \ldots, a_{(1-c)N}$ at time N. Since no information passed through the channel until time cN, at that time, the sender must possess (at least) the information $a_1, \ldots, a_{cN}$, thus his memory is lower bounded by $cN = \Omega(N)$.

As for the decoder, technically, the size of the output memory must be $\Omega(N)$ in order to decode a prefix of the window. However, if we wish to obtain a lower bound on the decoder's "work" memory, we note the following. In order for the decoder to guarantee at most a negligible probability of failure, the decoder's "work" memory must be $\omega(\log N)$. Otherwise, the decoder can save at most $O(\log N)$ transmissions, where a proportion c of them is corrupt in expectation. Regardless of the coding used for these elements, it would fail with non-negligible probability. Closing the memory-gap for the decoder is left for further research.

Effective Window. Since the decoder is able to decode a $(1 - c - \varepsilon)$-prefix of the current window, we can think of him as effectively decoding, at each time t a sliding window of size N', where $N' = (1 - c - \varepsilon)N$, whose start point is the same as the real window. The effective window ends cN time steps before the real window of the stream, which is precisely why the delay of our scheme is cN.

4   Conclusions and Open Questions

We have shown an efficient (polylogarithmic-time) coding scheme for data streams in the sliding window model. Somewhat surprisingly, while solving a problem in the sliding window model is usually a more difficult task than in the unbounded model, the case of communicating a stream in the sliding window model is simpler than the unbounded case. This allows us to improve on the methods built for the unbounded case [1] and achieve a more simple and efficient scheme for sliding windows.

While standard error-correcting schemes can resist a noise rate of at most $1/2$ (in one window), our scheme allows any error rate less than one at the cost of requiring the parties to pre-share some randomness before running the scheme. Our scheme is also advantageous in terms of delay.

While we aimed at achieving a constant-rate scheme, we have not analyzed the minimal possible rate as a function of the noise (R is clearly lower bounded by $1/(1 - c)$). Achieving an efficient scheme with optimal rate remains as an open question.

Finally, we mention that we can obtain a scheme that assumes no shared randomness, if we restrict the error rate up to $1/2$, and assume an *additive* (a.k.a. *oblivious*) adversary. Here, the channel fixes an error string to be added to the transmitted codeword, without seeing the transmitted codeword (though it can depend on the message and the coding scheme). This is done by employing the techniques of Guruswami and Smith [4]. Although this construction is somewhat technically involved, there are no novel techniques. We omit the details due to lack of space. Since the error rate is bounded by $1/2$, the only advantage this scheme has over the naïve approach described in Section 1, is reducing the delay.

References

1. Franklin, M., Gelles, R., Ostrovsky, R., Schulman, L.J.: Optimal coding for streaming authentication and interactive communication. In: Canetti, R., Garay, J.A. (eds.) CRYPTO 2013, Part II. LNCS, vol. 8043, pp. 258–276. Springer, Heidelberg (2013)
2. Gelles, R., Moitra, A., Sahai, A.: Efficient and explicit coding for interactive communication. In: FOCS 2011, pp. 768–777 (2011)
3. Guruswami, V., Indyk, P.: Linear-time encodable/decodable codes with near-optimal rate. IEEE Trans. on Information Theory 51(10), 3393–3400 (2005)
4. Guruswami, V., Smith, A.: Codes for computationally simple channels: Explicit constructions with optimal rate. In: FOCS 2010, pp. 723–732 (2010)

5. Kienzler, R., Bruggmann, R., Ranganathan, A., Tatbul, N.: Large-scale DNA sequence analysis in the cloud: a stream-based approach. In: Alexander, M., et al. (eds.) Euro-Par 2011, Part II. LNCS, vol. 7156, pp. 467–476. Springer, Heidelberg (2012)
6. Langberg, M.: Private codes or succinct random codes that are (almost) perfect. In: FOCS 2004, pp. 325–334. IEEE Computer Society, Washington, DC (2004)
7. Lipton, R.: A new approach to information theory. In: Enjalbert, P., Mayr, E.W., Wagner, K.W. (eds.) STACS 1994. LNCS, vol. 775, pp. 699–708. Springer, Heidelberg (1994)
8. Micali, S., Peikert, C., Sudan, M., Wilson, D.A.: Optimal error correction against computationally bounded noise. In: Kilian, J. (ed.) TCC 2005. LNCS, vol. 3378, pp. 1–16. Springer, Heidelberg (2005)
9. Munir, S., Lin, S., Hoque, E., Nirjon, S., Stankovic, J., Whitehouse, K.: Addressing burstiness for reliable communication and latency bound generation in wireless sensor networks. In: IPSN 2010, pp. 303–314 (2010)
10. Schulman, L.J.: Deterministic coding for interactive communication. In: STOC 1993, pp. 747–756 (1993)
11. Schulman, L.J.: Coding for interactive communication. IEEE Transactions on Information Theory 42(6), 1745–1756 (1996)
12. Shannon, C.E.: A mathematical theory of communication. ACM SIGMOBILE Mobile Computing and Communications Review 5(1), 3–55 (2001), originally appeared in Bell System Tech. J. 27, 379–423, 623–656 (1948)
13. Shannon, C.E.: A note on a partial ordering for communication channels. Information and Control 1(4), 390–397 (1958)
14. Smith, A.: Scrambling adversarial errors using few random bits, optimal information reconciliation, and better private codes. In: SODA 2007, pp. 395–404 (2007)
15. Spielman, D.: Computationally efficient error-correcting codes and holographic proofs. Ph.D. thesis, Massachusetts Institute of Technology (1995)
16. Spielman, D.: Linear-time encodable and decodable error-correcting codes. IEEE Transactions on Information Theory 42(6), 1723–1731 (1996)

Integrating UML Composite Structures and fUML

Alessandro Gerlinger Romero[1,*], Klaus Schneider[2],
and Maurício Gonçalves Vieira Ferreira[3]

[1] Brazilian National Institute for Space Research, São Paulo, Brazil
romgerale@yahoo.com.br
[2] Department of Computer Science, University of Kaiserslautern, Germany
schneider@cs.uni-kl.de
[3] Brazilian National Institute for Space Research, São Paulo, Brazil
mauricio@ccs.inpe.br

Abstract. To cope with the complexity of large systems, one usually makes use of hierarchical structures in their models. To detect and to remove design errors as soon as possible, these models must be analyzed in early stages of the development process. For example, UML models can be analyzed through simulation using the semantics of a foundational subset for executable UML models (fUML). However, the composite structures used to describe the hierarchy of systems in UML is not covered by fUML. In this paper, we therefore propose a complementary meta-model for fUML covering parts of UML's composite structures, and elaborate the rules previously defined in the literature for static semantics. These rules are described in an axiomatic way using first-order logic so that a large set of tools can be used for analysis. Our preliminary evaluation provides results about the applicability of the meta-model and the soundness of the rules.

Keywords: composite structures, static semantics, formal analysis.

1 Introduction

The modeling of today's systems requires to cope with their enormous complexity, which has at least two factors: disorder, and variety [12]. For the first factor, hierarchical structures are frequently used, while analysis is one of the techniques used to deal with the variety [12]. Models based on the Unified Modeling language (UML) use composite structures [7] to describe the structure of the systems in a hierarchical way. Furthermore, these models can be analyzed by simulation using the semantics of a foundational subset for executable UML models (fUML) [8].

UML composite structure is a fundamental technique to describe systems of systems with boundaries and connections between them [5]. This notion is

---

* This work was supported by the Brazilian Coordination for Enhancement of Higher Education Personnel (CAPES).

V. Geffert et al. (Eds.): SOFSEM 2014, LNCS 8327, pp. 269–280, 2014.

well suited for the definition of the components, which should have an explicit separation between internal elements, ports (a connection point) describing the provided features, and ports describing the required features of the environment [2]. However, fUML excluded the UML composite structures arguing that they are moderately used, and a straightforward translation is possible from them into the foundational subset ([8];pp. 20)([10]; pp.19). Moreover, the literature [2,4,5,6] recognizes the large number of ambiguities that emerges from a use of the composite structures without a thorough static semantics (context-sensitive constraints [8]). Although there are indications of how to deal with the integration of fUML and UML composite structures (see Section 2), the static semantics of composite structures (using fUML) still remains an open issue.

In fact, precise semantics for composite structures based on fUML is a request for proposal (RFP) from Object Management Group (OMG; [10]), which solicits specifications containing precise semantics for UML composite structures to enable execution and reduce ambiguities[10].

In this paper, we propose a complementary meta-model for fUML covering parts of UML composite structures, and elaborate the rules defined in the literature [2,4] for static semantics. These rules are described using axioms in first-order logic, what is in accordance with the RFP, which states: new axioms shall have explicit relationship with fUML base semantics ([10]; pp.22). The current research paper is not intended to satisfy all requirements from the RFP [10], and the major contributions of this work are: (1) integration of UML composite structure and fUML; and, (2) the static semantic rules. The latter is formally defined in first-order logic allowing advanced analysis and can be applied in any context where the assumptions and constraints described in Section 4 are fulfilled.

The remainder of this paper is organized as follows: in Section 2, we consider related work; in Section 3, the necessary background is presented; in Section 4, we define the CompositeStructure4fUML meta-model, explore the semantics, and declare the static semantics rules; in Section 5, we explore an example and discuss the results. Finally, conclusions are shared in the last section.

2   Related Works

Bock [1] described UML composite structures diagrams and their informal semantics. Equally important, it is stated in that paper that the goal of composite structures is to describe usages of classes (instances) and associations (links) in a given context, instead of associating classes globally. Similarly, Oliver and Luukkala [6] defined that the state space admitted by a composite structure is smaller than or equal to the state space admitted by classes and their associations.

Cuccuru et. al. [2] presented an evaluation of semantics for composite structures to support the request propagation across ports. Ober and Dragomir [4] refined the evaluation from [2] proposing that ports should be uni-directional because the bi-directionality raises typing problems.

Ober et. al. [5] discussed the gap between the expressiveness of UML and the requirements of the engineers. It is stated that the hierarchical decomposition, enabled by composite structures, proved to be a central technique for system modeling. Also, it recognized that ports are used to define simple connection points, where an incoming request is dispatched to a concrete handler. Nevertheless, UML defines a large number of options for port behavior modification.

Action language for foundational UML (Alf; [9]; see Section 3) has an informative annex defining a semantic integration with composite structure because *"executable behaviors will often be nested in some way within a component"* ([9]; pp. 365). This annex states that associations should be used by internal elements to access required services on ports, and it does not need associations to support provided services, which is an issue considering the concept of structure.

In conclusion, to the best of our knowledge, we have not found related work about the integration of composite structures and fUML on the previous static semantics rules studied in the literature [2,4,5]. Therefore, we elaborate the previous works covering issues that emerge from the integration.

3 UML Composite Structures and fUML

UML Composite Structures are *"a composition of interconnected elements, representing run-time instances collaborating over communication links to achieve some common objective"* ([7]; pp. 167). The main concepts in a composite structure are: internal elements, ports, and connectors. Internal elements are classifiers owned by an encapsulated classifier that can play a role in connections. Ports define the boundaries for a given classifier; they are defined to act as a connection point (at boundary) between the encapsulated classifier, its internal elements, and environment. Connectors are used to connect internal elements, and ports; a delegation connector is used to connect a port to internal elements, whereas an assembly connector is used to connect internal elements. In addition, connectors can be typed by an association, which provides the semantic channel.

fUML is a selection from a part of actions defined in UML (to model behavior), and part of expressiveness of classes (to model structure). The specification defines four elements for the language: (1) abstract syntax; (2) model library; (3) execution model; and, (4) base semantics [8]. It does not define a concrete syntax, so the only notation available for defining user models is the graphical notation provided by UML, namely activity diagrams, and class diagrams [8].

The abstract syntax is a subset of UML with complementary constraints. The model library defines primitive types and primitive functions. The execution model is an interpreter written in fUML (circular definition), which is defined using core elements from fUML that together form the base UML (bUML). Base semantics breaks the circular definition of fUML providing a set of axioms and inference rules that constrains the allowable execution of fUML models. Base semantics covers elements in bUML, and is defined using Common Logic Interchange Format (CLIF; [3]).

Alf is defined by providing a textual concrete syntax for fUML [9]. It is an action language that includes primitive types, primitive actions, and control flow mechanisms. Further, the execution semantics of Alf is given by mapping the Alf abstract syntax to the abstract syntax of fUML.

4   Metamodel - CompositeStructure4fUML

The integration of UML composite structures and fUML can be achieved through two techniques: (1) Translational or (2) Extensional. Focusing on the abstract syntax, Fig. 1 shows the relationship between these techniques and meta-models.

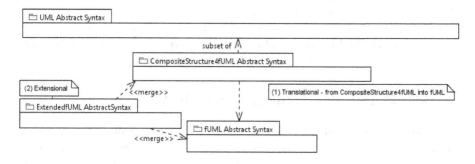

Fig. 1. Techniques to integrate UML Composite Structures and fUML

In the translational technique, the complementary meta-elements are translated into the fUML foundational elements. Therefore, there is no change of the fUML abstract syntax, and consequently, no change of the semantics. Moreover, this technique is constrained by the expressiveness of fUML. In contrast, a third meta-model (ExtendedfUML) is defined in the extensional technique by merging the models in it. As a result, the fUML abstract syntax is extended, therefore, its semantics must be reviewed.

Taken CompositeStructure4fUML into account, it can be used with these two techniques because its semantics is constrained by expressiveness of fUML. The constraints, requirements and assumptions for CompositeStructure4fUML are the following:

Constraint 1: One active object cannot access data that is managed by another active object (shared data between processes are forbidden). The reason for this constraint is that shared data can easily make systems inconsistent, and pose challenges to composability.

Constraint 2: The communication between objects cannot be bi-directional. The reason for this constraint is that the communication is best understood when the channel is uni-directional. This simplifies the static, and behavioral analyses, and there is no expressivity loss because a bi-directional channel can be replaced by two uni-directional channels [4].

Requirement 1: The semantics of CompositeStructure4fUML must be defined by fUML. As a result, a translation from CompositeStructure4fUML to fUML is not needed for simulation purposes.

Requirement 2: A translation from a UML surface (constrained by 1 and 2) to the CompositeStructure4fUML must be able to automatically generate predictable behaviors for ports and others elements needed for the semantics.

Assumption 1: Active objects (processes) are solely objects that can exchange messages asynchronously through signals because of the use of fUML [8].

Assumption 2: Connectors have two end points because connectors with more than two end points are rarely used ("A connector has two end points"; pp. 258; [5]; pp. 420; [4]) and introduce unnecessary complexity in the static semantics.

These constraints, requirements and assumptions lead to the definition of the abstract syntax, the semantics, and formal rules for static semantics.

4.1   Abstract Syntax

The abstract syntax for CompositeStructure4fUML is presented in Fig. 2, where meta-elements (classes, attributes, and relationships) from UML are included in the CompositeStructure4fUML through copy (as fUML [8]). The included elements are marked with part of their qualified name (*CompositeStructures*). The following attributes and associations are removed during the definition:

- From Port
  - isService - rationale: the goal of the ports is to establish connections between internal elements and the environment;
  - redefinedPort - rationale: port's redefinitions add significant complexity and are rarely used by engineers [5];
- From Connector
  - contract - rationale: the valid interaction patterns are defined by the features of the internal elements or connected ports;
  - redefinedConnector - rationale: connector's redefinitions add significant complexity and are rarely used by engineers [5].

Ports (from UML) are changed to compute the required and provided features based on abstract classes instead of interfaces (excluded from [8]). Further, required and provided features are mutually exclusive, which means: a port defines provided features through the abstract classes realized by its type[1], and the attribute is Conjugated equals to false; or, a port declares required features through the abstract classes[1] realized by its type, and the attribute is Conjugated equals to true (recall "Constraint 2"). If more than one independently defined feature have to be exposed by a given port, an abstract class that specializes all desired features must be defined[1].

---

[1] Features defined by abstract classes without receptions and operations are not considered.

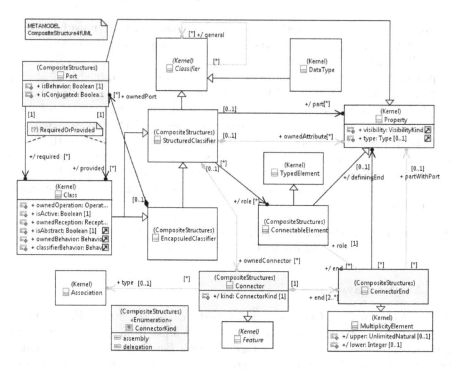

Fig. 2. Abstract syntax for CompositeStructure4fUML

Embedded Abstract Syntax. The CompositeStructure4fUML's abstract syntax is formally available through a technique called embedding, where relations in the abstract syntax are directly embedded in first-order logic [13]. This is in accordance with the base semantics of fUML [8]. However, the base semantics does not formalize the static semantics rules because it considers that a given model is compliant with all constraints imposed by UML and fUML [8]. As the static semantic rules for CompositeStructure4fUML are not part of fUML or UML, they can be defined using CLIF through the embedding technique.

Table. 1 shows a part of the abstract syntax representation in CLIF. These relations are embedded in CLIF using unary and binary relations. Instances of elements (individuals) use unary relations, e.g. a connector called *Connector1* is declared using *(cbuml:Connector Connector1)*. Attribute definitions are achieved using binary relations, e.g. to describe that the class *Class1* has a connector called *Connector1*, the predicate *(cbuml:ownedConnector Class1 Connector1)* is used. In fact, the meta-elements in Fig. 2 given in gray are embedded in CLIF. The prefix *"buml:"* is used to indicate relations that are defined in base semantics [8] while *"cbuml:"* is used to indicate relations defined by CompositeStructure4fUML. The prefix *"form:"* is used to indicate relations introduced for the formalization.

In summary, CLIF offers the logic syntax, the static semantics rules provide a set of inference rules and axioms that together with user axioms (describing com-

Table 1. Part of the embedded abstract syntax for CompositeStructure4fUML

Set	Set operations and relations	CLIF representation	Meta-elements
T			Type
F			Feature
C	$C \subseteq T$		Classifier
CL	$CL \subseteq C$	(buml:Class cl)	Class
P	$P \subseteq F$	(buml:Property p)	Property
CO	$CO \subseteq F$	(cbuml:Connector co)	Connector
G	$G = C \sqsubseteq C$	(buml:general c c)	general
OWA	$OWA = CL \sqsubseteq P$	(buml:ownedAttribute cl p)	ownedAttribute
PT	$PT = P \sqsubseteq T$	(buml:type p t)	type
CLAB	$CLAB \subseteq CL$	(cbuml:isAbstract cl form:true)	isAbstract
OWCO	$OWCO = CL \sqsubseteq CO$	(cbuml:ownedConnector cl co)	ownedConnector

posite structures) form a mathematical theory, which can be used for analysis. In this paper, the goal is to check if all axioms are consistent.

4.2 Semantics and Formal Rules

The constraints considered to define CompositeStructure4fUML are stated explicitly if they are not covered by fUML or UML. They are used to reject a model that is not compliant with them. "Constraint 1" is not covered because fUML allows the use of operation calls and value accesses between any active objects. Thus, the following rule is defined:

Rule Constraint 1. It is not allowed to define receptions, and public operations or public values in same class.

Although all the rules are available [11], only one formalized rule is presented due to lack of space. The following excerpt shows the formalized version of the above constraint:

```
(cl-text Constraint1
(forall (c)
   (if
      (buml:Class c)
      (not (and (exists (r)
                    (form:owned-reception-general c r))
             (or
                (exists(a)
                    (form:owned-attribute-general-visible c a))
                (exists(o)
                    (form:owned-operation-general-visible c o))
             ))))))
```

"Constraint 2" is covered by the definition of required and provided features for ports (subsection 4.1) and its derivatives are discussed below.

"Requirement 1" leads to the necessity of providing semantics for channels (connectors in abstract syntax) using the elements in fUML. Furthermore, Alf [9] suggests that associations should be used to integrate fUML and composite structures. Thus, associations are the mechanism used to provide semantics for connectors.

"Requirement 2" demands rules that enables automatic behavior generation according to the defined semantics as well as generation of the associations defined in the previous paragraph. These rules are grouped as follows: (1) rules for the constraints and the assumptions, (2) rules about ports, (3) rules about connectors, (4) rules about type compatibility, and (5) rules about associations. Groups 1, 2, 3, and 4 are designed to evaluate models in a UML surface (rejecting it in case of failure), whereas the group 5 is defined to check the results from a given translation into CompositeStructure4fUML. Group 1 is defined by the constraint presented above. In the following, the semantics considered for the groups 2, 3 and 4 are briefly presented. Afterwards, the rules for each of these groups are presented and explained, and, finally, the rules for the fifth group are listed. In the following, brackets are used to indicate values of the attributes.

The evaluation of semantics starts reviewing the concept of uni-directionality, which is applied to signals perfectly, while is not applicable for operations that have return values. Indeed, signals can be copied and sent to multiple targets, whereas operation calls cannot, because the definition of how to deal with multiple returns is not straightforward. Therefore, connectors that have at least one end as a port (isBehavior=false, and owning classifier (isActive=false)) must have multiplicity equals to one in the both ends. The reason is to allow a straightforward generation from the port's behavior. This semantics allows multiplicity greater than one for active objects or ports (isBehavior=true), and allows multicast for active objects. The behavior that should be generated for port (isBehavior=false) is: (a) active ports implement a classifier behavior that concurrently awaits for all signals, and for each signal received, a copy is dispatched for all the associations that connector matches the feature from the signal; (b) for passive classes, all operations are implemented with a synchronous call to the other end that satisfies the feature. Ports (isBehavior=false) are transformed in concrete classes accessed through associations. Public values are not supported in ports. Accordingly, ports (isBehavior=true) are innocuous because behaviors and associations are defined and pointed to the owning classifier.

Ports. The concept of required and provided features as well as the related use of the attribute isConjugated are defined above, so it remains to discuss the definition of the attribute isBehavior, which is defined by UML ([7]; pp. 186) as a mark specifying if the requests arriving at this port should be sent to classifier owning the port, or to internal elements. Under this semantics, and considering the previous definitions, the following set of rules is defined:

Rule Port 1: All ports (isBehavior=true) must have an abstract class as type;

- *Rule Port 1.1.* All ports (isBehavior=true) cannot have assembly connectors to its internal elements;
- *Rule Port 1.2.* All ports (isBehavior=true, isConjugated=false), a provided feature, the owning classifier must specialize the port's type.
- *Rule Port 1.3.* All ports (isBehavior=true, isConjugated=true), a required feature, at least, one assembly connector must reach the Port coming from external elements;

Rule Port 2: All ports (isBehavior=false) must have a concrete class as type;
- *Rule Port 2.1.* Port (isBehavior=false) must be a specialization from just one abstract class;
- *Rule Port 2.2.* Port (isBehavior=false), at least, one delegation connector must reach it coming from internal elements.

Rule Port 3: All ports in active classes must contain, at least, one reception;

Rule Port 4: All ports in passive classes must contain, at least, one public operation and no public values.

The last two rules for ports are consequences of "Constraint 1".

Connectors. Considering the elements that can be linked by connectors, it is not possible to connect DataTypes because such a connection would mean a constraint stating that the value would be always equal to the other end of the connection. For that reason, the following rule is defined:

Rule Connector 1. Connectors cannot be connected to a property whose type is a kind of DataType;

Due to "Constraint 1", the connectors relating passive classes and active classes are not allowed (and vice-versa), which generates the Rule Connector 2.

Rule Connector 2. Connectors are not allowed to establish relationships between classes with different value for the attribute isActive;

The semantics defined restricts the multiplicity of connectors in passive classes, which generates the rule below.

Rule Connector 3. Connectors relating one or more ports (isBehavior=false) from an owning classifier (isActive=false) must have multiplicity one as lower and upper in the both connector ends;

The rest of the rules in this group covers how connectors can link two ports.

Rule Connector 4. Assembly connectors between two ports must be defined using complement values for the attribute isConjugated;

Rule Connector 5. Delegation connectors between two ports must be defined using the same value for the attribute isConjugated.

Connectors and Type Compatibility. The definition of type compatibility depends on the coverability and predictability of each port:

Definition 1 (Coverability and Predicability). *Coverability is satisfied by a port p if p can communicate with the features in the other ends from its connectors. Predicability is satisfied by a port p, if it is possible to automatically* *generate a predictable behavior to dispatch all requests that can come to p evaluating its properties, its connectors, and the other ends.*

For the following theorem, we make use of the definitions below:

- P, the set of ports;
- PF_p, the set of features for a given port $p \in P$;
- $OEA_{pi}, i \in \mathbb{N}, p \in P$, set of features for each other connector end$_i$ owned by assembly connectors connected to p, when the other end(oe) is a port ($oe \in P$);
- $OEAWI_{pi}, i \in \mathbb{N}, p \in P$, set of features for each other connector end$_i$ owned by assembly connectors connected to p, when the other end(oe) is not a port ($oe \notin P$);
- $UOEAWI_p = \cup_i OEAWI_{pi}$;
- $UOEA_p = (\cup_i OEA_{pi}) \cup UOEAWI_p$;
- $OED_{pi}, i \in \mathbb{N}, p \in P$, set of features for each other connector end$_i$ owned by delegation connectors connected to p, when the other end(oe) is a port ($oe \in P$);
- $OEDWI_{pi}, i \in \mathbb{N}, p \in P$, set of features for each other connector end$_i$ owned by delegation connectors connected to p, when the other end(oe) is not a port ($oe \notin P$);
- $UOEDWI_p = \cup_i OEDWI_{pi}$;
- $UOED_p = (\cup_i OED_{pi}) \cup UOEDWI_p$;

Based on these, we can prove the following rules that express the intuition that assembly and delegation have a counterpart effect:

Theorem 1. *Coverability and Predictability for a given $p \in P$*

- Rule Type 1. *Coverability is satisfied if p(isConjugated=false) then ($PF_p \supseteq UOEAWI_p$ and $PF_p \subseteq UOED_p$).*
- Rule Type 2. *Coverability is satisfied if p (isConjugated=true) then ($PF_p \subseteq UOEA_p$ and $PF_p \supseteq UOEDWI_p$).*
- Rule Type 3. *Predicability is satisfied if p (isConjugated=false) then (OED_p is pairwise disjoint or type of p(isActive=true)).*
- Rule Type 4. *Predicability is satisfied if p (isConjugated=true) then (OEA_p is pairwise disjoint or type of p(isActive=true)).*

Connectors and Associations. The rules related with associations are defined to verify if the translation process from the UML surface to CompositeStructure4fUML generates a model that is compliant with the defined semantics, i.e., (1) the type of the connector end is a subtype (or the same type) of the association end, (2) the lower from the connector end is greater or equal than the association end, (3) the upper from the connector end is less or equal to the association end, (4) navigability is compliant with the ports involved, and (5) a port (isBehavior=true) cannot be used by the association (in this case, the association should have as end the owner of the port).

These rules state that the translation should generate associations that must cover all possible connections between the classes described by the composite structures assuring that fUML semantics can handle the channels.

5   Evaluation and Discussion

Fig. 3 shows one of the models used to evaluate the meta-model, and the translated axioms. It is a component for a player that can be used to define PingPong systems with two or more players. This model is defined considering the rules defined by CompositeStructure4fUML, and it satisfies the static semantics rules without translations. In this case, the internal behavior defined by the classifier behavior from the class Arm remains untouched when the system of players are mounted. However, the behavioral definition from the ports can change, e.g. the topology from players changed so the behavior from ports must be reviewed. This is the goal of the rules to guarantee that under the above semantics, port's behavior generation can be performed, and the resulting behavior is predictable.

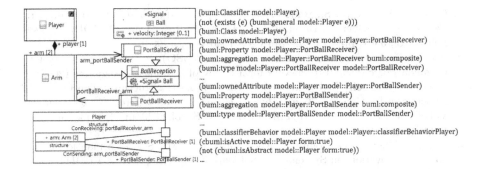

Fig. 3. Example of an analyzed Model (user-defined diagrams and translated axioms)

The formalized version of the model (user axioms) is generated using a MOF2Text (M2T) transformation with some degree of Object Constraint Language (OCL) embedded. There are some situations where the transformation M2T does preprocessing, e.g. counting the connectors end, and comparing numbers for multiplicity check.

These models provide empirical evidence indicating that the rules are sound. Besides, the techniques used to define the rules allow the use of advanced formal analysis of their completeness and soundness, as well as their redundancy.

The set of rules expresses the definition from composite structures that show classes and associations in context [1], and it is in accordance with fUML [8]. One of the benefits of these rules is supporting the evaluation of multiple scenarios described by composite structures [6].

The current work extends [2] exploring how to handle requests in behavioral ports for passive objects, and the impacts on the structure of the owning classifier. Concerning the work [4], the rules "Type 1, 2, 3, and 4" are less restrictive for connectors: it is not mandatory that all provided features should be used by some other end (Rule 8; [4]; pp. 427). Besides, ports for active objects use copies to distribute signals to possible multiple ports requiring the same feature, what is not considered a source of ambiguity in the semantics defined above.

6 Conclusion

We have presented a meta-model for the integration of composite structure, a fundamental technique to handle complexity, and fUML, namely CompositeStructure4fUML. This meta-model enables analysis through simulation (execution) without compromising the structure. Moreover, we have presented formal modular rules for static semantics, which can be applied in any context where the given constraints and assumptions hold.

References

1. Bock, C.: UML 2 Composition Model. Journal of Object Technology 3(10), 47–73 (2004)
2. Cuccuru, A., Gérard, S., Radermacher, A.: Meaningful composite structures. In: Czarnecki, K., Ober, I., Bruel, J.-M., Uhl, A., Völter, M. (eds.) MODELS 2008. LNCS, vol. 5301, pp. 828–842. Springer, Heidelberg (2008)
3. International Organization for Standardization (ISO): Information technology - Common Logic (CL): a framework for a family of logic-based languages (2007)
4. Ober, I., Dragomir, I.: Unambiguos UML Composite Structures: The OMEGA2 Experience. In: Černá, I., Gyimóthy, T., Hromkovič, J., Jefferey, K., Královič, R., Vukolić, M., Wolf, S. (eds.) SOFSEM 2011. LNCS, vol. 6543, pp. 418–430. Springer, Heidelberg (2011)
5. Ober, I., Ober, I., Dragomir, I., Aboussoror, E.: UML/SysML semantic tunings. Journal Innovations in Systems and Software Engineering, 257–264 (2011)
6. Oliver, I., Luukkala, V.: On UMLs Composite Structure Diagram. In: 5th Workshop on System Analysis and Modelling, SAM (2006)
7. Object Management Group (OMG): Unified Modeling Language Superstructure: Version: 2.4.1. USA: OMG, 2011 (2011), http://www.omg.org/spec/UML/2.4.1/ (access: April 14, 2013)
8. Object Management Group (OMG): Semantics of a Foundational Subset for Executable UML Models: Version 1.1 RTF Beta. USA: OMG, 2012 (2012), http://www.omg.org/spec/FUML/ (access: April 24, 2013)
9. Object Management Group (OMG): Concrete Syntax for UML Action Language (Action Language for Foundational UML - ALF): Version: 1.0.1 - Beta. USA: OMG, 2013 (2013), http://www.omg.org/spec/ALF/ (access: April 27, 2013)
10. Object Management Group (OMG): Precise Semantics of UML Composite Structures - Request For Proposal - OMG Document: ad/2011-12-07. USA: OMG, 2013 (2013), http://www.omg.org/cgi-bin/doc?ad/11-12-07/ (access: August 25, 2013)
11. Romero, A., Schneider, K., Ferreira, M.G.V.: Support files (2013), http://es.cs.uni-kl.de/people/romero/sofsem2014.zip (access: October 14, 2013)
12. Warfield, J.N., Staley, M.M.: Structural thinking: Organizing complexity through disciplined activity. Journal Systems Research 13, 47–67 (1996)
13. World Wide Web Consortium (W3C): An Axiomatic Semantics for RDF, RDF-S, and DAML+OIL (March 2001). W3C Note December 18, 2001, http://www.w3.org/TR/daml+oil-axioms (access: June 23, 2013)

Deciding the Value 1 Problem for ♯-acyclic Partially Observable Markov Decision Processes*

Hugo Gimbert[1] and Youssouf Oualhadj[2,3]

[1] LaBRI, CNRS, France
[2] Aix-Marseille Université, CNRS, LIF, Marseille, France
[3] Université de Mons, Belgium

Abstract. The value 1 problem is a natural decision problem in algorithmic game theory. For partially observable Markov decision processes with reachability objective, this problem is defined as follows: are there observational strategies that achieve the reachability objective with probability arbitrarily close to 1? This problem was shown undecidable recently. Our contribution is to introduce a class of partially observable Markov decision processes, namely *♯-acyclic* partially observable Markov decision processes, for which the value 1 problem is decidable. Our algorithm is based on the construction of a two-player perfect information game, called the knowledge game, abstracting the behaviour of a ♯-acyclic partially observable Markov decision process $\mathcal{M}$ such that the first player has a winning strategy in the knowledge game if and only if the value of $\mathcal{M}$ is 1.

1 Introduction

Partially Observable Markov Decision Processes (POMDP for short). Markov decision processes (MDPs) are well established tool for modelling systems that mix both probabilistic and nondeterministic behaviours. The nondeterminism models the choices of the system supervisor (the controller) and the probabilities describe the environment behaviour. When the system offers full information, it is rather easy for the controller to make the best choice, this follows from the fact that fully observable MDPs enjoy good algorithmic properties. For instance ω-regular objectives such as parity objective can be solved in polynomial time [10,8], as well as quantitative objectives such as average and discounted reward criterions [11,17]. Moreover, optimal strategies always exist for any tail winning condition [5,14]. Unfortunately, the assumption that a real life system offers a full observability is not realistic. Indeed, an everyday system cannot be made fully monitored because it is either too large (e.g. information system), or implementing full monitoring is too costly (e.g. subway system), or even not possible (e.g. electronic components of an embedded system). That is why partially observable Markov decision processes are a better suited theoretical tool for modelling real life system. In a POMDP, the state space is partitioned and the decision maker cannot observe the states themselves but only the partition they belong to also called the *observation*. Therefore, two executions that carry the same observations and the same actions are undistinguishable for the controller and hence its choice after both execution is going to be the same.

---

* This work has been supported by the ANR project ECSPER (JC09_472677) and the ARC project Game Theory for the Automatic Synthesis of Computer Systems.

V. Geffert et al. (Eds.): SOFSEM 2014, LNCS 8327, pp. 281–292, 2014.

In other words the strategies for the controller are mappings from sequences of observation and actions to actions.

Value 1 Problem. This problem is relevant for controller synthesis: given a discrete event system whose evolution is influenced by both random events and controllable actions, it is natural to look for controllers as efficient as possible, i.e. to compute strategies which guarantee a probability to win as high as possible. As opposed to the almost-sure problem where the controller is asked to find a strategy that ensures the win with probability exactly 1. There are toy examples in which an almost-sure controller does not exist but still there exists controllers arbitrarily efficient, and the system can be considered as safe. Moreover, in fully observable setting, the value 1 and the almost-sure winning coincide, this is actually the case for any tail winning condition for simple stochastic games. This property makes the study of fully observable models way easier and leads in most cases to decidability. But as we will see later, almost-sure winning and the value 1 problem do not coincide for POMDPs. Actually, the former problem is decidable [3,7] while the latter is not.

Related Work. The value one problem has been left open by Bertoni since the 1970's [1,2]. Recently, we showed that this problem is undecidable for probabilistic automata [15], this result extends to POMDP because they subsume probabilistic automata. Since then, much efforts were put into identifying nontrivial decidable families of probabilistic automata for the value 1 problem. For instance, $\sharp$-acyclic automata [15], structurally simple automata [9], and leaktight automata [12]. The common point between those subclasses is the use of two crucial notions. The first one is the iteration of actions, this operation introduced in [15] for probabilistic automata and inspired by automata-theoretic works, describes the long term effects of a given behaviour. The second one is the limit-reachability. Broadly speaking, limit-reachability, formalises the desired behaviour of a probabilistic automaton that has value 1. Therefore, the technical effort in the previously cited papers consists in relating the operation of iteration with the limit-reachability in a complete and consistent manner. Even though the consistency can be obtained rather easily, the completeness requires restrictions on the model considered. This is not surprising since the general case is not decidable. In this work, we consider POMDP, and identify a subclass for which the value 1 problem is decidable.

Contribution and Result. We extend the decidability result of [15] to the case of POMDPs. We define a class of POMDPs called $\sharp$-*acyclic* POMDPs and we show that the value 1 problem is decidable for this class.

The techniques we use are new compared to [15]. While in [15] the value 1 problem for $\sharp$-acyclic automata is reduced to a reachability problem in a graph, in the present paper, the value 1 problem for POMDPs is reduced to the computation of a winner in a two-player game: the two-player game is won by the first player if and only if the value of the POMDP is 1. While for $\sharp$-acyclic probabilistic automata the value 1 problem can be decided in PSPACE, the algorithm for the value 1 problem for $\sharp$-acyclic POMDP runs in exponential time. This algorithm is fixed-parameter tractable (FPT) when the parameter is the number of states per observation.

Even though the class may seem contrived, as the results on probabilistic automata show, this class is useful from a theoretical point of view in the sense that it allows the definition of appropriate formal tools. The main technical challenge was to extend both the notions of iteration and limit-reachability; While in a probabilistic automaton the behaviour of the controller can be described by a finite word, because there is no feed back that the controller could use to change its behaviour. This is not anymore true for a POMDP. The behaviour of the controller is described by a (possibly infinite) tree, in this case the choice of the next action actually depends on the sequence of observations received. Generalisation from words to trees is in general a nontrivial step and leads to both algorithmic blowups and technical issues. In our case, the effect of generalisation is mostly notable in the definition of limit-reachability. As one can see in Definition 2 limit-reachability expresses two level of uncertainty as opposed to probabilistic automata where one level is sufficient. The notion of limit-reachability is carefully chosen so that it is transitive in the sense of Lemma 2 and can be algorithmically used thanks to Lemma 3. We believe that this definition can be kept unchanged for handling further more general decidable families of POMDPs.

Outline of the Paper. in Section 2 we introduce POMDPs and related notations. In Section 3 we introduce the class of ♯-acyclic POMDPs and state the decidability of the value 1 problem for ♯-acyclic POMDPS which is our main theorem, namely Theorem 2. In Section 4 we define the *knowledge game* and prove the main result.

See [16] for a full version of the paper.

2   Notations

Given S a finite set, let $\Delta(S)$ denote the set of distributions over S, that is the set of functions $\delta : S \to [0,1]$ such that $\sum_{s \in S} \delta(s) = 1$. For a distribution $\delta \in \Delta(S)$, the support of δ denoted $\mathsf{Supp}(\delta)$ is the set of states $s \in S$ such that $\delta(s) > 0$. We denote by δ_Q the uniform distribution over a finite set Q.

2.1   Partially Observable Markov Decision Process

Intuitively, to play in a POMDP, the controller receives an observation according to the initial distribution then it chooses an action then it receives another observation and chooses another action and so on. The goal of the controller is to maximize the probability to reach the set of target states T. A POMDP is a tuple $\mathcal{M} = (S, A, \mathcal{O}, \mathsf{p}, \delta_0)$ where S is a finite set of states, A is a finite set of actions, $\mathcal{O}$ is a partition of S called the observations, $\mathsf{p} : S \times A \to \Delta(S)$ is a transition function, and δ_0 is an initial distribution in $\Delta(S)$.

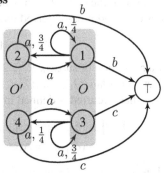

Fig. 1. Partially observable Markov decision process

We assume that for every state $s \in S$ and every action $a \in A$ the function $\mathsf{p}(s, a)$ is defined, i.e. every action can be played from every state. When the partition described by $O \in \mathcal{O}$ is a singleton $\{s\}$, we refer to state s as observable. An infinite play in a POMDP is an infinite word in $(\mathcal{O}A)^\omega$, and a finite play is a finite word in $\mathcal{O}(A\mathcal{O})^*$. We denote by Plays the set of finite plays.

Consider the POMDP $\mathcal{M}$ depicted in Fig 1. The initial distribution is at random between states 1 and 3, the play is winning if it reaches $\top$, and the observations are $\mathcal{O} = \{O, O', \{\top\}\}$ where $O = \{1, 3\}$ and $O' = \{2, 4\}$. State $\top$ is observable. The missing transitions lead to a sink and are omitted for the sake of clarity. A possible play in $\mathcal{M}$ is $\rho = OaOaO'(aO)^\omega$.

2.2 Strategies and Measure

To play the controller chooses the next action to apply in function of the initial distribution, the sequence of actions played, and the sequence of observations received along the play. Such strategies are said to be *observational*. Formally, an observational strategy for the controller is a function $\sigma : \mathsf{Plays} \to A$.

Notice that we consider only pure strategies, this is actually enough since in POMDPs randomized strategies are not more powerful than the pure ones [13,6].

Once an initial distribution δ_0 and a strategy σ are fixed, this defines uniquely a probability measure $\mathbb{P}_{\delta_0}^\sigma$ on $S(AS)^\omega$ as the probability measure of infinite trajectories of the Markov chain whose transition probabilities are fixed by δ_0, σ and $\mathsf{p} : S \times A \to \Delta(S)$. Using the natural projection $\pi : S(AS)^\omega \to \mathcal{O}(A\mathcal{O})^\omega$ we extend the probability measure $\mathbb{P}_{\delta_0}^\sigma$ to $\mathcal{O}(A\mathcal{O})^\omega$.

We define the random variables S_n, A_n, and O_n with values in S, A, and $\mathcal{O}$ respectively that maps an infinite play $w = s_0 a_1 s_1 a_2 s_2 \cdots$ to respectively the n-th state S_n, the n-th action A_n, and the n-th observation $O_n \in \mathcal{O}$ such that $S_n \in O_n$.

2.3 Outcome and Knowledge

Let $Q \subseteq S$ be a subset and a be a letter, we define $\mathsf{Acc}(Q, a)$ as the set of states $s \in S$ such that there exists $q \in Q$ and $\mathsf{p}(q, a)(s) > 0$.

The outcome of an action a given a subset of states Q is the collection $Q \cdot a$ of subsets of states that the controller may believe it is in after it has played action a in one of the states of Q and it has received its observation. Formally,

$$Q \cdot a = \{\mathsf{Acc}(Q, a) \cap O \mid O \in \mathcal{O}\} \ .$$

For a collection of subsets $\mathcal{R} \subseteq 2^S$ we write: $\mathcal{R} \cdot a = \bigcup_{R \in \mathcal{R}} R \cdot a$.

Let $w = O_0 a_1 O_1 \cdots a_n O_n \in \mathsf{Plays}$ be finite play. The knowledge of the controller after w has occurred is defined inductively as follows:

$$\begin{cases} K(\delta_0, O_0) = \mathsf{Supp}(\delta_0) \cap O_0 \ , \\ K(\delta_0, w) = \mathsf{Acc}(K(\delta_0, O_0 \cdots a_{n-1} O_{n-1}), a_n) \cap O_n \ . \end{cases}$$

It is a an elementary exercise to show that for every strategy σ, the following holds:

$$\mathbb{P}_{\delta_0}^\sigma(\forall n \in \mathbb{N}, S_n \in K(\delta_0, w)) = 1 \ . \tag{1}$$

2.4 Value 1 Problem

In the sequel we will concentrate on reachability objectives, hence when referring to the value of a POMDP it is implied that the objective is a reachability objective.

Definition 1 (Value). *Let $\mathcal{M}$ be a POMDP, $\delta_0 \in \Delta(S)$ be an initial distribution, and $T \subseteq S$ be a subset of target states. Then, the value of δ_0 in $\mathcal{M}$ is:*

$$\mathsf{Val}_{\mathcal{M}}(\delta_0) = \sup_{\sigma} \mathbb{P}^{\sigma}_{\delta_0}(\exists n \in \mathbb{N},\ S_n \in T) \ .$$

The value 1 problem consists in deciding whether $\mathsf{Val}_{\mathcal{M}}(\delta_0) = 1$ for given $\mathcal{M}$ and δ_0.

Example 1. The value of the POMDP of Fig 1 is 1 when the initial distribution is uniform over the set $\{1, 3\}$. Remember that missing edges (for example action c in state 1) go to a losing sink $\perp$, hence the goal of the controller is to determine whether the play is in the upper or the lower part of the game and to play b or c accordingly. Consider the strategy that plays long sequences of a^2 then compares the frequencies of observing O and O'; If O' was observed more than O then with high probability the initial state is 1 and by playing b state $\top$ is reached. Otherwise, with high probability the play is in 3 and by playing c again the play is winning. Note that the controller can make the correct guess with arbitrarily high probability by playing longer sequences of a^2, but it cannot win with probability 1 since it always has to take a risk when choosing between actions b and c. This example shows that the strategies ensuring the value 1 can be quite elaborated: the choice not only depends on the time and the sequence of observations observed, but also depends on the empirical frequency of the observations received.

The value 1 problem is undecidable in general, our goal is to extend the result of [15] and show that the value 1 problem is decidable for $\sharp$-*acyclic POMDP*. The idea is to abstract limit behaviours of finite plays using a finite two-player reachability game on a graph, so that limit-reachability in the POMDP in the sense of Definition 2 coincides with winning the reachability game on the finite graph.

The definition of limit reachability relies on the random variable that gives the probability to be in a set of states $T \subseteq S$ at step $n \in \mathbb{N}$ given the observations received along the $n - 1$ previous steps:

$$\phi_n(\delta, \sigma, T) = \mathbb{P}^{\sigma}_{\delta}(S_n \in T \mid O_0 A_1 \cdots A_n O_n) \ .$$

Definition 2 (Limit-reachability). *Let $Q \subseteq S$ be a subset of states and $\mathcal{T}$ be a nonempty collection of subsets of states. We say that $\mathcal{T}$ is limit-reachable from Q if for every $\varepsilon > 0$ there exists a strategy σ such that:*

$$\mathbb{P}^{\sigma}_{\delta_Q}\left(\exists n \in \mathbb{N}, \exists T \in \mathcal{T},\ \phi_n(\delta_Q, \sigma, T) \geq 1 - \varepsilon\right) \geq 1 - \varepsilon \ ,$$

where δ_Q is the uniform distribution on Q.

The intuition behind this definition is that when $\mathcal{T}$ is limit-reachable from Q, then if the play starts from a state randomly chosen in Q the controller has strategies so that with probability arbitrarily close to 1 it will know someday according to its observations that the play is in one of the sets $T \in \mathcal{T}$ and which set T it is.

The following lemma shows that the definition of limit-reachability is robust to a change of initial distribution as long as the support of the initial distribution is the same.

Lemma 1. *Let $\delta \in \Delta(S)$ be a distribution, $Q \subseteq S$ its support, $\mathcal{R}$ be a nonempty collection of subsets of states. Assume that for every $\varepsilon > 0$ there exists σ such that:*

$$\mathbb{P}_\delta^\sigma \left(\exists n \in \mathbb{N}, \ \exists R \in \mathcal{R}, \ \phi_n(\delta, \sigma, R) \geq 1 - \varepsilon \right) \geq 1 - \varepsilon \ ,$$

then $\mathcal{R}$ is limit-reachable from δ_Q.

The above lemma implies that the decision of the value 1 problem depends only on the support of the initial distribution.

We say that T is observable if it is formed by sets taken from the partition $\mathcal{O}$, i.e.

$$T = \bigcup_{\substack{O \in \mathcal{O} \\ O \cap T \neq \emptyset}} O \ .$$

Limit-reachability enjoys two nice properties. First the value 1 problem can be rephrased using limit-reachability, second limit-reachability is transitive.

Proposition 1. *Assume that T is observable, then $\mathsf{Val}_\mathcal{M}(\delta_0) = 1$ if and only if $2^T \setminus \emptyset$ is limit-reachable from $\mathsf{Supp}(\delta_0)$.*

Proposition 1 does not hold in the case where the set of target states is not observable. However there is a computable linear time transformation from a POMDP $\mathcal{M}$ to a POMDP $\mathcal{M}'$ with a larger set of states whose set of target states is observable and such that a distribution has value 1 in $\mathcal{M}$ if and only if it has value 1 in $\mathcal{M}'$. Therefore, our decidability result holds whether the target states are observable or not.

The proof of Proposition 1 and the above construction are available in the full version of the paper [16].

Limit-reachability is a transitive relation in the following sense.

Lemma 2 (Limit-reachability is transitive). *Let Q be a subset of states and $\mathcal{R}$ be a nonempty collection of subsets. Assume that $\mathcal{R}$ is limit-reachable from Q. Let $\mathcal{T}$ be a nonempty collection of subsets of states such that $\mathcal{T}$ is limit-reachable from every subset $R \in \mathcal{R}$. Then $\mathcal{T}$ is limit-reachable from Q.*

See [16] for a detailed proof.

3   The ♯-acyclic Partially Observable Markov Decision Processes

In this section we associate with every POMDP $\mathcal{M}$ a two-player zero-sum game on a graph $\mathcal{G}_\mathcal{M}$. The construction of the graph is based on a classical subset construction [4] extended with an iteration operation.

3.1   Iteration of Actions

Definition 3 (Stability). *Let $Q \subseteq S$ be a subset of states and $a \in A$ be an action, then Q is a-stable if $Q \subseteq \mathsf{Acc}(Q, a)$.*

Definition 4 (a-recurrence). *Let $Q \subseteq S$ be a subset of states and $a \in A$ be an action such that Q is a-stable. We denote by $\mathcal{M}[Q, a]$ the Markov chain with states Q and probabilities induced by a: the probability to go from a state $s \in Q$ to a state $t \in Q$ is $\mathsf{p}(s, a)(t)$. A state s is said to be a-recurrent if it is recurrent in $\mathcal{M}[S, a]$.*

The key notion in the definition of $\sharp$-acyclic POMDPs is *iteration of actions*. Intuitively, if the controller knows that the play is in Q then either someday it will receive an observation which informs him that the play is no more in Q or it will have more and more certainty that the play is trapped in the set of recurrent states of a stable subset of Q. Formally,

Definition 5 (Iteration). *Let Q be a subset of states, a be an action such that $Q \in Q \cdot a$ and R be the largest a-stable subset of Q. We define*

$$Q \cdot a^{\sharp} = \begin{cases} \{\{a\text{-recurrent states of } R\}\} \cup (Q \cdot a \setminus \{Q\}) & \text{if } R \text{ is not empty} \\ Q \cdot a \setminus \{Q\} & \text{otherwise .} \end{cases}$$

If $Q \cdot a^{\sharp} = \{Q\}$ then Q is said to be $a^{\sharp}$-stable, equivalently Q is a-stable and all states of Q are a-recurrent.

We will denote by $a^{\sharp}$ the iteration of a and by $A^{\sharp}$ the set $\{a^{\sharp} \mid a \in A\}$.

The action of letters and iterated letters is related to limit-reachability:

Proposition 2. *Let $Q \subseteq S$ and $a \in A$. Assume $Q \subseteq O$ for some $O \in \mathcal{O}$. Then $Q \cdot a$ is limit-reachable from Q. Moreover if $Q \in Q \cdot a$, then $Q \cdot a^{\sharp}$ is also limit-reachable from Q.*

Proof. Let $\varepsilon > 0$ and σ be the strategy that plays only a's. Since $Q \subseteq O$, $\mathbb{P}^{\sigma}_{\delta_Q}(O_0 = 0) = 1$. By definition of the knowledge $K(\delta_Q, O_0) = Q$ thus by definition of $Q \cdot a$,

$$\mathbb{P}^{\sigma}_{\delta_Q}(K(\delta_Q, O_0 a O_1) \in Q \cdot a) = 1 ,$$

and according to (1), $\mathbb{P}^{\sigma}_{\delta_Q}(S_1 \in K(\delta_Q, O_0 a O_1) \mid O_0 A_1 O_1) = 1$ thus

$$\mathbb{P}^{\sigma}_{\delta_Q}(\phi_1(\delta_Q, \sigma, K(\delta_Q, O_0 a O_1)) = 1) = 1 ,$$

and altogether we get

$$\mathbb{P}^{\sigma}_{\delta_Q}(\exists T \in Q \cdot a, \phi_1(\delta_Q, \sigma, T) = 1) = 1 ,$$

which proves that $Q \cdot a$ is limit-reachable from Q.

Assume that $Q \in Q \cdot a$. By definition of limit-reachability, to prove that $Q \cdot a^{\sharp}$ is limit-reachable from Q, it is enough to show for every $\varepsilon > 0$,

$$\mathbb{P}^{\sigma}_{\delta_Q}(\exists n \in \mathbb{N}, \exists T \in Q \cdot a^{\sharp}, \phi_n(\delta_Q, \sigma, T) \geq 1 - \varepsilon) \geq 1 - \varepsilon . \tag{2}$$

Let R the (possibly empty) largest a-stable subset of Q, and R' the set of a-recurrent states in R. Let $\mathsf{Stay}^n(O)$ be the event

$$\mathsf{Stay}^n(O) = \{\forall k \leq n, O_k = O\} \ .$$

The strategy σ plays only a's thus $\mathbb{P}^\sigma_{\delta_Q}$ coincides with the probability measure of the Markov chain $\mathcal{M}[S, a]$. Almost-surely the play will stay trapped in the set of a-recurrent states. Thus by definition of R',

$$(R' = \emptyset) \implies \left(\mathbb{P}^\sigma_{\delta_Q} (\mathsf{Stay}^n(O)) \xrightarrow[n \to \infty]{} 0 \right) \tag{3}$$

$$(R' \neq \emptyset) \implies \mathbb{P}^\sigma_{\delta_Q} (S_n \in R' \mid \mathsf{Stay}^n(O)) \xrightarrow[n \to \infty]{} 1 \ . \tag{4}$$

Now we complete the proof of (2). According to (4) if $R' \neq \emptyset$ there exists $N \in \mathbb{N}$ such that $\mathbb{P}^\sigma_{\delta_Q} \left(S_N \in R' \mid \mathsf{Stay}^N(O) \right) \geq 1 - \varepsilon$, thus

$$(R' \neq \emptyset) \implies \mathbb{P}^\sigma_{\delta_Q} \left(\phi_N(\delta_Q, \sigma, R') \geq 1 - \varepsilon \mid \mathsf{Stay}^N(O) \right) = 1 \ . \tag{5}$$

On the other hand if the play is in $\mathsf{Stay}^n(O)$ and not in $\mathsf{Stay}^{n+1}(O)$ it means the controller receives for the first time at step $n + 1$ an observation O_{n+1} which is not O. Since $Q \subseteq O$ it means the play has left Q thus $K(\delta_Q, O_0 a O_1 \cdots O_n) = Q$ and $K(\delta_Q, O_0 a O_1 \cdots O_n a O_{n+1}) = K(\delta_Q, Q a O_{n+1}) \in Q \cdot a \setminus \{Q\}$, thus for every $n \in \mathbb{N}$,

$$\mathbb{P}^\sigma_{\delta_Q} \left(\exists T \in Q \cdot a \setminus \{Q\}, \phi_n(\delta_Q, \sigma, T) = 1 \mid \mathsf{Stay}^n(O) \wedge \neg\mathsf{Stay}^{n+1}(O) \right) = 1. \tag{6}$$

Taking (5) and (6) together with the definition of $Q \cdot a^\sharp$ proves (2). $\qquad\square$

3.2 $\sharp$-acyclic POMDP

The construction of the knowledge graph is based on a classical subset construction (see e.g. [4]) extended with the iteration operation.

Definition 6 (Knowledge graph). *Let $\mathcal{M}$ be a POMDP, the knowledge graph $\mathcal{G}_\mathcal{M}$ of $\mathcal{M}$ is the labelled graph obtained as follows:*

- *The states are the nonempty subsets of the observations: $\bigcup_{O \in \mathcal{O}} 2^O \setminus \emptyset$.*
- *The triple (Q, a, T) is an edge if $T \in Q \cdot a$ and the triple $(Q, a^\sharp, T)$ is an edge if $Q \in Q \cdot a$ and $T \in Q \cdot a^\sharp$.*

Example 2. In Fig 2(a) is depicted a POMDP where the initial distribution is at random between states s and q. The states $\top, \bot, t$ are observable and $O = \{s, q\}$. In Fig 2(b) is the knowledge graph associated to it.

Definition 7 ($\sharp$-acyclic POMDP). *Let $\mathcal{M}$ be a POMDP and $\mathcal{G}_\mathcal{M}$ the associated knowledge graph. $\mathcal{M}$ is $\sharp$-acyclic if the only cycles in $\mathcal{G}_\mathcal{M}$ are self loops.*

This condition may seem very restrictive, nevertheless, it does not forbid cycles in the transition graph see e.g. [15] for an example. Of course one can check whether a POMDP is $\sharp$-acyclic or not in exponential time. The main result of the paper is:

Theorem 1. *The value 1 problem is decidable for $\sharp$-acyclic POMDPs. The complexity is polynomial in the size of the knowledge graph, thus exponential in the number of states of the POMDP and fix-parameter tractable with parameter $\max_{O \in \mathcal{O}} |O|$.*

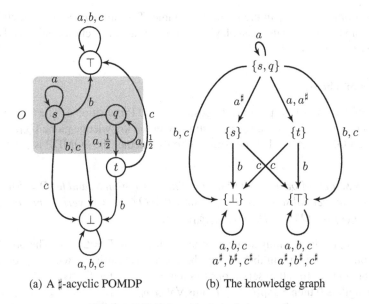

(a) A ♯-acyclic POMDP

(b) The knowledge graph

Fig. 2. A POMDP and its knowledge graph

4 Deciding the Value 1

In this section we show that given a POMDP $\mathcal{M}$ and its knowledge graph $\mathcal{G}_{\mathcal{M}}$ there exists a two-player (verifier and falsifier) perfect information game played on $\mathcal{G}_{\mathcal{M}}$ where verifier wins if and only if $\mathsf{Val}_{\mathcal{M}} = 1$.

4.1 The Knowledge Game

We first explain how to construct the game and how it is played. Let $\mathcal{M}$ be a POMDP and $\mathcal{G}_{\mathcal{M}}$ be the knowledge graph associated to $\mathcal{M}$. Starting from a vertex Q, the knowledge game is played on $\mathcal{G}_{\mathcal{M}}$ as follows:

- Verifier either chooses an action $a \in A$ or if $Q \in \mathcal{Q} \cdot a$ she can also choose an action $a \in A^{\sharp}$,
- falsifier chooses a successor $R \in Q \cdot a$ and $R \in Q \cdot a^{\sharp}$ in the second case,
- the play continues from the new state R.

Verifier wins if the game reaches a subset $R \subseteq T$ of target states.

Definition 8 (♯-**reachability**). *A nonempty collection of subsets $\mathcal{R}$ is ♯-reachable from a subset Q if there exists a strategy for verifier to reach one of the subsets $R \in \mathcal{R}$ against any strategy of falsifier in the knowledge game.*

Example 3. In the POMDP of Fig 2, assume that the initial distribution δ_0 is at random between state s and q. The value of the initial distribution is 1 because the controller can use the following strategy. Play long sequences of a and if the only observation received is O, with probability arbitrarily close to 1 the play is in state s otherwise with high probability the play would have moved to state q. On the other hand, verifier has a strategy to win from $\{s, q\}$ in the knowledge game. This strategy consists in choosing action $a^\sharp$ from the initial state, then playing action c if falsifier's choice is $\{t\}$ and action b if falsifier's choice is $\{s\}$.

4.2   Proof of Theorem 1

The proof of Theorem 1 is split into Proposition 3 and Proposition 4. The first proposition shows that if verifier has a winning strategy in the knowledge game $\mathcal{G}_\mathcal{M}$, then the value of the POMDP $\mathcal{M}$ is 1. Proposition 3 holds whether the POMDP is $\sharp$-acyclic or not.

Proposition 3. *Let $\mathcal{M}$ be a POMDP with initial distribution δ_0 and let $Q = \mathsf{Supp}(\delta_0)$. Assume that for every observation $O \in \mathcal{O}$ such that $O \cap Q \neq \emptyset$, verifier has a winning strategy in $\mathcal{G}_\mathcal{M}$ from $O \cap Q$. Then $\mathsf{Val}_\mathcal{M}(\delta_0) = 1$.*

Proof. Let $\sigma_\mathcal{M}$ be the winning strategy of the verifier and $\mathcal{T} = 2^T \setminus \emptyset$. The proof is by induction on the maximal number of steps before a play consistent with $\sigma_\mathcal{M}$ reaches $\mathcal{T}$ starting from $Q \cap O$ for all observations O such that $Q \cap O \neq \emptyset$.

If this length is 0 then $\mathsf{Supp}(\delta_0) \subseteq T$ thus $\mathsf{Val}_\mathcal{M}(\delta_0) = 1$.

Otherwise for every observation O such that $Q \cap O \neq \emptyset$, let $a_O = \sigma_\mathcal{M}(Q \cap O)$. Then by induction hypothesis, from every $R \in \mathsf{Supp}(Q \cap O) \cdot a_O$, $\mathsf{Val}_\mathcal{M}(\delta_R) = 1$. Given $\varepsilon > 0$, for every $R \in \mathsf{Supp}((Q \cap O) \cdot a_O)$ let σ_R a be a strategy in the POMDP to reach T from δ_R with probability at least $1 - \varepsilon$. Let σ be the strategy in the POMDP that receives the first observation O, plays a_O, receives the second observation O_1 then switches to $\sigma_{K(\delta_0, O_0 a_O O_1)}$.

By choice of σ_R, for every state $r \in R$, the strategy σ_R guarantees to reach T from $\delta_{\{r\}}$ with probability at least $1 - |R| \cdot \varepsilon$ thus σ guarantees to reach T from δ_0 with probability at least $1 - |Q| \cdot \varepsilon$. Since this holds for every ε, $\mathsf{Val}_\mathcal{M}(\delta_0) = 1$. $\square$

While it is not too difficult to prove that if verifier wins $\mathcal{G}_\mathcal{M}$ then $\mathsf{Val}_\mathcal{M} = 1$, the converse is much harder to establish, and holds only for $\sharp$-acyclic POMDPs.

Proposition 4. *Let $\mathcal{M}$ be a $\sharp$-acyclic POMDP and δ_0 be an initial distribution and denote $Q = \mathsf{Supp}(\delta_0)$. Assume that $\mathsf{Val}_\mathcal{M}(\delta_0) = 1$ then for every observation $O \in \mathcal{O}$ such that $O \cap Q \neq \emptyset$, verifier has a winning strategy in $\mathcal{G}_\mathcal{M}$ from $O \cap Q$.*

Lemma 3. *Let Q be a subset such that $Q \subseteq O$ for some observation $O \in \mathcal{O}$. Assume that a nonempty collection of subsets of states $\mathcal{T}$ is limit-reachable from Q, then either $Q \in \mathcal{T}$ or there exists a nonempty collection of subsets of states $\mathcal{R}$ such that:*

  i) $Q \notin \mathcal{R}$,
  ii) $\mathcal{R}$ is $\sharp$-reachable from Q,
  iii) $\mathcal{T}$ is limit-reachable from every subset in $\mathcal{R}$.

Proof. If $Q \subseteq T$ for some $T \in \mathcal{T}$, then there is nothing to prove. Assume the contrary. Since $\mathcal{T}$ is limit-reachable from Q, for every $n \in \mathbb{N}$ there exists a strategy σ_n such that:

$$\mathbb{P}^{\sigma_n}_{\delta_Q}\left(\exists m \in \mathbb{N}, \exists T \in \mathcal{T}, \phi_m(\delta_Q, \sigma_n, T) \geq 1 - \frac{1}{n}\right) \geq 1 - \frac{1}{n}. \tag{7}$$

Let $\pi_n = O a_1^n O a_2^n O \cdots$ the unique play consistent with the strategy σ_n such that the observation received all along π_n is O and let $\pi_n^m = O a_1^{(n)} O \cdots a_m^{(n)} O$. Let $A_Q = \{a \in A \mid (Q \in Q \cdot a) \wedge (Q \cdot a^\sharp = \{Q\})\}$ and let $d_n = \min \{k \mid \sigma_n(\pi_n^k) \notin A_Q\}$ with values in $\mathbb{N} \cup \{\infty\}$ and denote $(u_n)_{n \in \mathbb{N}}$ the sequence of words in A^* such that: $u_n = a_1^{(n)} \cdots a_{d_n-1}^{(n)}$.

We need the following preliminary result: there exists $\eta > 0$ such that for every $n \geq 0$

$$\mathbb{P}_{\delta_Q}^{\sigma_n} \left(\forall m < d_n, \; \forall T \in \mathcal{T}, \; \phi_m(\delta_Q, \sigma_n, T) \leq 1 - \eta \right) = 1 \; . \tag{8}$$

As a consequence of (8), it is not possible that for infinitely many n, $d_n = \infty$ otherwise (8) would contradict (7). We assume wlog (simply extract the corresponding subsequence from $(\sigma_n)_n$) that $d_n < \infty$ for every n thus all words u_n and plays $\pi_n^{d_n}$ are finite Since A is finite we also assume wlog that $\sigma_n(\pi_n^{d_n})$ is constant equal to some action $a \in A \setminus A_Q$. Since $a \notin A_Q$ then either $Q \notin Q \cdot a$ or $Q \in Q \cdot a$ and $Q \cdot a^\sharp \neq \{Q\}$. In the first case let $\mathcal{R} = Q \cdot a$ and in the second case let $\mathcal{R} = Q \cdot a^\sharp$.

We show that $\mathcal{R}$ satisfies the constraints of the lemma.

$i)$ holds because $a \notin A_Q$ and by definition of A_Q, $ii)$ holds because either $\mathcal{R} = Q \cdot a$ or $\mathcal{R} = Q \cdot a^\sharp$ hence playing a or $a^\sharp$ is a winning strategy for Verifier.

The proof of $iii)$ is fairly technical and is omitted due to pages restriction, see [16] for a complete proof. □

Proof (Proposition 4). Let $\mathcal{M}$ be a ♯-acyclic POMDP and δ_0 be an initial distribution. Assume that $\mathrm{Val}(\delta_0) = 1$ then by Proposition 1 we know that $\mathcal{T} = 2^T \setminus \emptyset$ is limit-reachable from $\mathrm{Supp}(\delta_0)$, using the sequence of strategies $(\sigma_n)_{n \in \mathbb{N}}$. Thanks to Lemma 3, we construct a winning strategy for verifier from $\mathrm{Supp}(\delta_0)$: when the current vertex Q is not in $\mathcal{T}$, compute $\mathcal{R}$ given by Lemma 3 and play a strategy to reach one of the vertices $R \in \mathcal{R}$. Because of condition i) of Lemma 3, a play consistent with this strategy will not reach twice in a row the same vertex until the play reaches some vertes $T \in \mathcal{T}$. Since $\mathcal{M}$ is ♯-acyclic, the only loops in $\mathcal{G}_{\mathcal{M}}$ are self loops and as a consequence the play will necessarilly end up in $\mathcal{T}$. □

Proposition 3 and Proposition 4 lead the following theorem:

Theorem 2. *Given a ♯-acyclic POMDP $\mathcal{M}$ and an initial distribution δ_0. Verifier has a winning strategy in the knowledge game $\mathcal{G}_{\mathcal{M}}$ if and only if $\mathrm{Val}_{\mathcal{M}}(\delta_0) = 1$.*

Theorem 1 follows directly from Theorem 2 and from the fact that the winner of a perfect information reachability game can be computed in quadratic time.

5   Conclusion

We have identified the class of ♯-acyclic POMDP and shown that for this class the value 1 problem is decidable. As a future research, we aim at identifying larger decidable classes such that the answer to the value 1 problem depends quantitatively on the transition probabilities as opposed to ♯-acyclic POMDPs. This would imply an improvement in the definition of the iteration operation.

References

1. Bertoni, A.: The solution of problems relative to probabilistic automata in the frame of the formal languages theory. In: Siefkes, D. (ed.) GI 1974. LNCS, vol. 26, pp. 107–112. Springer, Heidelberg (1975)
2. Bertoni, A., Mauri, G., Torelli, M.: Some recursive unsolvable problems relating to isolated cutpoints in probabilistic automata. In: Proceedings of the Fourth Colloquium on Automata, Languages and Programming, pp. 87–94. Springer, London (1977)
3. Bertrand, N., Genest, B., Gimbert, H.: Qualitative determinacy and decidability of stochastic games with signals. In: LICS, pp. 319–328 (2009)
4. Chatterjee, K., Doyen, L., Henzinger, T.A., Raskin, J.F.: Algorithms for omega-regular games of incomplete information. LMCS 3(3) (2007)
5. Chatterjee, K.: Concurrent games with tail objectives. Theor. Comput. Sci. 388(1-3), 181–198 (2007)
6. Chatterjee, K., Doyen, L., Gimbert, H., Henzinger, T.A.: Randomness for free. In: Hliněný, P., Kučera, A. (eds.) MFCS 2010. LNCS, vol. 6281, pp. 246–257. Springer, Heidelberg (2010)
7. Chatterjee, K., Doyen, L., Henzinger, T.A.: Qualitative analysis of partially-observable markov decision processes. In: Hliněný, P., Kučera, A. (eds.) MFCS 2010. LNCS, vol. 6281, pp. 258–269. Springer, Heidelberg (2010)
8. Chatterjee, K., Jurdzinski, M., Henzinger, T.A.: Quantitative stochastic parity games. In: SODA, pp. 121–130 (2004)
9. Chatterjee, K., Tracol, M.: Decidable problems for probabilistic automata on infinite words. In: LICS, pp. 185–194 (2012)
10. Courcoubetis, C., Yannakakis, M.: The complexity of probabilistic verification. J. ACM 42(4), 857–907 (1995)
11. Derman, C.: Finite State Markovian Decision Processes. Academic Press, Inc., Orlando (1970)
12. Fijalkow, N., Gimbert, H., Oualhadj, Y.: Deciding the value 1 problem for probabilistic leak-tight automata. In: LICS, pp. 295–304 (2012)
13. Gimbert, H.: Randomized Strategies are Useless in Markov Decision Processes. Technical report, Laboratoire Bordelais de Recherche en Informatique - LaBRI (July 2009), http://hal.archives-ouvertes.fr/hal-00403463
14. Gimbert, H., Horn, F.: Solving simple stochastic tail games. In: SODA, pp. 847–862 (2010)
15. Gimbert, H., Oualhadj, Y.: Probabilistic automata on finite words: Decidable and undecidable problems. In: ICALP (2), pp. 527–538 (2010)
16. Gimbert, H., Oualhadj, Y.: Deciding the Value 1 Problem for #-acyclic Partially Observable Markov Decision Processes. Technical report, Laboratoire Bordelais de Recherche en Informatique - LaBRI, Laboratoire d'Informatique Fondamentale de Marseille - LIF (July 2013), http://hal.archives-ouvertes.fr/hal-00743137
17. Putterman, M.L.: Markov Decision Processes: Discrete Stochastic Dynamic Programming. John Wiley and Sons, New York (1994)

Bidimensionality of Geometric Intersection Graphs

Alexander Grigoriev[1], Athanassios Koutsonas[2],
and Dimitrios M. Thilikos[2,3,*]

[1] School of Business and Economics Department of Quantitative Economics,
Maastricht University
[2] Department of Mathematics, National and Kapodistrian University of Athens
[3] AlGCo Project-Team, CNRS, LIRMM

Abstract. Let $\mathcal{B}$ be a finite collection of geometric (not necessarily convex) bodies in the plane. Clearly, this class of geometric objects naturally generalizes the class of disks, lines, ellipsoids, and even convex polygons. We consider geometric intersection graphs $G_{\mathcal{B}}$ where each body of the collection $\mathcal{B}$ is represented by a vertex, and two vertices of $G_{\mathcal{B}}$ are adjacent if the intersection of the corresponding bodies is non-empty. For such graph classes and under natural restrictions on their maximum degree or subgraph exclusion, we prove that the relation between their treewidth and the maximum size of a grid minor is linear. These combinatorial results vastly extend the applicability of all the meta-algorithmic results of the bidimensionality theory to geometrically defined graph classes.

Keywords: geometric intersection graphs, grid exlusion theorem, bidimensionality.

1 Introduction

Parameterized complexity treats problems as subsets of $\Sigma^* \times \mathbb{N}$, for some alphabet Σ. An instance of a parameterized problem is a pair (I, k) where I is the main part of the problem description and k is a, typically small, parameter. A parameterized problem Π is called *fixed parameter tractable*, if it admits an FPT-*algorithm*, namely one that runs in $f(k) \cdot n^{O(1)}$ time, where $n = |I|$. A central issue in parameterized complexity is to find which parameterized problems admit FPT-algorithms and, when this is the case, to reduce as much as possible the contribution of the function $f(\cdot)$, i.e., their *parametric dependance*. FPT-algorithms where $f(k) = 2^{o(k)}$ are called *sub-exponential parameterized algorithms*. It is known that such an algorithm where $f(k) = 2^{o(\sqrt{k})}$ is unlikely to

* The work of the last author is co-financed by the European Union (European Social Fund - ESF) and Greek national funds through the Operational Program "Education and Lifelong Learning" of the National Strategic Reference Framework (NSRF) - Research Funding Program: "Thales. Investing in knowledge society" through the European Social Fund.

V. Geffert et al. (Eds.): SOFSEM 2014, LNCS 8327, pp. 293–305, 2014.

exist for several problems on graphs, even when restricted to sparse graph classes such as planar graphs [2]. Therefore, a parametric dependance $f(k) = 2^{O(\sqrt{k})}$ is the best we may expect and this is what we may aim for.

A kernelization algorithm for a parameterized problem Π is one that, in polynomial time, can replace any instance (I, k) with a new equivalent one whose size depends exclusively on the parameter k. If such an algorithm exists and the size of the new instance is linear in k, then we say that Π *admits a linear kernel*. While the existence of an FTP-algorithm implies the existence of a kernel it is a challenge to find for which problems such a kernel can be polynomial [1].

Bidimensionality Theory. This theory was initially introduced in [4] as a general framework for designing parameterized algorithms with sub-exponential parametric dependance. Moreover, it also provided meta-algorithmic results in approximation algorithms [5,8] and kernelization [10] (for a survey on bidimensionality, see [3]). To present the consequences and the motivation of our results let us first give some brief description of the meta-algorithmic consequences of Bidimensionality theory. For this, we need first some definitions.

A *graph invariant* is a function **p** mapping graphs to non-negative integers. For an graph invariant **p**, we denote as $\Pi_{\mathbf{p}}$ the associated parameterized problem, that has as input a pair (G, k), where G is a graph and k is a non-negative integer, and asks whether $\mathbf{p}(G) \leq k$ (or, alternatively whether $\mathbf{p}(G) \geq k$). We also define the graph invariant **bg** such that given a graph G,

$$\mathbf{bg}(G) = \max\{k \mid G \text{ contains the } (k \times k)\text{-grid as a minor}\}.$$

Definition 1. *Given a graph invariant* **p** *we say that* $\Pi_{\mathbf{p}}$ *is minor-bidimensional if the following conditions hold:*

- **p** *is closed under taking of minors, i.e., for every graph G, if H is a minor of G, then* $\mathbf{p}(H) \leq \mathbf{p}(G)$.
- *If L_k is the $(k \times k)$-grid, then* $\mathbf{p}(L_k) = \Omega(k^2)$.

The main consequences of bidimensionality theory for minor closed invariants are summarized by the following: Suppose that **p** is a graph invariant such that $\Pi_{\mathbf{p}}$ is a minor bidimensional problem. Let also $\mathcal{G}$ be a graph class that satisfies the following property (for a definition of treewidth, $\mathbf{tw}(G)$, see Section 2)

$$\forall_{G \in \mathcal{G}} \ \mathbf{tw}(G) = O(\mathbf{bg}(G)) \tag{1}$$

and let $\Pi_{\mathbf{p}}^{\mathcal{G}}$ be the restriction of $\Pi_{\mathbf{p}}$ to the graphs in $\mathcal{G}$, i.e., the problem occurring if we alter all YES-instance of $\Pi_{\mathbf{p}}^{\mathcal{G}}$ whose graph is not in $\mathcal{G}$ to NO-instances. Then the following hold

1. if $\mathbf{p}(G)$ can be computed in $2^{O(\mathbf{tw}(G))} \cdot n^{O(1)}$ steps, then $\Pi_{\mathbf{p}}^{\mathcal{G}}$ can be solved by a sub-exponential parameterized algorithm that runs in $2^{O(\sqrt{k})} \cdot n^{O(1)}$ steps.
2. if **p** satisfies some separability property (see [5,8,10] for the precise definition) and $\Pi_{\mathbf{p}} = \{(G, k) \mid \exists S \subseteq V(G) : |S| \geq k \text{ and } (G, S) \models \psi\}$ where ψ is a sentence in Counting Monadic Second Order logic, then $\Pi_{\mathbf{p}}$ admits a linear kernel, i.e., there exists a polynomial algorithm reducing each instance (G, k) of $\Pi_{\mathbf{p}}$ to an equivalence instance (G', k') where $|V(G')| = O(k)$ and $k' \leq k$.

3. If **p** satisfies some separability property and is reducible (in the sense this is defined in [8]), then there is an EPTAS for computing $\mathbf{p}(G)$ on the graphs in $\mathcal{G}$.

According to the current state of the art all above meta-algorithmic results hold when $\mathcal{G}$ excludes graphs with some fixed graph H as a minor. This is due to the combinatorial result of Demaine and Hajiaghayi in [6], who proved (1) for every graph G excluding some fixed graph H as a minor. While such graphs are of a topological nature it remained an interesting question whether the applicability of the above theory can be extended for geometrically (rather than topologically) restricted graphs classes.

Our Results. Clearly, any extension of the applicability of bidimensionality theory on some class $\mathcal{G}$ requires a proof that it satisfies property (1). Recently, a first step to extend meta-algorithmic results for graph classes that are not topologically restricted was done in [9], where the bidimensionality condition was used to derive sub-exponential algorithms for H-free unit-disk intersection graphs and H-free map graphs, where a graph class is H-free if none of its graphs contains H as a subgraph. However, no meta-algorithmic results were known so far for more generic classes of geometric intersection graphs, for instance for intersection graphs of polygons. In this paper we vastly extend the combinatorial results of [9] to more general families of geometric intersection graphs. In particular, we prove that property (1) holds for several classes of geometric intersection graphs and open a new direction of the applicability of bidimensionality theory. Let $\mathcal{B}$ be a collection of geometrical objects. We denote by $G_{\mathcal{B}}$ the corresponding intersection graph (for the precise definition, see Section 3). Our results are the following.

1. Let $\mathcal{B}$ be a set of (not necessarily straight) lines in the plane such that for each $C_1, C_2 \in \mathcal{B}$ with $C_1 \neq C_2$, the set $C_1 \cap C_2$ is a finite set of points and at most two lines intersect in the same point. Assume also that each line is intersected at most ξ times. Then $\mathbf{tw}(G_{\mathcal{B}}) = O(\xi \cdot \mathbf{bg}(G_{\mathcal{B}}))$.

2. Let $\mathcal{B}$ be a set of ρ-convex bodies (where any two of their points can be joined by a polysegment of at most $\rho - 1$ bends that is entirely inside the body) such that for each $B_1, B_2 \in \mathcal{B}$, if $B_1 \cap B_2 \neq \emptyset$, then the set $B_1 \cap B_2$ has non-empty interior. Let $G_{\mathcal{B}}$ be the intersection graph of $\mathcal{B}$ and let Δ be the maximum degree of $G_{\mathcal{B}}$. Then $\mathbf{tw}(G_{\mathcal{B}}) = O(\rho^2 \cdot \Delta^3 \cdot \mathbf{bg}(G_{\mathcal{B}}))$.

3. Let H be a graph on h vertices, and let $\mathcal{B}$ be a collection of convex bodies in the plane such that for each $B_1, B_2 \in \mathcal{B}$, if $B_1 \cap B_2 \neq \emptyset$, then the set $B_1 \cap B_2$ has non-empty interior. If the intersection graph $G_{\mathcal{B}}$ of $\mathcal{B}$ is α-fat and does not contain H as a subgraph, then $\mathbf{tw}(G_{\mathcal{B}}) = O(\alpha^6 \cdot h^3 \cdot \mathbf{bg}(G_{\mathcal{B}}))$. (Given a real number α, we call the intersection graph of a collection of convex bodies α-fat, if the ratio between the minimum and the maximum radius of a circle where all bodies of the collection can be inscribed, and circumscribed respectively, is upper bounded by α.)

Notice that for H-subgraph free unit-disk intersection graphs treated in [9] is a very special case of the third result (unit-disk graphs are 1-convex and 1-fat).

The paper is organized as follows: In Section 2, we give some basic definitions and results. In Section 3 we prove the main technical results that are used in Section 4 for the derivation of its implications in a variety of geometric graph classes. Section 5 discusses extensions and conclusions of this work. The omitted proofs are deferred to the complete version of the paper.

2   Definitions and Preliminaries

All graphs in this paper are undirected and may have loops or multiple edges. If a graph has no multiple edges or loops we call it *simple*. Given a graph G, we denote by $V(G)$ its vertex set and by $E(G)$ its edge set. Let x be a vertex or an edge of a graph G and likewise for y; their distance in G, denoted by $\mathbf{dist}_G(x, y)$ is the smallest length of a path in G that contains them both. We call *part of a path* any sequence of adjacent edges in a given path. For any set of vertices $S \subseteq V(G)$, we denote by $G[S]$ the subgraph of G induced by the vertices from S.

Graph Embeddings. We use the term *graph embedding* to denote a drawing of a graph G in the plane, where each vertex is associated to a distinct point of the plane and each edge to a simple open Jordan curve, such that its endpoints are the two points of the plane associated with the endvertices of this edge. To simplify the presentation, when not necessary, we do not distinguish between a vertex of G and the point in the plane representing the vertex; likewise for an edge of G. Roughly speaking, we often do not distinguish between G and its embedding. Two edges of an embedding of a graph in the plane *cross*, if they share a non-vertex point of the plane. We use the term *plane graph* for an embedding of a graph without crossings. A graph is *planar* if it admits a plane embedding.

Geometric Bodies, Lines and Polysegments. We call a set of points in the plane a *2-dimensional geometric body*, or simply a *2-dimensional body*, if it is homeomorphic to the closed disk $\{(x, y)| \; x^2 + y^2 \leq 1\}$. Also a *line* is a subset of the plane that is homeomorphic to the interval $[0, 1]$. A *polysegment* C is a line that is the union of a sequence of straight lines $\overline{p_1 p_2}, \overline{p_2 p_3}, \cdots, \overline{p_{k-1} p_k}$ in the plane, where p_1 and p_k are the endpoints of C. We say that a polysegment C *contains* a point p_i and *joins* the endpoints p_1, p_k, and we refer to the rest points $p_2, p_3, \cdots, p_{k-1}$ as *bend points* of C. The length of a polysegment is defined as equal to the number of straight lines it contains (i.e., one more than the number of its bend points). Throughout the paper we assume that a polysegment is not self-crossing.

Minors and Distance Minors. Given two simple graphs H and G, we write $H \preccurlyeq G$ and call H a *minor* of G, if H can be obtained from a subgraph of G by edge contractions (the *contraction* of an edge $e = \{x, y\}$ in a graph G is the operation of replacing x and y by a new vertex x_e that is made adjacent with all the neighbors of x and y in G that are different from x and y). Moreover, we say that H is a *contraction* of G, if H can be obtained from G by contracting edges.

Let G be a simple graph. We denote as G^ℓ the graph obtained from G by adding a loop on each of its vertices. We also say that a subset F of $E(G^\ell)$ is *solid*, if for every $v_1, v_2 \in \bigcup_{e \in F} e$ there is a walk in G^ℓ from v_1 to v_2 consisting of edges in F and where each second edge is a loop. We define the relation $\preccurlyeq_\phi$ between two graphs as follows.

Let H and G be simple graphs. Then we write $H \preccurlyeq_\phi G$, if there is a function $\phi : E(G^\ell) \to V(H) \cup E(H) \cup \{\star\}$, such that

(1) for every vertex $v \in V(H)$, $\phi^{-1}(v)$ is a non-empty solid set,
(2) for every two distinct vertices $v_1, v_2 \in V(H)$, an edge in $\phi^{-1}(v_1)$ does not share a common endpoint with an edge in $\phi^{-1}(v_2)$.
(3) for every edge $e = \{v_1, v_2\} \in E(H)$ and every edge e' in $\phi^{-1}(e)$, e' is not a loop and shares its one endpoint with an edge in $\phi^{-1}(v_1)$ and the other with an edge in $\phi^{-1}(v_2)$.
(4) for every $e \in E(H)$, $|\phi^{-1}(e)| = 1$.

The following lemma reveals the equivalence between the relation defined previously and the minor relation.

Lemma 1. *If G and H are graphs, then $H \preccurlyeq_\phi G$ if and only if H is a minor of G.*

Given the existence of a function ϕ as in the definition above, we say H *is a ϕ-generated minor of G*. Moreover, H *is a distance minor of G* if H is a ϕ-generated minor of G and the following additional condition holds:

(5) for every $e_1, e_2 \in E(G) \setminus \phi^{-1}(\star)$, $\mathrm{dist}_H(\phi(e_1), \phi(e_2)) \leq \mathrm{dist}_G(e_1, e_2)$.

Contractions and c-Contractions. If the definition of the relation $\preccurlyeq_\phi$ is modified by omitting condition (4) and demanding that $\phi^{-1}(\star) = \emptyset$, then we deal with the contraction relation and we say that H *is a ϕ-generated contraction of G*. (Note, that condition (4) is not a requirement of the equivalence to the minor relation – see also the proof of Lemma 1.) Let c be a non negative integer. We say that H is a *c-contraction of G* if H is a ϕ-generated contraction of G and for all $v \in V(H)$, $G[\phi^{-1}(v)]$ is a graph of at most c edges.

In this paper we use the alternative, more complicated, definitions of minors and contractions as they are necessary for the proofs of our results.

Tree-Decompositions and Treewidth. A *tree-decomposition* of a graph G, is a pair $(T, \mathcal{X})$, where T is a tree and $\mathcal{X} = \{X_t : t \in V(T)\}$ is a family of subsets of $V(G)$, called *bags*, such that the following three properties are satisfied:

(1) $\bigcup_{t \in V(T)} X_t = V(G)$,
(2) for every edge $e \in E(G)$ there exists $t \in V(T)$ such that X_t contains both ends of e, and

(3) $\forall v \in V(G)$, the set $T_v = \{t \in V(T) \mid v \in X_t\}$ induces a tree in T.

The *width* of a tree-decomposition is the cardinality of the maximum size bag minus 1 and the *treewidth* of a graph G is the minimum width of a tree-decomposition of G. We denote the treewidth of G by $\mathbf{tw}(G)$. We say that a graph H is a *partial triangulation* of a plane graph G if G is a spanning subgraph of H and H is plane. The following result follows from [12].

Proposition 1. *Let r be an integer. Then, any planar graph with treewidth at least $4.5 \cdot r$ contains a partial triangulation of the $(r \times r)$-grid as a contraction.*

Lemma 2. *Let G be a planar graph and k an integer. If $\mathbf{tw}(G) \geq 18 \cdot k$ then G contains a $(k \times k)$-grid as a distance minor.*

3   Bidimensionality of Line Intersection Graphs

Let $\mathcal{B} = \{B_1, \dots, B_k\}$ be a collection of lines in the plane. The *intersection graph* $G_{\mathcal{B}}$ of $\mathcal{B}$, is a graph whose vertex set is $\mathcal{B}$, and that has an edge $\{B_i, B_j\}$ (for $i \neq j$) if and only if B_i and B_j intersect, namely $B_i \cap B_j \neq \emptyset$.

The following theorem states our main technical result.

Theorem 1. *Let $\mathcal{B}$ be a set of lines in the plane such that for each $C_1, C_2 \in \mathcal{B}$ with $C_1 \neq C_2$, the set $C_1 \cap C_2$ is a finite set of points and at most two lines intersect in the same point. Let also $G_{\mathcal{B}}$ be the intersection graph of $\mathcal{B}$ and let $\xi = \max_{C \in \mathcal{B}} |C \cap \bigcup_{C' \in \mathcal{B} \setminus C} C'|$. Then $\mathbf{tw}(G_{\mathcal{B}}) = O(\xi \cdot \mathbf{bg}(G_{\mathcal{B}}))$.*

For related results on intersection graphs of collections of lines (also called string graphs), see [11]. To prove Theorem 1 we will need a series of lemmata.

Lemma 3. *Let G be a graph and let H be a c-contraction of G. Then $\mathbf{tw}(G) \leq (c+1) \cdot (\mathbf{tw}(H) + 1) - 1$.*

Lemma 4. *Let G be a graph and let $V_1, \dots, V_r$ be a partition of the vertices of G, such that for each $i \in \{1, \dots, r\}$, $G[V_i]$ is a connected graph, and for each $i \in \{1, \dots, r-1\}$ there exist an edge of G with one endpoint in V_i and one endpoint in V_{i+1}. Let also $s \in V_1$ and $t \in V_r$. Then G has a path from s to t with a part P of length at least $\beta - \alpha + 2$, where $1 \leq \alpha < \beta \leq r$, so that P does not contain any edge in $G[V_i]$ for $i \in \{1, \dots, \alpha - 1\} \cup \{\beta + 1, \dots, r\}$.*

Lemma 5. *Let A, B, and C be graphs such that B is a ψ_1-generated contraction of A and C is a ψ_2-generated minor of A for some functions $\psi_1 : E(A^\ell) \to V(B) \cup E(B)$ and $\psi_2 : E(A^\ell) \to V(C) \cup E(C) \cup \{\star\}$. If*

$$\forall_{e \in E(C)} \ |\psi_2^{-1}(e) \cap \psi_1^{-1}(E(B))| = 1 \qquad (2)$$

then C is also a minor of B.

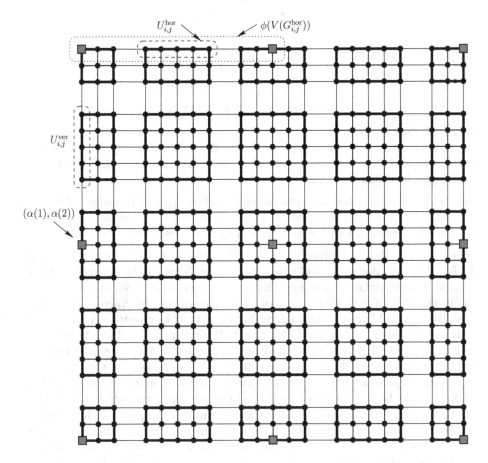

Fig. 1. An example of the proof of Lemma 6 for $c = 5$, $k = 21$, and $k' = 3$

Lemma 6. *Let G be a connected graph and let H be a c-contraction of G. If G contains a $(k \times k)$-grid as a distance minor, then H contains a (k', k')-grid as a minor, where $k' = \lfloor \frac{k-1}{2(c+1)} \rfloor + 1$.*

Proof. We assume that c is an odd number and equivalently prove the lemma for $k' = \lfloor \frac{k-1}{2c} \rfloor + 1$.

Let H be a σ-generated contraction of G for some $\sigma : E(G^\ell) \to V(H) \cup E(H)$ such that $G[\sigma^{-1}(v)]$ is a graph of at most c edges for all $v \in V(H)$. Suppose also that G contains a $(k \times k)$-grid L_k as a distance minor via a function $\phi : E(G^\ell) \to V(L_k) \cup E(L_k) \cup \{\star\}$.

We assume that $V(L_k) = \{1, \ldots, k\}^2$ where each (i, j) corresponds to its grid coordinates. Our target is to prove that the $(k' \times k')$ grid $L_{k'}$ is a minor of H. We define $\alpha : \{1, \ldots, k'\} \to \{1, \ldots, k\}$ such that $\alpha(i) = 2(i-1)c+1$. Notice that this definition is possible as $2(k'-1)c+1 \leq k$. For each $(i, j) \in \{1, \ldots, k'\}^2$, we define a *horizontal* and a *vertical* set of vertices in $V(L_k)$,

$$U_{i,j}^{\text{hor}} = \bigcup_{r \in \{\alpha(i)+(c+1)/2,\ldots,\alpha(i+1)-(c+1)/2\}} (r, \alpha(j)), \qquad (3)$$

$$U_{i,j}^{\text{ver}} = \bigcup_{r \in \{\alpha(j)+(c+1)/2,\ldots,\alpha(j+1)-(c+1)/2\}} (\alpha(i), r) \qquad (4)$$

and let $\mathcal{U}$ be the collection of all sets $U_{i,j}^{\text{hor}}$ or $U_{i,j}^{\text{ver}}$ defined in (3) and (4). For every horizontal (resp. vertical) $U \in \mathcal{U}$, we denote by $\mathcal{E}(U) \subseteq E(L_k)$ the set containing all horizontal (resp. vertical) edges of L_k with an endpoint in U. We will prove the following claim:

(∗)   Let U_1 and U_2 be two different sets of $\mathcal{U}$ and let e_1, e_2 be two edges of G such that $\phi(e_i) \in \mathcal{E}(U_i) \cup U_i$, for $i = \{1,2\}$. Then, there are no disjoint paths of length at most c from the endpoints of e_1 to the endpoints of e_2 in G.

Since L_k is a distance minor of G, it suffices to show that there is no cycle in L_k, that contains $\phi(e_1)$ and $\phi(e_2)$ together with two paths between them of length at most c. Let us suppose that such a cycle exists. Notice that by the definition of $\mathcal{U}$, if two vertices x, y of $V(L_k)$ belong to two different sets of $\mathcal{U}$, then $\mathbf{dist}_{L_k}(x, y) \geq c+1$. This implies that $\phi(e_i)$ must be an edge $v_i u_i$ of L_k with only one endpoint, say u_i, in U_i, for $i = \{1,2\}$, or else we are done. Likewise, it holds that $\mathbf{dist}_{L_k}(u_1, u_2) \geq c+1$ and hence one path of length at most c of the cycle must be from v_1 to u_2, the other from v_2 to u_1. It follows, that the edges $v_1 u_1$ and $v_2 u_2$ cannot be both vertical nor horizontal, and all vertices of the two disjoint paths lie inside the square part of the grid these two edges define. This contradicts the planarity of the grid, which completes the proof of the claim.

For every $i, j \in \{1,\ldots,k'\}^2$ we choose arbitrarily a vertex $v_{i,j}$ from the graph $G[\phi^{-1}(\alpha(i),\alpha(j))]$. This selection creates a collection of $k' \times k'$ vertices of G.

For each pair $\{(i,j),(i+1,j)\}$ where $(i,j) \in \{1,\ldots,k'-1\} \times \{1,\ldots,k'\}$, we observe that the graph

$$G_{i,j}^{\text{hor}} = G[\bigcup_{i' \in \{\alpha(i),\ldots,\alpha(i+1)\}} \phi^{-1}(i',\alpha(j))]$$

is connected, and for every $i' \in \{\alpha(i),\ldots,\alpha(i+1)\}$ the sets $\phi^{-1}(i',\alpha(j))$ form a partition of $V(G_{i,j}^{\text{hor}})$ and there is an edge of G between $\phi^{-1}(i',\alpha(j))$ and $\phi^{-1}(i'+1,\alpha(j))$. From Lemma 4, $G_{i,j}^{\text{hor}}$ contains a path $P_{i,j}^{\text{hor}}$ from $v_{i,j}$ to $v_{i+1,j}$ with a part of length at least $c+1$ in $\phi^{-1}(\mathcal{E}(U_{i,j}^{\text{hor}})) \cup \phi^{-1}(U_{i,j}^{\text{hor}})$. Clearly, one of the edges in this part of the path, say $e_{i,j}^{\text{hor}}$, is an edge of $\sigma^{-1}(E(H))$. We denote by $\overrightarrow{P}_{i,j}$ (resp. $\overleftarrow{P}_{i+1,y}$) the part of $P_{i,j}^{\text{hor}}$ starting from $v_{i,y}$ (resp. $v_{i+1,y}$) and containing only one endpoint of $e_{i,j}^{\text{hor}}$.

Working in the same way as before but following the "vertical" instead of "horizontal" direction, for each pair $\{(i,j),(i,j+1)\}$ where $(i,j) \in \{1,\ldots,k'\} \times \{1,\ldots,k'-1\}$, we define the graph

$$G_{i,j}^{\text{ver}} = G[\bigcup_{j' \in \{\alpha(j),\ldots,\alpha(j+1)\}} \phi^{-1}(\alpha(i),j')]$$

and we find the path $P_{i,j}^{\mathrm{ver}}$ in it starting from $v_{i,j}$ finishing in $v_{i,j+1}$ and containing an edge $e_{i,j}^{\mathrm{ver}}$ of $\sigma^{-1}(E(H))$ that belongs in $\phi^{-1}(\mathcal{E}(U_{i,j}^{\mathrm{ver}})) \cup \phi^{-1}(U_{i,j}^{\mathrm{ver}})$. As before, $P_{i,j}^{\mathrm{ver}}$ is decomposed to a path $\downarrow P_{i,j}$ (containing $v_{i,j}$), the edge $e_{i,j}^{\mathrm{ver}}$, and the path $\uparrow P_{i,j+1}$ (containing $v_{i,j+1}$). Let, finally, E^* be the set containing each $e_{x,y}^{\mathrm{hor}}$ and each $e_{x,y}^{\mathrm{ver}}$.

From Lemma 5, to prove that $L_{k'}$ is a minor of H, it is enough to define a function $\tau : E(G^\ell) \to V(L_{k'}) \cup E(L_{k'}) \cup \{\star\}$ certifying that $L_{k'}$ is a minor of G in a way that $\forall f \in E(L_{k'}) \; |\tau^{-1}(f) \cap \sigma^{-1}(E(H))| = 1$. For this, for every $(x, y) \in \{1, \ldots, k'\}$ we define $E_{x,y}$ as the union of the edges and the loops of the vertices of every path that exists in the set $\{\overleftarrow{P}_{x,y}, \overrightarrow{P}_{x,y}, \downarrow P_{x,y}, \uparrow P_{x,y}\}$ and for each $e \in E_{x,y}$ we set $\tau(e) = (x, y)$. Notice that for every $(x, y) \in \{1, \ldots, k'\}^2$, $G[\tau^{-1}(x, y)]$ is the union of a set of paths of G having a vertex in common, thus it induces a connected subgraph of G. Let now e be an edge of $L_{k'}$. In case $e = \{(x, y), (x+1, y)\}$ (resp. $e = \{(x, y), (x, y+1)\}$), then, by its definition, the edge $e_{x,y}^{\mathrm{hor}}$ (reps. $e_{x,y}^{\mathrm{ver}}$) connects an endpoint v_1 of an edge in $\tau^{-1}(x, y)$ (resp. $\tau^{-1}(x, y)$) with an endpoint v_2 of an edge in $\tau^{-1}(x+1, y)$ (resp. $\tau^{-1}(x, y+1)$). In any case, we set $\tau(v_1 v_2) = e$. It follows that $\tau(E^*) = E(L_{k'})$. For all edges of G^ℓ whose image has not been defined so far, we set $\tau(e) = \star$. It is now easy to verify that τ is a well-defined function and that $L_{k'}$ is a τ-generated minor of G.

Next we prove that $\forall f \in E(L_{k'}) \; |\tau^{-1}(f) \cap \sigma^{-1}(E(H))| = 1$. For this, first of all notice that, by the definition of τ, all edges in $\tau^{-1}(E(L_{k'})) = E^*$ are edges of $\sigma^{-1}(E(H))$. Therefore, it suffices to prove that for each $e \in E(H)$, $\sigma^{-1}(e)$ contains no more than one edge from E^*. Suppose in contrary that $e_1, e_2 \in \sigma^{-1}(e) \cap E^*$ and $e_1 \neq e_2$. As $\sigma(e_1) = \sigma(e_2) = e$, it follows that each e_i has an endpoint w_i in $\sigma^{-1}(w)$ and an endpoint z_i in $\sigma^{-1}(z)$, where $wz = e$. Since each subgraph $G[\sigma^{-1}(w)]$ and $G[\sigma^{-1}(z)]$ is connected, has at most c edges and both are disjoint, there are two disjoint paths of length at most c in G from w_1 to w_2 and from z_1 to z_2, a contradiction to (∗) as $e_1, e_2 \in E^*$. $\square$

Lemma 7. *Let H_1 and H_2 be two graphs. Consider a graph G such that H_1 is a c_1-contraction of G and H_2 is a c_2-contraction of G. If H_1 is planar then* $\mathbf{tw}(H_2) = O(c_1 \cdot c_2 \cdot \mathbf{bg}(H_2)) = 36 \cdot (c_1 + 1) \cdot (c_2 + 1) \cdot [\mathbf{bg}(H_2) - 1] + O(c_1).$

Proof. Let H_1, H_2 be two contractions of G generated by some $\sigma_i : E(G^\ell) \to V(H_i) \cup E(H_i)$, $i = 1, 2$ respectively. Let $r = \mathbf{tw}(H_2)$. As G contains H_2 as a contraction, it follows that $\mathbf{tw}(G) \geq r$. By Lemma 3, $\mathbf{tw}(H_1) \geq (r+1)/(c_1 + 1) - 1$. Since H_1 is planar, by Lemma 2, H_1 contains $L_{r'}$ as a distance minor, where $r' = \lfloor \frac{1}{18} \cdot (\frac{r+1}{c_1+1} - 1) \rfloor$. As H_1 is a contraction of G, then also G contains $L_{r'}$ as a distance minor. By Lemma 6, H_2 contains as a minor an (r'', r'')-grid, where $r'' = \lfloor \frac{r'-1}{2(c_2+1)} \rfloor + 1$, as claimed. $\square$

Proof of Theorem 1. Given a planar drawing of the lines of $\mathcal{B}$, consider any crossing p, that is a point of the plane that belongs to more than one line. By the assumptions, p belongs to exactly two lines, say L_1, L_2 and there is an open disc D of the plane containing p, but no lines other than L_1 and L_2 and no other point that belongs to more than one line. In addition, w.l.o.g. we can always assume that p is not an endpoint of L_1 or L_2; or else we can stretch inside D the line that ends in p without further altering the setting.

Then, let G_1 be the embedding in the plane of a simple graph, in which all endpoints of lines in $\mathcal{B}$ are vertices of G_1 and every line $L \in \mathcal{B}$ is an edge of G_1 joining the two vertices, which are endpoints of L. Note that G_1 is not necessarily plane – in fact, any crossing of two lines in $\mathcal{B}$ is as well a crossing of the corresponding edges of G_1.

For every crossing p of two lines L_1, L_2 in $\mathcal{B}$ and hence of the corresponding edges e_1, e_2 of G_1, we can consider as above an open disc D of the plane in a way, so that $D \cap e = \emptyset$ for any edge $e \in E(G_1) \setminus \{e_1, e_2\}$, the only point in D that belongs to both edges is p, no vertex of G_1 lies in D and all considered discs are pairwise disjoint. Then, we subdivide e_1 and e_2 by adding two new vertices x, y in $D \setminus \{p\}$ and we join x and y with a new edge f that lies entirely in the disc D and meets L_1 and L_2 only at its endpoints. We denote as M the set of these new edges. Notice that we can contract the edge f inside the disc D so that the resulting vertex is the point p, leaving the embedding of the graph outside D untouched. By doing so for every edge in M, we obtain a planar embedding of a graph. Let H be this graph and let G be the graph before contracting the edges in M, i.e., $G/M = H$. Clearly, H is an 1-contraction of G. Moreover, if we contract all edges of G that are not in M, we obtain the intersection graph $G_\mathcal{B}$. Since every edge of G_1 was subdivided into at most $\xi + 1$ edges of G, the graph $G_\mathcal{B}$ is a $(\xi + 1)$-contraction of G and the result follows from Lemma 7.   □

4   Modeling Body Intersections by Intersection of Polysegments

Let $\mathcal{B} = \{B_1, \ldots, B_k\}$ be a collection of 2-dimensional geometric bodies in the plane. We assume that if two bodies do intersect each other, then every connected component of the intersection has a non-empty interior. Our goal is to associate each geometric body $B \in \mathcal{B}$ with a polysegment C such that the resulting set $\mathcal{B}'$ of polysegments conveys all necessary information regarding the disposition of the bodies in the plane and their intersections.

For every body B_i let us pick a point of the sphere p_i that lies in B_i and for every body B_j intersecting B_i ($i \neq j$) in $\mathcal{B}$, a point p_{ij} that lies in $B_i \cap B_j$. We can assume without loss of generality, that these points are pairwise distinct and that any three of them are not co-linear. We stress that since $\mathcal{B}$ is finite this assumption is safe, because we can always consider an open disc D_p of small radius around any given point p of the sphere, such that if p lies in a body B then we can replace B with the possibly expanded body $B' = B \cup D_p$ without altering the intersection graph of $\mathcal{B}$. Let now $\mathcal{P}_i$ be the set of all points that contain the index i in the assigned subscript.

A geometric body B of the sphere is ρ-convex, if for any two points of B there exists a polysegment of length ρ that lies entirely inside B and its endpoints are the given two points. Notice, that the definition of a ρ-convex body naturally extends the standard definition of a convex body, which under this new perspective is also called 1-convex.

Lemma 8. *For any collection of ρ-convex bodies $\mathcal{B}$ on the sphere, there exists a collection of polysegments $\mathcal{B}'$ and a bijection $\phi : \mathcal{B} \rightarrow \mathcal{B}'$, such that two bodies in $\mathcal{B}$ intersect if and only if the corresponding polysegments in $\mathcal{B}'$ intersect. Moreover, each polysegment $C \in \mathcal{B}'$ is crossed by the polysegments from $\mathcal{B}' \setminus C$ at most $\xi = O(\rho^2 \cdot \Delta^3)$ times, where Δ is the maximum degree in the intersection graph $G_{\mathcal{B}}$ of $\mathcal{B}$.*

Straightforwardly applying Theorem 1 to the sets of polysegments constructed in Lemma 8, results to the following theorem for ρ-convex geometric bodies.

Theorem 2. *Let $\mathcal{B}$ be a set of ρ-convex bodies such that for each $B_1, B_2 \in \mathcal{B}$, if $B_1 \cap B_2 \neq \emptyset$, then the set $B_1 \cap B_2$ has non-empty interior. Let $G_{\mathcal{B}}$ be the intersection graph of $\mathcal{B}$ and let Δ be the maximum degree of $G_{\mathcal{B}}$. Then $\mathbf{tw}(G_{\mathcal{B}}) = O(\rho^2 \cdot \Delta^3 \cdot \mathbf{bg}(G_{\mathcal{B}}))$.*

Given a positive real number α, we define the class of α-fat convex intersection graphs as the class containing an intersection graph $G_{\mathcal{B}}$ of a collection·$\mathcal{B}$ of convex bodies, if the ratio between the minimum and the maximum radius of a circle where all objects in $\mathcal{B}$ can be inscribed, and circumscribed respectively, is upper bounded by α. The following lemma describes the manner in which the convex bodies of such a collection are being modeled by polysegments.

Lemma 9. *Let H be a graph on h vertices and let $\mathcal{B}$ be a collection of convex bodies on the sphere. If the intersection graph of $\mathcal{B}$ is α-fat and does not contain graph H as a subgraph, then there exists a collection of polysegments $\mathcal{B}'$ and a bijection $\phi : \mathcal{B} \rightarrow \mathcal{B}'$ such that two bodies in $\mathcal{B}$ intersect if and only if the corresponding polysegments in $\mathcal{B}'$ intersect. Moreover, each polysegment $C \in \mathcal{B}'$ is crossed by the polysegments from $\mathcal{B}' \setminus C$ at most $\xi = O(\alpha^6 \cdot h^3)$ times.*

Again, by straightforwardly applying Theorem 1 to the sets of polysegments constructed in Lemma 9, we derive an improved theorem for H-free α-fat convex intersection graphs of geometric bodies.

Theorem 3. *Let H be a graph on h vertices, and let $\mathcal{B}$ be a collection of convex bodies on the sphere such that for each $B_1, B_2 \in \mathcal{B}$, if $B_1 \cap B_2 \neq \emptyset$, then the set $B_1 \cap B_2$ has non-empty interior. If the intersection graph $G_{\mathcal{B}}$ of $\mathcal{B}$ is α-fat and does not contain H as a subgraph, then $\mathbf{tw}(G_{\mathcal{B}}) = O(\alpha^6 \cdot h^3 \cdot \mathbf{bg}(G_{\mathcal{B}}))$.*

5   Conclusions and Further Research

We believe that the applicability of our combinatorial results is even wider than what is explained in the previous section. The main combinatorial engine of this paper is Lemma 7 that essentially induces an edit-distance notion between graphs under contractibility. This is materialized by the following definition.

Definition 2. *Let G_1 and G_2 be graphs. We define the contraction-edit distance between G_1 and G_2, denoted by $\mathbf{cdist}(G_1, G_2)$, as the minimum c for which there* *exists a graph that contains both G_1 and G_2 as c-contractions. Given a graph G we define $\mathcal{B}^c(G) = \{H \mid \mathbf{cdist}(G, H) \leq c\}$. Finally, given a graph class $\mathcal{G}$, we define $\mathcal{B}^c(\mathcal{G}) = \bigcup_{G \in \mathcal{G}} \mathcal{B}^c(G)$. We refer to the class $\mathcal{B}^c(\mathcal{G})$ as the c-contraction extension of the class $\mathcal{G}$.*

A direct consequence of Lemma 7 is the following:

Corollary 1. *Let $\mathcal{P}$ be the class of planar graphs, then for every fixed constant c, $\mathcal{B}^c(\mathcal{P})$ satisfies (1).*

Actually, Corollary 1 can be extended much further than planar graphs. For this, the only we need analogues of Lemma 2 for more general graph classes. Using the main result of [7], it follows that Lemma 2 is qualitatively correct for every graph class that excludes an apex graph as a minor (an apex graph is a graph that can become planar after the removal of a vertex). By plugging this more general version of Lemma 2 to the proofs of the previous section we obtain the following.

Theorem 4. *Let H be an apex-minor free graph and let $\mathcal{G}_H$ be the class of graphs excluding H as a minor. Then, for every fixed constant c, the class $\mathcal{B}^c(\mathcal{G}_H)$ satisfies (1).*

All the algorithmic applications of this paper follow by the fact that all geometric intersection graph classes considered in this paper are subsets of $\mathcal{B}^c(\mathcal{P})$ for some choice of c. Clearly, Theorem 4 offers a much more wide framework for this, including graphs of bounded genus (including intersection graphs of lines or polygons on surfaces), graphs excluding a single-crossing graph, and $K_{3,r}$-minor free graphs. We believe that Theorem 4, that is the most general combinatorial extension of our results may have applications to more general combinatorial objects than just intersection graph classes. We leave this question open for further research.

References

1. Bodlaender, H.L., Downey, R.G., Fellows, M.R., Hermelin, D.: On problems without polynomial kernels. J. Comput. Syst. Sci. 75, 423–434 (2009)
2. Cai, L., Juedes, D.: On the existence of subexponential parameterized algorithms. J. Comput. System Sci. 67(4), 789–807 (2003)
3. Demaine, E., Hajiaghayi, M.: The bidimensionality theory and its algorithmic applications. The Computer Journal 51(3), 292–302 (2007)
4. Demaine, E.D., Fomin, F.V., Hajiaghayi, M., Thilikos, D.M.: Subexponential parameterized algorithms on graphs of bounded genus and H-minor-free graphs. Journal of the ACM 52(6), 866–893 (2005)
5. Demaine, E.D., Hajiaghayi, M.: Bidimensionality: new connections between FPT algorithms and PTASs. In: Sixteenth Annual ACM-SIAM Symposium on Discrete Algorithms, pp. 590–601. (electronic). ACM, New York (2005)
6. Demaine, E.D., Hajiaghayi, M.: Linearity of grid minors in treewidth with applications through bidimensionality. Combinatorica 28(1), 19–36 (2008)
7. Fomin, F.V., Golovach, P.A., Thilikos, D.M.: Contraction obstructions for treewidth. J. Comb. Theory, Ser. B 101(5), 302–314 (2011)

8. Fomin, F.V., Lokshtanov, D., Raman, V., Saurabh, S.: Bidimensionality and EP-TAS. In: 22nd ACM–SIAM Symposium on Discrete Algorithms (SODA 2011), pp. 748–759. ACM-SIAM, San Francisco, California (2011)
9. Fomin, F.V., Lokshtanov, D., Saurabh, S.: Bidimensionality and geometric graphs. In: 23rd Annual ACM-SIAM Symposium on Discrete Algorithms (SODA 2012), pp. 1563–1575 (2012)
10. Fomin, F.V., Lokshtanov, D., Saurabh, S., Thilikos, D.M.: Bidimensionality and kernels. In: 21st Annual ACM-SIAM Symposium on Discrete Algorithms (SODA 2010), pp. 503–510 (2010)
11. Fox, J., Pach, J.: Applications of a new separator theorem for string graphs. CoRR, abs/1302.7228 (2013)
12. Gu, Q.-P., Tamaki, H.: Improved bounds on the planar branchwidth with respect to the largest grid minor size. Algorithmica 64(3), 416–453 (2012)

Attack against a Pairing Based Anonymous Authentication Protocol

Lucjan Hanzlik and Kamil Kluczniak

Faculty of Fundamental Problems of Technology,
Wrocław University of Technology
{firstname.secondname}@pwr.wroc.pl

Abstract. Anonymous authentication protocols aim to provide means to anonymously prove membership in a group. Moreover, the membership should not be transferable i.e. a subgroup of members should not be able to help an outsider to gain access on behalf of a group. In this note we present two attacks on a recently published protocol of this kind (ICUIMC '11 Proceedings of the 5th International Conference on Ubiquitous Information Management and Communication, article no. 32) and thereby we show that it failed the security targets for an anonymous authentication protocol.

Keywords: anonymous authentication, attack, pairing.

1 Introduction

With the growth of technology many new kinds of information systems were created. There exist systems which can be accessed by anyone and there exist systems which require the user to authenticate before granting access. There are many ways of authentication such as login and password, asymmetric keys etc. One thing all those methods have in common is that the system always knows which user is trying to gain access. This is very convenient when the user accesses his account in the system. However, there are systems which only verify the users membership in the system e.g. access to buildings, company resources. Obviously, in such cases, some privacy issues arise when using those standard authentication methods. Some users do not want the system to know, how many times and when they accessed it. This issue inspired cryptographers to create anonymous authentication protocols. The idea is simple. We consider a set of users (called provers), a group manager and verifiers which verify the membership of users to a particular group (system), created by the group manager. So, a group manager is responsible for generating the system parameters, publishing a group verification key and issuing secret keys. Each prover, having a secret key, may use it to create a proof for a verifier such that he is certain that the prover is a member of a group which has access to the system. In addition, a verifier should not be able to tell if two attempts to access the system were made using the same keys (ie the same prover).

The notion of anonymous authentication protocols, or sometimes called group identification protocols were first proposed in [1]. Since that time, researchers have studied the problem of authenticating as member of a group. There are many approaches to

V. Geffert et al. (Eds.): SOFSEM 2014, LNCS 8327, pp. 306–314, 2014.

achieve this goals. One of it is using a group signature scheme. In adversity to anonymous authentication schemes, a signer can anonymously sign a message on behalf of the group. The problem of anonymously sign messages were first suggested by Chaum and van Heyst [2]. By now, group signatures come by a very rich literature, [3,4,5,6,7] to name only a few, and already cover many interesting problems.

To construct an authentication scheme from any group signature scheme it suffices, that a prover signs a challenge from a verifier. Another approach is to design a „clean" authentication scheme. Such schemes appeared in the literature in [8,9,10,11,12]. Unfortunately, in all of these schemes the execution time is linear dependent by the number of group members. This property make these schemes highly impractical, especially for large groups, or if the proving algorithm has to be implemented on devices with limited resources (e.g. smart cards). An interesting scheme was proposed in [13] which execution time and message size is independent of the size of the group. However, the schemes introduced [13] are in a bit different notion than group signatures or group authentication schemes, namely it introduces ad hoc anonymous identification. An ad hoc anonymous identification allows to create new groups in an ad hoc formation in the sense of ring signatures. The main contribution of [13] was introducing constant size ring and group signatures supporting efficient identity escrow capabilities. The main practical problem in [13] is that, when a new user joins a group any group member has to update their secret keys.

An anonymous authentication protocol have to fulfill some basic requirements. Such schemes must provide soundness (or unforgeability in case of group signatures), anonymity and correctness. Roughly speaking, by soundness we mean that the verifier rejects with overwhelming probability when the prover is not a legitimate group member. By anonymity we mean that no information about the users identity is revealed to the verifier besides that the prover is a group member. To be more specific, suppose we have two executions of an anonymous authentication protocol, then determining whether the executions were produced by a single prover or two distinct provers should be hard. Correctness of an anonymous authentication scheme guaranties that, if a prover is a member of the group, then a verifier accepts with overwhelming probability. Other desired features include revocation of a single user, and identity escrow. An very important property of anonymous authentication protocols is that only a group manager should be able to add new users to a group. So, we require untransferability of membership. Practically, this means that a coalition of users (possibly malicious) should not be able to create secret keys for a new group member. In some articles like [11] this is called „resilience" or „k-resilience" were k is a threshold number of corrupted users for which a scheme stays still resilient (i.e. more than k legitimate users are needed to create a new group member).

Contribution: In this paper we examine the recent work from [14]. The proposed protocol has the execution time and message size independent of the size of the group. Moreover, unlike the scheme from [13], group members do not have to update their keys when a new user joins the group. However, we encountered some design flaws which allows to perform attacks against the protocol. The first one is a collusion attack in which two cooperating group members, using their secret keys, are able to compute the group manager's secret keys. The second attack is against anonymity, where a verifier (or an eavesdropper) can easily determine whether two transcripts were produced by the same prover or by distinct provers.

Organization of the Paper: First, we give a full description of the protocol from [14] in section 2. In section 3 we describe a collusion attack against this scheme. Then, we show how to link two protocol executions of the protocol i.e. we describe an attack against anonymity. Finally, we give some conclusions and design suggestions.

2   Protocol Description

Designations: We denote as GM the group manager who's role is to issue secret keys to group members and public group verification keys. In the original protocol description in [14] the role of GM is denoted as TTP (Trusted Third Party). However, we want to stay consistent with other work in this field. We assume also that H is a secure hash function and $\|$ means concatenation of bit strings. The scheme from [14] describes also a certificate authority denoted as CA. The role of the certificate authority is to issue certificates on user public keys in order to verify the relevance of a user by the GM. Suppose that G is a point on an elliptic curve, then by $R_x(G)$ we denote the x coordinate of that point and by $R_y(G)$ the y coordinate.

Bilinear maps: Let $\mathbb{G}_1$, $\mathbb{G}_2$ be cyclic groups of prime order q and let $e : \mathbb{G}_1 \times \mathbb{G}_1 \to \mathbb{G}_2$ be a function with the following properties:

- (bilinearity) For all $P \in \mathbb{G}_1, Q \in \mathbb{G}_1$ and $a, b, c \in \mathbb{Z}_q$, we have

$$e(P^a + P^b, Q^c) = e(P, Q)^{a \cdot c} e(P, Q)^{b \cdot c}$$

  and

$$e(P^a, Q^b + Q^c) = e(P, Q)^{a \cdot b} e(P, Q)^{a \cdot c},$$

- (non-degeneracy) For all $P, Q \in \mathbb{G}_1, P \neq Q$ we have $e(P, Q) \neq 1$,
- (computability) e can be efficiently computed.

2.1   Setup

The GM chooses a group generator $G \in \mathbb{G}_1$. Then GM chooses two secret keys $s, k_{GM} \in \mathbb{Z}_q$ at random, and computes sG, $Y_{GM} = k_{GM}G$. The values G, sG and Y_{GM} are public, while s and k_{GM} are GM's secrets. We will denote $H_0 = H((R_x(Y_{GM}\|R_y(Y_{GM})))$. The value H_0 don't has to be published since any party in the system can compute it by itself.

Each user has a private key k_U and a corresponding public key $Y_U = k_U G$. User public key Y_U are bounded to a unique identifier ID_U with a certificate from CA. We denote the certificate as $cert_{CA}(Y_U, ID_U)$.

2.2 Registration

First, a user U establishes a secure communication channel with the GM. After this step U sends his certificate $cert_{CA}(Y_U, ID_U)$ to the GM. The GM verifies the certificate and extracts the users public key and identifier. Then it performs the following steps:

1. Selects $\omega_{U,1} \in \mathbb{Z}_q$ randomly,
2. Computes $A_U := \omega_{U,1}G - Y_U = \omega_{U,1}G - k_UG$,
3. Computes $\omega_{U,2} := s\omega_{U,1}$,
4. Computes $\gamma_{U,1} := H_0\omega_{U,1} + R_x(A_U)k_{GM}$,
5. Computes $\gamma_{U,2} := H_0\omega_{U,2} + k_{GM} = H_0s\omega_{U,1} + k_{GM}$,
6. Stores $\gamma_{U,1}, \gamma_{U,2}, A_U, Y_U, \omega_{U,1}, \omega_{U,2}$ and ID_U in GM's internal database,
7. Sends values $A_U, \gamma_{U,1}$ and $\gamma_{U,2}$ to user U.

User U can then verify the validity of these values by checking whether the equations:

$$\gamma_{U,1}G = H_0\omega_{U,1}G + R_x(A_U)Y_{GM} = H_0(A_U + Y_U) + R_x(A_U)Y_{GM}$$

and

$$\gamma_{U,2}G = H_0s\omega_{U,1}G + Y_{GM} = H_0(sA_U + sY_U) + Y_{GM}$$

hold.

Note that the user is able to compute sA_U by himself. He computes $\gamma_{U,2}G = H_0s\omega_{U,1}G + Y_{GM} = H_0sA_U + H_0sY_U + Y_{GM}$. Since $H_0sY_U = k_uH_0(sG)$ and sG is public, we have that $sA_U = H_0^{-1}(\gamma_{U,2}G) - Y_{GM} - H_0sY_U$.

2.3 Computing of Membership Proof

A verifier sends a random nonce N_s and the user U chooses five random values x, t, t_1, x_1 and $\psi \in \mathbb{Z}_q$. Then U computes the following values:

$$
\begin{aligned}
&t_1G, && t_2G, \\
&zG, && x_2 := x \cdot x_1^{-1}, \\
&L := xG, && sL := x(sG), \\
&L_1 := x_1G, && L_2 := x_2G, \\
&sL_1 := x_1(sG), && sL_2 := x_2(sG), \\
&T := \psi G, && z := t \cdot R_x(A_U), \\
&B := t(A_U + T), && t_2 := t \cdot t^{-1}, \\
&B_1 := t_1(A_U + T), && B_2 := t_2(A_U + T), \\
&stT := t\psi(sG), && sB := tsA_U + stT, \\
&sB_1 := t_1sA_U + st_1T, && sB_2 := t_2sA_U + st_2T.
\end{aligned}
$$

Let M be a message or set of instructions, which user U will send to the verifier. We define $\mathcal{D} = (M, L, sL, L_1, sL_1, L_2, sL_2, zG, t_1G, t_2G, B, B_1, B_2, sB, sB_1, sB_2)$. and the hash of $\mathcal{D}$ is computed as:

$$
\begin{aligned}
H(\mathcal{D}) := H(&H(M)\|H(L)\|H(sL)\|H(L_1)\|H(sL_1)\|H(L_2) \\
&\|H(sL_2)\|H(zG)\|H(t_1G)\|H(t_2G)\|H(B)\|H(B_1) \\
&\|H(B_2)\|H(sB)\|H(sB_1)\|H(sB_2)\|H(N_s))
\end{aligned}
$$

User U computes the last five values $\lambda_1, \lambda_{2,1}, \lambda_{2,2}, \lambda_{3,1}, \lambda_{3,2}$ as follows:

$$\lambda_1 := \gamma_{U,1}t + H_0(\psi t - k_U t + xH(\mathcal{D})),$$
$$\lambda_{2,1} := \gamma_{U,2}t_1 + H(\mathcal{D}),$$
$$\lambda_{2,2} := \gamma_{U,2}t_2 + H_0 x_2 H(\mathcal{D}),$$
$$\lambda_{3,1} := t_1(\psi - k_U) + x,$$
$$\lambda_{3,2} := t_2(\psi - k_U) + x.$$

Finally U sends the values $\lambda_1, \lambda_{2,1}, \lambda_{2,2}, \lambda_{3,1}, \lambda_{3,2}, L, sL, L_1, sL_1, L_2, sL_2, zG,$ $t_1 G, t_2 G, B, B_1, B_2, sB, sB_1, sB_2$ along with the message M to the verifier.

2.4   Verification of the Proof

The verifier, holding the values sent by a user U, performs the following steps:

- Computes $zY_{GM} := \lambda_1 G - H_0(B + H(\mathcal{D})L)$.
- Computes $t_1 Y_{GM}$ and $t_2 Y_{GM}$ as follows:

$$t_1 Y_{GM} := \lambda_{2,1}G + \lambda_{3,1}sG - H_0(sB_1 + H(\mathcal{D})L_1) - sL$$
$$t_2 Y_{GM} := \lambda_{2,2}G + \lambda_{3,2}sG - H_0(sB_2 + H(\mathcal{D})L_2) - sL$$

- For $j = \{1, 2\}$ checks whether the discrete logarithm of $t_j G$ equals the discrete logarithm of $t_j Y_{GM}$ with respect to G:

$$e(t_1 Y_{GM}, G) \stackrel{?}{=} e(Y_{GM}, t_1 G) \qquad e(t_2 Y_{GM}, G) \stackrel{?}{=} e(Y_{GM}, t_2 G)$$

- Similarly, verifies the relation between the computed value zY_{GM} and zG sent by the user.

$$e(zY_{GM}, G) \stackrel{?}{=} e(Y_{GM}, zG)$$

- Then, he verifies whether the discrete logarithm of L is the multiplication of the discrete logarithms of L_1 and L_2 with respect to G.

$$e(L_1, L_2) \stackrel{?}{=} e(L, G)$$

- Checks the relations:

$$e(sB, G) \stackrel{?}{=} e(B, sG) \qquad\qquad e(sB_1, G) \stackrel{?}{=} e(B_1, sG)$$
$$e(sB_1, t_2 G) \stackrel{?}{=} e(sB, G) \qquad\qquad e(sB_2, t_1 G) \stackrel{?}{=} e(sB, G)$$
$$e(sL, G) \stackrel{?}{=} e(sG, L) \qquad\qquad e(sL, G) \stackrel{?}{=} e(sG, L)$$
$$e(sL_1, G) \stackrel{?}{=} e(sG, L_1) \qquad\qquad e(sL_2, G) \stackrel{?}{=} e(sG, L_2)$$

- He accepts only if every of the above equations hold and having $\lambda_{3,1}$ and $\lambda_{3,2}$ the verifier checks whether the following equations hold

$$\lambda_{3,1}sG \stackrel{?}{=} sB_1 + sL \qquad\qquad \lambda_{3,1}sG \stackrel{?}{=} L_1$$
$$\lambda_{3,2}sG \stackrel{?}{=} sB_2 + sL \qquad\qquad \lambda_{3,2}sG \stackrel{?}{=} L_2$$

3 The Flaws and Attacks

In this section we give details about the flaws we found in the design of the examined protocol. We argue, that those flaws are so significant, that the protocol should never be implemented since it do not fulfill the basic security requirements for anonymous authentication schemes namely anonymity and untransferability of group membership.

3.1 Collusion Attack

We investigate the case when two users cooperate. Although, the authors of [14] assume that only one user can become malicious. This is a very strong assumption and may be impractical in real world. Note that the below attack also works if one user registers twice, possible under different identifiers.

Let U_1 and U_2 be two users registered by the GM. Each of them obtains three values from GM. Suppose U_1 received:

$$\gamma_{U_1,1} = H_0 \cdot \omega_{U_1,1} + R_x(A_{U_1}) \cdot k_{GM},$$
$$\gamma_{U_1,2} = H_0 \cdot s \cdot \omega_{U_1,1} + k_{GM}$$
$$\text{and} \quad A_{U_1}.$$

Note that H_0 is a publicly known value, thus we got two equations with three unknown variables $\omega_{U_1,1}$, s and k_{GM}. This gives us an indefinite system of linear equations. However, if another user registers into the GM, we get two additional equations

$$\gamma_{U_2,1} = H_0 \cdot \omega_{U_2,1} + R_x(A_{U_2}) \cdot k_{GM},$$
$$\gamma_{U_2,2} = H_0 \cdot s \cdot \omega_{U_2,1} + k_{GM},$$

but only one new unknown variable, namely $\omega_{U_2,1}$. Note that A_{U_2} is given to the user U_2 in the registration phase.

So, if these two users collude, then they have enough information to compute the GM's secrets s and k_{GM}, since we get a definitive system of four linear equations where there are four unknown variables s, k_{GM}, $\omega_{U_1,1}$ and $\omega_{U_2,1}$.

Computing the private keys of GM obviously corrupts the system totally. Two cooperating users having these keys can create new identities at will, so even tracking these „false" identities is highly impractical because of the possible number of identities which can be created.

3.2 Attack against Anonymity

In this section we explore the possibility to break the anonymity of the scheme presented in [14]. We will show that there exists an algorithm $\mathcal{A}$ that given two transcripts of a protocol execution can distinguish if they were produced by the same prover. The idea behind algorithm $\mathcal{A}$ is presented in the proof of lemma 1.

Lemma 1. *Let*

$$T_1 = (\lambda_1, \lambda_{2,1}, \lambda_{2,2}, \lambda_{3,1}, \lambda_{3,2}, L, sL, sL_1, L_2, sL_2, t_1G,$$
$$t_2G, zG, B, sB, B_1, sB_1, B_2, sB_2, N_s, H(D))$$
$$T_2 = (\lambda'_1, \lambda'_{2,1}, \lambda'_{2,2}, \lambda'_{3,1}, \lambda'_{3,2}, L', sL', sL'_1, L'_2, sL'_2, t'_1G,$$
$$t'_2G, z'G, B', sB', B'_1, sB'_1, B'_2, sB'_2, N'_s, H(D'))$$

be two transcripts of executions of the protocol from [14]. There exists an PPT algorithm $\mathcal{A}$ that on input:

$$(H_0, G, sG, (\lambda_{2,1}, sL, t_1 G, H(D)), (\lambda'_{2,1}, sL', t'_1 G, H(D')))$$

outputs, with overwhelming probability, 1 if T_1 and T_2 are transcript of communication between a verifier and the same prover and 0 otherwise.

Proof. We will show the construction of algorithm $\mathcal{A}$. First, it computes:

$$\lambda_{2,1} \cdot sG - H_0 \cdot H(D) \cdot sL =$$
$$= \gamma_{U,2} t_1 sG + H_0 x_1 H(D) sG - H_0 H(D) s x_1 G$$
$$= \gamma_{U,2} t_1 sG$$

and

$$\lambda'_{2,1} \cdot sG - H_0 \cdot H(D') \cdot sL' =$$
$$= \gamma_{U',2} t'_1 sG + H_0 x'_1 H(D') sG - H_0 H(D') s x'_1 G$$
$$= \gamma_{U',2} t'_1 sG.$$

Note that $\mathcal{A}$ can compute those values using only the received input. Finally, the algorithm outputs 1 if

$$e(\gamma_{U,2} t_1 sG, t'_1 G) \stackrel{?}{=} e(\gamma_{U',2} t'_1 sG, t_1 G)$$

and 0 otherwise. By the bilinearity of the pairing function e this final verification is equal to:

$$e(G, G)^{\gamma_{U,2} t_1 s t'_1} \stackrel{?}{=} e(G, G)^{\gamma_{U',2} t'_1 s t_1}$$

Note that this equation is only valid if $\gamma_{U,2} \stackrel{?}{=} \gamma_{U',2}$, which implies that $U = U'$.

Now, since by lemma 1 there exists such algorithm $\mathcal{A}$, so each verifier can run $\mathcal{A}$ and break the anonymity of any prover in the system. Note that a prover sends the required, by algorithm $\mathcal{A}$, data while sending the membership proof. In addition, imagine that there is an adversary that eavesdrops on the secure communication between provers and a verifier, using for example a man-in-the-middle attack. Such an adversary is therefore able to run algorithm $\mathcal{A}$, since it eavesdrops all necessary input data. Therefore, such attack cannot only be run by an active adversary in form of the verifier, but also by a passive adversary which eavesdrops on the communication between provers and verifiers.

4 Final Comments

4.1 Design Suggestions

The easiest way to secure the scheme against a collusion attack is to store the secret keys is secure memory (e.g. a Hardware Security Module) and assume that users cannot access these keys and the authentication algorithm is also executed in a separated environment. These however, are very strong assumptions which in practice costs additional infrastructure.

If a secret key issuance procedure, issues keys or part of keys which are linear equations, then one has to take into account the number of unknown variables issued to new users. Basically, for each new equation introduced to the system, in order to keep the linear system indefinite, a new unknown variable has to be introduced as well. There are security assumption which cover this problem e.g. one-more Diffe-Hellman assumptions, [15] for instance. The idea behind such type of assumptions is the following: even if the adversary is given l distributions of a given type, he is unable to produce a $l + 1$ distribution. Note that this simulates the collusion of l users who are trying to create a new user.

The protection of user identity is a more complex task. Especially, when we use pairing friendly groups in the system. The use of pairings allows to create new and interesting protocols which overcome some design problems difficult to solve without them. However we must remember that in some cases (e.g. type 1 pairing) the Decisional Diffe-Hellman is easy in the underlying group. The authors of [14] used a type 1 pairing and we exploited the symmetry of e (i.e. $e(P, Q) = e(Q, P)$ for all $P, Q \in \mathbb{G}_1$) to perform a cross attack against anonymity. We encourage them to look at type 3 pairing functions and the XDH (External Diffe-Hellman) assumption (see [16]). In such a scenario the pairing function e takes arguments not from one group $\mathbb{G}_1$ but from two different groups $\mathbb{G}_1$ and $\mathbb{G}_2$. Thus, pairing of type 3 is asymmetric. In addition, if we assume that the XDH assumption holds (it is suspected that it holds for certain MNT curves [17]) we may use the assumption that the DDH problem is intractable in $\mathbb{G}_1$.

4.2   Conclusion

We have shown flaws in the design of the scheme from [14] and we presented attacks exploiting them. It follows that the proposed scheme should not be implemented since it do not fulfill the security requirements for anonymous authentication protocols. We also described some design suggestions which may help the authors of [14] to fix the flaws in their construction and maybe produce a secure version of the proposed anonymous authentication protocol.

Acknowledgments. This work was done as part of the Ventures/2012- 9/4 project financed by the Foundation for Polish Science.

References

1. De Santis, A., Di Crescenzo, G., Persiano, G., Yung, M.: On monotone formula closure of SZK. In: Proceedings of the 35th Annual Symposium on Foundations of Computer Science, SFCS 1994, pp. 454–465. IEEE Computer Society, Washington, DC (1994)
2. Chaum, D., van Heyst, E.: Group signatures. In: Davies, D.W. (ed.) EUROCRYPT 1991. LNCS, vol. 547, pp. 257–265. Springer, Heidelberg (1991)
3. Ateniese, G., Camenisch, J., Hohenberger, S., de Medeiros, B.: Practical group signatures without random oracles. Cryptology ePrint Archive, Report 2005/385 (2005), http://eprint.iacr.org/

4. Bellare, M., Micciancio, D., Warinschi, B.: Foundations of group signatures: formal definitions, simplified requirements, and a construction based on general assumptions. In: Biham, E. (ed.) EUROCRYPT 2003. LNCS, vol. 2656, pp. 614–629. Springer, Heidelberg (2003)
5. Boneh, D., Boyen, X., Shacham, H.: Short group signatures. In: Franklin, M. (ed.) CRYPTO 2004. LNCS, vol. 3152, pp. 41–55. Springer, Heidelberg (2004)
6. Boyen, X., Waters, B.: Compact group signatures without random oracles. In: Vaudenay, S. (ed.) EUROCRYPT 2006. LNCS, vol. 4004, pp. 427–444. Springer, Heidelberg (2006)
7. Groth, J.: Fully anonymous group signatures without random oracles. In: Kurosawa, K. (ed.) ASIACRYPT 2007. LNCS, vol. 4833, pp. 164–180. Springer, Heidelberg (2007)
8. Handley, B.: Resource-efficient anonymous group identification. In: Frankel, Y. (ed.) FC 2000. LNCS, vol. 1962, pp. 295–312. Springer, Heidelberg (2001)
9. Jaulmes, É., Poupard, G.: On the security of homage group authentication protocol. In: Syverson, P. (ed.) FC 2001. LNCS, vol. 2339, pp. 106–116. Springer, Heidelberg (2002)
10. De Santis, A., Di Crescenzo, G., Persiano, G.: Communication-efficient anonymous group identification. In: Proceedings of the 5th ACM Conference on Computer and Communications Security, CCS 1998, pp. 73–82. ACM, New York (1998)
11. Boneh, D., Franklin, M.: Anonymous authentication with subset queries (extended abstract). In: Proceedings of the 6th ACM Conference on Computer and Communications Security, CCS 1999, pp. 113–119. ACM, New York (1999)
12. Schechter, S., Parnell, T., Hartemink, A.: Anonymous authentication of membership in dynamic groups. In: Franklin, M. (ed.) FC 1999. LNCS, vol. 1648, pp. 184–195. Springer, Heidelberg (1999)
13. Dodis, Y., Kiayias, A., Nicolosi, A., Shoup, V.: Anonymous identification in ad hoc groups. In: Cachin, C., Camenisch, J.L. (eds.) EUROCRYPT 2004. LNCS, vol. 3027, pp. 609–626. Springer, Heidelberg (2004)
14. Jalal, S., King, B.: A pairing based cryptographic anonymous authentication scheme. In: Proceedings of the 5th International Conference on Ubiquitous Information Management and Communication, ICUIMC 2011, pp. 32:1–32:8. ACM, New York (2011)
15. Boneh, D., Boyen, X.: Short signatures without random oracles and the sdh assumption in bilinear groups. J. Cryptology 21, 149–177 (2008)
16. Ballard, L., Green, M., de Medeiros, B., Monrose, F.: Correlation-resistant storage via keyword-searchable encryption. IACR Cryptology ePrint Archive 2005, 417 (2005)
17. Scott, M., Barreto, P.S.L.M.: Generating more mnt elliptic curves. Des. Codes Cryptography 38(2), 209–217 (2006)

Finding Disjoint Paths in Split Graphs[*]

Pinar Heggernes[1], Pim van 't Hof[1], Erik Jan van Leeuwen[2], and Reza Saei[1]

[1] Department of Informatics, University of Bergen, Norway
{pinar.heggernes,pim.vanthof,reza.saeidinvar}@ii.uib.no
[2] MPI für Informatik, Saarbrücken, Germany
erikjan@mpi-inf.mpg.de

Abstract. The well-known DISJOINT PATHS problem takes as input a graph G and a set of k pairs of terminals in G, and the task is to decide whether there exists a collection of k pairwise vertex-disjoint paths in G such that the vertices in each terminal pair are connected to each other by one of the paths. This problem is known to NP-complete, even when restricted to planar graphs or interval graphs. Moreover, although the problem is fixed-parameter tractable when parameterized by k due to a celebrated result by Robertson and Seymour, it is known not to admit a polynomial kernel unless NP $\subseteq$ coNP/poly. We prove that DISJOINT PATHS remains NP-complete on split graphs, and show that the problem admits a kernel with $O(k^2)$ vertices when restricted to this graph class. We furthermore prove that, on split graphs, the edge-disjoint variant of the problem is also NP-complete and admits a kernel with $O(k^3)$ vertices. To the best of our knowledge, our kernelization results are the first non-trivial kernelization results for these problems on graph classes.

1   Introduction

Finding vertex-disjoint or edge-disjoint paths with specified endpoints is one of the most studied classical and fundamental problems in algorithmic graph theory and combinatorial optimization, with many applications in such areas as VLSI layout, transportation networks, and network reliability; see, for example, the surveys by Frank [9] and by Vygen [24]. An instance of the VERTEX-DISJOINT PATHS problem consists of a graph G with n vertices and m edges, and a set $\mathcal{X} = \{(s_1, t_1), \ldots, (s_k, t_k)\}$ of k pairs of vertices in G, called the *terminals*. The question is whether there exists a collection $\mathcal{P} = \{P_1, \ldots, P_k\}$ of k pairwise vertex-disjoint paths in G such that P_i connects s_i to t_i for every $i \in \{1, \ldots, k\}$. The EDGE-DISJOINT PATHS problem is defined analagously, but here the task is to decide whether there exist k pairwise edge-disjoint paths instead of vertex-disjoint paths.

The VERTEX-DISJOINT PATHS problem was shown to be NP-complete by Karp [13], one year before Even et al. [8] proved that the same holds for EDGE-DISJOINT PATHS. A celebrated result by Robertson and Seymour [22], obtained

[*] This research is supported by the Research Council of Norway.

as part of their groundbreaking graph minors theory, states that the VERTEX-DISJOINT PATHS problem can be solved in $O(n^3)$ time for every fixed k. This implies that EDGE-DISJOINT PATHS can be solved in $O(m^3)$ time for every fixed k. As a recent development, an $O(n^2)$-time algorithm for each of the problems, for every fixed k, was obtained by Kawarabayashi, Kobayashi and Reed [14]. The above results show that both problems are fixed-parameter tractable when parameterized by the number of terminal pairs. On the negative side, Bodlaender, Thomassé and Yeo [3] showed that, under the same parameterization, the VERTEX-DISJOINT PATHS problem does not admit a polynomial kernel, i.e., an equivalent instance whose size is bounded by a polynomial in k, unless NP $\subseteq$ coNP/poly.

Due to their evident importance, both problems have been intensively studied on graph classes. A trivial reduction from EDGE-DISJOINT PATHS to VERTEX-DISJOINT PATHS implies that the latter problem is NP-complete on line graphs. It is known that both problems remain NP-complete when restricted to planar graphs [16,17]. On the positive side, VERTEX-DISJOINT PATHS can be solved in linear time for every fixed k on planar graphs [21], or more generally, on graphs of bounded genus [7,15]. Interestingly, VERTEX-DISJOINT PATHS is polynomial-time solvable on graphs of bounded treewidth [20], while EDGE-DISJOINT PATHS is NP-complete on series-parallel graphs [19], and hence on graphs of treewidth at most 2. Gurski and Wanke [11] proved that VERTEX-DISJOINT PATHS is NP-complete on graphs of clique-width at most 6, but can be solved in linear time on graphs of clique-width at most 2. Natarayan and Sprague [18] proved the NP-completeness of VERTEX-DISJOINT PATHS on interval graphs, and hence also on all superclasses of interval graphs such as circular-arc graphs and chordal graphs. On chordal graphs, VERTEX-DISJOINT PATHS is linear-time solvable for each fixed k [12].

Given the fact that the VERTEX-DISJOINT PATHS problem is unlikely to admit a polynomial kernel on general graphs, and the amount of known results for both problems on graph classes, it is surprising that no kernelization result has been known on either problem when restricted to graph classes. Interestingly, even the classical complexity status of both problems has been open on split graphs, i.e., graphs whose vertex set can be partitioned into a clique and an independent set, which form a well-studied graph class and a famous subclass of chordal graphs [4,10].

We present the first non-trivial kernelization results for VERTEX-DISJOINT PATHS and EDGE DISJOINT PATHS on graph classes, by showing that the problems admit kernels with $O(k^2)$ and $O(k^3)$ vertices, respectively, on split graphs. To complement these results, we prove that both problems remain NP-complete on this graph class. In this extended abstract, the proofs of results marked with a star are omitted due to page restrictions.

2 Preliminaries

All the graphs considered in this paper are finite, simple and undirected. We refer to the monograph by Diestel [5] for graph terminology and notation not defined below. Let G be a graph. For any vertex v in G, we write $N_G(v)$ to denote the neighborhood of v, and $d_G(v) = |N_G(v)|$ to denote the degree of v. A *split graph* is a graph whose vertex set can be partitioned into a clique C and an independent set I, either of which may be empty; such a partition (C, I) is called a *split partition*. Note that, in general, a split graph can have more than one split partition.

In any instance of VERTEX-DISJOINT PATHS or EDGE-DISJOINT PATHS, we allow different terminals to coincide. For this reason, by slight abuse of terminology, we define two paths to be *vertex-disjoint* if neither path contains an inner vertex of the other. This implies that in any solution $\mathcal{P}$ for an instance of VERTEX-DISJOINT PATHS, a terminal might be an endpoint of several paths in $\mathcal{P}$, but none of the paths in $\mathcal{P}$ contains a terminal as an inner vertex. Note that edge-disjoint paths are allowed to share vertices by definition. Hence, in any solution $\mathcal{P}$ for an instance of EDGE-DISJOINT PATHS, terminals may appear as inner vertices of paths in $\mathcal{P}$.

Let $(G, \mathcal{X})$ be an instance of the VERTEX-DISJOINT PATHS problem, where $\mathcal{X} = \{(s_1, t_1), \ldots, (s_k, t_k)\}$. A solution $\mathcal{P} = \{P_1, \ldots, P_k\}$ for the instance $(G, \mathcal{X})$ is *minimum* if there is no solution $\mathcal{Q} = \{Q_1, \ldots, Q_k\}$ for $(G, \mathcal{X})$ such that $\sum_{i=1}^{k} |E(Q_i)| < \sum_{i=1}^{k} |E(P_i)|$. Note that every path in a minimum solution for $(G, \mathcal{X})$ is an induced path in G.

For any problem Π, two instances I_1, I_2 of Π are *equivalent* if I_1 is a yes-instance of Π if and only if I_2 is a yes-instance of Π.

A *parameterized problem* is a subset $Q \subseteq \Sigma^* \times \mathbb{N}$ for some finite alphabet Σ, where the second part of the input is called the *parameter*. A parameterized problem $Q \subseteq \Sigma^* \times \mathbb{N}$ is said to be *fixed-parameter tractable* if for each pair $(x, k) \in \Sigma^* \times \mathbb{N}$ it can be decided in time $f(k)|x|^{O(1)}$ whether $(x, k) \in Q$, for some function f that only depends on k; here, $|x|$ denotes the length of input x. We say that a parameterized problem Q has a *kernel* if there is an algorithm that transforms each instance (x, k) in time $(|x| + k)^{O(1)}$ into an instance (x', k'), such that $(x, k) \in Q$ if and only if $(x', k') \in Q$ and $|x'| + k' \leq g(k)$ for some function g. Here, g is typically an exponential function of k. If g is a polynomial or a linear function of k, then we say that the problem has a *polynomial* kernel or a *linear* kernel, respectively. We refer the interested reader to the monograph by Downey and Fellows [6] for more background on parameterized complexity. It is known that a parameterized problem is fixed-parameter tractable if and only if it is decidable and has a kernel, and several fixed-parameter tractable problems are known to have polynomial or even linear kernels. Recently, methods have been developed for proving non-existence of polynomial kernels, under some complexity theoretical assumptions [1,2,3].

In the NP-completeness proofs in Section 3, we will reduce from a restricted variant of the SATISFIABILITY (SAT) problem. In order to define this variant, we need to introduce some terminology. Let x be a variable and c a clause of a Boolean formula φ in conjunctive normal form (CNF). We say that x *appears* in c if either x or $\neg x$ is a literal of c. If x is a literal of clause c, then we say that x *appears positively* in c. Similarly, if $\neg x$ is a literal of c, then x *appears negatively* in c. Given a Boolean formula φ, we say that a variable x appears positively (respectively negatively) if there is a clause c in φ in which x appears positively (respectively negatively). The following result, which we will use to prove that VERTEX-DISJOINT PATHS is NP-complete on split graphs, is due to Tovey [23].

Theorem 1 ([23]). *The* SAT *problem is NP-complete when restricted to CNF formulas satisfying the following three conditions:*

- *every clause contains two or three literals;*
- *every variable appears in two or three clauses;*
- *every variable appears at least once positively and at least once negatively.*

3   Finding Disjoint Paths in Split Graphs Is NP-Hard

Lynch [17] gave a polynomial-time reduction from SAT to VERTEX-DISJOINT PATHS, thereby proving the latter problem to be NP-complete in general. By modifying his reduction, he then strengthened his result and proved that VERTEX-DISJOINT PATHS remains NP-complete when restricted to planar graphs. In this section, we first show that the reduction of Lynch can also be modified to prove that VERTEX-DISJOINT PATHS is NP-complete on split graphs. We then prove that the EDGE-DISJOINT PATHS problem is NP-complete on split graphs as well, using a reduction from the EDGE-DISJOINT PATHS problem on general graphs.

We first describe the reduction from SAT to VERTEX-DISJOINT PATHS due to Lynch [17]. Let $\varphi = c_1 \vee c_2 \vee \ldots \vee c_m$ be a CNF formula, and let $v_1, \ldots, v_n$ be the variables that appear in φ. We assume that every variable appears at least once positively and at least once negatively; if this is not the case, then we can trivially reduce the instance to an equivalent instance that satisfies this property. Given the formula φ, we create an instance $(G_\varphi, \mathcal{X}_\varphi)$ of VERTEX-DISJOINT PATHS as follows.

The vertex set of the graph G_φ consists of three types of vertices: variable vertices, clause vertices, and literal vertices. For each variable v_i in φ, we create two *variable vertices* v_i and w_i; we call (v_i, w_i) a *variable pair*. For each clause c_j, we add two *clause vertices* c_j and d_j and call (c_j, d_j) a *clause pair*. For each clause c_j, we also add a *literal vertex* for each literal that appears in c_j as follows. If c_j contains a literal v_i, that is, if variable v_i appears positively in clause c_j, then we add a vertex $\ell_{i,j}^+$ to the graph, and we make this vertex adjacent to vertices c_j and d_j. Similarly, if c_j contains a literal $\neg v_i$, then we add a vertex $\ell_{i,j}^-$ and make it adjacent to both c_j and d_j. This way, we create $|c_j|$ paths of length exactly 2 between each clause pair (c_j, d_j), where $|c_j|$ is the number of literals in clause c_j.

For each $i \in \{1, \ldots, n\}$, we add edges to the graph in order to create exactly two vertex-disjoint paths between the variable pair (v_i, w_i) as follows. Let $c_{j_1}, c_{j_2}, \ldots, c_{j_p}$ be the clauses in which v_i appears positively, where $j_1 < j_2 < \cdots < j_p$. Similarly, let $c_{k_1}, c_{k_2}, \ldots, c_{k_q}$ be the clauses in which v_i appears negatively, where $k_1 < k_2 < \cdots < k_p$. Note that $p \geq 1$ and $q \geq 1$ due to the assumption that every variable appears at least once positively and at least once negatively. We now add the edges $v_i \ell_{i,j_1}^+$ and $\ell_{i,j_p}^+ w_i$, as well as the edges $\ell_{i,j_1}^+ \ell_{i,j_2}^+$, $\ell_{i,j_2}^+ \ell_{i,j_3}^+, \ldots, \ell_{i,j_{p-1}}^+ \ell_{i,j_p}^+$. Let $L_i^+ = v_i \ell_{i,j_1}^+ \ell_{i,j_2}^+ \cdots \ell_{i,j_{p-1}}^+ \ell_{i,j_p}^+ w_i$ denote the path between v_i and w_i that is created this way. Similarly, we add exactly those edges needed to create the path $L_i^- = v_i \ell_{i,k_1}^- \ell_{i,k_2}^- \cdots \ell_{i,j_{q-1}}^- \ell_{i,j_q}^- w_i$. This completes the construction of the graph G_φ.

Let $\mathcal{X}_\varphi$ be the set consisting of all the variable pairs and all the clause pairs in G_φ, i.e., $\mathcal{X}_\varphi = \{(v_i, w_i) \mid 1 \leq i \leq n\} \cup \{(c_j, d_j) \mid 1 \leq j \leq m\}$. The pair $(G_\varphi, \mathcal{X}_\varphi)$ is the instance of VERTEX-DISJOINT PATHS corresponding to the instance φ of SAT.

Theorem 2 ([17]). *Let φ be a CNF formula. Then φ is satisfiable if and only if $(G_\varphi, \mathcal{X}_\varphi)$ is a yes-instance of the VERTEX-DISJOINT PATHS problem.*

We are now ready to prove our first result.

Theorem 3. *The VERTEX-DISJOINT PATHS problem is NP-complete on split graphs.*

Proof. We reduce from the NP-complete variant of SAT defined in Theorem 1. Let $\varphi = c_1 \vee c_2 \vee \ldots \vee c_m$ be a CNF formula that satisfies the three conditions mentioned in Theorem 1, and let $v_1, \ldots, v_n$ be the variables that appear in φ. Let $(G_\varphi, \mathcal{X}_\varphi)$ be the instance of VERTEX-DISJOINT PATHS constructed from φ in the way described at the beginning of this section. Now let G be the graph obtained from G_φ by adding an edge between each pair of distinct literal vertices, i.e., by adding all the edges needed to make the literal vertices form a clique. The graph G clearly is a split graph.

We will show that $(G, \mathcal{X}_\varphi)$ is a yes-instance of VERTEX-DISJOINT PATHS if and only if $(G_\varphi, \mathcal{X}_\varphi)$ is a yes-instance of VERTEX-DISJOINT PATHS. Since $(G_\varphi, \mathcal{X}_\varphi)$ is a yes-instance of VERTEX-DISJOINT PATHS if and only if the formula φ is satisfiable due to Theorem 2, this suffices to prove the theorem.

If $(G_\varphi, \mathcal{X}_\varphi)$ is a yes-instance of VERTEX-DISJOINT PATHS, then so is $(G, \mathcal{X}_\varphi)$ due to the fact that G is a supergraph of G_φ. Hence it remains to prove the reverse direction. Suppose $(G, \mathcal{X}_\varphi)$ is a yes-instance of VERTEX-DISJOINT PATHS. Let $\mathcal{P} = \{P_1, \ldots, P_n, Q_1, \ldots, Q_m\}$ be a minimum solution, where each path P_i connects the two terminal vertices in the variable pair (v_i, w_i), and each path Q_j connects the terminals in the clause pair (c_j, d_j). We will show that all the paths in $\mathcal{P}$ exist also in the graph G_φ, implying that $\mathcal{P}$ is a solution for the instance $(G_\varphi, \mathcal{X}_\varphi)$.

The assumption that $\mathcal{P}$ is a minimum solution implies that every path in $\mathcal{P}$ is an induced path in G. By the construction of G, this implies that all the inner vertices of every path in $\mathcal{P}$ are literal vertices. Moreover, since the literal vertices form a clique in G, every path in $\mathcal{P}$ has at most two inner vertices.

Let $j \in \{1, \ldots, m\}$. Since $N_G(c_j) = N_G(d_j)$, the vertices c_j and d_j are non-adjacent, and Q_j is an induced path between c_j and d_j, the path Q_j must have length 2, and its only inner vertex is a literal vertex. Recall that we only added edges between distinct literal vertices when constructing the graph G from G_φ. Hence the path Q_j exists in G_φ.

Now let $i \in \{1, \ldots, n\}$. We consider the path P_i between v_i and w_i. As we observed earlier, the path P_i contains at most two inner vertices, and all inner vertices of P_i are literal vertices. If P_i has exactly one inner vertex, then P_i exists in G_φ for the same reason as why the path Q_j from the previous paragraph exists in G_φ. Suppose P_i has two inner vertices. Recall the two vertex-disjoint paths L_i^+ and L_i^- between v_i and w_i, respectively, that were defined just above Theorem 2. Since v_i appears in at most three different clauses, at least once positively and at least once negatively, one of these paths has length 2, while the other path has length 2 or 3. Without loss of generality, suppose L_i^+ has length 2, and let ℓ denote the only inner vertex of L_i^+. Note that both v_i and w_i are adjacent to ℓ. Since P_i is an induced path from v_i to w_i with exactly two inner vertices, P_i cannot contain the vertex ℓ. From the construction of G, it is then clear that both inner vertices of P_i must lie on the path L_i^-. This implies that L_i^- must have length 3, and that $P_i = L_i^-$. We conclude that the path P_i exists in G_φ. $\square$

We now prove the analogue of Theorem 3 for EDGE-DISJOINT PATHS.

Theorem 4. *The* EDGE-DISJOINT PATHS *problem is NP-complete on split graphs.*

Proof. We reduce from EDGE-DISJOINT PATHS on general graphs, which is well-known to be NP-complete [16]. Let $(G, \mathcal{X})$ be an instance of EDGE-DISJOINT PATHS, where $\mathcal{X} = \{(s_1, t_1), \ldots, (s_k, t_k)\}$. Let G' be the graph obtained from G by adding, for every $i \in \{1, \ldots, k\}$, two new vertices s_i' and t_i' as well as two edges $s_i' s_i$ and $t_i' t_i$. Let $\mathcal{X}' = \{(s_1', t_1'), \ldots, (s_k', t_k')\}$. Clearly, $(G, \mathcal{X})$ is a yes-instance of EDGE-DISJOINT PATHS if and only if $(G', \mathcal{X}')$ is a yes-instance of EDGE-DISJOINT PATHS. From G', we create a split graph G'' as follows. For every pair of vertices $u, v \in V(G)$ such that $uv \notin E(G)$, we add to G' the edge uv as well as two new terminals s_{uv}, t_{uv}. Let G'' be the resulting graph, let $\mathcal{Q} = \{(s_{uv}, t_{uv}) \mid u, v \in V(G), uv \notin E(G)\}$ be the terminal pairs that were added to G' to create G'', and let $\mathcal{X}'' = \mathcal{X}' \cup \mathcal{Q}$. We claim that $(G'', \mathcal{X}'')$ and $(G', \mathcal{X}')$ are equivalent instances of EDGE-DISJOINT PATHS.

Since G'' is a supergraph of G', it is clear that $(G'', \mathcal{X}'')$ is a yes-instance of EDGE-DISJOINT PATHS if $(G', \mathcal{X}')$ is. For the reverse direction, suppose that $(G'', \mathcal{X}'')$ is a yes-instance. For every pair $(s_{uv}, t_{uv}) \in \mathcal{Q}$, let P_{uv} be unique path of length 3 in G'' between s_{uv} and t_{uv}, and let $\mathcal{P}'$ be the set consisting of these paths. It can be shown that there is a solution $\mathcal{P}$ for $(G'', \mathcal{X}'')$ such that $\mathcal{P}' \subseteq \mathcal{P}$. Note that the paths in $\mathcal{P}'$ contain all the edges that were added between non-adjacent vertices in G' in the construction of G''. This implies that for every $(s, t) \in \mathcal{X}'$, the path in $\mathcal{P}$ connecting s to t contains only edges that already existed in G'. Hence $\mathcal{P} \setminus \mathcal{P}'$ is a solution for the instance $(G', \mathcal{X}')$. $\square$

4   Two Polynomial Kernels

In this section, we present polynomial kernels for VERTEX-DISJOINT PATHS and EDGE-DISJOINT PATHS on split graphs, parameterized by the number of terminal pairs.

Before we present the kernels, we introduce some additional terminology. Let $(G, \mathcal{X}, k)$ be an instance of either the VERTEX-DISJOINT PATHS problem or the EDGE-DISJOINT PATHS problem, where $\mathcal{X} = \{(s_1, t_1), \ldots, (s_k, t_k)\}$. Every vertex in the set $\{s_1, \ldots, s_k, t_1, \ldots, t_k\}$ is called a *terminal*. If $s_i = v$ (resp. $t_i = v$) for some $v \in V(G)$, then we say that s_i (resp. t_i) is a *terminal on* v; note that, in general, there can be more than one terminal on v. A vertex $v \in V(G)$ is a *terminal vertex* if there is at least one terminal on v, and v is a *non-terminal vertex* otherwise. Given a path P in G and a vertex $v \in V(G)$, we say that P *visits* v if $v \in V(P)$.

4.1   Polynomial Kernel for VERTEX-DISJOINT PATHS on Split Graphs

Our kernelization algorithm for VERTEX-DISJOINT PATHS on split graphs consists of four reduction rules. In each of the rules below, we let $(G, \mathcal{X}, k)$ denote the instance of VERTEX-DISJOINT PATHS on which the rule is applied, where we fix a split partition (C, I) of G. The instance that is obtained after the application of the rule on $(G, \mathcal{X}, k)$ is denoted by $(G', \mathcal{X}', k')$. We say that a reduction rule is *safe* if $(G, \mathcal{X}, k)$ is a yes-instance of VERTEX-DISJOINT PATHS if and only if $(G', \mathcal{X}', k')$ is a yes-instance of this problem. A reduction rule is only applied if none of the previous rules can be applied, i.e., for every $i \in \{2, 3, 4\}$, Rule i is applied only if Rule j cannot be applied for any $j \in \{1, \ldots, i-1\}$.

Rule 1. *If there exists a terminal vertex $v \in V(G)$ such that $v = s_i = t_i$ for some terminal pair $(s_i, t_i) \in \mathcal{X}$, then we set $\mathcal{X}' = \mathcal{X} \setminus \{(s_i, t_i)\}$ and $k' = k - 1$. If v becomes a non-terminal vertex, then we set $G' = G - v$; otherwise, we set $G' = G$.*

Rule 2. *If there exists a non-terminal vertex $v \in I$, then we set $G' = G - v$, $\mathcal{X}' = \mathcal{X}$, and $k' = k$.*

Lemma 1. *Both Rule 1 and Rule 2 are safe.*

Proof. Rule 1 is safe since there is no need to find a path between s_i and t_i, and we make sure that v cannot serve as an inner vertex of another path. To see why Rule 2 is safe, suppose there exists a non-terminal vertex $v \in I$. It is clear that if $(G', \mathcal{X}', k')$ is a yes-instance of VERTEX-DISJOINT PATHS, then $(G, \mathcal{X}, k)$ is also a yes-instance of VERTEX-DISJOINT PATHS, as G is a supergraph of G'. For the reverse direction, suppose $(G, \mathcal{X}, k)$ is a yes-instance of VERTEX-DISJOINT PATHS, and let $\mathcal{P}$ be a minimum solution for this instance. Since all the paths in $\mathcal{P}$ are induced and v is not a terminal vertex, v is not visited by any of the paths in $\mathcal{P}$. Hence $\mathcal{P}$ is also a solution for the instance $(G', \mathcal{X}', k')$. □

Rule 3. *If there exists a terminal vertex $v \in I$ with $d_G(v) \geq 2k - p$, where $p \geq 1$ is the number of terminals on v, then we set G' to be the graph obtained from G by deleting all edges incident with v, adding p new vertices $\{x_1, \ldots, x_p\}$ to C, and making these new vertices adjacent to v, to each other, and to all the other vertices in C. We also set $\mathcal{X}' = \mathcal{X}$ and $k' = k$.*

Lemma 2. *Rule 3 is safe.*

Proof. Suppose there exists a terminal vertex $v \in I$ with $d_G(v) \geq 2k - p$, where $p \geq 1$ is the number of terminals on v. Let $X = \{x_1, \ldots, x_p\}$ be the set of vertices that were added to C during the execution of the rule. Hence, after the execution of the rule, $X \subseteq C$. Let $Y = \{y_1, \ldots, y_p\}$ be the set of terminals on v.

First suppose $(G, \mathcal{X}, k)$ is a yes-instance of VERTEX-DISJOINT PATHS, and let $\mathcal{P} = \{P_1, \ldots, P_k\}$ be an arbitrary solution for this instance. We construct a solution $\mathcal{P}' = \{P_1', \ldots, P_k'\}$ for $(G', \mathcal{X}', k')$ as follows. Let $i \in \{1, \ldots, k\}$. First suppose that neither of the terminals in the pair (s_i, t_i) belongs to the set Y. Since the paths in $\mathcal{P}$ are pairwise vertex-disjoint and v is a terminal vertex, the path P_i does not contain an edge incident with v. Hence P_i exists in G', and we set $P_i' = P_i$. Now suppose $v \in \{s_i, t_i\}$. The assumption that Rule 1 cannot be applied implies that $s_i \neq t_i$. Suppose, without loss of generality, that $v = s_i$. Then $s_i \in Y$, so $s_i = y_r$ for some $r \in \{1, \ldots, p\}$. Let vw be the first edge of the path P_i in G. We define P_i' to be the path in G' obtained from P_i by deleting the edge vw and adding the vertex x_r as well as the edges vx_r and $x_r w$. Let $\mathcal{P}' = \{P_1', \ldots, P_k'\}$ denote the collection of paths in G' obtained this way. Since the paths in $\mathcal{P}$ are pairwise vertex-disjoint in G, and every vertex in $\{x_1, \ldots, x_p\}$ is visited by exactly one path in $\mathcal{P}'$, it holds that the paths in $\mathcal{P}'$ are pairwise vertex-disjoint in G'. Hence $\mathcal{P}'$ is a solution for the instance $(G', \mathcal{X}', k')$.

For the reverse direction, suppose $(G', \mathcal{X}', k')$ is a yes-instance of VERTEX-DISJOINT PATHS, and let $\mathcal{Q} = \{Q_1, \ldots, Q_k\}$ be a minimum solution. Then each of the paths in $\mathcal{Q}$ is an induced path in G'. Let $\mathcal{Q}^* \subseteq \mathcal{Q}$ be the set of paths in $\mathcal{Q}$ that visit a vertex in the set $X = \{x_1, \ldots, x_p\}$. Since there are p terminals on v, and v has exactly p neighbors in G' (namely, the vertices of X), every path in $\mathcal{Q}^*$ has v as one of its endpoints and $|\mathcal{Q}^*| = p$. Moreover, as no vertex in X is a terminal vertex, and the only neighbors of a vertex $x_i \in X$ are v and the vertices of $C \setminus \{x_i\}$, every path in $\mathcal{Q}^*$ visits exactly one vertex of $C \setminus X$. Finally, we observe that each of the $k - p$ paths in $\mathcal{Q} \setminus \mathcal{Q}^*$ visits at most two vertices of C and none of X, as C is a clique and every vertex in X is a non-terminal vertex that is already visited by some path in $\mathcal{Q}^*$. Recall that $d_G(v) \geq 2k - p$. Therefore, at least p vertices of $N_G(v)$, say $z_1, \ldots, z_p$, are not visited by any path in $\mathcal{Q}$.

Armed with the above observations, we construct a solution $\mathcal{P} = (P_1, \ldots, P_k)$ for $(G, \mathcal{X}, k)$ as follows. For every path $Q_i \in \mathcal{Q} \setminus \mathcal{Q}^*$, we define $P_i = Q_i$. Now let $Q_i \in \mathcal{Q}^*$. The path Q_i visits v, one vertex $x_\ell \in X$, and one vertex $z \in C \setminus X$. If $z \in N_G(v)$, then we define P_i to be the path in G whose single edge is vz. If $z \notin N_G(v)$, then we define P_i to be the path obtained from Q_i by replacing the vertex x_ℓ by z_ℓ. It is easy to verify that $\mathcal{P}$ is a solution for the instance $(G, \mathcal{X}, k)$. □

Rule 4. *If there exists a non-terminal vertex $v \in C$ that has no neighbors in I, then we set $G' = G - v$, $\mathcal{X}' = \mathcal{X}$, and $k' = k$.*

Lemma 3. (★) *Rule 4 is safe.*

We now prove that the above four reduction rules yield a quadratic vertex kernel for VERTEX-DISJOINT PATHS on split graphs.

Theorem 5. *The* VERTEX-DISJOINT PATHS *problem on split graphs has a kernel with at most* $4k^2$ *vertices, where* k *is the number of terminal pairs.*

Proof. We describe a kernelization algorithm for VERTEX-DISJOINT PATHS on split graphs. Let $(G, \mathcal{X}, k)$ be an instance of VERTEX-DISJOINT PATHS, where G is a split graph. We fix a split partition (C, I) of G. We then exhaustively apply the four reduction rules, making sure that whenever we apply Rule i for some $i \in \{2, 3, 4\}$, Rule j is not applicable for any $j \in \{1, \ldots, i-1\}$. Let $(G', \mathcal{X}', k')$ be the resulting instance on which none of the reduction rules can be applied. From the description of the reduction rules it is clear that G' is a split graph, and that $\mathcal{X}' = \mathcal{X}$ and $k' \leq k$. By Lemmas 1, 2 and 3, $(G', \mathcal{X}', k')$ is a yes-instance of VERTEX-DISJOINT PATHS if and only if $(G, \mathcal{X}, k)$ is a yes-instance. Hence, the algorithm indeed reduces any instance of VERTEX-DISJOINT PATHS to an equivalent instance.

We now determine an upper bound on the number of vertices in G'. Let (C', I') be the unique partition of $V(G')$ into a clique C' and an independent set I' such that $I' = V(G') \cap I$, i.e., the independent set I' contains exactly those vertices of I that were not deleted during any application of the reduction rules. Since Rule 2 cannot be applied, every vertex in I' is a terminal vertex, so $|I'| \leq 2k$. Similarly, since Rules 3 and 4 cannot be applied, every vertex in I' has degree at most $2k - 2$ and every vertex in C' has at least one neighbor in I', implying that $|C'| \leq 2k(2k - 2)$. This shows that $|V(G')| \leq 4k^2 - 2k \leq 4k^2$.

It remains to argue that the above algorithm runs in polynomial time. Rule 1 is applied at most k times. Rules 2 and 3 together are applied at most $|I|$ times in total, as each vertex in I is considered only once. Since every vertex x_i that is created in an application of Rule 2 has exactly one neighbor in I, Rule 4 is never applied on such a vertex. Consequently, Rule 4 is applied at most $|C|$ times. This means that the algorithm executes all the reduction rules no more than $k + |I| + |C| = k + |V(G)|$ times in total. Since each of the reduction rules can trivially be executed in polynomial time, the overall running time of the kernelization algorithm is polynomial. □

4.2   Polynomial Kernel for EDGE-DISJOINT PATHS on Split Graphs

In this section, we present a kernel with $O(k^3)$ vertices for the EDGE-DISJOINT PATHS problem on split graphs. We need the following two structural lemmas.

Lemma 4. (★) *Let* $(G, \mathcal{X}, k)$ *be an instance of* EDGE-DISJOINT PATHS *such that* G *is a complete graph. If* $|V(G)| \geq 2k$, *then* $(G, \mathcal{X}, k)$ *is a yes-instance.*

Lemma 5. *Let* $(G, \mathcal{X}, k)$ *be an instance of* EDGE-DISJOINT PATHS *such that* G *is a split graph with split partition* (C, I), $\mathcal{X} = \{(s_1, t_1), \ldots, (s_k, t_k)\}$ *and* $s_i \neq t_i$ *for every* $i \in \{1, \ldots, k\}$. *If the degree of every terminal vertex is at least the number of terminals on it and* $|C| \geq 2k$, *then* $(G, \mathcal{X}, k)$ *is a yes-instance.*

Proof. The proof of this lemma consists of two steps: project to C, and route within C. In the first step, we project the terminals to C. Consider any terminal vertex $x \in I$. For each terminal on x, we project it to a neighbor of x in such a way that no two terminals on x are projected to the same vertex; if the terminal is s_i, denote this neighbor by s_i', and if the terminal is t_i, denote this neighbor by t_i'. Since the degree of every terminal vertex is at least the number of terminals on it, this is indeed possible. For any terminal s_i that is on a terminal vertex in C, let $s_i' = s_i$, and for any terminal t_i that is on a terminal vertex in C, let $t_i' = t_i$. Let $\mathcal{X}' = \{(s_i', t_i') \mid i = 1, \ldots, k\}$, and let $G' = G[V(G) \setminus I]$.

Since G' is a complete graph and $|V(G')| = |C| \geq 2k$, there exists a solution $\mathcal{P}' = (P_1', \ldots, P_k')$ for the instance $(G', \mathcal{X}', k)$ due to Lemma 4. We now show that we can extend the paths in $\mathcal{P}'$ to obtain a solution $\mathcal{P}$ for the instance $(G, \mathcal{X}, k)$. For every $i \in \{1, \ldots, k\}$, we extend the path P_i' using the edges $s_i s_i'$ (if $s_i \neq s_i'$) and $t_i t_i'$ (if $t_i \neq t_i'$); let the resulting path be P_i. Since for every terminal vertex $x \in I$, no two terminals on x were projected to the same neighbor of x, the paths in $\mathcal{P}$ are pairwise edge-disjoint. We conclude that $(G, \mathcal{X}, k)$ is a yes-instance. □

Our kernelization algorithm for EDGE-DISJOINT PATHS on split graphs consists of two reduction rules. In each of the two reduction rules below, we let $(G, \mathcal{X}, k)$ denote the instance on which the rule is applied, where we fix a split partition (C, I) of G, and assume that $\mathcal{X} = \{(s_i, t_i) \mid i = 1, \ldots, k\}$. The instance that is obtained after the application of a rule is denoted by $(G', \mathcal{X}', k')$. A reduction rule is *safe* if $(G, \mathcal{X}, k)$ is a yes-instance of EDGE-DISJOINT PATHS if and only if $(G', \mathcal{X}', k')$ is a yes-instance of this problem. Reduction Rule B is only applied if Rule A cannot be applied on the same instance.

Rule A. *If $s_i = t_i$ for some terminal pair $(s_i, t_i) \in \mathcal{X}$, then we set $G' = G$, $\mathcal{X}' = \mathcal{X} \setminus \{(s_i, t_i)\}$, and $k' = k - 1$.*

Rule A is trivially safe. Suppose now that $(G, \mathcal{X}, k)$ is an instance on which Rule A cannot be applied, that is, $s_i \neq t_i$ for every $i \in \{1, \ldots, k\}$. Let (C, I) be an arbitrary split partition of G. If $|C| \geq 2k$, then Lemma 5 ensures that $(G, \mathcal{X}, k)$ is a yes-instance, so our kernelization algorithm will output a trivial yes-instance. If $|C| \leq 2k - 1$, then we still need to upper bound the size of I by a polynomial in k in order to obtain a polynomial kernel. This is exactly what the second reduction rule will achieve. Before we describe the rule, we need to define an auxiliary graph.

Let T be the set of all terminal vertices in G. We construct an auxiliary bipartite graph $H = (I \setminus T, A, F)$, where $I \setminus T$ and A are the two sides of the bipartition and F is the set of edges. Here, the set A is defined as follows: for each pair v, w of vertices of C, we add vertices $a_1^{vw}, \ldots, a_{4k+1}^{vw}$. The set F is then constructed by, for each $x \in I \setminus T$, adding an edge from x to all $a_1^{vw}, \ldots, a_{4k+1}^{vw}$ if and only if x is adjacent to both v and w in G.

Using the graph H, we can now define our second rule. Here, given a matching M of H, we say that $x \in I$ is *covered* by M if x is an endpoint of an edge in M.

Rule B. *Let M be any maximal matching of H, and let R be the set of vertices of $I \setminus T$ that are not covered by M. We set $G' = G - R$, $\mathcal{X}' = \mathcal{X}$, and $k' = k$.*

Lemma 6. *Rule B is safe.*

Proof. It is clear that if $(G', \mathcal{X}', k')$ is a yes-instance of EDGE-DISJOINT PATHS, then $(G, \mathcal{X}, k)$ is also a yes-instance of EDGE-DISJOINT PATHS, as G is a super-graph of G'. For the reverse direction, suppose that $(G, \mathcal{X}, k)$ is a yes-instance of EDGE-DISJOINT PATHS. Note that there exists a solution for $(G, \mathcal{X}, k)$ such that no path in the solution visits a vertex more than once. Among all such solutions, let $\mathcal{P} = (P_1, \ldots, P_k)$ be one for which the total number of visits by all paths combined to vertices from R is minimized. We claim that no path in $\mathcal{P}$ visits a vertex in R.

For contradiction, suppose that some path $P_j \in \mathcal{P}$ visits some vertex $r \in R$. Since $r \notin T$, there are two vertices $v, w \in C$ such that the edges vr and wr appear consecutively in the path P_j. As $r \in R$, it is not covered by the maximal matching M used in Rule B. Since r is adjacent to all the vertices in $\{a_1^{vw}, \ldots, a_{4k+1}^{vw}\}$ and M is a maximal matching, M covers all the vertices in $\{a_1^{vw}, \ldots, a_{4k+1}^{vw}\}$, and consequently at least $4k + 1$ vertices of $I \setminus T$ that are adjacent to both v and w. Let Z denote this set of vertices. Note that all vertices in Z are adjacent to both v and w by the construction of H. By the choice of $\mathcal{P}$, no path of $\mathcal{P}$ visits a vertex twice. Hence, there are at most $4k$ edges of $\bigcup_{i=1}^{k} E(P_i)$ incident with v or w in G. Therefore, there exists a vertex $z \in Z$ such that $\bigcup_{i=1}^{k} E(P_i)$ contains neither the edge vz nor the edge wz. Let P_j' be the path obtained from P_j by replacing r with z and shortcutting it if necessary (i.e., if $z \in V(P_j)$). Then, $\mathcal{P}' = (P_1, \ldots, P_{j-1}, P_j', P_{j+1}, \ldots, P_k)$ is a solution for $(G, \mathcal{X}, k)$ where each path visits each vertex at most once, and where the total number of visits by all paths combined to vertices from R is at least one smaller than $\mathcal{P}$, contradicting the choice of $\mathcal{P}$. Therefore, no path of $\mathcal{P}$ visits a vertex of R. Hence, $\mathcal{P}$ is also a solution for $(G', \mathcal{X}', k')$, and thus it is a yes-instance. $\square$

Rules A and B, together with Lemmas 4 and 5, yield the following result.

Theorem 6. (★) *The* EDGE-DISJOINT PATHS *problem on split graphs has a kernel with at most $8k^3$ vertices, where k is the number of terminal pairs.*

5   Conclusion

It would be interesting to investigate whether or not VERTEX-DISJOINT PATHS or EDGE-DISJOINT PATHS admits a *linear* kernel on split graphs. Another interesting open question is whether either problem admits a polynomial kernel on chordal graphs, a well-known superclass of split graphs.

Bodlaender et al. [3] asked whether or not VERTEX-DISJOINT PATHS admits a polynomial kernel when restricted to planar graphs. What about the EDGE-DISJOINT PATHS problem on planar graphs?

References

1. Bodlaender, H.L., Downey, R.G., Fellows, M.R., Hermelin, D.: On problems without polynomial kernels. J. Comp. Syst. Sci. 75(8), 423–434 (2009)

2. Bodlaender, H.L., Fomin, F.V., Lokshtanov, D., Penninkx, E., Saurabh, S., Thilikos, D.M. (Meta) kernelization. In: 50th Annual IEEE Symposium on Foundations of Computer Science, FOCS 2009, pp. 629–638. IEEE Computer Society (2009)
3. Bodlaender, H.L., Thomasse, S., Yeo, A.: Kernel bounds for disjoint cycles and disjoint paths. Theor. Comp. Sci. 412(35), 4570–4578 (2011)
4. Brandstädt, A., Le, V.B., Spinrad, J.: Graph Classes: A Survey. SIAM Monographs on Discrete Mathematics and Applications (1999)
5. Diestel, R.: Graph Theory, Electronic edn. Springer (2005)
6. Downey, R.G., Fellows, M.R.: Parameterized Complexity. Monographs in Computer Science. Springer (1999)
7. Dvorak, Z., Král', D., Thomas, R.: Three-coloring triangle-free planar graphs in linear time. In: Mathieu, C. (ed.) SODA 2009, pp. 1176–1182. ACM-SIAM (2009)
8. Even, S., Itai, A., Shamir, A.: On the complexity of timetable and multicommodity flow problems. SIAM J. Comp. 5, 691–703 (1976)
9. Frank, A.: Packing paths, circuits, and cuts – a survey. In: Korte, B., Lovász, L., Prömel, H.J., Schrijver, A. (eds.) Paths, Flows, and VLSI-Layout, pp. 47–100. Springer, Berlin (1990)
10. Golumbic, M.C.: Algorithmic Graph Theory and Perfect Graphs. Annals of Disc. Math. 57 (2004)
11. Gurski, F., Wanke, E.: Vertex disjoint paths on clique-width bounded graphs. Theor. Comput. Sci. 359, 188–199 (2006)
12. Kammer, F., Tholey, T.: The k-disjoint paths problem on chordal graphs. In: Paul, C., Habib, M. (eds.) WG 2009. LNCS, vol. 5911, pp. 190–201. Springer, Heidelberg (2010)
13. Karp, R.M.: On the complexity of combinatorial problems. Networks 5, 45–68 (1975)
14. Kawarabayashi, K., Kobayashi, Y., Reed, B.A.: The disjoint paths problem in quadratic time. J. Comb. Theory B 102, 424–435 (2012)
15. Kobayashi, Y., Kawarabayashi, K.: Algorithms for finding an induced cycle in planar graphs and bounded genus graphs. In: Mathieu, C. (ed.) SODA 2009, pp. 1146–1155. ACM-SIAM (2009)
16. Kramer, M., van Leeuwen, J.: The complexity of wirerouting and finding minimum area layouts for arbitrary VLSI circuits. Adv. Comput. Res. 2, 129–146 (1984)
17. Lynch, J.F.: The equivalence of theorem proving and the interconnection problem. ACM SIGDA Newsletter 5(3), 31–36 (1975)
18. Natarajan, S., Sprague, A.P.: Disjoint paths in circular arc graphs. Nordic J. Comp. 3, 256–270 (1996)
19. Nishizeki, T., Vygen, J., Zhou, X.: The edge-disjoint paths problem is NP-complete for series-parallel graphs. Discrete Applied Math. 115, 177–186 (2001)
20. Reed, B.A.: Tree width and tangles: A new connectivity measure and some applications. In: Bailey, R.A. (ed.) Surveys in Combinatorics, pp. 87–162. Cambridge University Press (1997)
21. Reed, B.A., Robertson, N., Schrijver, A., Seymour, P.D.: Finding disjoint trees in planar graphs in linear time. In: Contemp. Math., vol. 147, pp. 295–301. Amer. Math. Soc., Providence (1993)
22. Robertson, N., Seymour, P.D.: Graph minors XIII. The disjoint paths problem. J. Comb. Theory B 63(1), 65–110 (1995)
23. Tovey, C.A.: A simplified NP-complete satisfiability problem. Discrete Applied Math. 8, 85–89 (1984)
24. Vygen, J.: Disjoint paths. Technical report 94816, Research Institute for Discrete Mathematics, University of Bonn (1998)

A New Asymptotic Approximation Algorithm for 3-Dimensional Strip Packing*

Klaus Jansen and Lars Prädel

Universität Kiel, Institut für Informatik, Christian-Albrechts-Platz 4, 24118 Kiel, Germany
{kj,lap}@informatik.uni-kiel.de

Abstract. We study the 3-dimensional Strip Packing problem: Given a list of n boxes $b_1, \ldots, b_n$ of the width $w_i \leq 1$, depth $d_i \leq 1$ and an arbitrary length ℓ_i. The objective is to pack all boxes into a strip of the width and depth 1 and infinite length, so that the packing length is minimized. The boxes may not overlap or be rotated. We present an improvement of the current best asymptotic approximation ratio of 1.692 by Bansal et al. [2] with an asymptotic $3/2 + \varepsilon$-approximation for any $\varepsilon > 0$.

Keywords: Strip Packing, Packing Boxes, Approximation Algorithms.

1 Introduction

We study the 3-dimensional Strip Packing problem: Given a list of n boxes $b_1, \ldots, b_n$ of the width $w_i \leq 1$, depth $d_i \leq 1$ and an arbitrary length ℓ_i. The objective is to pack all boxes into a strip of the width and depth 1 and infinite length, so that the packing length is minimized. The boxes may not overlap or be rotated.

3-dimensional Strip Packing is known to be $\mathcal{NP}$-hard as it is the 2-dimensional counterpart. Thus, unless $\mathcal{P} = \mathcal{NP}$, there will be no polynomial time approximation algorithm that computes a packing with the optimal packing length. Therefore, we study approximation algorithms that have polynomial running time. An asymptotic approximation algorithm A for a minimization problem X with approximation ratio α and additive constant β is a polynomial-time algorithm, that computes for any instance I of the problem X a solution with $A(I) \leq \alpha \cdot \mathrm{OPT}(I) + \beta$, where $\mathrm{OPT}(I)$ is the optimal value of the instance and $A(I)$ is the value of the output. If $\beta = 0$, we call α also absolute approximation ratio. A family of asymptotic approximation algorithms with ratio $1 + \varepsilon$, for any $\varepsilon > 0$ is called an $\mathcal{APTAS}$.

Known results 3-dimensional Strip Packing is a generalization of the 2-dimensional Bin Packing Problem: Given is a list of rectangles $r_1, \ldots, r_n$ of the widths w_i and the heights h_i and an infinite set of 2-dimensional unit-squares, called bins. The objective is to pack all rectangles axis-parallel and non-overlapping into the bins in order to minimize the bins used. Rotations of the rectangles are not allowed. This problem

---

* Research supported by German Research Foundation (DFG) project JA612/12-2, "Approximation algorithms for two- and three-dimensional packing problems and related scheduling problems".

V. Geffert et al. (Eds.): SOFSEM 2014, LNCS 8327, pp. 327–338, 2014.

is a special case of the 3-dimensional Strip Packing problem, where the lengths of all boxes are 1. Thus, the lower bounds for 2-dimensional Bin Packing hold also for our problem. In the non-asymptotic setting, there is no approximation algorithm strictly better than 2, otherwise the problem Partition can be solved in polynomial time. In the asymptotic setting it is proven that there is no $\mathcal{APTAS}$ for this problem, unless $P = NP$ by Bansal et al. [1]. This lower bound was further improved by Chlebík & Chlebíková [3] to the value $1 + 1/2196$. On the positive side there is an asymptotic 3.25 [9], 2.89 [10], and 2.67 [11] approximation for our problem. More recently an asymptotic 2-approximation was given by Jansen and Solis-Oba [6] that was improved by Bansal et al. [2] to an asymptotic 1.692-approximation.

New results We present a significant improvement of the current best asymptotic approximation ratio:

Theorem 1. *For any* $\varepsilon > 0$ *and any instance* I *of the 3-dimensional Strip Packing problem that fits into a strip of length* $\mathrm{OPT}_{3D}(I)$, *we produce a packing of the length* $A(I)$ *such that*

$$A(I) \leq (3/2 + \varepsilon) \cdot \mathrm{OPT}_{3D}(I) + \varepsilon + f(\varepsilon, \ell_{\max}),$$

where $\ell_{\max}$ *is the length of the largest box in* I *and* $f(\varepsilon, \ell_{\max})$ *is a function in* ε *and* $\ell_{\max}$. *The running time is polynomial in the input length.*

Techniques In our work, we use a new result of the 2-dimensional Bin Packing Problem. In [5,12], there is the following result given:

Theorem 2. *For any* $\varepsilon > 0$, *there is an approximation algorithm* A *which produces a packing of a list* I *of* n *rectangles in* $A(I)$ *bins such that*

$$A(I) \leq (3/2 + \varepsilon) \cdot \mathrm{OPT}_{2D}(I) + 69,$$

where $\mathrm{OPT}_{2D}(I)$ *is the optimal number of bins. The running time of* A *is polynomial in* n.

There it is proven, that it is possible to round/enlarge some rectangles, so that there are only a constant number of different types and an optimal packing of the enlarged rectangles fit into roughly $(3/2 + \varepsilon) \cdot \mathrm{OPT}_{2D}(I)$ bins. Since there are only a constant number of different types of rectangles, a solution of them can be computed by solving an (Integer) Linear Program. In our work we present a non-trivial method to use these results for the 3-dimensional Strip Packing problem. Therefore, we adopt also some techniques from [6] to transform an instance of the 3-dimensional Strip Packing problem to an instance of the 2-dimensional Bin Packing problem. The main difficulty is to obtain a solution of our problem from the solution of the 2-dimensional Bin Packing problem.

2   2-Dimensional Bin Packing

As mentioned above, we use the results of the work [5,12] of the 2-dimensional Bin Packing problem. Thus, we give here a brief overview over the results obtained in this work.

2.1 Modifying Packings

We assume that we have an optimal solution in OPT_{2D} bins of an arbitrary instance of the 2-dimensional Bin Packing problem given. We denote by $a(X)$, $w(X)$ and $h(X)$ the total area, width and height of a set X of rectangles. In the first step we find a value δ and divide the instance into big, wide, long, small and medium rectangles. We use therefore the following result given in [5,12], where ε' is in dependency of the precision of the algorithm specified later. A formal proof is given in the full version.

Lemma 1. *We find a value δ, so that $\varepsilon'^{2^{2/\varepsilon'}} < \delta \leq \varepsilon'$ and $1/\delta$ is a multiple of 24 holds and all rectangles r_i of the width $w_i \in [\delta^4, \delta)$ or the height $h_i \in [\delta^4, \delta]$ have a total area of at most $\varepsilon' \cdot OPT_{2D}$.*

The value $1/\delta$ has to be a multiple of 24 for technical reasons, which we will not discuss further. A rectangle is *big* when the width $w_i \geq \delta$ and height $h_i \geq \delta$ holds, it is *wide* when the width $w_i \geq \delta$ and the height $h_i < \delta^4$ holds, when the width $w_i < \delta^4$ and the height $h_i \geq \delta$ holds it is *long*, when the width $w_i < \delta^4$ and height $h_i < \delta^4$ holds it is *small*. If none of these conditions holds, i.e. at least one side is within $[\delta^4, \delta)$ it is a *medium* rectangle. These medium rectangles are packed separately.

Our optimal solution can be transformed so that the widths and the heights of the big rectangles are rounded up to at most $2/\delta^4$ different types of rectangles (cf. Figure 1(a)). We denote the types by $B_1, \ldots, B_{2/\delta^4}$. The wide rectangles are cut in the height. The widths of the resulting slices of the wide rectangles are rounded up to at most $4/\delta^2$ values. The set of the slices of the wide rectangles of the different widths are denoted by $W_1, \ldots, W_{4/\delta^2}$. Vice versa, the long rectangles are cut in the width and the heights of the resulting slices are rounded up to at most $4/\delta^2$ different values. The sets of these slices of different heights are denoted by $L_1, \ldots, L_{4/\delta^2}$. The slices are packed into wide and long containers. There are at most $6/\delta^3$ wide and long containers in each bin. The wide containers have at most $4/\delta^2$ different widths and the containers of one certain width have at most $1/\delta^4$ different heights. Thus, there are at most $4/\delta^6$ different types. The same holds vice versa for the long containers. We denote the sets of wide containers of the different widths by $CW_1, \ldots, CW_{4/\delta^2}$, each set CW_i is separated in sets $CW_{i,1}, \ldots, CW_{i,1/\delta^4}$ of containers of different heights. The sets for the long containers of the different heights are denoted by $CL_1, \ldots, CL_{4/\delta^2}$ and each set CL_i is also separated into sets $CL_{i,1}, \ldots, CL_{i,1/\delta^4}$ of different widths. The small rectangles are cut in the width and height and are packed into the wide and long containers. The total number of bins without the medium rectangles is increased with these steps to at most $(3/2 + 22\delta)OPT_{2D} + 53$ bins.

2.2 2-Dimensional Bin Packing Algorithm

After showing the above described modification steps of any optimal solution, it is possible to compute a solution with only a small increased number of bins. Therefore, all necessary values are guessed via an enumeration step. It is possible to find the value OPT_{2D}, the widths and the heights of the different types of big rectangles and long and wide containers, the different widths of the wide rectangles and the different heights

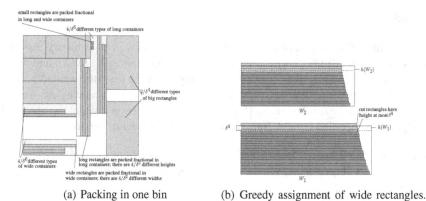

(a) Packing in one bin

(b) Greedy assignment of wide rectangles. Grey rectangles are packed separately.

Fig. 1. Structure of a packing in one bin and assignment of wide rectangles

of the long rectangles in polynomial time in the input length. Note that it is not beforehand clear to which values a rectangle is rounded, i.e. to which set it belongs. We only know that it is possible to enlarge/round it to one of the types. Thus, we have to assign each big, wide and long rectangle to one of the sets. The cardinality of the sets of big rectangles and containers are also guessed via an enumeration step. Whereas we guess approximately with an error of δ^4 the total heights of each set of wide rectangles $h(W_i)$ and the total width of each set of long rectangles $w(L_i)$. Since there are only fractions of these rectangles in these sets, it is not possible to enumerate the cardinality of them. In the following we assume that we are in the iteration of the right guess of all above described values. The big rectangles are assigned via a network flow algorithm to the sets. The wide rectangles are sorted by their widths and greedily assigned to one of these groups, beginning with the widest group, so that the total heights $h(W_i)$ of each set is strictly exceeded (cf. Figure 1(b)). With the right approximate guess of the total heights of each group, we can ensure that all rectangles are assigned to one group. While removing the rectangles of the total height at most $3\delta^4$, we secure that the total height of the wide rectangles is below $h(W_i)$ and thus they fit fractionally into the containers. We denote the set of rectangles that are rounded to the ith width by $\overline{W_i}$. We need 1 additional bin for the removed rectangles of all sets. We do the analogous steps for assigning the long rectangles to the groups. At this moment, the big, long and wide rectangles are assigned to one group and are rounded. The wide rectangles are packed into the wide containers via a linear program, that is similar to the linear program in [7]. We pack the wide rectangles fractionally into the containers of a certain width. Here $\mathfrak{C}_j^{(\ell)}$ represents a configuration of wide rectangles that fit into a wide container of the set CW_ℓ. The configurations are all possible multi-sets of wide rectangles that have a total width of at most the width of the container. There is only a bounded number q of possible configurations. $a(i, \mathfrak{C}_j^{(\ell)})$ represents the number of wide rectangles in the set W_i that are in configuration $\mathfrak{C}_j^{(\ell)}$. The variable $x_j^{(\ell)}$ gives the total height of one configuration in the container. $LP(1)$:

$$\sum_{j=1}^{q^{(\ell)}} x_j^{(\ell)} = h(CW_\ell) \qquad\qquad \ell \in \{1,\ldots,4/\delta^2\}$$

$$\sum_{\ell=1}^{t} \sum_{j=1}^{q^{(\ell)}} a(i, \mathfrak{C}_j^{(\ell)}) \cdot x_j^{(\ell)} \geq h(\overline{W}_i) \qquad\qquad i \in \{1,\ldots,4/\delta^2\}$$

$$x_j^{(\ell)} \geq 0 \qquad\qquad j \in \{1,\ldots,q^{(\ell)}\}, \ell \in \{1,\ldots,4/\delta^2\}$$

The first line secures, that the total height of the containers of a certain width is not exceeded by the configurations, while the second line secures, that there is enough area to occupy all wide rectangles. Since we have at most $2 \cdot 4/\delta^2$ conditions, a basic solution has also at most $8/\delta^2$ non-zero variables, i.e. there are only $8/\delta^2$ different configurations in the solution. This and the fact that there is a bounded number of containers allows us to generate a non-fractional packing of the wide rectangles into the containers. The same is done with the long rectangles that are packed in the long containers. The small rectangles are packed with Next-Fit-Decreasing-Height by Coffman et al. [4] in the remaining gaps, by losing only a small amount of additional bins.

Finally, the big rectangles and wide and long containers are packed with an Integer Linear Program $ILP(1)$:

$$\min \sum_{k=1}^{q} x_k$$

$$s.t. \sum_{k=1}^{q} b(i, C_k) \cdot x_k \geq n_i^b \qquad\qquad i \in \{1,\ldots,2/\delta^4\}$$

$$\sum_{k=1}^{q} w(i, j, C_k) \cdot x_k \geq n_{i,j}^w \qquad i \in \{1,\ldots,4/\delta^2\}, j \in \{1,\ldots,4/\delta^6\}$$

$$\sum_{k=1}^{q} \ell(i, j, C_k) \cdot x_k \geq n_{i,j}^\ell \qquad i \in \{1,\ldots,4/\delta^2\}, j \in \{1,\ldots,4/\delta^6\}$$

$$x_k \in \mathbb{N} \qquad\qquad k \in \{1,\ldots,q\}$$

Therefore, we build also configurations C_k of rectangles that fit into one sole bin. Since we have a constant number of rectangles/containers in one bin and only a constant number of different types, the number of configurations q is also only a constant. n_i^b represents the total number of big rectangles in the set B_i, and $b(i, C_k)$ gives the number of big rectangles in the set B_i that are in configuration C_k. The analogous values for the containers are represented by $n_{i,j}^w$ and $w(i, j, C_k)$ for the wide containers of type $CW_{i,j}$ and $n_{i,j}^\ell$ and $\ell(i, j, C_k)$ for the long containers of the type $CL_{i,j}$. This Integer Linear Program computes a value of at most $(3/2 + 22\delta)\text{OPT}_{2D} + 53$ bins.

3   3-Dimensional Strip Packing

After giving the overview of the 2-dimensional Bin Packing algorithm we start with the presentation of our 3-dimensional Strip Packing algorithm. We start also with an optimal solution of an arbitrary given instance and show how to modify this. Some of the techniques used here are also used in [6]. Afterwards, we present our algorithm. We denote by $vol(X)$ the total volume of a set X of boxes. Furthermore, we call the rectangle of the width w_i and height d_i of a box b_i by the base of b_i.

3.1   Modifying Packings

We first modify an optimal solution of an instance for our problem that fits into a strip of the length OPT_{3D}. We scale the lengths of the whole instance by the value OPT_{3D}, the total length of the optimal packing is thus 1. Afterwards, we extend the lengths of the boxes to the next multiple of ε'/n for a given $\varepsilon' > 0$. This enlarges the strip by at most ε', since each box is enlarged by at most ε'/n and there are at most n boxes on top of each other. The length of the strip is $1 + \varepsilon'$. Furthermore, this allows us to place the boxes on z-coordinates that are multiples of ε'/n. A formal proof of this fact is already given in [6]. In the next step, we cut the strip horizontally on each z-coordinate that is a multiple of ε'/n. Each slice of length ε'/n of the packing is treated as one 2-dimensional bin. Note that each box intersects a slice of the solution completely, or not at all. Each slice of a box b_i is a copy of its base, i.e. a rectangle of the width w_i and height d_i. It follows that we obtain from the optimal solution of the 3-dimensional Strip Packing problem a solution of a 2-dimensional Bin Packing instance in $(1 + \varepsilon') \cdot n/\varepsilon' = n/\varepsilon' + n$ bins. We denote by $\mathrm{OPT}_{2D} \leq n/\varepsilon' + n$ the minimal number of bins used in an optimal packing of this 2-dimensional Bin Packing instance.

We use the modification steps of the 2-dimensional Bin Packing as described above. The medium rectangles/boxes are discarded. Thus we have a packing into one strip of the total length:

$$((3/2 + 22\delta)\mathrm{OPT}_{2D} + 53) \cdot \varepsilon'/n \leq (3/2 + 22\delta) \cdot (n/\varepsilon' + n) + 53) \cdot \varepsilon'/n$$
$$\leq (3/2 + 22\delta) \cdot (1 + \varepsilon') + 53\varepsilon'/n$$
$$\leq 3/2 + 22\delta + 3/2\varepsilon' + 22\delta\varepsilon' + 53\varepsilon'$$

After rescaling the lengths of the boxes by OPT_{3D}, we obtain a packing length of $(3/2 + 22\delta + 3/2\varepsilon' + 22\delta\varepsilon' + 53\varepsilon')\mathrm{OPT}_{3D}$.

3.2   Algorithm

In the first step we set ε' as the largest value so that $1/\varepsilon'$ is a multiple of 24 and $\varepsilon' \leq \min\{\varepsilon/236, 1/48\}$ holds. Thus, $\varepsilon' \geq \varepsilon/260$.

For dual approximation we approximately guess the optimal length $L_{\mathrm{OPT}_{3D}}$ so that $\mathrm{OPT}_{3D} \leq L_{\mathrm{OPT}_{3D}} < \mathrm{OPT}_{3D} + \varepsilon'$ holds. To do this we use a naïve approach. It holds $\mathrm{OPT}_{3D} \in [\ell_{\max}, n \cdot \ell_{\max}]$. Thus, we test less than $n \cdot \ell_{\max}/\varepsilon'$ values with binary search. This takes time at most $\mathcal{O}(\log(n \cdot \ell_{\max}/\varepsilon'))$ and is thus polynomial in the encoding length of the input. In the following we assume that we are in the iteration where we

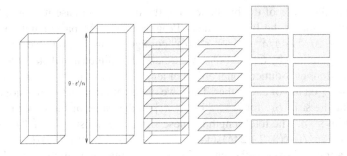

Fig. 2. Transform a box into rectangles

found the correct value $L_{\text{OPT}_{3D}}$. We scale the lengths of the boxes in the input by the value $L_{\text{OPT}_{3D}}$. An optimal packing fits now into a strip of length 1.

Our algorithm rounds afterwards the lengths of the boxes to the next multiple of ε'/n and we cut each box at each multiple of ε'/n. Each slice of a box b_i is treated as a 2-dimensional rectangle of the width w_i and the height d_i. There are now at most $n \cdot n/\varepsilon' = n^2/\varepsilon'$ rectangles that fit in an optimal packing into at most $\text{OPT}_{2D} \le (1 + \varepsilon') \cdot n/\varepsilon'$ bins.

Gap-Creation and Medium Boxes. We find a value δ with Lemma 1, and partition the instance into big, wide, long, small and medium rectangles. We also guess all necessary values that are needed to run the 2-dimensional Bin Packing algorithm. We assume that we are in the iteration, where all values are guessed correctly (cf. also Algorithm 1).

The medium rectangles are divided into two sets $M_{w\delta}$ and $M_{h\delta}$ of rectangles of the width within $[\delta^4, \delta)$ and the remaining rectangles of the height within $[\delta^4, \delta)$. The total area $a(M_{w\delta} \cup M_{h\delta})$ is bounded by $\varepsilon'\text{OPT}_{2D} \le \varepsilon'(n/\varepsilon' + n) = n + \varepsilon'n$. Thus, the total volume of the corresponding 3-dimensional (medium, scaled) boxes is bounded by $\varepsilon'/n \cdot (n + \varepsilon'n) = \varepsilon'(1 + \varepsilon')$. After rescaling by $L_{\text{OPT}_{3D}}$, the total volume is increased to $vol(M_{w\delta} \cup M_{h\delta}) \le L_{\text{OPT}_{3D}}\varepsilon'(1 + \varepsilon') \le (\text{OPT}_{3D} + \varepsilon')\varepsilon'(1 + \varepsilon') \le 2\varepsilon'(\text{OPT}_{3D} + \varepsilon')$. We pack the medium boxes into a strip S_0 with the following Lemma. Furthermore, we assign the wide and long rectangles non-fractional to the groups $W_1, \ldots, W_{4/\delta^2}$ and $L_1, \ldots, L_{4/\delta^2}$, so that all slices of one box belong to one group. This is done similarly as assigning the wide and long rectangles in the 2-dimensional Bin Packing algorithm (cf. Figure 1(b)). Therefore, we have to pack some wide and long boxes that cannot be assigned into S_0. The proof of the following Lemma is given in the full version.

Lemma 2. *We need a strip S_0 of the total length $6\varepsilon'\text{OPT}_{3D} + \varepsilon' + 6\ell_{\max}$ to*

1. *pack the medium boxes and*
2. *assign the wide and long rectangles into the groups $W_1, \ldots, W_{4/\delta^2}$ and $L_1, \ldots, L_{4/\delta^2}$ so that all slices of one box belong to one group.*

Packing the Containers and Big-Slices In the end of the 2-dimensional Bin Packing algorithm, an Integer Linear Programs $(ILP(1))$ is solved to pack the 2-dimensional containers and the slices of the big boxes into the bins. In our case it is an advantage to use the relaxation of the Integer Linear Program, since the basic solution consists of at most $m \leq 2/\delta^4 + 4/\delta^6 + 4/\delta^6 \leq 9/\delta^6$ configurations. W.l.o.g. we denote these non-zero configurations by $C_1, \ldots, C_m$. We treat each 2-dimensional object in the non-zero configurations as 3-dimensional object of length ε'/n and pack the objects of each configuration C_k on top of each other. Thus, we obtain at most m 3-dimensional strips. The length of the strip S_k, for $k \in \{1, \ldots, m\}$ is the value of the configuration $\overline{x}_k$ multiplied with ε'/n. The total length of these strips is at most $\varepsilon'/n \cdot ((3/2 + 24\delta) \cdot$ $\mathrm{OPT}_{2D} + 53) \leq \varepsilon'/n \cdot ((3/2 + 24\delta) \cdot (1 + \varepsilon') \cdot n/\varepsilon' + 53) \leq (3/2 + 24\delta) \cdot (1 + \varepsilon') + 53\varepsilon'$.

After rescaling the boxes by the length $L_{\mathrm{OPT}_{3D}}$ we obtain the following total packing length:

$$
\begin{aligned}
L &\leq L_{\mathrm{OPT}_{3D}}((3/2 + 24\delta) \cdot (1 + \varepsilon') + 53\varepsilon') \\
&\leq (\mathrm{OPT}_{3D} + \varepsilon')((3/2 + 24\delta) \cdot (1 + \varepsilon') + 53\varepsilon') \\
&\leq (3/2 + 24\delta) \cdot (\mathrm{OPT}_{3D} + \varepsilon' + \varepsilon'\mathrm{OPT}_{3D} + \varepsilon'^2) + 53\varepsilon'\mathrm{OPT}_{3D} + 53\varepsilon'^2 \\
&= (3/2 + 24\delta + 3/2\varepsilon' + 24\delta\varepsilon' + 53\varepsilon')\mathrm{OPT}_{3D} + (3/2 + 24\delta) \cdot (\varepsilon' + \varepsilon'^2) + 53\varepsilon'^2 \\
&\leq (3/2 + 80\varepsilon')\mathrm{OPT}_{3D} + 6\varepsilon',
\end{aligned}
$$

since $\delta \leq \varepsilon' \leq 1/48$. It is left to pack the big boxes into the strip at the places of their placeholders and to pack the wide, long and small boxes into the containers. Remember that we assume that we have guessed all values correctly, so there is a fractional packing of the big boxes into the strips. We show in the full version how to use a result by Lenstra et al. [8] for scheduling jobs on unrelated machines to pack the big boxes into the strips.

3.3 3-Dimensional Containers

We describe in this section how to pack the wide and long boxes into the corresponding containers. We will focus on the wide boxes, since the steps for the long boxes are analogous.

Packing the Wide and Long Boxes into the Containers We remain in the 2-dimensional representation to pack the slices into the 2-dimensional containers. We use the linear program $LP(1)$ to select at most $8/\delta^2$ configurations in the containers so that all wide slices are fractionally covered.

Afterwards, we transform the slices of the containers to 3-dimensional objects by adding lengths of the value $\varepsilon'/n \cdot L_{\mathrm{OPT}_{3D}}$. We keep the configurations of the linear program, that forms slots in the 3-dimensional containers (cf. Figure 3). Each 3-dimensional container is divided into slots and possibly an empty space on the right side for the small boxes. If a 3-dimensional container consists of different configurations, we split the entire strip at this length. This increases the number of strips by less than $8/\delta^2$. By doing the analogous steps for the long slices, the number of strips grows to at most $\overline{m} \leq m + 16/\delta^2 = 9/\delta^6 + 16/\delta^2$ strips. At this moment there is only one configuration in each strip and each container. Furthermore, all wide boxes fit fractionally (cut in the depth and length) into the 3-dimensional slots inside the wide containers. We increase all strips by the length $\ell_{\max}$. The total length of all $\overline{m}$ strips is

$$
\overline{L} \leq L + \overline{m} \cdot \ell_{\max} \leq (3/2 + 80\varepsilon')\mathrm{OPT}_{3D} + 6\varepsilon' + (9/\delta^6 + 16/\delta^2) \cdot \ell_{\max}.
$$

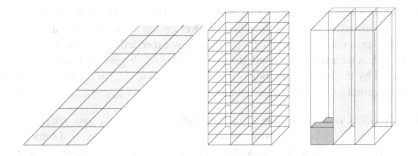

Fig. 3. Configurations of the linear program build slots in the wide container

We pack the wide boxes into the 3-dimensional wide containers, therefore we focus one 3-dimensional wide container C of the width w_C, depth d_C and length ℓ_C. We increased the length by ℓ_{max}, thus the length is $\ell_C + \ell_{max}$. Since all wide boxes fit fractionally into the slots inside the wide containers (when all values are guessed correctly) it holds that the total volume of the wide boxes is at most the total volume of these slots. The same holds with the long boxes and the slots in the long containers. All small boxes fit also fractionally into the wide and long containers in the remaining gaps. Thus, if we find a packing of the boxes that occupies the total volume of the containers, then we know that all boxes are packed. We prove that we either occupy the total volume of one container, or we are running out of boxes. Therefore, we have to extend the side-lengths of each container. We already increased the length of each container by ℓ_{max}. Now we extend also the depth by δ^4 (cf. Figure 4) and the width by δ^4. The side-lengths of C is now $w_C + \delta^4$, $d_C + \delta^4$ and $\ell_C + \ell_{max}$.

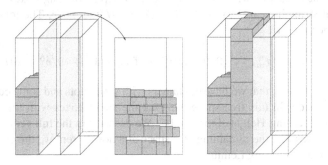

Fig. 4. Next-Fit heuristic

The left side of the container is parted by the slots of the linear program $LP(1)$. We focus one of these slots S of the width w_S, depth $d_S = d_C + \delta^4$ and length $\ell_C + \ell_{max}$. In this slot we pack only boxes b_i of the (rounded) width $w_i = w_S$. We sort the boxes by their lengths and pack them with a Next-Fit heuristic into the slots. When the next box does not fit into the slot we form a new level on top of the first box and continue to pack the boxes until the length of the container is exceeded. Each box has a depth of at most δ^4, thus we exceed the depth d_C in each level (cf. Figure 4). It follows that the area covered by the bases of the boxes is at least $a := w_S \cdot d_C$ in each level. With the following Lemma 3 we cover a total volume of $w_S \cdot d_C \cdot (\ell_C + \ell_{\max} - \ell_{\max}) = w_S \cdot d_C \cdot \ell_C$.

Lemma 3. *Given a target region X of the width w_X, depth d_X and length ℓ_X and k levels $\mathcal{L}_1, \ldots, \mathcal{L}_k$, where the bases of the boxes in each level covers a total area of at least a. Let w.l.o.g. b_i be the largest box in level $\mathcal{L}_i$ and $b_{i'}$ be the smallest box. If $\ell_i \geq \ell_{i+1}$ for all $i \in \{1, \ldots, k-1\}$, then we are able to pack boxes into X with a total volume of at least $a \cdot (\ell_X - \ell_{\max})$.*

Proof. We pack the levels on top of each other until the next level does not fit into the target region. Let $\ell_{k+1} := \ell_X - \sum_{i=1}^{k} \ell_i$ be the length on top of the uppermost target region. It holds $\ell_{i'} \geq \ell_{i+1}$ for all $i \in \{1, \ldots, k\}$. Furthermore, we have $vol(\mathcal{L}_i) \geq a \cdot \ell_{i'}$. Thus, $\sum_{i=1}^{k} vol(\mathcal{L}_i) \geq \sum_{i=1}^{k} a \cdot \ell_{i'} \geq \sum_{i=1}^{k} a \cdot \ell_{i+1} = a \cdot (\sum_{i=2}^{k+1} \ell_i) \geq a \cdot (\ell_X - \ell_{\max})$
□

We do this with all slots in all configurations. Since the total volume of the boxes of one width w is at most the total volume of all slots of the same width w, we are able to pack all boxes. It happens for each slot of one specified width exactly once that the boxes running out and that there is some free space in the slot that we have to use for the small boxes. In this case, we change the order of the slots and exchange this slot with the rightmost slot. If there are several slots in one strip where this happens, then we sort the slots by non-increasing packing lengths (cf. Figure 4). Each time when this happens, we split the entire strip into two strips. The length of the lower part is extended by the length $\ell_{\max}$, so that the cut boxes still fit in the lower strip. There are $2\delta^2$ different widths of wide and $2/\delta^2$ different depths of long boxes. Therefore, the number of strips increases to $\overline{\overline{m}} \leq \overline{m} + 4\delta^2 \leq 9/\delta^6 + 20\delta^2$. The total length of all strips is increased to

$$\overline{\overline{L}} \leq \overline{L} + 4/\delta^2 \cdot \ell_{\max} \leq (3/2 + 80\varepsilon')\text{OPT}_{3D} + 6\varepsilon' + (9/\delta^6 + 20/\delta^2) \cdot \ell_{\max}.$$

The advantage is, that we have containers with some slots and some cubic free space at the right side. In this free space we pack the small boxes with the 2-dimensional Next-Fit-Decreasing-Height algorithm [4]. We describe in the full version how to pack the small boxes into the remaining free space and how to remove the extensions of the boxes with the following Lemma:

Lemma 4. *We are able to remove the extensions of the containers and to pack the intersecting boxes into a strip $S_{\overline{\overline{m}}+1}$ of the length $48\delta\overline{\overline{L}} \leq 152\varepsilon'\text{OPT}_{3D} + 6\varepsilon' + (9/\delta^6 + 20/\delta^2) \cdot \ell_{\max}$.*

3.4 Summary

To summarize our results, we state the entire algorithm in Algorithm 1.

Algorithm 1. Algorithm for 3-dimensional Strip Packing

1: Set $\varepsilon' := \min\{\varepsilon/236, 1/48\}$, so that $1/\varepsilon'$ is a multiple of 24
2: **Find** $L_{\mathrm{OPT}_{3D}}$, so that $\mathrm{OPT}_{3D} \le L_{\mathrm{OPT}_{3D}} < \mathrm{OPT}_{3D} + \varepsilon'$ holds with binary search **for each guess do**
3:   Scale length of boxes by $1/L_{\mathrm{OPT}_{3D}}$ and round them to next multiple of ε'/n
4:   Split boxes at multiple of ε'/n and obtain instance for 2-dim. Bin Packing
5:                    \\begin 2-dim. Bin Packing algorithm
6:   **Find** OPT_{2D} **for each guess do**
7:     Compute δ and partition the rectangles
8:     **Find** widths and heights and number of $2/\delta^4$ different types of big rectangles;
       widths and approx. total height of $4/\delta^2$ different types of wide rectangles;
       heights and approx. total width of $4/\delta^2$ different types of long rectangles;
       widths and heights and number of $4/\delta^6$ different types of long and wide containers **for each guess do**
9:       Greedy assignment of wide and long rectangles to different types
10:      Solve $LP(1)$ to find fractional packing of wide rectangles in wide containers and long rectangles in long containers
11:      Solve relaxation of $ILP(1)$ to find fractional packing of big rectangles and wide and long containers
12:                   \\end 2-dim. Bin Packing algorithm
13:      Add lengths of value $\varepsilon' L_{\mathrm{OPT}_{3D}}/n$ to 2-dim. bins and build strip for each configuration in the basic solution of relaxation of $ILP(1)$
14:      Extend each strip by $\ell_{\max}$
15:      Pack big boxes into strip with result by Lenstra et al. [8]
16:      Pack wide, long and small boxes into extended 3-dimensional containers and remove the extensions
17:      Pack medium boxes with use of Steinbergs algorithm [13]

We have packed the boxes into the strips $S_0, \ldots, S_{\overline{\overline{m}}+1}$. We simply stack them on top of each other and obtain one strip of the total length:

$$4\varepsilon'\mathrm{OPT}_{3D} + \varepsilon' + 6\ell_{\max} + \overline{\overline{L}} + 48\delta\overline{\overline{L}}$$

$$\le (3/2 + 84\varepsilon')\mathrm{OPT}_{3D} + 7\varepsilon' + (6/\delta^6 + 20/\delta^2 + 6) \cdot \ell_{\max} + 48\delta\overline{\overline{L}}$$

$$\le (3/2 + 236\varepsilon')\mathrm{OPT}_{3D} + 13\varepsilon' + (15/\delta^6 + 40/\delta^2 + 6) \cdot \ell_{\max}$$

$$\le (3/2 + 236\varepsilon')\mathrm{OPT}_{3D} + 13\varepsilon' + (16/\delta^6) \cdot \ell_{\max}$$

$$\le (3/2 + \varepsilon)\mathrm{OPT}_{3D} + \varepsilon + 16/\varepsilon'^{12/\varepsilon'}\ell_{\max}$$

$$\le (3/2 + \varepsilon)\mathrm{OPT}_{3D} + \varepsilon + 4160/\varepsilon^{3120/\varepsilon}\ell_{\max}.$$

By stacking the strips on top of each other it follows Theorem 1 with $f(\varepsilon, \ell_{\max}) = 4160/\varepsilon^{3120/\varepsilon}\ell_{\max}$.

4 Conclusion

We presented an asymptotic $3/2 + \varepsilon$-approximation for the 3-dimensional Strip Packing problem. This is a significant improvement over the previous best known asymptotic approximation ratio of 1.692 by Bansal et al. [2]. It is of interest, if it is possible to improve our new upper bound or to show that the lower bound of $1 + 1/2196$ by Chlebík & Chlebíková [3] can be lifted.

References

1. Bansal, N., Correa, J.R., Kenyon, C., Sviridenko, M.: Bin packing in multiple dimensions: Inapproximability results and approximation schemes. Mathematics of Operations Research 31(1), 31–49 (2006)
2. Bansal, N., Han, X., Iwama, K., Sviridenko, M., Zhang, G.: A harmonic algorithm for the 3d strip packing problem. SIAM Journal on Computing 42(2), 579–592 (2013)
3. Chlebík, M., Chlebíková, J.: Inapproximability results for orthogonal rectangle packing problems with rotations. In: Calamoneri, T., Finocchi, I., Italiano, G.F. (eds.) CIAC 2006. LNCS, vol. 3998, pp. 199–210. Springer, Heidelberg (2006)
4. Coffman Jr., E.G., Garey, M.R., Johnson, D.S., Tarjan, R.E.: Performance bounds for level-oriented two-dimensional packing algorithms. SIAM Journal on Computing 9(4), 808–826 (1980)
5. Jansen, K., Prädel, L.: New approximability results for two-dimensional bin packing. In: Proceedings of the 24th Annual ACM-SIAM Symposium on Discrete Algorithms (SODA 2013), pp. 919–936 (2013)
6. Jansen, K., Solis-Oba, R.: An asymptotic approximation algorithm for 3d-strip packing. In: Proceedings of the Seventeenth Annual ACM-SIAM Symposium on Discrete Algorithm (SODA 2006), pp. 143–152. ACM Press (2006)
7. Kenyon, C., Rémila, E.: A near-optimal solution to a two-dimensional cutting stock problem. Mathematics of Operations Research 25(4), 645–656 (2000)
8. Lenstra, J.K., Shmoys, D.B., Tardos, É.: Approximation algorithms for scheduling unrelated parallel machines. Mathematical Programming 46, 259–271 (1990)
9. Li, K., Cheng, K.-H.: On three-dimensional packing. SIAM Journal on Computing 19(5), 847–867 (1990)
10. Li, K., Cheng, K.-H.: Heuristic algorithms for on-line packing in three dimensions. Journal of Algorithms 13(4), 589–605 (1992)
11. Miyazawa, F.K., Wakabayashi, Y.: An algorithm for the three-dimensional packing problem with asymptotic performance analysis. Algorithmica 18(1), 122–144 (1997)
12. Prädel, L.: Approximation Algorithms for Geometric Packing Problems. PhD thesis, University of Kiel (2012), http://eldiss.uni-kiel.de/macau/receive/dissertation_diss_00010802
13. Steinberg, A.: A strip-packing algorithm with absolute performance bound 2. SIAM Journal on Computing 26(2), 401–409 (1997)

A Stronger Square Conjecture on Binary Words

Nataša Jonoska[1], Florin Manea[2], and Shinnosuke Seki[3,4]

[1] Department of Mathematics and Statistics, University of South Florida, 4202 East
Fowler Avenue, Tampa, FL 33620, USA
jonoska@math.usf.edu
[2] Institut für Informatik, Christian-Albrechts-Universität zu Kiel, Kiel, Germany
flm@informatik.uni-kiel.de
[3] Helsinki Institute for Information Technology (HIIT)
[4] Department of Information and Computer Science, Aalto University,
P. O. Box 15400, FI-00076, Aalto, Finland
shinnosuke.seki@aalto.fi

Abstract. We propose a stronger conjecture regarding the number of
distinct squares in a binary word. Fraenkel and Simpson conjectured in
1998 that the number of distinct squares in a word is upper bounded by
the length of the word. Here, we conjecture that in the case of a word
of length n over the alphabet $\{a, b\}$, the number of distinct squares is
upper bounded by $\frac{2k-1}{2k+2}n$, where k is the least of the number of a's and
the number of b's. We support the conjecture by showing its validity for
several classes of binary words. We also prove that the bound is tight.

1   Conjectures

Let Σ be an alphabet and Σ^* be the set of all words over Σ. Let $w \in \Sigma^*$. By $|w|$,
we denote its length. For a letter $a \in \Sigma$, we denote the number of occurrences
of a's in w by $|w|_a$. In this paper, n exclusively denotes the length of a word in
which squares are to be counted.

Let $\mathrm{Sq}(w) = \{uu \mid w = xuuy \text{ for some } x, y \in \Sigma^* \text{ with } w \neq xy\}$ be the set of
all squares occurring in w. Its size, denoted by $\#\mathrm{Sq}(w)$, has been conjectured to
be bounded from above by the length of w [1].

That is to say, $\#\mathrm{Sq}(w) \leq n$ for any word w of length n; a slightly stronger
conjecture is $\#\mathrm{Sq}(w) \leq n - |\Sigma|$, given in [2]. Notable upper bounds shown so
far are $\#\mathrm{Sq}(w) \leq 2n$ [1], further improved by Ilie to $\#\mathrm{Sq}(w) \leq 2n - \log n$ [3],
this being the best bound known so far.

An infinite word, over the binary alphabet $\Sigma_2 = \{a, b\}$, whose finite factors
have a relatively large number of distinct squares compared to their length was
given by Fraenkel and Simpson [1]:

$$w_{\mathrm{fs}} = a^1 b a^2 b a^3 b a^2 b a^3 b a^4 b a^3 b a^4 b a^5 b a^4 b a^5 b a^6 b \cdots . \tag{1}$$

None of its factors of length n with k letters b contain more than $\frac{2k-1}{2k+2}n$ distinct squares (Corollary 2). In fact, we propose (Conjecture 1) that this upper
bound holds not only for the factors of w_{fs} but for all binary words. A computer

V. Geffert et al. (Eds.): SOFSEM 2014, LNCS 8327, pp. 339–350, 2014.

program verified the conjecture for all binary words of length less than 30, as well as for randomly generated binary words of length up to 500 without any counterexample found. Due to this, we propose, as our first contribution, the following stronger conjecture regarding the number of squares.

Conjecture 1. Let $k \geq 2$. For any binary word $w \in \Sigma_2^+$ of length n with k b's where $k \leq \lfloor \frac{n}{2} \rfloor$,

$$\#\mathrm{Sq}(w) \leq \frac{2k-1}{2k+2}n.$$

The bound is defined here as a function of the number of b's. However, Conjecture 1 gives an upper bound on number of squares more generally by redefining k as $\min\{|w|_a, |w|_b\}$, as the number of distinct squares in a binary word is invariant under the isomorphism swapping the letters a and b. Another conjecture proposed in [2] states that for a binary word w we have $\#\mathrm{Sq}(w) \leq n - 2$. Our conjecture is, however, stronger, because $\frac{2k-1}{2k+2}n \leq n - 2$ whenever $2 \geq k \leq \lfloor \frac{n}{2} \rfloor$.

Conjecture 1 doesn't consider words with at most one b because they are square sparse. It is clear that $\mathrm{Sq}(a^n) = \{(a)^2, (aa)^2, \ldots, (a^{\lfloor n/2 \rfloor})^2\}$, and hence, $\#\mathrm{Sq}(a^n) = \lfloor n/2 \rfloor$. The sole b in a word cannot be a part of any square, so its presence cannot increase the number of squares. Thus, the upper bound $\lfloor n/2 \rfloor$ holds canonically for any binary word with at most one b.

Note that our conjecture not only strengthens the conjecture that $\#\mathrm{Sq}(w) \leq n$, but its dependency on the number of b's suggests that a possible proof might be obtained by induction on this number. We show here that it holds when at most nine b's are present in the word.

Parenthesizing the sequence of positive integers representing the powers of a's in the word w_{fs}, in a convenient manner gives the sequence $(1, 2, 3)$, $(2, 3, 4)$, $(3, 4, 5), \ldots$. This reveals the structure of w_{fs} as catenation of simpler words $a^i b a^{i+1} b a^{i+2} b$, $i = 1, 2, \ldots$. As another contribution, we propose a structurally simpler infinite word, whose coefficients just increment:

$$w_{\mathrm{jms}} = a^1 b a^2 b a^3 b a^4 b a^5 b a^6 b \cdots, \tag{2}$$

and prove that it is quite rich with respect to the number of squares its factors contain. Indeed, we show that its factors achieve the upper bound in Conjecture 1 asymptotically.

The word w_{jms} points out that a word does not necessarily need a complicated structure in order to have many squares. Thus, we further prove that for any word w of length n with k letters b, whose coefficient sequence is sorted (incrementing or decrementing), Conjecture 1 holds (see Theorem 2). This result follows by induction on the number of b's on the word.

As an important technical tool, our analysis is not based on combinatorial properties that the word itself has, but rather on the combinatorial properties of the sequence of powers of the letters a (called here "coefficient sequence"). This allows us to define more general classes of words for which the conjecture holds (e.g., Theorem 3).

2 Preliminaries

Let Σ be an alphabet; for this section this alphabet can even be infinite (for instance, the set of positive integers). For words $u, v \in \Sigma^*$, v is a *prefix* (*suffix*) of u if $u = vy$ (resp. $u = xv$) for some word $y \in \Sigma^*$ (resp. $x \in \Sigma^*$). If $u \neq v$, v is called a *proper prefix* (resp. *proper suffix*). The prefix and proper prefix relations are denoted by $v \leq_p u$ and $v <_p u$, respectively. The suffix and proper suffix relations, $\geq_s$ and $>_s$, are defined analogously. If $u = xvy$, then v is a *factor* of u. A factor that is not a prefix or suffix is said to be *proper*.

Three square lemmas concern the occurrence of two squares at the same location in the word, with another square there, or "nearby" (see, e.g., [4,5,3,6]). We give an analogous lemma, not on squares, but on words of the form uau as Lemma 2, which plays an important part in our inductive analysis. Its proof is a modification of the proof of Theorem 1 in [5], but based on the variant of synchronization lemma below.

Lemma 1. *Let $x, y \in \Sigma^*$ and $a \in \Sigma$ be such that xay is primitive of length at least 2. If $(xay)^2 = z_1 y x z_2$ for some $z_1, z_2 \in \Sigma^*$, then $z_1 = xa$ and $z_2 = ay$.*

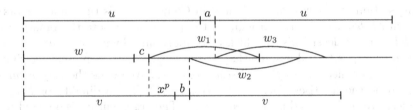

Fig. 1. Three words uau, vbv, wcw starting at the same position

Lemma 2. *For words $u, v, w \in \Sigma^*$ and letters $a, b, c \in \Sigma$, if $u \leq_p wcw <_p vbv <_p uau$ holds, then the word $cwcwc$ occurs as a factor on vbv.*

Proof. Note that the prefix relations imply $|w| \geq 1$ (hence, $|v| \geq 2$ and $|u| \geq 3$). As illustrated in Fig. 1, we denote the second occurrence of w in wcw by w_1, the prefix w of the second v by w_2, and the prefix w of the second u by w_3.

Let $v = wcx^p$ for some primitive word x and $p \geq 0$. If $p = 0$, then $w = b^i$ and $v = b^i c$ for some $i \geq 1$. Note that $|ua| < |vbv| < |ua| + |v|$ and $v = b^i c <_p u$. These mean that the rightmost letter c of the second v is on the prefix b^i of the second u, and hence, $b = c$. Now we have $w = c^i$ and $vbv = c^{2i+3} = cwcwc$.

The case of p being positive is considered below. Note that w_2 certainly overlaps with w_3, while it does not overlap with w_1 if and only if $u = wcw = vb$ holds. We handle the non-overlapping subcase first. In this subcase, $u = vb$ implies $v = a^j$ and $u = a^j b$ for some $j \geq 2$. With $u = wcw$, they give $c = a$ and

w is a power of a, and hence, $a = b = c$. Thus, all of wcw, vbv, uau are a power of a. Now we examine the other case when w_1 and w_2 overlap. Let $x^p b = y^s$ for some primitive word y and $s \geq 1$. The overlap gives $w = y^q y'$ for some $q \geq s$, where y' is a proper prefix of y such that $y = y' y''$ for some $y'' \in \Sigma^+$. If $|y| = 1$, then $w = a^q$, $v = a^q ca^p$, and $p < s \leq q$. Since w_3 is a proper factor of v, we can conclude that $c = a$. If $|y| \geq 2$, on the other hand, we can apply Lemma 1 to the overlap between w_2 and w_3, which is of length at least $|y| - 1$, and obtain that the overlap is $y^r y'$ for some $s \leq r \leq q$. The remaining suffix of w_3, which is $y'' y^{q-r-1} y'$, and $(w^2)^{-1} v = cy^s b^{-1}$ begin at the same position. Then we have $y'' y' = cyb^{-1}$, and this means $b = c$ because $y'' y'$ is a conjugate of y. Now synchronization gives us $y'' = c$ and $y' = yc^{-1}$. Then $w = y^{q+1} c^{-1}$, and hence, $wcw = y^{2(q+1)} c^{-1}$. We also have $v = wbx^p = y^{q+1+s} c^{-1}$, and this gives $vbv = y^{2(s-1)} yc^{-1} cy^{2(q+1)} c^{-1} cyc^{-1} = y^{2(s-1)} yc^{-1} \underline{cwcwc}\, yc^{-1}$. $\square$

The results of this section will not be applied for binary words, in which we count squares, but rather for their coefficient sequences, which are words over integer alphabets. These two lemmas enable us to develop a series of technical tools, which are important to our analysis, e.g., Lemmas 6 and 7.

3   Counting Squares

In this section we show that the bound in Conjecture 1 is tight and factors of w_{jms} with k b's achieve it. Throughout the paper, we denote the binary word with k b's (and at least k a's) in which we count squares by $w_k = a^{i_0} ba^{i_1} b \cdots ba^{i_k}$, where $i_0, \ldots, i_k \geq 0$, and assume that it is of length n. We represent w_k simply as $\langle i_0, \ldots, i_k \rangle$ called the *coefficient sequence* of w_k. We define the *coefficient set* of w_k to be the multiset $I(w_k) = \{i_0, i_1, \ldots, i_k\}$. The cardinality of $I(w_k)$ is considered to be $k+1$ and is denoted $|I(w_k)|$. The argument w_k is omitted from $I(w_k)$ when it is understood in the context. For $j \leq k+1$, by $I[j]$ we denote the *j-th smallest element* of I. Since its maximum element $I[k+1]$ is often referred to, it is more convenient to denote it by $I[\max]$. More generally, by $I[\max - (j-1)]$ we mean the j-th largest element of I.

Squares in w_k that are free from b can be counted simply by checking the largest coefficient in $I(w_k)$ as:

$$\#(\mathrm{Sq}(w_k) \cap a^+) = \left\lfloor \frac{I(w_k)[\max]}{2} \right\rfloor. \tag{3}$$

In counting squares including b's, we first classify them with respect to the equivalence relation "cyclic shift of a's", and then do counting per class. For instance, $x = a^3 baba^3 bab$ and $ababa^3 baba^2$ are members of the same equivalence class because cyclically shifting the first two a's transforms the former into the latter. The class of x contains other two words $a^2 baba^3 baba$ and $baba^3 baba^3$. In contrast, $aba^3 baba^3 b$ does not belong to the same class as x because one has to shift a b in order to obtain this word from x.

In general, a binary square uu with $2m$ b's (m b's per u) has a coefficient sequence $\langle i_0, i_1, \ldots, i_{2m} \rangle$ ($i_0, \ldots, i_{2m} \geq 0$) such that $i_0 + i_{2m} = i_m$ and $i_j = i_{m+j}$

for $1 \leq j \leq m-1$. Let $c = i_m$. The first property lets $i_{2m} = c-i_0$ and by replacing the sequences $i_1, \ldots, i_{m-1}$ and $i_{m+1}, \ldots, i_{2m-1}$ by μ, the square can be written as $\langle i_0, \mu, c, \mu, c-i_0 \rangle$. The squares that result from applying cyclic shift of a's to uu are those written as $\langle c, \mu, c, \mu, 0 \rangle$, $\langle c-1, \mu, c, \mu, 1 \rangle$, $\langle c-2, \mu, c, \mu, 2 \rangle$, $\ldots$, $\langle 0, \mu, c, \mu, c \rangle$, and they compose one equivalence class. We denote the equivalence class with $\langle \mu, c, \mu \rangle$. Its cardinality is $c+1$.

By $\#\mathrm{Sq}_{\langle \mu, c, \mu \rangle}(w)$, we denote the number of squares in the class $\langle \mu, c, \mu \rangle$ that occur in a binary word w. Clearly, $\#\mathrm{Sq}_{\langle \mu, c, \mu \rangle}(w) \leq c+1$. When the equality holds, we say that the class $\langle \mu, c, \mu \rangle$ is *saturated in* w.

Example 1. The class $\langle 1, 3, 1 \rangle$ consists of the squares a^3baba^3bab, a^2baba^3baba, $ababa^3baba^2$, and $baba^3baba^3$. It is not saturated in $a^2baba^3baba^3$ as the first square is missing, while it is saturated in $a^3baba^3baba^3$ or in $a^2baba^3baba^3bab$.

Now, we count the squares in the class $\langle \mu, c, \mu \rangle$ of the word w_k. First of all, the coefficient sequence $\langle i_0, \ldots, i_k \rangle$ of w_k must contain the class identifier $\langle \mu, c, \mu \rangle$ as its *proper* factor in order for such a square to occur in w_k. When $\langle \mu, c, \mu \rangle$ occurs exactly once, that is, there are unique coefficients $\ell, r \geq 0$ such that $\langle \ell, \mu, c, \mu, r \rangle$ is a factor of the coefficient sequence, the count is

$$\#\mathrm{Sq}_{\langle \mu, c, \mu \rangle}(w_k) = \begin{cases} \min\{\ell, c\} + \min\{c, r\} - c + 1 & \text{if } \ell + r \geq c \\ 0 & \text{otherwise} \end{cases}$$

$$\leq \min\{\ell, c, r\} + 1 \qquad (4)$$

where 4 follows from $\min\{i, k\} + \min\{k, j\} - k \leq \min\{i, j, k\}$ for $i, j, k \geq 0$.

It must be noted that (4) does not depend on μ.

We verify Conjecture 1 for a word $w_2 = a^{i_0}ba^{i_1}ba^{i_2}$. The sole class whose square can occur in w_2 is $\langle i_1 \rangle$. Using (4), squares in this class are counted in w_k as $\#\mathrm{Sq}_{\langle i_1 \rangle} \leq \min\{i_0, i_1, i_2\} + 1 = I(w_2)[\max - 2] + 1$. Summing this and (3) gives

$$\#\mathrm{Sq}(w_2) \leq \frac{I(w_2)[\max]}{2} + I(w_2)[\max - 2] + 1$$

$$\leq \frac{I(w_2)[\max]}{2} + \frac{1}{2}(n - 2 - I(w_2)[\max]) + 1 = \frac{1}{2}n.$$

Proposition 1. $\#\mathrm{Sq}(w_2) \leq \frac{1}{2}n$ *for any binary word w_2 of length n with 2 b's.*

Double-counting is a significant issue. When the coefficient sequence of a word w_k includes the factor $\langle \mu, c, \mu \rangle$ exactly twice as $\langle u \rangle = \langle \ell_1, \mu, c, \mu, r_1 \rangle$ and $\langle v \rangle = \langle \ell_2, \mu, c, \mu, r_2 \rangle$, we have

$$\#\mathrm{Sq}_{\langle \mu, c, \mu \rangle}(w_k) = \#\mathrm{Sq}_{\langle \mu, c, \mu \rangle}(u) + \#\mathrm{Sq}_{\langle \mu, c, \mu \rangle}(v)$$

$$- \max\{\min\{\ell_1, \ell_2, c\} + \min\{c, r_1, r_2\} - c + 1, 0\}. \qquad (5)$$

The subtracted term accounts for double-counting. It is 0 (i.e., u does not share any square in the class $\langle \mu, c, \mu \rangle$ with v) if and only if $\min\{\ell_1, \ell_2\} + \min\{r_1, r_2\} < c$.

Before proceeding, we note that Lemmas 1 and 2 deal with words of the form $\mu c \mu$, so, by extension, with generative classes $\langle \mu, c, \mu \rangle$. The two lemmas offer a better understanding of the combinatorial properties of such generative classes, and provide the fundamentals needed to use them in our proofs.

The tightness of the bound $\frac{2k-1}{2k+2}n$ for $k \geq 2$ follows by considering the factors of w_{jms}, defined in (2) . We parameterize by m the largest factor of w_{jms} with k b's as $w_{\mathrm{jms},k}(m) = a^m b a^{m+1} b \cdots b a^{m+k}$. As the coefficients of such a factor are pairwise distinct, squares in any class $\langle \mu, c, \mu \rangle$ with μ being nonempty do not occur in the factor. In fact, the classes whose squares are capable of occurring are $\langle m+1 \rangle, \langle m+2 \rangle, \ldots, \langle m+k-1 \rangle$ (not being a proper factor, neither $\langle m \rangle$ nor $\langle m+k \rangle$ can find its square in $w_{\mathrm{jms},k}$). Moreover, $w_{\mathrm{jms},k}(m)$ contains all but one squares of each class. Due to (3) and (4), we have

$$\#\mathrm{Sq}(w_{\mathrm{jms},k}(m)) = \left\lfloor \frac{m+k}{2} \right\rfloor + \sum_{i=1}^{k-1}(m+i) = \left\lfloor \frac{m+k}{2} \right\rfloor + \frac{(k-1)(2m+k)}{2}.$$

Removing the floor function gives two formulae sandwiching $\#\mathrm{Sq}(w_{\mathrm{jms},k}(m))$ and dividing them by $|w_{\mathrm{jms},k}(m)| = \frac{m(2k+2)+k(k+3)}{2}$ yields

$$\frac{m(2k-1)+k^2-2}{m(2k+2)+k(k+3)} \leq \frac{\#\mathrm{Sq}(w_{\mathrm{jms},k}(m))}{|w_{\mathrm{jms},k}(m)|} \leq \frac{m(2k-1)+k^2}{m(2k+2)+k(k+3)}.$$

The sandwiching functions are monotonically-increasing in m for $k \geq 2$ and their limit as m approaches infinity is $(2k-1)/(2k+2)$.

4   Towards the Inductive Proof

The main aim of this section is to propose an inductive approach to Conjecture 1. The next inequality is of principal significance for this purpose.

Lemma 3. *The inequality*

$$\#\mathrm{Sq}(w_k) \leq \left\lfloor \frac{I(w_k)[\max]}{2} \right\rfloor + \sum_{j=1}^{|I(w_k)|-2}(I(w_k)[j]+1) \tag{6}$$

implies $\#\mathrm{Sq}(w_k) \leq \frac{2k-1}{2k+2}n$.

Proof. The inequality (6) is expanded as:

$$\#\mathrm{Sq}(w_k) \leq \left\lfloor \frac{I[\max]}{2} \right\rfloor + \left\lfloor \frac{k-1}{k}(n-I[\max]) \right\rfloor \leq \left\lfloor \frac{k-1}{k}n - \frac{k-2}{2k}\left\lceil \frac{n-k}{k+1} \right\rceil \right\rfloor$$

$$= \left\lfloor \frac{2k-1}{2k+2}n + \frac{k-2}{2k+2} \right\rfloor = \left\lceil \frac{2k-1}{2k+2}n + \frac{k-1}{2k+2} \right\rceil - 1,$$

where the first inequality follows from the fact that each term in the sum in (6) is at most $\lfloor (n-I[\max])/k \rfloor$. The second inequality is due to $I[\max] \geq \lceil (n-k)/(k+1) \rceil$, and at the end, we employ a standard conversion of floors to ceilings. Since $\#\mathrm{Sq}(w_k)$ is an integer, this implies $\#\mathrm{Sq}(w_k) \leq \frac{2k-1}{2k+2}n$.   □

Lemma 4. *Let* $I = \{i_0, i_1, \ldots, i_{k-1}\}$ *and* $J = I \cup \{i_k\}$ *be multisets. Then* $\sum_{j=1}^{|I|-2} I[j] + \min\{J[\max -2], i_k\} = \sum_{\ell=1}^{|J|-2} J[\ell]$.

With the case of $k = 2$ as the basis (Proposition 1), induction proceeds as: choose an operation that yields w_k from another word w' with less b's. Use the induction hypothesis that w' satisfies (6) to prove that the operation does not create too many squares, and conclude that (6) holds for w_k.

One such operation is catenation. Catenating ba^{i_k} to the end of $w_{k-1} = a^{i_0}b \cdots ba^{i_{k-1}}$ yields w_k. By saying that the catenation *creates* a square, we mean that the square does not occur in w_{k-1} but occurs in w_k. Let $\langle \mu, c, \mu \rangle$ be a class of squares. In order for the catenation to create a square in this class, $\langle \mu, c, \mu \rangle$ must be a *proper suffix* of the coefficient sequence $\langle i_0, \ldots, i_{k-1} \rangle$ of w_{k-1}. When $\langle i_0, \ldots, i_{k-1} \rangle$ contains a proper suffix $\langle \mu, c, \mu \rangle$ which creates new squares we say that the class $\langle \mu, c, \mu \rangle$ is *generative* for the catenation. Observe that any saturated class in w_{k-1} cannot be generative.

4.1 Induction Based on Catenation with Single Generative Class

We further show how induction would work by verification of Conjecture 1 for words whose all coefficients, but the leftmost and rightmost, are pairwise-distinct. Proposition 1 allows us to only consider the induction step ($k \geq 3$). Let $w_{k-1} = a^{i_0}b \cdots ba^{i_{k-1}}$, and we assume (6) holds for it as an induction hypothesis. The catenation of ba^{i_k} to w_{k-1} yields $w_k = w_{k-1}ba^{i_k}$. The pairwise-distinctiveness of the coefficient sequence makes $\langle i_{k-1} \rangle$ the sole generative class for the catenation, and $\min\{i_{k-2}, i_{k-1}, i_k\} + 1$ squares are thus created due to (4). With $\min\{i_{k-2}, i_{k-1}, i_k\} \leq I(w_k)[\max -2]$, Lemma 4 verifies (6) for w_k as follows:

$$
\begin{aligned}
\#\mathrm{Sq}(w_k) &\leq \left\lfloor \frac{\max\{I(w_{k-1})[\max], i_k\}}{2} \right\rfloor \\
&\quad + \sum_{j=1}^{|I(w_{k-1})|-2} (I(w_{k-1})[j] + 1) + (\min\{I(w_k)[\max -2], i_k\} + 1) \\
&\leq \frac{I(w_k)[\max]}{2} + \sum_{j=1}^{|I(w_k)|-2} (I(w_k)[j] + 1).
\end{aligned}
$$

Theorem 1. *For* $k \geq 2$, *let* $w_k = a^{i_0}ba^{i_1}b \cdots ba^{i_{k-1}}ba^{i_k}$ *be a binary word of length* n *with* k *b's. If* $i_1, \ldots, i_{k-1}$ *are pairwise-distinct, then* $\#\mathrm{Sq}(w_k) \leq \frac{2k-1}{2k+2}n$.

Corollary 1. *For any factor* w_k *of* w_{jms} *with* k *b's,* $\#\mathrm{Sq}(w_k) \leq \frac{2k-1}{2k+2}n$, *where* n *is the length of* w_k.

Before proceeding to the analysis of multiple generative classes, let us introduce and examine a class of words for which the bound can be verified inductively based on catenation with single generative class. A word $w_k = a^{i_0}ba^{i_1}b \cdots ba^{i_{k-1}}ba^{i_k}$ is an *ascending (descending) slope* if $i_0 \leq i_1 \leq \cdots \leq i_k$

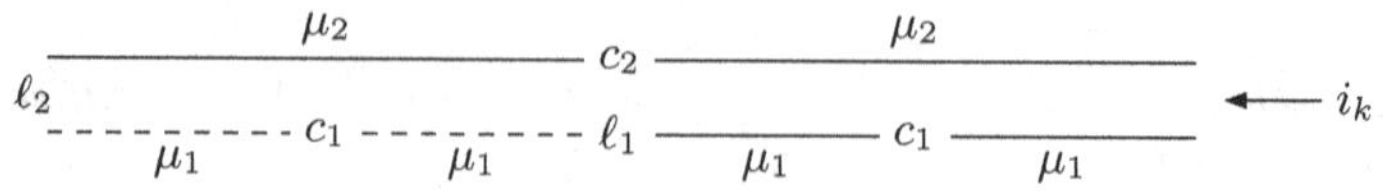

Fig. 2. Two classes $\langle \mu_2, c_2, \mu_2 \rangle, \langle \mu_1, c_1, \mu_1 \rangle$, which occur as suffixes, and hence, can be generative for the catenation of ba^{i_k} at the end

(resp. $i_0 \geq i_1 \geq \cdots \geq i_k$) holds. This notion is generalized as: w_k is a *padded slope* if its factor $a^{i_1}b \cdots ba^{i_{k-1}}$ is a slope.

Theorem 2. *For $k \geq 2$, if a binary word w_k of length n with k b's is a padded slope, then $\#\mathrm{Sq}(w_k) \leq \frac{2k-1}{2k+2}n$.*

Proof. Let $w_k = a^{i_0}ba^{i_1}b \cdots ba^{i_{k-1}}ba^{i_k}$. As induction hypothesis, assume that $w_{k-1} = a^{i_0}b \cdots ba^{i_{k-1}}$ fulfills inequality (6). Invariance of the number of squares under reversal allows us to proceed with the assumption that w_k is ascending.

Let $\langle \mu, c, \mu \rangle$ be a generative class for the catenation of ba^{i_k} to w_{k-1}. Let i_m be such that $m \geq 1$ and $i_{m-1} < i_m = \cdots = i_{k-1}$. Due to the ascending property, $\langle i_m, \ldots, i_{k-1} \rangle \geq_s \langle \mu, c, \mu \rangle$ must hold. Let $d = i_m = \cdots = i_{k-1}$. The sole class that can be generative for the catenation is $\langle d^{\lfloor (k-m)/2 \rfloor}, d, d^{\lfloor (k-m)/2 \rfloor} \rangle$ because for any $j < \lfloor (k-m)/2 \rfloor$, the class $\langle d^j, d, d^j \rangle$ has been already saturated in w_{k-1}. At most $\min\{i_\ell, d, i_k\} + 1$ squares in this class can be created due to (4), where $\ell = k - 2\lfloor (k-m)/2 \rfloor - 2$. This is clearly at most $\min\{I(w_k)[\max -2], i_k\} + 1$. Lemma 4 now concludes that the inequality (6) holds for w_k. □

Remark. We note that the proofs of Theorems 1 and 2 can be adjusted such that whenever the catenation of ba^{i_k} to w_{k-1} yields a single generative class, the Conjecture 1 holds.

4.2 Induction Based on Catenation with Multiple Generative Classes

However, catenation may involve more than one generative class. For instance, in extending the prefix $a^1ba^2ba^3ba^2$ of w_{fs} (see (1)) by ba^3, squares in the two classes $\langle 2 \rangle$ and $\langle 2, 3, 2 \rangle$ are created.

We begin with examining catenation with two generative classes. Consider two generative classes $\langle \mu_1, c_1, \mu_1 \rangle, \langle \mu_2, c_2, \mu_2 \rangle$ for the catenation of ba^{i_k} to $w_{k-1} = a^{i_0}ba^{i_1}b \cdots ba^{i_{k-1}}$, which yields $w_k = w_{k-1}ba^{i_k}$. From (4), we get that the catenation creates at most $\min\{I(w_k)[\max -3], i_k\} + \min\{I(w_k)[\max -2], i_k\} + 2$ squares in these classes. When $\langle \mu_2 \rangle \geq_s \langle \mu_1, c_1, \mu_1 \rangle$, the number turns out to be bounded by $\min\{I(w_k)[\max -2], i_k\} + 1$, as shown in the next lemma.

Lemma 5. *Let $\langle \mu_1, c_1, \mu_1 \rangle, \langle \mu_2, c_2, \mu_2 \rangle$ be generative classes for the catenation of ba^{i_k} to w_{k-1} to yield $w_k = w_{k-1}ba^{i_k}$. If $\langle \mu_2 \rangle \geq_s \langle \mu_1, c_1, \mu_1 \rangle$, then the number of squares in the classes created by the catenation is at most $\min\{c_1, i_k\} + 1 \leq \min\{I(w_k)[\max -2], i_k\} + 1$. Moreover, if $\langle \mu_2 \rangle >_s \langle \mu_1, c_1, \mu_1 \rangle$, then $c_2 < c_1$.*

Proof. Let $w_{k-1} = a^{i_0} b \cdots b a^{i_{k-1}}$, and we have $\langle i_0, \ldots, i_{k-1} \rangle \geq_s \langle \ell_2, \mu_2, c_2, \mu_2 \rangle$ and $\langle c_2, \mu_2 \rangle \geq_s \langle \ell_1, \mu_1, c_1, \mu_1 \rangle$ for some ℓ_1, ℓ_2. The catenation creates at most[1]

$$\min\{\ell_2, c_2\} + \min\{c_2, i_k\} - c_2 + 1 \tag{7}$$

squares in the class $\langle \mu_2, c_2, \mu_2 \rangle$ due to (4).

We first consider the case when $\langle \mu_1, c_1, \mu_1 \rangle$ is a *proper* suffix of $\langle \mu_2 \rangle$. In counting the number of distinct squares to be created in the class $\langle \mu_1, c_1, \mu_1 \rangle$, we should take into account the factor $\langle \ell_1, \mu_1, c_1, \mu_1, c_2 \rangle$ of w_{k-1}. In order for the class to be generative, we have $c_2 < \min\{c_1, i_k\}$. Then at most $\min\{c_1, i_k\} - c_2$ squares in the class are created, and the subtraction term "$-c_2$" cancels (7). As a result, the catenation creates at most $\min\{c_1, i_k\} + 1$ squares in these two classes. Moreover, this is upper bounded by $\min\{I(w_k)[\max -2], i_k\} + 1$ since $I(w_k)$ contains two c_1's.

Next we consider the case of $\langle c_2, \mu_2 \rangle = \langle \ell_1, \mu_1, c_1, \mu_1 \rangle$ (see Fig. 2), that is, $c_2 = \ell_1$ and $\langle \mu_2 \rangle = \langle \mu_1, c_1, \mu_1 \rangle$. It creates at most the following number of distinct squares in the class $\langle \mu_1, c_1, \mu_1 \rangle$:

$$\min\{c_2, c_1\} + \min\{c_1, i_k\} - c_1 + 1$$
$$- \max\{\min\{\ell_2, c_2, c_1\} + \min\{c_1, c_2, i_k\} - c_1 + 1, 0\}. \tag{8}$$

The subtraction term, due to (5), takes into account that $\langle \ell_2, \mu_1, c_1, \mu_1, c_2 \rangle$ already appears in $\langle i_0, \ldots, i_{k-1} \rangle$. The number of distinct squares in these classes created by the catenation is given as the sum (7) + (8). The last subtraction term in (8) is 0 if and only if $\min\{\ell_2, c_2\} + \min\{c_2, i_k\} < c_1$. Then,

$$(7) + (8) = (\min\{\ell_2, c_2\} + \min\{c_2, i_k\} + 1 - c_1)$$
$$+ \min\{c_1, i_k\} + (\min\{c_2, c_1\} - c_1) + 1 \leq \min\{c_1, i_k\} + 1.$$

If the term is positive, on the other hand, then we obtain

$$(7) + (8) = (\min\{\ell_2, c_2\} + \min\{c_2, c_1\} - \min\{\ell_2, c_2, c_1\} - c_2)$$
$$+ (\min\{c_2, i_k\} + \min\{c_1, i_k\} - \min\{c_2, c_1, i_k\}) + 1 \leq i_k + 1.$$

Now we prove that the sum is at most $c_1 + 1$, and it suffices to do so under the condition $c_1 < i_k$. Then the sum is $(\min\{\ell_2, c_2\} + \min\{c_2, i_k\} - c_2 - \min\{\ell_2, c_2, c_1\}) + c_1 + 1$. If $\min\{\ell_2, c_2, c_1\} = \min\{\ell_2, c_2\}$, then the terms inside the parentheses amount to 0 and hence the sum is at most $c_1 + 1$. This condition must hold because if $\min\{\ell_2, c_2, c_1\} = c_1$, then the class $\langle \mu_1, c_1, \mu_1 \rangle$ would have been already saturated in w_{k-1} so that it could not be generative.   □

Now we develop the previous argument for arbitrary number of generative classes: $\langle \mu_m, c_m, \mu_m \rangle, \ldots, \langle \mu_2, c_2, \mu_2 \rangle, \langle \mu_1, c_1, \mu_1 \rangle$ with $m \geq 3$ such that $\langle \mu_m, c_m, \mu_m \rangle >_s \cdots >_s \langle \mu_1, c_1, \mu_1 \rangle$. Interestingly, *no matter how many generative classes are involved,* catenation creates at most $(\min\{I(w_k)[\max -3], i_k\} + 1) + (\min\{I(w_k)[\max -2], i_k\} + 1)$ squares. The next lemma enables us to divide the classes into two groups so that the classes in one group are responsible for the first term and those in the other are for the second term.

---

[1] Here we say "at most" because w_{k-1} may contain some squares in this class already.

Lemma 6. *Let* $\langle \mu_1, c_1, \mu_1 \rangle, \langle \mu_2, c_2, \mu_2 \rangle, \langle \mu_3, c_3, \mu_3 \rangle$ *be three generative classes of the catenation of* ba^{i_k} *to* w_{k-1} *to yield* $w_k = w_{k-1}ba^{i_k}$ *such that* $\langle \mu_3, c_3, \mu_3 \rangle >_s$ $\langle \mu_2, c_2, \mu_2 \rangle >_s \langle \mu_1, c_1, \mu_1 \rangle$. *Then* $\langle \mu_3 \rangle >_s \langle \mu_1, c_1, \mu_1 \rangle$ *and* $c_3 < c_1$ *hold, and the number of squares in the classes* $\langle \mu_3, c_3, \mu_3 \rangle$ *and* $\langle \mu_1, c_1, \mu_1 \rangle$ *created by the catenation is at most* $\min\{c_1, i_k\} + 1 \leq \min\{I(w_k)[\max -2], i_k\} + 1.$

Proof. If $\langle \mu_3 \rangle >_s \langle \mu_1, c_1, \mu_1 \rangle$ did not hold, then, by Lemma 2, $\langle c_1, \mu_1, c_1, \mu_1, c_1 \rangle$ would be a factor of the coefficient sequence of w_{k-1}, that is, the class $\langle \mu_1, c_1, \mu_1 \rangle$ would be saturated in w_{k-1}, a contradiction. Thus, $\langle \mu_3 \rangle >_s \langle \mu_1, c_1, \mu_1 \rangle$ must hold. The other two results derive from this due to Lemma 5.   □

Consider the catenation of ba^{i_k} to w_{k-1} from the right, and let $\langle \mu_{i_\ell}, c_{i_\ell}, \mu_{i_\ell} \rangle$, $\ldots, \langle \mu_{i_1}, c_{i_1}, \mu_{i_1} \rangle$ be its generative classes with $i_m > \cdots > i_1$. We say that they form a *(length-halving) chain* if for any $1 < j \leq \ell$, $\langle \mu_{i_j} \rangle \geq_s \langle \mu_{i_{j-1}}, c_{i_{j-1}}, \mu_{i_{j-1}} \rangle$. Lemmas 5 and 6 imply:

Lemma 7. *For any* $\ell \geq 1$, *if the catenation of* ba^{i_k} *to* w_{k-1} *involves* ℓ *generative classes* $\langle \mu_{i_\ell}, c_{i_\ell}, \mu_{i_\ell} \rangle, \ldots, \langle \mu_{i_1}, c_{i_1}, \mu_{i_1} \rangle$ *that satisfy* $\langle \mu_{i_\ell}, c_{i_\ell}, \mu_{i_\ell} \rangle >_s$ $\cdots >_s \langle \mu_{i_1}, c_{i_1}, \mu_{i_1} \rangle$ *and also form a chain, then the catenation creates at most* $\min\{c_{i_1}, i_k\} + 1$ *squares in these classes.*

Lemma 6 enables us to divide the classes into (at most) two groups so as for the classes in each group to form a chain. The index-parity-based division: $\ldots, \langle \mu_4, c_4, \mu_4 \rangle, \langle \mu_2, c_2, \mu_2 \rangle$ and $\ldots, \langle \mu_3, c_3, \mu_3 \rangle, \langle \mu_1, c_1, \mu_1 \rangle$ is such a division. With Lemma 7, now we complete the proof that the catenation cannot create more than $\min\{I(w_k)[\max -3], i_k\} + \min\{I(w_k)[\max -2], i_k\} + 2$ squares.

4.3   Towards an Inductive Proof for General Words

Any word can be factorized into slopes. Given a word, a *proper factor* $\langle i_\ell, i_{\ell+1}, \ldots, i_{r-1}, i_r \rangle$ ($l > 0$ and $r < k$) of its coefficient sequence is called a (local) *minimum* (*maximum*) if $i_\ell > i_{\ell+1} = i_{\ell+2} = \cdots = i_{r-1} < i_r$ (resp. $i_\ell < i_{\ell+1} = \cdots = i_{r-1} > i_r$). Minima and maxima are collectively called *extrema*. It must be noted that by definition extrema are a *proper* factor so that the leftmost or rightmost coefficient of the given word cannot be a part of them. For $m \geq 0$, we say that a word is an *m*-extrema word if it contains *at most* m extrema. By $\mathcal{E}_m$, we denote the class of all *m*-extrema words.

We identify two minima $\langle i_{\ell_1}, i_{\ell_1+1}, \ldots, i_{r_1-1}, i_{r_1} \rangle, \langle i_{\ell_2}, i_{\ell_2+1}, \ldots, i_{r_2-1}, i_{r_2} \rangle$ if $r_1 - \ell_1 = r_2 - \ell_2$ and $i_{\ell_1+j} = i_{\ell_2+j}$ for any $0 \leq j \leq r_1 - \ell_1$; otherwise, we say they are *distinct*.

Although μ_i is a subsequence of integers, we consider it a word where each integer is a symbol. This notation is applied in the following lemma.

Lemma 8. *If a catenation involves two generative classes* $\langle \mu_1, c_1, \mu_1 \rangle, \langle \mu_2, c_2, \mu_2 \rangle$ *in different chains, and all minima of the resulting word are pairwise-distinct, then one of the following holds:*

1. $c_1 > c_2$, $\mu_2 = c_2^m c_1 c_2^j$, and $\mu_1 = c_2^j$ for some $j \geq 1$ and $m < j$;

2. $c_1 < c_2$, $\mu_2 = c_1^{2j+m+1}c_2c_1^{j+m}$, and $\mu_1 = c_1^j c_2 c_1^{j+m}$ for some $j \geq 1$ and $m \geq 0$;

3. $c_1 \neq c_2$, $\mu_2 = d^j c_1 c_2 d^j$, and $\mu_1 = c_2 d^j$ for some $j \geq 0$ and coefficient d with $d \leq c_1$ and $d < c_2$.

Theorem 3. *If all minima of a word w_k of length n with k b's are pairwise-distinct, then $\#\mathrm{Sq}(w_k) \leq \frac{2k-1}{2k+2}n$.*

Proof. Let $w_k = a^{i_0}b \cdots ba^{i_k}$ and consider the catenation of ba^{i_k} to $w_{k-1} = a^{i_0}b \cdots ba^{i_{k-1}}$ to yield w_k. Assume two generative classes $\langle \mu_1, c_1, \mu_1 \rangle$, $\langle \mu_2, c_2, \mu_2 \rangle$ are involved in it, and moreover, they are in different chains. To them, Lemma 8 is applicable to represent these classes in three ways. Proofs for all these representations take the same strategy: spotting a coefficient i_j such that catenating ba^{i_j} creates so small number of squares that offsets the number of squares to be created by the catenation of ba^{i_k}. Therefore, in the following, we just examine the first representation.

We have that $\langle \ell, \mu_2, c_2, \mu_2 \rangle$ is a suffix of the coefficient sequence $\langle i_0, \ldots, i_{k-1} \rangle$ of w_{k-1} and $\mu_2 = c_2^m c_1 c_2^j$ for some coefficients ℓ, c_1, c_2 with $c_1 > c_2$ and $j \geq 1$, $m \geq 0$ with $j > m$. The right μ_2 is actually the sequence $\langle i_{k-j}, \ldots, i_{k-j+m-1}, i_{k-j+m}, \ldots, i_{k-1} \rangle$. Consider the successive catenations of $ba^{i_{k-j+m}}, \ldots, ba^{i_{k-1}}$ to $w_{k-j+m-1} = a^{i_0}b \cdots ba^{i_{k-j+m-1}}$. If $\ell \neq c_2$, then the first catenation creates $\max\{c_2 - \ell, 0\}$ squares in the class $\langle c_2^m, c_1, c_2^m \rangle$, which is its sole generative class. The catenation of i_k creates at most $\min\{c_2, i_k\} + \min\{\ell, c_2, i_k\} + 2$ squares. As a result, they introduce two additive terms. Moreover, due to $\ell \neq c_2$, each of other catenations involves just one chain. If $\ell = c_2$, then $\langle c_2^m, c_1, c_2^m \rangle$ is not generative any more, but instead the class $\langle c_2^j c_1 c_2^m, c_2, c_2^j c_1 c_2^m \rangle$ can be. If it is not, then no square is created, and this offsets one term brought by the catenation of ba^{i_k}. Otherwise, $\langle i_0, \ldots, i_{k-1} \rangle \geq_s \langle \ell', c_2^j c_1 c_2^m, c_2, c_2^j c_1 c_2^m \rangle$ for some $\ell' \geq 0$. The catenation of i_k creates at most $\min\{\ell', c_2\} + \min\{c_2, i_k\} - c_2 + 1$ squares in the class $\langle \mu_2, c_2, \mu_2 \rangle$ and $c_2 + \min\{c_1, i_k\} - c_1 + 1 - (\min\{\ell', c_2\} + \min\{c_2, i_k\} - c_1 + 1)$, where the subtraction term is to avoid the double-counting (note $\langle \ell', \mu_1, c_1, \mu_1, c_2 \rangle$ is in $\langle i_0, \ldots, i_{k-1} \rangle$). Thus, it creates at most $\min\{c_1, i_k\} + 1$ squares. □

As its corollary, we can verify the bound $\frac{2k-1}{2k+2}n$ for the word (1) by Fraenkel and Simpson, or more precisely, for its factors with k b's, since all of their minima are pairwise-distinct.

Corollary 2. *For any factor $w_{\mathrm{fs},k}$ of w_{fs} with k b's, $\#\mathrm{Sq}(w_{\mathrm{fs},k}) \leq \frac{2k-1}{2k+2}|w_{fs,k}|$.*

Maxima-pairwise-distinct variants of Lemma 8 and Theorem 3 hold. As for the variant of the lemma, all inequalities must be inverted. From them, the next result holds.

Corollary 3. *For any word $w_k \in \mathcal{E}_3$ with k b's, $\#\mathrm{Sq}(w_k) \leq \frac{2k-1}{2k+2}|w_k|$.*

Corollary 4. *For any $k \leq 6$ and word w_k with k b's, $\#\mathrm{Sq}(w_k) \leq \frac{2k-1}{2k+2}|w_k|$.*

Proof. It suffices to observe that, for any $k \geq 3$, words with k b's can contain at most $k - 3$ extrema. Then this immediately follows from Corollary 3. □

The more classes are involved, the more strictly the structure of w_{k-1}, to which we catenate ba^{i_k}, is restricted. In fact, we can easily show that for $k \leq 9$, either the catenation involves just one chain or all minima (or maxima) of the resulting word are pairwise-distinct.

Proposition 2. *For any $k \leq 9$ and word w_k with k b's, $\#\mathrm{Sq}(w_k) \leq \frac{2k-1}{2k+2}|w_k|$.*

5   Conclusions

Our results are partial steps in showing Conjecture 1. However, we identified several ways to approach this conjecture. For instance, one may follow the technique in which we examine a word as a sequence of slopes, and try to identify how the number of squares increases when the words have non pairwise distinct minima (maxima). Nevertheless, it may be the case that a direct inductive proof with respect to the number of b's would validate the conjecture; using the generative classes with respect to catenation we only analyzed the cases when this number is at most 9, but it is our hope that our method can be generalized.

Finally, we discussed only the case of binary words. It seems unlikely that the tools we developed could be used directly to obtain upper bounds on the number of squares in words over larger alphabets.

Acknowledgement. We gratefully acknowledge helpful discussions with Florence Linez, Robert Mercaş, and Mike Müller. Mike Müller kindly implemented a computer program for the experimental verification of Conjecture 1. The research was partially supported by the NSF grants No. DMS-0900671 and CCF-1117254 and the NIH grant R01 GM109459-01 to N. J., by the DFG grant 596676 to F. M., and by the HIIT Pump Priming Project Grant 902184/T30606 and the Academy of Finland, Postdoctoral Researcher Grant 13266670/T30606 to S. S.

References

1. Fraenkel, A.S., Simpson, J.: How many squares can a string contain? Journal of Combinatorial Theory, Series A 82, 112–120 (1998)
2. Deza, A., Franek, F., Jiang, M.: A d-step approach for distinct squares in strings. In: Giancarlo, R., Manzini, G. (eds.) CPM 2011. LNCS, vol. 6661, pp. 77–89. Springer, Heidelberg (2011)
3. Ilie, L.: A note on the number of squares in a word. Theoretical Computer Science 380, 373–376 (2007)
4. Fan, K., Puglisi, S.J., Smyth, W.F., Turpin, A.: A new periodicity lemma. SIAM Journal of Discrete Mathematics 20(3), 656–668 (2005)
5. Ilie, L.: A simple proof that a word of length n has at most $2n$ distinct squares. Journal of Combinatorial Theory, Series A 112(1), 163–164 (2005)
6. Kopylova, E., Smyth, W.F.: The three squares lemma revisited. Journal of Discrete Algorithms 11, 3–14 (2012)

DSL Based Platform for Business Process Management

Audris Kalnins, Lelde Lace, Elina Kalnina, and Agris Sostaks

Institute of Mathematics and Computer Science, University of Latvia
{Audris.Kalnnins,Lelde.Lace,Elina.Kalnina,
Agris.Sostaks}@lumii.lv

Abstract. Currently nearly all commercial and open source BPMS are based on BPMN as a process notation. In contrast, the paper proposes to build a BPMS based on a domain specific language (DSL) as a process notation – DSBPMS. In such a DSBPMS a specific business process support could be created by business analysts. A platform for creating such DSBPMS with feasible efforts is described. This platform contains a Configurator for easy creation of graphical editors for the chosen DSL and a simple mapping language for transforming processes in this DSL to a language directly executable by the execution engine of this platform. The engine includes also all typical execution support functions so no other tools are required.

Keywords: Business process management systems, Domain specific languages.

1 Introduction

Currently nearly all commercial and open source Business Process Management Systems (BPMS) are based on BPMN [1] as a process notation. The main rationale is the standardization and potential model exchange, nevertheless the process notation is only a part of a complete system definition, data model and form definitions are important as well. These BPMS aspects are not covered by BPMN, each BPMS offers its own solution there. Taking into account the complexity of the full BPMN 2.0 language it is clear that standardization does not outweigh the enhanced efforts of using BPMN for every simple process definition [2].

With the advance of domain specific languages (DSLs) in all domains of modeling and development, it is worth to revitalize also the use of DSLs for BPMS. The given paper proposes a DSL-based solution for BPMS – for domains where really a domain specific notation provides a significant gain in development speed. Since the whole development becomes domain specific, we can call the approach *DSBPMS*.

Using the DSBPMS approach it is possible to create a process definition language based on concepts and notations typical for the given domain. Then domain experts can not only read the process definitions but also create and modify them. Typical examples of such domains are insurance, healthcare, logistics etc. For example, the insurance domain could contain actions: Client Action, Broker Action, Employee Action; start event kinds: Claim Received, Risk Level Reached and domain elements used by actions: Claim, Risk etc. Similarly, processes in a healthcare institution should be based on terms understandable by doctors and healthcare personnel.

V. Geffert et al. (Eds.): SOFSEM 2014, LNCS 8327, pp. 351–362, 2014.

The approach is applicable also to domains where simple and flexible business processes dominate, such as internal document processing in various government institutions, for example education. There BPMN with its intricate control structures would make the notation unnecessary complicated. A simple process language based on UML activity basics would be much more suitable (see Section 4).

In this paper we propose the platform named GraDe3, by means of which the implementation of a DSBPMS even for quite a narrow domain would become feasible in the sense of efforts needed and would pay off shortly. In addition, processes in such a DSL could be easily modified to meet the goal of agile process management.

In this approach the first step is to choose a relevant domain and define an adequate process specification DSL for this domain. The platform is then used to create an advanced graphical editor set ready for building a complete process definition on the basis of the chosen domain specific process language. In addition to the editor for the process language the editor set contains editors for data model and form definitions. The next step supported by the platform is the creation of a transformation in a simple domain specific mapping language from the chosen process DSL to the language directly supported by the execution engine. In addition, this transformation defines the semantics of the created DSL in a simple and precise way. The platform contains a complete runtime support for the developed DSBPMS including user management, process execution monitoring etc., thus no other tools are required to build and execute a specific business process support in this DSBPMS. It should be noted that all steps in this development – the DSL definition, the editors for creating a concrete process management system, the transformation of the system definition to its runtime form and even the execution – are completely model-driven, with the corresponding metamodels precisely defined.

2   Related Work

There are a lot of tools, frequently named also Business Process Management Suites [3], available for the development of business process support systems. They are provided by software industry heavyweights such as IBM [4], Oracle [5], SAP [6] and others, and smaller vendors such as BizAgi [7]. In addition, a large number of open source solutions are also available – BonitaSoft [8], ProcessMaker [9] et al. Nearly all of the BPMSs use BPMN as a process modeling notation, only some use custom process languages (e.g., ProcessMaker [9]). The Gartner report 2010 on commercial BPMS [3] considers BPMN support as one of the key features in its tool evaluation.

None of the popular BPMSs are based on the idea of a DSL for a process definition. Most of the suites mentioned here are very complicated to use, with a large number of service features included – they are intended to be applied in large companies with complicated business processes and with high runtime performance in mind.

One of suites most oriented to building simple systems is BizAgi [7]. Process modeling there is based on a relatively large subset of BPMN. A sort of simple E-R model is used for data modeling, there is a relatively advanced form editor and an expression language for specifying guard conditions on flows, display lists for data selection controls etc. The main difference is that the process language is fixed to BPMN, while our approach is based on a DSL having a notation adapted to the chosen kind of processes and concepts.

The open source BPMS, one of the most usable between them being BonitaSoft [8], are also nearly all based on BPMN. In addition, they typically require at least some development in an OOP language (mostly Java, including BonitaSoft).

There are very few approaches explicitly using a DSL for BPMS. One of such is the approach based on Karlsruhe's Integrated Information Management (KIM) [10], however there only a choice from existing standard process languages (BPMN, UML activity, Petri nets) is offered as a process DSL. The approach closest to ours is DSLs4BPM [11], there new graphical DSLs can be defined using the Eclipse framework, however the possible diagram structure must remain very constrained and close to the very specific PICTURE language [12] used as the base.

3   Languages and Platform for DSBPMS

3.1   General Principles of the Approach

The goal of the approach is to enable the building of a DSBPMS based on a DSL for process design with as little effort as possible. Here we want to briefly explain how the building of a DSBPMS and its usage would look like from the viewpoint of the involved stakeholders and what steps are to be done. The first step would be the choice of an appropriate domain and the conceptual design of a process DSL for this domain, with special emphasis on finding the typical kinds of custom actions. Certainly, a modeling expert is required for this task. The next task is to formalize the graphical syntax of the DSL and create a graphical editor for it using the Graphical language definition environment of the platform. A DSL developer with some skills in graphical editor building is required here. The process editor is coupled with two predefined graphical editors in the platform – for building a data model and screen forms. The next task in DSBPMS definition is to map the defined process DSL to the language directly executed by the runtime engine of the platform (the Base language, see the next section). Thus both a very simple "compiler" in the DSBPMS is built and the precise execution semantics of the DSL is defined. The skills required for this task are similar to the previous one. In some cases the execution engine has to be extended by custom libraries built using an OOP language. Now the DSBPMS is completely built and ready for use. In order to build an executable system for a business process using this DSBPMS, the process must be precisely defined in the given DSL (typically, by several related process diagrams), this task is best performed by business analysts who now all the process details. Then the process definition must be extended by the data model, form definitions for user actions in the process and expressions showing how form elements and constraints in the process are linked to the data model. This step (to be performed by a system analyst) requires some IT skills though explicit programming is not required here. After applying the mapping the system is ready to use. Fig 1 shows an overview of all these activities.

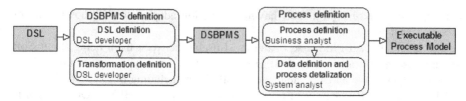

Fig. 1. Overview of the approach

3.2   Language Components for Building a Process DSL

Base Language. The proposed platform includes a process Execution engine which directly executes a simple (but functionally complete) process language, named the *Base* language in the approach. The rationale for the selection of the Base language, on the one hand, has been the simplicity of its implementation, but on the other hand, the ease of mapping the elements of a simple process definition DSL to it.

The process execution features of the Base language are chosen as a very basic subset of the UML 2.x Activity notation [13]. It includes the most used kinds of actions – general Action and CallProcessAction. Some more kinds of actions are not exactly from the UML standard, but are included because most of BPM languages (including BPMN) contain them. There are two such general subcases of action – UserAction and SystemAction. The most used kind of UserAction is ShowFormAction, and SystemAction also has several predefined sub-kinds – SendMailAction, DataAssignmentAction, CallServiceAction and a generic pattern-based AsignAllUserActions. One more important kind of elements is Custom action – CustomUserAction and CustomSystemAction. Custom actions are not directly implemented by the engine – their implementation must be supplied by the developer defining the DSL. Control structures include Start and End nodes, Decision and Merge nodes and Fork and Join Nodes, all flow control is performed explicitly by these nodes (only one flow can enter/exit an action). Only control flow edges are used, possibly with guard conditions. One more aspect is Roles and Users – each UserAction in a process has one or more Roles specified who can perform this action. The assignment of a specific user to this action is done at runtime according to the Roles.

Selection of a DSL for Process Notation. The Base language is used as a foundation for defining a specific DSL for the process notation in the given DSBPMS. It should be noted that the simplest DSL to be built is that directly coinciding with the Base language, but normally the DSL is modified to fit the chosen domain in the best way. Typically the modified elements can be directly mapped to the Base language elements or their groups. One of the goals of such mappings is to introduce derived notations for typical language constructs in the DSL. Completely new action kinds can be introduced in the DSL by mapping them to appropriate Custom actions to be implemented by the language developer.

The semantics of the defined DSL is precisely defined by its mapping to the Base language.

Data Modeling Language. Besides the process component of the DSL, each specific process management system built via the DSBPMS contains the persistent data model. The data model consists of the fixed part used directly by the execution engine and a specific data model for a given system. This specific model is to be built within a simple fixed subset of UML class diagrams. The data model is interpreted as a typical ORM image of a data base schema defining the persistent data for the given system. The fixed system runtime data can also be referenced in this custom model (using the <<system>> stereotype). Data models of existing partner systems the given system has to communicate with can be referenced as well, by the <<external>> stereotype (see a data model example in Fig. 4). It should be noted that BPMN based BPMSs (except BizAgi) typically use persistent process variables to model data (as required by the BPMN standard) which is less natural in practice (the real data are persisted in databases anyway).

Form Definition Language. Each ShowFormAction uses a specific form to be defined in a simple Form definition language. Forms in this language can contain all basic kinds of controls (textboxes, read-only fields, listboxes, checkboxes, tables as nested forms etc.). This language is also fixed in the platform. Typically one form is used in several actions, but with small modifications – some elements hidden, some made read-only etc. To support this situation in a simple way, a form with maximum details (a "main" form) can be defined and a "clone" of this form with small modifications can be easily specified for a user action. The logical structure of the form is uniquely defined by the form definition language but its style can be customized.

Expression Language. Both the process sublanguage and Form sublanguage use one fixed common element – a textual Expression language relying on the defined data model. Expressions in this language are used for binding form elements to data classes, defining the selection lists for listboxes etc. Another use of expression language is for guard conditions on control flows exiting a decision. The expression language is reused, e.g., for explicit assignment of values to process data element properties in DataAssignmentActions. Yet another use of expressions is to define parameter values for actions. Namely the last feature contributes a lot to the easy extension of functionality of the DSL. The expression language is OCL-like, however with some syntax simplification taken from OO languages, in order to support easy navigation inside the data model. Each process definition in the supported DSLs must have a base class chosen from the data model (denoted by the *self* keyword), this significantly simplifies specifying the expressions linking form and data elements. All the navigation in the Expression language is specified by using the "." symbol.

3.3   Platform Components

The proposed GraDe3 platform provides support both for the development of a DSBPMS and for its usage for building specific process management systems. Some initial ideas of such platform and mappings from the defined DSL to the execution language have been presented in [14]. In this paper the main emphasis will be on the DSBPMS development aspects.

Graphical Language Definition Environment. The graphical definition environment is based on the Transformation Based Graphical Tool Building Platform GrTP [15], which in turn is based on the TDA [16] platform. The GrTP component directly used for defining a DSL is the Configurator [17]. The definition of a diagram graphical syntax (diagram type) – its node types, edge types, their styles and their related text elements (compartments) – is also a simple diagram itself to be built in the Configurator. For compartments the relevant property dialogs can also be easily defined. Thus a completely specified graphical editor is obtained for the given diagram type.

The result of a language definition is a set of graphical editors based on TDA platform which includes the editor for the defined process DSL and the predefined editors for the data model definition and form definition. These editors constitute a workplace for developing a specific system in this DSBPMS. The generated process editor provides two views – the Business view with all data related details hidden (to be used by business analysts) and the Detailed view with all expressions visible.

Transformation/Mapping Definition Environment. A system model in the defined DSL must be converted to a model in the Base language before it can be executed. Therefore the DSL definition environment has one more component – a tool for defining a transformation (mapping) from the DSL to the Base language. In addition, this transformation defines the precise semantics of the DSL – in fact it is a very simple "compiler". Only the process sublanguage has to be mapped, the data and form sublanguages are predefined. The transformation definition is the second task for the DSL developer (see Fig. 1). We treat the DSL definition in the Configurator as a DSL metamodel, with element types corresponding to classes and compartment types to their attributes. Now a classical model transformation language could be applied. Fig. 2 shows a fragment of such a metamodel for the example DSL of this paper.

However, in most cases the transformation is so straightforward, that a simple domain specific mapping language provided in the platform is the best solution. Typically a class instance together with its direct environment in the source model (the DSL) fully determines which transformation rule is to be applied. This means that the source pattern is very simple – a class (node) instance with its attributes and incoming/outgoing edges. The corresponding target pattern normally consists of one class (node) instance in the Base language, with edges connected in a way isomorphic to the source model. In some cases an additional node instance must be added before or after the direct map target, with an edge connecting them in an obvious way (and external edges reconnected accordingly). Other elements of the source model can be transformed in a fixed way. An edge in the source maps to a target edge in the simplest way, with attributes, if any, copied. Only the predefined guard "*else*" is automatically substituted by a relevant not-based expression. A source process is mapped one-to-one to a target process and all node/edge mapping is localized inside a process.

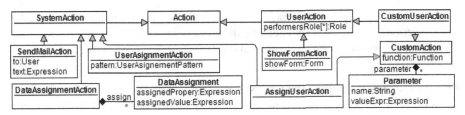

Fig. 2. Metamodel of the example DSL (fragment)

The mapping rules in the platform are defined in a tabular way, with three elements defining a mapping: the class to be mapped, the filter condition and the target class in the Base language (instance of which is to be created). If additional before/after instances must be created, their classes are specified as well. The filter condition must be specified as a Lua/lQuery expression [18]. We remind that Lua is a functional language having a collection (map) of arbitrary objects as the main data type. lQuery adds powerful expressions for filtering such collections or creating derived collections via navigating the model.

For the most typical cases predefined functions are offered, namely, the function *outLine* returning the collection of outgoing edges of a node and *inLine* returning that of incoming edges. The *size* function can be applied to a collection. The function *attr* *(<attribute_name>)* returns the value of the given attribute for the node (class). Expressions can be used to set the target class attributes.

The simplest mappings are for DSL elements coinciding with the Base language elements, there only the mapped class and the target class must be specified.

The two non-trivial mappings for the DSL used in this paper are described now. A simpler one using only a filter condition is that for inserting an explicit Merge node in the target when more than one flow enters an action in the source model:

```
MappedClass: Action, FilterCondition: inLine.size > 1
TargetClass: - , TargetClassBefore: Merge
```

The *Action* class is an abstract superclass therefore this rule is combined with the rule defined for each specific action class, the target class being defined in that rule.

Another more complicated mapping is used for processing the specific AssignUserAction action by mapping it to a CustomSystemAction based on the implemented function *AssignActionUser* with two parameters – a string (the action name) and a *User* class instance to be assigned. This mapping creates three class instances in the target model – a *CustomSystemAction* instance (having the chosen implementation) and two *Parameter* instances linked to the action:

```
MappedClass: AssignUserAction
TargetClass: CustomSystemAction (function="AssignUser")
    Parameter (name="action",valueExpr=attr("actionName"))
    Parameter (name="user",valueExpr=attr("userExpr"))
```

Since only one association links the used target model classes in the metamodel there is no need to specify it explicitly here.

The same mapping language facilities, especially the filter conditions and predefined functions, can be used for checking the consistency of a model in the DSL, thus a syntax checker can be easily created and only valid models need to be mapped.

Process Execution Engine. The process execution engine directly executes the Base language, with all form- and data-related actions included. It is based on the runtime metamodel implemented via a database. The engine maintains the state of all active process instances by means of tokens in a way inspired by UML activity semantics. It involves maintaining user action lists assigned to a user and automatic invoking of system actions. When an action is complete the tokens are moved, taking into account the control nodes. Due to the subset chosen, this token management is quite straightforward. All the execution is logged, in order to provide data for process monitoring.

The user management is provided via the administrator portal where users can be registered and linked to roles and process execution monitored. Regular users access the system via the user portal. There they can start a new process instance when their roles permit this. In another tab a user sees the user tasks assigned to him and ready to execute; when a task is selected the corresponding form opens.

The current version of Engine is implemented in MS.NET, with forms using the ASP.NET and data access based on Entity framework. A very valuable component of the engine is the Expression language interpreter which evaluates any valid expression over the defined data model including also the system runtime data.

4   Example – A DSL for Internal Document Processing

The example represents a simplified business trip management system in a university. Such a system should be built for the University of Latvia, but since the requirements are not finalized yet, some similar systems with available descriptions from universities in USA [19, 20] are used as a prototype for the example. The provided diagrams describe the initial part the business trip process – preparing the trip request and approving it, the whole process description would contain two more diagrams of a similar size. Though the basic path of a document in this system is quite straightforward, it contains some subtle moments related to who can actually perform the given action. In such institutions it is typical that an administrator can delegate its approval rights to another employee. The delegation rules are defined via the specific AssignUserAction included in the DSL. This is an aspect not so easily definable within the BPMN 2.0.

4.1   Description of the Example DSL

The process language of the proposed DSL reuses many of the Base language elements. However, there are some simplifications of the control structure and a new specific action is added in this DSL. The chosen DSL is well adjusted for internal document management systems; certainly, the language is slightly simplified in order to a have a complete description in the paper.

The given DSL contains the following actions – ShowFormAction, CallProcessAction, DataAssignmentAction, SendMailAction and two kinds of user assignment action – the pattern-based AssignAllUserActions taken from the Base language and one specific for this DSL – AssignUserAction. Except for the last kind, the actions are one-to-one with the Base language. The specific AssignUserAction assigns at runtime a user to the selected UserAction (by its name) in the given process instance on the basis of an expression (which must return an instance of the type *User*). It is defined by mapping it to the CustomSystemAction, a specific implementation of which is provided for this DSL (see the description of this mapping in section 3.3). According to the parameter mechanism of the Base language, both the relevant action name and the user expression are evaluated at runtime by the engine and passed as parameters to the function implementing this kind of CustomSystemAction.

The following control nodes are included in the DSL: Start, End and Decision. For simplicity we omit the concurrent flow management (Fork and Join), concurrent actions are not so frequent in internal document processing. The Merge node is substituted by the possibility to have more than one flow entering an action – certainly, all such situations are mapped to explicit Merge nodes. The mapping inserting the Merge node where required is also provided in 3.3. Thus there are only two non-trivial mappings to the Base language for this DSL, the other ones are one-to-one.

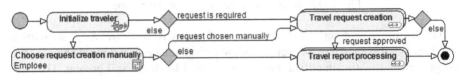

Fig. 3. Main process – University Travel System

4.2   The Example in Brief

In totality the example contains four process diagrams, one data model diagram and 5 "main" form diagrams in this system. The main process (marked as an entry point to the whole system) is the University Travel System (see Fig. 3). This process is shown in the Business view – with all data related expressions hidden. Only instances of this process can be directly created by authorized users of the system. The Role of a user who can start a new instance of this process is *Employee*. The base data class for this process is Travel (see the data model in Fig. 4), this class will be denoted by *self* in all expressions related to this process and a new instance of this class will be created at the process start. The first action to be executed in the main process is the DataAssignmentAction *Initialize Travel*. The main process invokes two subprocesses – Travel Request creation and Travel Report processing. According to the rules of University, not always the travel request has to be created before the travel (it depends on the unit where the employee works). The employee can also choose to create the request manually using the form in the ShowFormAction Choose Request creation manually.

All the expressions in the example are based on its data model, therefore we now briefly describe this model (Fig. 4). The classes without stereotypes are those built for the Travel System. The classes with the stereotype <<external>> are for some existing systems – here it is the Human Resources system. Classes with the stereotype <<system>> are those for the workflow engine runtime (only one of them is shown). We assume here that the system links each *User* instance (via the *user-emp* association) to the relevant *Employee* instance. Classes with the <<codifier>> stereotype represent instance sets for selection. Note that all classes used for the Travel System should be somehow linked to its root – the *Travel* class.

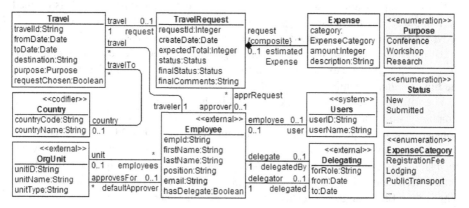

Fig. 4. Data model (the fragment used in expressions)

The process to be described in the most detailed way is the *Travel Request creation* (see Fig. 5) – it is shown in the Detailed view with all data expressions visible. In fact, the process is slightly simplified, but all used action kinds are still present. The base data class for this process is *TravelRequest*. The CallProcessAction (here the one in the main process in Fig. 3) invoking the process can specify the path in the model from the base class of the caller to the base class of the callee (here it is *self.request*).

The first action in the process is the pattern based user assignment. The pattern is "ToStarter" – the user starting the main process instance would be assigned to all actions where the filter is true – here to actions where the specified role is *Employee* (it was the starter role). The next action is the DataAssignmentAction *Initialize Request*. Two attributes of the newly created *TravelRequest* instance are set by means of built-in functions. The link to default approver (the *approver* link) is set by a more complicated expresssion (*self.travel.traveler.unit.defaultApprover*). The action *Assign Approver* assigns a user for the action *Approve Request*, by checking whether a delegate is currently set for the default approver – the setting of a delegate is done in another system (Human Resources). The SendMailAction *Notify Traveler on Final status* uses the keyword *me* for the To expression – it specifies the starter user of the whole process. The other parameter for this action is the message text. An essential ShowFormAction here is *Create Travel Request*.

To complete the example, the forms must be defined and bound to the data model. Then the defined mapping to the Base language must be run and the Travel system is ready to use. It should be reminded that no low-level programming is needed for building such a system. The example confirms the usability of the approach and the suitability of the chosen DSL for internal document processing systems.

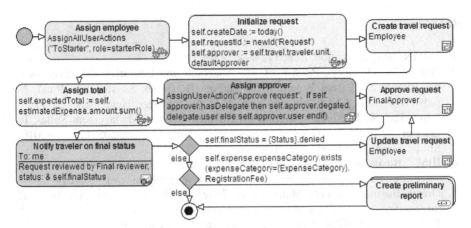

Fig. 5. Process diagram *Travel request creation*

5   Conclusions and Future Work

A new approach to building BPMS has been proposed in the paper. Instead of using the standard BPMN notation for business process behavior description the usage of a DSL best suited for the given process domain is recommended, thus yielding a DSBPMS for the domain. The goal of the approach is to simplify the development of a specific business process support system in such a DSBPMS so that the process development could be performed by domain experts. Certainly, all this makes sense only when the development of the DSBPMS itself can be done with relatively little effort. Therefore a new GraDe3 platform is proposed in the paper. The key components of this platform are the Configurator for easy definition of a graphical editor for the chosen DSL and a new simple mapping language for defining a transformation, by means of which process descriptions in the DSL are transformed to the language directly executable by the execution engine of the platform. An example of such a DSL is given containing two typical use cases – a domain specific control structure simplification and adding a new domain specific action kind.

The implementation of the GraDe3 platform prototype is nearly complete. The DSBPMS based on the example DSL for this paper has been built and several process examples implemented in it. Another DSL for study process management is being built and evaluated on the bachelor study management at the University of Latvia. The experiments confirm the usability of the approach for building DSLs by language designers and user-friendliness of these DSLs for business analysts. One of the key factors enabling the usability of DSLs is the included domain specific actions.

The close integration of DSL definition, transformation definition and process execution engine in the platform permits to add new features for process management. A process execution monitoring based on diagrams in the original DSL notation will be provided, as well as various queries on execution status based on such diagrams.

References

1. BPMN 2.0 specification, `http://www.bpmn.org/`
2. Genon, N., Heymans, P., Amyot, D.: Analysing the Cognitive Effectiveness of the BPMN 2.0 Visual Notation. In: Malloy, B., Staab, S., van den Brand, M. (eds.) SLE 2010. LNCS, vol. 6563, pp. 377–396. Springer, Heidelberg (2011)
3. Sinur, J., Hill, J.: Magic Quadrant for Business Process Management Suites. In: Gartner RAS Core Research Note G00205212 (2010), `http://www.gartner.com/id=1453527`
4. IBM Business Process Manager, v 8.5, `http://www-03.ibm.com/software/products/us/en/business-process-manager-family`
5. Oracle Business Process Management Suite 11g, `http://www.oracle.com/us/technologies/bpm/suite/overview/inex.html`
6. SAP NetWeaver BPM, `http://scn.sap.com/community/bpm`
7. Bizagi BPM suite 10.1, `http://www.bizagi.com/index.php`
8. BonitaSoft - Bonita Open Solution, Open Source BPM, `http://www.bonitasoft.com`
9. ProcessMaker Workflow management and BPM, `http://www.processmaker.com`
10. Freudenstein, P.: Web Engineering for Workflow-based Applications: Models, Systems and Methodologies. KIT Scientific Publishing, Karlsruhe (2009)
11. Heitkötter, H.: A Framework for Creating Domain-specific Process Modeling Languages. In: Proceedings of the ICSOFT 2012, pp. 127–136 (2012)
12. Becker, J., Pfeiffer, D., Räckers, M.: Domain specific process modelling in public administrations–the PICTURE-approach. In: Wimmer, M.A., Scholl, J., Grönlund, Å. (eds.) EGOV 2007. LNCS, vol. 4656, pp. 68–79. Springer, Heidelberg (2007)
13. UML specification v.2.4.1, `http://www.omg.org/spec/UML`
14. Lace, L., Liepiņš, R., Rencis, E.: Architecture and Language for Semantic Reduction of Domain-Specific Models in BPMS. In: Aseeva, N., Babkin, E., Kozyrev, O. (eds.) BIR 2012. LNBIP, vol. 128, pp. 70–84. Springer, Heidelberg (2012)
15. Barzdins, J., Zarins, A., Cerans, K., Rencis, E., et al.: GrTP: Transformation Based Graphical Tool Building Platform. In: Proc. of MDDAUI 2007 Workshop of MODELS 2007, Nashville, Tennessee, USA, CEUR Workshop Proceedings, vol. 297 (2007), `http://ceur-ws.org`
16. Kozlovics, S., Barzdins, J.: The Transformation-Driven Architecture for interactive systems. In: Automatic Control and Computer Sciences, vol. 47(1/2013), pp. 28–37. Allerton Press, Inc (2013)
17. Sprogis, A.: The Configurator in DSL Tool Building. In: Scientific Papers, vol. 756, pp. 173–192. University of Latvia (2010)
18. Liepiņš, R.: Library for model querying: IQuery. In: Proceedings of the 12th Workshop on OCL and Textual Modelling (OCL 2012), pp. 31–36. ACM, New York (2012)
19. Welcome to e-Expense Travel & Business Expense System. A User Guide, Tufts University, `http://finance.tufts.edu/accpay/files/eExpenseGuide.pdf`
20. The Ohio State University, eTravel ASSIST, `https://assist-erp.osu.edu/assistTravel/index.htm`

Bounded Occurrence Edit Distance:
A New Metric for String Similarity Joins
with Edit Distance Constraints

Tomoki Komatsu, Ryosuke Okuta, Kazuyuki Narisawa, and Ayumi Shinohara

Graduate School of Information Science, Tohoku University, Japan
{tomoki_komatsu@shino.,ryousuke_okuta@shino.,
narisawa@,ayumi@}ecei.tohoku.ac.jp

Abstract. Given two sets of strings and a similarity function on strings, *similarity joins* attempt to find all similar pairs of strings from each respective set. In this paper, we focus on similarity joins with respect to the edit distance, and propose a new metric called the bounded occurrence edit distance and a filter based on the metric. Using the filter, we can reduce the total time required to solve similarity joins because the metric can be computed faster than the edit distance by bitwise operations. We demonstrate the effectiveness of the filter through experiments.

Keywords: Edit distance, Similarity join problem, Similarity search, Data integration.

1 Introduction

Given two sets of strings and a similarity function, similarity join problems attempt to find all pairs of strings from each set such that each pair's similarity value satisfies a given criterion. This problem has attracted significant research attention recently because it has a variety of applications. Query reinforcement for Web search [7], coalition detection [4], typographical error checking, and data integration are typical applications.

In this study, we focus on similarity joins with respect to edit distance, or *edit similarity join*. The edit distance between two strings is the minimum number of edit operations (insertion, deletion, and substitution) required to transform one string into the other. Edit similarity join returns all pairs of strings whose edit distance is less than or equal to a given threshold. As is well known, edit distance can be calculated easily by dynamic programming in quadratic time [8]. However, calculating the edit distance for all pairs of strings from each set still incurs high cost.

Various approaches have been proposed to reduce the cost of similarity joins. These approaches generate candidate pairs by filtering the string pair from each given set and validate candidate pairs by calculating the edit distance. Gravano *et al.* [3] proposed a filter that uses a q-gram. Bayardo *et al.* [1] proposed an inverted index approach. Xiao *et al.* [11] focused on the lower bounds of edit distance as

V. Geffert et al. (Eds.): SOFSEM 2014, LNCS 8327, pp. 363–374, 2014.

obtained by analysing the locations and contents of mismatching q-grams. Narita *et al.* [5] proposed a filter by the hash-join approach. Wang *et al.* [9] used a trie structure to efficiently find similar string pairs. These approaches substantially reduced candidate sizes, but the filtering cost is large.

In this paper, we propose a new distance metric called the bounded occurrence edit distance (BOED for short). This metric focuses on character occurrence. The L_1 distance, which is used in the existing algorithm [11], also focuses on character occurrence but cannot be calculated efficiently. The proposed distance metric can be calculated efficiently by bit operations without an array. We derive several theorems on relations between the standard edit distance and BOED. In order to reduce the cost of the edit similarity join, we propose new filtering methods based on BOED. Because these bit-based filters are simple, they can be combined with other algorithms easily and effectively. We present some experiments to verify that our filters eliminate many fault candidate pairs very efficiently. We also analyze several techniques that improve performance in these experiments.

2   Preliminaries

Let $\mathbb{N}$ be the set of all non-negative integers. Let Σ be a finite set of symbols, called an *alphabet*. An element of Σ^* is called a *string*. The length of string s is denoted by $|s|$. ε denotes the empty string. The i-th character of string s is denoted by $s[i]$ for $0 \leq i < |s|$. Let $occ(s, c)$ be the number of occurrences of symbol $c \in \Sigma$ in string s, and $difflen(x, y) = \big| |x| - |y| \big|$ for strings $x, y \in \Sigma^*$.

Definition 1. *For two strings $x, y \in \Sigma^*$, the edit distance between x and y, denoted by $ED(x, y)$, is the minimum cost of edit operations (insertions, deletions, and substitutions) required to change x into y.*

Definition 2. *Given two string sets X and Y, and threshold τ for edit distance, an edit similarity join returns all string pairs (x, y) in $X \times Y$ with $ED(x, y) \leq \tau$.*

Existing approaches use a distance defined by the total sum of the differences between the number of characters in two strings, which is called the L_1 *distance*.

Definition 3 ([11,2,6]). *For two strings $x, y \in \Sigma^*$, the L_1 distance between x and y is defined by*

$$L_1(x, y) = \sum_{c \in \Sigma} \big| occ(x, c) - occ(y, c) \big|.$$

3   Bit-Vector-Based Join

We propose several filters in order to reduce the number of calculations for the edit distance, based on several *weak* edit distances. Given two strings $x, y \in \Sigma^*$ and threshold τ, these filters give several sufficient conditions for $ED(x, y) > \tau$,

---

Algorithm 1. algorithm to solve similarity joins with filter

Input: two sets X, Y of strings, and threshold τ
Output: set S of string pairs
1 $S \leftarrow \emptyset$;
2 **foreach** $x \in X$ **do**
3   **foreach** $y \in Y$ **do**
4   **if** $Filter(x, y, \tau)$ =false **then**
5   **if** $ED(x, y) \leq \tau$ **then**
6                  $S \leftarrow S \cup (x, y)$;

7 **return** S;

---

and these filters can be computed significantly faster than the original edit distance. Therefore, we have to only compute $ED(x, y)$ for a small fraction of pairs (x, y) that pass the filters. Algorithm 1 shows the pseudo-code for edit similarity join with a filter. Function $Filter(x, y, \tau)$ returns false if it has confidence that $ED(x, y) > \tau$. Existing filters such as ED-Join [11] and Trie Join [9] are enough precise to reduce candidate pairs substantially. ED-Join uses q-gram approach and Trie Join uses trie structure to filter the candidate pairs. However, these filters are costly because many operations related to q-gram are required and constructing trie structure needs calculation cost. Thus, we propose a new metric which can be calculated efficiently and applied to similarity joins filter easily. This metric can be calculated by bit operations efficiently. We begin by introducing an (unbounded) occurrence edit distance, which does not use special bit operations.

3.1 Occurrence Edit Distance

In this section, we define the occurrence edit distance, which can be computed at lower cost than the edit distance, and propose a filter using this distance. First, we define a distance by using L_1 distance, which is called the occurrence edit distance. This distance can be applied to similarity joins filter.

Definition 4. *For two strings* $x, y \in \Sigma^*$, *the* occurrence edit distance *(OED for short) between* x *and* y *is defined as*

$$OED(x, y) = \frac{L_1(x, y) + difflen(x, y)}{2}.$$

For instance, $OED(\mathsf{AABC}, \mathsf{ABBCCC}) = ((1 + 1 + 2) + 2)/2 = 3$.

Lemma 1. *For any strings* $x \in \Sigma^*$ *and* $y = y_1 y_1' y_2 y_2' \cdots y_n y_n'$ *satisfying that* $|y_1 y_2 \cdots y_n| = |x|$ *and* $y_i, y_i' \in \Sigma^*$ *for* $1 \leq i \leq n$, *let* $z = y_1 y_2 \cdots y_n$ *and* $z' = y_1' y_2' \cdots y_n'$. *Then* $L_1(x, y) \leq L_1(x, z) + L_1(\varepsilon, z')$ *holds.*

Proof. We prove it by showing $L_1^c(x, y) \leq L_1^c(x, z) + L_1^c(\varepsilon, z')$ for each $c \in \Sigma$, where we define $L_1^c(x, y) = \big|occ(x, c) - occ(y, c)\big|$. At first, $L_1^c(\varepsilon, z') =$

$occ(z', c) = occ(y, c) - occ(z, c)$, because $occ(\varepsilon, c) = 0$. If $occ(x, c) \leq occ(y, c)$, then $L_1^c(x, z) + L_1^c(\varepsilon, z') - L_1^c(x, y) = |occ(x, c) - occ(z, c)| + (occ(y, c) - occ(z, c)) - (occ(y, c) - occ(x, c)) = |occ(x, c) - occ(z, c)| + (occ(x, c) - occ(z, c)) \geq 0$. Otherwise, $occ(z, c) \leq occ(y, c) \leq occ(x, c)$ and $L_1^c(x, z) + L_1^c(\varepsilon, z') - L_1^c(x, y) = (occ(x, c) - occ(z, c)) + (occ(y, c) - occ(z, c)) - (occ(x, c) - occ(y, c)) = 2(occ(y, c) - occ(z, c)) \geq 0$. □

The next theorem shows the inequality between the edit distance and the OED.

Theorem 1. *For any $x, y \in \Sigma^*$, $ED(x, y) \geq OED(x, y)$.*

Proof. Let $S(x) = \{s \in \Sigma^* \mid L_1(x, s) = 0\}$. We prove the theorem by showing that $ED(x, y) \geq \min_{s \in S(x)} ED(s, y) \geq OED(x, y)$. The left inequality immediately holds because $L_1(x, x) = 0$, so $x \in S(x)$. Next, we show that the right inequality also holds. Let $x' \in S(x)$ be a string satisfying $ED(x', y) = \min_{s \in S(x)} ED(s, y)$. It is obtained from x by transpositions so that the longest common subsequence between it and y is maximised.

For any two distinct characters $a, b \in \Sigma$, we have $ED(a, b) = L_1(a, b)/2$ because $L_1(a, b) = 2$ and $ED(\varepsilon, b) = L_1(\varepsilon, b) = \textit{difflen}(\varepsilon, b) = 1$. This implies that $ED(s, t) = L_1(s, t)/2$ and $ED(\varepsilon, t) = L_1(\varepsilon, t) = \textit{difflen}(\varepsilon, t)$ for any $s, t \in \Sigma^n$.

Now we assume $|x'| \leq |y|$ without loss of generality. There exists a string y' of length $|x|$ such that y' is a subsequence of y and can be transformed from x' with substitution operations only. This implies that y can be transformed from y' by insertion operations only. Let y'' be the string deleted from y' to create y. Hence,

$$\min_{s \in S(x)} ED(s, y) = ED(x', y)$$

$$= ED(x', y') + ED(y', y) = ED(x', y') + ED(\varepsilon, y'')$$

$$\geq \frac{L_1(x', y) - L_1(\varepsilon, y'')}{2} + L_1(\varepsilon, y'') \qquad \text{(by Lemma 1)}$$

$$= \frac{L_1(x', y) + L_1(\varepsilon, y'')}{2} = \frac{L_1(x', y) + \textit{difflen}(x', y)}{2}$$

$$= \frac{L_1(x, y) + \textit{difflen}(x, y)}{2}.$$

□

The above theorem ensures that any pair (x, y) satisfying $OED(x, y) > \tau$ also satisfies $ED(x, y) > \tau$. Algorithm 2 shows the pseudo-code of the OED filter. The filter can be computed in $O(|x| + |y| + |\Sigma|)$ time and $O(|\Sigma|)$ space, most of which is required for computation of the L_1 distance. This computation is significantly faster than the computation of the original edit distance $ED(x, y)$, which requires $O(|x| |y|)$ time and space.

3.2   Bounded Occurrence Edit Distance

In this section, we propose a new metric which can be calculated efficiently by bitwise operations. For that purpose, we place a restriction on the L_1 distance,

---

Algorithm 2. OED filter

Input: two strings $x, y \in \Sigma^*$ and threshold τ

Output: a boolean value that indicates whether (x, y) can be removed from the candidates

1 **if** $difflen(x, y) > \tau$ **then return** true;
2 $d \leftarrow L_1(x, y)$;
3 **if** $(d + difflen(x, y))/2 > \tau$ **then return** true;
4 **return** false;

Table 1. Examples of occurrence bit vectors for $\Sigma = \{A, B, C, D\}$ (bits are arranged from right to left)

(a) Occurrence Bit Vector

a_i		A	B	C	D
$f(a_i)$		2	1	2	1
b	$pos(a_i)$	4	3	1	0
	l	2 1	1	2 1	1
$OBV(\mathtt{AACC}; f)$		1 1	0	1 1	0
$OBV(\mathtt{AACCCC}; f)$		1 1	0	1 1	0
$OBV(\mathtt{ACCCD}; f)$		0 1	0	1 1	1
$OBV(\mathtt{ABBCCCD}; f)$		0 1	1	1 1	1

(b) Freq-OBV

$A(a_i)$		B,D	A	C
$g(a_i)$		1	2	3
b	$pos'(a_i)$	5	3	0
	l	1	2 1	3 2 1
$Freq\text{-}OBV(\mathtt{AACC}; g)$		0	1 1	0 1 1
$Freq\text{-}OBV(\mathtt{AACCCC}; g)$		0	1 1	1 1 1
$Freq\text{-}OBV(\mathtt{ACCCD}; g)$		1	0 1	1 1 1
$Freq\text{-}OBV(\mathtt{ABBCCCD}; g)$		1	0 1	1 1 1

which is the bound of counting characters in the bit width. We explain the bounded L_1 distance before the bounded occurrence edit distance.

Let $f : \Sigma \to \mathbb{N}$ be a mapping for the character-counting bounds.

Definition 5. *A bound function for Σ is a mapping $f : \Sigma \to \mathbb{N}$. For two strings $x, y \in \Sigma^*$, the bounded L_1 distance between x and y is defined by*

$$BL_1(x, y; f) = \sum_{c \in \Sigma} |\min\{occ(x, c), f(c)\} - \min\{occ(y, c), f(c)\}| .$$

In order to compute $BL_1(x, y; f)$ efficiently, we introduce a bit-vector representation of counters.

Let $pos : \Sigma \to \mathbb{N}$ be a mapping called a *position function for Σ*, which indicates the position of the right-most bit associated with each character.

Definition 6. *Let f be a bound function for $\Sigma = \{a_1, a_2, \cdots a_k\}$, and pos be a position function for Σ. For string $s \in \Sigma^*$, $OBV(s; f)$ is a bit vector b of length $\sum_{i=1}^{k} f(a_i)$, consisting of*

$$b[pos(a_i) + l] = \begin{cases} 1 & l \leq occ(s, a_i) \\ 0 & otherwise \end{cases}, \quad for\ 1 \leq i \leq k\ and\ 1 \leq l \leq f(a_i).$$

Table 1(a) presents examples of OBV for $\Sigma = \{A, B, C, D\}$ with $f(A) = 2$, $f(B) = 1$, $f(C) = 2$, $f(D) = 1$, $pos(A) = 4$, $pos(B) = 3$, $pos(C) = 1$, and $pos(D) = 0$. The bits corresponding to $b[pos(A) + 1]$, $b[pos(A) + 2]$, $b[pos(C) + 1]$, and

---

Algorithm 3. Compute $OBV(s; f)$

    Input: string s
    Output: $OBV(s; f)$ as an integer
1   $b \leftarrow 0$;
2   **for** $i = 1$ to $|s|$ **do**
3 | $mask \leftarrow ((1 \ll f(s[i])) - 1) \ll pos(s[i])$;
4 | $b \leftarrow b \mid (((b \ll 1) \mid (1 \ll pos(s[i]))) \mathbin{\&} mask)$;

5   **return** b;

$b[pos(\mathtt{C}) + 2]$ are 1 in $OBV(\mathtt{AACC}; f)$ because $\mathtt{AACC}$ contains two $\mathtt{A}$'s and two $\mathtt{C}$'s. It does not have a sufficient number of bits to record occurrences of $\mathtt{C}$ more than twice, because $f(\mathtt{C}) = 2$. Therefore, $OBV(\mathtt{AACC}; f)$ is equal to $OBV(\mathtt{AACCCC}; f)$.

Algorithm 3 shows the pseudo-code for constructing $OBV(s; f)$ as an integer. Unary counters are efficiently implemented using bit operations. We can compute $OBV(s; f)$ in $O(|s|)$ time and $O(1)$ space. Now, we are ready to calculate bounded L_1 distance by bit operations. Given $x, y \in \Sigma^*$ and bound function f, we can calculate $BL_1(x, y; f)$ by

$$BL_1(x, y; f) = bitcount(OBV(x; f) \oplus OBV(y; f)),$$

where $\oplus$ denotes the XOR operation, and $bitcount(w)$ is a function that returns the number of bits that are 1 in w. Using an XOR operation and $bitcount$ function computed by either a special instruction in the target machine or a smart combination of bit operations (e.g., [10]), we can keep the costs of the calculation for $BL_1(x, y; f)$ low. For example, in Table 1(a), $BL_1(\mathtt{AACC}, \mathtt{ABBCCCD}; f) = bitcount(\mathtt{110110} \oplus \mathtt{011111}) = bitcount(\mathtt{101001}) = 3$.

Definition 7. *Let f be a bound function for Σ. For two strings $x, y \in \Sigma^*$, the bounded occurrence edit distance between x and y is defined by*

$$BOED(x, y; f) = \frac{BL_1(x, y; f) + \textit{difflen}(x, y)}{2}.$$

We show some lemmas for bounded occurrence edit distance.

Lemma 2. *Let f and g be bound functions satisfying that $f(c) \geq g(c)$ for any $c \in \Sigma$. Then, $BOED(x, y; f) \geq BOED(x, y; g)$ for any $x, y \in \Sigma^*$.*

Proof. Because $BOED(x, y; f) - BOED(x, y; g) = (BL_1(x, y; f) - BL_1(x, y; g))/2$, we only have to prove that $BL_1(x, y; f) \geq BL_1(x, y; g)$. We assume that $occ(x, c) \geq occ(y, c)$ without loss of generality. Let $occ(x, c; f) = \min\{occ(x, c), f(c)\}$ and $docc(x, y, c; f) = |occ(x, c; f) - occ(y, c; f)|$. Then

$$docc_f(x, y, c) - docc_g(x, y, c)$$

$$= \begin{cases} 0 & (f(c) \geq g(c) \geq occ(x, c) \geq occ(y, c)) \\ occ(x, c) - g(c) & (f(c) \geq occ(x, c) \geq g(c) \geq occ(y, c)) \\ occ(x, c) - occ(y, c) & (f(c) \geq occ(x, c) \geq occ(y, c) \geq g(c)) \\ f(c) - g(c) & (occ(x, c) \geq f(c) \geq g(c) \geq occ(y, c)) \\ f(c) - occ(y, c) & (occ(x, c) \geq f(c) \geq occ(y, c) \geq g(c)) \\ 0 & (occ(x, c) \geq occ(y, c) \geq f(c) \geq g(c)) \end{cases}.$$

Hence, we can obtain $BL_1(x, y; f) \geq BL_1(x, y; g)$ because $BL_1(x, y; f)$ is the summation of $docc(x, y, c; f)$ for all c. □

The following lemma shows the relation between OED and the bounded occurrence edit distance.

Lemma 3. *Let f be any bound function for Σ. Then, $OED(x, y) \geq BOED(x, y; f)$ for any $x, y \in \Sigma^*$.*

Proof. $OED(x, y) = BOED(x, y; f)$ if $f(c) \geq \max\{occ(x, c), occ(y, c)\}$ for any $c \in \Sigma$. By Lemma 2, we have $OED(x, y) \geq BOED(x, y; f)$. □

By Lemma 3 and Theorem 1, we obtain the following theorem.

Theorem 2. *Let f be any bound function for Σ. Then, $ED(x, y) \geq BOED(x, y; f)$ for any $x, y \in \Sigma^*$.*

We can construct a new filter based on Theorem 2 by changing line 2 in Algorithm 2 to $d \leftarrow bitcount(OBV(x; f) \oplus OBV(y; f))$. Lemma 2 shows that assigning more bits to frequent characters improves the performance of the filter.

3.3   Frequency-Based Bounded Occurrence Edit Distance

Lemma 2 shows that assigning more bits to frequent characters would improve the performance of filters. Thus, assigning bits to characters which seldom occur is wasteful. We consider assigning the same bits to multiple (infrequent) characters, so that we can assign more bits to frequent characters. We realize this extension by simply setting $pos(c_1) = pos(c_2) = \cdots = pos(c_k)$, where $c_1, c_2, \ldots c_k$ are the characters that share the same bits.

Now, we propose an effective bit assignment method depending on character frequency in Algorithm 4. The integer $width$ represents the number of bits, and $C[c]$ denotes the frequency of character $c \in \Sigma$ in the dataset. The function $round$ returns the nearest integral value to a given value (Line 4). This algorithm first assigns some bits according to character frequencies (Lines 4-4). Next, the remaining bits are assigned to the characters that are not assigned yet (Lines 4-4). These characters share the same bits.

Algorithm 4. Bit Assignment

Input: integer *width* and mapping C from characters to integers
Output: a pair of mappings of the bound function and the position function

1   $total \leftarrow \sum_{c \in \Sigma} C[c]$;
2   $rest \leftarrow width$;
3   $f, pos \leftarrow$ empty mapping from characters to integers;
4   **foreach** $c \in \Sigma$ **do**
5      $n \leftarrow \min(rest, \text{round}(width * C[c]/total))$;
6      $f[c] \leftarrow n$;
7      $pos[c] \leftarrow width - rest$;
8      $rest \leftarrow rest - n$;

9   **foreach** $c \in \Sigma$ **do**
10   **if** $C[c] > 0$ *and* $f[c] = 0$ **then**
11        $f[c] \leftarrow rest$;
12        $pos[c] \leftarrow width - rest$;

13 **return** (f, pos);

Definition 8. *Let* f *be a bound function for* $\Sigma = \{a_1, a_2, \cdots a_k\}$. *Let pos be a position function for* Σ, *and* $A(a) = \{x \in \Sigma \mid pos(a) = pos(x)\}$. *For string* $s \in \Sigma^*$, *Freq-OBV*$(s; f)$ *is a bit vector* **b** *consisting of*

$$b[pos(a_i) + l] = \begin{cases} 1 & l \leq \sum_{x \in A(a_i)} occ(s, x) \\ 0 & otherwise \end{cases}, \ for \ 1 \leq i \leq k \ and \ 1 \leq l \leq f(a_i).$$

We can use Freq-OBV instead of OBV in the bounded L_1 distance. Freq-OBV is superior to OBV, as we will see in experiments. For example, in Table 1, $BOED(\texttt{AACC}, \texttt{ACCCD}; g) = (3 + 1)/2 = 2$ and $BOED(\texttt{AACC}, \texttt{ACCCD}; f) = (2 + 1)/2 = 1.5$, because $BL_1(\texttt{AACC}, \texttt{ACCCD}; g) = 3$ and $BL_1(\texttt{AACC}, \texttt{ACCCD}; f) = 2$, respectively.

4 Performance Evaluation

In this section, we present comparison experiments conducted on proposed filters using both artificial data and real-world data. All experiments address *self join*, which attempts to find all string pairs $(x, y) \in S \times S$ satisfying $ED(x, y) \leq \tau$ for given a string set S and threshold τ. We conduct all experiments on a PC with an Intel(R) Xeon(R) CPU X5660 @ 2.80 GHz and 48 GB of RAM. All algorithms are implemented in C++ and compiled using GCC 4.4.7. We use POPCNT in SSE4.2 to count the number of bits that are 1 in given bit vector.

4.1 Experiments on Artificial Data

First, we conduct several experiments using artificial data. For our performance evaluation, uniform random data is unsuitable because our proposals mostly

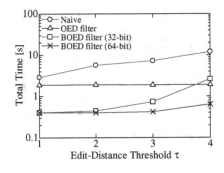
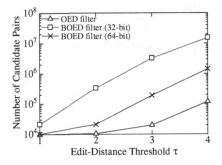

Fig. 1. Total running time required to solve self join on artificial data

Fig. 2. Number of candidate pairs that passed the filters on artificial data

target various documents written in natural languages. Therefore, in these experiments, we generate artificial data according to the Zipfian distribution. This distribution connects a character frequency *freq* and its ranking k by $freq(k) = \frac{1/k}{\sum_{i=1}^{|\Sigma|} 1/i}$. Let $|\Sigma| = 32$, and let S be a set of 10,000 strings of length 20 over Σ generated from this distribution. In this experiment, we compare proposed filters with respect to running time.

Comparisons of Filters. We evaluate our three filters, (OED filter, 32-bit BOED filter, and 64-bit BOED filter), and compare them with a naive method.

- **Naive**: Computes the edit distance for all pairs without any filters.
- **OED Filter**: This is shown as Algorithm 2 (no special bit operations).
- **BOED Filter (32-bit)**: This uses 32 bits for OBV and $f(c) = 1$ for any character $c \in \Sigma$.
- **BOED Filter (64-bit)**: This uses 64 bits for OBV and $f(c) = 2$ for any character $c \in \Sigma$.

Fig. 1 shows the relation between threshold τ and the total running time. The total running time includes the time required to filter queries and validate candidate pairs by calculating the edit distance. Fig. 2 shows the relation between threshold τ and the number of pairs that the filter passed, which must be validated by calculating the edit distance. In this figure, we omitted the result of the naive method because it was 10^8 constantly. The total time of OED filter in Fig. 1 is almost independent of the threshold τ. The reason is that the cost of OED filter is much higher than that of validation by calculating the edit distance. Fig. 2 shows that OED filter is more precise than other filters. However, BOED filter is superior to OED filter in Fig. 1 because the BOED filter is calculated more efficiently than OED. BOED filter (64-bit) outperformes BOED filter (32-bit) because 64-bit version is more precise than 32-bit version. These results clearly show that the proposed metric can be applied to similarity joins filter effectively.

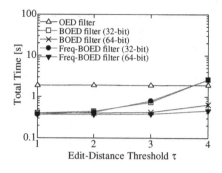
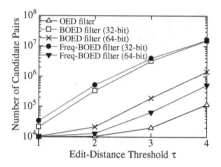

Fig. 3. Total running time required to solve self join on real-world data

Fig. 4. Number of candidate pairs that passed the filters on real-world data

Comparisons of Bit Assignment Methods. Lemma 2 implies that the number of pairs filtered depends on the bound function, that is, bit assignments. We experimentally investigate the relation between performance and bit assignment methods. Added to **BOED filter (32-bit)** and **BOED filter (64-bit)**, we examine **Freq-BOED filter (32-bit)** and **Freq-BOED filter (64-bit)**.

Fig. 3 shows the total running time for each assignment, and Fig. 4 shows the number of pairs that the filter passed. When the given string is too long relative to the number of bits, OBV is occupied almost entirely by 1-bits and loses its ability to keep track of character occurrence.

4.2 Experiments on Real-World Data

Next, we experimentally demonstrate the performance of our filter on real-world data. We use three datasets [1]: English Dict, DBLP Author, and AOL Query Log. The distributions of words included in each dataset are plotted in Fig. 5.

We use the following algorithms in this experiment.

- **Ed-Join-1 Only**: This was proposed in [11] and is a method with a filter based on a q-gram. This experiment uses the most time efficient q for each dataset, i.e. $q = 2$ for an English Dict, $q = 3$ for DBLP Author, and $q = 4$ for AOL Query Log.
- **Ed-Join-1 + OED Filter**: After Ed-Join-1, we run the OED filter.
- **Ed-Join-1 + Freq-BOED Filter (64-bit)**: After Ed-Join-1, we run the Freq BOED filter (64-bit).

Calculating filters for all candidates in similarity joins still requires huge time although our proposed filter can run fast. Hence, we use Ed-Join-1 algorithm as a preprocess in order to reduce candidates.

The results are shown in Figs. 6 and 7 and demonstrate that our proposed filters are effective for real-world data. These results show that proposed metric can be combined with other algorithms effectively.

---

[1] http://dbgroup.cs.tsinghua.edu.cn/wangjn/projects/triejoin/

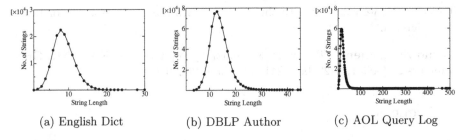

(a) English Dict   (b) DBLP Author   (c) AOL Query Log

Fig. 5. Distributions of string length for each dataset

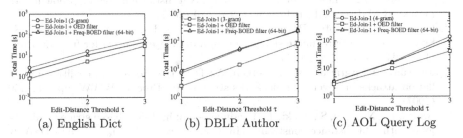

(a) English Dict   (b) DBLP Author   (c) AOL Query Log

Fig. 6. Running time required to solve *self join* for each algorithm

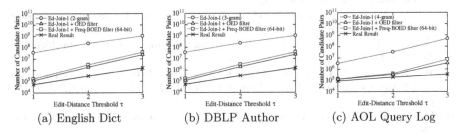

(a) English Dict   (b) DBLP Author   (c) AOL Query Log

Fig. 7. Number of candidate pairs verified by calculating edit distance

5 Conclusion and Future Work

We proposed new distance metrics which can be applied to filtering for the edit similarity join. These metrics can be calculated rapidly by bit operations. We proved several theorems on the standard edit distance and these metrics to perform bit-based filtering. Bit-based filtering can be combined easily with other algorithms. In our experiments, we combined our technique with Ed-Join-1 [11] and demonstrated the efficiency of our approach. We also showed that assigning bits depending on character frequencies improves performance.

Our future work includes dealing with longer strings. The efficiency of our approach degrades with increasing string length because OBV uses a limited number of bits. When a given string becomes too long, OBV is occupied almost entirely by 1-bits and loses its ability to keep track of the character occurrence.

It seems that we can deal with this problem with a q-gram approach or better bit assignment.

Acknowledgements. This work was partially supported by KAKENHI Grant Numbers 23300051, 23220001, 24106010, and 25240003.

References

1. Bayardo, R.J., Ma, Y., Srikant, R.: Scaling up all pairs similarity search. In: Proc. of WWW, pp. 131–140 (2007)
2. Cormode, G., Muthukrishnan, S.: The string edit distance matching problem with moves. ACM Trans. Algorithms 3(1), 2:1–2:19 (2007)
3. Gravano, L., Ipeirotis, P.G., Jagadish, H.V., Koudas, N., Muthukrishnan, S., Srivastava, D.: Approximate string joins in a database (almost) for free. In: Proc. of VLDB, pp. 491–500 (2001)
4. Metwally, A., Agrawal, D., El Abbadi, A.: Detectives: detecting coalition hit in-flation attacks in advertising networks streams. In: Proc. of WWW, pp. 241–250 (2007)
5. Narita, K., Nakadai, S., Araki, T.: Landmark-join: hash-join based string similarity joins with edit distance constraints. In: Cuzzocrea, A., Dayal, U. (eds.) DaWaK 2012. LNCS, vol. 7448, pp. 180–191. Springer, Heidelberg (2012)
6. Ohad, L., Ely, P.: Approximate pattern matching with the l_1, l_2 and l; metrics. Algorithmica 60(2), 335–348 (2011)
7. Sahami, M., Heilman, T.D.: A web-based kernel function for measuring the simi-larity of short text snippets. In: Proc. of WWW, pp. 377–386 (2006)
8. Wagner, R.A., Fischer, M.J.: The string-to-string correction problem. J. ACM 21(1), 168–173 (1974)
9. Wang, J., Feng, J., Li, G.: Trie-join: efficient trie-based string similarity joins with edit-distance constraints. Proceedings of the VLDB Endowment 3(1-2), 1219–1230 (2010)
10. Warren, H.S.: Hacker's Delight. Addison-Wesley Longman Publishing Co., Inc. (2002)
11. Xiao, C., Wang, W., Lin, X.: Ed-join: an efficient algorithm for similarity joins with edit distance constraints. Proceedings of the VLDB Endowment 1(1), 933–944 (2008)

Deterministic Verification of Integer Matrix Multiplication in Quadratic Time[*]

Ivan Korec[1] and Jiří Wiedermann[2]

[1] Mathematical Institute, Slovak Academy of Sciences
Štefánikova 49, 814 73 Bratislava, Slovakia
[2] Institute of Computer Science, Academy of Sciences of the Czech Republic
Pod Vodárenskou věží 2, 182 07 Prague 8, Czech Republic
jiri.wiedermann@cs.cas.cz

Abstract. Let **A**, **B** and **C** be $n \times n$ matrices of integer numbers. We show that there is a deterministic algorithm of quadratic time complexity (w.r.t. the number of arithmetical operations) verifying whether **AB** = **C**. For the integer matrices this result improves upon the best known result by Freivalds from 1977 that only holds for a randomized (Monte Carlo) algorithm. As a consequence, we design a quadratic time nondeterministic integer and rational matrix multiplication algorithm whose time complexity cannot be further improved. This indicates that any technique for proving a super-quadratic lower bound for deterministic matrix multiplication must exploit methods which would not work for the non-deterministic case.

1 Introduction

Matrix multiplication admittedly belongs among the most studied problems in computer science. There are at least two main reasons for this. First, it is the immense practical importance of the problem. Second, it is a deep algorithmic beauty of the problem. The run after efficient matrix multiplication algorithms began in 1968 when Volker Strassen discovered an algorithm for matrix multiplication of complexity $O(n^{2.807})$ [12] (in terms of the number of arithmetical operations). This has been a significant improvement over the classical algorithm of complexity $O(n^3)$. The search for more efficient algorithms continued steadily through a series of incremental improvements and the recent champion in matrix multiplication is the algorithm of Vasilevska Williams from 2011 achieving the performance of $O(n^{2.3727})$ arithmetical operations [13]. Unfortunately, in order to overcome the influence of the constant hidden in the Big O notation, in practice the asymptotically fast multiplication algorithms appear to be an improvement over the classical $O(n^3)$ solution only for very large matrices.

Suppose that we have implemented an asymptotically fast matrix multiplication algorithm and we want to verify the correctness of its implementation. The

[*] This research was partially supported by RVO 67985807 and the GA ČR grant No. P202/10/1333. The paper is based on the joint research of both authors which started shortly before the untimely death of Ivan Korec in 1998.

V. Geffert et al. (Eds.): SOFSEM 2014, LNCS 8327, pp. 375–382, 2014.
© Springer International Publishing Switzerland 2014

most direct approach would be to execute the fast algorithm on some concrete instances of matrices and to verify the correctness of the result. In our case, given three $n \times n$ matrices $\mathbf{A}$, $\mathbf{B}$, and $\mathbf{C}$ of real numbers, we want to verify whether $\mathbf{AB} = \mathbf{C}$. Obviously, it does not make sense to exploit the classical $O(n^3)$ algorithm for such a purpose since the verification would last longer than the original computation and for sufficiently large matrices might even be infeasible. Is there a faster way to do the verification?

In 1977 Rusins Freivalds designed a randomized algorithm with a bounded error probability for verifying matrix multiplication in quadratic randomized time [5], [6]. Since then his algorithm has become a standard textbook example illustrating the power of randomized computations over the deterministic ones (cf. [3],[9]). In order to verify whether $\mathbf{AB} = \mathbf{C}$ for any three matrices $\mathbf{A}$, $\mathbf{B}$ and $\mathbf{C}$ of real numbers of size $n \times n$, Freivalds' algorithm chooses a specific (column) vector $\mathbf{x}$ of length n and compares the product $\mathbf{ABx}$ with the product $\mathbf{Cx}$. Both products can be computed using $O(n^2)$ arithmetical operations (the former product thanks to the associativity of the matrix products: $(\mathbf{AB})\mathbf{x} = \mathbf{A}(\mathbf{Bx})$). The entries of vector $\mathbf{x}$ are uniformly chosen from the set $\{0, 1\}$. It can be shown that if $\mathbf{AB} \neq \mathbf{C}$, then the Freivalds algorithm returns a wrong answer with the probability at most $1/2$ (cf. [3]). The probability of error can be reduced to $1/2^k$ by performing k independent iterations of the algorithm.

In our paper we propose an alternative algorithm of quadratic time complexity for verification of matrix multiplication. Our approach differs from Freivalds' solution in several aspects.

First, unlike the Freivalds algorithm which makes use of randomization, for the case of integer matrices our algorithm is a truly deterministic algorithm. Thus, our verification algorithm always returns a correct answer — there is no margin for an error. Second, our algorithm is faster (albeit not asymptotically) since there is no need to iterate the verification process in order to decrease the error margin — a single run of the algorithm is enough to obtain a definitive and correct answer. Last but not least, the deterministic verification opens the possibility for designing a non-deterministic algorithm for matrix multiplication (that is, not merely for the verification of the matrix multiplication) still in quadratic time. This is the first known matrix multiplication algorithm whose performance is linear in the number of inputs. Of course, this algorithm is mainly of academic interest since the underlying computation exploits a non–deterministic machine model which is not realistic. Nevertheless, this result is of a methodological value because it shows that any technique for proving a super–quadratic lower bound for deterministic matrix multiplication must exploit methods which would not work for the non–deterministic case.

The structure of the paper is as follows. In Section 2 we introduce the main ideas leading first to the proposal of probabilistic and then also to the deterministic verification algorithm for matrix multiplication of quadratic time complexity. Based on the previous results in Section 3 we present our final result — a non-deterministic algorithm for matrix multiplication in quadratic time. Section 4 contains conclusions.

Before continuing we briefly describe the computational model used in the sequel on which our algorithms are implemented. It will be the Blum-Shub-Smale machine, or *BSS machine* [2]. This model has been designed in order to capture (idealized, i.e., exact) computations over the real numbers. Essentially, a BSS machine is a random access machine with registers that can store arbitrary real numbers, can compute rational functions over reals at unit cost and can compare the reals using "$\geq$" operator. A *non-deterministic variant of the BSS model* has the instruction for guessing and storing an arbitrary real number. Since the use of such an instruction produces an uncountable number of guesses there is an uncountable number of computational paths following each nondeterministic guessing instruction. The set of all guesses done along a particular computational path in a computation on a given input is called a *witness* for that input. A witness is used to pin down a computational path that satisfies the required input-output relation. A witness then enters into this relation as the third parameter. More formally, a computation of a nondeterministic BSS machine will be described by the so-called *input-output function* Φ from a subset of the input space and the witness space to the output space. An element of the input space x is given to the machine at the beginning of its computation, the witness w is guessed during the computation on input x, and the output $z = \Phi(x, w)$ is printed at the end of the computation. For more details, cf. [2].

2   A Simple Probabilistic Verification Algorithm

Our algorithm has a similar structure like the original Freivalds' algorithm. Thus, in order to verify, whether $\mathbf{AB} = \mathbf{C}$ for any three matrices $\mathbf{A}$, $\mathbf{B}$ and $\mathbf{C}$ of real numbers of size $n \times n$, we choose a specific (column) vector $\mathbf{x}$ of length n and compare the product $\mathbf{A}(\mathbf{Bx})$ with the product $\mathbf{Cx}$. Both products can be computed using $O(n^2)$ arithmetical operations. However, whereas Freivalds has made use of vector $\mathbf{x}$ with entries uniformly chosen from $\{0, 1\}$ (or $\{-1, 1\}$), we will make use of a vector $\mathbf{x}$ of form $\mathbf{x} = (1, r, r^2, \ldots, r^{n-1})^T$ (here operator T denotes the transposition of a row vector into a column vector), for any real number r appropriately chosen. The details are given in the sequel.

Lemma 1. *If $\mathbf{D} \neq \mathbf{0}$ is a real matrix of size $n \times n$ then there are at most $n-1$ real numbers r such that*

$$\mathbf{D} \begin{pmatrix} 1 \\ r \\ r^2 \\ \ldots \\ r^{n-1} \end{pmatrix} = \mathbf{0} \tag{$*$}$$

Proof: Since $\mathbf{D} \neq \mathbf{0}$ at least one row of $\mathbf{D}$ is non-zero. Corresponding to this row in the resulting matrix-vector product $(*)$ there is one algebraic equation in indeterminate r of degree less than n. This equation has at most $n-1$ real roots.

□

Based on the previous lemma we first design a probabilistic algorithm for verifying the product of two matrices of quadratic time complexity.

To verify $\mathbf{AB} = \mathbf{C}$, we pick a "random" real number r, create vector $\mathbf{x} = (1, r, r^2, \ldots, r^{n-1})^T$ and compute $\mathbf{Y} = \{y_{i,j}\} = \mathbf{A}(\mathbf{Bx}) - \mathbf{Cx}$.

If $\mathbf{Y} = \mathbf{0}$ then $\mathbf{AB} = \mathbf{C}$ with probability 1 (because there are at most $n-1$ "bad" numbers r causing $(\mathbf{AB} - \mathbf{C})\mathbf{x} = \mathbf{0}$ even if $\mathbf{AB} \neq \mathbf{C}$).

If $\mathbf{Y} \neq \mathbf{0}$ then $\mathbf{AB} \neq \mathbf{C}$ "for sure".

This leads to the following probabilistic algorithm for verification of matrix multiplication:

Probabilistic verification of matrix multiplication:

1. Input matrices $\mathbf{A}$, $\mathbf{B}$ and $\mathbf{C}$;
2. Pick randomly a real number r and compute vector $\mathbf{x} = (1, r, r^2, \ldots, r^{n-1})^T$;
3. If $\mathbf{A}(\mathbf{Bx}) = \mathbf{Cx}$ then output $Prob(\mathbf{AB} = \mathbf{C}) = 1$ else output $\mathbf{AB} \neq \mathbf{C}$;

The required computation can be done using $O(n^2)$ operations $+$, $-$, $\times$ and n comparisons (with zero). The latter number can be diminished to 1 by first computing the norm $\|\mathbf{Y}\| = y_1^2, \ldots, y_n^2$ of $\mathbf{Y}$ and comparing it to zero.

Now the following proposition is obvious:

Proposition 1. *Let $\mathbf{A}$, $\mathbf{B}$, $\mathbf{C}$, r, and $\mathbf{x}$ be as before, let $\mathbf{Y} = \mathbf{AB} - \mathbf{C} \neq \mathbf{0}$. Then the previous algorithm returns a wrong answer if and only if $\mathbf{Yx} = \mathbf{0}$, i.e., if and only if $P_i(r) = \sum_{j=1}^{n} y_{i,j} r^{j-1} = 0$ for $i = 1, 2, \ldots, n$.*

Thus, we get a wrong answer only in an extremely unlikely case that we randomly select r which turns out to be a root of all polynomials $P_i(r) = 0$ for $i = 1, 2, \ldots, n$.

Remark: Incidentally, the previous verification algorithm is formally similar to that proposed in [8]. However, the similarity is superficial. Namely, a single run of our algorithm with a randomly chosen real number leads to a correct verification with probability one. On the other hand, a single run of the algorithm from [8] delivers a correct answer with probability greater than $1/2$. Although the reliability of the latter algorithm can be increased iteratively it will never achieve a correct answer with probability 1. The correctness and complexity analysis of both algorithms are different.

3   A Deterministic Algorithm for Verification of Integer Matrices Product

In order to turn the previous probabilistic algorithm into a purely deterministic one we must show that without the necessity of computing $\mathbf{Y}$ we can deterministically find number r for which not all polynomials $P_i(r)$ defined in Proposition

1 are zeroed. Such r can be found for the case of integer matrices. Therefore, in the sequel we will only consider matrices with integer entries.

For determining r we can make use of any of the known theorems giving an upper bound on the magnitude of roots of a polynomial. Probably the simplest of such theorems is due to Cauchy [4] — the so–called *Cauchy's bound* (for a proof, cf. the textbook [7], p. 82) :

Theorem 1. *Let* $P(x) = a_k x^k + a_{k-1} x^{k-1} + \ldots + a_1 x + a_0$ *be a polynomial with real coefficients. If x is a root of $P(x) = 0$ then $|x| < 1 + A/|a_k|$, with $A = \max_{i=0}^{k-1}\{|a_i|\}$.*

It is seen that in order to upper-bound the magnitude of the roots we have to know coefficient a_k and the maximum of the absolute value of all coefficients in a polynomial. In our case, $k = n$ in the previous theorem and if $\mathbf{Y} = \{y_{i,j}\}$, then $y_{i,j} = \sum_{j=1}^{n} a_{i,j} b_{j,i} - c_{i,j}$ and $P_i(r) = \sum_{j=1}^{n} y_{i,j} r^{j-1} = 0$ for $i = 1, 2, \ldots, n$.

Let $c_{max} = \max\{|a_{i,j}|, |b_{i,j}|, |c_{i,j}|\}$. Then the maximal coefficient in any polynomial — the value of A — can be upper-bounded by $n c_{max}^2 + c_{max}$. Further, for any i, the absolute value of the leading coefficient in front of the highest power of r in $P_i(r)$ can be lower-bounded by 1 since we deal with integer matrices. (Note that it is here where we had to restrict ourselves to the integer matrices since for the real valued matrices a lower bound on $|a_k|$ cannot be found without computing $y_{i,j}$.)

From Cauchy's bound it follows then that for any polynomial $P_i(r)$ the absolute value of its roots are upper-bounded by $\alpha = n c_{max}^2 + c_{max} + 1$.

Thus, choosing any $r \geq \alpha$ in the previous proposition will guarantee that $P_i(r) \neq 0$ and hence $\mathbf{Y x} = \mathbf{0}$ can only hold for $\mathbf{Y} = \mathbf{0}$. The deterministic algorithm for matrix multiplication verification follows easily:

A deterministic verification of matrix multiplication for integer matrices:

1. Input matrices $\mathbf{A}$, $\mathbf{B}$ and $\mathbf{C}$, $\mathbf{C} \neq \mathbf{0}$;
2. Compute α and set $r := \alpha$;
3. Compute vector $\mathbf{x} = (1, r, r^2, \ldots, r^{n-1})^T$;
4. If $\mathbf{A}(\mathbf{B x}) = \mathbf{C x}$ then output YES else output NO;

From the definition of α it is seen that it can be computed in $O(n^2)$ operations since the entries of all matrices must be inspected. Vector $\mathbf{x}$ can be computed in time $O(n)$ and hence the entire deterministic verification algorithm is of quadratic time complexity.

The previous algorithm can be seen as a derandomized version of Freivald's algorithm for the verification of the product of integer matrices.

4   The Non-deterministic Algorithm for Integer Matrix Multiplication in Quadratic Time

The lastly considered deterministic algorithm for matrix multiplication verification can be turned into a non-deterministic algorithm for matrix multiplication working in quadratic time as follows. The idea is that the algorithm guesses matrix $\mathbf{C}$, deterministically computes number r (using Theorem 1) and verifies, whether $\mathbf{AB} = \mathbf{C}$ similarly as in Proposition 1. The guessing space can be bounded since the size of entries in (so-far unknown) matrix $\mathbf{C}$ depends on entries in matrices $\mathbf{A}$ and $\mathbf{B}$. If $e_{max} = \max\{|a_{i,j}|, |b_{i,j}|\}$ than the absolute value of any entry in $\mathbf{C}$ cannot be greater than $\beta = ne_{max}^2$. This means that it is enough to guess matrices $\mathbf{C}$ with entries of size, in absolute value, at most β.

The resulting algorithm to be implemented on a non-deterministic BSS machine works as follows:

A non-deterministic integer matrix multiplication algorithm:

1. Input matrices $\mathbf{A}$ and $\mathbf{B}$;
2. Guess the witness — matrix $\mathbf{C} \neq \mathbf{0}$ with the absolute values of entries bounded by β;
3. Compute $r := \beta + e_{max}$;
4. Compute vector $\mathbf{x} = (1, r, r^2, \ldots, r^{n-1})^T$;
5. Deterministically verify $\mathbf{A}(\mathbf{Bx}) = \mathbf{Cx}$;
6. Output $\mathbf{C}$.

It is clear that the time complexity of this algorithm is $O(n^2)$. Its correctness follows from the fact that after Step 2, there exist infinitely many computational paths (one for each guess of $\mathbf{C}$) in the computation, but with the help of witness $\mathbf{C}$ the verification in Step 5 selects exactly one of them satisfying $\mathbf{A}(\mathbf{Bx}) = \mathbf{Cx}$ which is equivalent to verifying $\mathbf{AB} = \mathbf{C}$ thanks to Theorem 1. Note that the seemingly similar idea of guessing $\mathbf{C}$ and its subsequent verification à la Freivalds cannot work since through the error margin of the respective algorithm, however small, matrices $\mathbf{C}$ can come through for which $\mathbf{AB} \neq \mathbf{C}$. Thus, getting rid of probabilistic features of a verification algorithm for matrix multiplication turns out to be a crucial ingredient for the success of our nondeterministic algorithm.

Note that the previous algorithm can be generalized to the case of matrices of rational numbers given as numerator and denominator pairs. The idea is to transform all entries in matrices to the common denominator and to extract it, as a scalar, in front of matrices. By this we get the integer matrices and can proceed accordingly. The transformation of entries to a common denominator can be straightforwardly done using $O(n^2)$ arithmetical operations. If d_{max} is the maximal absolute value of some denominator in $\mathbf{A}$ or $\mathbf{B}$ then the size of the common denominator would be of order d_{max}^n. This, however, does not play any role in our BSS model that computes with the unit cost measure independently of the size of arguments of arithmetical operations.

Theorem 2. *For a BSS computer there exists a non-deterministic integer or rational matrix multiplication algorithm of quadratic time complexity.*

Since any algorithm for multiplying two $n \times n$ matrices has to process all $2n^2$ entries, $\Omega(n^2)$ is an asymptotic lower bound of the number of operations needed for matrix multiplication. For bounded coefficient arithmetic circuits over the real or complex numbers Raz [11] proved a lower bound of order $\Omega(n^2 \log n)$. This bound does not apply to our case since we are using a different model than Raz — namely the BSS model where computations with arbitrary large reals and equality tests are allowed. The result from the last theorem indicates that any technique for proving a super-quadratic lower bound for deterministic matrix multiplication on a BSS machine must exploit methods which would not work for the non-deterministic case.

5   Conclusions

The presented results bring a further shift in our understanding of matrix multiplication algorithms.

The Freivald's idea of replacing direct checking of matrix–matrix product by checking the matrix product with a randomly chosen vector is a great one. During its more than thirty years long history the Freivalds' algorithm has gained a firm place among both practical and theoretically important algorithms. It has also served as a source of inspiration for using similar ideas in other contexts of algebra and in checking approximate computations over the reals (cf. [1]).

The present paper comes with yet an other variation of the initial Freivalds' idea. There have been efforts to diminish the amount of randomness in the algorithm (cf. [8],[10]). For the case of integer matrices we have succeeded in getting rid of randomized steps entirely in the algorithm for no extra cost (still assuming a unit cost of operation), but at the expense of computing with large numbers. On the BSS model this has no effect on complexity but on more realistic models of computations (like on uniform cost RAMs or Turing machines) the use of numbers whose size substantially exceeds the size of the matrix elements presents a bottleneck in practical considerations. On the other hand, the entirely deterministic algorithm for matrix multiplication verification has enabled the design of a nondeterministic algorithm for integer matrix multiplication in quadratic time. This seems to be the first known algorithm for such a task achieving this theoretically best possible performance. Whether there exists a nondeterministic algorithm for real matrices multiplication of quadratic time complexity on a BSS machine remains an open problem.

Although the results are not of immediate practical value they contain a clear methodological message. First, the results strengthen the hope that on the BSS machines also deterministic matrix multiplication of quadratic time complexity algorithms might exist (as many computer scientists in this field believe), and second, any effort for proving super–quadratic lower bounds for this task for the BSS machines must avoid considerations of non-deterministic algorithms.

It would be a great surprise if on a BSS machine matrices could be multiplied faster nondeterministically than deterministically.

Acknowledgment. The authors wish to thank to the anonymous reviewers for their comments on the previous draft of the paper.

References

1. Ar, S., Blum, M., Codenotti, B., Gemmel, P.: Checking approximate computations over the reals. In: Proc. 25th ACM STOC, pp. 786–795 (1993)
2. Blum, L., Cucker, F., Shub, M., Smale, S.: Complexity and Real Computation, 452 p. Springer (1998)
3. Brassard, G., Bratley, P.: Fundamentals of Algorithmics, 524 p. Prentice Hall, Englewood Cliffs (1996)
4. Cauchy, A.L.: Exercises de Mathematique, Oeuvres (2), vol. 9, 122 p. (1829)
5. Freivalds, R.: Probabilistic machines can use less running time. In: Proceedings of the IFIP Congress 1977 Information Processing, pp. 839–842 (1977)
6. Freivalds, R.: Fast Probabilistic Algorithms. In: Bečvář, J. (ed.) Proc. Mathematical Foundations of Computer Science MFCS 1979. LNCS, vol. 74, pp. 57–69. Springer (1979)
7. Householder, A.: The Numerical Treatment of a Single Nonlinear Equation. McGraw-Hill, New York (1970)
8. Kimbrel, T., Sinha, R.K.: A probabilistic algorithm for verifying matrix products using $O(n^2)$ time and $\log_2 n + O(1)$ random bits. Inf. Process. Lett. 45(2), 107–110 (1993)
9. Motwani, R., Raghavan, P.: Randomized Algorithms. Cambridge University Press, New York (1995)
10. Naor, J., Naor, M.: Small-bias probability spaces: efficient constructions and applications. SIAM J. Comput. 22(4), 838–856 (1993)
11. Raz, R.: On the complexity of matrix product. In: Proc. of the 34th Annual ACM Symposium on Theory of Computing. ACM Press (2002)
12. Strassen, V.: Gaussian elimination is not optimal. Numer. Math. 13, 354–356 (1969)
13. Vassilevska Williams, V.: Multiplying matrices faster than Coppersmith-Winograd. ACM STOC, 887–898 (2012)

Comparison of Genetic Algorithms for Trading Strategies

Petr Kroha[1] and Matthias Friedrich[2]

[1] Czech Technical University in Prague,
Faculty of Information Technology, Department of Software Engineering,
Thakurova 9, 160 00 Praha 6, Czech Republic
kroha@informatik.tu-chemnitz.de
[2] Chemnitz University of Technology, Strasse der Nationen 62, 09111 Chemnitz, Germany
matthias_friedrich@ymail.com

Abstract. In this contribution, we describe and compare two genetic systems which create trading strategies. The first system is based on the idea that the connection weight matrix of a neural network represents the genotype of an individual and can be changed by genetic algorithm. The second system uses genetic programming to derive trading strategies. As input data in our experiments, we used technical indicators of NASDAQ stocks. As output, the algorithms generate trading strategies, i.e. buy, hold, and sell signals. Our hypothesis that strategies obtained by genetic programming bring better results than buy-and-hold strategy has been proven as statistically significant. We discuss our results and compare them to our previous experiments with fuzzy technology, fractal approach, and with simple technical indicator strategy.

Keywords: Genetic algorithms, neurogenetic approach, neuroevolutionary system, genetic programming, neural network, investment, forecast, trading, financial modeling, technical analysis.

1 Introduction

Analyzing financial markets is a very interesting and popular field. Especially, forecasting is a hot topic. However, the question is how successfully and reliable a market behavior can be predicted. There is no consensus in the expert community because two main, contradictory, competing hypotheses on market processes have been formulated.

Efficient market hypothesis [6], [11] states that markets are efficient in the sense that current stock prices reflect completely all currently known information that could anticipate future market, i.e. there is no information hidden that could be used to predict future market development.

Later, inefficient market hypothesis [15] was formulated because some anomalies in market development have been found that cannot be explained as being caused by efficient markets. More or less, market trading need buyers and sellers at the same time. So, a consensus would stop trading.

Compared to systems in physics, reflexivity of markets and investors is a very important factor. It states that investors influence the market by changing their biases, mind,

V. Geffert et al. (Eds.): SOFSEM 2014, LNCS 8327, pp. 383–394, 2014.

interest, and trading rules. Similarly, market changes influence behavior of investors. It will be investigated by crowd psychology.

Market processes are driven by events and by trends. Events happen and are represented by news. Predictable events have usually no influence on stock prices because investors presume them and prices adapt to them before. Unpredictable events cause big changes in stock prices but they are unpredictable like earthquake. Trends are given by investor's behavior. The chance to predict trends seems to be slightly better than to predict earthquake. So, the effort to forecast markets is more or less the effort to predict investors' behavior. It can bring good results in time periods when trends are dominating and only expected events happen. We did not investigate possibilities of short term prediction, e.g. for day trading, because the influence of noise is stronger than in the case of long term prediction. Because of that we compare the performance of our prototypes with buy-and-hold method that will be used as a standard in such investigations.

Genetic algorithms can be used in many ways to optimize systems. The important parts of the algorithms are: how to specify what will be coded as genotype, and how the fitness will be calculated.

The first possibility of using genetic algorithms is that neural network (Fig. 1) connection weights can be optimized by a genetic algorithm instead of the commonly used back propagation method. Elements of the weight matrix describing the neural network topology are coded as real values and mapped into genotype. Such a system will be called neurogenetic or neuroevolutionary system. We implemented such system as our prototype A - Section 3.1.

The second possibility how to use genetic algorithms in financial application is that system parameters (e.g. open and close stock value, technical indicators) are coded as tree leaves (operands), and operators working with them (e.g. or, and, if, less-or-equal) are coded as tree nodes (Fig. 2). This method will be called genetic programming (GP). We start using a random placement of operators and operands into a tree that will represent the genotype (in our prototype B). As published in [1], we used the swapping of subtrees from both parents for the crossover. The following mutation selects a subtree of a parent and replaces it by a randomly created tree - Section 3.2.

All genetic algorithms follow the same procedure. After a initial population of genotypes is constructed, the fitness of each individual is calculated. For the recombination operation, two parents are selected according to their fitness (in our prototypes with the linear ranking method), and their genotypes are crossed. The new created individual is subsequently modified by the mutation operation. This process will continue until the new population is created, which is consistently followed by applying the fitness calculation and the genetic operations.

In this paper, we describe the trading system A based on neurogenetic approach and the genetic programming trading system B. Our original contribution is that we compounded methods of [2], [9], and [14] for the neurogenetic approach in our prototype A, and also improved the genetic programming method (tree swapping) presented in [1] in our prototype B. We added technical indicators as tree nodes and we modified the probability of the creation of tree nodes. We specified our own fitness functions for both prototypes. Both methods are described in detail in Section 3.

Additionally, we successfully tested statistical significance of the hypothesis that the genetic programming prototype B brings better results than buy-and-hold strategy used in comparisons as a standard. Further, we compared all achieved results, and results obtained using fuzzy and fractal technology that we used in our previous works [8].

The rest of the paper is organized as follows. In Section 2, we discuss related work. In Section 3, we introduce the developed prototypes. Fitness function is described in Section 4. Then, we present data used in Section 5. The following section 6 describes the implementation, experiments, and results. Statistical significance proving is given in Section 7. The comparison of genetic with fuzzy and fractal methods is mentioned in Section 8. In the last section, we conclude our work.

2   Related Work

There are many interesting works investigating similar problems as our research.

In [9], a neural network having one hidden-layer has been used. It was trained using back propagation to find a local optimum. Compared to this work, we did not use back propagation in the prototype A.

In [2], a multi-layer network was used. Genetic algorithm mutation changes weights but also the topology of a cycle-free neural network. It was used for day traders, and after a planned day profit was reached, the system generated a sell signal. Compared to this work, we did not optimize the topology, but we used recombination of individuals and applied the moving window technique in the prototype A. Our output was not day trader oriented.

The authors of [14] suppose that there is no optimum of buy and sell signals, and because of that supervised learning (e.g. back propagation) cannot be used. To optimize weights, only genetic algorithm was applied using the moving window technique.

In [5], back propagation is replaced by simulated annealing, the input vector represents 5 technical indicators. The output is a stock value predicted for the next day, i.e. buy-, hold-, and sell signals are not generated.

Fuzzy technology instead of genetic algorithm to optimize network topology is used in [10].

In [7], authors focus on optimization of technical indicator parameters but differently to our approach they do not use tree swapping.

A method to evaluate individuals which were proposed to be applied in automated trading is described in [12]. Instead, we used our own fitness function described below.

It is very difficult to compare the methods mentioned above because of the large variety of approaches, parameters, and used data. So, we compared tests of our prototypes running on the same data.

3   Our Prototypes

3.1   Our Neurogenetic Prototype A

Our goal was to investigate how a trading strategy in terms of buy, sell, and hold signals can be represented and generated from output values of a neurogenetic system. Furthermore we wanted to prospect whether using a neurogenetic system can bring better

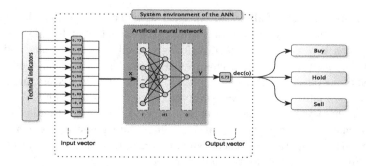

Fig. 1. Instance of a neural network as an individuum of genetic algorithm

results than genetic programming, fuzzy and fractal technology that we investigated in our previous work [8].

For the basis of our prototype A we used ideas from [2], [9], and [14] in a specific composition with some additional improvements.

The important aspect of genetic optimization is how to map the problem into the genotype. In this case, the input vector is represented by system parameters. They are combined using an activation function and weight parameters which are stored in a matrix correspondingly to the network topology. The output vector is finally decoded into buy, hold and sell signals. To recombine genotype of two parents (i.e. their matrices) the 2D-recombination method will be used. For the mutation, we developed the noise-layer-mutation, which added small randomly created values to each element of the matrix of a specific layer using the gaussian distribution $\mathcal{N}(0, 1)$.

The weight optimization was implemented using a genetic algorithm in periods with moving window technique. We abstained from the back propagation algorithm because we agree with the assumption in [14] saying that buy and sell decisions for middle and long term trading strategy cannot be predicted using supervised learning algorithms.

3.2   Our Prototype B Based on Genetic Programming

Our genetic programming prototype B is an improved algorithm based on the method published in [1]. As described in Section 1, the trading rule is represented by a tree using system parameters as leaves and operators as tree nodes Fig. 2.

Our main improvement is that we used different selection probabilities for different kinds of nodes. This means, when creating a random tree (e.g. for initial population or mutation), the selection probability is not uniform distributed over nodes in one category, as given in Table 1. The effect is that system values like technical indicators have a higher probability of occurrence within the tree, and therefore they influence the trading strategy more than fixed parameters. Furthermore, it reduced the number of combinations that have semantically only very limited or improbable occurrence but cause a tree explosion. The probabilities we used are our estimations based on our experience with description of many strategies.

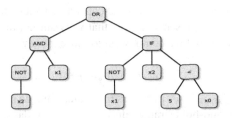

Fig. 2. An instance of a tree in genetic programming

Table 1. Probabilities of rule application

	Parameter	Value	P
Boolean	Basic functions	if-then-else, and, or, not	12.5 % each
	Enhanced functions	<, >	25.0 % each
Numeric	Basic functions	+, -, *, :, norm	2.0 % each
	Close value	price	30 %
	Enhanced function	number, average, maximum, minimum, lag	4.0 % each
	Indicators	ROC, MACD, SO, TCI	10 % each

4 Fitness Function

Fitness function evaluates behavior of phenotype, i.e. behavior of the individual that was generated from genotype. Next generation individuals will be constructed from two individuals that have usually a high fitness function value, since all common selection methods (i.e. linear ranking) are fitness-oriented.

In our applications, individuals represent trading strategies. An important factor of the fitness function is the money earned by each strategy m_{strat}. An individual obtains a start money amount m_{init} and uses it following its strategy s_t during n days. The money m_{strat} of an individual, i.e. of each strategy, achieved in a period is calculated using close stock value x_t and the market position pos_t of day t corresponding to the trading strategy:

$$m_{strat} = m_{init} \cdot \prod_{t=i-n+1}^{i} \left(\frac{x_t}{x_{t-1}}\right)^{pos_t}, \quad pos_t = \begin{cases} 1 & \text{, if } s_{t-1} \Leftrightarrow \text{buy} \\ 0 & \text{, if } s_{t-1} \Leftrightarrow \text{sell} \\ pos_{t-1} & \text{, if } s_{t-1} \Leftrightarrow \text{hold} \end{cases} \quad (1)$$

The calculation of the fitness function in our neurogenetic system contained several components as described below:

– profit of each strategy f_{strat} is calculated in relation to the profit of the buy-and-hold strategy (m_{bah} denotes money obtained by the buy-and-hold strategy)

$$f_{strat} = m_{strat} - m_{bah} = m_{strat} - \left(m_{init} \cdot \frac{x_i}{x_{i-n+1}}\right) \quad (2)$$

– absolute value of profit - $f_{strat-abs}$ - indicates the profit or loss of money caused by the strategy, i.e. it represents the rule that we want not only to be better as the buy-and-hold strategy, but we do not want to loose money

$$f_{strat-abs} = m_{strat} - m_{init} \tag{3}$$

– penalty function - $f_{strat-pen}$ - penalizes individuals that do not change position and simply follow buy-and-hold strategy during $\tilde{n}$ coherent days, using an adjustment factor α

$$f_{strat-pen} = \alpha \cdot m_{init} \cdot \left(\frac{\tilde{n}}{n}\right)^3 \tag{4}$$

– absolute relation between profit and loss - $f_{strat-pl}$ - respects the risk β resulted from the using of the trading strategy

$$f_{strat-pl} = \beta \cdot \bar{p} + (1 - \beta) \cdot \bar{l} \tag{5}$$

The average profit $\bar{p}$ is given by positive return during a period of days running consecutively

$$\bar{p} = \frac{1}{\sum\limits_{t=i-n+1}^{i} p_t} \cdot \sum_{t=i-n+1}^{i} \left(\frac{x_t}{x_{t-1}} \cdot m_{t-1}\right) \cdot p_t, \quad p_t = \begin{cases} 1 & \text{, if } x_t > x_{t-1} \\ 0 & \text{, otherwise} \end{cases} \tag{6}$$

The average loss $\bar{l}$ is calculated similarly:

$$\bar{l} = \frac{1}{\sum\limits_{t=i-n+1}^{i} l_t} \cdot \sum_{t=i-n+1}^{i} \left(\frac{x_t}{x_{t-1}} \cdot m_{t-1}\right) \cdot l_t, \quad l_t = \begin{cases} 1 & \text{, if } x_t < x_{t-1} \\ 0 & \text{, otherwise} \end{cases} \tag{7}$$

The components described above are combined in the fitness value $\mathcal{F}$:

$$\mathcal{F}_{strat} = (1 - \gamma) \cdot f_{strat-bah} + \gamma \cdot f_{strat-abs} + f_{strat-pen} + f_{strat-pl} \tag{8}$$

The parameter γ represents the weight of f_{bah} and f_{abs}. In our preliminary study, we found that $\gamma = 0.6$ was the most suitable value. Other authors construct different fitness functions.

5 Data Used

In the first part of our experiments called preliminary study, we tried to specify suitable parameters and their values that could be used as fixed during the optimization problem. Since we used connection weights to be optimized in our neurogenetic prototype A, there are many methods which could be applied for the genetic algorithm. Because of the complexity, it is practically impossible to optimize all the parameters.

During the preliminary study we evaluated different parameter combinations for the neurogenetic system, i.e. we altered the topology, recombination and mutation methods. Starting with the configuration given in Table 3, we picked the parameters in succession and evaluated different values and methods. The description of all experiments and evaluations of our preliminary study is out of scope of this paper. Of course, we know that the parameter combination we fixed does not guarantee the global optimum but computational complexity of other approach were immense. The final configuration used for the test is given in Table 4.

Table 2. Input vector variables of our neural network

Position	Variable	Position	Variable	Position	Variable
1	$close_t$	14	$MACD_t(20,40)$	27	$SO_t(40)$
2	$SMA_t(5)$	15	$ROC_t(20)$	28	$TCI_t(40,80)$
3	$EMA_t(5)$	16	$SO_t(20)$	29	$BB_t(40,2.0)$
4	$MACD_t(10,20)$	17	$TCI_t(20,40)$	30	$RAVI_t(10,100)$
5	$ROC_t(10)$	18	$BB_t(20,2.0)$	31	$RSI_t(14)$
6	$SO_t(10)$	19	$RAVI_t(6,60)$	32	$SMA_t(50)$
7	$TCI_t(10,20)$	20	$RSI_t(9)$	33	$EMA_t(50)$
8	$BB_t(10,2.0)$	21	$high_t$	34	$SMA_t(100)$
9	$RAVI_t(3,30)$	22	low_t	35	$EMA_t(100)$
10	$RSI_t(3)$	23	$SMA_t(20)$	36	$SMA_t(200)$
11	$open_t$	24	$EMA_t(20)$	37	$EMA_t(200)$
12	$SMA_t(10)$	25	$MACD_t(40,80)$		
13	$EMA_t(10)$	26	$ROC_t(40)$		

As 37 input values, we used technical indicators (SMA means Simple Moving Average, EMA means Exponential Moving Average etc. - all acronyms are given in [13]) and stock values (open, high, low, close) as given in Table 2.

Both prototypes have been tested at stocks of companies listed in NASDAQ-100 in June 2009. As time period for the evaluation we considered the 01.01.2003 start and the 01.10.2009 as end point. A test case is represented by the values of a time series of a stock in one year, which we used as test period. The training and selection period contained the values of 12 respectively 6 months directly before start of the test period. Since some companies do not have stock values in the whole periods, there were 621 test cases.

6   Experiments and Results

To run our experiments with prototype A, we used an Apple Mac Pro 4,1 with two Intel Xeon E5520 processors, 4 cores / 8 threads each, clock rate 2,26 GHz. The experiments of our prototype B have been executed on a computer with processor Intel Core 2 Duo E6300, clock rate 1,86 GHz.

Table 3. The configuration of the neurogenetic system in the preliminary study

Category	Parameter	Value
Topology	Input-layer	37
	Hidden-layer 1	20
	Hidden-layer 2	8
	Output-layer	1
Activation function	Function type	Logistic Function
	Parameter α	2
	Parameter β	3
	Parameter γ	3
Threshold parameters	Activation potential θ	1.0
	Buy signal t_{buy}	0.66
	Sell signal t_{sell}	−0.66
Population	Initialization	Gaussian distribution
	Number of generations	100
Selection	Selection method	Linear ranking
	Elitism	0
	Minimum	0.5
	Maximum	1.5
Crossover	Recombination method	Layer-crossover
	Probability p_c	0.8
Mutation	Mutation method	Noise-layer
	Probability p_m	0.2
	Distance	0.1
Capital	Seed money	10000 $
	Transaction costs	Relative: 0.25 %
Fitness measurement	Computation strategy of fitness	See section 4
Moving windows	Number of windows	6
	Number of overlapping days	30
Population size	Training period	1000
	Selection period	50
	Test period	1

Experiments were very time consuming. Each algorithm execution took about 27 minutes and 18 seconds, i.e. the tests would need altogether approximately 282 hours and 38 minutes. Since we could parallelize our prototype A, we only needed 156 hours 49 minutes for the evaluation.

Both methods achieved better results than the strategy buy-and-hold. In the case of genetic programming, the prototype B makes 2.72 % a higher profit towards the buy-and-hold strategy.

Compared to buy-and-hold strategy, the strategies of the neurogenetic prototype A earned in average 91.82 % more during the training period, but only 0.65 % more during the test period. It achieved in 99.19 % better results as buy-and-hold strategy during the training period, but only 44.28 % during the test period. It seems to be overfitted.

The strategies generated by the genetic programming system (prototype B) earned in average only 55.20 % more than buy-and-hold strategy in training period, but 2.72 %

Table 4. The configuration of the final tests

Parameter	Neurogenetic system	Genetic programming
Attributes of Individuals	Input-layer: 37 Hidden-layer: 19 Output-layer: 1 Tangens hyperbolicus $t_{buy} = 0.5$; $t_{sell} = -0.33$; $\theta = 0.0$	Non-uniform distribution, see Table 1 Maximum tree depth: 7
Population	Initialization: Gaussian distribution Generations: 150	Generations: 75
Selection	Linear ranking Minimum: 0.5 Maximum: 1.5 Individuals for elitism: 0	
Crossover	2D-recombination $p_c = 0.9$	Tree swapping $p_c = 0.7$
Mutation	Noise-layer $p_m = 0.7$ Distance: 0.1 (Gaussian distribution)	Tree switching $p_m = 0.7$
Capital	Seed money: 10.000 \$ Transaction costs: 10 \$ for each transaction	
Fitness measurement	See section 4	
Moving windows	Number of windows: 2 Number of overlapping days: 75	–
Population size	Training period: 100 Selection period: 50 Test period: 1	Training period: 50 Selection period: 20 Test period: 1

more during the test period. During the training period, it was better in 90.82 % of cases; during the test period, it was better in 44.61 % of cases.

If we consider only stocks that provide a better performance using the generated strategies compared to buy-and-hold strategy then strategies generated by prototype A (neurogenetic) give 26.98 % and strategies generated by prototype B (genetic programming) give 28.73 % more earning during the test period. The results are summarized in Table 5.

7 Statistical Hypothesis Testing

Because of the results described above, we used statistical hypothesis testing (Z-test because the sample size is large and the population variance known) and proved both prototypes using the following hypotheses with the parameters $\alpha = 5$ % and $\mu_0 = 1.0$ which leads to the parameter $\Phi(z_{1-\alpha}) = 1.6449$.

H_0 = the expected earning of the generated strategy is equal or less compared to the earning of the buy-and-hold strategy

H_1 = the expected earning of the generated strategy is greater compared to the earning of the buy-and-hold strategy

Table 5. Results of neurogenetic and genetic programming system compared to buy-and-hold strategy

	Neurogenetic system	Genetic programming
Average (Training)	1.9182	1.5520
Average (Test)	1.0065	1.0272
Number of better cases (Training)	616	564
Number of worse cases (Training)	5	57
Number of better cases (Test)	275	277
Number of worse cases (Test)	346	344
Average of better cases(Training)	1.9392	1.6160
Average of better cases (Test)	1.2698	1.2873

For the neurogenetic system we obtained $z = 0.4728 < \Phi(z_{1-\alpha})$. It means that the hypothesis H_0 can neither be rejected nor accepted.

In contrast, we obtained $z = 1.9910 > \Phi(z_{1-\alpha})$ for the genetic programming. This means that the hypothesis H_0 can be rejected. Therefore we can state that the strategies generated by genetic programming method earns more than the strategy buy-and-hold.

8   Comparison to Fuzzy and Fractal Technology

In our previous work [8], we investigated how fuzzy technology, fractal technology, and using of technical indicators can be used to generate trading strategies. Now, we used the same time series to prove the generated strategies as in [8], which allows the comparison of all five methods. The results obtained are shown sorted in Table 6. We can see that the system based on genetic programming implemented in prototype B brings the best results.

Table 6. Comparison of neurogenetic and genetic programming systems to results of fuzzy technology, fractal technology, and simple application of technical indicators

	Average	Standard deviation
Genetic programming	1.1149	0.8872
Neurogenetic system	0.9851	0.6954
Fractal analysis	0.9664	0.7247
Fuzzy control	0.8392	0.3566
Technical indicators	0.6473	0.3118

The small difference between Table 5 and Table 6 occurred because we investigated only 82 stocks of NASDAQ-100 in our previous work but all 100 stocks in this work.

There is a question concerning the size of data necessary for the training and selection process. To use our results in practice with several stocks, the time consumed by the execution must be reduced. A modest trader in Middle Europe respects the stock exchange in New York closing at 22:30 (CET) and Frankfurt stock exchange opening at

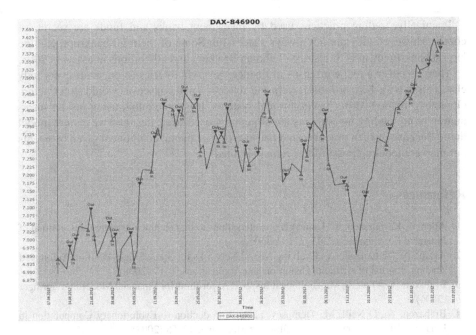

Fig. 3. Buy and Sell Signals for the DAX generated by genetic programming, separated in training, selection and test periods

9:00 (CET) the next day. There are 10.5 hours available to compute the most promising trading strategy for the next day influenced by the last data.

In subsequent experiments, we investigated a time series of one stock only and reduced the training, selection, and test to 1 month each and used 150 generations. The process running in 2 threads took 2 minutes only and gave signals shown in Fig. 3.

9 Conclusion

We implemented, improved, and tested two prototypes of methods based on genetic algorithms that generate trading strategies. In difference to other works, we took transaction costs into account (Table 4).

We expected that the neurogenetic prototype A based on genetic optimization of neural network weights brings the best results. However, we found that the method of genetic programming generates a better trading strategy, and that it brings more profit than the buy-and-hold strategy which is usually used for comparison. We proved statistical significance of this result. As we mentioned above, the optimization process is very time consuming when large data is used. We recognized that prior behavior of markets has a limited information content relative to current trading behavior. The difference between performance of trading decisions during a training period and during a test period is very big (Table 5).

In common, complex system like markets are influenced by very many parameters which can be further investigated.

Practically, trading practices used by investment banks and funds are secret, of course. However, they are not always successful. Some of them go bankrupt, e.g. the largest bankruptcy in U.S. history - Lehman Brothers loss 600 bilions in assets in 2008.

There is never a real consensus in finance - everybody tries to outsmart everybody else. This is why there are still buyers and sellers - a real consensus would stop trading. Obviously, markets differ from physical systems. It is known that using insider information or pretending investments are commonly used strategies, and we cannot model them. Because of such practices and because of the chaotic component given by events our market modeling possibilities remain limited.

References

1. Allen, F., Karjalainen, R.: Using genetic algorithms to find technical trading rules. Journal of Financial Economics 51, 245–271 (1999)
2. Azzini, A., Tettamanzi, A.: Evolving Neural Networks for Static Single-Position Automated Trading. Journal of Artificial Evolution and Applications, 1–17 (2008)
3. Brabazon, A., O'Neill, M.: Biological Inspired Algorithms for Financial Modelling. Springer (2006)
4. Brabazon, A., O'Neill, M., Dempsey, I.: An Introduction to Evolutionary Computation in Finance. IEEE Computational Intelligence Magazine, 42–55 (2008)
5. El-Henawy, I.M., Kamal, A.H., Abdelbary, H.A., Abas, A.R.: Predicting Stock Index Using Neural Network Combined with Evolutionary Computation Methods. In: The 7th International Conference on Informatics and Systems (INFOS), pp. 1–6 (2010)
6. Fama, E.: Efficient capital markets: A review of theory and empirical work. Journal of Finance 25, 383–417 (1970)
7. Kapoor, V., Dey, S., Khurana, A.P.: Genetic Algorithm: An Application to Technical Trading System Design. International Journal of Computer Applications 36(5) (2011)
8. Kroha, P., Lauschke, M.: Using Fuzzy and Fractal Methods for Analyzing Market Time Series. In: Proceedings of the International Conference on Fuzzy Computation and International Conference on Neural Computation ICFC 2010 and ICNC 2010, pp. 85–92 (2010)
9. Kwon, Y.-K., Moon, B.-R.: A Hybrid Neurogenetic Approach for Stock Forecasting. IEEE Transactions on Neural Networks 18, 851–864 (2007)
10. Li, R., Xiong, Z.: A Modified Genetic Fuzzy Neural Network with Application to Financial Distress Analysis. In: International Conference on Computational Intelligence for Modeling, Control and Automation and International Conference on Intelligent Agents, Web Technologies and Internet Commerce (2006)
11. Malkiel, B.: A Random Walk Down Wall Street. W.W. Norton, New York (1996)
12. Matsui, K., Sato, H.: Neighborhood Evaluation in Acquiring Stock Trading Strategy Using Genetic Algorithms. International Journal of Computer Information Systems and Industrial Management Applications 4, 366–373 (2012)
13. Murphy, J.J.: Technical Analysis of the Financial Markets. Prentice Hall (1999)
14. Skabar, A., Cloete, I.: Neural networks, Financial Trading and the Efficient Markets Hypothesis. In: Proceedings of the Twenty-Fifth Australasian Conference on Computer Science ACSC 2002, vol. 4, pp. 241–249 (2002)
15. Shleifer, A.: Inefficient Markets – An Introduction to Behavioral Finance. Oxford University Press (2000)

Probabilistic Admissible Encoding on Elliptic Curves - Towards PACE with Generalized Integrated Mapping*

Łukasz Krzywiecki, Przemysław Kubiak**, and Mirosław Kutyłowski

Wrocław University of Technology, Institute of Mathematics and Computer Science

Abstract. We consider *admissible encodings on an elliptic curve*, that is, the hash functions that map bitstrings to points of the curve. We extend the framework of admissible encodings, known from CRYPTO 2010 paper, to some class of non-deterministic mapping algorithms. Using Siguna Müller's probabilistic square root algorithm we show a mapping that works efficiently for *any* finite field $\mathbb{F}_q$ of characteristic greater than 3, and that is *immune to timing attacks*. Thereby we remove limitations of the mappings analyzed in the CRYPTO 2010 paper. Consequently, we remove limitations of a so called PACE Integrated Mapping protocol, which has recently been standardized by ICAO, and is used to protect contactless identity documents against unauthorized access.

Keywords: indifferentiability, admissible encoding, non-deterministic square root algorithm, finite field, elliptic curve.

1 Introduction

Many cryptographic protocols use efficient algebraic structures, such as elliptic curves, but at the same time operate on ordinary objects such as binary strings. In such a case we frequently need a *mapping* that converts the binary strings to points of the algebraic structure. This problem has profound practical implications, since an inefficient mapping may outweigh computational advantages of using the target algebraic structure.

The above issue is very important for smart card protocols, and in particular for electronic identity documents (e-ID), which are issued on a large scale. In fact, the problem concerned in this paper emerged while constructing authentication protocols for e-ID. Note that efficiency and security issues are critical for the issuer of e-IDs – less efficient protocols require more expensive smart cards on which the e-ID could be implemented, while a security flaw might cause exchange of identity documents with enormous organizational costs.

For the rest of the paper we focus on one fundamental issue for contactless e-IDs, namely preventing activation of an e-ID via the contactless interface without consent of the document owner. This problem is solved by cryptographic protocols based on a

---

* The paper has been supported by the Polish Ministry of Science and Higher Education: during the initial stage (i.e., numerical experiments) by project O R00 0015 07, later by project N N206 369 839; the third author is supported by the Foundation for Polish Science, "Mistrz" Programme.
** Corresponding author: `przemyslaw.kubiak@pwr.wroc.pl`

V. Geffert et al. (Eds.): SOFSEM 2014, LNCS 8327, pp. 395–406, 2014.

secret (password) shared by the document owner and the e-ID card. There are many password-based authentication protocols, however we focus on a standardized solution called PACE (Password Authenticated Connection Establishment) introduced by the German federal authority BSI ([7]). The aim of PACE is to establish a secure channel between a contactless smart card and a reader based on a password memorized by the owner or engraved on a surface of the card. Due to PACE, an active but non-invasive adversary can only guess the password and try it in an interaction with the card. A PACE execution for a password π consists of three stages:

1. So called domain parameters and the result of encryption of a nonce s are sent from the smart card to the reader. The domain parameters consist of a definition of an elliptic curve over a finite field $\mathbb{F}_q$, point G of prime order belonging to the curve, and the co-factor ℓ, i.e., the integer such that $(\text{ord}G) \cdot \ell$ equals the number of points on the curve. The encryption of the nonce is performed with a key derived from the password π stored inside the card. On the other hand, the reader learns the password π via another channel (e.g., the owner of the card enters it manually) and decrypts the message s.

2. Secondly, with help of the nonce s (and some other data exchanged between the reader and the card) a point $\hat{G}$ is calculated locally both on the side of the reader and on the side of the card.

3. $\hat{G}$ is used as a base point in the Elliptic Curve Diffie-Hellman key agreement protocol (ECDH). The resulting key is used to generate session keys needed to establish a secure channel between the reader and the card.

The second step is called *mapping*, since it maps the nonce s to the elliptic curve indicated in the domain parameters of the smart card. The mapping implemented on German e-ID documents is called the Generic Mapping. The Generic Mapping includes a separate execution of ECDH, therefore two ECDH executions are included in this variant of PACE.

In [11], [9] another mapping for PACE is proposed – the so called Integrated Mapping is more efficient than the Generic Mapping both in terms of communication and computational costs. The resulting variant of PACE is denoted as PACE IM.

The main building block of the Integrated Mapping is an algorithm that encodes a bit string as a point of the curve. Two encoding algorithms were proposed: the simplified Shallue-Woestijne-Ulas (SWU) algorithm or Icart's encoding (see [5]). Both encoding algorithms work in time $O((\log_2 q)^3)$, where $\mathbb{F}_q$ is the field over which the elliptic curve has been defined.[1]

The SWU simplified algorithm works for any field $\mathbb{F}_q$ with $\text{char}(\mathbb{F}_q) > 3$, however, it is used by PACE IM only for $q \equiv 3 \mod 4$. The condition $q \equiv 3 \mod 4$ arises from focusing on deterministic square root algorithms in finite fields. Icart's encoding method is also deterministic and works for any field $\mathbb{F}_q$, where $\text{char}(\mathbb{F}_q) > 3$ and $q \equiv 2 \mod 3$.

The choice of deterministic encoding algorithms is a consequence of utilizing the notion of *indifferentiability* in the proof of quality of the encoding (cf. [6] and the full version [5]). Proving the security of hash based cryptosystems in the Random Oracle

---

[1] PACE uses only prime fields $\mathbb{F}_q$. but encoding algorithms work also for extension fields.

Model (ROM) requires that a hash function is modeled as an ideal functionality accessible by all parties. However real designs impose that hashes do not have a *monolithic* construction, but are rather iteratively built from some smaller blocks. Indifferentiability property for deterministic algorithms guarantees that such a compound construction can replace the monolithic ROM model in cryptosystems with security scenarios related to a single, stateful adversary (for multi-stage protocols and resource-restriction models see [15], [10]). However, indifferentiability assumes that the algorithm implementing function f is deterministic. Therefore the security of PACE IM was analyzed only for deterministic encoding functions.

Our Contribution. We extend the framework of indifferentiable hashing given in [6], [5] to some class of non-deterministic algorithms. The extended framework justifies the use of certain probabilistic algorithms in the simplified SWU method. The resulting encoding protocol works for any field $\mathbb{F}_q$ with $\mathrm{char}(\mathbb{F}_q) > 3$, has expected running time $O((\log_2 q)^3)$ (like the deterministic simplified SWU method and Icart's encoding) and enjoys small variance. In particular, our generalization allows to use the NIST P-224 curve in the probabilistic variant of PACE IM (the standardization report [11] excludes this curve).

Note that removing constraints imposed by the mapping algorithm on the fields $\mathbb{F}_q$ of characteristic greater than 3 gives more flexibility in other applications (an example could be the signature scheme [4]) and leaves more room for future designs. Moreover, having replaced specialized Integrated Mappings with a single general standard we could reduce deployment costs.

Timing attacks. Apart from the high quality randomness ensured by the mapping calculated in the second step of PACE, the protocol must protect confidentiality of the password π. In particular, execution time of the protocol must not depend on the *value* of π. This condition is satisfied by both versions of PACE and our extension meets this condition as well.

Notation. The following notation is used in the rest of the paper.

Definition 1 (parity operator $\mathrm{parity}_\mathcal{V}() : \mathbb{F}_q \to \{0, 1\}$**).** *Let $\mathbb{F}_q$ be a finite field of characteristic $p > 2$. Let $\mathcal{V} = \{v_0, v_1, \ldots, v_{d-1}\}$ be a base of the vector space $\mathbb{F}_q$ defined over the field $\mathbb{F}_p$. Accordingly, d is the extension degree $[\mathbb{F}_q : \mathbb{F}_p]$. Let $a \in \mathbb{F}_q$, and let $a = \sum_{i=0}^{d-1} a_i v_i$ be the representation of a in base $\mathcal{V}$ with $a_i \in \mathbb{F}_p$ for $i = 0, \ldots, d-1$. Define $\mathrm{parity}_\mathcal{V}(a)$ as the parity of a_0.*

Definition 2 (mapping $\tilde{m}_\mathcal{V} : \mathbb{Z}_q \to \mathbb{F}_q$**).** *Let $\mathbb{F}_q$ be a finite field of characteristic $p > 2$, let $\mathcal{V}$ be a base from Def. 1. We define the mapping $\tilde{m}_\mathcal{V}$ as follows: for any $t \in \mathbb{Z}_q$ represent t as a nonnegative integer written in base p, that is $t = (t_{d-1} \ldots t_1 t_0)_p$, where each t_i belongs to the set $\{0, 1, \ldots, p-1\}$. Then $\tilde{m}_\mathcal{V}(t) = \sum_{i=0}^{d-1} t_i v_i$.*

It is easy to see that $\tilde{m}_\mathcal{V}()$ is a one-to-one mapping.

For $a \in \mathbb{F}_q$, let $\eta(a)$ denote the quadratic character of a, i.e., $\eta(a) = 0$ for $a = 0$, $\eta(a) = 1$ if $a = b^2$ for some $b \in \mathbb{F}_q^*$, and $\eta(a) = -1$ otherwise. Recall that $\eta(a)$ can be calculated in time $O\left((\log q)^2\right)$ (cf. [2]). By $\nu_p(n)$ we denote the greatest exponent α such that p^α divides n. By $\oplus$ we denote *exclusive or*. The expression $x \leftarrow_\$ X$ denotes the random choice of an element x from the set X with uniform probability distribution. $[k]G$ denotes the multiplication of an element G of some additive group by the scalar k.

2 Siguna Müller's Square Root Algorithm

We recall the square root algorithm from [13] constructed for finite fields $\mathbb{F}_q$ with odd characteristic and based on Lucas sequences (cf. [1, Annex I.1.3]). Its expected running time is $O((\log_2 q)^3)$. An alternative to [13] may be Peralta's algorithm (see [2, Sect. 3]) with the same expected running time. However, Peralta's algorithm works only for $q \equiv 1 \bmod 4$.

The square root algorithm from [13] uses elements of the Lucas sequence $V_k(\tilde{P}, 1)$, where $V_0(\tilde{P}, \tilde{Q}) = 2, V_1(\tilde{P}, \tilde{Q}) = \tilde{P}, V_{2n}(\tilde{P}, \tilde{Q}) = (V_n(\tilde{P}, \tilde{Q}))^2 - 2\tilde{Q}^n, V_{2n+1}(\tilde{P}, \tilde{Q}) = V_{n+1}(\tilde{P}, \tilde{Q}) \cdot V_n(\tilde{P}, \tilde{Q}) - \tilde{P} \cdot \tilde{Q}^n$. These formulas constitute a base for a left-to-right binary algorithm [14] and allow to compute $V_k(\tilde{P}, 1)$ at a cost of at most $2 \cdot \lfloor \log_2 k \rfloor$ multiplications and squarings in $\mathbb{F}_q$.

The base $\mathcal{V}$ being the input of Algorithm 1 is usually defined as a polynomial or as a normal basis. Recall that $\mathcal{V}$ and the modular polynomial defining the field $\mathbb{F}_q$ both determine arithmetic in the field.

Algorithm 1. SM_sqrt – Siguna Müller's square root algorithm from [13]

Input: definition of $\mathbb{F}_q$ and basis $\mathcal{V}$, value Q being a square in $\mathbb{F}_q^*$, *parity_bit* $\in \{0, 1\}$
Output: $\tilde{a} \in \mathbb{F}_q^*$ such that $\tilde{a}^2 = Q$ in $\mathbb{F}_q$ and that $\mathrm{parity}_\mathcal{V}(\tilde{a}) = parity\_bit$,
1: $\delta \leftarrow -1$ for $q \equiv 1 \bmod 4$, and $\delta \leftarrow 1$ for $q \equiv 3 \bmod 4$
2: **repeat**
3:      $r \leftarrow_\$ \mathbb{F}_q^*$
4:      $P \leftarrow Q \cdot r^2 - 4$
5: **until** $\eta(P) = \delta$
6: $P \leftarrow P + 2$
7: $a[0] \leftarrow V_{(q+\delta)/4}(P, 1)$
8: $a[0] \leftarrow a[0] \cdot r^{-1}$ {now we have $(a[0])^2 = Q$}
9: $a[1] \leftarrow -a[0]$ {only one of the elements $a[0], a[1] \in \mathbb{F}_q^*$ has the required *parity_bit*}
10: **return** $a[\mathrm{parity}_\mathcal{V}(a[0]) \oplus parity\_bit]$

With help of the reasoning from Sect. 2 of [3] one may show that both for $q \equiv 1 \bmod 4$ and $q \equiv 3 \bmod 4$, the probability that for a single choice of $r \in \mathbb{F}_q^*$ (i.e., for a single iteration of the loop) the condition in line 5 of Algorithm 1 is not satisfied is equal to $\frac{1}{2} + \frac{1}{q-1}$.

Note that randomization by r completely equalizes the chances of all squares Q to reach line 6 of the algorithm in some given number of iterations of the **repeat...until** loop. Timing attacks at this place are therefore mitigated.

Since complexity of the computation of the quadratic character $\eta(P)$ may be bounded by $O\left((\log_2 q)^2\right)$, it is easy to see that with an overwhelming probability the algorithm will return the required square root in time $O\left((\log_2 q)^3\right)$. Thus we have some level of uncertainty that the main loop of Algorithm 1 will take no more than $\log_2 q$ iterations. However, the probability that the loop takes more iterations than $\log_2 q$ equals $(\frac{1}{2} + \frac{1}{q-1})^{\log_2 q}$. Note that some small level of uncertainty turned out to be acceptable in the case of the Miller-Rabin primality test deployed on smart cards.

To summarize, the number of iterations of the **repeat...until** loop of Algorithm 1 is described by a random variable X_{loop} having geometric distribution with parameter $\hat{p} = \frac{1}{2} - \frac{1}{q-1}$. Hence $E[X_{loop}] = 1/\hat{p}$ and $Var[X_{loop}] = (1 - \hat{p})/(\hat{p})^2$.

3   Probabilistic Variant of the Simplified SWU Method

Let n be a non-square in the field $\mathbb{F}_q$. If the mapping is to be utilized for establishing a secure channel between two parties, then both parties should use the same n. The simplest way to achieve this is to define the basepoint G included in the set of domain parameters in such a way that the x-coordinate of G is a non-square in $\mathbb{F}_q$.

Henceforth we assume that $\mathbb{F}_q$ is a field of characteristic greater than 3. Let $\mathcal{V}$ be a base from Def. 1, and let $\tilde{m}_{\mathcal{V}}()$ be the mapping defined by Def. 2. By Algorithm 2 – a probabilistic variant of the simplified SWU algorithm — we define a mapping $f : \mathbb{Z}_{2q} \to E_{a,b}(\mathbb{F}_q)$, where $E_{a,b}(\mathbb{F}_q)$ is an elliptic curve defined by equation $y^2 = x^3 + ax + b$ for $a, b, x, y \in \mathbb{F}_q$.

Algorithm 2. Encoding $f : \mathbb{Z}_{2q} \to E_{a,b}(\mathbb{F}_q)$

Input: $a, b, n, \mathbb{F}_q, \mathcal{V}$, and argument $t \in \mathbb{Z}_{2q}$ to be mapped on $E_{a,b}(\mathbb{F}_q)$
Output: deterministic result of the mapping in non-deterministic time
1: $parity \leftarrow t \bmod 2$
2: $t' \leftarrow t \bmod q$ {from the Chinese Remainder Theorem (CRT) the pair $(parity, t')$ uniquely represents $t \in \mathbb{Z}_{2q}$}
3: $S \leftarrow n \cdot (\tilde{m}_{\mathcal{V}}(t'))^2$
4: $X_2 \leftarrow \frac{-b}{a}(1 + \frac{1}{S^2+S})$
5: $X_3 \leftarrow S \cdot X_2$
6: $rndbit \leftarrow_\$ \{0, 1\}$ {we shall save some computations from time to time}
7: $h \leftarrow \left((X_{2+rndbit})^2 + a\right) \cdot X_{2+rndbit} + b$
8: **if** $\eta(h) = 1$ **then**
9:     **return** $(X_{2+rndbit}, \texttt{SM\_sqrt}(\mathbb{F}_q, \mathcal{V}, h, parity))$
10: **else**
11:     $h \leftarrow \left((X_{3-rndbit})^2 + a\right) \cdot X_{3-rndbit} + b$
12:     **return** $(X_{3-rndbit}, \texttt{SM\_sqrt}(\mathbb{F}_q, \mathcal{V}, h, parity))$
13: **end if**

Note that the random choice made in line 6 of Algorithm 2 is independent of the input. Moreover, the execution time of the main loop of Algorithm 1 is independent from that choice (recall the randomization by r in the main loop of Algorithm 1). Consequently, for fixed parameters $a, b, n, \mathbb{F}_q, \mathcal{V}$ the execution time of the whole Algorithm 2 is described by a random variable $Y_1 + Y_2$ that is independent of Algorithm's 2 input argument t: the random variable Y_1 describes the execution time of the code included in the listing of Algorithm 2, excluding SM_sqrt subroutine, whereas the random variable Y_2 describes the execution time of the SM_sqrt subroutine. Y_1 is independent from Y_2.

Tests. Our tests confirm that Algorithm 2 is pretty efficient. E.g., for the NIST P-224 curve, during 10^5 calls of Algorithm 2, we measured that for the number of multiplications, squaring and Jacobi symbol computations: a sample mean equals ≈ 135.503, ≈ 226.503, ≈ 3.002 correspondingly, square root of a sample variance equals ≈ 1.506, ≈ 1.506, ≈ 1.424 respectively, The worst case measured during these 10^5 calls equals $155, 246, 22$ correspondingly. There are exactly 3 inversions computed during each call of the Algorithm 2.

4 Indifferentiability for a Non-deterministic Case

The "indifferentiability" notion was introduced in [12] and used also in [8]. In both papers only deterministic algorithms are concerned.

To formally justify Algorithm 2 we introduce the following notion of a *Time-oblivious Turing machine*:

Definition 3 (Time-oblivious Turing machine). *Let C be a non-deterministic Turing machine that computes a function. For each input Q, belonging to the domain $\mathcal{D}_C$ of allowable arguments, the time when C delivers the result is unknown a priori, but is described by the random variable X_Q. Let F_{X_Q} denote the probability distribution of X_Q.*

The machine C is called a time-oblivious *if for any $Q, Q' \in \mathcal{D}_C$ the probability distributions F_{X_Q}, $F_{X_{Q'}}$ are exactly the same, (thus we have a single distribution F_X), and the expected value $E[X_Q]$ is finite for each Q.*

Note that X_Q takes only non-negative values, so $E[X_Q] < \infty$ implies that the probability of the event that C will not deliver the result in finite time for an argument from $\mathcal{D}_C$ equals 0. We denote the event by $C = \bot$.

An example of a time-oblivious Turing machine is a Turing machine implementing directly the simplified SWU method. More generally, we may consider a Turing machine C' implementing a function, and possessing a single, distinguished state S such that starting in S the machine might take one of two sequences of actions: the first sequence is deterministic and terminates the execution, the second one is also deterministic but terminates in the state S (hence we have a loop). The choice of the sequence is made by tossing a coin that might be asymmetric but the same coin is used for all arguments and all iterations of the loop. We assume that state S is the only possibility allowing the machine C' to enter the infinite loop. However, we do not exclude the possibility that before the state S is reached for the first time machine C' makes some other random choices independent of the input. However, since C' implements a function none of the random choices (including those made at state S) influences the final output.

Now we recall the definition of a *random oracle* and *indifferentiability* from Sect. 2.2 of [5]. According to [5], an *ideal primitive* is an algorithmic entity which receives inputs from one of the parties and delivers its output *immediately* to the querrying party.

A *random oracle* into a finite set S is an ideal primitive which provides a random output in S for each new querry; identical input queries are given the same answer.

Let $\mathcal{D}^{R_1,R_2}$ denote a Turing machine $\mathcal{D}$ that uses two different oracles R_1 and R_2. During its execution $\mathcal{D}^{R_1,R_2}$ can state queries to R_1 and R_2 and get immediately an answer.

Definition 4 (indifferentiability [12], [8], [5]). *A Turing machine C with oracle access to an ideal primitive h is said to be $(t_D, t_S, q_D, \varepsilon)$-indifferentiable from an ideal primitive H if there exists a simulator S with oracle access to H and running in time at most t_S, such that for any distinguisher $\mathcal{D}$ running in time at most t_D and making at most q_D queries, it holds that:*

$$\left| \Pr\left[\mathcal{D}^{C^h,h} = 1 \right] - \Pr\left[\mathcal{D}^{H,S^H} = 1 \right] \right| < \varepsilon.$$

C^h is said to be indifferentiable from H, if ε is a negligible function of the security parameter k, for polynomially bounded q_D, t_D and t_S.

Let us explain the above definition. One can try to emulate the primitive H with a Turing machine C using another primitive h as a subroutine. Intuitively, the machine C transforms the results of primitive h so that the final results mimic the primitive H. If we try to distinguish the results of H and C^h, then we not only have to show that the results of C^h are as good as the results of H (i.e., might have been obtained by H). A more difficult part is to convince an observer that a result given by H might have been obtained by C^h. For this purpose we need a simulator S: it provides "an output of h" that would be used by C to provide the same result as obtained by H. The machine $\mathcal{D}^{R_1,R_2}$ is supposed to output 1 if $(R_1, R_2) = (C^h, h)$ and 0 if $(R_1, R_2) = (H, S^H)$, hence Def. 4 means that $\mathcal{D}$ is unable to distinguish between these two cases.

Note that if the construction C^h is indifferentiable from an ideal primitive H, then C^h can replace H in a cryptosystem with security scenarios related to a single stateful adversary, and the resulting cryptosystem is almost as secure as before.

Recall that the *ideal primitive* is an algorithmic entity which receives inputs from one of the parties and delivers its output *immediately* to the querying party. Moreover, the simulator S might include some probabilistic algorithm, like e.g., the sampling algorithm from the proof of Theorem 3 in [5], yielding non-constant execution time. These remarks suggest that time does not matter for the distinguisher $\mathcal{D}$, otherwise $\mathcal{D}$ could distinguish S^H from h simply by time measurements. Indeed, it is not necessary that an observer cannot say whether he is *observing* H or C^h (in the real world we will always observe C^h). The point is that the *set of results* should have essentially the same properties no matter where it comes from and that the execution time for C^h must not leak any additional side channel information. However, the last condition is satisfied automatically if C^h is a deterministic or a time-oblivious Turing machine. Therefore we generalize Def. 4 in the following way:

Definition 5 (indifferentiability for a non-deterministic case). *A non-deterministic Turing machine $\tilde{C}$ with oracle access to an ideal primitive h is $(t_{\tilde{C}}, t_D, t_S, q_D, \varepsilon, \varepsilon_{\tilde{C}})$-indifferentiable from an ideal primitive H, if there exists a simulator S with oracle access to H and running in time at most t_S, such that for any distinguisher $\mathcal{D}$ running in time at most t_D and making at most q_D queries, the following conditions hold:*

- $\tilde{C}^h$ is a time-oblivious Turing machine,
- the probability that in time $t_{\tilde{C}}$ the machine $\tilde{C}^h$ will not return an answer to the query is less than $\varepsilon_{\tilde{C}}$,
- for all events $\tilde{C}^h \neq \perp$

$$\left| \Pr\left[\mathcal{D}^{\tilde{C}^h, h} = 1 \right] - \Pr\left[\mathcal{D}^{H, S^H} = 1 \right] \right| < \varepsilon \ ,$$

$\tilde{C}^h$ is said to be indifferentiable from H, if $\tilde{C}^h$ is $(t_{\tilde{C}}, t_D, t_S, q_D, \varepsilon, \varepsilon_{\tilde{C}})$-indifferentiable from an ideal primitive H, for $\varepsilon, \varepsilon_{\tilde{C}}$ being negligible functions of the security parameter k, and polynomially bounded $t_{\tilde{C}}, t_D, t_S, q_D$.

Note that the constraints on t_S, t_D are exactly the same in Def. 4 and 5 . As we shall see imposing additional constraints on t_S is unnecessary when extending the notion of admissible encoding to the case we investigate (in fact, almost the same simulator as the one defined in [5] could be used – the only difference lies in inversion of the mapping f, which must now take *parity* into account). On the other hand, imposing additional constraints on t_D would weaken the property defined by Def. 5 .

5   Probabilistic Admissible and Weak Encodings

Below we extend the notions of *admissible encoding* and *weak encoding* to some non-deterministic cases. Both notions are known from [6], [5] and we recall them below. But first we recall definition of statistically indistinguishable distributions.

Definition 6 (statistically indistinguishable distributions). *Let X and Y be two random variables over a set S. The distributions of X and Y are ε-statistically indistinguishable if: $\sum_{s \in S} |\Pr[X = s] - \Pr[Y = s]| \leq \varepsilon$. The distributions X and Y are statistically indistinguishable, if ε is a negligible function of the security parameter.*

Definition 7 (admissible encoding from [5]). *A function $F : S \to R$ between finite sets is an ε-admissible encoding if it satisfies the following properties:*

1. *Computable: F is computable in deterministic polynomial time.*
2. *Regular: for s uniformly distributed in S, the distribution of $F(s)$ is ε-statistically indistinguishable from the uniform distribution in R.*
3. *samplable: There is an efficient randomized algorithm $\mathcal{I}$ such that for any $r \in R$, $\mathcal{I}(r)$ induces a distribution that is ε-statistically indistinguishable from the uniform distribution in $F^{-1}(r)$.*

F is an admissible encoding if ε is a negligible function of the security parameter.

Definition 8 (admissible probabilistic encoding). *A function $F : S \to R$ between finite sets is an $(\varepsilon, t_F, \varepsilon_F)$-admissible probabilistic encoding, if*

1. *F can be implemented on some time-oblivious Turing machine C such that probability that C will not return the result in time t_F is smaller than ε_F;*
2. *properties 2. and 3. from Def. 7 hold for F.*

F is an admissible encoding if ε, ε_F are negligible functions of the security parameter k, for polynomially bounded t_F.

By analogy to [6], [5], we define the function $\tilde{H} : \{0,1\}^* \rightarrow R$ as $\tilde{H}(m) := F(h(m))$, where $F : S \rightarrow R$ is an admissible probabilistic encoding, and $h : \{0,1\}^* \rightarrow S$ is a function whose output is seen as an output of a random oracle. To justify the latter condition consider the following thought experiment: we have two black-boxes, each containing the same implementation of hash function h. That is, each box include the same code for h and is built from the same hardware components. For each argument m each black-box actually calculates the value $h(m)$, but in one of the black-boxes at the very end of the computations the value $h(m)$ is replaced by a value returned by some random oracle assigned to the given black-box. If after a series of queries an external observer not knowing the code for h cannot indicate the black-box returning the values of h, then we say that the output of h might be seen as the output of a random oracle.

Theorem 1 (Theorem 1 from [5]). *Let $F : S \rightarrow R$ be an ε-admissible encoding. The construction $H(m) := F(h(m))$ is $(t_D, t_S, q_D, \varepsilon')$-indifferentiable from a random oracle, in the random oracle model for $h : \{0,1\}^* \rightarrow S$, with $\varepsilon' = 4q_D \cdot \varepsilon$ and $t_S = 2q_D \cdot t_I$, where t_I is the maximum running time of F's sampling algorithm.*

Theorem 2 (analogous to Theorem 1 from [5]). *Let $F : S \rightarrow R$ be an $(\varepsilon, t_F, \varepsilon_F)$-admissible probabilistic encoding, let maximum running time of $h : \{0,1\}^* \rightarrow S$ be t_h, where t_h is bounded by a polynomial in the security parameter. The construction $\tilde{H}(m) := F(h(m))$ is $(t_F + t_h, t_D, t_S, q_D, \varepsilon', \varepsilon_F)$-indifferentiable from a random oracle, in the random oracle model for the output of $h : \{0,1\}^* \rightarrow S$, with $\varepsilon' = 4q_D \cdot \varepsilon$ and $t_S = 2q_D \cdot t_I$, where t_I is the maximum running time of F's sampling algorithm.*

Proof (Sketch). Since h is implemented on some deterministic Turing machine $\tilde{h}$ running in time t_h which is not affected even if $\tilde{h}$ includes a call to some random oracle, then this Turing machine may be incorporated into time-oblivious Turing machine C implementing function F. In this way the time-oblivious Turing machine $\tilde{C}^h$ from Def. 5 is constructed, and from assumptions on the function F we get that the probability that in time $t_F + t_h$ the machine $\tilde{C}^h$ will not return an answer to the query is less than ε_F (note that $\tilde{h}$ is called once and output of $\tilde{h}$ constitutes input to F). The rest of the proof follows the proof of Theorem 1 from [5]. □

5.1 Weak Probabilistic Encoding

Definition 9 (weak encoding from [5]). *A function $f : S \rightarrow R$ between finite sets is said to be an α-weak encoding if it satisfies the following properties:*

1. *Computable: f is computable in deterministic polynomial time.*
2. *α-bounded: for s uniformly distributed in S, the distribution of $f(s)$ is α-bounded in R, i.e., the inequality $\Pr_s[f(s) = r] \leq \alpha/|R|$ holds for any $r \in R$.*
3. *Samplable: there is an efficient randomized algorithm $\mathcal{I}_f$ such that $\mathcal{I}_f(r)$ induces the uniform distribution in $f^{-1}(r)$ for any $r \in R$. Additionally $\mathcal{I}_f(r)$ returns $N_r = |f^{-1}\{r\}|$ for $r \in R$.*

f is a weak encoding if α is a polynomial function of the security parameter.

Definition 10 (weak probabilistic encoding). *A function $f : S \to R$ between finite sets is said to be $(\alpha, t_f, \varepsilon_f)$-weak probabilistic encoding if*

1. *f can be implemented on some time-oblivious Turing machine C such that the probability that C will not return the result in time t_f is smaller than ε_f.*
2. *Properties 2. and 3. from Def. 9 hold for f.*

The function f is a weak probabilistic encoding if α and t_f are polynomial functions of the security parameter, and ε_f is a negligible function of the security parameter.

Note that properties 2 and 3 enumerated in Def. 10 concern conditions *for the function f.* Therefore they do not refer to the event $C = \perp$ concerning an *implementation of f* (recall that the event $C = \perp$ is *independent* from the input of f). The event $C = \perp$ is implicitly served by the first property enumerated in Def. 10.

Proposition 1. f *is an $(\alpha, c \cdot (\log_2 q)^3, (\frac{1}{2} + \frac{1}{q-1})^{\log_2 q})$-weak probabilistic encoding with some constant c and $\alpha = 8N/(2q)$, where N is the elliptic curve order.*

Proof (Sketch). It is easy to see that the encoding f defined by Algorithm 2 satisfies Lemma 6 [5]. Namely, the y-coordinate of the resulting point uniquely determines $t \bmod 2$, and the x-coordinate has preimage size at most 8 elements $\tilde{m}(t')$ from $\mathbb{F}_q$ and all the elements can be found in polynomial time. The proof for the x-coordinate follows exactly the proof of Lemma 6 [5]. Note that inversion of the mapping $\tilde{m}()$ from Def. 2 is easily computable, hence given an element $\tilde{t}$ from $\mathbb{F}_q$ it is easy to find $t \in \mathbb{Z}_q$ such that $\tilde{m}(t) = \tilde{t}$. All in all, from the CRT we conclude that for the mapping f the pre-image of any point $P \in E_{a,b}(\mathbb{F}_q)$ contains at most 8 elements from $\mathbb{Z}_{2q}$, and the pre-image can be computed in polynomial time. We assume that the constant c is chosen so that probability that the time-oblivious Turing machine C from Def. 10 will not return the result in time $c \cdot (\log_2 q)^3$ is *strictly smaller* than $(\frac{1}{2} + \frac{1}{q-1})^{\log_2 q}$, so the first condition from Def. 10 is also satisfied. □

Note that by Hasse Theorem $\alpha \leq \frac{8(\sqrt{q}+1)^2}{2q}$. Hence $\alpha < 8$ for $q \geq 7$. The same value $\alpha = 8N/(2q)$ applies, if we multiply the result (x, y) of the encoding f by a co-factor ℓ: let the order N of the group $E_{a,b}(\mathbb{F}_q)$ satisfy the condition $N = \ell \cdot N'$, where N' is prime and $\gcd(\ell, N') = 1$. The integer ℓ is called the co-factor (of N'). Hence $E_{a,b}(\mathbb{F}_q)$ has only one cyclic subgroup of order N'. Let G be a generator of this subgroup. If we take the result (x, y) of the encoding f and multiply by the co-factor ℓ, then we obtain an element of group the $\langle G \rangle$. Consequently, we have a map $f' = [\ell] \circ f$ such that $f' : \mathbb{Z}_{2q} \to \langle G \rangle$. Now we will find all pre-images of a given $P \in \langle G \rangle$ with respect to the map f'. For each element $P \in \langle G \rangle$ it is easy to obtain the unique element $P' \in \langle G \rangle$ being the inverse of P with respect to the scalar multiplication with $[\ell]$ in $\langle G \rangle$. Namely, to obtain P' it suffices to multiply P by the scalar $\ell^{-1} \bmod \mathrm{ord}\, G$. Since $\gcd(\ell, N') = 1$, we have $E_{a,b}(\mathbb{F}_q) = \langle G \rangle \times H$ where H is the subgroup of $E_{a,b}(\mathbb{F}_q)$ of order ℓ. To obtain all candidates in $E_{a,b}(\mathbb{F}_q)$ for the pre-image of P with respect to the scalar multiplication with $[\ell]$, we must collect all results of point addition $P' + P''$, where $P'' \in H$. There are ℓ such results. For each of them there are at most 8 elements in $\mathbb{Z}_{2q}$ being its preimage with respect to f. Altogether, for each of N' elements of $\langle G \rangle$ we have preimage of size at most $8 \cdot \ell$, hence $\alpha = (8 \cdot \ell) \cdot N'/(2q) = 8N/(2q)$.

5.2 The Resulting Admissible Probabilistic Encoding

Theorem 3 (Theorem 3 from [5], weak → admissible encoding). *Let $\mathbb{G}$ be cyclic additive group of order N, and let G be a generator of $\mathbb{G}$. Let $f : S \to \mathbb{G}$ be an α-weak encoding. Then the function $F : S \times \mathbb{Z}_N \to \mathbb{G}$ with $F(s,x) := f(s) + [x]G$ is an ε-admissible encoding into $\mathbb{G}$, with $\varepsilon = (1 - 1/\alpha)^t$ for any t being a polynomial in the security parameter k, and with $\varepsilon = 2^{-k}$ for $t = \alpha \cdot k$.*

The above theorem can be generalized as follows:

Theorem 4 (weak probabilistic → admissible probabilistic encoding). *Let $\mathbb{G}$ be an additive cyclic group of order N, and let G be a generator of $\mathbb{G}$. Let $f : S \to \mathbb{G}$ be an $(\alpha, t_f, \varepsilon_f)$-weak probabilistic encoding. Then the function $F : S \times \mathbb{Z}_N \to \mathbb{G}$ with $F(s,x) := f(s) + [x]G$ is an $(\varepsilon, t_f + t_x, \varepsilon_f)$-admissible probabilistic encoding into $\mathbb{G}$, where t_x is the maximum running time of the scalar multiplication $[x]G$ together with addition of resulting elements (i.e. elements $f(s), [x]G$), and $\varepsilon = (1 - 1/\alpha)^t$ for any t polynomially bounded in the security parameter k, and with $\varepsilon = 2^{-k}$ for $t = \alpha \cdot k$.*

Proof (Sketch). The *deterministic* Turing machine implementing the scalar multiplication $[x]G$ and the addition of the resulting elements may be incorporated into the time-oblivious Turing machine C implementing f. In this way execution time of the time-oblivious Turing machine grows at most by t_x (if $\mathbb{G}$ is a subgroup of some elliptic curve defined over a field $\mathbb{F}_q$, then time t_x of scalar multiplication $[x]G$ together with two points addition $f(s) + [x]G$ can be bounded by $c' \cdot (\log_2 q)^3$ for some constant c'). The rest of the proof follows exactly the proof of Theorem 3 from [5]. □

Consequently, in order to obtain an Admissible Probabilistic Encoding we should apply Theorem 4 to the encoding $f' = [\ell] \circ f$, that is, to the composition of the encoding f defined by Algorithm 2 and the scalar multiplication by the co-factor ℓ.

We also obtain an extension of the result from [5]: Let $h_1 : \{0,1\}^* \to \mathbb{Z}_{2q}$ and $h_2 : \{0,1\}^* \to \mathbb{Z}_N$ be two hash functions of running time bounded by polynomial in the security parameter k. Then the function $H : \{0,1\}^* \to \mathbb{G}$ defined by:

$$H(m) := f'(h_1(m)) + [h_2(m)]G$$

is (according to Def. 5) indifferentiable from a random oracle in the random oracle model for outputs generated by h_1 and h_2.

6 Conclusions

Security analysis of PACE IM utilizes admissible encodings (cf. [9]). By extending the framework from [5] we have shown that the Admissible Probabilistic Encoding defined in Subsect. 5.2 preserves the level of security of its predecessor. Thus Algorithm 1 may be used in PACE IM in place of the currently used simplified SWU mapping.

Note that complexity of evaluation of $V_{(q+\delta)/4}(P,1)$ in Algorithm 1 is comparable with the cost of the worst case exponentiation in field $\mathbb{F}_q$ via square and multiply. Moreover, the running time of the Algorithm 1 has small variation. Consequently, the difference between the running time of Algorithm 2 and the simplified SWU mapping will be negligible (or even unnoticeable) for the owner of a smart card. At the expense of a small decrease of efficiency we have gained more flexibility in choice of a field for a definition of the elliptic curve, hence we have more freedom in other applications like e.g., BLS signatures [4]. What is more, a standardized, general mapping procedure decreases deployment costs of the infrastructure supporting the protocols, especially in reference to the Common Criteria certification process.

Acknowledgements. We would like to thank Bart Preneel for many helpful comments.

References

1. Accredited Standards Committee X9, Inc., Financial Industry Standards: ANS X9.62-2005, Public Key Cryptography for the Financial Services Industry: The Elliptic Curve Digital Signature Algorithm (ECDSA). American National Standard for Financial Services (2005)
2. Bach, E.: A note on square roots in finite fields. IEEE Transactions on Information Theory 36(6), 1494–1498 (1990)
3. Bernstein, D.J.: Faster square roots in annoying finite fields. Note: to be incorporated into author's High-speed cryptography book (November 2001)
4. Boneh, D., Lynn, B., Shacham, H.: Short signatures from the Weil pairing. J. Cryptology 17(4), 297–319 (2004)
5. Brier, E., Coron, J.S., Icart, T., Madore, D., Randriam, H., Tibouchi, M.: Efficient indifferentiable hashing into ordinary elliptic curves. Cryptology ePrint Archive, Report 2009/340 (2009)
6. Brier, E., Coron, J.-S., Icart, T., Madore, D., Randriam, H., Tibouchi, M.: Efficient indifferentiable hashing into ordinary elliptic curves. In: Rabin, T. (ed.) CRYPTO 2010. LNCS, vol. 6223, pp. 237–254. Springer, Heidelberg (2010)
7. BSI: Advanced Security Mechanisms for Machine Readable Travel Documents 2.11. Technische Richtlinie TR-03110-3 (2013)
8. Coron, J.-S., Dodis, Y., Malinaud, C., Puniya, P.: Merkle-Damgård revisited: How to construct a hash function. In: Shoup, V. (ed.) CRYPTO 2005. LNCS, vol. 3621, pp. 430–448. Springer, Heidelberg (2005)
9. Coron, J.-S., Gouget, A., Icart, T., Paillier, P.: Supplemental Access Control (PACE v2): Security analysis of PACE Integrated Mapping. In: Naccache, D. (ed.) Quisquater Festschrift. LNCS, vol. 6805, pp. 207–232. Springer, Heidelberg (2012)
10. Demay, G., Gaži, P., Hirt, M., Maurer, U.: Resource-restricted indifferentiability. In: Johansson, T., Nguyen, P.Q. (eds.) EUROCRYPT 2013. LNCS, vol. 7881, pp. 664–683. Springer, Heidelberg (2013)
11. ISO/IEC JTC1 SC17 WG3/TF5 for the International Civil Aviation Organization: Supplemental access control for machine readable travel documents. Technical Report (2011) version 1.02 (March 2008)
12. Maurer, U., Renner, R., Holenstein, C.: Indifferentiability, impossibility results on reductions, and applications to the random oracle methodology. In: Naor, M. (ed.) TCC 2004. LNCS, vol. 2951, pp. 21–39. Springer, Heidelberg (2004)
13. Müller, S.: On the computation of square roots in finite fields. Des. Codes Cryptography 31(3), 301–312 (2004)
14. Postl, H.: Fast evaluation of Dickson polynomials. Contrib. to General Algebra 6, 223–225 (1988)
15. Ristenpart, T., Shacham, H., Shrimpton, T.: Careful with composition: Limitations of the indifferentiability framework. In: Paterson, K.G. (ed.) EUROCRYPT 2011. LNCS, vol. 6632, pp. 487–506. Springer, Heidelberg (2011)

An Algebraic Framework for Modeling of Reactive Rule-Based Intelligent Agents

Katerina Ksystra, Petros Stefaneas, and Panayiotis Frangos

National Technical University of Athens
Iroon Polytexneiou 9, 15780 Zografou, Athens, Greece
katksy@central.ntua.gr, petros@math.ntua.gr, pfrangos@central.ntua.gr

Abstract. As the use of intelligent agents in critical domains increases, the need for verifying their behavior becomes stronger. Reactive rules are the main reasoning formalism for intelligent agents. For this reason, we propose the use of the OTS/CafeOBJ method for the specification of reactive rules, which will permit the verification of safety properties for reactive rule-based intelligent agents.

Keywords: intelligent agents, reactive rules, CafeOBJ, Observational Transition Systems, rule-based system.

1 Introduction

Intelligent agents are a new paradigm for developing software applications. An intelligent agent is defined either as anything that can be viewed as perceiving its environment through sensors and acting upon that environment through effectors [1], or as a software that carries out some set of operations and acts on behalf of a user [2], or finally as a computational process that implements the autonomous functionality of an application [3]. Agent-based systems usually consist of many agents that communicate with each other and are known as multi-agent systems.

The use of rule-based systems as the main reasoning model of agents that are part of a multi-agent system has been proposed in early attempts. In this approach each agent includes a rule engine and is able to perform rule-based inference [4]. Thus, an intelligent agent is called rule-based, if its behavior and its knowledge are expressed by means of rules.

The task of verifying the behavior of rule-based agents is difficult because rules can interact during execution and this interaction can cause undesirable results [5]. For example, one rule may trigger another rule and cause a chain of rule triggerings. Also, changes to the rule base (add, remove, change rules) can introduce errors in the behavior of the system if the effects of the changes are not examined beforehand. Thus, using rules in critical systems implies that the system's behavior must be extensively analyzed. Formal methods provide powerful means for analyzing system's behavior and can prove really helpful for preventing design errors at an early stage of development. In this paper, we address the problem of formally analyzing reactive rule-based agents as follows:

V. Geffert et al. (Eds.): SOFSEM 2014, LNCS 8327, pp. 407–418, 2014.

- We present an algebraic framework for formally expressing Production and Event-Condition-Action rules (Section 3).
- We use the OTS/CafeOBJ method for the specification of reactive rule-based intelligent agents and the verification of their behavior (Section 3).
- We apply the framework to a supply chain management system and prove security properties in order to demonstrate its effectiveness (Section 4).

The proposed framework offers the ability to formally specify an intelligent agent whose behavior is expressed in terms of reactive rules, to verify its behavior and thus ensure its correctness. This work is in continuation of [6], where Observational Transition System (OTS) semantics were provided for reactive rules.

1.1   Related Work

A lot of research concerning analysis of rule-based systems exists in the area of active databases. For example, in [7] authors present an overview of processing rules in production systems, deductive and active databases. A larger survey on the different approaches of reaction rules can be found in [18]. Most of the approaches addressing formal analysis of such systems however deal with checking properties such as termination, confluence and completeness. One such attempt to verify rule-based systems can be found in [8], where authors use Petri-nets to analyze various types of structure errors such as inconsistency, incompleteness, redundancy and circularity of rules. Also, ECA-LP [15] which is based on a labeled transaction logic semantics, supports state based knowledge updates including a test case based verification and validation for transactional updates.

Few papers targeting the verification of the behavior of active rule-based systems/agents exist. More precisely, in [9] authors describe a reasoning framework for Ambient Intelligence that uses the Event Calculus formalism for reasoning about actions and causality. Also, an approach to verify the behavior of Event-Condition-Action rules is presented in [5] where a tool that transforms such rules to timed automata is developed. Then the Uppaal tool is used to prove desired properties for a rule-based application. This last work is the closest to ours with the difference that in [5] authors use model checking, while our approach uses theorem proving techniques.

Our framework focuses on verifying the behavior of rule-based agents, rather than proving correctness properties or handling problems with negations, mainly for two reasons; first, the verification of such properties has been studied in many other approaches [8] and second the OTS/CafeOBJ method does not study properties about the transitions of the system but analyzes their effects in the system's behavior. We believe that our framework has the following advantages over existing approaches; it can be used for the verification of complex systems due to the simplicity of the CafeOBJ language and its natural affinity for abstraction. Also, it has the ability to specify systems with infinite states (in contrast with approaches that use model-checking techniques) and it allows the reusability not only of the specification code but also of the proofs [17].

2 Observational Transition Systems and CafeOBJ

An Observational Transition System (OTS) is a transition system written in terms of equations [10]. Assuming that there exists a universal state space Y and that each data type we need to use, including their equivalence relationship, has been declared in advance, an OTS S is defined as the triplet $S = \langle O, I, T \rangle$ where:

1. O is a finite set of observers. Each $o \in O$ is a function $o : Y \to D$, where D is a data type that may differ from observer to observer. Given an OTS S and two states u_1, $u_2 \in Y$ the equivalence $u_1 =_s u_2$ between them with respect to S is defined as; $\forall o \in O, o(u_1) = o(u_2)$ i.e. two states are considered behaviorally equivalent if all the observers return for these states the same data values.
2. I is the set of initial states, such that $I \subseteq Y$.
3. T is a set of conditional transitions. Each $\tau \in T$ is a function $\tau : Y \to Y$ and preserves the equivalence between two states; if $u_1 =_s u_2$ then $\tau(u_1) =_s \tau(u_2)$. For each $u \in Y$, $\tau(u)$ is called the successor state of u wrt τ. The condition c-τ is called the effective condition of τ. Also, for each $u \in Y$, c-$\tau(u) = false$ $\Rightarrow u =_s \tau(u)$. Finally, observers and transitions may be parameterized by data type values.

Observational transition systems can be described as behavioral specifications in CafeOBJ, an algebraic specification language and processor [11],[21]. In a CafeOBJ module we can declare sorts, operators, variables and equations. There exists two kinds of sorts; a visible sort denotes an abstract data type and a hidden sort denotes the state space of an abstract machine. Two kinds of behavioral operators can be applied to hidden sorts: action and observation operators. An observation operator can only be used to observe the inside of an abstract machine while an action operator can change its state. Finally, CafeOBJ system rewrites a given term by regarding equations as left-to-right rewrite rules.

CafeOBJ is used to specify OTSs [10]. The universal state space Y of an OTS is denoted in CafeOBJ by a hidden sort and an observer by an observation operator. Any initial state in I is denoted by a constant and a transition by an action operator. The transitions are defined by describing what the value, returned by each observer in the successor state, becomes when the transitions are applied in an arbitrary state u. Finally, for expressing the effective conditions, conditional equations are used.

3 An Algebraic Framework for Reactive Rules

The proposed framework aims to enhance reactive rules with verification capabilities. More precisely, it supports Event Condition Action and Production rules. This will allow proving desired safety properties about intelligent agents/systems whose behavior is expressed in terms of such rules. Because we are interested in proving application specific properties, additional characteristics (observers and/or transitions) about the specific system will be required in order to specify its behavior. These characteristics will differ from application to application and thus the specification cannot become fully automated. This framework however will serve as the basis for specifying and verifying reactive rule based systems and most importantly for capturing the semantics of their rules.

3.1 Production Rules in CafeOBJ

A Production rule is a statement of rule programming logic, which specifies the execution of an action in case its conditions are satisfied, i.e. production rules react to states changes. Their essential syntax is *if Condition do Action*. Some usual predefined actions supported by Rule Markup languages are: add, retract, update knowledge and generic actions with external effects [12].

A Production rule can be naturally expressed in our framework if we map the action of the rule to a transition which has as effective condition the condition of the rule. Also, since most of the actions correspond to changes of the knowledge base in order to describe their effects we need an observer that will observe the knowledge base (KB) at any given time. Thus, the observer *knowledge* : $Y \to$ *Set of Bool* which returns the set of boolean elements that belong to the knowledge base is needed. For expressing the functionalities of the KB, the following operators are required; /in which returns true if an element belongs to the knowledge base, | which denotes that an element is added to the KB and / which denotes that an element is removed from the KB. Formally, the definition of a set of Production rules as an OTS is presented below.

Definition 1. *Assume the universal state space Y and the following set of Production rules; $\{if\ C_i\ do\ A_i,\ i = 1, \ldots, n \in \mathbb{N}\}$, where without harm of generality we also assume that the conditions of the rules are disjoint. We define an OTS $S = \langle O, I, T \rangle$ from this set of rules as follows:*

- $O = \{O' \cup knowledge\}$
- $T = \{A_i\}$
- $I =$ *the set of initial states, such that $I \subseteq Y$*

In the above definition, O' denotes the rest of the system's observers. Transitions are the actions of the rules, $A_i : Y D_1 \ldots D_l \to Y$. They can be generic actions (with external changes) or the usual predefined actions *assert* : $Y Bool \to Y$ (add a fact to KB), *retract* : $Y Bool \to Y$ (remove a fact from KB), *update* : $Y Bool\ Bool \to Y$ (remove/add a fact) [20]. Facts are denoted by boolean-sorted CafeOBJ terms. Formally, the actions of Production rules are defined as transitions through the following steps;

1. The effective condition of an action A_i is defined as; eq c-Ai(u,d1,...,dn) = Ci(d1,...,dn) /in knowledge(u).
2. If A_i is an assert action its effect on the knowledge observer is defined as; knowledge(assert(u,ki(d1,...,dn))) = ki(d1,..,dn)|knowledge(u) if c-assert(u,ki(d1,...,dn)).

3. If A_i is a retract action its effect on the knowledge observer is defined as;
   `knowledge(retract(u,ki(d1,..,dn))) = knowledge(u)/ki(d1,..,dn)`
   `if c-retract(u,ki(d1,..,dn))`.
4. If A_i is an update action its effect on the knowledge observer is defined
   as; `knowledge(update(ki(d1,..,dn),kj(d1,..,dn))) = (knowledge(u)`
   `/ki(d1,..,dn))|kj(d1,..,dn) if c-update(ki(d1,..,dn),kj(d1,...`
   `,dn))`.
5. If A_i is a generic action, we define; `knowledge(ai(u,d1,...,dn)) = ai(d1`
   `,...,dn)|knowledge(u) if c-ai(u,d1,...,dn)` and
   `oi(ai(u,d1,...,dn)) = vi if c-ai(u,d1,..,dn)`.

Step 1 declares that an action a_i can be successfully applied if the condition
of the rule holds, i.e. belongs to the knowledge base[1]. Step 2 states that when
a transition $assert(u, k_i(d_1, \ldots, d_n))$ is applied successfully in an arbitrary state
u, k_i is added to the knowledge base. Where k_i is the fact being asserted. In
step 3 it is stated that when the transition $retract(u, k_i(d_1, \ldots, d_n))$ is applied
successfully in an arbitrary state u, k_i is removed from the knowledge base. When
the transition $update(u, k_i(d_1, \ldots, d_n), k_j(d_1, \ldots, d_n))$ is applied successfully in
an arbitrary state u, ki is removed and kj is added, as step 4 defines. Finally,
step 5 states that when we have the application of a generic action we add to our
KB the information that this action occurred. But generic actions may have side
effects and in order to describe them we may have to use additional observers
$o_i \in O$ and define how their values change when the action is applied successfully.

3.2 Event Condition Action Rules in CafeOBJ

In contrast to Production rules, Event Condition Action (ECA) rules define an
explicit event part which is separated from the conditions and actions of the rule.
Their essential syntax is; *on Event if Condition do Action*. The ECA paradigm
states that a rule autonomously reacts to actively or passively detected simple or
complex events by evaluating a condition or a set of conditions and by executing
a reaction whenever the event happens and the condition(s) is true [13].

In order to express ECA rules in our framework we need an observer that will
remember the occurred events. For this reason, in each event we assign a natural
number and when an event is detected its number is stored in the observer *event-memory* : $Y \to Nat$. Using *event-memory* we can map events to transitions.
The actions of ECA rules are assert, retract, update, or generic actions and are
mapped to transitions, as before. However, their semantics differs as the actions
of ECA rules can be applied only if their triggering event has been detected first.
Formally, the definition of a set of ECA rules as an OTS is presented below.

Definition 2. *Assume the universal state space Y and a finite set of ECA rules*
$\{on\ E_i\ if\ C_i\ do\ A_i,\ i = 1, \ldots, n \in \mathbb{N}\}$, *where without harm of generality we*

---

[1] If we have negation-as-failure in the condition of the rule, i.e. if the condition cannot
be proved, this is expressed in our framework as; if $c_i \notin knowledge(u)$, since this
basically means that there is no information (in our knowledge base) about the
condition.

also assume that for $i \neq j$; $E_i, A_i, C_i \neq E_j, A_j, C_j$, respectively. The OTS $S = \langle O, I, T \rangle$ modeling these rules is defined as:

- $O = \{O' \cup knowledge, event - memory\}$
- $T = \{E_i, A_i\}$
- $I = the\ set\ of\ initial\ states\ such\ that\ I \subseteq Y$

Here, O' is the same as in definition 1. Transitions are the events, E_i : $YD_1 \ldots D_n \to Y$ and the actions, $A_i : YD_1 \ldots D_n \to Y$. Formally, the rule on E_i if C_i do A_i is defined in CafeOBJ terms through the following steps:

1. The effective condition of an event E_i is denoted as $c\text{-}ei(u, d1, \ldots, dn)$ and states the conditions under which the system is able to detect the event.
2. The effects of the application of the event E_i in an arbitrary system state u is the following;

```
knowledge(Ei(u,d1,...,dn)) = ei(d1,...,dn)|knowledge(u) if
c-ei(u,d1,...,dn) and event-memory(u) = null .
event-memory(Ei(u,d1,...,dn)) = i if c-ei(u,d1,...,dn) and
event-memory(u) = null .
```

3. The effective condition of the action A_i, is defined as; eq `c-Ai(u,d1,..,dn)` = `Ci(d1,...,dn)` /in `knowledge(u)` and `ei(d1,...,dn)` /in `knowledge(u)`.
4. The effects of the action A_i, if it is an assert action, is described through the following equations;

```
knowledge(assert(u,ki(d1,..,dn))) = ki(d1,..,dn)|knowledge(u)
/ei(d1,...,dn) if c-assert(u,ki(d1,...,dn)) .
event-memory((u,ki(d1,...,dn))) = null if
c-assert(u,ki(d1,...,dn)) .
```

The effects of the rest of the actions are defined in a similar way. Step 2 states that when the transition/event E_i is applied, the name of the occurred event (e_i) is added to the knowledge base as a fact if in the previous state the detection conditions of the event were true and event-memory was null (denoting that no events had occurred). Also, when the event is applied, event-memory stores the identification number of the event (here i). Step 3 declares that the action will be applied successfully, if the condition of the rule belongs to the KB and the triggering event of the action has been detected. In step 4 it is stated that when the action $assert(u, ki(d1, \ldots, dn))$ is applied, the fact k_i is added to the knowledge base and its triggering event is consumed, i.e. its name is removed from the knowledge observer. Also, *event-memory* becomes null.

We must mention here that as we will see in the following section, sometimes the names of the events are not removed from the observer event-memory if they are required for the detection of complex events. Also, if many rules (either Production or ECA) can be executed at the same time, a selection function is used from the inference engine of the system such as those presented in [14], [15]. It is quite straightforward to include this characteristic in our framework but is out of the scope of this paper.

One of the challenges we met while expressing these rules into our framework was the difference between events and actions, i.e. while events *can* occur at anytime and can be straightforwardly mapped to transitions, actions *must* be executed after the detection of their triggering events. To capture this difference we used the observer event-memory. Initially it returns the value null (meaning that no events have been detected) denoting that any event can occur, but when an event is detected then the only applicable transition in the system is the action of the detected event.

3.3 Complex Events Definition

Sometimes ECA rules react to the detection of complex events. Complex events are created by primitive event(s) and event operator(s). A typical set of event operators for defining complex events include the following; Xor (Mutually Exclusive), Disjunction (Or), Conjunction (And), Any, Concurrent (Parallel), Sequence (Ordered), Aperiodic, Periodic. In [14] definitions of such operators are presented in more details. In this section we will present how the basic event operators can be expressed in our framework.

Assume primitive events A_i and B_j defined as transitions with effective conditions $c\text{-}A_i$ and $c\text{-}B_j$ respectively. Complex event $Xor(Ai, Bj)$ means that either event Ai happens or Bj, but not both. The application of the complex event/transition $e_k : xor(u, Ai, Bj)$ to an arbitrary system state is defined as:

```
knowledge(xor(u,Ai,Bj)) = xor(Ai,Bj)|knowledge(u) if
Ai /in knowledge(u) xor Bj /in knowledge(u) .
event-memory(xor(u,Ai,Bj)) = k if Ai /in knowledge(u) xor
Bj /in knowledge(u) .
```

The above equations state that the complex event is detected (its occurrence is added to the KB) if its detection conditions are fulfilled, i.e. if we have detected either the primitive event Ai or event Bj. Also, the observer event-memory stores the id number k of the event (where xor is a built-in operator) if the same conditions hold. $Disjunction(Ai, Bj)$ means that either event Ai happens or Bj (or both). In a similar way, the application of the event $disjunction(u, Ai, Bj)$ is defined as; `knowledge(disjunction(u,Ai,Bj)) = disjunction(Ai,Bj)| knowledge(u)` if `Ai /in knowledge(u) or Bj /in knowledge(u)`.

$Conjunction(Ai, Bj)$ means that both events Ai and Bj occur in any order. The application of the event $conjunction(u, Ai, Bj)$ is defined as; `knowledge (conjuction(u,Ai,Bj)) = conjuction(Ai,Bj)|knowledge(u)` if `Ai /in knowledge(u) and Bj /in knowledge(u)`.

$Sequence(Ai, Bj)$ corresponds to the ordered execution of events Ai and Bj. The application of $sequence(u, Ai, Bj)$ is defined as; `knowledge(Bj(u))= sequence(Ai,Bj)|knowledge(u)` if `Ai /in knowledge(u) and event-`

`memory(u) = i`. This complex event is detected (its occurrence is added to KB) during the occurrence of event Bj, which can occur if in the previous state Ai had occurred, i.e. event-memory had stored i (and not if the memory is equal to null). By using the observer event-memory and declaring which event had occurred before we can avoid the unintended semantics these operators can have, which are caused because the events, in the active database sense, are treated as if they occur at an atomic instant. This problem is discussed in [15], [19] where also a solution is proposed by defining an interval-based effect semantics in terms of an interval-based event calculus formalization. The alternative interval-based semantics could be implemented in our framework by extending the definition of an event with the time of its occurrence and introducing the notions of event and time intervals. The rest event operators (Concurrent, Aperiodic and Periodic), which are used less often, cannot be straightforwardly expressed in our framework and an extension is required in order to include them as well.

4 Case Study: A Supply Chain Management System

To demonstrate the expressiveness of our framework we applied it to an industrial case study that uses Event Condition Action rules to control the activities of its agents. These activities are inter-enterprise business processes and thus their verification is an important task. In [16] authors present an integrated workflow-supported supply chain management system that was developed so that Nanjing Jin Cheng Motorcycle Corporation in China and its suppliers could handle better their inner processes. The proposed system consists of a set of business function agents whose tasks are to deal with outsourcing, production planning, sales, customer service, inventory, and so on. Each agent is an autonomic and independent entity. ECA rules are used to control the execution sequence of agents' activities. These rules are presented in table 1. A more detailed description of the system is presented in appendix A.

Table 1. ECA rules controlling the activities of the manufacturer

R1 On end(sales)
     Do st(charge)
	R5 On end(ManufacturePlan)
	     if isMaterialsEnough

R2 On end(sales) and end(charge)
     if payment $\geq$ totalprice
     Do st(QueryInventory)
	Do st(Manufacture)
	R6 On end(ManufacturePlan)
	     if not isMaterialsEnough

R3 On end(queryinventory)
     if IsGoodsEnough
     Do st(DeliverGoods)
	Do st(Outsource)
	R7 On end(Outsource)
	     if ArrivedMaterials

R4 On end(queryinventory)
     if not IsGoodsEnough
     Do st(ManufacturePlan)
	Do st(Manufacture)
	R8 On end(Manufacture)
	     Do st(DeliverGoods)

4.1 Formal Specification and Verification of the System

Rules R1-R8 were expressed in our framework according to the previous definitions. For example, the first rule was defined in CafeOBJ using the transitions
endsales and stcharge. The first transition represents the event part of the
rule and the second the action. The definition of the transition endsales can be
seen below:

```
-- endsales
op c-endsales : Sys department customer Nat -> Bool
eq c-endsales(S,Sales,C,N) = (order(S,Sales,C) = N)
and (event-memory(S) = null) .
ceq knowledge(endsales(S,Sales,C,N)) = (endsales|knowledge(S))
if c-endsales(S,Sales,C,N) .
ceq event-memory(endsales(S,Sales,C,N)) = 1
if c-endsales(S,Sales,C,N) .
```

The effective condition c-endsales denotes that the event endsales can be
detected when the sales department receives an order from a customer and if no
other event had been detected in the previous state. The observer order returns
the cost of the order a department receives from a customer. When the event
is successfully detected its name enters the knowledge base and event-memory
stores its identification number. The transition stcharge is defined as follows:

```
-- stcharge
op c-stcharge : Sys Nat customer -> Bool
eq c-stcharge(S,N,C1) = (endsales /in knowledge(S)) and
(event-memory(S) = 1) .
ceq event-memory(stcharge(S,N,C1)) = null if c-stcharge(S,N,C1) .
eq knowledge(stcharge(S,N,C1)) = knowledge(S) .
ceq payment(stcharge(S,N,C1),C2) = pending if c-stcharge(S,N,C1)
and (C1 = C2) .
```

The effective condition c-stcharge denotes that the action stcharge will occur
if endsales belongs to the KB and event memory contains the id number of the
event. After the execution of the action, the observer event-memory becomes
null, knowledge base stays the same (because the occurrence of endsales event
is needed for the detection of the complex event end(sales) *and* end(charge) of
R2) and the payment of the customer is pending until a receipt is received.
The sixth rule was defined in CafeOBJ using the transitions stoutsource and
endmanufactureplan. The definition of the transition endmanufactureplan is
presented below;

```
-- endmanufactureplan
op c-endmanufactureplan : Sys bill inventory -> Bool
eq c-endmanufactureplan(S,B,I) = (materials(S,B,I) = computed)
and (event-memory(S) = null) .
```

```
ceq knowledge(endmanufactureplan(S,B,I)) = (endmanufactureplan|
knowledge(S)) if c-endmanufactureplan(S,B,I) .
ceq event-memory(endmanufactureplan(S,B,I)) = 5 if
c-endmanufactureplan (S,B,I) .
```

The effective condition `c-endmanufactureplan` denotes that the event can be detected when it is computed if there are enough materials to produce goods for the order and if event-memory is null. When the event is detected the name of the event enters the knowledge base and the observer event-memory stores the number of the event, i.e. 5. The transition `stoutsource` is defined as follows;

```
-- stoutsource
op c-stoutsource : Sys bill inventory agent -> Bool
eq c-stoutsource(S,B,I,A) = endmanufactureplan /in knowledge(S)
and (event-memory(S) = 5) and (materials(S,B,I) < enough) .
ceq knowledge(stoutsource(S,B,I,A)) = (knowledge(S)/
endmanufactureplan) if c-stoutsource(S,B,I,A) .
ceq event-memory(stoutsource(S,B,I,A)) = null if
c-stoutsource(S,B,I,A) .
ceq list(stoutsource(S,B,I,A),A) = true if
c-stoutsource(S,B,I,A) .
```

The effective condition `c-stoutsource` declares that the action can be successfully applied if the event endmanufactureplan has been detected and the condition of the action holds, i.e. the materials are not enough. When the action occurs, the observer event-memory becomes null, the occurrence of the event is removed from the knowledge base and a list is sent to the outsourcing agent.

In a similar way we expressed all the rules in our framework. We also defined the transitions whose occurrence makes the detection conditions of the events true. For example, in order to detect the event endmanufacture, the products for the order must have been produced. Thus, we defined the transition `produceproducts`. When this transition is successfully applied, the value of the observer products becomes "produced", indicating that the event endmanufacture can be detected; `ceq products(produceproducts(S)) = produced if c-produceproducts(S) .`

In the above case study, the events may seem as simple propositional representations, or similar in format, but in the context of the whole specification they fully express the functionalities of the system (appendix A). In order to specify this manufacturer agent, 18 transitions (12 that correspond to events and actions and 6 external transitions) and 14 observers were needed.

The most important feature of the proposed framework is the ability to verify the behavior of reactive rule-based intelligent agents using the proof score methodology [10,17]. The type of properties that can be proved with the framework are safety properties, that hold in any reachable state of the system (called invariant properties), and liveness properties, which denote that something will eventually happen. For the supply chain system of the previous section, we proved that the process of delivering the goods to the customer must not be activated if the payment of the customer does not cover the total cost of the order. This is an invariant property, important for the purpose of the system. Invariant 1, is defined in CafeOBJ terms as; `eq inv1(S,C) = not(not(payment(S,C) >= cost(S,C)) and (delivered(S,C) = true))`. Following the CafeOBJ/ OTS method [10,17] we successfully verified invariant 1 and two more invariants that were needed to conclude the proof (for more details see appendix B). The full specification, the proofs and the appendices can be found at http://cafeobjntua. wordpress.com.

5   Conclusions and Future Work

We believe that due to the fact that reactive rule-based intelligent agents are increasingly used in critical systems, there is a strong need for ensuring their intended behavior. This task is difficult because rules interact during execution and thus can have complex and unpredictable behavior. For this reason we have presented a framework for formally specifying reactive rules with the help of the OTS/CafeOBJ method. This framework can express complex systems while capturing the semantics of the underlying reactive rules, and can be used for the verification of safety properties reactive rule-based agents should meet. In order to demonstrate its effectiveness, we have applied it to a case study of a manufacturer business agent. In the future, we intend to develop a tool that will automatically translate a set of reactive rules, written in a Rule Markup language, to CafeOBJ and that will support online verification. Finally, this framework could be extended for modeling operational reactive systems that need to define an optimized proof-theoretic and operational semantics.

Acknowledgments. This research has been co-financed by the European Union (European Social Fund ESF) and Greek national funds through the Operational Program "Education and Lifelong Learning" of the National Strategic Reference Framework (NSRF) - Research Funding Program: THALIS

References

1. Russell, S., Norvig, P.: Artificial Intelligence: A Modern Approach, 1st edn. Prentice Hall (1995)
2. Gilbert, D.: Intelligent Agents: The Right Information at the Right Time. IBM Intelligent Agent White Paper
3. FIPA (Foundation for Intelligent Physical Agents), www.fipa.org

4. Badica, C., Braubach, L., Paschke, A.: Rule-based Distributed and Agent Systems. In: 5th International Conference on Rule-Based Reasoning, Programming, and Applications, RuleML 2011, pp. 3–28. Springer (2011)
5. Ericsson, A., Berndtsson, M., Pettersson, P.: Verification of an industrial rule-based manufacturing system using REX. In: 1st International Workshop on Complex Event Processing for Future Internet, iCEP-FIS (2008)
6. Ksystra, K., Triantafyllou, N., Stefaneas, P.: On the Algebraic Semantics of Reactive Rules. In: Bikakis, A., Giurca, A. (eds.) RuleML 2012. LNCS, vol. 7438, pp. 136–150. Springer, Heidelberg (2012)
7. Vlahavas, I., Bassiliades, N.: Parallel, object-oriented, and active knowledge base systems. Kluwer Academic Publishers, Norwell (1998)
8. Xudong, H., Chu, C., Yang, H., Yang, S.J.H.: A New Approach to Verify Rule-Based Systems Using Petri Nets. Information and Software Technology 45(10), 663–669 (2003)
9. Patkos, T., Chrysakis, I., Bikakis, A., Plexousakis, D., Antoniou, G.: A Reasoning Framework for Ambient Intelligence. In: Konstantopoulos, S., Perantonis, S., Karkaletsis, V., Spyropoulos, C.D., Vouros, G. (eds.) SETN 2010. LNCS (LNAI), vol. 6040, pp. 213–222. Springer, Heidelberg (2010)
10. Ogata, K., Futatsugi, K.: Proof scores in the OTS/CafeOBJ method. In: Najm, E., Nestmann, U., Stevens, P. (eds.) FMOODS 2003. LNCS, vol. 2884, pp. 170–184. Springer, Heidelberg (2003)
11. Diaconescu, R., Futatsugi, K.: CafeOBJ report: the language, proof techniques, and methodologies for object-oriented algebraic specification. AMAST series in computing. World Scientific, Singapore (1998)
12. Paschke, A., Boley, H., Zhao, Z., Teymourian, K., Athan, T.: Reaction RuleML 1.0: Standardized Semantic Reaction Rules. In: Bikakis, A., Giurca, A. (eds.) RuleML 2012. LNCS, vol. 7438, pp. 100–119. Springer, Heidelberg (2012)
13. Paschke, A.: ECA-RuleML: An Approach combining ECA Rules with temporal interval-based KR Event/Action Logics and Transactional Update Logics. ECA-RuleML Proposal for RuleML Reaction Rules Technical Goup (2005)
14. Paschke, A., Boley, H.: Rules Capturing Events and Reactivity. In: Giurca, A., Gasevic, D., Taveter, K. (eds.) Handbook of Research on Emerging Rule-Based Languages and Technologies: Open Solutions and Approaches, pp. 215–252. IGI Publishing (2009)
15. Paschke, A.: ECA-LP / ECA-RuleML: A Homogeneous Event-Condition-Action Logic Programming Language. In: Int. Conf. on Rules and Rule Markup Languages for the Semantic Web, Athens, Georgia, USA (2006)
16. Liua, J., Zhangb, J., Hub, J.: A case study of an inter-enterprise workflow-supported supply chain management system. Information and Management 42, 441–454 (2005)
17. Futatsugi, K., Goguen, J.A., Ogata, K.: Verifying Design with Proof Scores. Verified Software: Theories, Tools, Experiments 4171, 277–290 (2005)
18. Paschke, A., Kozlenkov, A.: Rule-Based Event Processing and Reaction Rules. In: Governatori, G., Hall, J., Paschke, A. (eds.) RuleML 2009. LNCS, vol. 5858, pp. 53–66. Springer, Heidelberg (2009)
19. Teymourian, K., Paschke, A.: Semantic Rule-Based Complex Event Processing. In: Governatori, G., Hall, J., Paschke, A. (eds.) RuleML 2009. LNCS, vol. 5858, pp. 82–92. Springer, Heidelberg (2009)
20. Reaction RuleML, http://ruleml.org/reaction
21. Diaconescu, R., Futatsugi, K., Ogata, K.: CafeOBJ: Logical Foundations and Methodologies. Computing and Informatics 22, 257–283 (2003)

Parameterized Prefix Distance
between Regular Languages

Martin Kutrib, Katja Meckel, and Matthias Wendlandt

Institut für Informatik, Universität Giessen
Arndtstr. 2, 35392 Giessen, Germany
{kutrib,meckel,matthias.wendlandt}@informatik.uni-giessen.de

Abstract. We investigate the parameterized prefix distance between regular languages. The prefix distance between words is extended to languages in such a way that the distances of all words up to length n to the mutual other language are summed up. Tight upper bounds for the distance between unary as well as non-unary regular languages are derived. It is shown that there are pairs of languages having a constant, degree k polynomial, and exponential distance. Moreover, for every constant and every polynomial, languages over a binary alphabet are constructed that have exactly that distance. From the density and census functions of regular languages the orders of possible distances between languages are derived and are shown to be decidable.

1 Introduction

Finite state devices are used in several applications and implementations in software engineering, programming languages and other practical areas in computer science. They are one of the first and most intensely investigated computational models. Due to several applications and implementations of transducers in theoretical and practical areas of computer science, their fault-tolerance or even usability in the presence of failures is a natural question of crucial importance. The applications are widely spread. For example, finite state transducers are currently used for compiler constructions [1], language and speech processing [7], and even for the design of controllability systems in aircraft design [9]. Much of the underlying theory has originated from linguistics. In natural language and speech processing transducers with more than one hundred million states may be used [8]. All of the components involved may be subject to failure. However, not all faults necessarily incapacitate the automaton entirely. In several applications small aberrations are tolerable. From this point of view the questions of what are tolerable aberrations arise immediately. We consider the distance between the languages accepted by the original and the faulty machine as measure for this purpose. So, even in the case of transducers we regard the accepting part of the computation only.

Inspired by these considerations, here we start to investigate the parameterized prefix distance between regular languages. In [4] several notions of distances

V. Geffert et al. (Eds.): SOFSEM 2014, LNCS 8327, pp. 419–430, 2014.

have been extended from distances between strings to distances between languages (see also [6]). To this end, a relative distance between a language L_1 and a language L_2 is defined to be the supremum of the minimal distances of all words from L_1 to L_2. The distance between L_1 and L_2 is defined as the maximum of their mutual relative distances. Since here we are interested in computations of faulty finite state devices that are still tolerable, we stick with the prefix distance and consider a parameterized extension. For words w_1 and w_2 the prefix distance sums up the number of all letters of w_1 and w_2 that do not belong to a common prefix of these words. One can suppose that on the common prefixes the computations of both machines are the same until a faulty component comes into play and the computations diverge. The parameterized prefix distance between languages sums up the distances of all words up to length n from one language to their closest words from the other language, and vice versa.

Since the distance between identical words should always be 0, for the distances between languages, the number of words in their symmetric difference plays a crucial role. In this connection we utilize the density and census functions that count the number of words in a language. The study of densities of regular languages has a long history (see, for example, [3,5,10,11,12,13]). Restricted to the number of unary words in a binary language the census function has been shown to be log-space many-one complete for #L in [2].

In particular, in the present paper tight upper bounds for the parameterized prefix distance between unary as well as non-unary regular languages are derived. It is shown that there are pairs of languages having a constant, degree k polynomial, and exponential distance. Moreover, for every constant and every polynomial, languages over a binary alphabet are constructed that have exactly that distance. From practical as well as theoretical point of view, it is important to decide this order. Here, the orders of possible distances between regular languages are derived and are shown to be decidable.

2   Preliminaries

We write Σ^* for the set of all words over the finite alphabet Σ, and $\mathbb{N}$ for the set $\{0, 1, 2, \dots\}$ of non-negative integers. The *empty word* is denoted by λ and the *reversal* of a word w by w^R. For the *length* of w we write $|w|$ and for the number of occurrences of a symbol a in w we use the notation $|w|_a$. We use $\subseteq$ for *inclusions* and $\subset$ for *strict inclusions*, and write 2^S for the powerset of a set S.

In general, a *distance* over Σ^* is a function $d : \Sigma^* \times \Sigma^* \to \mathbb{N} \cup \{\infty\}$ satisfying, for all $x, y, z \in \Sigma^*$, the conditions $d(x, y) = 0$ if and only if $x = y$, $d(x, y) = d(y, x)$, and $d(x, y) \le d(x, z) + d(z, y)$.

For example, words w_1 and w_2 over Σ^* can be compared by summing up the number of all letters of w_1 and w_2 that do not belong to a common prefix of these words. This so-called prefix distance $d_{\mathrm{pref}} : \Sigma^* \times \Sigma^* \to \mathbb{N}$ between words is defined to be $d_{\mathrm{pref}}(w_1, w_2) = |w_1| + |w_2| - 2\max\{\,|v| \mid w_1, w_2 \in v\Sigma^*\,\}$. Clearly, $d_{\mathrm{pref}}(w_1, w_2) = 0$ if and only if $w_1 = w_2$, and $d_{\mathrm{pref}}(w_1, w_2) = |w_1| + |w_2|$ if and only if the first letters of w_1 and w_2 are different. Moreover, the prefix distance between two words can be large if their length difference is large.

Distances over Σ are extended to distances between a word and a language by taking the minimum of the distances between the word and the words belonging to the language. For the prefix distance we obtain pref-$d : \Sigma^* \times 2^{\Sigma^*} \to \mathbb{N} \cup \{\infty\}$ which is defined to be

$$\text{pref-}d(w, L) = \begin{cases} \min\{ d_{\text{pref}}(w, w') \mid w' \in L \} & \text{if } L \neq \emptyset \\ \infty & \text{otherwise} \end{cases} .$$

Clearly, pref-$d(w, L) = 0$ if $w \in L$.

The next step is to extend the distance between a word and a language to a distance between two languages $L_1, L_2 \subseteq \Sigma^*$. This can be done by taking the maximum of the suprema of the distances of all words from L_1 to L_2 and vice versa. However, here we are interested in a parameterized definition, where the distance additionally depends on the length of the words. So, the *parameterized prefix distance* between languages pref-$D : \mathbb{N} \times 2^{\Sigma^*} \times 2^{\Sigma^*} \to \mathbb{N} \cup \{\infty\}$ is defined by

$$\text{pref-}D(n, L_1, L_2) = \sum_{\substack{w \in L_1, \\ 0 \leq |w| \leq n}} \text{pref-}d(w, L_2) + \sum_{\substack{w \in L_2, \\ 0 \leq |w| \leq n}} \text{pref-}d(w, L_1).$$

In general, one cannot expect to obtain a convenient description of the parameterized prefix distance for *all* n. So, in the following, if not stated otherwise, it is understood that pref-$D(n, L_1, L_2) = f(n)$ means pref-$D(n, L_1, L_2) = f(n)$, for all n greater than some constant n_0.

The following technical proposition is a useful tool for the analysis and construction of regular languages having a certain distance.

Proposition 1. *Let $L_1, L_2 \subseteq \Sigma^*$ be two languages so that $L_1 \subseteq L_2$.*

1. *For a word $v \in L_2 \setminus L_1$, let $L_1' = L_1 \cup \{v\}$ and $L_2' = L_2 \setminus \{v\}$. Then*

$$\text{pref-}D(n, L_1, L_2) > \text{pref-}D(n, L_1', L_2) \text{ and}$$
$$\text{pref-}D(n, L_1, L_2) > \text{pref-}D(n, L_1, L_2').$$

2. *For a word $v \in \Sigma^* \setminus L_2$, let $L_2' = L_2 \cup \{v\}$. Then*

$$\text{pref-}D(n, L_1, L_2) < \text{pref-}D(n, L_1, L_2').$$

3. *For a word $v \in L_1$, let $L_1' = L_1 \setminus \{v\}$. Then*

$$\text{pref-}D(n, L_1, L_2) < \text{pref-}D(n, L_1', L_2).$$

Example 2. We consider the two regular languages $L_1 = \{a, b\}^* \{ab, ba\} \{a, b\}^*$, that is, the language of all words over $\{a, b\}$ containing the factor ab or ba, and $L_2 = \{a, b\}^* b \{a, b\}^* b \{a, b\}^*$, that is, the language of all words over $\{a, b\}$ containing at least two symbols b.

In order to compute their parameterized prefix distance, first the distances of all words from L_1 to L_2 are determined. All words of L_1 that belong to L_2 are of the forms $\{a, b\}^* \{ab, ba\} \{a, b\}^* b \{a, b\}^*$ or $\{a, b\}^* b \{a, b\}^* \{ab, ba\} \{a, b\}^*$. So, we only have to compute the prefix distances of words w from $a^*\{ab, ba\}a^*$ to L_2, which is 1 since $wb \in L_2$ and pref-$d(w, L_2) = |w| + |wb| - 2|w|$.

Second, all words w from L_2 that are not included in L_1 are of the form b^2b^*. Their prefix distance to L_1 is also always 1, since $wa \in L_1$.

Together, the prefix distance between L_1 and L_2 is

$$\text{pref-}D(n, L_1, L_2) = \sum_{\substack{w \in a^*ba^*, \\ 2 \leq |w| \leq n}} \text{pref-}d(w, L_2) + \sum_{\substack{w \in b^2b^*, \\ 2 \leq |w| \leq n}} \text{pref-}d(w, L_1).$$

These sums can be reformulated by summing up over the sizes of the words and multiplying by their prefix distance to the languages they are not contained in. So, we obtain

$$\text{pref-}D(n, L_1, L_2) = \sum_{i=2}^{n} \left|\{w \in a^*ba^* \mid |w| = i\}\right| \cdot 1 + \sum_{i=2}^{n} \left|\{w \in b^2b^* \mid |w| = i\}\right| \cdot 1.$$

The set $\{w \in a^*ba^* \mid |w| = i\}$ of the first sum contains i words. The set $\{w \in b^2b^* \mid |w| = i\}$ of the third sum has a size of 1. Therefore, the result is

$$\text{pref-}D(n, L_1, L_2) = \sum_{i=2}^{n} i + \sum_{i=2}^{n} 1 = \frac{(n+1)n}{2} - 1 + n - 1 = \frac{n^2}{2} + \frac{3}{2}n - 2.$$

$\square$

3   Upper and Lower Bounds for the Prefix Distance

We turn to investigate the range of possible parameterized distances between regular languages. We are interested in upper bounds and whether these upper bounds are tight, that is, whether there are witness languages showing that the upper bound is, in fact, the best possible.

To determine the upper bound of the prefix distance between two languages $L_1, L_2 \subseteq \Sigma^*$ we consider some word $w \in L_1$ and the shortest word $s \in L_2$. In any case we have pref-$d(w, L_2) \leq |w| + |s|$ and, thus, the word w contributes in a maximal way to the distance if it does not have a common prefix with s. In this case, we have pref-$d(w, L_2) = |w| + |s|$. This observation leads to a general upper bound as follows.

Proposition 3. *Let* $L_1, L_2 \subseteq \Sigma^*$ *be two non-empty languages,* $m_1 = \min\{|w| \mid w \in L_1\}$ *and* $m_2 = \min\{|w| \mid w \in L_2\}$ *be the lengths of the shortest words of* L_1 *and* L_2, *respectively,* $m = \min\{m_1, m_2\}$, *and* $M = \max\{m_1, m_2\}$. *Then*

$$\text{pref-}D(n, L_1, L_2) \leq \sum_{i=m}^{n} |\Sigma|^i \cdot (i + M).$$

The next lemma identifies properties that are necessary for two languages to match the upper bound.

Lemma 4. *Let L_1, L_2 be languages with*

$$m = \min\{\min\{\,|w| \mid w \in L_1\,\}, \min\{\,|w| \mid w \in L_2\,\}\}\ and$$

$$M = \max\{\min\{\,|w| \mid w \in L_1\,\}, \min\{\,|w| \mid w \in L_2\,\}\}.$$

Then the upper bound of Proposition 3 is met only if (i) each word $w \in L_1 \cup L_2$ contributes $|w| + M$ to the prefix distance, (ii) $L_1 \cap L_2 = \emptyset$ if $m \geq 1$, and $L_1 \cap L_2 \subseteq \{\lambda\}$ if $m = 0$, and (iii) $L_1 \cup L_2 = \{\, w \in \Sigma^* \mid |w| \geq m \,\}$.

Proof. We assume $m \geq 1$ and $L_1 \cap L_2 \neq \emptyset$, or $m = 0$ and $L_1 \cap L_2$ is not a subset of $\{\lambda\}$. In both cases there exists at least one word w of length greater than or equal to $\max\{1, m\}$ that does not contribute to pref-$D(|w|, L_1, L_2)$. So there must be a word in $L_1 \cup L_2$ that contributes more than $|w| + M$ to the prefix distance. However, this is a contradiction to the choice of M to be the maximum of the sizes of the shortest words. So, (ii) is a necessary condition.

If $L_1 \cup L_2 \neq \{\, w \in \Sigma^* \mid |w| \geq m \,\}$, then there is a word not in $L_1 \cup L_2$ whose length is at least $\max\{1, m\}$. This word can be added to both languages L_1 and L_2 without affecting pref-$D(n, L_1, L_2)$. Since in this case the intersection $L_1 \cap L_2$ contains a non-empty word, we have a contradiction to (ii). This shows (iii).

At last we assume that there exists a word $w \in L_1 \cup L_2$ that contributes less than $|w| + M$ to pref-$D(|w|, L_1, L_2)$. Then there must be a word in $L_1 \cup L_2$ that contributes more than $|w| + M$ to the prefix distance. The same contradiction as for (ii) shows case (i). □

Lemma 4 particularly shows that the upper bound cannot be reached if $m < M$. Let in this case w with $|w| = M$ be a shortest word in its language, say L_2. Then pref-$d(w, L_1) \leq |w| + m < |w| + M$. So, condition (i) of the lemma is violated. Next we turn to show that the upper bound of Proposition 3 is the best possible, in the sense that there are worst case languages for which it is matched. These languages necessarily satisfy the conditions of Lemma 4.

Proposition 5. *For any $M = m \geq 0$, there are binary regular languages $L_1, L_2 \subseteq \{a, b\}^*$ so that pref-$D(n, L_1, L_2) = \sum_{i=m}^{n} |\Sigma|^i \cdot (i + M)$, where m is the minimum and M is the maximum of the lengths of the shortest words in L_1 and L_2.*

Proof. For any $m \geq 1$, we use the disjoint regular witness languages $L_1 = a\{a, b\}^{m-1}\{a, b\}^*$ and $L_2 = b\{a, b\}^{m-1}\{a, b\}^*$. In particular, no two words of L_1 and L_2 have a common prefix.

Let $w \in L_1$ be some word. Its prefix distance to L_2 is $|w| + m$. Similarly, the prefix distance of every word $w \in L_2$ to the language L_1 is $|w| + m$. So we have

$$\text{pref-}D(n, L_1, L_2) = \sum_{\substack{w \in L_1 \setminus L_2, \\ m \leq |w| \leq n}} |w| + m + \sum_{\substack{w \in L_2 \setminus L_1, \\ m \leq |w| \leq n}} |w| + m.$$

Since $L_1 \cup L_2 = \{ w \in \Sigma^* \mid |w| \geq m \}$ and $L_1 \cap L_2 = \emptyset$ this in turn is

$$\text{pref-}D(n, L_1, L_2) = \sum_{i=m}^{n} |\Sigma|^i \cdot (i + m).$$

If $m = 0$, then the empty word belongs to both languages. In this case we set $L_1 = \{\lambda\} \cup a\{a, b\}^*$ and $L_2 = \{\lambda\} \cup b\{a, b\}^*$ and obtain

$$\text{pref-}D(n, L_1, L_2) = \sum_{i=1}^{n} |\Sigma|^i \cdot i = \sum_{i=0}^{n} |\Sigma|^i \cdot (i + 0) = \sum_{i=m}^{n} |\Sigma|^i \cdot (i + m).$$

$\square$

So far, we considered languages over alphabets with at least two letters. For unary languages the situation changes significantly. An immediate observation is, that every two words have a distance to each other which is given by their length difference only.

Proposition 6. *Let $L_1 \subseteq \{a\}^*$ and $L_2 \subseteq \{a\}^*$ be two non-empty unary languages. Then $\text{pref-}D(n, L_1, L_2) \leq \frac{n(n+1)}{2} + 1$.*

As for the general case, the upper bound for the parameterized prefix distance of unary regular languages is tight. However, the witness languages of the following proof are the only ones whose distance meets the upper bound.

Proposition 7. *There are unary regular languages $L_1, L_2 \subseteq \{a\}^*$ so that their prefix distance is $\text{pref-}D(n, L_1, L_2) = \frac{n(n+1)}{2} + 1$.*

Proof. Let $L_1 = aa^*$ and $L_2 = \{\lambda\}$. These languages are unary, regular, and disjoint. Therefore, the prefix distance of each word in $w \in L_1$ to L_2 is $|w|$. For the only word λ in L_2 its distance to L_1 is $d_{\text{pref}}(\lambda, a) = 1$. So we have $\text{pref-}D(n, L_1, L_2) = 1 + \sum_{i=1}^{n} i = \frac{n(n+1)}{2} + 1$. $\square$

4   Distances Below the Upper Bound

So far, we have explored the upper and lower bounds for parameterized prefix distances. Here we are interested in the question which functions are possible to obtain by considering the prefix distance of two regular languages. The next proposition gives an example for regular languages whose parameterized prefix distance is superpolynomial.

Proposition 8. *There are regular languages L_1 and L_2 even over a binary alphabet so that $\text{pref-}D(n, L_1, L_2) \in \Theta(n2^n)$.*

Proof. Here we can use the witness languages L_1 and L_2 from the case $m = 0$ in the proof of Proposition 5. There,

$$\text{pref-}D(n, L_1, L_2) = \sum_{i=1}^{n} |\Sigma|^i \cdot i = \sum_{i=1}^{n} 2^i \cdot i.$$

has been shown. This sum is equal to $n2^{n+2} - (n+1)2^{n+1} + 2 \in \Theta(n2^n)$. $\square$

Next we give evidence that, for any constant $c \geq 1$, there are regular languages having parameterized prefix distance c.

Proposition 9. *Let* $c \geq 1$ *be an integer. Then there are unary regular languages* L_1 *and* L_2 *so that* pref-$D(n, L_1, L_2) = c$, *for all* $n \geq c$.

Proof. We use the languages $L_1 = \{\lambda\}$ and $L_2 = \{\lambda, a^c\}$ as witnesses. Since $\lambda \in L_1 \cap L_2$ the empty word in L_1 and L_2 does not contribute to the distance between L_1 and L_2. Clearly, pref-$d(a^c, L_1) = c$ and, thus, pref-$D(n, L_1, L_2) = c$, for all $n \geq c$. $\square$

Now we turn to the main part of this section. Given an arbitrary polynomial p with integer coefficients whose leading coefficient is positive, we show how to construct two regular languages over a binary alphabet having exactly the parameterized prefix distance p. Clearly, a negative leading coefficient does not make sense since it would yield a negative distance.

Theorem 10. *Let* $p(n) = x_k \cdot n^k + x_{k-1} \cdot n^{k-1} + \cdots + x_0$ *be a polynomial of degree* $k \geq 0$ *with integer coefficients* x_i, $0 \leq i \leq k$, *and* $x_k \geq 1$. *Then two regular languages* L_1 *and* L_2 *over the alphabet* $\{a, b\}$ *can effectively be constructed so that* pref-$D(n, L_1, L_2) = p(n)$, *for all* $n \geq n_0$, *where* n_0 *is some constant.*

Proof. Proposition 9 already shows the special case $k = 0$. Therefore, we assume $k \geq 1$. The basic idea of the construction is to start with two languages whose distance is already a polynomial of degree k, but its coefficients may be incorrect. Subsequently, the coefficients are corrected one after the other, from x_k to x_0. When coefficient x_i is corrected, the coefficients x_k to x_{i+1} are not affected while the coefficients x_{i-1} to x_0 may be changed.

In general, language L_1 will always be a subset of L_2. In this way, the words from L_1 never contribute to the distance.

For the corrections of the coefficients a set of equally long prefixes is used. So, we define $P \subseteq \{a, b\}^l$, for some constant l, with $P = \{p_0, p_1, \ldots, p_m\}$. Assume for a moment that $l \geq k$ is large enough to perform the following constructions. Later we will give evidence that it always can be chosen appropriately.

We consider auxiliary languages

$$L_{r,-1} = \{p_r\} \text{ and } L_{r,-1,b} = L_{r,-1} \cup L_{r,-1}b,$$
$$L_{r,s} = \{p_r v \mid v \in \{a, b\}^*, |v|_b = s\} \text{ and } L_{r,s,b} = L_{r,s} \cup L_{r,s}b$$

for $s \geq 0$ and $p_r \in P$. Clearly, there are $\binom{n-|p_r|}{s} = \binom{n-l}{s} \in \Theta(n^s)$ many words of length n in the languages $L_{r,s}$. Considering the distance between $L_{r,-1}$ and $L_{r,-1,b}$ we obtain pref-$D(n, L_{r,-1}, L_{r,-1,b}) = 1$. For the distance between $L_{r,s}$ and $L_{r,s,b}$, all words from $L_{r,s}b$ contribute 1 while the words from $L_{r,s}$ contribute nothing. For $s \geq 1$, we obtain

$$\text{pref-}D(n, L_{r,s-1}, L_{r,s-1,b}) = \sum_{i=1}^{n} \binom{i-l}{s-1} = \binom{n-l+1}{s}$$
$$= \frac{(n-l+1) \cdot (n-l) \cdot (n-l-1) \cdots (n-l-s+2)}{s!}$$

which gives us a term of the form $\frac{n^s + y_{s-1} \cdot n^{s-1} + y_{s-2} n^{s-2} + y_{s-3} n^{s-3} + \cdots + y_0}{s!}$, where a rough and simple estimation yields $|y_i| \leq 3^s \cdot l^s$, $0 \leq i \leq s - 1$.

We start the construction by using the union of auxiliary languages with $x_k \cdot k!$ many different prefixes, that is,

$$L_1 = \bigcup_{i=0}^{x_k \cdot k! - 1} L_{i,k-1} \text{ and } L_2 = \bigcup_{i=0}^{x_k \cdot k! - 1} L_{i,k-1,b}.$$

So, we start with a distance of the form

$$x_k n^k + z_{k-1} n^{k-1} + z_{k-2} n^{k-2} + z_{k-3} n^{k-3} + \cdots + z_0,$$

where x_k is already the correct coefficient and $|z_i| \leq x_k \cdot 3^k \cdot l^k$, $0 \leq i \leq k - 1$.

Next we correct the remaining coefficients. Let $x_{max} = \max\{ x_i \mid 0 \leq i \leq k \}$. Concluding inductively, we assume that currently

$$\text{pref-}D(n, L_1, L_2) = x_k n^k + x_{k-1} n^{k-1} + \cdots + x_{k-i+1} n^{k-i+1} + z_{k-i} n^{k-i} + \cdots + z_0,$$

where the coefficients $x_k, x_{k-1}, \ldots, x_{k-i+1}$ are already correct and, moreover, $|z_{k-i}|, |z_{k-i-1}|, \ldots, |z_0| \leq 3^{i-1} \cdot x_{max} \cdot (3^k \cdot l^k)^i$.

In order to obtain the correct coefficient x_{k-i}, we set $d = z_{k-i} - x_{k-i}$ and distinguish the two cases, where d is negative or positive. Clearly, if $d = 0$ the coefficient x_{k-i} is already correct and nothing has to be done.

If $d < 0$, the distance has to be increased. To this end, the auxiliary languages $L_{j,k-i-1}$ and $L_{j,k-i-1,b}$ are used. We add their unions with $|d| \cdot (k - i)!$ many new different prefixes to L_1 and L_2, that is,

$$\bigcup_{j=0}^{|d| \cdot (k-i)! - 1} L_{j,k-i-1} \text{ is added to } L_1 \text{ and } \bigcup_{j=0}^{|d| \cdot (k-i)! - 1} L_{j,k-i-1,b} \text{ is added to } L_2.$$

Since all the prefixes p_j are new and $L_1 \subseteq L_2$, again all words from L_2 contribute 1 to the distance while the words in L_1 contribute nothing. In particular, we have added $|d| n^{k-i} + z'_{k-i-1} n^{k-i-1} + z'_{k-i-2} n^{k-i-2} + \cdots + z'_0$ words up to length n to L_2, where $|z'_{k-i-1}|, |z'_{k-i-2}|, \ldots, |z'_0| \leq |d| \cdot 3^{k-i} \cdot l^{k-i} \leq |d| \cdot 3^k \cdot l^k$. This implies

$$\text{pref-}D(n, L_1, L_2) = x_k n^k + x_{k-1} n^{k-1} + \cdots + x_{k-i} n^{k-i} + z_{k-i-1} n^{k-i-1} + \cdots + z_0,$$

where $x_k, x_{k-1}, \ldots, x_{k-i}$ are already correct and $|z_{k-i-1}|, |z_{k-i-2}|, \ldots, |z_0|$ are at most

$$3^{i-1} \cdot x_{max} \cdot (3^k \cdot l^k)^i + |d| \cdot 3^k \cdot l^k$$
$$= 3^{i-1} \cdot x_{max} \cdot (3^k \cdot l^k)^i + (3^{i-1} \cdot x_{max} \cdot (3^k \cdot l^k)^i + x_{max}) \cdot 3^k \cdot l^k$$
$$= 3^{i-1} \cdot x_{max} \cdot (3^k \cdot l^k)^i + 3^{i-1} \cdot x_{max} \cdot (3^k \cdot l^k)^i \cdot 3^k \cdot l^k + x_{max} \cdot 3^k \cdot l^k$$
$$\leq 3^i \cdot x_{max} \cdot (3^k \cdot l^k)^{i+1}.$$

This concludes the first case.

If $d > 0$, the distance has to be decreased. To this end, words from L_2 are added to L_1 so that they do not contribute to the distance anymore. Let $\tilde{p}$ be one of the $x_k \cdot k!$ prefixes used at the beginning of the induction to establish a polynomial distance of degree k. Moreover, we may assume that $\tilde{p}$ has not been used for the current purpose before.

Then, for $r, t \geq 0$ and $s \geq r$, another auxiliary language is defined as $\tilde{L}_{\tilde{p},r,s,t} = \{ \tilde{p}ub^r v \mid uv \in \{a,b\}^*, |u| = t, |uv|_b = s - r \}$. Here, we set $\tilde{L}_{\tilde{p},r,r-1,t} = \{ \tilde{p}b^r \}$. In these languages the position of the block b^r is fixed, so that the union $\bigcup_{j=0}^{d \cdot (k-i)!-1} \tilde{L}_{\tilde{p},i,k-1,j}$ contains $dn^{k-i} + z'_{k-i-1}n^{k-i-1} + z'_{k-i-2}n^{k-i-2} + \cdots + z'_0$ words up to length n, for $n \geq d \cdot (k-i)! + l + i$, where $|z'_{k-i-1}|, |z'_{k-i-2}|, \ldots, |z'_0| \leq d \cdot 3^{k-i} \cdot (l+i)^{k-i} \leq d \cdot 3^k \cdot l^k$. Now all these words are concatenated with a symbol b and are added to L_1. Since all words do belong to L_2 as well, we obtain

$$\text{pref-}D(n, L_1, L_2) = x_k n^k + x_{k-1} n^{k-1} + \cdots + x_{k-i} n^{k-i} + z_{k-i-1} n^{k-i-1} + \cdots + z_0,$$

where the coefficients $x_k, \ldots, x_{k-i}$ are already correct and analogously to the first case $|z_{k-i-1}|, |z_{k-i-2}|, \ldots, |z_0|$ are at most $3^i \cdot x_{max} \cdot (3^k \cdot l^k)^{i+1}$. This concludes the second case.

The construction is concluded by the observation that choosing $n_0 > l + k$ is sufficient for the auxiliary languages applied in the initial step and the correction steps in the first case. For the corrections in the second case

$$d \cdot (k - i)! + l + i \leq 3^{k+1} \cdot x_{max} \cdot (3^{k^2} \cdot l^{k^2}) \cdot k! \leq n_0$$

is sufficient.

Finally, it has to be shown that the prefix length l always can be chosen appropriately. In the first step, $x_k \cdot k!$ many prefixes are used. For the correction steps, no additional prefix is used in the second case, and $|d| \cdot (k - i)!$ prefixes in the first case. The latter is less than

$$(3^{i-1} \cdot x_{max} \cdot (3^k \cdot l^k)^i + x_{max}) \cdot (k - i)! \leq 3^k \cdot x_{max} \cdot 3^{k^2} \cdot l^{k^2} \cdot k!.$$

Therefore, altogether less than $3^k \cdot x_{max} \cdot 3^{k^2} \cdot l^{k^2} \cdot (k + 1)!$ many prefixes are necessary. On the other hand, there are 2^l prefixes of length l. So it is sufficient to choose l large enough so that $2^l \geq 3^k \cdot x_{max} \cdot 3^{k^2} \cdot l^{k^2} \cdot (k + 1)!$ which is always possible since k and x_{max} are constants and on the right-hand side there is only a polynomial in l. □

5 Decidability of the Order of the Distances

From a practical as well as from a theoretical point of view, it is interesting to decide the order of magnitude of the distance between regular languages. In the definition of the distances, the number of words in the symmetric difference of the languages plays a crucial role. Summing up the distance of each of these words gives the distance of two languages. So, the question arises of how many words up to a certain length are in a given language. The function that counts the number of words of a fixed length n is called the *density function* (see, for example, [12,13] and the references therein). The function that counts the number of words up to a given length n is called *census function*. Clearly, both are closely related. So, first we deduce some decidability results for census functions from results on density functions shown in [12]. From these we derive the orders of possible distances between regular languages and show that the orders are decidable.

More formally, let L be a language over some alphabet Σ. Then its *density function* $\varrho_L : \mathbb{N} \to \mathbb{N}$ is defined as $\varrho_L(n) = |L \cap \Sigma^n| = |\{\, w \in L \mid |w| = n \,\}|$ and its census function $\mathrm{cens}_L : \mathbb{N} \to \mathbb{N}$ as $\mathrm{cens}_L(n) = \sum_{i=0}^{n} \varrho_L(i) = |\{\, w \in L \mid |w| \leq n \,\}|$.

The regular languages often are given in terms of minimal deterministic finite automata (DFA). For simplicity, in the following we write cens_A for $\mathrm{cens}_{L(A)}$, where A is a DFA.

Proposition 11. *Let A be a minimal DFA. Then it is decidable whether cens_A is ultimately constant.*

Proof. The function cens_A is ultimately constant if and only if A accepts a finite language. The finiteness of a regular language is decidable by checking whether each accepting path of A is acyclic.   □

In [12] the following gaps for the density of regular languages have been shown: (i) For any $k \geq 0$, there is no regular language whose density is in $\omega(n^k) \cap o(n^{k+1})$, and (ii) there is no regular language whose density is in $\omega(n^\ell)$ for all $\ell \geq 0$, and in $2^{o(n)}$. So, there is no density function of order $\Theta(\sqrt{n})$, $\Theta(n \log(n))$, or $\Theta(2^{\sqrt{n}})$. But note, the density of, say, the regular language $R_k = \{\, w \in \{a, b\}^* \mid |w|_a = k+1 \text{ and } |w| \text{ is even} \,\}$ is $\varrho_{R_k}(n) \in \Theta(n^{k+1})$ if n is even, and $\varrho_{R_k}(n) = 0$ if n is odd. So, it is neither in $O(n^k)$ nor in $\Omega(n^{k+1})$.

In the following, we say that the density is *polynomial* if the function mapping n to $\max\{\, \varrho(i) \mid 0 \leq i \leq n \,\}$ is of order $\Theta(n^k)$, for some $k \geq 1$. It is *exponential*, if it is neither constant nor polynomial. In the latter case it has to be of the form $2^{\Omega(n)}$.

Since the density function of every regular language is either bounded by a constant, polynomial, or exponential, the next corollary follows.

Corollary 12. *The census function of every regular language is either ultimately constant, polynomial, or exponential.*

Proof. By definition we obtain the census function $\mathrm{cens}(n)$ by summing up the densities up to n. Summing up polynomials of degree $k \geq 0$ gives a polynomial at most of degree $k + 1$. Similarly, summing up exponential functions of the form $2^{\Omega(n)}$ gives again an exponential function of that form.   □

Though not explicitly stated, from the results in [12] it follows that it is decidable whether the density function of a regular language has an upper bound that is constant, polynomial, or exponential.

Moreover, the results in [12] imply a decision procedure for the question whether the census function of a regular language is polynomial or exponential, and for the former cases, whether it is of a certain degree.

Theorem 13. *Let A be a DFA. Then it is decidable whether* cens_A *is exponential or a polynomial. If it is a polynomial, the degree can be computed.*

Proof. If $L(A)$ is a unary language, then cens_A is either ultimately constant or linear. By Theorem 11 we can decide whether it is ultimately constant. If not by the results in [12] it can be decided whether ϱ_A is exponential or polynomial, where in the latter case the degree of the polynomial is computable. From the orders of the density we can derive the order of cens_A.   □

Now we turn to the classes of parameterized prefix distances between regular languages. As mentioned before, for their computation the words in their symmetric difference are central, since only these contribute to the distance.

Let L_1 and L_2 be two languages. By $L_1 \oplus L_2$ we denote their symmetric difference. Let us recall briefly the observation $1 \leq \mathrm{pref}\text{-}d(w, L_1) \leq |w| + |s|$, for $w \notin L_1$ and s being a shortest word in L_1.

Theorem 14. *Let L_1 and L_2 be two regular languages. Then it is decidable whether the parameterized prefix distance* $\mathrm{pref}\text{-}D(n, L_1, L_2)$ *is ultimately constant.*

Proof. The family of regular languages is effectively closed under symmetric difference. So, a representation, say a DFA A, accepting $L_1 \oplus L_2$ can effectively be constructed from DFA accepting L_1 and L_2. Clearly, if $L_1 \oplus L_2$ is finite, then $\mathrm{pref}\text{-}D(n, L_1, L_2)$ is ultimately constant. Conversely, if $L_1 \oplus L_2$ is infinite, then $\mathrm{pref}\text{-}D(n, L_1, L_2)$ cannot be bounded by a constant, since all the infinitely many words in the symmetric difference contribute at least 1 to the distance. Now the theorem follows from the decidability of finiteness of regular languages.   □

Theorem 15. *Let L_1 and L_2 be two regular languages. Then it is decidable whether the parameterized prefix distance* $\mathrm{pref}\text{-}D(n, L_1, L_2)$ *is exponential.*

Proof. As in the proof of Theorem 14 we may assume without loss of generality that a DFA A accepting $L_1 \oplus L_2$ can effectively be constructed from L_1 and L_2. Moreover, one can decide whether $\mathrm{pref}\text{-}D(n, L_1, L_2)$ is ultimately constant. So, assume that it is not.

Any word $|w|$ in the symmetric difference contributes at least 1 and at most $|w| + |s|$ to the distance, where s is the shortest word in the language w does not belong to. Therefore, we know $\mathrm{cens}_A(n) \leq \mathrm{pref}\text{-}D(n, L_1, L_2) \leq (c + n) \cdot \mathrm{cens}_A(n)$, where c is the maximum of the lengths of the shortest words in L_1 and L_2. Since cens_A can only be ultimately constant, polynomial, or exponential, $\mathrm{pref}\text{-}D(n, L_1, L_2)$ is exponential if and only if cens_A is exponential. Now the theorem follows from the possibility to decide whether cens_A is exponential.   □

Theorem 16. *Let L_1 and L_2 be two regular languages and $k \geq 1$ be a constant. Then it is decidable whether the parameterized prefix distance* $\mathrm{pref}\text{-}D(n, L_1, L_2)$ *belongs to* $\Omega(n^k) \cap O(n^{k+1})$.

Proof. It is decidable whether $\mathrm{pref}\text{-}D(n, L_1, L_2)$ is ultimately constant or exponential. If it is neither of these, both census functions $\mathrm{cens}_{L_1 \setminus L_2}$ and $\mathrm{cens}_{L_2 \setminus L_1}$

are ultimately constant or polynomial. Theorem 13 shows that the degree k of the polynomial can be computed. With the fact, that each word $|w|$ contributes at least 1 and at most $|w| + |s|$ to the distance, where s is the shortest word in the language to which w does not belong, we derive pref-$D(n, L_1, L_2) \in \Omega(n^k) \cap O(n^{k+1})$.   □

References

1. Aho, A.V., Sethi, R., Ullman, J.D.: Compilers: Principles, Techniques, and Tools. Addison-Wesley (1986)
2. Àlvarez, C., Jenner, B.: A very hard log space counting class. In: Structure in Complexity Theory Conference, pp. 154–168. IEEE Computer Society (1990)
3. Berstel, J., Reutenauer, C.: Rational Series and Their Languages. EATCS Monographs on Theoretical Computer Science. Springer (1988)
4. Choffrut, C., Pighizzini, G.: Distances between languages and reflexivity of relations. Theoret. Comput. Sci. 286, 117–138 (2002)
5. Eilenberg, S.: Automata, Languages, and Machines. Academic Press (1974)
6. Kruskal, J.B.: An overview of sequence comparison. In: Time Warps, String Edits, and Macromolecules: The Theory and Practice of Sequence Comparison, pp. 1–44. Addison-Wesley (1983)
7. Mohri, M.: Finite-state transducers in language and speech processing. Computational Linguistics 23, 269–311 (1997)
8. Mohri, M.: On the use of sequential transducers in natural language processing. In: Finite-State Language Processing, pp. 355–381. MIT Press (1997)
9. Nerode, A., Kohn, W.: Models for hybrid systems: Automata, topologies, controllability, observability. In: Grossman, R.L., Ravn, A.P., Rischel, H., Nerode, A. (eds.) HS 1991 and HS 1992. LNCS, vol. 736, pp. 317–356. Springer, Heidelberg (1993)
10. Salomaa, A., Soittola, M.: Automata-Theoretic Aspects of Formal Power Series. Texts and monographs in computer science. Springer (1978)
11. Schützenberger, M.P.: Finite counting automata. Inform. Control 5, 91–107 (1962)
12. Szilard, A., Yu, S., Zhang, K., Shallit, J.: Characterizing regular languages with polynomial densities. In: Havel, I.M., Koubek, V. (eds.) MFCS 1992. LNCS, vol. 629, pp. 494–503. Springer, Heidelberg (1992)
13. Yu, S.: Regular languages. In: Handbook of Formal Languages, vol. 1, ch. 2, pp. 41–110. Springer, Berlin (1997)

Ordered Restarting Automata
for Picture Languages[*]

František Mráz[1] and Friedrich Otto[2]

[1] Charles University, Faculty of Mathematics and Physics
Malostranské nám. 25, 118 25 Prague 1, Czech Republic
frantisek.mraz@mff.cuni.cz
[2] Fachbereich Elektrotechnik/Informatik, Universität Kassel
34109 Kassel, Germany
otto@theory.informatik.uni-kassel.de

Abstract. We introduce a two-dimensional variant of the restarting automaton with window size three-by-three for processing rectangular pictures. In each rewrite step such an automaton can only replace the symbol in the middle position of its window by a symbol that is smaller with respect to a fixed ordering on the tape alphabet. When restricted to one-dimensional inputs (that is, words) the deterministic variant of these *ordered restarting automata* only accepts regular languages, while the nondeterministic one can accept some languages that are not even context-free. We then concentrate on the deterministic two-dimensional ordered restarting automaton, showing that it is quite expressive as it can simulate the deterministic sgraffito automaton, and we present some closure and non-closure properties for the class of picture languages accepted by these automata.

Keywords: restarting automaton, ordered rewriting, picture language.

1 Introduction

The restarting automaton was introduced in [5] as a formal device to model the linguistic technique of *analysis by reduction*. Since then many variants and extensions of the basic model have been introduced and studied (for an overview, see, e.g., [8]), and several classical families of formal languages, like the regular languages, the deterministic context-free languages, and the context-free languages, have been characterized by certain types of restarting automata. Recently, the restarting automaton has even been extended to a model that processes two-dimensional inputs, that is, *rectangular pictures* [9]. This model, called *restarting tiling automaton*, is a stateless device with a two-by-two window. In each cycle it scans the current picture based on a given scanning strategy until, at some place, it performs a rewrite step in which it rewrites a single symbol from the current content of its window by a symbol with smaller weight and restarts. If

---

[*] The first author was supported by the Grant Agency of the Czech Republic under the project P103/10/0783.

V. Geffert et al. (Eds.): SOFSEM 2014, LNCS 8327, pp. 431–442, 2014.

no rewrite operation can be performed, then the automaton halts after scanning the current picture completely. It is said to accept if at that point the current picture satisfies certain local conditions at each position. In this it is similar to a tiling automaton (see, e.g., [3]).

Here we introduce and study a type of two-dimensional restarting automaton, the so-called *deterministic 2-dimensional 3-way ordered restarting automaton* (or det-2D-3W-ORWW-automaton, for short) that works more in the spirit of the restarting automata as introduced in [5]. Such an automaton has a window of size 3-by-3, and it scans a given rectangular input picture starting at the top left corner. Based on the current state and the content of its window, it can change its state and move either to the right, down, or up, but not to the left. That's why it is called a *3-way* automaton. It keeps on moving until it either halts, accepting or rejecting, or until it performs a rewrite step, in which it replaces the symbol in the middle of its window by a symbol that is strictly smaller with respect to a given ordering on its tape alphabet. After performing such a rewrite, the automaton restarts immediately, that is, it jumps back to the top left corner, and its internal state is reset to the initial state.

When restricted to one-dimensional inputs (that is, words), then this device just accepts the regular languages. For two-dimensional inputs, however, it is quite powerful, as it can simulate the deterministic sgraffito automaton of [10], which in turn is known to be more expressive than the *four-way alternating automaton* [7] and the deterministic four-way one-marker automaton of Blum and Hewitt [1], and which accepts the *sudoku-deterministically recognizable* picture languages of Borchert and Reinhardt [2] (see [11] and [12]).

This paper is structured as follows. In Section 2 we introduce the *ordered restarting automaton* (ORWW-automaton) for processing words and establish its main properties. In Section 3 we restate the some basic notions on picture languages, we define the two-dimensional extension of the ORWW-automaton, the det-2D-3W-ORWW-automaton, and we illustrate its definition by a detailed example. In the next section we show that this automaton can simulate the deterministic sgraffito automaton, and we prove that the class of picture languages accepted by our devices is incomparable to the class of picture languages that are accepted by nondeterministic sgraffito automata. In the concluding section we present a few closure and non-closure properties, and we conclude with a number of open problems for future research.

2   Ordered Restarting Automaton for Words

An *ordered restarting automaton*, an ORWW-automaton for short, is a one-tape machine that is described by an 8-tuple $M = (Q, \Sigma, \Gamma, \vdash, \dashv, q_0, \delta, >)$, where Q is a finite set of states, Σ is a finite input alphabet, Γ is a finite tape alphabet containing Σ, the symbols $\vdash, \dashv \notin \Gamma$ serve as markers for the left and right border of the work space, respectively, $q_0 \in Q$ is the initial state,

$$\delta : Q \times (((\Gamma \cup \{\vdash\}) \cdot \Gamma \cdot (\Gamma \cup \{\dashv\})) \cup \{\vdash\dashv\}) \to 2^{(Q \times \{\mathsf{MVR}\}) \cup \Gamma \cup \{\mathsf{Accept}\}}$$

is the *transition relation*, where 2^S denotes the powerset of the set S, and $>$ is a *partial ordering* on Γ. The transition relation describes three different types of transition steps:

(1) A *move-right step* has the form $(q', \mathsf{MVR}) \in \delta(q, a_1 a_2 a_3)$, where $q, q' \in Q$, $a_1 \in \Gamma \cup \{\vdash\}$ and $a_2, a_3 \in \Gamma$. If M is in state q and sees the word $a_1 a_2 a_3$ in its read/write window, then this move-right step causes M to shift the read/write window one position to the right and to enter state q'. Observe that no move-right step is possible, if the content of the read/write window ends with the symbol $\dashv$.

(2) A *rewrite step* has the form $b \in \delta(q, a_1 a_2 a_3)$, where $q \in Q$, $a_1 \in \Gamma \cup \{\vdash\}$, $a_2, b \in \Gamma$, and $a_3 \in \Gamma \cup \{\dashv\}$ such that $a_2 > b$ holds. It causes M to replace the symbol a_2 in the middle of its read/write window by the symbol b and to restart, that is, M moves its read/write window to the left end of the tape, so that it contains the left delimiter $\vdash$ and the first two letters of the current tape content, and to reenter the initial state q_0.

(3) An *accept step* has the form $\mathsf{Accept} \in \delta(q, a_1 a_2 a_3)$, where $q \in Q$, $a_1 \in \Gamma \cup \{\vdash\}$, $a_2 \in \Gamma$, and $a_3 \in \Gamma \cup \{\dashv\}$. It causes M to halt and accept. In addition, we allow an accept step of the form $\mathsf{Accept} \in \delta(q_0, \vdash\dashv)$.

If $\delta(q, u) = \emptyset$ for some $q \in Q$ and $u \in ((\Gamma \cup \{\vdash\}) \cdot \Gamma \cdot (\Gamma \cup \{\dashv\})) \cup \{\vdash\dashv\}$, then M necessarily halts, when it is in state q seeing u in its read/write window, and we say that M *rejects* in this situation. Further, the letters in $\Gamma \smallsetminus \Sigma$ are called *auxiliary symbols*.

A *configuration* of M is a word $\alpha q \beta$, where $q \in Q$ and $|\beta| \geq 3$, and either $\alpha = \lambda$ (the empty word) and $\beta \in \{\vdash\} \cdot \Gamma^+ \cdot \{\dashv\}$ or $\alpha \in \{\vdash\} \cdot \Gamma^*$ and $\beta \in \Gamma \cdot \Gamma^+ \cdot \{\dashv\}$; here $q \in Q$ represents the current state, $\alpha \beta$ is the current content of the tape, and it is understood that the read/write window contains the first three symbols of β. In addition, we admit the configuration $q_0 \vdash\dashv$. A *restarting configuration* has the form $q_0 \vdash w \dashv$; if $w \in \Sigma^*$, then $q_0 \vdash w \dashv$ is an *initial configuration*. Further, we use Accept to denote the *accepting configurations*, which are those configurations that M reaches by executing an Accept instruction. A configuration of the form $\alpha q \beta$ such that $\delta(q, \beta_1) = \emptyset$, where β_1 is the current content of the read/write window, is a *rejecting configuration*. A *halting configuration* is either an accepting or a rejecting configuration.

In general, the automaton M is *nondeterministic*, that is, there can be two or more instructions with the same left-hand side (q, u), and thus, there can be more than one computation for an input word. If this is not the case, the automaton is *deterministic*. By det-ORWW we denote the deterministic ordered restarting automata.

We observe that any computation of an ordered restarting automaton M consists of certain phases. A phase, called a *cycle*, starts in a restarting configuration, the head moves along the tape performing MVR operations until a rewrite operation is performed and thus a new restarting configuration is reached. If no further rewrite operation is performed, any computation necessarily finishes in a halting configuration – such a phase is called a *tail*. By $\vdash_M^c$ we denote the execution of a complete cycle, and $\vdash_M^{c*}$ is the reflexive transitive closure of $\vdash_M^c$.

An input $w \in \Sigma^*$ is accepted by M, if there exists a computation of M which starts with the initial configuration $q_0 \vdash w \dashv$, and which finally ends with executing an Accept instruction. The language consisting of all words that are accepted by M is denoted by $L(M)$.

As each cycle ends with a rewrite operation, which replaces a symbol a by a symbol b that is strictly smaller than a with respect to the given ordering $>$, we see that each computation of M on an input of length n consists of at most $n \cdot (|\Gamma| - 1)$ many cycles. Thus, M can be simulated by a nondeterministic single-tape Turing machine in time $O(n^2)$.

The following example illustrates the way in which an ORWW-automaton works. To simplify the presentation we use *meta-instructions* (see, e.g., [8]) to describe the behaviour of this ORWW-automaton. A meta-instruction of the form $(E, u \rightarrow v)$ means that when the current tape contains u preceded by a word from the regular language E, then M can rewrite u into v and restart. A meta-instruction of the form (E, Accept) means that if the tape content is a word from the regular language E, then M can accept it without restart.

Example 1. Let M be the nondeterministic ORWW-automaton on $\Sigma = \{a, b, \#\}$ and $\Gamma = \{a, a_1, a_2, b, b_1, b_2, \#\}$ that is given by the following meta-instructions using the linear ordering $\# > a > b > a_1 > b_1 > a_2 > b_2$, where $c, d, e \in \{a, b\}$:

(1) $(\lambda, \vdash cd \rightarrow \vdash c_1 d)$,
(2) $(\lambda, \vdash c_1 d \rightarrow \vdash c_2 d)$,
(3) $(\vdash \cdot \{a_2, b_2\}^*, c_2 de \rightarrow c_2 d_1 e)$,
(4) $(\vdash \cdot \{a_2, b_2\}^*, c_2 d_1 e \rightarrow c_2 d_2 e)$,
(5) $(\vdash \cdot \{a_2, b_2\}^*, c_2 d\# \rightarrow c_2 d_1 \#)$,
(6) $(\vdash \cdot \{a_2, b_2\}^*, c_2 d_1 \# \rightarrow c_2 d_2 \#)$,
(7) $(\vdash \cdot \{a_2, b_2\}^* \cdot d_1 \cdot \{a, b\}^+ \cdot \# \cdot \{a, b\}^*, cd \dashv \rightarrow cd_1 \dashv)$,
(8) $(\vdash \cdot \{a_2, b_2\}^* \cdot d_2 \cdot \{a, b\}^+ \cdot \# \cdot \{a, b\}^*, cd_1 \dashv \rightarrow cd_2 \dashv)$,
(9) $(\vdash \cdot \{a_2, b_2\}^* \cdot d_1 \cdot \{a, b\}^+ \cdot \# \cdot \{a, b\}^*, cde_2 \rightarrow cd_1 e_2)$,
(10) $(\vdash \cdot \{a_2, b_2\}^* \cdot d_2 \cdot \{a, b\}^+ \cdot \# \cdot \{a, b\}^*, cd_1 e_2 \rightarrow cd_2 e_2)$,
(11) $(\vdash \cdot \{a_2, b_2\}^* \cdot d_1 \cdot \{a, b\}^*, \#de_2 \rightarrow \#d_1 e_2)$,
(12) $(\vdash \cdot \{a_2, b_2\}^* \cdot d_2 \cdot \{a, b\}^*, \#d_1 e_2 \rightarrow \#d_2 e_2)$,
(13) $(\vdash \cdot \{a_2, b_2\}^+ \cdot \# \cdot \{a_2, b_2\}^+ \cdot \dashv, \mathsf{Accept})$.

Then M accepts the following language L on Σ:

$$L = \{ w\#u \mid w, u \in \{a, b\}^*, |w|, |u| \geq 2, u \text{ is a scattered subsequence of } w^R \},$$

which is context-free, but not regular.

Actually, the construction of Example 1 can easily be changed to obtain an ORWW-automaton for the language

$$L'_{\text{copy}} = \{ w\#u \mid w, u \in \{a, b\}^*, |w|, |u| \geq 2, u \text{ is a scattered subsequence of } w \},$$

which is not even context-free. However, while nondeterministic ORWW-automata are quite expressive, it turns out that their deterministic variants are fairly weak.

Theorem 2. REG $= \mathcal{L}(\text{det-ORWW}) \subsetneq \mathcal{L}(\text{ORWW})$.

Proof. Obviously each regular language is accepted by a det-ORWW-automaton. Conversely, let $M = (Q, \Sigma, \Gamma, \vdash, \dashv, q_0, \delta, >)$ be a det-ORWW-automaton, and let $L = L(M)$. A one-tape Turing machine T can simulate M as follows.

When simulating the first sweep to the right, T stores in each tape field the current letter together with the letter from the previous field and the state in which the read/write window of M reaches the position in which the letter under consideration is in the middle of the read/write window. When M rewrites the letter at position i by some smaller letter, then T does the same. In the next step, T moves one position to the left, and from the information stored at that position and from the letter written at position i, it can now determine the operation that M will perform in the next cycle at position $i - 1$. Observe that, since M is deterministic, it must perform MVR-steps until this very position. If M executes another MVR-step at position $i - 1$, then so does T, and it will then store the state in which M reaches position i together with the symbol at position $i - 1$ and the current symbol at position i. If, however, M performs a rewrite step at position $i-1$, then T simulates this rewrite and moves to position $i - 2$. It should be clear that in this way T correctly simulates the computation of M, that is, we see that $L(T) = L(M)$ holds.

At each position M performs at most $\gamma := |\Gamma| - 1$ many rewrite steps. We claim that T visits every tape field at most $2\gamma + 1$ many times. In fact, consider tape position i such that $1 \leq i \leq n = |w|$, where w is the given input. At some point T visits this position for the first time. After each rewrite step executed at position i, T moves left, and hence, it may return to position i again. Thus, it enters position i at most $\gamma + 1$ times from the left. Further, each time a rewrite step is executed at position $i + 1$, T moves to position i from the right. Thus, position i is entered at most γ times from the right. Together this means that T moves to the tape field at position i at most $2\gamma + 1$ many times. Now Hennie has shown in [4] that a Turing machine that visits each of its tape fields at most a constant number of times can only accept a regular language. It follows that the language $L(M) = L(T)$ is regular.

Finally, Example 1 shows that REG is properly contained in $\mathcal{L}(\text{ORWW})$. $\square$

However, it remains open at this point whether every context-free language is accepted by some ORWW-automaton.

3   Picture Languages

Here we use the common notation and terms on pictures and picture languages (see, e.g., [3]). Let Σ be a finite alphabet, and let $P \in \Sigma^{*,*}$ be a *picture* over Σ, that is, P is a two-dimensional array of symbols from Σ. If P is of size $m \times n$, then we write $P \in \Sigma^{m,n}$, $\text{row}(P) = m$ denotes the number of rows of P, and $\text{col}(P) = n$ denotes the number of columns of P. Further, $P(i, j)$ denotes the symbol at row i and in column j for all $1 \leq i \leq m$ and $1 \leq j \leq n$. We introduce a set of five special markers (*sentinels*) $\mathcal{S} = \{\vdash, \dashv, \top, \bot, \#\}$, and we assume that

$\Sigma \cap \mathcal{S} = \emptyset$ for any alphabet Σ considered. In order to enable an automaton to detect the border of P easily, we define the *boundary picture* $\widehat{P}$ over $\Sigma \cup \mathcal{S}$ of size $(m+2) \times (n+2)$. It is illustrated by the following schema:

#	⊤	⊤	$\cdots$	⊤	⊤	#
⊢						⊣
⋮			P			⋮
⊢						⊣
#	⊥	⊥	$\cdots$	⊥	⊥	#

We now extend the det-ORWW-automaton to a model that processes two-dimensional input. This automaton will have a read/write window of size 3×3, which it can move across a given bordered picture $\widehat{P}$ while changing its state. It uses the set $\mathcal{H} = \{R, D, U\}$ of possible *window movements*, where R denotes a step to the *right*, D denotes a step *down*, and U denotes a step *up*. Observe that we do not allow a movement to the left.

Definition 3. *A deterministic two-dimensional three-way ordered RWW-automaton, a* det-2D-3W-ORWW-*automaton for short, is given through a 7-tuple* $M = (Q, \Sigma, \Gamma, \mathcal{S}, q_0, \delta, >)$, *where*

- Q *is a finite set of states containing the initial state* q_0,
- Σ *is a finite input alphabet,* Γ *is a finite tape alphabet containing* Σ *such that* $\Gamma \cap \mathcal{S} = \emptyset$, *and* $>$ *is a partial ordering on* Γ, *and*
- $\delta : Q \times (\Gamma \cup \mathcal{S})^{3,3} \to (Q \times \mathcal{H}) \cup \Gamma \cup \{\mathsf{Accept}\}$ *is the transition function that satisfies the following four restrictions for all* $q \in Q$ *and all* $C \in (\Gamma \cup \mathcal{S})^{3,3}$:
  1. *if* $C(1,2) = \top$, *then* $\delta(q, C) \neq (q', U)$ *for all* $q' \in Q$,
  2. *if* $C(2,3) = \dashv$, *then* $\delta(q, C) \neq (q', R)$ *for all* $q' \in Q$,
  3. *if* $C(3,2) = \bot$, *then* $\delta(q, C) \neq (q', D)$ *for all* $q' \in Q$,
  4. *if* $\delta(q, C) = b \in \Gamma$, *then* $C(2,2) > b$ *with respect to the ordering* $>$.

In addition, we admit the possible transition $\delta\left(\begin{pmatrix} \# & \# \\ \# & \# \end{pmatrix}\right) = \mathsf{Accept}$, *which means that* M *may accept the empty picture.*

Given a picture $P \in \Sigma^{m,n}$ as input, M begins its computation in state q_0 with its read/write window reading the subpicture of size 3×3 of $\widehat{P}$ at the upper left corner. Thus, M sees the subpicture $\begin{pmatrix} \# & \top & \top \\ \vdash & P(1,1) & P(1,2) \\ \vdash & P(2,1) & P(2,2) \end{pmatrix}$. Applying its transition function, M now moves through $\widehat{P}$ until it reaches a state q and a position such that

- either $\delta(p, C)$ is undefined, where C denotes the current content of the read/write window, or
- $\delta(p, C) = \mathsf{Accept}$, or
- $\delta(p, C) = b$ for some letter $b \in \Gamma$ such that $C(2,2) > b$.

In the first case, M gets stuck, and so the current computation ends without accepting, in the second case, M halts and accepts, and in the third case, M replaces the symbol $C(2,2)$ by the symbol b, moves its read/write window back to the upper left corner, and reenters its initial state q_0. This latter step is therefore called a *combined rewrite/restart step*.

In principle it could happen that M does not terminate on some input picture, as it may get stuck on a column, just moving up and down. To avoid this, we *require explicitly* that M halts on all input pictures! This could be realized by either providing a simple pattern, e.g., up* − down* − up* − down*, such that on each column, the sequence of up and down movements must fit this pattern, or one could use an external counter that, for each column entered in the course of a computation, counts the number of uninterrupted up and down movements, making sure that the computation fails as soon as more than $(m \cdot |Q|)$- many such steps are encountered on a column of height m. Actually, for all our examples termination follows easily from the fact that within a column, our automata are just looking for a specific occurrence of some symbol, and if that is not found, then the computation fails anyway.

A picture $P \in \Sigma^{*,*}$ is *accepted* by M, if the computation of M on input P ends with an **Accept** instruction. By $L(M)$ we denote the picture language that consists of all pictures over Σ that M accepts.

When restricted to one-row pictures $P \in \Sigma^{1,*}$, then the det-2D-3W-ORWW-automaton coincides with the det-ORWW-automaton. Thus, Theorem 2 implies that the det-2D-3W-ORWW-automaton only accepts regular word languages.

Example 4. Let $\Sigma = \{0,1\}$, and let $L_{\mathrm{perm}} \subseteq \Sigma^{*,*}$ be the picture language

$$L_{\mathrm{perm}} = \{\, P \in \Sigma^{*,*} \mid row(P) = col(P) \geq 1,$$
$$\text{each row and column contains exactly one symbol 1}\,\}.$$

We describe a det-2D-3W-ORWW-automaton M_{perm} that accepts this language. Obviously, M_{perm} can easily check whether each column of the given input picture P contains exactly one occurrence of the symbol 1 by traversing P column by column from left to right. However, the task of checking that each row contains a unique occurrence of the symbol 1 is much more difficult for M_{perm}, as it cannot perform any move-left steps. Thus, it must use its ability to perform rewrite operations for making this check. As M_{perm} is reset to the initial state and the initial position after each rewrite step, it cannot remember which rows it has checked already. Thus, it must also use the rewrite operations to keep track of these rows.

Let $\Gamma = \Sigma \cup \{0',1'\}$, and let $1 > 0 > 1' > 0'$ be the ordering on Γ to be used. The automaton M_{perm} proceeds as follows:

1. M_{perm} moves down the first column until it finds a row the first symbol of which belongs to Σ. If no such row is found, then M_{perm} checks whether each column contains a unique occurrence of the symbol $1'$, by traversing all columns from left to right. In the affirmative, it halts and accepts, and in the negative, it just halts without accepting.

2. If the first symbol in the current row is from Σ, then M_{perm} moves across this row from left to right. If all symbols in this row are from Σ, then M_{perm} checks that there is a unique occurrence of the symbol 1 in this row. In the negative, it just halts without accepting, and in the affirmative, it rewrites the last symbol 0 or 1 in that row into $0'$ or $1'$, respectively. If, however, M_{perm} enounters a symbol $a' \in \Gamma \setminus \Sigma$ in the current row, then it rewrites the previous symbol from Σ also into its primed variant.

Obviously, M_{perm} rewrites the symbols in each row from right to left into their primed variants, provided each row contains a unique occurrence of the symbol 1. Thereafter, M_{perm} checks that also each column contains a unique occurrence of the symbol 1. In the affirmative, the given input P is a square that belongs to the language L_{perm}. Hence, we see that $L(M_{\text{perm}}) = L_{\text{perm}}$.

Given an input picture P over Σ of size $m \times n$, a det-2D-3W-ORWW-automaton $M = (Q, \Sigma, \Gamma, \mathcal{S}, q_0, \delta, >)$ can execute at most $m \cdot n \cdot (|\Gamma| - 1)$ many cycles, as in each cycle it rewrites one of the $m \cdot n$ many symbols of the current picture by a symbol that is strictly smaller. As each cycle takes at most $m \cdot n \cdot |Q|$ many steps without getting into an infinite loop, we see that for accepting P, M executes at most $m^2 \cdot n^2 \cdot (|\Gamma| - 1) \cdot |Q|$ many steps. Thus, a two-dimensional Turing machine can simulate M in time $O(m^2 \cdot n^2)$. A multi-tape Turing machine that stores P column by column needs m steps to simulate a single move-right step of M. However, during each cycle M can execute at most $n - 1$ such steps. As P is of size $m \cdot n$, we obtain the following upper bound for the time complexity.

Theorem 5. $\mathcal{L}(\text{det-2D-3W-ORWW}) \subseteq \text{DTIME}((\text{size}(P))^2)$.

4 Simulating Sgraffito Automata

Now the question arises about the expressive power of det-2D-3W-ORWW-automata. As a first step towards answering this question, we compare the det-2D-3W-ORWW-automaton to the deterministic sgaffito automaton of [10].

Definition 6. A two-dimensional sgraffito automaton (2SA) is given by a 7-tuple $\mathcal{A} = (Q, \Sigma, \Gamma, \delta, q_0, Q_F, \mu)$, where Σ is an input alphabet and Γ is a working alphabet such that $\Sigma \subseteq \Gamma$, Q is a set of states containing the initial state q_0 and the set Q_F of final states, $\mu : \Gamma \to \mathbb{N}$ is a weight function, and

$$\delta : (Q \setminus Q_F) \times (\Gamma \cup \mathcal{S}) \to 2^{Q \times (\Gamma \cup \mathcal{S}) \times \mathcal{H}}$$

is a transition relation, where $\mathcal{H} = \{R, L, D, U, Z\}$ is the set of possible head movements (the first four elements denote directions (right, left, down, up), while Z represents no movement), such that the following two properties are satisfied:

1. $\mathcal{A}$ is bounded, that is, whenever it scans a symbol from $\mathcal{S}$, then it immediately moves to the nearest field of P without changing this symbol,
2. $\mathcal{A}$ is weight-reducing, that is, for all $q, q' \in Q$, $d \in \mathcal{H}$, and $a, a' \in \Gamma$, if $(q', a', d) \in \delta(q, a)$, then $\mu(a') < \mu(a)$.

Finally, $\mathcal{A}$ is deterministic (a 2DSA), *if $|\delta(q,a)| \leq 1$ for all $q \in Q$ and $a \in \Gamma \cup \mathcal{S}$.*

The notions of configuration and computation are defined as usual. In the initial configuration on input P, the tape contains $\widehat{P}$, $\mathcal{A}$ is in state q_0, and its head scans the top-left corner of P (or the bottom right corner of $\widehat{P}$ when P is empty). The automaton $\mathcal{A}$ accepts P iff there is a computation of $\mathcal{A}$ on input P that finishes in a state from Q_F.

In [10] some closure and non-closure properties are shown for the language classes $\mathcal{L}(2SA)$ and $\mathcal{L}(2DSA)$, and it is proved in [10,11,12] that deterministic sgraffito automata can simulate the 4-way alternating automata of Kari and Moore [7], the deterministic 4-way one-marker automata of Blum and Hewitt [1], and that they accept the sudoku-deterministically recognizable picture languages of Borchert and Reinhardt [2], but that they are strictly weaker than the two-dimensional deterministic forgetting automata of Jiřička and Král [6]. Also the two-dimensional language $L_! = \{ \square^{n,n!} \mid n \in \mathbb{N} \}$, which is not an element of the class REC of recognizable picture languages [3], is accepted by a deterministic sgraffito automaton [12]. Here we show that each deterministic sgraffito automaton can be simulated by a det-2D-3W-ORWW-automaton.

Theorem 7. $\mathcal{L}(2DSA) \subseteq \mathcal{L}(\text{det-2D-3W-ORWW})$.

Proof. Let $\mathcal{A} = (Q, \Sigma, \Gamma, \delta, q_0, Q_F, \mu)$ be a 2DSA accepting $L \subseteq \Sigma^{*,*}$. We define a det-2D-3W-ORWW-automaton $M = (Q', \Sigma, \Omega, \mathcal{S}, q_0', \delta', >)$ by taking $\Omega = \Gamma \cup \Gamma' \cup \{ a_q^{(1)}, a_q^{(2)} \mid a \in \Gamma, q \in Q \}$, where $\Gamma' = \{ a' \mid a \in \Gamma \}$ is a marked copy of Γ, and by letting $>$ be any ordering on Ω that satisfies the following conditions:

1. for all $a \in \Gamma$ and all $q \in Q$, $a > a' > a_q^{(1)} > a_q^{(2)}$, and
2. for all $a, b \in \Gamma$ and all $q \in Q$, if $\mu(a) > \mu(b)$, then $a_q^{(2)} > b$.

These conditions imply that, for all $a, b \in \Gamma$, if $\mu(a) > \mu(b)$, then $a > b$. Thus, as each rewrite step of $\mathcal{A}$ is weight-reducing with respect to μ, it is also ordered with respect to $>$. It remains to describe the transition function δ' of M.

Let $P \in \Sigma^{m,n}$ be the given input picture. If $m = n = 0$ or $m = n = 1$, then M can accept immediately, if $P \in L$. Thus, we assume in what follows that at least one of m, n is larger than 1. In this case, M proceeds as follows:

1. In state q_0', if the current content at tape field $(2,2)$ (which contains the top-left corner of P) is a symbol $a \in \Sigma$, then M rewrites a into $a_{q_0}^{(2)}$. Now the tape of M contains an encoding of the initial configuration of $\mathcal{A}$, where the actual state is encoded together with the current content in the field that is currently visited by the head of $\mathcal{A}$.
2. In state q_0', if the current content of tape field $(2,2)$ is not a symbol from Σ, then M knows that it is already within the simulation of a computation of $\mathcal{A}$. Accordingly, M proceeds to step 3.
3. M scans its tape column by column, from left to right, until it detects a tape field that contains a symbol of the form $c_p^{(1)}$ or $a_q^{(2)}$.
   (a) If it detects a symbol $a_q^{(2)}$, then it realizes that $\mathcal{A}$ has been scanning the symbol a while being in state q. If $q \in Q_F$, that is, q is a final state of $\mathcal{A}$,

then M halts and accepts. Otherwise, assume that $\delta(q, a) = (p, b, d)$. Then M determines the direction d in which the head of $\mathcal{A}$ would move, and it checks whether the tape field reached by d contains a symbol of the form $c_p^{(1)}$. If not, then M rewrites the current content of that field, which is either c or c' for some $c \in \Gamma$, by the symbol $c_p^{(1)}$; otherwise, M rewrites the symbol $a_q^{(2)}$ by the symbol b which would be produced by $\mathcal{A}$ at that position. Actually, if this happens to be the initial position $(2, 2)$, then instead of b, the symbol b' will be written there (see step 2).

(b) If it detects a symbol $c_p^{(1)}$, then it checks whether there is a neighboring tape field that contains a symbol of the form $a_q^{(2)}$. If so, then it rewrites the symbol $a_q^{(2)}$ by the symbol b (or b', see above), which would be produced by $\mathcal{A}$ at that position. Otherwise, M just replaces $c_p^{(1)}$ by the symbol $c_p^{(2)}$.

Thus, for simulating a single step of $\mathcal{A}$, the det-2D-3W-ORWW-automaton M executes three cycles. In the first cycle it searches for the current position of the head of $\mathcal{A}$ by looking for a symbol of the form $a_q^{(2)}$, it determines the next step $\delta(q, a) = (p, b, d)$ of $\mathcal{A}$ from the content of that field, and it encodes the information about this step into the tape field that $\mathcal{A}$ will reach next by replacing the content c or c' of that field by the symbol $c_p^{(1)}$. In the second cycle, M replaces the symbol $a_q^{(2)}$ by the symbol b (or b') that $\mathcal{A}$ would write there, and in the third cycle it replaces the symbol $c_p^{(1)}$ by the symbol $c_p^{(2)}$. Observe how the exponents (1) and (2) are used to distinguish between the tape field that will be the next position of the head of $\mathcal{A}$ and the tape field that is the current position of the head of $\mathcal{A}$. From this description it should be rather clear that $L(M) = L$. □

To show that the above inclusion is proper, we consider another example.

Example 8. Let $L_{1\text{col}}$ be the following picture language over $\Sigma = \{a, b\}$:

$$L_{1\text{col}} = \{\, P \in \Sigma^{2n,1} \mid n \geq 1, \, P(1, 1) \ldots P(n, 1) = (P(n + 1, 1) \ldots P(2n, 1))^R \,\},$$

that is, $L_{1\text{col}}$ consists of all pictures with a single column of even length such that the content of this column read from top to bottom is a palindrome. Using the strategy of Example 1, and noting the fact that a det-2D-2W-ORWW-automaton can freely move up and down a column, it can be shown that this language is accepted by some det-2D-2W-ORWW-automaton.

According to [10] the class $\mathcal{L}(2\text{SA})$ is closed under the operation of *rotation*. This operation turns the language $L_{1\text{col}}$ into the word language L_{pal} of palindromes of even length, which is a non-regular language. As sgraffito automata only accept regular word languages, it follows that $L_{1\text{col}}$ is not accepted by any sgraffito automaton. Thus, we obtain the following.

Corollary 9. $\mathcal{L}(2\text{DSA}) \subsetneq \mathcal{L}(\text{det-2D-3W-ORWW})$.

From Theorem 2 and Example 8 we also obtain the following negative result.

Corollary 10. $\mathcal{L}(\text{det-2D-3W-ORWW})$ *is not closed under rotation.*

From Example 8 we know that det-2D-3W-ORWW-automata accept some languages that are not even accepted by non-deterministic sgraffito automata. Is there a picture language that is accepted by a 2SA, but not by any det-2D-3W-ORWW-automaton? Before we try to answer this question we establish an easy closure property for $\mathcal{L}(\text{det-2D-3W-ORWW})$.

Proposition 11. $\mathcal{L}(\text{det-2D-3W-ORWW})$ *is closed under complement.*

Proof. Let M be a det-2D-3W-ORWW-automaton on Σ that accepts a language $L \subseteq \Sigma^{*,*}$. As M halts on all inputs, we obtain a det-2D-3W-ORWW-automaton for the language $L^c = \Sigma^{*,*} \smallsetminus L$ simply by interchanging Accept transitions with undefined transitions. □

Let $\Sigma = \{0,1\}$, and let L_{dub} denote the language of *duplicates* that consists of all pictures $P \oplus P$, where P is any quadratic picture over Σ and $\oplus$ denotes the operation of *column concatenation* (see [3]). For example,

1	0	1	0	1	0	1	0
0	1	0	0	0	1	0	0
0	0	1	0	0	0	1	0
1	1	1	0	1	1	1	0

is an element of L_{dub}. It is shown in [10] that $L_{\text{dub}} \notin \mathcal{L}(\text{2SA})$, while its complement $(L_{\text{dub}})^c$ is accepted by a sgraffito automaton. Concerning this language we have the following negative result (proof omitted).

Proposition 12. $L_{\text{dub}} \notin \mathcal{L}(\text{det-2D-3W-ORWW})$.

As $\mathcal{L}(\text{det-2D-3W-ORWW})$ is closed under complement, Proposition 12 implies that $(L_{\text{dub}})^c \in \mathcal{L}(\text{2SA}) \smallsetminus \mathcal{L}(\text{det-2D-3W-ORWW})$. Together with the fact that $L_{\text{1col}} \in \mathcal{L}(\text{det-2D-3W-ORWW}) \smallsetminus \mathcal{L}(\text{2SA})$, this yields the following incomparability results.

Corollary 13. *The class of picture languages* $\mathcal{L}(\text{det-2D-3W-ORWW})$ *is incomparable to the classes* $\mathcal{L}(\text{2SA})$ *and* REC *with respect to inclusion.*

Concerning the incomparability to the class REC of recognizable picture languages, it is known $L_{\text{1col}} \notin$ REC, while $(L_{\text{dub}})^c \in$ REC [3,7].

5   Concluding Remarks

We have introduced a class of two-dimensional restarting automata, the det-2D-3W-ORWW-automata. These automata are a direct generalization of the (ordered) restarting automata from words to pictures, and conceptually they are much closer to the underlying ideas of restarting automata as the tiling restarting automata of [9]. However, it can be shown that each deterministic tiling restarting automaton that reads its input pictures column by column, from left to right, can be simulated by a det-2D-3W-ORWW-automaton. Here we have seen that the det-2D-3W-ORWW-automata are even more expressive than deterministic sgraffito automata, although they still only accept regular word languages. In addition, we have seen above that the class of picture languages $\mathcal{L}$(det-2D-3W-ORWW) is closed under complement, but not under the operation of rotation. In addition, it can be shown that this class is closed under union and intersection, but unfortunately it is neither closed under projection nor under horizontal product. However, it is still open whether it is closed under vertical product. We used the explicit requirement that our automata halt on all input pictures. Is there a way to transform a det-2D-3W-ORWW-automaton that does not halt for all input pictures into an equivalent one that does?

References

1. Blum, M., Hewitt, C.: Automata on a 2-dimensional tape. In: Proc. IEEE Computer Society, SWAT 1967, Washington, DC, USA, pp. 155–160 (1967)
2. Borchert, B., Reinhardt, K.: Deterministically and sudoku-deterministically recognizable picture languages. In: Loos, R., Fazekas, S., Martin-Vide, C. (eds.) LATA 2007, Preproc. Report 35/07. Research Group on Mathematical Linguistics, pp. 175–186. Universitat Rovira i Virgili, Tarragona (2007)
3. Giammarresi, D., Restivo, A.: Two-dimensional languages. In: Rozenberg, G., Salomaa, A. (eds.) Handbook of Formal Languages, vol. 3, pp. 215–267. Springer, New York (1997)
4. Hennie, F.: One-tape, off-line Turing machine computations. Inform. Contr. 8, 553–578 (1965)
5. Jančar, P., Mráz, F., Plátek, M., Vogel, J.: Restarting automata. In: Reichel, H. (ed.) FCT 1995. LNCS, vol. 965, pp. 283–292. Springer, Heidelberg (1995)
6. Jiřička, P., Král, J.: Deterministic forgetting planar automata are more powerful than non-deterministic finite-state planar automata. In: Rozenberg, G., Thomas, W. (eds.) DLT 1999, pp. 71–80. World Scientific, Singapore (2000)
7. Kari, J., Moore, C.: New results on alternating and non-deterministic two-dimensional finite-state automata. In: Ferreira, A., Reichel, H. (eds.) STACS 2001. LNCS, vol. 2010, pp. 396–406. Springer, Heidelberg (2001)
8. Otto, F.: Restarting automata. In: Ésik, Z., Martin-Vide, C., Mitrana, V. (eds.) Recent Advances in Formal Languages and Applications. SCI, vol. 25, pp. 269–303. Springer, Berlin (2006)
9. Průša, D., Mráz, F.: Restarting tiling automata. In: Moreira, N., Reis, R. (eds.) CIAA 2012. LNCS, vol. 7381, pp. 289–300. Springer, Heidelberg (2012)
10. Průša, D., Mráz, F.: Two-dimensional sgraffito automata. In: Yen, H.-C., Ibarra, O.H. (eds.) DLT 2012. LNCS, vol. 7410, pp. 251–262. Springer, Heidelberg (2012)
11. Průša, D., Mráz, F., Otto, F.: Comparing two-dimensional one-marker automata to sgraffito automata. In: Konstantinidis, S. (ed.) CIAA 2013. LNCS, vol. 7982, pp. 268–279. Springer, Heidelberg (2013)
12. Průša, D., Mráz, F., Otto, F.: New results on deterministic sgraffito automata. In: Béal, M.-P., Carton, O. (eds.) DLT 2013. LNCS, vol. 7907, pp. 409–419. Springer, Heidelberg (2013)

Unary NFAs with Limited Nondeterminism

Alexandros Palioudakis, Kai Salomaa, and Selim G. Akl

School of Computing, Queen's University, Kingston, Ontario K7L 3N6, Canada
{alex,ksalomaa,akl}@cs.queensu.ca

Abstract. We consider unary finite automata employing limited nondeterminism. We show that for a unary regular language, a minimal finite tree width nondeterministic finite automaton (NFA) can always be found in Chrobak normal form. A similar property holds with respect to other measures of nondeterminism. The latter observation is used to establish relationships between classes of unary regular languages recognized by NFAs of given size where the nondeterminism is limited in various ways. Finally, we show that the branching measure of a unary NFA is always either bounded by a constant or has an exponential growth rate.

Keywords: finite automata, limited nondeterminism, state complexity, unary regular languages.

1 Introduction

The descriptive complexity of finite automata has been studied for over half a century, and there has been particularly much work done over the last two decades. Good general surveys on the topic include [8,9] and as examples of early papers on state complexity of finite automata we mention [17,18,19].

Motivated by the well known exponential trade-off in the NFA (nondeterministic finite automaton) to DFA (deterministic finite automaton) conversion, the literature has considered various ways of quantifying the amount of nondeterminism in finite automata. The *degree of ambiguity* of an NFA refers to the number of accepting computations on a given input [15,23]. The *guessing measure*, roughly, counts the number of advice bits used by an accepting computation on a given input [7,11]. The *branching* of an NFA is the product of the degrees of nondeterministic choices on the best accepting computation [7,14] and the *trace* of an NFA is the corresponding worst-case measure [21]. The *tree width measure* [20] counts the total number of computation paths corresponding to a given input. This measure is called *leaf size* in [10,11], see also [1]. The reader is referred to [6] for more information and references on NFAs employing limited nondeterminism.

With a few exceptions, little is known about the interrelationships of the different nondeterminism measures from a descriptional complexity point of view. Directly based on the definitions it follows that the branching and guessing measure are exponentially related [7] and some further results can be found in [10,11,21]. The size trade-off between NFAs of finite branching and DFAs with multiple initial states has been considered in [13,22].

V. Geffert et al. (Eds.): SOFSEM 2014, LNCS 8327, pp. 443–454, 2014.

In this paper we study the interrelationships of the different nondeterminism measures for the special case of unary NFAs. We show that for a given $k \in \mathbb{N}$ and a unary regular language, a minimal NFA with tree width k or trace k can always be found in Chrobak normal form. An analogous result for unary NFAs with finite ambiguity is known from [12]. The above normal form result is used to show that the state complexity classes defined by bounded tree width and by bounded trace, respectively, coincide in the case of unary regular languages and a similar correspondence, with certain limitations, holds for state complexity classes of unary regular languages defined by bounded ambiguity. The situation is different for the branching measure. In contrast with the measures of tree width, trace and ambiguity, it remains open whether unary NFAs with finite branching could have a normal form with a simple nice structure.

In the literature it is known that the growth rate of the degree of ambiguity and of tree width can be either constant, polynomial or exponential and that the growth rate of the trace measure is always either constant or exponential [16,11,21]. As our main result in Section 4, we show that the branching function of a unary NFA is either constant or grows exponentially, and in the latter case give a lower bound for the exponential growth rate that depends only on the number of states. It remains open whether for an NFA defined over an arbitrary alphabet that has unbounded branching, the branching growth rate is always exponential.

2   Preliminaries

We assume that the reader is familiar with the basic definitions concerning finite automata [24,25] and descriptional complexity [6,9]. Here we just fix some notation needed in the following.

The set of strings, or words, over a finite alphabet Σ is Σ^*, the length of $w \in \Sigma^*$ is $|w|$ and ε is the empty string. The set of positive integers is denoted by $\mathbb{N}$. The cardinality of a finite set S is $\#S$.

A nondeterministic finite automaton (NFA) is a 5-tuple $A = (Q, \Sigma, \delta, q_0, F)$, where Q is a finite set of states, Σ is a finite alphabet, $\delta : Q \times \Sigma \to 2^Q$ is the transition function, q_0 is the initial state and $F \subseteq Q$ is the set of accepting states. The function δ is extended in the usual way as a function $Q \times \Sigma^* \to 2^Q$ and the language recognized by A, $L(A)$, consists of strings $w \in \Sigma^*$ such that $\delta(q_0, w) \cap F \neq \emptyset$. An NFA A is called deterministic finite automaton (DFA) if for every state q of A and letter a of the input alphabet of A, the transition function goes to at most one state, i.e. $\#\delta(q, a) \leq 1$. Unless otherwise mentioned, we assume that any state q of an NFA A is reachable from the start state and some computation originating from q reaches a final state. The *size of A* is the number of states of A, i.e. $\text{size}(A) = \#Q$.

A special case of an NFA $A = (Q, \Sigma, \delta, q_0, F)$ is when the alphabet Σ has a unique letter. In this case we call the NFA A *unary* and we omit the alphabet of its tuple notation. Similarly, the transition function δ of an unary NFA has one argument, i.e. $\delta : Q \to 2^Q$. For a unary NFA $A = (Q, \delta, q_0, F)$ over an alphabet

$\Sigma = \{a\}$ we say that the number m is accepted by the NFA A instead of the word a^m, when $a^m \in L(A)$. To avoid confusion between operations on numbers and strings we use the symbols $+, \times, \cup, \cdot$ for the operations addition, multiplication, union, and concatenation respectively.

Every unary regular language L has a period and a preperiod. The period and preperiod of a regular language L are natural numbers m and n_0, respectively, where for all $n > n_0$ we have $n \in L$ if and only if $n + m \in L$.

The minimal size of a DFA (respectively, an NFA) recognizing a regular language is called the state complexity (respectively, the nondeterministic state complexity) of L and denoted $\mathrm{sc}(L)$ (respectively, $\mathrm{nsc}(L)$). Note that we allow DFAs to be incomplete and, consequently, the deterministic state complexity of L may differ by one from a definition using complete DFAs.

A *computation* of an NFA A from a state s_1 to a state s_2 is a sequence of transitions (q_i, a_i, p_i), $1 \le i \le k$, where $q_{i+1} = p_i$, $i = 1, \ldots, k-1$, and $s_1 = q_1$, $s_2 = p_k$. The underlying word of a computation $(q_1, a_1, q_2) (q_2, a_2, q_3)$ $\cdots (q_m, a_m, q_{m+1})$ is $a_1 a_2 \cdots a_m$. For $x \in \Sigma^*$, $\mathrm{comp}_A(x)$ denotes the set of all computation of A with underlying word x, starting from the initial state of A. We call a computation of A accepting if it starts from the initial state and it finishes at a final state. For $x \in \Sigma^*$, $\mathrm{acc\_comp}_A(x)$ denotes the set of all accepting computations of A with underlying word x.

We say that the computations C and C' on word w are equivalent if C and C' begin in the same state and they both end in the same state.

The *branching of a transition* (q, a, p) of an NFA A, denoted by $\beta_A((q, a, p))$, is the number $\#\delta(q, a)$ and the *branching of a computation* C, denoted by $\beta_A(C)$, is the product of the branching of each transition in C. The *branching of a word* $x \in L(A)$ is the minimum branching among all accepting computations by reading the word x, the branching of a word x is given by the formula $\beta_A(x) = \min\{\beta_A(C) \mid C \in \mathrm{acc\_comp}_A(x)\}$. The branching of an NFA A, denoted by $\beta(A)$, is the maximum branching of A on any string, assuming this quantity is bounded. More details on the branching measure can be found in [7].

We have also considered a worst-case variant of the above measure, so called *trace* [21]. The trace of an NFA A on a string x is the maximum branching among all computations reading the word x (accepting or not). The trace of a word x is given by the formula $\tau_A(x) = \max\{\beta_A(C) \mid C \in \mathrm{comp}_A(y), y \text{ is a prefix of } x\}$ (the prefixes of the word x are in the given formula to emphasize that we include also computation reading only an initial part of the word x). The trace of an NFA A, denoted by $\tau(A)$, is the maximum trace of A on any string, assuming this quantity is bounded.

The computation tree of an NFA A on string w is defined in the natural way and denoted as $T_{A,w}$. The tree width of A on w, $\mathrm{tw}_A(w)$, is the number of leaves of $T_{A,w}$ and the tree width of A, $\mathrm{tw}(A)$ (if it is finite) is the maximum tree width of A on any string w. The formal definitions associated with computation trees and tree width of an NFA can be found in [20,21].[1] The ambiguity of A

---

[1] Note that the tree width of an NFA is unrelated to the notion of tree width as used in graph theory [2].

on w, $\mathrm{amb}_A(w)$, is the number of accepting leaves of $T_{A,w}$ and the ambiguity of A, $\mathrm{amb}(A)$ (if it is finite) is the maximum ambiguity of A on any string w. Ambiguity is a well studied measure of nondeterminism, more details on ambiguity in NFAs can be found in [6].

Next we want to consider questions that involve the state complexity of classes of NFAs of limited nondeterminism. To formalize such question we have to define the following notation, where $sNFA$ is the set of all NFAs, α is a measure of nondeterminism, and c a constant.

$$nsc_{\alpha \leq c}(L) = \min_{A \in sNFA} \{size(A) \mid L = L(A) \text{ and } \alpha(A) \leq c\}$$

Now the numbers $nsc_{\beta \leq k}(L)$ and $nsc_{\tau \leq k}(L)$ have a meaning. The number $nsc_{\beta \leq k}(L)$ is the size of a smallest NFA A such that $L = L(A)$ and $\beta(A) \leq k$. The number $nsc_{\tau \leq k}(L)$ is the smallest number of states required from an automaton B such that $\tau(B) \leq k$. It is easy to see that $nsc_{\beta \leq k}(L) \leq nsc_{\tau \leq k}(L)$ by the definitions of the measures branching and trace.

Let us remind to the reader the Chrobak normal form [3]. A unary NFA A is in Chrobak normal form if initially the states of A form a 'tail' and later, at the end of the tail, are followed nondeterministically by disjoint deterministic cycles. Note, that the only state with nondeterministic choices is the last state of the tail. Formally, the NFA $M = (Q, \delta, q_0, F)$ is in Chrobak normal form if it has the following properties:

(i) $Q = \{q_0, \ldots, q_{t-1}\} \cup C_1 \cup \cdots \cup C_k$, where $C_i = \{p_{i,0}, p_{i,1}, \ldots, p_{i,y_i-1}\}$ for $i \in \{1, \ldots, k\}$,

(ii) $\delta = \{(q_i, q_{i+1}) \mid 0 \leq i \leq t-2\} \cup \{(q_{t-1}, p_{i,0}) \mid 1 \leq i \leq k\} \cup \{(p_{i,j}, p_{i,j+1}) \mid 1 \leq i \leq k, 1 \leq j \leq y_i - 2\} \cup \{(p_{i,y_i-1}, p_{i,0}) \mid 1 \leq i \leq k\}$.

We will use also a more relaxed normal form for unary NFAs which we call a *semi-Chrobak normal form*. A semi-Chrobak normal form NFA consists of a tail and a finite number of disjoint cycles. The only nondeterministic transitions are from the last state of the tail to the cycles, however, as opposed to the usual Chrobak normal form now there may be more than one transition from the last state of the tail to the same cycle. An example of a unary NFA in semi-Chrobak normal can be found in Figure 1.

3 Finite Tree Width and Chrobak Normal Form

For an NFA A in Chrobak normal form it is easy to determine the various nondeterminism measures of A.

Lemma 3.1. *Let A be a Chrobak normal form NFA with k cycles. Then $\beta(A) = \tau(A) = \mathrm{tw}(A) = k$.*

Furthermore, if A is a minimal NFA for $L(A)$ and $m \geq k$, then

$$size(A) = nsc_{\beta \leq m}(L(A)) = nsc_{\tau \leq m}(L(A)) = nsc_{\mathrm{tw} \leq m}(L(A)) = nsc(L(A)).$$

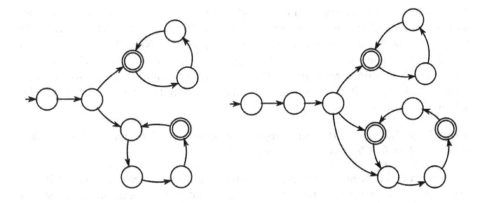

Fig. 1. Two unary NFAs in Chrobak normal form and semi-Chrobak normal form respectively

Proof. The first claim follows directly from the definition of Chrobak normal form. If A is minimal, $\text{size}(A) = \text{nsc}(L(A))$ and the chain of equalities follows because for any $m \geq k$ and $\varphi \in \{\beta, \tau, \text{tw}\}$, $\text{nsc}(L(A)) \leq \text{nsc}_{\varphi \leq m}(L(A)) \leq \text{size}(A)$. □

The semi-Chrobak normal form is a less restrictive variant of the Chrobak normal form. Lemma 3.2 shows that a semi-Chrobak NFA can be transformed to a Chrobak normal form NFA of the same size.

Lemma 3.2. *Every semi-Chrobak normal form NFA has an equivalent Chrobak normal form NFA of same size.*

Proof outline. A semi-Chrobak normal form NFA A with tail T and cycles $C_1, \ldots, C_k$ can be transformed to a Chrobak normal form NFA B with the same tail T and cycles $C_1, \ldots, C_k$. The NFA B has only one transition from the last state of T to each cycle C_i and to compensate for the omitted transitions we add new final states to the cycles. □

Chrobak showed in [3] that every unary NFA can be transformed into an equivalent NFA in Chrobak normal form with losing in efficiency in terms of the number of states. In the following lemma we show that we can transform any NFA with finite tree width into an NFA in Chrobak normal form without losing in efficiency.

Theorem 3.1. *Let A be a unary n-state NFA with tree width k. Then there exists an equivalent Chrobak normal form NFA B with at most n states and tree width k.*

Proof outline. Since the NFA A has finite tree width, we can divide its states into two groups. The states of the first group are the ones belonging in a cycle of A and the second group has the rest. The first group can have only disjoint cycles and its states can have only deterministic choices. We can replace the states of the second group with a chain (the number of the new states is at most as the number of states in the second group). The tail of the NFA B is made from the new states and its cycles are the cycles of the NFA A.   □

Theorem 3.1 speaks about NFAs with finite tree width, not only minimal automata. Sometimes we may use the minimality of an automaton so we want to emphazise that it also holds for minimal automata. We do that with the following corollary.

Corollary 3.1. *For any unary regular language, a state minimal finite tree width NFA is in Chrobak normal form.*

Moreover, Theorem 3.1 suggests a better comparison between NFA with finite tree width and a deterministic finite automaton with multiple initial states (MDFA) [5]. Recall that in [22] we have seen that the size of an MDFA can be exponentially larger than the size of a finite tree width NFA as a function of the degree of its tree width. In the next corollary we show that this is not the case for unary languages.

Corollary 3.2. *Let B be an n-state unary NFA with tree width $k \geq 2$. Then, there is an MDFA B' equivalent with B such that it has at most $k \times n - 5 \times (k-1)$ states.*

Corollary 3.2 gives an upper bound on the size of MDFAs in terms of the size of equivalent NFAs with finite tree width. The size of an MDFA can be linearly more than the size of an equivalent finite tree width NFA. However, note that the limited state complexity of a regular language L for finite tree width k is at most the limited state complexity of MDFAs having k initial states plus one. We can not do better than this, since there are languages that make these quantities equal. Such a language is $L = (p_1)^* \cup \ldots \cup (p_k)^*$, where the numbers p_j are prime, for $1 \leq j \leq k$.

In [21] we have seen that every NFA has finite tree width if and only if it has finite trace. In that paper we have also seen that the trace of an NFA can be as small as its tree width, but the trace can also be exponentially larger than the tree width. In the next corollary, we show that these two measures are equivalent for unary minimal NFAs. Its proof comes from Theorem 3.1 and Lemma 3.1.

Corollary 3.3. *For every unary regular language L and every natural number k, we have the following equality,*

$$\mathrm{nsc}_{\tau \leq k}(L) = \mathrm{nsc}_{\mathrm{tw} \leq k}(L)$$

Moreover, there is an NFA A with tree width k and trace k such that $L = L(A)$ and $size(A) = \mathrm{nsc}_{\tau \leq k}(L) = \mathrm{nsc}_{\mathrm{tw} \leq k}(L)$.

Corollary 3.3 compares minimal NFAs with finite tree width and NFAs with finite trace. The corollary says that the size of a minimal NFA with tree width k is the same as the size of a minimal NFA with trace k, and vice versa. The question here is whether this is true comparing NFAs with finite branching with

NFAs with finite trace or finite tree width. This question seems more difficult since the branching and tree width measures are not comparable, in general. For example take the automaton A of Figure 2, then, for all $m \in \mathbb{N}$, we have that $\mathrm{tw}_A(m) = m + 1$ and $\beta_A(m) = 2$. For the automaton B of the same figure, for all $m \in \mathbb{N}$, we have $\mathrm{tw}_B(m) = m + 1$ and $\beta_B(m) = 2^m$. Both NFAs A and B are minimal for the language $L = \{i \in \mathbb{N} \mid i \geq 1\}$. However, from Corollary 3.3 we have that for bounded branching and bounded tree width we can show an inequality between $\mathrm{nsc}_{\mathrm{tw} \leq k}(L)$ and $\mathrm{nsc}_{\beta \leq k}(L)$. Since for any NFA A, $\beta(A) \leq \tau(A)$, as a consequence of Corollary 3.3 we have Corollary 3.4.

Fig. 2. The NFA A is on the left, the NFA B is on the right

Corollary 3.4. *For every unary regular language L and every natural number k, we have that* $\mathrm{nsc}_{\beta \leq k}(L) \leq \mathrm{nsc}_{\mathrm{tw} \leq k}(L)$

In contrast to Corollary 3.3, the inequality for a unary language L, $\mathrm{nsc}_{\beta \leq k}(L) \leq \mathrm{nsc}_{\mathrm{tw} \leq k}(L)$ of Corollary 3.4 cannot be replaced by an equality. Let A be the unary NFA depicted in Figure 3. On any accepted input, the NFA A needs to go through the first cycle at most two times and, hence, $\beta_A = 4$. On the other hand, it is easy to verify that any NFA with finite tree width for the language $L(A)$ needs at least 6 states.

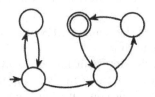

Fig. 3. A unary NFA recognizing the language $2^* \cdot 3^+$

Another consequence of Theorem 3.1 is a connection between finite tree width NFAs and finite ambiguous NFAs. Recall in [20] we have seen that the limited state complexity of finite tree width NFAs can be exponentially larger than the size of unambiguous equivalent NFAs. The next theorem, from [12], is a similar with Theorem 3.1 for finite ambiguity.

Theorem 3.2 ([12]). *Let A be a unary n-state NFA with ambiguity k. There is an equivalent Chrobak normal form NFA B with at most n states and ambiguity k.*

As a result of Theorem 3.1 and Theorem 3.2 we have the following corollary.

Corollary 3.5. *Let A be a unary NFA with n states, for every $k \geq \frac{n}{2}$ we have* $\mathrm{nsc}_{\mathrm{amb} \leq k}(L(A)) = \mathrm{nsc}_{\mathrm{tw} \leq k}(L(A))$.

Our final result for this section shows that there is a strict hierarchy for the state complexity of NFA with finite tree width and respectively with finite trace.

Lemma 3.3. *For any $k \in \mathbb{N}$ there exists a unary regular language L such that*

$$\mathrm{nsc}_{\tau \leq k+1}(L) < nsc_{\tau \leq k}(L) \quad and \quad \mathrm{nsc}_{\mathrm{tw} \leq k+1}(L) < nsc_{\mathrm{tw} \leq k}(L).$$

4   Growth Rate of the Branching Measure

In this section we study the growth rate of the branching function for unary automata. We show that the β-function of a unary NFA is either bounded by a constant or grows exponentially.

Before we show our result on the growth rate of the β-function, we will give two lemmas which we are going to use in the proof of the main theorem of this section. The first lemma shows that every computation going through a deterministic cycle S can be transformed into an equivalent computation that repeats cycles outside of S at most a fixed number of times.

Lemma 4.1. *Let A be a unary NFA and consider a computation C of A that contains a deterministic cycle S of length k. Then the computation C has an equivalent computation C' containing the deterministic cycle S, such that every state of A appearing in C' and not in S appears at most k times.*

Proof. Let $A = (Q, \Sigma, q_0, F)$ be a unary NFA and a computation C containing a deterministic cycle S of length k.

Consider a state q appearing in C but not in the cycle S. Let us assume that the state q appears at least $k + 1$ times in the computation C, notice here that since cycle S is deterministic if the computation C enter the cycle S then it stays inside the cycle S.

Let d_i be the length of the computation until i-th occurrence of the state q in the computation C, for $i = 1, \ldots, k + 1$. Two of the d_i numbers must be in same congruence class modulo k. Then, the steps of the computation C between these two occurrences of q can be shifted into the cycle S.

Continuing similar for all the states of A appearing in the computation C but not in the cycle S, we end up with a computation C', equivalent with the computation C, such that every state before the cycle S appears at most k times.

$\square$

In [3] Chrobak showed that for every unary NFA A with n states, there is a unary NFA A' in Chrobak normal form with n states participating in cycles and $O(n^2)$ states in its tail. For our purposes, we need a more accurate estimation on the size of the tail, which is due to Gawrychowski in [4].

Lemma 4.2 ([4]). *For each unary NFA A with n states, there is a unary NFA A' in Chrobak normal form with at most n states participating in cycles and with a tail with at most $n \times (n-1)$ states.*

Now we are ready to give the main result of this section.

Theorem 4.1. *Let A be a unary NFA with n states. Then either for every natural number m, $\beta_A(m) \leq n^{n \times (n-1)}$, or for every natural number $m > n \times (n-1)$,*

$$\beta_A(m) \geq 2^{\lfloor \frac{m}{e\sqrt{n \times \log n}} \rfloor}$$

Proof. Let $A = (Q, \delta, q_0, F)$ be a unary NFA with n states. Let the sets S_N and S_D be

$S_D = \{i \in \mathbb{N} \mid i \text{ is accepted by a path that enters a deterministic cycle}\}$
$S_N = \{i \in \mathbb{N} \mid i \text{ is accepted by a path that does not enter a deterministic cycle}\}$

We have that $L(A) = S_D \cup S_N$, note here that the sets S_D and S_N do not need to be disjoint.

Let us have now the NFA $D = (Q, \delta, q_0, F')$ which is exactly like the NFA A except the final states. The final states of D are the final states of A which are in a deterministic cycle. Since, if a computation enters a deterministic cyclic cannot exit this cycle, we have that $L(D) = S_D$. Similarly, we define the NFA $N = (Q, \delta, q_0, F/F')$ which we get by the NFA A by changing the final states appearing in deterministic cycles to non-final states. The NFA N recognizes the language S_N.

Since the unary languages S_N and S_D are regular, they both have a period and a preperiod. From Lemma 4.2 there are unary NFAs N' and D' in Chrobak normal form, respectively equivalent to the NFAs N and D, with tails of size at most $n \times (n-1)$. Then, the number $n \times (n-1)$ is a preperiod for both of the sets S_D and S_N.

Now we are interested in the relationship between S_D and S_N after their preperiod $n \times (n-1)$. To simplify things, we denote $S^{(k)} = \{x \in S \mid x > k\}$. Then, we are interested in the sets $S_N^{(n \times (n-1))}$ and $S_D^{(n \times (n-1))}$. Here we can have two cases, in the first case we have $S_N^{(n \times (n-1))} \subseteq S_D^{(n \times (n-1))}$, and in the second case we have $S_N^{(n \times (n-1))} \setminus S_D^{(n \times (n-1))} \neq \emptyset$.

In the former case, where $S_N^{(n \times (n-1))} \subseteq S_D^{(n \times (n-1))}$, we have that for every number k in $L(A)$ greater than $n \times (n-1)$ there is a computation C_k which enters a deterministic cycle and accepts k. In this case we argue that for every $m \in \mathbb{N}$, we have $\beta_A(m) \leq n^{n \times (n-1)}$. For every computation C with length at most $n \times (n-1)$ the maximum branching that C can have is $n^{n \times (n-1)}$. For any other accepting computation C with length greater than $n \times (n-1)$ there is an equivalent computation C' which enters in a deterministic cycle. From Lemma 4.1, we can safely assume that the computation C' enters a deterministic cycle and the number of states before this cycle is at most $\frac{(n)^2}{4}$, which implies that the branching of C' is at most $n^{n \times (n-1)}$.

The latter case is a bit more complicated. In this case we have that $S_N^{(n \times (n-1))} \setminus S_D^{(n \times (n-1))} \neq \emptyset$ which means that there is an accepting computation C which does not enter a deterministic cycle and there is not an equivalent computation with C such that enters a deterministic cycle. In the rest of this proof we will argue that since the computation C exists, there are infinitely many such computations and additionally the distance between two consecutive computations is at most $e^{\sqrt{n \times \log n}}$.

From Lemma 4.2 both of the NFAs N' and D', which are in Chrobak normal form, have disjoint cycles with the sum of sizes at most n. Consider the least common multiple of the sizes of the cycles of the NFAs N' and D' combined, denote this least common multiple by z. Then the number z is a period for both of the languages S_D and S_N (which implies also the same for the language $L(A)$). To justify this, consider that there is an accepting computation C_1 which has finished in a final state q inside of one of these cycles, call it S_1, in one of these NFAs (same argument applies to all the cycles of N' and D'). From the definition of the number z the size of the cycle S_1 divides z, then by continuing the computation C_1 after it has reached state q for z more steps, we end up again at the state q. This is true for all final states that are in any cycle of these NFAs. Vice versa, if there is an accepting computation C_2 greater in length than $n \times (n-1)+z$, then it finishes at a final state p which appears in one of these cycles, call it S_2. Considering the computation C_2' which follows computation C_2 and stops exactly z steps before. The computation C_2 is greater that $n \times (n-1)$ which implies that the computation C_2' finishes also in the same cycle S_2. Since the size of S_2 divides z the computation C_2' finishes at the state p as the computation C_2. From the Landau's function we know that z is at most $e^{\sqrt{n \times \log n}}$.

Since the number z is a period for the sets S_N and S_D and there is a number i in $S_N^{(n \times (n-1))} \setminus S_D^{(n \times (n-1))} \neq \emptyset$, we have that for every integer m, such that $i + m \times z > n \times (n-1)$, the number $i + m \times z$ is in S_N but not in S_D. Since we know that $z \leq e^{\sqrt{n \times \log n}}$ we get that for every $e^{\sqrt{n \times \log n}}$ consecutive values (or even more frequently) we have an integer in $S_N \setminus S_D$. The computation on such an integer has at least one nondeterministic step (i.e. the transition function has a choice of at least 2) for every n steps of the computation. We know that the β function is monotone, but we do not know if there are numbers in S_D between two consecutive numbers in $S_N \setminus S_D$. Hence, we have that for every number $m > n \times (n-1)$ the function $\beta_A(m)$ is at least as the function

$$f(m) = \begin{cases} 2^{\frac{m}{n}} & \text{if } m \equiv 0 \bmod \lceil e^{\sqrt{n \times \log n}} \rceil, \\ f(m-1) & \text{otherwise .} \end{cases}$$

It is easy to see that for every value m the function $f(m)$ is at least $2^{\lfloor \frac{m}{e^{\sqrt{n \times \log n}}} \rfloor}$, which implies that the branching $\beta_A(m)$ for every number $m \geq n \times (n-1)$ is at least $2^{\lfloor \frac{m}{e^{\sqrt{n \times \log n}}} \rfloor}$. □

In Theorem 4.1 the values, given as a function of the number of states, are not intended to be the best possible. With a more careful analysis, especially the constant upper bound for $\beta_A(m)$ in the first case, $n^{n \times (n-1)}$, could be significantly improved. It is somewhat less clear whether, in the second case, the factor $e^{\sqrt{n \times \log n}}$ in the exponent can be improved.

5 Conclusion and Open Problems

We have seen that for a unary regular languages, a minimal finite tree width NFA can always be found in Chrobak normal form. It remains open whether there is a similar simple structure for a minimal finite branching unary NFA.

We also studied the growth rate of branching for unary NFAs. We have seen that the branching function of unary NFAs is either bounded by a constant or grows exponentially. The characterization of possible growth rates of the branching function of an NFA defined over an arbitrary alphabet remains open. Here techniques used for our result dealing with the unary case seem not directly applicable.

References

1. Björklund, H., Martens, W.: The tractability frontier for nfa minimization. In: Aceto, L., Damgård, I., Goldberg, L.A., Halldórsson, M.M., Ingólfsdóttir, A., Walukiewicz, I. (eds.) ICALP 2008, Part II. LNCS, vol. 5126, pp. 27–38. Springer, Heidelberg (2008)
2. Bondy, J., Murty, U.: Graph theory. Graduate texts in mathematics, vol. 244. Springer (2008)
3. Chrobak, M.: Finite automata and unary languages. Theor. Comput. Sci. 47(3), 149–158 (1986)
4. Gawrychowski, P.: Chrobak normal form revisited, with applications. In: Bouchou-Markhoff, B., Caron, P., Champarnaud, J.-M., Maurel, D. (eds.) CIAA 2011. LNCS, vol. 6807, pp. 142–153. Springer, Heidelberg (2011)
5. Gill, A., Kou, L.T.: Multiple-entry finite automata. J. Comput. Syst. Sci. 9(1), 1–19 (1974)
6. Goldstine, J., Kappes, M., Kintala, C.M.R., Leung, H., Malcher, A., Wotschke, D.: Descriptional complexity of machines with limited resources. J. UCS 8(2), 193–234 (2002)
7. Goldstine, J., Kintala, C.M.R., Wotschke, D.: On measuring nondeterminism in regular languages. Inf. Comput. 86(2), 179–194 (1990)
8. Holzer, M., Kutrib, M.: Nondeterministic finite automata - recent results on the descriptional and computational complexity. Int. J. Found. Comput. Sci. 20(4), 563–580 (2009)
9. Holzer, M., Kutrib, M.: Descriptional and computational complexity of finite automata - a survey. Inf. Comput. 209(3), 456–470 (2011)

10. Hromkovič, J., Karhumäki, J., Klauck, H., Schnitger, G., Seibert, S.: Measures of nondeterminism in finite automata. In: Montanari, U., Rolim, J.D.P., Welzl, E. (eds.) ICALP 2000. LNCS, vol. 1853, pp. 199–210. Springer, Heidelberg (2000)
11. Hromkovic, J., Seibert, S., Karhumäki, J., Klauck, H., Schnitger, G.: Communication complexity method for measuring nondeterminism in finite automata. Inf. Comput. 172(2), 202–217 (2002)
12. Jiang, T., McDowell, E., Ravikumar, B.: The structure and complexity of minimal nfa's over a unary alphabet. In: Biswas, S., Nori, K.V. (eds.) FSTTCS 1991. LNCS, vol. 560, pp. 152–171. Springer, Heidelberg (1991)
13. Kappes, M.: Descriptional complexity of deterministic finite automata with multiple initial states. Journal of Automata, Languages and Combinatorics 5(3), 269–278 (2000)
14. Kintala, C.M.R., Wotschke, D.: Amounts of nondeterminism in finite automata. Acta Inf. 13, 199–204 (1980)
15. Leung, H.: On finite automata with limited nondeterminism. Acta Inf. 35(7), 595–624 (1998)
16. Leung, H.: Separating exponentially ambiguous finite automata from polynomially ambiguous finite automata. SIAM J. Comput. 27(4), 1073–1082 (1998)
17. Lupanov, O.B.: A comparison of two types of finite sources. Problemy Kibernetiki 9, 328–335 (1963)
18. Meyer, A.R., Fischer, M.J.: Economy of description by automata, grammars, and formal systems. In: SWAT (FOCS), pp. 188–191. IEEE Computer Society (1971)
19. Moore, F.R.: On the bounds for state-set size in the proofs of equivalence between deterministic, nondeterministic, and two-way finite automata. IEEE Transactions on Computers C-20(10), 1211–1214 (1971)
20. Palioudakis, A., Salomaa, K., Akl, S.G.: State complexity and limited nondeterminism. In: Kutrib, M., Moreira, N., Reis, R. (eds.) DCFS 2012. LNCS, vol. 7386, pp. 252–265. Springer, Heidelberg (2012)
21. Palioudakis, A., Salomaa, K., Akl, S.G.: Comparisons between measures of nondeterminism on finite automata. In: Jurgensen, H., Reis, R. (eds.) DCFS 2013. LNCS, vol. 8031, pp. 217–228. Springer, Heidelberg (2013)
22. Palioudakis, A., Salomaa, K., Akl, S.G.: Finite nondeterminism vs. dfas with multiple initial states. In: Jurgensen, H., Reis, R. (eds.) DCFS 2013. LNCS, vol. 8031, pp. 229–240. Springer, Heidelberg (2013)
23. Ravikumar, B., Ibarra, O.H.: Relating the type of ambiguity of finite automata to the succinctness of their representation. SIAM J. Comput. 18(6), 1263–1282 (1989)
24. Shallit, J.O.: A Second Course in Formal Languages and Automata Theory. Cambridge University Press (2008)
25. Yu, S.: Regular Languages. In: Handbook of Formal Languages, vol. 1, pp. 41–110. Springer (1998)

Recommending for Disloyal Customers with Low Consumption Rate

Ladislav Peska and Peter Vojtas

Faculty of Mathematics and Physics
Charles University in Prague
Malostranske namesti 25, Prague, Czech Republic
{Peska,vojtas}@ksi.mff.cuni.cz

Abstract. In this paper, we focus on small or medium-sized e-commerce portals. Due to high competition, users of these portals are not too loyal and e.g. refuse to register or provide any/enough explicit feedback. Furthermore, products such as tours, cars or furniture have very low average consumption rate preventing us from tracking unregistered user between two consecutive purchases. Recommending on such domains proves to be very challenging, yet interesting research task. For this task, we propose a model coupling various implicit feedbacks and object attributes in matrix factorization. We report on promising results of our initial off-line experiments on travel agency dataset. Our experiments corroborate benefits of using object attributes; however we are yet to decide about usefulness of some implicit feedback data.

Keywords: Recommender systems, implicit feedback, content-based attributes, e-commerce, matrix factorization.

1 Introduction

Recommending on the web is both an important commercial application and popular research topic. The amount of data on the web grows continuously and it is nearly impossible to process it directly by a human. The keyword search engines were adopted to fight information overload but despite their undoubted successes, they have certain limitations. Recommender systems can complement onsite search engines especially when user does not know exactly what he/she wants.

1.1 Our Motivation

Recently, a lot of attention was attracted by the NetFlix prize[1] aiming to predict future user rating based on previously rated objects. Our scenario is however substantially different.

First, we want to recommend on e-commerce applications, where users are generally less willing to provide explicit feedback (besides they are usually not

---

[1] http://www.netflixprize.com/

V. Geffert et al. (Eds.): SOFSEM 2014, LNCS 8327, pp. 455–465, 2014.

capable to provide relevant feedback for products, they did not buy or test yet). The vast majority of e-shops do not force users to register at all, which makes it difficult to track them. Combination of unregistered user and low consumption rate on particular product domains (tours, cars, furniture, specialized sport goods etc.) prevents us in many cases from effectively tracking consecutive purchases of the user.[2]

Given the described preconditions, the vast majority of users appear to be new users exacerbating the cold start problem. We cannot hope for tens of ratings as in multimedia portals, but rather need to cope with a few visited pages.

On the other hand, we can monitor additional data to improve our prospects. It is possible to record various types of implicit feedback (page-views, time on page, mouse usage, scrolling etc.) or track user behavior on category pages. Objects of the e-shops also contain vast number of content-based attributes..

1.2   Main Contribution

The main contributions of this paper are:
- Identifying challenging domain for recommender systems.
- Proposed recommending methods based on matrix factorization incorporated with implicit feedback and object attributes.
- Off-line experiments on travel agency dataset.
- Travel agency dataset for further experiments.

The rest of the paper is organized as follows: review of some related work is in section 2. In section 3 we describe travel agency dataset and in section 4 methods how to incorporate it into matrix factorization. Section 5 contains results of our off-line experiment on a travel agency dataset. Finally section 6 concludes our paper and points to our future work.

2   Related Work

The area of recommender systems has been extensively studied recently and it is out of scope of this paper to provide more elaborated overview. We suggest Konstan and Riedl [6] paper as a good starting point.

Implicit user feedback did not draw too much attention during early research on recommender systems, but grows on importance recently mainly in commercial area. A well known attempt to categorize various types of implicit feedback was proposed by Kelly and Teevan [5]. The list of feedback types presented in this study is due to the age of the survey not exhaustive any more (e.g. commencement of social networks and related activities), however it still presents a good starting point.

---

[2] Unregistered users are usually tracked by cookies stored within the browser. The cookie however may be lost for various reasons e.g. flushing data stored in browser, switching browsers or updating hardware. For a fixed user, the average distance between two consecutive purchases of e.g. a tour is approximately one year.

An approach to categorize methods for collecting implicit feedback was made by Gauch et al. [3]. Authors however didn't mention one important method: tracking user directly by a subprogram, e.g. a JavaScript code, deployed on the current website. Such approach needs cooperation from the site owners and it is possible to track only within cooperating sites, but data are received about vast majority of the users and it is possible to monitor more different information than e.g. by analyzing weblog files. Such user-tracking is also well suited to form a knowledge base for site-wide recommender systems.

Although there is quite large variety in implicit feedback types, publicly available datasets e.g. Last.fm[3] usually provides only very limited (if any) implicit feedback. As result, many research papers consider only binary implicit feedback and only one or a few types of feedback [4], [10].

Our feedback collecting method is based on JavaScript tracking of multiple feedback types on a single site. We have tested usefulness of various implicit feedback types in an on-line experiment in our previous work [11] and described method for learning user preferences from implicit feedback in [12].

Matrix factorization techniques [8] are currently main-stream algorithm to learn user preferences gaining their popularity during NetFlix prize. We use content-boosted matrix factorization as proposed in Forbes and Zhu [2] as one of the preference learning methods. Content-based and hybrid recommender systems may benefit from using additional data sources such as LOD,[4] however not too much research was made in this area yet. One of the exception is work of Ostuni et al. [10] proposing graph based recommender system combining LOD knowledge and implicit feedback, or our own work on incorporating LOD directly as object attributes [13].

Interesting phenomenon is also development of user preferences over time. However no general conclusion was yet achieved: e.g. Xia et al.[14] uses time decay in item-to-item recommender systems, however e.g. Koren [7] argues that classical time-windows or instance-decay methods loses too much signal.

Among papers concerning recommending for e-commerce we would like to mention Linden et al. [9] describing Item-to-item collaborative filtering or from more recent work Belluf et al. [1] measuring business impact of recommender systems deployment in an on-line experiment.

3   Datasets

We have collected usage data from one of the major Czech travel agencies. Data were collected from December 2012 to April 2013. Travel agency is typical e-commerce enterprise, where customers buy products only once in a while (most typically once a year). The site does not force users to register and so we can track unique users only with cookies stored in the browser. User typically browses or searches through several

---

[3] http://lastfm.com, dataset available from http://ir.ii.uam.es/hetrec2011/datasets.html

[4] Linked Open Data, http://linkeddata.org/

categories, compares few objects (possibly on more websites) and eventually buys a single object. Buying more than one object at the time is very rare.

Our main objective while collecting dataset was to capture various possibly interesting data and thereafter decide about their usefulness in the experiments.

3.1 Implicit Feedback Data

In our previous work [11], only the user behavior on objects was monitored. Certain actions user committed was stored into database table to serve as implicit feedback. The table is in form of:

ImpFeedback(UID, OID, PageView, Mouse, Scroll, Time)

UID and *OID* are unique user and object identifiers and Table 1 contains full description of implicit feedbacks. Note that *UID* is based on cookie stored by browser, so we cannot e.g. distinguish between two persons using the same computer. Table contains approx. 39 000 records with 0.09% density of *UID*× *OID* matrix.

Table 1. Description of the considered implicit feedbacks for user visiting an object

Factor	Description
PageView	Count(*OnLoad()* event on object page)
Mouse	Count(*OnMouseOver()* events on object page)
Scroll	Count(OnScroll() events on object page)
Time	Sum(time spent on object page)

3.2 Click Stream Data

However pages showing detail of an object represents less than 50% of visited pages. The rest consists mostly from various category pages accessed either via site menu or attributes search. The depth and broadness of category tree depends on the current user interface and nature of the objects of the domain. The travel agency website used in the experiments allows user to more or less freely combine values/intervals of several attributes (not all attributes of objects), where some predefined combinations are accessible via site menu and others can be derived through binary search. In order to determine importance of category pages for computing user preference, we have collected dataset containing user's click-stream throughout the website. The table is in form of:

ClickStream(UID, PageID, SessionNo, Timestamp)

PageID serves as unique identifier of visited page. There is unique mapping from *OID* to *PageID*. **ClickStream** table contains approx. 121900 records and matrix *UID×PageID* has density of 0.17%.

3.3 Content Based Attributes

Finally each object and category page can be assigned with several content-based attributes. The information value of content-based attributes varies in different domains from very informative like e.g. laptops and computers to almost valueless like secondhand bookshops [13]. The travel agency dataset can be classified somewhere in between as there are some informative attributes, but a lot of important information is accessible only through textual description leaving some space for employing textual data mining techniques in the future. Table 2 contains list of available attributes for travel agency domain. In order to handle attributes properly in the experiments, they were transferred into the Boolean vector (Integer values e.g price was discretized equipotently into 10 intervals). The resulting **Attributes** matrix contains 2300 objects (and categories) with 925 features each.

Table 2. Description of content-based attributes and their cardinality per tour

Attribute	Description
TourType	Type of the tour (e.g. sightseeing)
Country	Destination country of the tour (e.g. Spain) [1..n]
Destination	More specific destination (e.g. Costa Brava) [0..n]
AccomodationType	Quality of the accommodation (e.g. 3*);
Accommodation	Specific accommodation for the tour [0..n]
Board	Type of board (e.g. breakfast, half-board)
Transport	Type of transport (coach, air…)
Price	Base price per person; integer
AdditionalInfo	IDs of information linked to the tour (e.g. about visited places, destination country etc.) [0..n]

4   Algorithms

Matrix factorization techniques are currently leading methods for learning user preferences, so we decided to adopt and slightly adjust them to suit our needs. We skip more elaborated introduction to the matrix factorization as it is quite well known technique and suggest Koren et al. [8] for more elaborated introduction.

Given the list of users $U = \{u_1,...,u_n\}$ and objects $O = \{o_1,...,o_m\}$, we can form the user-object rating matrix $\mathbf{R} = [r_{uo}]_{n \times m}$. With lack of explicit feedback, user-object rating r_{uo} in our case carries only Boolean information whether user u visited object o. For a given number of latent factors f, matrix factorization aims to decompose original $\mathbf{R}$ matrix into $\mathbf{U}\mathbf{O}^T$, where $\mathbf{U}$ is $n \times f$ matrix of user latent factors (μ_i^T stands for latent factors vector for particular user u_i) and $\mathbf{O}^T$ is $f \times m$ matrix of object latent factors (σ_i is vector of latent factors for particular object o_i).

$$\mathbf{R} \approx \mathbf{U}\mathbf{O}^T = \underbrace{\begin{bmatrix} \mu_1^T \\ \mu_2^T \\ \vdots \end{bmatrix}}_{n \times f} \times \underbrace{[\sigma_1 \quad \sigma_2 \quad ...]}_{f \times m} \tag{1}$$

Unknown rating for user i and object j is predicted as $\hat{r}_{ij} = \mu_i^T \sigma_j$. Our target is to learn matrixes $\mathbf{U}$ and $\mathbf{O}$ minimizing errors on known ratings. Regularization penalty is added to prevent overfitting. The optimization equation is defined as follows:

$$\min_{\mathbf{U},\mathbf{O}} \left\| \mathbf{R} - \mathbf{U}\mathbf{O}^T \right\|^2 + \lambda(\|\mathbf{U}\|^2 + \|\mathbf{O}\|^2) \tag{2}$$

This equation can be solved e.g. by Stochastic Gradient Descent (SGD) technique iterating for each object and user vectors:

$$\mu_i = \mu_i + \eta \left(\sum_{j \in K_{ui}} (r_{ij} - \mu_i^T \sigma_j)\sigma_j - \lambda \mu_i \right)$$

$$\sigma_j = \sigma_j + \eta \left(\sum_{i \in K_{oj}} (r_{ij} - \mu_i^T \sigma_j)\mu_i - \lambda \sigma_j \right) \tag{3}$$

Where η is learning rate, K_{ui} set of all objects rated by user u_i and K_{oj} set of all users, who rates object o_j. The described method represents **baseline** algorithm. We now present three extensions to this method. Those extensions are independent of each other and can be combined freely.

4.1   Category Extension

Category extension expands list of objects to include also category pages covered in **ClickStream** dataset (see section 3.2). The rest of the algorithm remains the same. This allows us to better track movement of the user over the website and thus his/her preferences (note that both objects and categories may share some content attributes and thus we can infer their similarity).

Categories however cannot form full-bodied objects. One problem is that categories usually do not carry all features of the regular objects (e.g. there are some object attributes not available for binary search); however this is strongly domain dependent. The other problem is that categories cannot be effectively recommended, except maybe some very simple cases, as it would be impossible to derive any explanation for such recommendations (e.g. recommending category of Family Holidays in Spain in 2*hotels with half-board etc.). So the category objects may only aid us in inferring user preferences, not in fulfilling them.

4.2   Implicit Feedback Extension

Implicit feedback (*ImpF* in experiments) extension involves deeper studying of user activity within the object and thus better estimating how much it is preferred. User activity is stored in **ImpFeedback** dataset (currently page views, time on page, mouse moves and scrolling events are monitored). Improved user-to-object rating r_{uo}^+ replaces original Boolean r_{uo} where applicable.

The algorithm for computing r_{uo}^+ was presented in our previous work [12]. It works in two steps, where various methods can be used in both steps. At first, it considers each feedback type (page views, time on page etc.) separately resulting into the vector of so called *local preferences* – real numbers from [0, 1] interval representing inferred user preference based on a single feedback type. In the second step, those values are aggregated together into a single value (r_{uo}^{+}).

In our previous work [12], we experimented with various procedures in both steps of the algorithm, but used rather simple recommending methods to test it. In our current work we focused more on usability of such information in up-to-date recommending methods and so used standard procedures here. We leave more detailed study of combination of both factors to our future work.

With the absence of any explicit feedback, we have adopted business-like approach and stated that buying an object represents full user preference. Preference based on other types of feedback is defined through its similarity with the purchasing behavior. Details of this approach and argumentation supporting it can be found in [12], in our current research was used weighted average to combine local preferences and discretized intervals of feedback type values in local preferences computation.

4.3   Attributes Extension

Attributes extension involves using content-based attributes of objects (and categories). We adopt approach of *Forbes and Zhu* [2] to deal with object attributes and implement their algorithm as PHP library. Their *content boosted matrix factorization method* is based on the assumption that each object's latent factors vector is a function of its attributes. Having $O_{m\times f}$ matrix of object latent factors, $A_{m\times a}$ matrix of object attributes and $B_{a\times f}$ matrix of latent factors for each attribute, the constraint can be formulated as:

$$O = AB \tag{4}$$

Under the constraint (4), we can reformulate both matrix factorization problem (1), its optimization equation (2) and gradient descend equations (3):

$$\mathbf{R} \approx \mathbf{U}\mathbf{O}^T = \mathbf{U}\mathbf{B}^T\mathbf{A}^T = \underbrace{\begin{bmatrix} \mu_1^T \\ \mu_2^T \\ \vdots \end{bmatrix}}_{n\times f} \times \underbrace{\mathbf{B}^T}_{f\times a} \times \underbrace{\begin{bmatrix} a_1 & a_2 & \ldots \end{bmatrix}}_{a\times m} \tag{1a}$$

$$\mu_i = \mu_i + \eta \left(\sum_{j\in K_{ui}} (r_{ij} - \mu_i^T\mathbf{B}^T a_j)\mathbf{B}^T a_j - \lambda\mu_i \right)$$

$$\sigma_j = \sigma_j + \eta \left(\sum_{(i,j)\in K} (r_{ij} - \mu_i^T\mathbf{B}^T a_j)a_j\mu_i^T - \lambda\mathbf{B} \right) \tag{3a}$$

5   Experiments

In order to determine usefulness of additional data, we have examined each of their combination in an off-line experiment. We set the off-line evaluation as close to the real setting as possible, because we intent to deploy the recommender system into the full operation in the future.

5.1   Experimental Settings

We first needed to set experiment goals and success metrics. As the datasets contains only implicit feedback, we cannot rely on user rating and related error metrics e.g. RMSE or MAE (no need to mention, that those metrics do not reflect well real-world success metrics anyway). Precision / Recall methods are also problematical as the nature of observed implicit feedback is only positive[5] and absence of any feedback also cannot be automatically interpreted as negative feedback (user might not be aware of the object). As result, for arbitrary fixed user, we only have some evidence of positive preference for the minority of objects and know nothing about the rest.

Typical usage of recommender systems in e-commerce is to present a list of top-k objects to the user. We let recommending methods to rank objects and denote as success if the algorithm manages to rank well enough those objects, we have some evidence of their positive preference.

As we lack any explicit feedback, we need to infer positive preference from the implicit data. For the purpose of this rather early work we consider that every object the user has *visited* is positively preferred by him/her. It is possible to use more selective meanings of positive preference e.g. to consider only purchased objects as positively preferred. Such assumption will however lead to insufficient amount of data in the test set so we leave the problem of finer grained preference to the future work and bigger experimental datasets.

Recommending method evaluation was carried out as follows: For each user, his/her click stream was divided into two halves according to its timestamp – earlier data serving as train set and following as test set. Note that only users with at least two visited objects qualify for the experiment. There are other ways to divide train/test set e.g. to apply cross-validation, but we rather took advantage of possibility to use and compare stream or time-aware algorithms on the same dataset in future. The resulting train set contains 32480 records from 2956 users (13124 concerning objects and 19356 category pages). The test set contains 12749 records (objects only, as we intent to recommend only objects to the user).

All learning methods were initialized and trained with the same train set, 10 latent factors and maximal 50 iterations.

Then, for arbitrary fixed user, we let each method to rate all objects, sort them according to the rating and look up positions of objects from the test set. In production recommender system, we should take into account also other metrics like diversity, novelty or serendipity and probably want to pre-select the list of candidate objects, but for purpose of our experiment, we will focus on rating only.

We adopt normalized distributed cumulative gain (*nDCG*) as our main success metric. The premise of nDCG is that relevant documents appeared low in the

---

[5] Although some experiments and/or models of negative implicit feedback can be found in the literature, They are yet to be more extensively tested and eventually accepted.

recommended list should be penalized (logarithmical penalty applied) as they are less likely to attract user attention. This fits well for the recommending scenario, where lower-ranked objects are presented on less desirable positions. It is also possible to restrict DCG to sum only up to *top-k*th position as only *top-k* objects are shown to the user. However there is no justification to set any particular *top-k* and the list of eligible objects for recommendation could be pre-filtered (e.g. keep only objects from certain category if user is browsing the category). As result objects on lower ranks can keep some value too.

The results of ***Presence@top-k*** metric are also presented as it has more intuitive meaning. ***Presence@top-k*** for arbitrary fixed user is defined as quantity of preferred objects (objects from test set) within the *top-k* best objects list according to the prediction of current recommending method for current user. ***Presence@top-k*** is then summed over all users. ***Presence@top-k*** can quite well depict twists in methods performance over various top-k sizes.

5.2   Results

Figure 1 displays results of recommending methods in ***nDCG*** aggregated by the train set size and Figure 2 shows distribution of ***Presence@top-k*** up to top-100. Although smaller top-k would be used in the real deployment, the list of objects eligible for recommendation would be probably pre-filtered too, leaving some influence also to objects beyond typical top-k boundary.

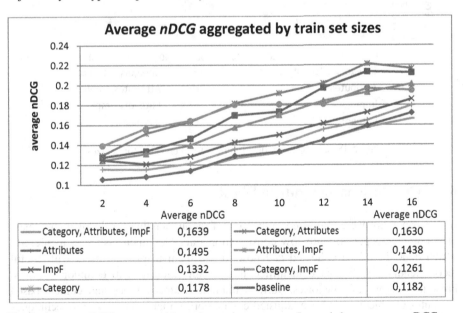

	Average nDCG		Average nDCG
Category, Attributes, ImpF	0,1639	Category, Attributes	0,1630
Attributes	0,1495	Attributes, ImpF	0,1438
ImpF	0,1332	Category, ImpF	0,1261
Category	0,1178	baseline	0,1182

Fig. 1. Average nDCG aggregated by train set sizes per user. Legend shows average nDCG per all users and train set sizes

From the three aspects taken into account (data from categories, additional implicit feedback and objects attributes), using attributes of the objects have proven to be the most crucial. All methods with *Attributes* performed significantly better in terms of average nDCG than methods without *Attributes* (p-value $< 10^{-6}$, TukeyHSD test[6]). The price for this improvement is time complexity rising with the number of attributes.

Using additional data from category pages also improves recommendation, but only if accompanied with object attributes (p-value $< 10^{-6}$). This is quite easy to justify as only object attributes can sufficiently describe similarity between categories and corresponding objects.

The results are rather inconclusive about usage of other implicit feedback data. If used without *Attributes*, they improved recommendations significantly (p-value $< 3.1*10^{-5}$), however using it together with *Attributes* did not improve results much further.

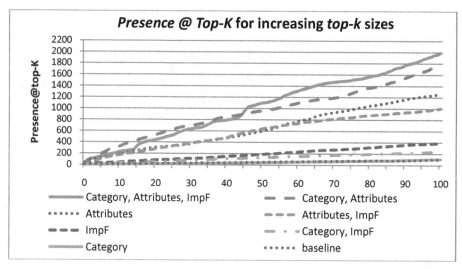

Fig. 2. Presence@top-k development for increasing top-k sizes

6   Conclusions and Future Work

In this paper, we were interested in the area of recommending for disloyal customers on e-commerce portals with low consumption rate. Our main task was to identify which data should be collected and how to use them to provide users with useful recommendations. We have identified three main extensions to common datasets: browsing history of categories, various implicit feedback types and content-based attributes of objects. We have adjusted contemporary matrix factorization techniques to handle such data and compare them in off-line experiment.

---

[6] Tukey Honest Significant Differences, http://stat.ethz.ch/R-manual/R-patched/library/stats/html/TukeyHSD.html

Both category browsing history and object attributes have proven to be worthy enhancement to the recommendation algorithms, however the results are not very conclusive about usage of more various implicit feedback features.

Future work involves e.g. experimenting with other recommendation methods or parameters of current ones. We would like to also consider other approaches to enhance matrix factorization with implicit feedback and object attributes and last but not least employing time-aware recommending algorithms. If experiments on other domains corroborate our ideas, our long-term goal is to deploy the system on real e-commerce portal and continue with on-line experimentation.

Acknowledgments. This work was supported by the grant SVV-2013-267312, P46 and GAUK-126313.

References

1. Belluf, T., Xavier, L., Giglio, R.: Case study on the business value impact of personalized recommendations on a large online retailer. In: RecSys 2012, pp. 277–280. ACM (2012)
2. Forbes, P., Zhu, M.: Content-boosted matrix factorization for recommender systems: experiments with recipe recommendation. In: RecSys 2011, pp. 261–264. ACM (2011)
3. Gauch, S., Speretta, M., Chandramouli, A., Micarelli, A.: User Profiles for Personalized Information Access. In: Brusilovsky, P., Kobsa, A., Nejdl, W. (eds.) The Adaptive Web 2007. LNCS, vol. 4321, pp. 54–89. Springer, Heidelberg (2007)
4. Jawaheer, G., Szomszor, M., Kostkova, P.: Comparison of implicit and explicit feedback from an online music recommendation service. In: HetRec 2010, pp. 47–51. ACM (2010)
5. Kelly, D., Teevan, J.: Implicit feedback for inferring user preference: a bibliography. SIGIR Forum 37, 18–28 (2003)
6. Konstan, J., Riedl, J.: Recommender systems: from algorithms to user experience. UMUAI 22, 101–123 (2012)
7. Koren, Y.: Collaborative filtering with temporal dynamics. In: ACM SIGKDD 2009, pp. 447–456. ACM (2009)
8. Koren, Y., Bell, R., Volinsky, C.: Matrix Factorization Techniques for Recommender Systems. Computer 42, 30–37 (2009)
9. Linden, G., Smith, B., York, J.: Amazon.com recommendations: item-to-item collaborative filtering. IEEE Internet Computing 7, 76–80 (2003)
10. Ostuni, V.C., Di Noia, T., Di Sciascio, E., Mirizzi, R.: Top-N recommendations from implicit feedback leveraging linked open data. In: RecSys 2013, pp. 85–92. ACM (2013)
11. Peska, L., Vojtas, P.: Evaluating Various Implicit Factors in E-commerce. In: RUE (RecSys) 2012. CEUR, vol. 910, pp. 51–55.
12. Peska, L., Vojtas, P.: Negative Implicit feedback in E-commerce Recommender Systems. In: Proc. of WIMS 2013, pp. 45:1-45:4. ACM (2013)
13. Peska, L., Vojtas, P.: Enhancing Recommender System with Linked Open Data. In: Larsen, H.L., Martin-Bautista, M.J., Vila, M.A., Andreasen, T., Christiansen, H. (eds.) FQAS 2013. LNCS (LNAI), vol. 8132, pp. 483–494. Springer, Heidelberg (2013)
14. Xia, C., Jiang, X., Liu, S., Luo, Z., Yu, Z.: Dynamic item-based recommendation algorithm with time decay. In: ICNC 2010, pp. 242–247. IEEE (2010)

Security Constraints in Modeling of Access Control Rules for Dynamic Information Systems

Aneta Poniszewska-Maranda

Institute of Information Technology, Lodz University of Technology, Poland
anetap@ics.p.lodz.pl

Abstract. Rapid development of new technologies brings with it a need for the new security solutions. Identifying, defining and implementing of security constraints is an important part of the process of modeling and developing of application/information systems and its administration.

The paper presents the issue of security constraints of information system from the point of view of Usage Role-based Access Control approach - it deals with the classification of constraints and their implementation in the process of modeling the access rules for dynamic information systems.

1 Introduction

The main idea of access control is to restrict and protect an access to some resource and ensure that only those allowed to use it can access it [10]. Resource can be any element of the system, like a file, folder, database or printer. Apart from controlling an access to a resource it also deals with how and when the resource can be used [11]. As an example, operating system controls access to the files and a certain user may have access to edit a given file, but only during working hours. The aim of access control is to prevent any unwanted or undesired access to resources. What is more, if properly managed, access control also promotes proper information sharing across users and applications [12].

Many models of access control are currently available and present in information systems, each having their advantages and disadvantages. The problem analyzed in the presented paper is how the approach of Usage Role-based Access Control [9], in particularly the security constraints of this approach, can deal with the issue of the logical security. Identifying, defining and implementing of security constraints is a very important part of the process of modeling and developing of an application or information systems and their later administration. *Security constraints* can be defined as an information assigned to the elements of a system that specifies some additional conditions to be fulfilled by these elements or elements related to them in order to ensure compliance with the security rules and ensure the global coherence of security schema.

In general the security constraints can be divided into two groups: the application constraints that are identified and defined at level of application's development by its developer and the organization constraints (i.e. global constraints)

V. Geffert et al. (Eds.): SOFSEM 2014, LNCS 8327, pp. 466–477, 2014.

that are determined at level of system exploitation by the global security administrator. Security constraints are very important during the specification and definition of access control rules for the users of application/information system and also during the execution of these rules. Moreover, security constraints play an important role in ensuring the global coherence of all elements of security schema (both at application and organization level) at the moment of attaching a new application with new elements, i.e. roles, functions and permissions, to the existing system.

The presented paper analysis the concept of security constraints and their impact on designing and execution of access control rules in the framework of dynamic information systems. The constraints are presented from the point of view of Usage Role-based Access Control approach [9], that introduces some improvements to the logical security of information systems. The paper is composed as follows: section 2 gives the outline of approach based on role concept and usage concept in aspect of access control (Usage Role Based Access Control, URBAC) and section 3 presents the concept of security constraints in URBAC approach - the classification of constraints with examples and the methods of their implementation.

2   Access Control Based on Role Concept and Usage Concept

It is very hard to design an access control model that is perfect and applicable to many types of systems and differing needs. Due to this, each of the traditional access control models or their extensions has some limitations. With rapid development of new systems and applications the needs for control of data are constantly changing with the new problems needing to be solved.

Mandatory Access Control (MAC) provides very strict and rigid control. It highly limits the user's possible actions and doesn't consider any dynamic alterations of underlying policies [13]. The main downfall of this model is difficulty to implement in the real-world applications and systems, which have to be rewritten in order to adhere to the model's labelling concept. Discretionary Access Control (DAC), unlike MAC gives more freedom to the users. It is left to their discretion to specify the access rules for files they are owners of. The main problem arising in this model is no protection from the copy operation. If a user can read another user's file, there is nothing stopping him from copying that file to a file that he owns and can freely share its contents [11]. Role-based Access Control (RBAC) provides a structure of access control tailored to the needs of enterprises. However it also creates a challenge between easy administration and strong security [12]. The role engineering may also pose a challenge as an access control may not always be compatible with organization's structure [1]. Usage Control concept [3,4,5] was introduced as an answer to the limitations of the above concepts. As all of them focused on the authorizations done before an access, this concept introduced a possibility to check them also during an access. It is a very abstract model, that does not provide a clear structure like

RBAC model and does not deal with who defines and modifies the rights that the subjects posses.

These disadvantages and the needs of present information systems caused the creation of unified approach that can encompass the use of traditional access control models and allow to define the dynamic rules of access control. Two access control concepts are chosen in our studies to develop the new approach for dynamic information systems: role concept [6] and usage concept [3,5]. The first one allows to represent the whole system organization in the complete, precise way while the second one allows to describe the usage control with authorizations, obligations, conditions, continuity (ongoing control) and mutability attributes.

2.1   Approach of Usage Role-Based Access Control

Usage Role-based Access Control (URBAC) [9,16] is based on role concept from extended Role-Based Access Control [6] and usage concept from Usage Control [5]. It uses a complete and precise way to represent the entire system organization with the use of roles and functions. URBAC also incorporates the control of usage in data access with authorizations, obligations and conditions that can be applied both before and during an access (Fig. 1).

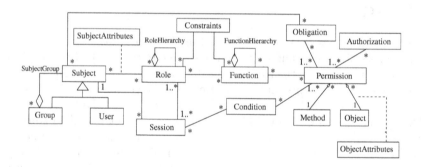

Fig. 1. Meta-model of URBAC approach

Subject can represent users and groups of users, that share the same rights as well as obligations and responsibilities. Session is the interval of time during which a user is actively logged into the system and may execute the actions in it that require the appropriate rights. User is logged in to the system in a single session, during which the roles can be activated.

A **Role** can be regarded as a reflection of position or job title in an organization, that holds with it the authority as well as responsibilities. It allows to accomplish certain tasks connected with processes in an organization. Users are assigned to them based on their competencies and qualifications. Therefore, role is associated with subjects, where user or group of users can take on different roles, but one role can also be shared among users. This association also contains *Subject Attributes*, like identity or credits, which are additional subject properties that can be considered in usage decision.

As each role specifies a possibility to perform specific tasks, it consists of many **functions**, which users may apply. Like with roles, function hierarchy can be defined with inheritance relations between specific functions. Function in turn, can be split to more atomic elements which are operations that are performed on objects. Those are granted by **permissions**. We therefore can view the functions as sets or sequences of permissions, that grant them right to perform the specified methods on a specified objects.

In the model when permissions are assigned to objects, the specification of security constraints is necessary. Those constraints are first of all authorizations, conditions and obligations. Constraint determines that some permission is valid only for a part of the object instances. We can denote a permission as a function: $p(o, m, Cst)$ where o represents an object, m a method that can be executed on the object and Cst is a set of constraints that determine this permission. Taking into consideration a concept of authorization (A), obligation (B) and condition (C), the set of constraints can take the following form $Cst = \{A, B, C\}$ and the permission can be presented as a function $p(o, m, \{A, B, C\})$. According to this, the permission is given to all instances of the object class except the contrary specification.

Objects are entities that can be accessed by users. They have a direct relationship with permissions that can be further described with the use of Object Attributes. Those represent additional object attributes specific to the relation, like for instance the security labels or ownerships. Attributes of both subjects and objects can be mutable, which means they can be updated by the system as consequences of subject usage on objects. Attributes can also be immutable and cannot be changed by the system, but only at administrator's discretion.

The **constraints** can be defined for each main element of the model presented above (i.e user, group, subject, session, role, function, permission, object and method), and also for the relationships among the elements. The concept of constraints was described widely in the literature [1,2,6,7,8,14,15] and it is presented in the following section.

2.2 Creation Process of Security Profiles for Users of Information System

Design and realization process of security schema for an information system can be divided into two stages [9,16]: *conception stage* (realized at local application level by application/system developer who knows its specification that should be realized) and *deployment stage* (realized at global level of information system by security administrator who knows the general security rules and constraints that should be taken into consideration at the whole company level).

Conception Stage. The realization process of information system or a simple application is provoked by a client's requests. Basing on client's needs and requirements the application/system developer creates the logical model of application/system and next the project of this system that will be the base for its implementation. The responsibilities of application developer based on URBAC are as follows:

− definition of permissions - identification of methods,
− definition of object attributes associated to objects according with access control rules,
− assignment of elements: permissions to functions and functions to roles,
− definition of security constraints assigned to application's elements, i.e. authorizations, obligations and conditions on the permissions and standard constraints on roles, functions and their relationships.

Deployment Stage. The security administrator defines the administration rules and company constraints according to the global security policy and application/system rules received from the developer. He should also check if these new security constraints remain in agreement with the security constraints defined for the elements of existing information system in order to guarantee the global coherence of the new system. The duties of security administrator are as follows:

− definition of users' rights basing on their responsibilities and their business functions in an organization - assignment of users to roles of information system,
− organize the users in groups and definition of access control rights for the groups of users that realize for example the same business functions - assignment of groups to the roles of information system,
− definition of subject attributes associated to certain users or groups of users that allows to determine the dynamic aspects of security constraints,
− definition of security constraints for the relationships between users and roles or group of users and roles.

Example for URBAC Approach. The concept used to illustrate URBAC approach is an online store. It allows to represent the access control in an extensive way. It also deals with an access to digital objects, which is a large part of Usage Control, that is a part of URBAC. There are different types of roles available for the users - *Regular*, *Premium* and *Seller* and a special role of the *Administrator*, who has access to all the application models and data to define the new elements of the access control in accordance with URBAC model directly without the need to modify the application's code.

Each role consists of a collection of functions. Therefore the administrator is able to assign different functions to various roles. Examples of the main available functions are: buy an ebook, download an ebook, edit an ebook, register, remove an ebook, upgrade an account, sell an ebook, get credits, give feedback, feature an ebook. For instance the function to *edit an ebook* is assigned only to the role *Seller*, while the function *Buy an Ebook* is assigned to all the roles available in the application. To perform a function a user has to make a sequence of activities. Some *permissions* are assigned to each of these activities.

When a permission is evaluated it consists of *authorizations, obligations* and *conditions*. All three have to be satisfied to give a user access and let him proceed to next step in the function. Examples of *authorizations* for the *purchase of the ebook* permission, that the administrator can define, presented in form of questions, can be as follows:

- Does a user own the ebook object?
- Does the user have enough credits to purchase the ebook?
- Has the user already bought the item?
- Does the user have any downloads of the ebook left?

Obligations for instance, can consider if the user has agreed to the terms of service and the copyright agreement. If he has not agreed his activities in the application will be limited. *Conditions*, in turn, can focus on allowing the users to execute the functions only during the business hours and setting a limit on the number of ebooks that can be featured in the store.

3   Security Constraints for URBAC Approach

The security administration process of an information system is a complex task. Many security constraints have to be specified in order to correctly define the security strategy of the system. The application developer is able in a relatively easy way to define the constraints that should be associated with this application. The security administrator from the other hand knows the global security policy and it is able to fix the constraints at global level.

The security constraints can be specified with the use of development tools that we chosen according to the specification of URBAC approach and creation process of security profiles:

1. *Object Constraint Language, OCL* [17,18] at level of application developer - is a part of UML and is used to define and specify the constraints formulated in model created during the analysis and design of information system.
2. *Role-based Constraint Language, RCL* [14,15] at level of security administrator - created to define and specify the constraints of RBAC model and we appropriately modified and extend it to meet the needs of URBAC model - *Usage and Role Constraint Language (uRCL)* [16].

3.1   Classification of Security Constraints of URBAC Model

Classification of security constraints for URBAC approach reflects the life-cycle of an application, which in the simplest form, is divided into two stages: *analysis-development*, realized by application developer and *exploitation-maintenance*, realized by security administrator, to guarantee the global coherence of system security schema. Therefore, the basic classification of security constraints distinguishes two groups of constraints:

- *constraints from developer point of view* - constraints at application level, determined in order to create a complex application with its access control rules and
- *constraints from security administrator point of view* - constraints at organization level, defined for global security of an organization.

The second classification level of security constraints represents the limitations associated with the URBAC approach:

- *constraints for permission* - limit the set of objects available for a method, including the authorization, obligations and conditions,
- *separation of duty, SOD, constraints* - represents the concept of mutually exclusive roles, mutually exclusive functions and mutually exclusive permissions [1,14,15],
- *prerequisite constraints* - based on the concept of prerequisite roles, prerequisite functions and prerequisite permissions,
- *cardinality constraints* - numerical limitations defined for classes in role-based system and numerical limitations for application elements,
- *session constraints*, that can be expressed by obligations,
- *role hierarchy constraints* and *function hierarchy constraints*.

The third classification level includes two categories of constraints:

- *static constraints*, that are constant and apply before the access to data in a system,
- *dynamic constraints*, that can be changeable during the system working and appear during the session activation by a system user.

General classification of constraints, based on above division and on scope of duties and responsibilities assigned to application/system developer and security administrator is presented in figure 2.

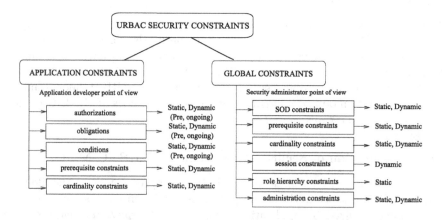

Fig. 2. Classification of security constraints based on URBAC approach

3.2   Constraints from Developer Point of View in URBAC Approach

Three main types of security constraints are distinguished at level of application/system developer: constraints for permission, including the authorization, obligations and conditions, prerequisite constraints and cardinality constraints.

Security constraints for permissions are:

1. *Pre-authorizations* or in other words *static authorizations* - limit the set of objects on which a user can execute the method, regardless of this user; based on the subject attributes and object attributes. For example "permission of reading the document files with *.sec* extension is accessible only for role *Manager* (in OCL):

   context File read()
   pre : Set (Objects) → select(obj | obj.oclIsTypeOf (File)) and
   obj.getName() = ∗.sec and actor.type = Manager

2. *Ongoing-authorizations*, i.e. *dynamic authorizations* - limit the set of objects accessible during the session (during the access realization); more based on subject attributes and object attributes that can change during a session that provoke the changes in the constraints. For example "permission of downloading the e-book files from the Internet shop server to read them by a user is dependent on current state of his account (state of account has to be at least equal to the value of requested document)":

   context File read()
   pre : Set (Objects) → select(obj | obj.oclIsTypeOf (eBookFile)) and
   user (s_i) .account.meter >= obj.value

   where *user (s_i)* returns a set of users (i.e. one user) of session s_i.

3. *Pre-obligations*, i.e. *static obligations* use the history of previous activities in order to check whether certain action have been taken; they can use the subject attributes or object attributes. E.g. "user has to accept the terms of service and copyright agreement before the access to certain digital data":

   context Data get()
   pre : Set (Objects) → select(obj | obj.oclIsTypeOf (Data)) and
   user.actor.accept(ServiceCopyrightAgreement) = true

4. *Ongoing-obligations*, i.e. *dynamic obligations* - have to be fulfill continuously or periodically during the given right are used; they can use the subject attributes or object attributes. For example "user has to agree to make the log information available the provider before reading an e-book file":

   context File read()
   Set (Objects) → select(obj | obj.oclIsTypeOf (eBookFile)) and
   user (s_i) .accept(LogsToProvider) = true

5. *Pre-conditions*, i.e. *static conditions* represent the features of an application or a system that are used to take the decision about the usage; a condition is evaluated before usage. For example "permission of read the data from the server files only during the working hours, from 8.00 am to 6.00 pm":

   context File read()
   pre : Set (Objects) → select(obj | obj.oclIsTypeOf (ServerFile)) and

$actor.timePeriodAccess >= 8.00am$ and
$actor.timePeriodAccess <= 6.00pm$

6. *Ongoing-conditions*, i.e. *dynamic conditions* - represent the current state or status of a system or environment during a user session; condition has to be fulfilled during a usage. For example "permission of reading the document with important data only one by the user during his current session":
context FileImportant read()
$pre : Set (Objects) \rightarrow select(obj \mid obj.oclIsTypeOf (File))$ and
$user (s_i) .timeAccess <= 1$

The other two types of security constraints at developer level are:

1. *Prerequisite constraints*, that are based on prerequisite concept that in the case of URBAC approach signifies:
   - permission $p_1 (m_1, o_1)$ can be assigned to function f if this function has already permission $p_2 (m_2, o_2)$ assigned before:
   context Function inv :
   $self.method \rightarrow includes (m_1)$ and $Set (Objects) \rightarrow$
   $select(obj \mid obj.oclIsTypeOf (o_1))$ *implies*
   $self.method \rightarrow includes (m_2)$ and $Set (Objects) \rightarrow$
   $select(obj \mid obj.oclIsTypeOf (o_2))$
   - function f_1 can be assigned to role r if this role has already function f_2 assigned before:
   context Role inv :
   $self.useCase \rightarrow includes (f_1)$ *implies* $self.useCase \rightarrow includes (f_2)$
   For example "function of downloading the document files from server needs the function of logging to this server":
   context Role inv :
   $self.useCase \rightarrow includes ('perform file transfer')$ *implies*
   $self.useCase \rightarrow includes ('login')$.

2. *Cardinality constraints* - numerical limitation concerning the application elements. For example "function of downloading a file containing the confidential information can be assigned only to one certain role":
   context Function inv :
   $self.permission.method \rightarrow includes ('get()')$ and $self.permission.object \rightarrow$
   $includes ('InfoConfidential')$ *implies* $self.actor \rightarrow size = 1$.

3.3 Constraints from Security Administrator Point of View in URBAC Approach

The security constraints at level of security administrator first of all concern directly the users and roles of a system. The basic task of these constraints is to guarantee the security rules for the whole organization and in consequence the preserving of coherence of global security schema.

The basic security constraints at level of exploiting an application or a whole system are the *separation of duty (SOD) constraints*, that are based on the concept of mutually exclusive roles [1,14,15]. It has as a purpose to split the tasks of each users and privileges assigned to them, needed to implement a task or a set of related tasks. SOD constraints can concern different concepts and therefore we can define: *constraints on conflicting users (CU)*, *constraints on conflicting roles (CR)*, *constraints on conflicting functions (CF)*, *constraints on conflicting permissions (CP)*.

The basic division of SOD constraints divides them into two groups:

1. *Static SOD* - constraints defined in static way prior to the execution of certain actions by the user in the system. For example "constraints that any user cannot be assigned to two conflicting roles - conflicting roles cannot have common users" with the use of uRCL:

$$|roles\,(OE\,(U)) \cap OE\,(CR)| \leq 1$$

where $(OE\,(X))$ signifies the choice of one element from a set X.

2. *Dynamic SOD* - constraints that respect the roles activated by a user during the session; based on the actions that cannot be executed simultaneously. For example "there are not the users with two conflicting roles - conflicting roles can have common users but they cannot activate simultaneously the conflicting roles":

$$|roles\,(OE\,(S)) \cap OE\,(CR)| \leq 1$$

The other types of constraints that are defined at level of security administrator are:

1. *Prerequisite constraints* - based on concept of *prerequisite roles*. For example "a user can be assigned to role r_1 only when he has role r_2 assigned previously":

$$r_1 \in roles\,(OE\,(U)) \;\Rightarrow\; r_2 \in roles\,(OE\,(U))$$

E.g. "a user can be assigned to role *Premium User* only when he was previously assigned to role *Regular User* (it guarantees that only the users assigned to role *Regular User* could be assigned to role *Premium User*)":

$$'Premium\,User' \in roles\,(OE\,(U)) \;\Rightarrow\; 'Regular\,User' \in roles\,(OE\,(U))$$

2. *Cardinality constraints* - numerical restrictions defined for the classes in a system based on role. For example "number of users assigned to the role is limited", e.g. "there is exists only one person who can be assigned to role *Managing Director*":

$$|user\,('Managing\,Director')| = 1$$

3. *Session constraints*, that have to be expressed by the obligations. For example "a users that can be a member of two roles r_1, r_2 cannot to activate them at the same time during one session":

$$roles\,(OE\,(U)) = \{r_1,\,r_2\} \;\wedge\; |sessions\,(r_1) \cap sessions\,(r_2)| = \emptyset$$

4. *Role hierarchy constraints.* For example "a user cannot be assigned to child roles that inherit the parent roles that are in conflict":

$$|roles^\star (OE\,(U)) \cap OE\,(CR)| \leq 1$$

3.4   Implementation of Security Constraints in URBAC Model

Definition and implementation of security constraints of URBAC model can be realized both in OCL (*Object Constraint Language*) - for constraints from level of application/system developer and in extended uRCL - for constraints from level of security administrator. OCL gives the possibility to define the *invariants* for model's classes, the invariants for stereotypes and *pre-conditions* or *post-conditions* for operations. OCL expressions can be used in different situations:

- for attributes or operations of the classes in class diagram,
- a condition can be attached to a message to specify a situation in which it may be sent to the specified object in interaction diagram (i.e. sequence or communication diagram),
- a message can have the attributes, which values can be specified with the use of OCL expressions in interaction diagram.

Security constraints can be defined in different UML diagrams. The class diagram and sequence diagram were chosen to specify the constraints of URBAC model. The constraints in class diagram are presented the most often in form of *invariants* and in sequence diagram - in form of *pre-conditions*.

Implementation of URBAC model is based on role engineering process, which needs the definition of appropriate functions, permissions and security constraints [9]. Definition of set of roles for an information system is realized with the use of UML diagrams and the result is presented in widely known format, e.g. in XML format [9]. In connection with this, the implementation of security constraints of URBAC approach consists in generating of XML files containing the set of constraints, provided in form legible and understandable both for developer and for security administrator. The constraints defined in OCL by application/system developer are next translated to uRCL for security administrator who defines also the global security constraints with the use of uRCL.

4   Conclusion

Usage Role-based Access Control introduces a new approach to the logical security of information systems. It is flexible model that can be tailored to various needs. It allows to support the security of dynamic information systems thanks to the concepts of mutability and continuity where the dynamic changes of security policy can be transformed into the changes of values of subject attributes or object attributes. The presented approach gives the possibilities to specify not only the permissions but also the prohibitions and duties in framework of information systems and the security rules dependent on the application context. It is possible thanks to different types of security constraints, presented in paper.

The possibility of defining the constraints both at developer level and administrator level allows to assure the global coherence of security schema for the whole information system, containing many components (i.e. applications) and supporting many users in their activities. Developer and administrator define the constraints with the use of two different tools, according to the specification of the whole process of modeling, creation and exploitation of information system and its security schema but the concepts of these tools can be transformed between these two levels to assure the global coherence.

References

1. Ferraiolo, D., Sandhu, R.S., Gavrila, S., Kuhn, D.R., Chandramouli, R.: Proposed NIST Role-Based Access control. ACM TISSEC (2001)
2. Park, J., Zhang, X., Sandhu, R.: Attribute Mutability in Usage Control. In: 18th IFIP WG 11.3 Working Conference on Data and Applications Security (2004)
3. Lazouski, A., Martinelli, F., Mori, P.: Usage control in computer security: A survey. Computer Science Review 4(2), 81–99 (2010)
4. Pretschner, A., Hilty, M., Basin, D.: Distributed usage control. Communications of the ACM 49(9) (2006)
5. Zhang, X., Parisi-Presicce, F., Sandhu, R., Park, J.: Formal Model and Policy Specification of Usage Control. ACM TISSEC 8(4), 351–387 (2005)
6. Poniszewska-Maranda, A.: Conception Approach of Access Control in Heterogeneous Information Systems using UML. Journal of Telecommunication Systems 45(2-3), 177–190 (2010)
7. Strembeck, M., Neumann, G.: An Integrated Approach to Engineer and Enforce Context Constraints in RBAC Environments. ACM TISSEC 7(3) (2004)
8. Bertino, E., Ferrari, E., Atluri, V.: The Specification and Enforcement of Authorization Constraints in Workflow Management Systems. ACM TISSEC 2(1)
9. Poniszewska-Maranda, A.: Modeling and design of role engineering in development of access control for dynamic information systems. Bulletin of the Polish Academy of Sciences, Technical Science 61(3) (2013)
10. Kim, D., Solomon, M.: Fundamentals of Information Systems Security. Jones & Bartlett Learning (2012)
11. Ferraiolo, D.F., Kuhn, D.R., Chandramouli, R.: Role-Based Access Control, 2nd edn. Artech House (2007)
12. Hu, V.C., Ferraiolo, D.F., Kuhn, D.R.: Assessment of Access Control Systems. Interagency Report 7316, NIST (2006)
13. Stewart, J.M., Chapple, M., Gibson, D.: CISSP: Certified Information Systems Security Professional Study Guide, 6th edn. John Wiley & Sons (2012)
14. Ahn, G.-J.: The RCL 2000 language for specifying role-based authorization constraints, Ph.D. thesis, George Mason University, USA (1999)
15. Ahn, G.-J., Sandhu, R.S.: Role-based authorization constraints specification. ACM Trans. on Information and Systems Security 3(4), 207–226 (2000)
16. Poniszewska-Maranda, A.: Logical security models and their implementations in information systems (in Polish). EXIT (2013)
17. Booch, G., Rumbaugh, J., Jacobson, I.: The Unified Modelling Language User Guide. Addison Wesley (1998)
18. OMG, OMG Unified Modeling Language Specification (2011)

A New Plane-Sweep Algorithm
for the K-Closest-Pairs Query

George Roumelis[1,*], Michael Vassilakopoulos[2,*], Antonio Corral[3,*,**],
and Yannis Manolopoulos[1,*]

[1] Dept. of Informatics, Aristotle University, GR-54124 Thessaloniki, Greece
{groumeli,manolopo}@csd.auth.gr
[2] Dept. of Computer Science and Biomedical Informatics,
University of Thessaly, 2-4 Papasiopoulou st., 35100 Lamia, Greece
mvasilako@dib.uth.gr
[3] Dept. of Informatics, University of Almeria, 04120 Almeria, Spain
acorral@ual.es

Abstract. One of the most representative and studied Distance-Based
Queries in Spatial Databases is the K-Closest-Pairs Query ($KCPQ$).
This query involves two spatial data sets and a distance function to measure the degree of closeness, along with a given number K of elements
of the result. The output is a set of pairs of objects (with one object element from each set), with the K lowest distances. In this paper, we study
the problem of processing KCPQs between RAM-based point sets, using
Plane-Sweep (*PS*) algorithms. We utilize two improvements that can be
applied to a *PS* algorithm and propose a new algorithm that minimizes
the number of distance computations, in comparison to the *classic PS*
algorithm. By extensive experimentation, using real and synthetic data
sets, we highlight the most efficient improvement and show that the new
PS algorithm outperforms the *classic* one, in most cases.

Keywords: Spatial Query Processing, Plane-Sweep, Closest-Pair Query,
Algorithms.

1 Introduction

Spatial database is a database that offers spatial data types (for example, types
for points, line segments, regions, etc.), a query language with spatial predicates,
spatial indexing techniques and efficient processing of spatial queries [11]. It has
grown in importance in several fields of application such as urban planning,
resource management, transportation planning, etc. Together with them come
various types of complex queries that need to be answered efficiently. While

---

* Work funded by the GENCENG project (SYNERGASIA 2011 action, supported
by the European Regional Development Fund and Greek National Funds); project
number 11SYN_ 8_1213.
** Supported by the Junta Andalucia research project [TIC-06114].

V. Geffert et al. (Eds.): SOFSEM 2014, LNCS 8327, pp. 478–490, 2014.
© Springer International Publishing Switzerland 2014

queries involving a single data set have been studied extensively in the literature, Distance Join Queries (DJQs) on spatial data like K-Closest-Pairs Query
(KCPQ) has not been paid similar attention. For this reason, in this work we
will improve this kind of DJQ for spatial data (points) in terms of execution
time using the *plane-sweep (PS)* technique.

One of the most important techniques in the computational geometry field
is the *PS* algorithm, which is a type of algorithm that uses a conceptual *sweep
line* to solve various problems in the Euclidean plane, E^2, [10]. The name of *PS*
is derived from the idea of sweeping the plane from left to right with a vertical
line (front) stopping at every transaction point of a geometric configuration to
update the front. All processing is done with respect to this moving front, without
any backtracking, with a look-ahead on only one point each time [6]. The *PS*
technique has been successfully applied in spatial query processing, mainly for
intersection joins [8]. In the context of DJQs, the *PS* technique has been used
to restrict all possible combinations of pairs of points from the two data sets.

In the context of computational geometry, in [6], the *PS* algorithm is applied
to find the closest pair in a set of points, in an elegant way. Two improvements
when a new pair can be formed are proposed. The first one examines only candidates which may form a new closest pair with the fixed point p on the sweep
line that lie in a half-circle centered at this point, with radius δ (the current distance threshold). The second one, since the use of a half-circle in a *PS* algorithm
is complex, examines only candidates within a boundary rectangle (a rectangle
with width δ in X-axis and, height $2 * \delta$ in Y-axis ($p + \delta$ and $p - \delta$ from p)). A
critical observation made in [6] is that, as the sweep line passes through a fixed
point, there is at most a constant number of points that need to be checked.
But this property does not hold in the KCPQ, which is essentially a generalization of the Bichromatic Closest-Pair problem and the number of points with
monochromatic color in this problem cannot be bounded (Section 5.1 of [13]).
Moreover, the algorithm of [6] uses an array and a balanced binary tree (e.g.
AVL-tree) to sort on both axes, while we will use one array for each data set,
sorting on one axis (e.g. X). Finally, our proposed *PS* algorithm can be easily
adapted to distance-based join query processing on disk resident data.

The contributions of this paper consist in the following:

1. We enhance the *classic PS* algorithm for KCPQ with two improvements
   (sliding window and sliding semi-circle), which were proposed in [6] for
   the closest-pair problem over one data set, and here have been adapted to
   KCPQ, where two data sets are involved.
2. We improve processing of the *classic PS algorithm* for KCPQ, with a new
   algorithm called Reverse Run Plane-Sweep, *RRPS* or *RR PS*, that minimizes
   Euclidean and sweeping axis distance computations.
3. We present results of an extensive experimentation, that compares the performance of the different algorithms and algorithmic improvements.

The paper is organized as follows. In Section 2, we review the related literature
and motivate the research reported here. In Section 3, we describe the *classic
PS* for KCPQ. In Section 4, a new *PS* algorithm for KCPQ is presented. In

Section 5, a comparative performance study is reported. Finally, in Section 6, conclusions on the contribution of this paper and future work are summarized.

2   Related Work and Motivation

There are numerous papers that study processing of join queries, and recently, an exhaustive analysis of techniques for spatial join taking into account a filter-and-refinement approach appeared in [8]. We can classify the spatial join methods depending on whether the sets of objects involved are indexed or not, but in all cases the *PS* technique plays an important role for reducing the CPU cost [8].

The *K*CPQ discovers the *K* pairs of data elements formed from two data sets that have the *K* smallest distances between them. The *K*CPQ is a combination of spatial join and nearest neighbor queries. Like a spatial join query, all pairs of objects are candidates for the result. Like a nearest neighbor query, proximity forms the basis for the final ordering. If both data sets are indexed by R-trees, the concept of synchronous tree traversal and Depth-First (DF) or Best-First (BF) traversal order can be combined for the query processing [3,4,7,12]. In [7], incremental and non-recursive algorithms based on BF traversal using R-trees and additional priority queues for DJQs are presented. In [12], additional techniques as sorting and application of *PS* during the expansion of node pairs, and the use of the estimation of the distance of the *K*-th closest pair to suspend unnecessary computations of MBR distances are included to improve [7]. In [3,4] non-incremental recursive (DF) and non-recursive (BF) algorithms are presented for solving the *K*CPQ, when the data sets are indexed by R*-trees. The main issue of the non-incremental variant is to separate the treatment of the terminal candidates (the elements of the leaf nodes) from the rest of the candidates (internal nodes) in the index data structures. In [4] the *PS* technique is also applied to limit the number of MBRs and points that must be paired up, thus reducing the number of distance computations. Finally, in [9] the *PS* technique has been used to answer the *K* Distance Join Query, another way to call to *K*CPQ, in order to find exactly *K* nearest pairs in non-incremental fashion, where R*-tree and Quadtree-like index structures are compared.

Recently, in [13] the *PS* technique is used to obtain the α-Distance for spatial query processing for fuzzy objects. Essentially, the computation of the α-Distance is to find the closest pair of qualified points of two fuzzy objects. The main property of this variant of the *PS* method is the use of two sweep lines to facility the search for the particular types of spatial queries with fuzzy objects, that has been presented in such research work.

Finally, the main motivation of this work is to improve the *classic PS algorithm* proposed in [4,12] for *K*CPQs, in order to reduce the number of distance computations, making the query processing faster in terms of execution time.

3   Plane-Sweep in *K*-Closest-Pairs Query Processing

A common approach to performing spatial joins when both data sets are stored on disk is to partition the data until it is of a size that can be processed using internal memory PS algorithm [8]. The *classic PS* algorithm for KCPQ sweeps a scan line (*sweepline*) that is vertical to one of the axes through two sorted (in relation to this axis) arrays of the two data sets. In general terms, this algorithm is an adaptation of the PS algorithm for intersection of MBRs presented in [2], considering two sets of points and a distance threshold δ (the distance of the K-th closest pair found so far) in the sweeping axis [4,12]. This *classic PS* algorithm can be considered a *greedy* variant of the algorithm for the closest-pair problem described in [10,6]. It is a greedy algorithm, because it always makes the choice that looks best at the moment. It also combines PS and *nested loop* techniques. Compared to the naive nested loop algorithm, except in unlikely situations, the *sweepline* limits the number of points that must be tested against one another [11].

In general, if we assume that the two point sets are P and Q, the *classic PS algorithm* consists of the following steps [4,12]:

1. Sorting the entries of the two point sets, based on the coordinates of one of the axes in increasing or decreasing order. The axis for the *sweepline* can be established based on sweeping axis criteria (e.g. X-axis) and the order can be fixed by sweeping direction criteria (e.g. *forward sweep* (increasing order) or *backward sweep* (decreasing order)); both criteria are presented in [12].

2. After that, two pointers are maintained initially pointing to the first entry for processing of each sorted set of points. Assuming that X-axis is the sweeping axis and the order is increasing (from left to right, i.e *forward sweep*), let *pivot* be the point with the smallest X-value pointed by one of these two pointers, e.g. P, then the *pivot* is initialized to this point, $p_{pivot} \in P$.

3. Afterwards, the *pivot* must be paired up with the points stored in the other set of points ($q_j \in Q$) from left to right, satisfying $dx = q_j.x - p_{pivot}.x \leq \delta$, processing all points as candidate pairs where the *pivot* is fixed. After all possible pairs of entries that contain *pivot* have been paired up (i.e., the forward lookup stops when $q_j.x - p_{pivot}.x > \delta$ is verified), the pointer to the set of the *pivot* is increased to the next entry, *pivot* is updated with the point of the next smallest X-value pointed by one of the two pointers, and the process is repeated until one of the set of points is fully processed.

Highlight that the PS *technique* applies the distance function over the sweeping axis (in this case, the X-axis) because in the PS *technique*, the sweep is only over one axis (the best axis according to the criteria suggested in [12]). Moreover, the search is only restricted to the closest points with respect to the *pivot* according to the current *distance threshold* (δ). No duplicated pairs are obtained, since the points are always checked over sorted sets. Note that this kind of processing is called *forward sweep*, since it scans from left to right (or from right to left, *backward sweep*) the sorted sets in order to obtain pairs of points that will have a distance smaller than or equal to δ.

Clearly, the application of this technique can be viewed as a *sliding strip* on the sweeping axis with a width equal to the δ value starting from the *pivot* (i.e., $[0, \delta]$ in the X-axis), where we only choose all possible pairs of points that can be formed using the *pivot* and the other points from the remainder entries of

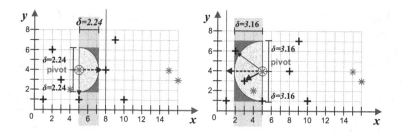

Fig. 1. *Classic* (left) and *RR* (right) *PS* algorithms, using sliding window or semi-circle

the other set of points that fall into the current *strip*. See in Figure 1[1] left, the strip in light grey color.

According to the ideas proposed in [6] to improve the *PS* applied to the closest-pair problem over one set of points, here we will propose two improvements of the previous *classic PS* algorithm over two data sets to reduce the number of point-point distance computations on KCPQ algorithms.

1. An intuitive way to save the number of point-point distance computations is to bound the other axis (not only the sweeping axis) by δ as is illustrated in Figure 1 left. In this case, the search space is now restricted to the closest points inside the *window* with width δ and a height $2 * \delta$ (i.e., $[0, \delta]$ in the X-axis and $[-\delta, \delta]$ in the Y-axis, from the *pivot*). Clearly, the application of this technique can be viewed as a *sliding window* on the sweeping axis with a width equal to δ and height equal to $2 * \delta$ starting from the *pivot*. And we only choose all possible pairs of points that can be formed using the *pivot* point and the remainder points of the other data set that fall into the current *window*. See in Figure 1 left, the *window* in dark grey color.

2. If we try to reduce even more the search space, we can only select those points inside the *semi-circle* (or half-circle) centered in the *pivot* with radius δ (remember that the equation of all points $t = (t.x, t.y) \in E^2$ that fall inside the circle, centered in the point $p_{pivot} = (p_{pivot}.x, p_{pivot}.y) \in P$ and radius δ is $(t.x - p_{pivot}.x)^2 + (t.y - p_{pivot}.y)^2 \leq \delta^2$). And the application of this new improvement can be viewed as a *sliding semi-circle* with radius δ along the sweeping axis and centered on the *pivot* point, choosing only the points that fall inside the current *semi-circle*. See in Figure 1 left, the *semi-circle* centered at *pivot*.

4   Reverse Run Plane-Sweep Algorithm for KCPQs

The *Reverse Run Plane-Sweep (RRPS, or RR PS)* algorithm is based on two concepts. First, every point that is used as a *reference point* (pivot) forms a *run*

---

[1] In both parts of Figure 1, a dotted arrow points to a point not included in the KCPQ result (paired with *pivot*), due to large dx; a thin arrow points to a point not included in the result, due to large distance to *pivot*; a thick arrow points to a point included in the KCPQ result (paired with *pivot*); the value of K is 3.

with other subsequent points of the same set. A *run* is a continuous sequence of points of the same set that doesn't contain any point from the other set. For each set, we keep a *left limit*, which is updated (moved to the right) every time that the algorithm concludes that it is only necessary to compare with points of this set that reside on the right of this limit. Each point of the *active run* (*reference point*) is compared with each point of the other set (*current point*) that is on the left of the first point of the *active run*, until the *left limit* of the other set is reached. Second, the *reference points* (and their *runs*) are processed in ascending X-order (the sets are X-sorted before the application of the *RRPS* algorithm). Each point of the *active run* is compared with the points of the other set (*current points*) in the opposite order (descending X-order).

A max binary heap (keyed by pair distances and called *MaxKHeap*) that keeps the K closest point pairs found so far is utilized. For each point of the *active run* being compared with a *current point*, there are 2 cases. *Case 1:* If the distance between this pair of points d is smaller than the distance of δ, then the pair will be inserted in the heap (rule 1). In case the heap is not full (it contains less than K pairs), the pair will be inserted in the heap, regardless of the pair distance. *Case 2:* If the distance between this pair of points in the sweeping axis dx is larger than or equal to δ, then there is no need to calculate the distance of the pair (rule 2). The *left limit* of the other set must be updated at the index value of the point being compared (a comparison with a point of the other set having an index value smaller than, or equal to the updated *left limit* will have X-distance larger than dx and is unnecessary).

Moreover, if the rightmost *current point* has an index value equal to the left limit of its set, then all the points of the *active run* will have larger dx from all the *current points* of the other set and the relevant pairs need not participate in computations (the algorithm advances to the start of the next run - rule 3). The *RRPS* algorithm is depicted in Algorithm 1. The following example illustrates its operation. Let's consider the points of the right part of Figure 1 (depicting a snapshot of the *RRPS* operation), presented, in commonly sorted X-order, in Table 1. To simplify the algorithm operation (the stopping conditions), a sentinel point with X-coordinate equal to ∞ is added to each set (line 2). In case the two point sets overlap in X-dimension, initialization sets i (j) equal to 0, since the first run of P (Q) set starts at $P[0]$ ($Q[0]$). Moreover, initialization sets the left limit *leftp* (*leftq*) equal to -1 (line 5), since the first P (Q) point to be used in comparisons is $P[leftp+1]$ ($Q[leftq+1]$).

Table 1. The points of Figure 1 in X-sorted order

i	0	1	2		3	4	5	6		7
$P[i]$	(1,1)	(2,6)	(3,3)		(5,1)	(8,4)	(9,7)	(10,1)		$(\infty,-)$
j				0	1				2	3	4
$Q[j]$			(4,2)	(5,4)				(15,4)	(16,3)	$(\infty,-)$

Algorithm 1. Reverse Run Plane-Sweep

Input: $P[0..N-1], Q[0..M-1]$: X-sorted arrays of points. K: positive int
Output: *MaxKHeap*: binary Max Heap storing the K closest pairs between P and Q
      // Initialization
1: $i \leftarrow 0$ $j \leftarrow 0$ *continue* $\leftarrow$ TRUE
2: $P[N].x \leftarrow \infty$ $Q[M].x \leftarrow \infty$ ⸴   // sentinel points for simpler stoping conditions
3: **if** $P[N-1].x \leq Q[0].x$ **then** $i \leftarrow N$ // the sets do not overlap
4: **if** $Q[M-1].x \leq P[0].x$ **then** $j \leftarrow M$ // the sets do not overlap
5: *leftp* $\leftarrow -1$ *leftq* $\leftarrow -1$ // comparisons start at $P[leftp+1]$, $Q[leftq+1]$
      // Main Algorithm. $P[i]$ ($Q[j]$): start of next P-run (Q-run)
6: **while** *continue* **do**
7:     **if** $P[i].x < Q[j].x$ **then**   // the active run is from the P set
8:         **while** $P[i].x < Q[j].x$ **do**   // while active run unfinished. $P[i]$: *ref_point*
9:             **if** $j - 1 = leftq$ **then**   // $Q[j-1]$: last *cur_point* - rule 3
10:                advance i to next P-run and break **while** l.8
11:             **for** $k = j - 1$ **downto** $leftq + 1$ **do**   // $Q[k]$: *cur_point*
12:                **if** *MaxKHeap* is not full **then**
13:                    calculate distance d b/t *ref_point* ($P[i]$) and *cur_point* ($Q[k]$)
14:                    insert (*ref_point, cur_point*) with key d into *MaxKHeap*
15:                **else**
16:                    calculate x-distance dx b/t *ref_point* ($P[i]$) and *cur_point* ($Q[k]$)
17:                    **if** $dx \geq$ key of *MaxKHeap* root **then**   // $dx \geq \delta$ - rule 2
18:                       *leftq* $\leftarrow k$ and break **for** l.11
19:                    calculate distance d b/t *ref_point* ($P[i]$) and *cur_point* ($Q[k]$)
20:                    **if** $d <$ key of *MaxKHeap* root **then**   // $d < \delta$ - rule 1
21:                       insert (*ref_point, cur_point*) with key d into *MaxKHeap*
22:         increment i // update *ref_point* $P[i]$
23:     **else if** $j < M$ **then**   // the active run is from the (unfinished) Q set
24:         $P[N].x \leftarrow Q[M-1].x + 1$ // $P[N]$ should be $< Q[M]$, since ...
        // ... **else** l.23 handles equal X-values b/t P and Q points
25:         **while** $Q[j].x \leq P[i].x$ **do**   // while active run unfinished. $Q[j]$: *ref_point*
26:             **if** $i - 1 = leftp$ **then**   // $P[i-1]$: last *cur_point* - rule 3
27:                advance j to the next Q-run start and break **while** l.25
28:             **for** $k = i - 1$ **downto** $leftp + 1$ **do**   // $P[k]$: *cur_point*
29:                **if** *MaxKHeap* is not full **then**
30:                    calculate distance d b/t *ref_point* ($Q[j]$) and *cur_point* ($P[k]$)
31:                    insert (*ref_point, cur_point*) with key d into *MaxKHeap*
32:                **else**
33:                    calculate x-distance dx b/t *ref_point* ($Q[j]$) and *cur_point* ($P[k]$)
34:                    **if** $dx \geq$ key of *MaxKHeap* root **then**   // $dx \geq \delta$ - rule 2
35:                       *leftp* $\leftarrow k$ and break **for** l.28
36:                    calculate distance d b/t *ref_point* ($Q[j]$) and *cur_point* ($P[k]$)
37:                    **if** $d <$ key of *MaxKHeap* root **then**   // $d < \delta$ - rule 1
38:                       insert (*ref_point, cur_point*) with key d into *MaxKHeap*
39:         increment j // update *ref_point* $Q[j]$
40:         $P[N].x \leftarrow Q[N].x$ // revert the P sentinel at the maximum real X-value
41:     **else** *continue* $\leftarrow$ FALSE   // the points of both sets have been processed

Since $P[0].x < Q[0].x$ (line 7), the *active run* is from the P set and consists of $P[0], P[1]$ and $P[2]$ ($P[2]$ is the last point of P before $Q[j]$). Each of the points of the *active run* should be compared with each of the *current points* of Q in reverse X-order which form the sequence $Q[j-1], \ldots, Q[leftq+1]$. However, since $j - 1 = leftq$ (line 9), i is advanced to 3 ($P[3]$ is the start of the next P run) and processing of the *active run* is broken (rule 3).

During the next iteration of the "while" loop at line 6, $P[3].x > Q[0].x$. Thus the *active run* is from the Q set (line 23[2]) and consists of $Q[0]$ and $Q[1]$ ($Q[1]$ is the last point of Q before $P[i]$) and each of these points will be compared with each of the *current points* of P in reverse X-order which form the sequence $P[i-1], \ldots, P[leftp+1]$ ($P[2], \ldots, P[0]$). The pairs $(P[2], Q[0]), (P[1], Q[0])$ and $(P[0], Q[0])$ are inserted in the non-full heap ($K = 3$). The pair $(P[2], Q[1])$ is inserted in the heap, replacing $(P[1], Q[0])$, since its distance is smaller than δ (rule 1). The pair $(P[1], Q[1])$ is not inserted in the heap, due to its distance. The pair $(P[0], Q[1])$ is not inserted in the heap, due to $dx \geq \delta$ (rule 2). *leftp* is advanced to 0 (only comparisons with P points after $P[0]$ are necessary) and "for" loop at line 28 is broken. The *active run* ends with $i = 3$ and $j = 2$.

During the next iteration of the while loop at line 6, the *active run* consists of $P[3], P[4], P[5]$ and $P[6]$. Each of these points will be compared with $Q[1]$ and $Q[0]$. The pair $(P[3], Q[1])$ is inserted in the heap, replacing $(P[0], Q[0])$ since its distance is smaller than δ (rule 1). Next, the pair $(P[3], Q[0])$ is inserted in the heap, replacing $(P[3], Q[1])$, since its distance is smaller than δ (rule 1). The pair $(P[4], Q[1])$ is not inserted in the heap, due to $dx \geq \delta$ (rule 2). *leftq* is advanced to 1 (only comparisons with Q points after $Q[1]$ are necessary) and "for" loop at line 28 is broken (so the comparison of $P[4]$ with $Q[0]$ is avoided). During the next iteration of the "while" loop at line 8, $j - 1 = leftq$ (line 9), meaning that the rest of the P run will be skipped, saving comparisons (rule 3). The *active run* ends with $i = 7$ and $j = 2$.

During the next iteration of the while loop at line 6, the *active run* consists of $Q[2]$ and $Q[3]$. Each of these points will be compared with each point of the sequence $P[6], \ldots, P[1]$. The pair $(P[6], Q[2])$ is not inserted in the heap, due to $dx \geq \delta$ (rule 2). *leftp* is advanced to 6 (only comparisons with P points after $P[0]$ are necessary) and "for" loop at line 28 is broken (so the comparisons of $Q[2]$ with $P[5], \ldots, P[1]$ are avoided).

During the next iteration of the "while" loop at line 25, $i - 1$ is equal to *leftp* (line 26), meaning that the rest of the Q run (in fact, only point $Q[3]$) will be skipped, saving comparisons (rule 3). The *active run* ends with $i = 7$ and $j = 3$.

During the next iteration of the while loop at line 6, since $P[i = 7] = Q[j = 3] = \infty$ and $j = M$, processing of the two sets is completed.

---

[2] Note that the extra check "$j < M$" at line 23 guarantees that we have not reached the sentinel of the Q set and is necessary, since the "else" part of the main loop (Lines 23-39) handles the case of equal X-values between points and, at the time of this check, both sentinels equal ∞. At line 24, only for the duration of this "else" part, we set the P sentinel to a value between the last Q point and the Q sentinel (to terminate the loop at line 24 when the last run belongs to the Q set).

Note that the *classic PS* algorithm always processes pairs from left to right, even when the distance of the *pivot* point to its closest point of the other set is large (this is likely, since, *runs* of the two sets are in general interleaved). On the contrary, *RRPS* processes pairs of points in opposite X-orders, starting from pairs consisting of points that are the closest possible, avoiding further processing of pairs that is guaranteed not to be part of the result and substituting distance computations by simpler dx computations, when possible. This way, δ is expected to be updated more fastly and the processing cost of *RRPS* to be lower. In the specific example described previously, the *classic PS* algorithm would perform 9 distance computations, 15 dx computations, 8 heap insertions and would examine 18 pairs. *RRPS* performed 7 distance computations, 7 dx computations, 6 heap insertions and examined 10 pairs.

5   Experimentation

In order to evaluate the behavior of the proposed algorithms, we have used 6 real spatial data sets of North America, representing cultural landmarks (NAcl with 9203 points) and populated places (NApp with 24493 points), roads (NArd with 569120 line-segments) and railroads (NArr with 191637 line-segments). To create sets of points, we have transformed the MBRs of line-segments from NArd and NArr into points by taking the center of each MBR (i.e., |NArd| = 569120 points, |NArr| = 191637 points). Moreover, in order to get the double amount of points from NArr and NArd, we chose the two points with min and max coordinates of the MBR of each line-segment (i.e., |NArdD| = 1138240 points, |NArrD| = 383274 points). The data of these 6 files were normalized in the range $[0, 1]$ and the files were combined in pairs, excluding the combinations of NArr with NArrD and NArd with NArdD (since these data sets are correlated, due to the way D versions were created), we made 13 combinations of input sets. We have also created synthetic clustered data sets of 125000, 250000, 500000 and 1000000 points, with 125 clusters in each data set (uniformly distributed in the range $[0 - 1]$), where for a set having N points, $N/125$ points were gathered around the center of each cluster, according to Gaussian distribution. We made 4 combinations of synthetic data sets by combining two separate instances of data sets, for each of the above 4 cardinalities. For each of these 17 (=13+4) combinations of data sets, we executed the *classic PS* algorithm and the *RRPS* algorithm, using a sliding strip, a sliding window and a sliding semi-circle, for K equal to 1, 10, 100, 1000 and 10000. This sums up to 510 experiments (17 combinations × 2 algorithms × 3 variations × 5 K values). All experiments were performed on a PC with Intel Core 2 Duo, 2.2 GHz CPU with 4 GB of RAM and several GBs of secondary storage, with Ubuntu Linux v. 14.04, using the GNU C/C++ compiler (gcc). The performance measurements were:

1. The response time (total query execution time) of processing the KCPQ, not counting reading from disk files to main memory and sorting.
2. The number of Euclidean distance computations (*dist*).
3. The number of X-axis distance computations (*dx*).

5.1 Performance Comparison of *PS* Algorithms for *K*CPQs

In the following, out of the large amount of results obtained from experimentation, some representative results are presented. In the upper (lower) part of Table 2, the execution time in milliseconds of the *classic* (*RRPS*) algorithm, for two real and two synthetic data set combinations, for the sliding strip, window and semi-circle variations and for all K values are depicted. First, it is observed that among the algorithmic variations, the sliding semi-circle is constantly the most efficient one in both algorithms. Second, it is observed that the *RRPS* algorithm outperforms the *classic* one in all cases for the NApp-NArdD combination, for $K > 1$ for the 250KC-250KC combination, for $K = 1$ / sliding strip and for $K > 1$ / sliding window or semi-circle for the 1000KC-1000KC combination and for $K > 100$ for the NArr-NArd combination. In Figure 2,

Table 2. Execution times of the *classic* (above) and *RR* (below) *PS* algorithms

PS	NApp-NArdD			NArr-NArd			250KC-250KC			1000KC-1000KC		
K	Strip	Window	sCircle	Strip	Window	sCircle	Strip	Window	sCircle	Strip	Window	sCircle
1	7.7	7.4	7.5	9.0	8.3	7.6	21.5	15.5	12.1	133.2	87.6	67.9
10	9.4	8.7	8.2	19.9	16.4	12.5	45.1	31.2	21.9	255.8	165.8	119.0
100	15.2	13.1	10.5	39.8	32.3	20.9	121.4	81.2	53.4	740.5	464.5	322.5
1000	34.3	28.1	19.6	98.1	76.9	46.1	329.2	221.0	140.9	2193.3	1378.3	936.7
10000	131.6	109.7	77.9	300.0	241.4	145.8	806.2	570.0	349.6	5353.3	3561.7	2281.1

RRPS	NApp-NArdD			NArr-NArd			250KC-250KC			1000KC-1000KC		
K	Strip	Window	sCircle	Strip	Window	sCircle	Strip	Window	sCircle	Strip	Window	sCircle
1	5.8	5.7	5.4	9.0	8.5	7.7	21.0	15.8	12.3	127.3	85.8	67.4
10	7.5	7.1	6.2	19.6	16.7	12.6	40.1	28.6	20.3	219.3	147.1	105.6
100	13.0	11.2	8.2	39.2	32.2	21.1	96.3	66.7	43.5	537.0	355.6	238.1
1000	30.4	24.8	15.5	94.7	74.2	44.2	247.6	172.0	106.4	1457.5	970.0	625.4
10000	108.9	88.5	56.2	276.1	214.6	124.8	667.5	475.5	288.5	3848.3	2615.0	1632.5

the relative performance of the two algorithms is depicted for the NApp-NArdD (upper-left diagram), NArr-NArd (upper-right diagram), 250KC-250KC (lower-left diagram) and 1000KC-1000KC (lower-right diagram) combinations, for all K values and all algorithmic variations. The percentages depicted express the fraction of the difference of the execution time of the *classic* minus the execution time of the *RRPS* algorithm, over the execution time of the *classic* algorithm (called gain). In other words, they express how much time is saved (positive values) or wasted (negative values) when the *RRPS* replaces the *classic* algorithm. These two figures were created by the same data that are depicted in Table 2 and they visualize the above conclusions about the relative performance of the two algorithms. Note that the variation of gain values depends on the distributions of the data sets and the value of K, both of which affect the number of computations each algorithm performs and how fast it approaches a good δ.

In Table 3 we summarize the results of relative performance for all the 255 (=510/2) cases of experimental comparisons performed. A "−" expresses gain $\leq -1.5\%$ (the *classic* algorithm is more efficient), a "×" expresses $-1.5\% <$ gain

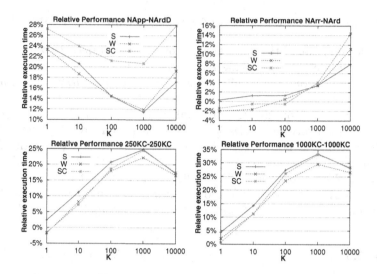

Fig. 2. Relative performance of the *classic* and *RR PS* algorithms

< 1.5% (the two algorithms are almost equal) and a "+" expresses gain ≥ 1.5% (the *RRPS* algorithm is more efficient). Moreover, in each row the minimum and maximum gain is shown. The *RRPS* algorithm is more efficient in 217 (or in 85% of the) cases (in fact, gain ≥ 5% in 76% of the cases), the two algorithms are equal in 22 cases, while the *classic* algorithm is more efficient in 16 cases. Moreover, in all experiments the sliding semi-circle was the most efficient variation for both algorithms.

Table 3. Summary of the relative performance of the *classic* and *RR PS* algorithms

Algorithmic variant: Set combinations K:	Sliding Strip 10^0	10^1	10^2	10^3	10^4	Sliding Window 10^0	10^1	10^2	10^3	10^4	Sliding Semi-Circle 10^0	10^1	10^2	10^3	10^4	gain % min	max
NAcl−NApp	+	×	+	+	+	−	×	+	+	+	−	×	+	+	+	-4.4	39.6
NArr−NArd	×	×	×	+	+	−	−	×	+	+	×	×	×	+	+	-2.0	14.4
NArrD−NArd	−	−	×	+	+	−	−	×	+	+	−	−	−	+	+	-9.9	11.3
NArrD−NArdD	+	+	×	+	+	−	×	×	+	+	−	×	×	+	+	-2.6	10.6
All other (9/13) real data combinations	+	+	+	+	+	+	+	+	+	+	+	+	+	+	+	3.5	36.4
125KC−125KC	×	+	+	+	+	−	+	+	+	+	×	+	+	+	+	-2.5	19.8
250KC−250KC	+	+	+	+	+	−	+	+	+	+	−	+	+	+	+	-1.8	24.8
500KC−500KC	+	+	+	+	+	×	+	+	+	+	×	+	+	+	+	-0.5	29.7
1000KC−1000KC	+	+	+	+	+	+	+	+	+	+	×	+	+	+	+	0.7	33.5

In Table 4, the relative gains in *dist* and *dx* computations of using the *RRPS* algorithm instead of the *classic* algorithm, utilizing only (due to space limitations) the semi-circle variant in both algorithms, for the same data set combinations and K values of Table 2, are depicted. It is obvious that the use of the *RRPS* algorithm saves both *dist* and *dx* computations. The percentages of gain varies significantly. Studying the rest of the results of the 510 experiments performed, we reach the same conclusion: the use of the *RRPS* algorithm always saves *dist* and *dx* computations, but the percentage of gain varies significantly and depends on the data sets combination, the algorithmic variation and K. Moreover, there is no linear or other obvious relation of these percentages to the execution time gain of using the *RRPS* algorithm instead of the *classic* algorithm. We plan to investigate this relation futher in future work.

Table 4. Relative gain in *dist* and *dx* computations of *RRPS* (semi-circle variant)

K	NApp-NArdD dist	dx	NArr-NArd dist	dx	250KC-250KC dist	dx	1000KC-1000KC dist	dx
1	80.5%	86.1%	91.6%	47.0%	36.4%	16.7%	38.2%	12.9%
10	56.3%	67.1%	83.2%	18.9%	33.7%	19.0%	38.4%	19.9%
100	37.9%	36.9%	75.7%	7.6%	37.8%	23.8%	40.3%	29.6%
1000	41.2%	16.9%	59.7%	3.0%	38.2%	25.4%	33.3%	34.0%
10000	47.2%	7.0%	52.5%	1.9%	25.6%	16.4%	26.6%	28.0%

6   Conclusions and Future Work

In this paper, we studied the problem of answering the KCPQ between two point sets that reside on RAM, using *PS* algorithms. We utilized two improvements (sliding window and sliding semi-circle) and proposed a new algorithm (*RRPS*) that minimizes the number of *dist* and *dx* computations, in comparison to the *classic PS* algorithm. By extensive experimentation using real and synthetic data sets, we concluded that the semi-circle improvement is the most efficient one, while the *RRPS* algorithm outperformes the *classic* one in 76% (85%) of the cases with a performance gain $\geq$ 5% (1.5%) and may approach 40%.

In future work, we plan to extend the *RRPS* algorithm for finding closest pairs between non point data sets, like MBR sets that are stored in nodes of two trees and are combined during processing of distance join queries.

The development of the *RRPS* algorithm is the first step in developing a *PS* algorithm for the K-Closest-Pairs query for data sets that cannot be completely transferred to RAM, due to their large cardinalities. For such cases, the *PS* algorithm should utilize the available RAM and process the data sets in parts, minimizing not only distance computations but disk accesses too.

References

1. Beckmann, N., Kriegel, H.P., Schneider, R., Seeger, B.: The R*-tree: An Efficient and Robust Access Method for Points and Rectangles. In: SIGMOD Conference, pp. 322–331 (1990)
2. Brinkhoff, T., Kriegel, H.P., Seeger, B.: Efficient Processing of Spatial Joins Using R-trees. In: SIGMOD Conference, pp. 237–246 (1993)
3. Corral, A., Manolopoulos, Y., Theodoridis, Y., Vassilakopoulos, M.: Closest Pair Queries in Spatial Databases. In: SIGMOD Conference, pp. 189–200 (2000)

4. Corral, A., Manolopoulos, Y., Theodoridis, Y., Vassilakopoulos, M.: Algorithms for Processing K-Closest-Pair Queries in Spatial Databases. Data & Knowledge Engineering 49(1), 67–104 (2004)
5. Guttman, A.: R-trees: A Dynamic Index Structure for Spatial Searching. In: SIGMOD Conference, pp. 47–57 (1984)
6. Hinrichs, K., Nievergelt, J., Schorn, P.: Plane-Sweep Solves the Closest Pair Problem Elegantly. Information Processing Letters 26(5), 255–261 (1988)
7. Hjaltason, G.R., Samet, H.: Incremental Distance Join Algorithms for Spatial Databases. In: SIGMOD Conference, pp. 237–248 (1998)
8. Jacox, E.H., Samet, H.: Spatial Join Techniques. TODS 32(1), article 7, 1–44 (2007)
9. Kim, Y.J., Patel, J.: Performance Comparison of the R*-tree and the Quadtree for kNN and Distance Join Queries. IEEE TKDE 22(7), 1014–1027 (2010)
10. Preparata, F.P., Shamos, M.I.: Computational Geometry: An Introduction. Springer (1985)
11. Rigaux, P., Scholl, M., Voisard, A.: Introduction to Spatial Databases: Applications to GIS. Morgan Kaufmann (2000)
12. Shin, H., Moon, B., Lee, S.: Adaptive and Incremental Processing for Distance Join Queries. IEEE TKDE 15(6), 1561–1578 (2003)
13. Zheng, K., Zhou, X., Fung, P.C., Xie, K.: Spatial query processing for fuzzy objects. VLDB Journal 21, 729–751 (2012)

Mastering Erosion of Software Architecture in Automotive Software Product Lines

Arthur Strasser[1], Benjamin Cool[1], Christoph Gernert[1], Christoph Knieke[1],
Marco Körner[1], Dirk Niebuhr[1], Henrik Peters[1], Andreas Rausch[1],
Oliver Brox[2], Stefanie Jauns-Seyfried[2], Hanno Jelden[2], Stefan Klie[2],
and Michael Krämer[2]

[1] TU Clausthal, Department of Computer Science, Software Systems Engineering
Julius-Albert-Straße 4, D-38678 Clausthal-Zellerfeld, Germany
[2] Volkswagen AG, Powertrain Electronics
P.O. 16870, D-38436 Wolfsburg, Germany

Abstract. Most automobile manufacturers maintain many vehicle types to keep a successful position on the market. Through the further development all vehicle types gain a diverse amount of new functionality. Additional features have to be supported by the car's software. For time efficient accomplishment, usually the existing electronic control unit (ECU) code is extended.

In the majority of cases this evolutionary development process is accompanied by a constant decay of the software architecture. This effect known as software erosion leads to an increasing deviation from the requirements specifications. To counteract the erosion it is necessary to continuously restore the architecture in respect of the specification.

Automobile manufacturers cope with the erosion of their ECU software with varying degree of success. Successfully we applied a methodical and structured approach of architecture restoration in the specific case of the brake servo unit (BSU). Software product lines from existing BSU variants were extracted by explicit projection of the architecture variability and decomposition of the original architecture. After initial application, this approach was capable to restore the BSU architecture recurrently.

Keywords: Architecture design, Reuse, Engineering methodologies, Model driven development, Software product lines, Software erosion, Automotive.

1 Introduction

In the automotive sector, global markets have to be served, in which different requirements for the vehicle exist, e.g. country and culturally specific. Thus, specific adjusted variants of the vehicle types have to be developed and produced. Due to high cost pressure, variants and vehicle types can not be developed independently. Instead, potential synergies have to be exploited.

V. Geffert et al. (Eds.): SOFSEM 2014, LNCS 8327, pp. 491–502, 2014.
© Springer International Publishing Switzerland 2014

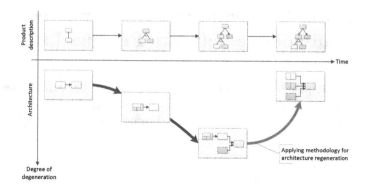

Fig. 1. Degeneration of the architecture based on iterative software development

As a result, components are built in multiple vehicle variants with different configurations. Hence, this approach also applies for the corresponding software modules.

The context, e.g. the available sensors and actuators, in which the software module is used, may differ between vehicle variants. Consequently, software modules have to be able to deal with this variability. For this, product line approaches have proven to be successful in the past.

As shown in Fig. 1, a product line is designed initially and developed over time. It makes no difference whether the product line is explicitly planned or exists only implicitly in the minds of the participants. This product line with its variance regarding the available sensors, actuators and software features provides the basis for the development of an appropriate architecture.

Over the lifetime of the product line new variation points and variants are added. For logical reasons functioning systems are not developed from scratch, but the current state of their architecture and implementation is consequently developed evolutionarily. Since the implementation of new features has to be as flawless, cost-effective and timely as possible, this type of development is often applied in cases where the risks for faults and an explosion of costs and time are the lowest. Apparently that is not always the ideal place in regard to the quality of the resulting software, thus it leads to the erosion of the originally planned architecture.

At some point structural difference between the product line and the architecture size reaches a level that it will be increasingly difficult to integrate new features into the system with the required quality, cost and time. Using an example from practice we show how a fundamental procedure may be designed to optain a renewed architecture in a methodical and strutured fashion, starting from a degenerated architecture.

[8] shows an approach to extract a product line from a user documentation. Our approach has a similar objective, but assumes bahavior models.

This paper is organized as follows: Section 2 gives an overview of the state of research and introduces our approach. The application example is introduced in Section 3. It outlines the main steps for the extraction of product lines from the existing variants of the brake servo unit. Section 4 summarizes the results of the proposed approach and provides an outlook on further work.

2   Background

In the iterative development process new software product variants of high quality are developed on existing software artifacts. Initial design concepts serve reuse and further development of product variants in each cycle of development. If the requirements specification no longer corresponds to the architecture and the implemented software products, the consistency between those artifacts must be restored.

The aim of our approach is the approximation between the architecture and the specification to minimize the so-called software erosion. In Fig. 2 this step is shown as activity for the architecture regeneration. Then the result artifacts can be reused and further developed in the phase of *long-term evolution*.

2.1   Erosion

The challenge in the development of software architecture is to minimize the effort for implementation of design changes, particularly if the architecture yet has been further developed. The fundamental problem is that all design decisions based on the requirements during the domain analysis are only implicitly included in the evolving architecture. If this information is not available while changing the existing architecture (knowledge vaporization), these changes can violate initial design decisions. This effect is called erosion and conclusively leads to architecture degeneration [10,2].

The implemented architecture is based on the requirements specification, resulting in a component-based architecture description [11]. This software architecture description provides the basis for the reuse of existing software components and their iterative development through the implementation of new ECU software functions.

The software components realize requirements specified by the stakeholders depending on the vehicle model. The resulting artifacts are the interfaces and the behavior model which are managed with tool support [11]. Implemented variances which arise from the vehicle model, functional specifications and the installed hardware are also specified in the result artifacts. By querying a condition of the behavior model e.g. the presence of a vehicle-specific sensor can be specified. The interfaces can be parameterized in this manner as well. This increases the complexity of the software component.

This approach for the software component implementing complicates component reuse and development for new and existing vehicle models. New dependencies arise by the specification of variances through conditions in several functionally distinct parts of the behavior model. If this kind of variant specification is implemented repeatedly, the behavioral model parts, originally specified as independent, can be difficult to identify. If the conditions also depend upon each other, the distinction between a valid and invalid variance gets complicated without initial design knowledge. This design freedom does not have to be documented explicitly [13]. The risk of architectural erosion increases. In order to ensure maintainability and extensibility, it is pursued to keep this risk in the long term to a manageable level.

2.2   Model Based Development of Automotive Software

There are several requirements in developing software architecture of powertrain systems. Higher reuse and extensibility reduce time-to-market during the lifetime of vehicles software and ECU generations. Separation of concerns is the necessary design concept to fulfill these requirements. It is handled by the well-known OSEK abstraction layers that are also applied in the AUTOSAR standard.

The models are used to separate the development of hardware related software services from customer related software functionality (basic and application software). Moreover it defines the architecture on a higher level of abstraction with standardized interfaces which is suitable for independent development of automotive software. However requirements and concepts for individual developed software architectures of the applications layer needs to be identified to explore design options [12]. Otherwise it is hardly possible to specify the applications' concerns of functionality. Further the specification of the applications' responsibility is needed to identify reusable development artifacts. The exploration step is challenging due to complex couplings between software functions like in powertrain systems where a specific sensor or actuator is controlled by a set of

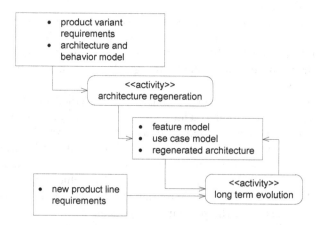

Fig. 2. Required activities for mastering erosion in software product line

frequently interacting functions [3] and due to the lack of separation between functional and architecture model during model based development [12].

In the automotive domain software architects and engineers of different disciplines model software with Matlab/Simulink or Ascet. Design decision are immediately implemented in a graphical language as behavior models [12]. The benefit of behavior models is the possibility of early integration which satisfies flexibility at design time [1,11]. But the implementation does not clearly specify the concern which is realized by a behavior model. However for further development an architect must consider the couplings to system interfaces (sensors, actuators) and the remaining distributed deployed applications in the application layer. Otherwise modularity of the affected application can suffer after completion of modifications.

Our approach consists of two essential steps that are called architecture regeneration and long term evolution to address the shortcomings. The first step aims to extract architectural significant concerns that led to design decisions in the eroded model artefacts. All extracted concerns are specified in the appropriate artefacts:

– feature model: To specify corresponding vehicle variants. Associations to variation points within the use case model and template architecture document the affect of a configured variant in the model.
– use case model: To specify functionality-concerns as use cases of the application which represent functional requirements.
– template architecture: To specify the components as resulting elements of the appropriate design decisions. The template architecture provides a reusable application design for an appropriate configuration in the feature model.

Associations between the resulting artefacts are used to track applied design decisions in the template architecture. These tracks are considered for further development in a long term as illustrated in Fig. 2.

2.3   Software Product Line Extraction

The aim of Software Product Line Extraction is to identify all the valid points of variation and the associated functional requirements of component diagrams. The correspoding steps *variant analyses* and *requirements documentation* are illustrated in Fig. 3 The diagrams represent the behavior models and their interfaces. For the variability extraction the requirements specification of the application and the specified vehicle variants (products that reuse the application) are analysed. This step is necessary to identify requirements that match the applications' functionality and the dependencies to the systems' architecture as described in section 2.2. In the setting of powertrain systems a functionality can be specified, for example, as the calculation of the injected fuel quantity [3]. It can be documented in our approach as a use case model.

In order to obtain the variability in identified product line requirements, we apply the use case modelling technique [7]. These application requirements are coupled with the characteristic vehicle variants that are common to all or a part of systems' applications by points of variation. In [4] variability of a system characteristic is described in a feature model as variable features that can be mapped to use cases. Feature Oriented analysis is a feasible approach for a software product line (SPL) in complex systems [5].

Our approach exploits the associated variation points between both models. A variation point is not only used to describe variability but also the applications extensibility. For example the powertrain system configuration is modified by adding sensors or subsystems. The modifications are specified in the feature model as features using existing or new variation points. Due to the mapping of features to use cases, associated use case variation points are influenced. These points must be taken into account for further development. Thereby valid points for applications' changes can be tracked.

The Product Line UML-based Software Engineering (PLUS) approach permits variability analysis based on use case scenarios and the specification of variable properties in a feature model [6]. To carry out the variability analysis, we use the PLUS approach to describe the variability models in a consistent syntax. By applying the software product line approach new software products are developed from reusable development artefacts. An alternative to SPL would be an concept, where for each product the whole set of functions is available for reuse [9]. Because of the limited amount of ECU memory this concept is not an option.

The use cases of the product line core are central components of the product line and thus present in every derived product. Optional use cases describe tasks that do not have to be carried out by all products. These may or may not have to be carried out by a particular product. On the other hand alternative use cases mutually exclude each other, if they are located within the same group. This means that two alternative use cases cannot simultaneously be part of a derived product. After the identification of all core and variable use cases these are grouped into features. Commonly reused scenarios are each assigned to a feature. This assignment may be stated in a table. In conclusion the dependencies between the features in the variability model are specified according to PLUS. Thereby the root node always groups all applications defined as the core scenarios. All remaining features, such as the use cases, carry the attributes *optional* or *alternative* and can also exhibit dependency relationships defined through PLUS. A validly selected feature set characterizes the set of requirements (use cases) of a particular software product.

The resulting artefacts contains the coarse architecture based on the use case and variability model providing the starting point for the product line re-engineering.

2.4   Software Architecture Re-engineering

Based on the result artifacts of the product line extraction, the architecture components are designed in the next step. This is illustrated as product line re-engineering in Fig. 3. Initial design decisions of the behavior and interface

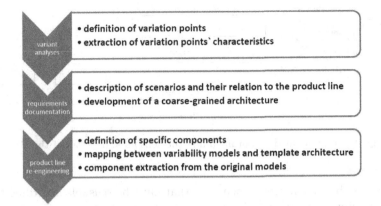

Fig. 3. Procedure of the architecture regeneration approach

model must be extracted with the aid of the result artefacts of the product line extraction. The goal is the fulfillment of the architecture quality requirements, such as modularity and reusability of components. In particular, the quality should be maintained on subsequent architectural changes.

A key component of the approach is the design of reusable software artefacts by the selection of features in the variability model. All valid variants and their requirements can be derived from features. The use cases are associated with the respective features. In this way the variability does not need to be explicitly described by the architecture.

Based on the existing architecture and variability model the coarse architecture is refined. For each use case, software components are described, taking into account eroded architectural elements. The variants were detected in the step of product line extraction. Based on the features and use cases, components are designed which are either reusable for each product or only in certain products. The extend and include relationships in the use case model characterize variation points for further development. This is relevant during the component design.

This approach allows further development of the product line. Dependencies between the variants are recorded in the extraction phase. A new vehicle variant can be introduced as a new feature of a variation. Existing features and their variation points provide their associated use cases that represent a part of the functional application requirements. Significant requirements to determine design decisions are carried out in the seperation of the use cases.

Each use case is associated to components that implement a set of requirements. The introduction of a variation point within a use case is significant to all components that are associated with that point. The components must be implemented as reusable application artefacts. For example differenten types of brake-force calculations are used depending on the vehicles' ECU configuation. The calculated value does not vary but the type of calculation is specified as parameterizable. A certain functionality (e.g. brake-force calculation) can be

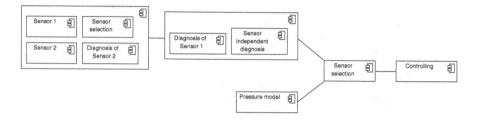

Fig. 4. Component diagram including the sensors and actors of the BSU

implemented by varying component sets that must be reusable according to the variation points. Thereby the design significant requirements can be carried out by tracking the asscociated variation points of each component.

In the re-engineering step seperatable concerns are derived from use case and feature models to redesign cohesion lowering elements. All decisions about allowed seperations are made by considering the extracted variability. In this way, the modularization during component design is supported in the long term evolution step. New variation points must be recorded in the use case, feature model and must be associated with the architecture. The subsequent architecture design can be verified against the re-engineered result artefacts.

After the design, the components are connected according to their interfaces. The connection depends on the mapped features and the variability model, resulting in a product line architecture.

3   Experimental Details

The purpose of the brake servo unit (BSU) in cars is to assist the driver by boosting his brake force. To achieve this, a vacuum (for example from the intake manifold) is used to generate a pressure difference within the BSU. The pressure difference results in an additional force inside the braking system.

The softwares' objective is to guarantee a sufficiently high pressure difference in order to achieve the required boost. Initially, only one sensor was available for pressure measurement. The development of the controller fulfilling requirements (e.g. by adjusting the engine speed) was model-based using behavior models.

Over time, new systems were added in cars, to which the BSU software had to be adapted. This concerned, for example, new sensors available after the introduction of electronic stability control systems (ESC), as well as new actuators (e.g. an electrical vacuum pump). In addition cross effects of new functions such as *Start-Stop* caused disturbances which the existing controllers could not handle.

As shown in Fig. 4 the continuous erosion of the software architecture led to wider component interfaces and a reduction of the components' cohesion.

We were able to identify two major variation points within the BSU software: *sensors* and methods for *vacuum generation*. After examination of the models, we found two versions of the variation point "methods for vacuum generation", specifically "Subsystem x" (intake manifold evacuation), and "vacuum-pump evacuation".

The calculation of the pressure model, selection of the sensor signal and pressure controller are located in the core of the BSU software. The use cases "sensor 1" and "sensor n" capture data from different sensors and extend the core use case "Choose sensor value". Each pressure control method is encapsulated in a separate use case. They extend the core use case "Control pressure difference".

For a more detailed description of the use cases, templates are used to document the variability adequately.

We were able to decompose the function into smaller components with higher cohesion with respect to these use cases (Fig. 6) and the feature model (Fig. 5) and developed an architecture template model (Fig. 7) to describe the possible connections of the components. This model can be used to derive an architecture for every configuration.

4   Results and Discussion

Originally there was a general variant of the software, which evacuated the vacuum chamber of the BSU through the intake manifold. Later on a variant of BSU software was added, that featured an electric vacuum pump for the evacuation. The software variance was constituted by the presence or absence of a mechanical vacuum pump. When implementing the variability into software the developers chose the simplest and fastest way: Since the mechanical vacuum pump was installed only in diesel vehicles, the variance was realized by a query whether there is a gasoline or diesel engine. This query was already used in other vehicle functions.

This solution established itself over time, but was insufficient with the introduction of hybrid vehicles, as they may have both a gasoline engine and an electric vacuum pump. Therefore, the developers extended the initial "gasoline

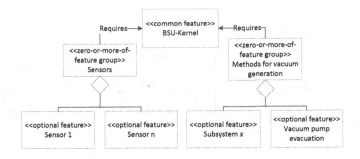

Fig. 5. Resulting feature model after the architecture regeneration activity

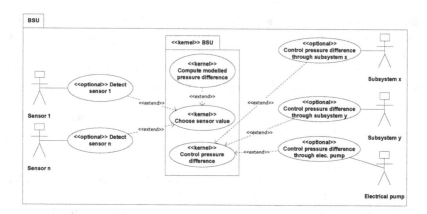

Fig. 6. Use case diagram of the BSU. Sensors and subsystems are displayed as UML actors.

or diesel engine" query by another query, whether it is a hybrid vehicle. While this was purposeful to allow quick implementation of the BSU software for hybrid vehicles, it no longer corresponded to the original motivation whether a vacuum pump is available. Multiple implementations of such quick solutions in both actuator and sensor variance, led to a difficult to overlooking, hardly maintainable and extensible BSU software system. An extensive analysis and de novo establishment of a product line within the BSU software, an *architecture regeneration* was required. In the extraction phase we specified two use case groups and two variation points that match the systems' functionality. One use case group considers requirements which must be fulfilled by all ECU configurations (uses core BSU control application) and another group that must only be realized by some ECU configurations. The variation points that extend (modify) a certain functionality at a defined point of bahavior are associated with two features of the feature model. The two feature-groups group another features that

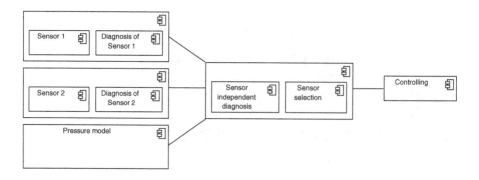

Fig. 7. Resulting domain specific template model of the BSU software

represent different variants of evacuation methods and sensors. The group is associated with a variation point within the use case "control pressure difference". Every grouped feature can be independently selected for a valid ECU configuration. Thereby we documented the extensibility requirements as variation points within the two use cases.

The extracted product line requirements include significant design constraints that must be considered by the architect. All components that are associated with the kernel use cases realize the functionality BSU pressure control. All components that are associated with extended use cases interact with the BSU pressure control through generic interfaces. Moreover a certain extended use case (e.g. sensor 1) is associated with components that realize the appropriate sensor concern. These design decision can be tracked and verified against further modifications to obtain a modular design. This enables the interconnect of all components according to feature selection. We call the final design "template architecture".

A design decision that would cause additional dependencies between components which are associated to different variation points of the BSU, violate the modularity. Our approach establishes a relation between the variability models (functionality model) and the template architecture (architecture model) by applying the extraction phase. In the BSU real world example all use cases are one-to-one mapped to variation point features. The verification of components that are affected by a more complex variability is still under research of our working group.

5   Conclusion

We proposed an approach to support effectively avoiding erosion of software architecture by a product line which is enhanced iteratively. By continuously applying the long term evolution step, a strong cohesion of the software components is achieved improving maintainability and extensibility of the software.

One important advantage of the template architecture is modularity. Evolution can be done on every single component without considering further BSU component dependencies. New functional requirements must be realized with respect to the existing use case model, variability model and template architecture, following our two staged methodology. This minimizes the risk to violate the requirements specification by making future design modifications. In addition, the effort for software configuration in certain hardware environments is decreased. By realizing variability on the level of software architecture, readability of all model components is improved.

Initially the BSU software was developed ECU independent. The corresponding behavior model does not include any memory or processor specific data types but was configurable by parameters depending on used actuator- and sensor-hardware. We recommend to focus on those features in automotive software architectures when applying our methodology.

Our approach was demonstrated by a real-world case study of a brake servo unit. The results of the case study are relevant for future automotive systems, too. The case study shows how the methodology is used to design modular and reusable software components, characterizing the variability of the system (ECU) configuration.

The BSU software based on the extracted product lines has already been partially implemented. Experiences concerning extendability of the software could be gained. Future work can refer to elaborating an integrated tool-chain supporting the modeling languages used in our approach adequately.

References

1. Andrews, D., Bate, I., Nolte, T., Otero-Perez, C., Petters, S.M.: Impact of embedded systems evolution on rtos use and design. In: 1st International Workshop Operating System Platforms for Embedded Real-Time Applications, OSPERT 2005 (2005)
2. Bosch, J.: Software architecture: The next step. In: Oquendo, F., Warboys, B.C., Morrison, R. (eds.) EWSA 2004. LNCS, vol. 3047, pp. 194–199. Springer, Heidelberg (2004)
3. Claraz, D., Kuntz, S., Margull, U., Niemetz, M., Wirrer, G.: Deterministic execution sequence in component based multi-contributor powertrain control systems. In: Embedded Real Time Software and Systems Conference (2012)
4. Clements, P., Northrop, L.: Software Product Lines: Practices and Patterns. Addison Wesley (2001)
5. Czarnecki, K., Grünbacher, P., Rabiser, R., Schmid, K., Wąsowski, A.: Cool features and tough decisions: a comparison of variability modeling approaches. In: Proceedings of the Sixth International Workshop on Variability Modeling of Software-Intensive Systems, VaMoS 2012, pp. 173–182. ACM, New York (2012)
6. Gomaa, H.: Designing Software Product Lines with UML: From Use Cases to Pattern-Based Software Architectures. Addison-Wesley Professional (2004)
7. Jacobson, I., Griss, M., Jonsson, P.: Software reuse: architecture, process and organization for business success. ACM Press/Addison-Wesley Publishing Co. (1997)
8. John, I., Dörr, J.: Elicitation of requirements from user documentation. In: Ninth International Workshop on Requirements Engineering: Foundation for Software Quality, REFSQ 2003, Klagenfurt/Velden (2003)
9. Krueger, C.W.: Introduction to the emerging practice software product line development. Methods & Tools 14, 3–15 (2006)
10. Perry, D.E., Wolf, A.L.: Foundations for the study of software architecture. ACM SIGSOFT Software Engineering Notes 17(4), 40–52 (1992)
11. Pretschner, A., Broy, M., Krüger, I.H., Stauner, T.: Software engineering for automotive systems: A roadmap. In: Future of Software Engineering, FOSE 2007 (2007)
12. Sangiovanni-Vincentelli, A., Di Natale, M.: Embedded system design for automotive applications. Computer 40(10), 42–51 (2007)
13. Weber, M., Weisbrod, J.: DaimlerChrysler-Research: Requirements engineering in automotive development: Experiences and challenges. In: Proceedings of the IEEE Joint International Conference on Requirements Engineering 2002 (2002)

Shortest Unique Substrings Queries in Optimal Time

Kazuya Tsuruta, Shunsuke Inenaga, Hideo Bannai, and Masayuki Takeda

Department of Informatics, Kyushu University, Japan
{inenaga,bannai,takeda}@inf.kyushu-u.ac.jp

Abstract. We present an optimal, linear time algorithm for the shortest unique substring problem, thus improving the algorithm by Pei et al. (ICDE 2013). Our implementation is simple and based on suffix arrays. Computational experiments show that our algorithm is much more efficient in practice, compared to that of Pei et al.

1 Introduction

The shortest unique substring problem was proposed by Pei et al. [4]. Given a string S and position p, the problem is to find a *shortest unique substring* (SUS) of S that contains position p, that is, a substring that only occurs once in S, and whose occurrence contains position p. They also consider a version of the problem where S may be preprocessed, and SUS queries for arbitrary positions may be answered efficiently.

For the first version of the problem, Pei et al. [4] presented an algorithm that computes the SUS for any given position p in $O(n)$ time and space, where n is the length of string S. For the second version, they present an $O(hn)$ time and $O(n)$ space preprocessing algorithm which allows queries to be answered in constant time, where h is a value depending on S. However, h is only bounded by $O(n)$, and in the worst case, this results in $O(n^2)$ time pre-processing.

The contributions of this paper is as follows: First, we give optimal time solutions for both problems and show that S can be preprocessed in $O(n)$ time so that a SUS for any query position can be answered in $O(1)$ time. This considerably improves the theoretical worst case running time compared to Pei et al. [4], allowing us to output a SUS for *all* positions in the string in $O(n)$ total time. Second, we consider the general problem of computing all SUSs that contain a given position. Although there can be multiple shortest substrings that contain a given query position, Pei et al. [4] only considered the problem of answering a single SUS that contains a position. We show that the same linear time preprocessing above also allows us to return *all* SUSs that contain a given query position in $O(k)$ time, where k is the size of the output. Finally, we implement our algorithm and show through computational experiments that our implementation is much more practical and scalable compared to an implementation of the algorithm by Pei et al. [4] made available by the authors.

V. Geffert et al. (Eds.): SOFSEM 2014, LNCS 8327, pp. 503–513, 2014.

2   Preliminaries

2.1   Strings

Let Σ be an ordered finite *alphabet*. An element of Σ^* is called a *string*. The length of a string w is denoted by $|w|$. The empty string ε is a string of length 0. Let Σ^+ be the set of non-empty strings, i.e., $\Sigma^+ = \Sigma^* - \{\varepsilon\}$. For a string $w = xyz$, x, y and z are called a *prefix*, *substring*, and *suffix* of w, respectively. A prefix (resp. substring, suffix) x of w is called a *proper prefix* (resp. substring, suffix) of w if $x \neq w$. The i-th character of a string w is denoted by $w[i]$, where $1 \leq i \leq |w|$. For any integers $i \leq j$, let $[i..j]$ denote an interval, i.e. the set of integers $\{i, \ldots, j\}$, and let $|[i..j]| = j - i + 1$ denote the length of the interval. For convenience, let $[i..j]$ denote the empty set when $i > j$. For a string w and interval $[i..j]$ where $1 \leq i \leq j \leq |w|$, let $w[i..j] = w[i] \cdots w[j]$ denote the substring of w that begins at position i and ends at position j. For convenience, let $w[i..j] = \varepsilon$ when $i > j$. For any string w and S, we call a position p an *occurrence* of w in S, if $S[p..p + |w| - 1] = w$.

Given two distinct positions i, j ($i < j$), we say that i is to the *left* and j the *right*. Two distinct intervals are *nested*, if one is a subset of the other. For two non-nested intervals $[i..j]$ and $[i'..j']$, we say that $[i..j]$ is to the left and $[i'..j']$ is to the right, if $i < j$. Since, for any interval $[i..j]$ ($1 \leq i \leq j \leq |S|$) there is a corresponding substring $S[i..j]$ of S, we abuse the language and will many times call an interval a substring.

2.2   Unique Substrings

We say that a substring w of S is *unique*, if there is exactly one occurence of w in S. When w is unique, the interval $[i..i + |w| - 1]$ such that $S[i..i + |w| - 1] = w$ is called a unique interval of S. We say that a unique substring w of S contains position p, if $w = S[i..i + |w| - 1]$ and $p \in [i..i + |w| - 1]$. It is easy to see that any string that contains a substring that is unique, is also unique, and any interval that contains a sub-interval that is unique, is also unique.

Definition 1 (Shortest Unique Substring). *A substring w is a shortest unique substring (SUS) of S that contains position p, if $w = S[i..j]$ is unique in S, $i \leq p \leq j$, and no other substring $w' = S[i'..j']$ such that $i' \leq p \leq j'$ and $j' - i' < j - i$ is unique in S.*

Note that there can be more than one SUS that contains position p as shown in the following example. Let $SUS_S(p)$ denote the set of intervals corresponding to SUSs of S that contains position p. Note that $SUS_S(p) \neq \emptyset$ for any position $1 \leq p \leq |S|$.

Example 1 (SUS). Let $S = $ aabaabcababbaabdbab. Then, $SUS_S(2) = \{[1..4],$ $[2..5]\}$, $SUS_S(4) = \{[1..4], [2..5], [4..7]\}$, $SUS_S(9) = \{[7..9]\}$, $SUS_S(10) = \{[10..12]\}$. (See Fig. 1)

In this paper, we focus on the following problems.

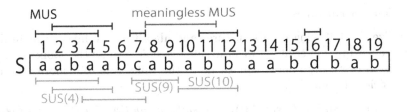

Fig. 1. Example of a string and its SUSs (see Definition 1) and MUSs (see Definition 2). Although all 6 MUSs are depicted, $SUS(p)$ is depicted only for positions $4, 9$ and 10. $MUS_S = \{[1..4], [2..5], [7..7], [8..11], [11..12], [16..16]\}$. $SUS_S(4) = \{[1..4], [2..5], [4..7]\}$, $SUS_S(9) = \{[7..9]\}$, $SUS_S(10) = \{[10..12]\}$. The MUS $[8..11]$ is meaningless since no SUS contains it, while the others are meaningful (see Definition 7).

Problem 1 (SUS query). Given string S of length n, compute for all positions p $(1 \leq p \leq n)$, a shortest unique substring that contains position p.

Problem 2 (All SUS query). Given string S of length n, compute for all positions p $(1 \leq p \leq n)$, all shortest unique substrings that contain position p.

Problem 1 was first considered by Pei et al. [4]. They first gave a simple $O(n)$ time algorithm for computing an SUS for a single p. However, this would result in $O(n^2)$ time for computing a SUS for each p. Thus, they further showed an improved algorithm which pre-process S in $O(hn)$ time, and allows queries for any p in $O(1)$ time, where h is a parameter that depends on S. This results in an $O(hn)$ time solution for computing the SUS for all positions $1 \leq p \leq n$. Although Pei et al. [4] gave empirical evidence that h is not very large in practice, they were not able to give a good theoretical bound on h, mentioning that h can be as large as $O(n)$, resulting in $O(n^2)$ time worst case pre-processing time.

In this paper, we give optimal time solutions for both problems, and show that S can be preprocessed in $O(n)$ time so that the queries can be answered in $O(k)$ time, for any query position p, where k is the size of the output. Noting that k is $O(1)$ for Problem 1, this results in an $O(n)$, i.e. a truly linear time solution for computing the SUS for all positions $1 \leq p \leq n$.

Like the algorithm by Pei et al. [4], our algorithm finds SUSs, based on the concept of Minimal Unique Substrings defined below.

Definition 2 (Minimal Unique Substring). *A substring w of S is a minimal unique substring (MUS) if w is unique in S, and no proper substring of w is unique in S.*

Let MUS_S denote the set of intervals corresponding to MUSs of string S. Notice that by definition, MUSs of S can overlap with each other, but cannot be nested. This implies that there can exist at most one MUS starting at a given position in S. Also, since there must exist at least one MUS, we have $0 < |MUS_S| \leq |S|$.

Example 2 (MUS). Let $S = \texttt{aabaabcababbaabdbab}$, the same string as in Example 1. $MUS_S = \{[1..4], [2..5], [7..7], [8..11], [11..12], [16..16]\}$. (See Fig. 1)

2.3 Data Structures

We utilize the following data structures and algorithms. While the main data structure used by Pei et al. [4] was the suffix tree [5], we use the suffix array [3], which is theoretically almost equivalent to the suffix tree, but more time and space efficient in practice.

Definition 3 (Suffix Array [3]). *The suffix array SA of a string S of length n is a permutation of integers* $\{1, \ldots, n\}$, *such that* $SA[i] = j$ *represents the ith lexicographically smallest suffix* $S[j..n]$ *of S.*

Theorem 1 ([1]). *Assuming an integer alphabet, the suffix array of a string S of length n can be constructed in* $O(n)$ *time.*

Definition 4 (Rank array). *The rank array* SA^{-1} *of a string S of length n, is a permutation of integers* $\{1, \ldots, n\}$, *such that* $SA^{-1}[SA[i]] = i$.

Given SA, SA^{-1} can be computed in $O(n)$ time by a simple loop over SA.

Definition 5 (LCP array). *The longest common prefix (lcp) array LCP of a string S of length n, is an array of integers where* $LCP[1] = 0$ *and* $LCP[i]$ *for* $1 < i \le n$ *holds the length of the longest common prefix between suffix* $S[SA[i-1]..n]$ *and* $S[SA[i]..n]$, *where SA is the suffix array of S.*

Theorem 2 ([2]). *Given string S of length n and its suffix array SA, the lcp array LCP of S can be computed in* $O(n)$ *time.*

3 Algorithm

3.1 Finding All MUSs

Here, we describe how to find all MUSs of a string S in linear time, using the suffix and lcp arrays of S.

Lemma 1. *All MUSs of a string S of length n can be found in* $O(n)$ *time and space.*

Proof. Let SA and LCP respectively be the suffix array and lcp array of S. For any suffix $S[j..n]$ where $SA[i] = j$ (or $SA^{-1}[j] = i$), the shortest prefix of $S[j..n]$ that is unique is given by $S[j..j + \ell_j]$ where

$$\ell_j = \begin{cases} \max\{LCP[i], LCP[i+1]\} & 1 \le i < n \\ LCP[i] & i = n. \end{cases}$$

The definition of ℓ_j implies that $S[j..j + \ell_j - 1]$ is not unique. Thus, $S[j..j + \ell_j]$ is the only candidate for a MUS starting at position j, and is a MUS if and only if $S[j + 1..j + \ell_j]$ is not unique. Since the definition of ℓ_{j+1} implies that $S[j+1..j+\ell_{j+1}]$ is not unique, $S[j..j+\ell_j]$ is a MUS if and only if $j+\ell_j \le j+\ell_{j+1}$. Once SA, SA^{-1}, and LCP are computed in $O(n)$ time, this can be checked in $O(1)$ time for each j. Therefore, the lemma follows since ℓ_j for all $1 \le j \le n$ can be computed in a total of $O(n)$ time. $\square$

3.2 SUSs from MUSs

Next, we consider the relation between MUSs and SUSs.

Definition 6. *For an interval $[i..j]$ and position p, let $cover([i..j], p)$ denote the smallest interval that contains $[i..j]$ and p, i.e. $cover([i..j], p) = [\min(i, p).. \max(j, p)]$. We say that $cover([i..j], p)$ is derived from interval $[i..j]$ and position p.*

We first show that any $SUS_S(p)$ is derived from an element in MUS_S. The following Lemma is essentially the same as Theorem 2 in [4], but the statement has been reworded for our purposes.

Lemma 2. *For any position p and interval $[i..j] \in SUS_S(p)$, there exists exactly one sub-interval $[i'..j'] \in MUS_S$ of $[i..j]$ such that $[i..j] = cover([i'..j'], p)$.*

Proof. Since $[i..j]$ is unique, it must contain at least one minimal unique sub-interval. Let $[i'..j']$ be any MUS contained in the SUS $[i..j]$. Since $i \le p \le j$, $cover([i'..j'], p)$ is unique, contains position p, and is a sub-interval of $[i..j]$. Thus, $[i..j] = cover([i'..j'], p)$ must hold, since otherwise, $cover([i'..j'], p)$ would be an interval shorter than $[i..j]$ containing position p, contradicting the assumption that $[i..j]$ is an SUS.

Next we show that there is exactly one MUS contained in a SUS. Suppose there are two distinct minimal unique sub-intervals $[i_1..j_1]$ and $[i_2..j_2]$ of $[i..j]$. From the above arguments, $[i..j] = cover([i_1..j_1], p) = cover([i_2..j_2], p)$ must hold. Since MUSs cannot be nested, both must be proper sub-intervals of $[i..j]$, and we assume without loss of generality that $i \le i_1 < i_2$ and $j_1 < j_2 \le j$. However, if $i \le p < j$, then $cover([i_1..j_1], p) \ne [i..j]$ since $\max\{p, j_1\} < j$, and if $i < p \le j$, then $cover([i_2..j_2], p) \ne [i..j]$ since $\min\{p, i_2\} > i$. Thus, there can only be one MUS that is contained in a given SUS. □

For the purpose of describing our algorithm, we define a generalization of SUSs with respect to a subset of MUSs, namely, MUSs that begin at or before a certain position. Let $MUS_S^k = \{[i..j] \in MUS_S \mid i \le k\}$. We define $SUS_S^k(p)$ to be the subset of intervals which are shortest, of the intervals that can be derived from intervals in MUS_S^k and position p, i.e., $[i..j] \in SUS_S^k(p)$ if $[i..j] = cover([i'..j'], p)$ for some $[i'..j'] \in MUS_S^k$, and $|[i..j]| \le |cover([i''..j''], p)|$ for any $[i''..j''] \in MUS_k$. Let $lmSUS_S^k(p)$ denote the leftmost interval of $SUS_S^k(p)$, and $lmMUS_S^k(p)$ the interval in MUS_k that derives $lmSUS_S^k(p)$.

Note that $MUS_S = MUS_S^n$, and $SUS_S(p) = SUS_S^n(p)$. Also note that although for any $k < k'$, $MUS_S^k \subseteq MUS_S^{k'}$, it is not necessarily the case that $SUS_S^k(p) \subseteq SUS_S^{k'}(p)$.

Next, we define the concept of *meaningful* and *meaningless* MUSs, which is the main difference of our algorithm with [4].

Definition 7 (Meaningful Minimal Unique Substring). *We say that an interval $[i..j] \in MUS_S^k$ is meaningful with respect to MUS_S^k, if, for some position p, $cover([i..j], p) \in SUS_S^k(p)$. Otherwise, we say that a minimal unique substring is meaningless with respect to MUS_S^k.*

Example 3 (Meaningful MUS). Let $S = $ aabaabcababbaabdbab, the same string as in Example 1. Then, the set of MUSs $\{[1..4], [2..5], [7..7], [11..12], [16..16]\}$ are meaningful, since they respectively derive SUSs corresponding to positions 4, 9 and 10. However, MUS $[8..11]$ is meaningless, it does not derive any SUS. (See Fig. 1)

Observation 1. *For any $k < k'$, if an interval $[i..j] \in MUS_S^k$ is meaningless with respect to MUS_S^k, then it is meaningless with respect to $MUS_S^{k'}$.*

Let $MMUS_S^k$ denote the set of all meaningful MUSs with respect to MUS_S^k. We first show that if we have an array $MMUS_S = MMUS_S^n$ of meaningful MUSs with respect to MUS_S, in order of their occurrence, and for each position p we hold an index $L[p]$ such that $MMUS_S[L[p]] = lmMUS_S^n(p)$, we can answer $SUS_S(p)$ in $O(|SUS_S(p)|)$ time.

To prove this, we give a more specific characterization of which MUSs can derive elements of $SUS_S(p)$. Let $MUS_S(p)$ denote the set of MUSs that contain position p, i.e.,

$$MUS_S(p) = \{S[i..j] \in MUS_S \mid i \le p \le j\}.$$

$MUS_S(p)$ can be empty. For any position p, let $pred_S(p) = [i..j]$ represent the rightmost MUS that occurs before position p if one exists, that is, $[i..j] \in MUS_S$, $j < p$, and there exists no $[i'..j'] \in MUS_S$ such that $j < j' < p$. Similarly, let $succ_S(p) = [i..j]$ represent the leftmost MUS that occurs after position p if one exists, that is, $[i..j] \in MUS_S$, $i > p$, and there exists no $[i'..j'] \in MUS_S$ such that $p < i' < i$. We say that the set $\{pred_S(p), succ_S(p)\} \cup MUS_S(p)$ is the MUSs in the *neighborhood* of position p.

The following lemma shows that $|cover([i..j], p)|$ for MUSs $[i..j]$ in the neighborhood of position p which are meaningful with respect to MUS_S^k and are to the right of $lmMUS_S^k(p)$ (including $lmMUS_S^k(p)$), form a monotonically increasing sequence.

Lemma 3. *Consider any position p and integer k, and let $[i..j] = lmMUS_S^k(p)$. Any two distinct intervals $[i_1..j_1], [i_2..j_2] \in \{\{pred_S(p), succ_S(p)\} \cup MUS_S(p)\} \cap MMUS_S^k$ such that $i \le i_1 < i_2$, satisfy $|cover([i_1..j_1], p)| \le |cover([i_2..j_2], p)|$.*

Proof. Suppose to the contrary that $|cover([i_1..j_1], p)| > |cover([i_2..j_2], p)|$. Since $cover([i..j], p) \in SUS_S^k(p)$, it holds that $|cover([i..j], p)| \le |cover([i_2..j_2], p)| < |cover([i_1..j_1], p)|$. For all positions $i \le p' < p$, it holds that $|cover([i..j], p')| \le |cover([i..j], p)| < |cover([i_1..j_1], p)|$. Since $[i..j] = lmMUS_S^k(p)$ and $i < i_1$, it holds that $[i_1..j_1] \ne pred_s(p)$ and $p' < p \le j_1$. Since $p' < p \le j_1$, it holds that $|cover([i_1..j_1], p)| \le |cover([i_1..j_1], p')|$. Similarly, for all positions $p < p' < j_2$, it holds that $|cover([i_2..j_2], p')| = |cover([i_2..j_2], p)| < |cover([i_1..j_1], p)|$. Since $|cover([i_1..j_1], p)| > |cover([i_2..j_2], p)|$, it holds that $[i_1..j_1] \ne succ_s(p)$ and $i_1 \le p < p'$. It holds that $|cover([i_1..j_1], p)| \le |cover([i_1..j_1], p')|$.

Also, for any position $p' < i$, $|cover([i..j], p)| < |cover([i_1..j_1], p)|$, and for any position $p' > j_2$, $|cover([i_2..j_2], p)| < |cover([i_1..j_1], p)|$. This implies that $[i_1..j_1]$

cannot be meaningful for all positions $1 \leq p' \leq n$, and must be meaningless with respect to MUS_S^k, contradicting the assumption that $[i_1..j_1] \in MMUS_S^k$. Thus, it must be that $|cover([i_1..j_1], p)| \leq |cover([i_2..j_2], p)|$. $\square$

The next lemma shows that intervals in $SUS_S^k(p)$ are the shortest ones derived from MUSs in the neighborhood of position p which are meaningful with respect to MUS_S^k.

Lemma 4. *Consider position p, integer k, interval $[i..j] \in MUS_S^k$, and let $Y = \{\{pred_S(p), succ_S(p)\} \cup MUS_S(p)\} \cap MMUS_S^k$. If $cover([i..j], p) \in SUS_S^k(p)$, then $[i..j] \in Y$ and $|cover([i..j], p)| \leq |cover([i'..j'], p)|$ for all intervals $[i'..j'] \in Y$.*

Proof. Assume $cover([i..j], p) \in SUS_S^k(p)$ holds. Since $Y \subseteq MUS_S^k$ and by the defintion of $SUS_S^k(p)$, $|cover([i..j], p)| \leq |cover([i'..j'], p)|$ holds for all $[i'..j'] \in Y$.

It is easy to see that $[i..j]$ cannot be to the left of $pred_S(p)$, since then, $|cover([i..j], p)| > |cover(pred_S(p), p)|$ and $[i..j]$ could not be in $SUS_S^k(p)$. Similarly, $[i..j]$ cannot be to the right of $succ_S(p)$, since then, $|cover([i..j], p)| > |cover(succ_S(p), p)|$ and again, $[i..j]$ could not be in $SUS_S^k(p)$.

Finally, by the definition of meaningful, $[i..j] \in MMUS_S^k$. $\square$

Algorithm 1. $SUS_S(p)$ from L and $MMUS_S$.

Input: position p, $MMUS_S$, L
Output: $SUS_S(p)$

```
1  t ← L[p];
2  l ← |cover(MMUS_S[t], p)| ;                    // length of SUS
3  while |cover(MMUS_S[t], p)| = l do
4  |    output cover(MMUS_S[t], p);
5  |    t ← t + 1;
6  end
```

Theorem 3. *Given an array $MMUS_S$ of all meaningful MUSs with respect to MUS_S in order of occurrence, and an array L of size n, where, for each position $1 \leq p \leq n$, $MMUS_S[L[p]] = lmMUS_S^n(p)$, we can compute $SUS_S(p)$ in $O(|SUS_S(p)|)$ time.*

Proof. The pseudo code of the algorithm is shown in Algorithm 1. By definition of $MMUS_S$ and L, it is clear that the first output is $lmSUS_S^n(p)$, i.e., the leftmost SUS that contains position p. From Lemma 2 and by the definition of a meaningful interval, it is easy to see that all MUSs that derive elements in $SUS_S(p)$ must be in $MMUS_S$.

It remains to prove that each element in $SUS_S(p)$ is derived from MUSs in a contiguous range in $MMUS_S$. This can be seen from Lemmas 3 and 4, which claim that all MUSs in $SUS_S(p)$ are in the neighborhood of position p that are meaningful with respect to MUS_S, and for all such meaningful MUSs $[i..j]$ to the right of $lmMUS_S^n(p)$ (including $lmMUS_S^n(p)$), $cover([i..j], p)$ forms a monotonically increasing sequence. $\square$

Algorithm 2. Create array $MMUS_S$ of meaningful MUSs and an array of pointers L to $lmMUS$ for each position of string S

Input: LCP and RANK array for string S.
Output: $MMUS[1..|MMUS.size()|]$: array of meaningful MUSs; $L[1..n]$: index in $MMUS$ of leftmost SUS for each position.

1   **for** $p \leftarrow 1$ **to** n **do**
2     $\ell \leftarrow MMUS.size()$;
    // lmMUS for position p wrt MUS_S^{p-1} is either the same as $p-1$, or the next one.
3   **if** $p = 1$ **then**
4       $\lfloor$ $L[1] \leftarrow 1$;   // Core MUS of position 1 is leftmost MUS.
5   **else if** $L[p-1] < \ell$ **and**
     $|cover(MMUS[L[p-1]+1], p)| < |cover(MMUS[L[p-1]], p)|$ **then**
6       $\lfloor$ $L[p] \leftarrow L[p-1] + 1$;
7   **else**
8       $\lfloor$ $L[p] \leftarrow L[p-1]$;
    // update $MMUS$ and L to values wrt MUS_S^p
9   **if** *exists MUS*: $newMUS = [p, e]$ *for some* $e \geq p$. **then** // $O(1)$ time using LCP and RANK array
10   **if** $\ell > 0$ **then**
        // j: rightmost position that doesn't need update
11         $j \leftarrow \max\{i \leq p \mid |cover(MMUS[L[i]], i)| \leq |cover(newMUS, i)|\}$;
12   **if** $j = p$ **then**   // No updates for L. Remove meaningless MUSs from $MMUS$
13           $MMUS \leftarrow MMUS[1..k]$ where
           $\lfloor$ $k = \max\{k' \leq \ell \mid |cover(MMUS[k'], p)| \leq |cover(newMUS, p)|\}$;
14   **else** // remove meaningless MUSs after the one pointed by j and $newMUS$
15           $MMUS \leftarrow MMUS[1..k]$ where $k = \max\{k' \leq \ell \mid$
          $|cover(MMUS[k'], j)| \leq |cover(MMUS[L[j]], j)|\}$;
16           $\lfloor$ **for** $j + 1 \leq i \leq p$ **do** $L[i] \leftarrow k + 1$;   // update L to new MUS
17       $MMUS.push\_back(newMUS)$;

Next we show that $MMUS_S$ and L can be constructed in linear time, by incrementally updating $MMUS_S^k$ and L. Let L^k denote an array of indices where $MMUS_S^k[L^k[p]] = lmMUS_S^k(p)$.

Lemma 5. $L^{p-1}[p]$ *is either the MUS* $[i..j]$ *pointed to by* $L^{p-1}[p-1]$, *or the next MUS* $[i'..j']$ *in* $MMUS_S^{p-1}$, *i.e., the one pointed to by* $L^{p-1}[p-1] + 1$.

Proof. By definition, $[i..j] = lmMUS_S^{p-1}(p-1)$. Let $[i''..j'']$ be an arbitrary interval in $MMUS_S^{p-1}$ to the right of $[i..j]$. Then, $[i''..j''] \in MUS_S^{p-1}(p-1) \cap MUS_S^{p-1}(p)$, since we have that $i < i'' \leq p-1$, and if $j < j'' < p$, then $|cover([i..j], p-1)| = |[i..p-1]| > |[i''..p-1]| = |cover([i''..j''], p-1)|$

contradicting the definition of $[i..j]$. Thus, we have that $|cover([i''..j''], p-1)| = |cover([i''..j''], p)|$, and from Lemma 3, these values are monotonically increasing. Therefore, the first one, which is $[i'..j'] = MMUS_S^{p-1}[L^{p-1}[p-1]+1]$, gives the smallest value. Note that $[i_p..j_p] = lmMUS_S^{p-1}(p)$ cannot be to the left of $[i..j]$; If $p \leq j_p$, then since $i_p < i < p \leq j_p < j$ and from the definition of $[i_p..j_p]$, we have $|cover([i_p..j_p], p-1)| = |cover([i_p..j_p], p)| \leq |cover([i..j], p)| = |cover([i..j], p-1)|$ which contradicts the definition of $[i..j]$ If $j_p \leq p-1$, then $cover([i_p..j_p], p-1) + 1 = cover([i_p..j_p], p) \leq cover([i..j], p) \leq cover([i..j], p-1) + 1$, again contradicting the definition of $[i..j]$. Thus, $lmMUS_S^{p-1}(p)$ must be either $[i..j]$ or $[i'..j']$.   □

Theorem 4. *$MMUS_S$ and L can be constructed in linear time.*

Proof. The pseudo code of the algorithm is shown in Algorithm 2. The algorithm computes $MMUS_S$ and L for increasing positions. For each value of p, we assume that $MMUS_S^{p-1}$ and $L^{p-1}[1..p-1]$ are correctly computed, and we update them to correct values of $MMUS_S^p$ and $L^p[1..p]$.

Lines 3-8 in Algorithm 2 compute $L^{p-1}[p]$ from $L^{p-1}[p-1]$, and $MMUS_S^{p-1}$. The correctness can be seen from Lemma 5. The calculation for updating L can be done in constant time for each position.

Next, we show how to compute $MMUS_S^p$ and $L^p[1..p]$ given $MMUS_S^{p-1}$ and $L^{p-1}[1..p]$. The existence of an MUS starting at position p can be checked in constant time with Lemma 1. If there exists no such MUS, then, since $MUS_S^{p-1} = MUS_S^p$, $MMUS_S^p = MMUS_S^{p-1}$ and $L^p[p] = L^{p-1}[p]$, and no update is required. If there does exist $[p..e] \in MUS_S$ for some $e \geq p$, we check previous positions $i \leq p$ to see if $L^{p-1}[i]$ needs to be updated to $L^p[i]$. Such positions i satisfy $|cover(MMUS_S^{p-1}[L^{p-1}[i]], i)| > |cover([p..e], i)|$, and if for some position j this does not hold, then it is easy to see that it does not hold for all $j' \leq j$. Let j be the rightmost position such that the condition does not hold, i.e., $L^{p-1}[1..j]$ does not need to be updated.

If $j = p$, this means that no values in $L^{p-1}[1..p]$ need to be updated, and $L^p[p] = L^{p-1}[p]$. Concerning updating $MMUS_S^{p-1}$, we can easily see that $cover([p..e], n) \in SUS_S^p(n)$, and thus $[p..e]$ will be the last element in $MMUS_S^p$. However, MUSs in $MMUS_S^{p-1}$ may become meaningless with respect to MUS_S^p, because of the addition of $[p..e]$. These are the ones to the right of $[i'..j'] = MMUS_S^{p-1}[L^p[p]]$. They can be found and removed in line 13, whose correctness can be seen from Lemma 3.

If $j < p$, MUSs in $MMUS_S^{p-1}[L[j]+1..\ell]$ such that $|cover(MMUS_S[k'], j)| > |cover(MMUS_S[j], j)|$, i.e., those that do not derive an interval in $SUS_S^p(j)$ become meaningless with respect to MUS_S^p, so are removed in line 15. The correctness can also be seen from Lemma 3.

Although there may be more than a constant number of positions and MUSs that need to be updated with the addition of $[p..e]$, the cost can be amortized. Such operations correspond to lines 11, 13, 15, and 16 of Algorithm 2.

The time required for lines 11 and 16 is linear in the number of updates required for L. We show that $L[p]$ for each p is updated only a constant number of times. $L^{p-1}[p]$ is first determined at lines 3-8, with respect to MUS_S^{p-1}, pointing to $pred_S(p)$ or the leftmost shortest element in $MUS_S(p) \cap MUS_S^{p-1}$. It can be seen from Lemma 4 that for all $p' \geq p$, $L^{p'}[p]$ can only point to $pred_S(p)$, the leftmost shortest element in $MUS_S(p)$ ($= MUS_S(p) \cap MUS_S^{p'}$), or $succ_S(p)$. There are only two possibly remaining MUSs that will be added to $MUS_S(p) \cap MUS_S^{p-1}$ and update $L[p]$; an MUS in $MUS_S(p)$ beginning at position p, or $succ_S(p)$. Thus, the total time for this is linear in the number of positions.

The time required for lines 13 and 15, is linear in the number of intervals added or deleted from $MMUS$. Since each interval in $MMUS$ is added or removed at most once, the total time for this update is linear in the total number of MUSs in S, which is $O(n)$. Thus, the total time of the algorithm is $O(n)$.   □

From Theorems 3 and 4, we obtain the following main theorem.

Theorem 5. *A string S of length n can be preprocessed in $O(n)$ time and space so that shortest unique substring queries can be answered in $O(k)$ time, where k is the number of shortest substrings returned. Notably, outputting a single SUS can be done in $O(1)$ time.*

4 Computational Experiments

We implemented our algorithm using the C++ language. All computational experiments were conducted on a MacPro (Early 2008) with two 3.2GHz Quad Core Xeon processors and 18GB Memory (DDR2 FB-DIMM 800MHz). We use libdivsufsort (http://code.google.com/p/libdivsufsort/) for construction of the suffix array.

Table 1. Comparison of Computation Time

n (MB)	english ($|\Sigma| = 239$) time (sec)		dna ($|\Sigma| = 16$) time (sec)		dblp.xml ($|\Sigma| = 97$) time (sec)		protein ($|\Sigma| = 27$) time (sec)	
	TSUS	RSUS	TSUS	RSUS	TSUS	RSUS	TSUS	RSUS
10	4.21	122.31	4.79	18.63	3.42	14.34	4.01	28.28
20	9.16	324.58	10.54	40.46	7.44	29.98	9.04	66.74
30	14.13	445.84	16.45	61.80	11.43	46.51	14.57	108.00
40	20.14	500.19	23.06	84.75	16.17	62.76	21.68	151.85
50	25.62	580.00	29.31	107.34	20.35	78.73	28.90	197.99
60	31.20	667.16	36.08	131.38	24.62	95.55	35.61	242.55
70	38.26	N/A	43.90	N/A	30.14	728.71	43.96	N/A
80	44.00	N/A	50.83	N/A	34.67	N/A	51.01	N/A
90	50.37	N/A	57.88	N/A	39.03	N/A	58.13	N/A
100	56.71	N/A	65.17	N/A	43.30	N/A	64.22	N/A

We used data taken from the Pizza & Chile corpus (http://pizzachili.dcc.uchile.cl/texts.html), namely, english texts, DNA sequences, XML, and protein sequences. We compared our algorithm with the implementation RSUS of [4] available at `https://bitbucket.org/wush_iis/rsus`. RSUS is actually a combination of an interface for the R language (`http://www.r-project.org`) and core routines written in C++. For comparison in our experiments, we modified the RSUS C++ routines to be called from a C++ program so that all programs utilize only the C++ language.

The results of experiments for the 4 data are shown in Table 1. We take a prefix of length n for each data, and measure the running times of RSUS [4], and TSUS (the implementation of the algorithm in this paper). The entries marked N/A for RSUS was when the time exceeded 1 hour, at which time the execution of the program was stopped. The cause for the sudden increase in running times for RSUS was due to the fact that RSUS consumed all of the available physical memory. The results show that our algorithm is much faster (as fast as 20 times) in preprocessing time compared to RSUS.

References

1. Kärkkäinen, J., Sanders, P.: Simple linear work suffix array construction. In: Baeten, J.C.M., Lenstra, J.K., Parrow, J., Woeginger, G.J. (eds.) ICALP 2003. LNCS, vol. 2719, pp. 943–955. Springer, Heidelberg (2003)
2. Kasai, T., Lee, G., Arimura, H., Arikawa, S., Park, K.: Linear-time Longest-Common-Prefix Computation in Suffix Arrays and Its Applications. In: Amir, A. (ed.) CPM 2001. LNCS, vol. 2089, pp. 181–192. Springer, Heidelberg (2001)
3. Manber, U., Myers, G.: Suffix arrays: A new method for on-line string searches. SIAM J. Computing 22(5), 935–948 (1993)
4. Pei, J., Wu, W.C.H., Yeh, M.Y.: On shortest unique substring queries. In: Proc. ICDE, pp. 937–948 (2013)
5. Weiner, P.: Linear pattern-matching algorithms. In: Proc. of 14th IEEE Ann. Symp. on Switching and Automata Theory, pp. 1–11. Institute of Electrical Electronics Engineers, New York (1973)

Oracle Pushdown Automata, Nondeterministic Reducibilities, and the Hierarchy over the Family of Context-Free Languages

Tomoyuki Yamakami

Department of Information Science, University of Fukui
3-9-1 Bunkyo, Fukui 910-8507, Japan

Abstract. We implement various oracle mechanisms on nondeterministic pushdown automata, which naturally induce nondeterministic reducibilities among formal languages in a theory of context-free languages. In particular, we examine a notion of nondeterministic many-one CFL-reducibility and carry out ground work of formulating a coherent framework for further expositions. Another more powerful reducibility—Turing CFL-reducibility—is also discussed in comparison. The Turing CFL-reducibility, in particular, makes it possible to induce a useful hierarchy (the CFL hierarchy) built over the family CFL of context-free languages. For each level of this hierarchy, basic structural properties are proven and three alternative characterizations are presented. We also show that the CFL hierarchy enjoys an upward collapse property. The first and second levels of the hierarchy are proven to be different. We argue that the CFL hierarchy coincides with a hierarchy over CFL built by applications of many-one CFL-reductions. Our goal is to provide a solid foundation for structural-complexity analyses in automata theory.

Keywords: regular language, context-free language, pushdown automaton, oracle, many-one reducibility, Turing reducibility, CFL hierarchy, polynomial hierarchy, Dyck language.

1 Backgrounds and Main Themes

A fundamental notion of *reducibility* has long played an essential role in the development of a theory of NP-completeness. In the 1970s, various forms of polynomial-time reducibility emerged, most of which were based on models of multi-tape oracle Turing machine, and they provided a technical means to study *relativizations* of associated families of languages. Most typical reducibilities in use today in computational complexity theory include many-one, truth-table, and Turing reducibilities obtained by imposing appropriate restrictions on the functionality of oracle mechanism of underlying Turing machines. Away from standard complexity-theoretical subjects, we will shift our attention to a theory of formal languages and automata. Within this theory, we wish to lay out a framework for a future extensive study on structural complexity issues by providing a solid foundation for various notions of reducibility and their associated relativizations.

V. Geffert et al. (Eds.): SOFSEM 2014, LNCS 8327, pp. 514–525, 2014.

Among many languages, we are particularly interested in *context-free languages*, which are characterized by context-free grammars or one-way nondeterministic pushdown automata (or npda's, hereafter). The context-free languages are inherently nondeterministic. In light of the fact that the notion of nondeterminism appears naturally in real life, this notion has become a key to many fields of computer science. The family CFL of context-free languages has proven to be a fascinating subject, simply because the languages in CFL behave quite differently from the languages in the corresponding nondeterministic polynomial-time class NP. For instance, whereas NP is closed under any Boolean operation (possibly) except for complementation, CFL is not even closed under intersection. This non-closure property is caused by the lack of flexibility in a use of its memory storage on an underlying model of npda. On the contrary, a restricted use of memory helps us prove a separation between the first and the second levels of the Boolean hierarchy $\{\text{CFL}_k \mid k \geq 1\}$ built over CFL by applying alternatingly two operations of intersection and union to CFL [13]. Moreover, we can prove that a family of languages $\text{CFL}(k)$ composed of intersections of k context-free languages truly forms an infinite hierarchy [7]. Such an architectural constraint sometimes becomes a crucial issue in certain applications of pushdown automata.

A most simple type of well-known reduction is probably *many-one reduction* and, by adopting the existing formulation of this reducibility, we intend to bring a notion of nondeterministic many-one reducibility into context-free languages under the name of *many-one CFL-reducibility*. We write CFL_m^A to denote the family of languages that are many-one CFL-reducible to oracle A. Notice that Reinhardt [8] earlier considered many-one reductions that are induced by nondeterministic finite automata (or nfa's), which use no memory space. We wish to build a hierarchy of language families over CFL using our new reducibility by immediate analogy with constructing the *polynomial(-time) hierarchy* over NP. For this purpose, we choose npda's rather than nfa's. Owing mostly to a unique architecture of npda's, our reducibility exhibits quite distinctive features; for instance, this reducibility in general does not admit a transitivity property. (For this reason, our reducibility might have been called a "quasi-reducibility" if the transitive property is a prerequisite for a reducibility notion.) As a consequence, the family CFL is not closed under the many-one CFL-reducibility (namely, $\text{CFL}_m^{\text{CFL}} \neq \text{CFL}$). This non-closure property allures us to study the family $\text{CFL}_{m[k]}^{\text{CFL}}$ whose elements are obtained by the k-fold application of many-one CFL-reductions to languages in CFL. As shown in Section 3.1, the language family $\text{CFL}_{m[k]}^{\text{CFL}}$ turns out to coincide with $\text{CFL}_m^{\text{CFL}(k)}$.

We further discuss another more powerful reducibility in use—Turing CFL-reducibility based on npda's. This reducibility introduces a hierarchy analogous to the polynomial hierarchy: the hierarchy $\{\Delta_k^{\text{CFL}}, \Sigma_k^{\text{CFL}}, \Pi_k^{\text{CFL}} \mid k \geq 1\}$ built over CFL, which we succinctly call the *CFL hierarchy*, and this hierarchy turns out to be quite useful in classifying the computational complexity of formal languages. As quick examples, two languages $Dup_2 = \{xx \mid x \in \{0,1\}^*\}$ and $Dup_3 = \{xxx \mid x \in \{0,1\}^*\}$, which are known to be outside of CFL, fall into the second level Σ_2^{CFL} of the CFL hierarchy. A simple matching language $Match = \{x\#w \mid \exists u, v\,[w = uxv]\}$ is also in Σ_2^{CFL}. Two more languages $Sq = \{0^n 1^{n^2} \mid n \geq 1\}$ and $Prim = \{0^n \mid n$ is a prime number $\}$ belong to Σ_2^{CFL} and Π_2^{CFL}, respectively. A slightly more complex language $MulPrim = \{0^{mn} \mid m$ and n are prime numbers $\}$ is a member of Σ_3^{CFL}. The first and second levels of the CFL hierarchy are easily proven to be different. Regarding the aforementioned language families $\mathrm{CFL}(k)$ and CFL_k, we can show later that the families $\mathrm{CFL}(\omega) = \bigcup_{k \geq 1} \mathrm{CFL}(k)$ and $\mathrm{BHCFL} = \bigcup_{k \geq 1} \mathrm{CFL}_k$ belong to $\Sigma_2^{\mathrm{CFL}} \cap \Pi_2^{\mathrm{CFL}}$ of the CFL hierarchy. In Section 4.1, we show that $\mathrm{CFL}_m^{\mathrm{CFL}(\omega)}$ is located within Σ_3^{CFL}. Despite obvious similarities between their definitions, the CFL hierarchy and the polynomial hierarchy are quite different in nature. Because of npda's architectural restrictions, "standard" techniques of simulating a two-way Turing machine, in general, do not apply; hence, we need to develop new simulation techniques for npda's.

In this paper, we employ three simulation techniques to obtain some of the aforementioned results. The first technique is of guessing and verifying a *stack history* to eliminate a use of stack, where a stack history roughly means a series of consecutive stack operations made by an underlying npda. The second technique is applied to the case of simulating two or more tape heads by a single tape head. To adjust the different head speeds, we intentionally insert extra dummy symbols to generate a single query word so that an oracle can eliminate them when it accesses the query word. The last technique is to generate a string that encodes a computation path generated by a nondeterministic machine. All the techniques are explained in details in Sections 3.1–3.2. Those simulation techniques actually make it possible to obtain three alternative characterizations of the CFL hierarchy in Section 4.2.

Topics excluded from this extended abstract are found in its full version available at arXiv:1303.1717.

2   Preparation

Given a finite set A, the notation $|A|$ expresses the number of elements in A. Let $\mathbb{N}$ be the set of all *natural numbers* (i.e., nonnegative integers) and set $\mathbb{N}^+ = \mathbb{N} - \{0\}$. For any number $n \in \mathbb{N}^+$, $[n]$ denotes the integer set $\{1, 2, \ldots, n\}$. The term "polynomial" always means a polynomial on $\mathbb{N}$ with coefficients of nonnegative integers. In particular, a *linear polynomial* is of the form $ax + b$ with $a, b \in \mathbb{N}$. The notation $A - B$ for two sets A and B indicates the *difference* $\{x \mid x \in A, x \notin B\}$ and $\mathcal{P}(A)$ denotes the *power set* of A. The *Kleene closure* Σ^* of Σ is the infinite union $\bigcup_{k \in \mathbb{N}} \Sigma^k$. Similarly, the notation $\Sigma^{\leq k}$ is used to mean $\bigcup_{i=1}^{k} \Sigma^i$. Given a language A over Σ, its *complement* is $\Sigma^* - A$, which is also denoted by $\overline{A}$. We use the following three class operations between two language families $\mathcal{C}_1$ and $\mathcal{C}_2$: $\mathcal{C}_1 \wedge \mathcal{C}_2 = \{A \cap B \mid A \in \mathcal{C}_1, B \in \mathcal{C}_2\}$, $\mathcal{C}_1 \vee \mathcal{C}_2 = \{A \cup B \mid A \in \mathcal{C}_1, B \in \mathcal{C}_2\}$, and $\mathcal{C}_1 - \mathcal{C}_2 = \{A - B \mid A \in \mathcal{C}_1, B \in \mathcal{C}_2\}$, where A and B must be defined over the same alphabet. For a use of *track notation* $[\begin{smallmatrix} x \\ y \end{smallmatrix}]$, see [9].

As our basic computation models, we use the following types of finite-state machines: *one-way deterministic finite automaton* (or dfa, in short) with λ-moves and *one-way nondeterministic pushdown automaton* (or npda) with λ-moves, where a λ-*move* (or a λ-*transition*) is a transition of the machine's configurations in which a target tape head stays still. Whenever we refer to a *write-only tape*, we always assume that (i) initially, all cells of the tape are blank, (ii) a tape head starts at the so-called *start cell*, (iii) the tape head steps forward whenever it writes down any non-blank symbol, and (iv) the tape head can stay still only in a blank cell. Therefore, all cells through which the tape head passes during a computation must contain no blank symbols. An *output* (or *outcome*) along a computation path is a string produced on the output tape after the computation path is terminated. We call an output string *valid* (or *legitimate*) if it is produced along a certain accepting computation path. When we refer to the machine's outputs, we normally disregard any invalid strings left on the output tape on a rejecting computation path. REG, CFL, and DCFL stand for the families of all regular languages, of all context-free languages, and of all deterministic context-free languages, respectively.

3   Natural Reducibilities

A typical way of comparing the computational complexity of two formal languages is various forms of *resource-bounded reducibility*. Such reducibility is also regarded as a *relativization* of its underlying language family. We refer the reader to [2] for basics of computational complexity theory.

3.1   Many-One Reductions by Npda's

Our exposition begins with an introduction of an appropriate form of nondeterministic many-one reducibility whose reductions are operated by npda's.

Our "reduction machine" is essentially a restricted version of "pushdown transducer" or "algebraic transduction" (see, e.g., [1]). Here, we define this notion in a style of "oracle machine." An *m-reduction npda* M is a standard npda equipped with an extra *query tape* on which the machine writes a string surrounded by blank cells starting at the designated *start cell* for the purpose of making a query to a given *oracle*. We treat this query tape as an output tape, and thus the query-tape head must move to a next blank cell whenever it writes a non-blank symbol. Formally, an m-reduction npda is a tuple $(Q, \Sigma, \{\mathbb{c}, \$\}, \Theta, \Gamma, \delta, q_0, Z_0, Q_{acc}, Q_{rej})$, where Θ is a query alphabet and δ is of the form: $\delta : (Q - Q_{halt}) \times (\check{\Sigma} \cup \{\lambda\}) \times \Gamma \to \mathcal{P}(Q \times \Gamma^* \times (\Theta \cup \{\lambda\}))$, where $Q_{halt} = Q_{acc} \cup Q_{rej}$ and $\check{\Sigma} = \Sigma \cup \{\mathbb{c}, \$\}$. There are two types of λ-moves. Assuming $(p, \tau, \xi) \in \delta(q, \sigma, \gamma)$, if $\sigma = \lambda$, then the input-tape head stays still (or makes a λ-move); in contrast, if $\tau = \lambda$, then the query-tape head stays still (or makes a λ-move). Since repetitions of λ-moves potentially produce extremely long output strings, we should require the following *termination condition* for M. Recall that, as a consequence of Greibach's normal form theorem, all context-free languages can be recognized by λ-*free* npda's (i.e., npda's with no λ-moves) whose computation paths have length $O(n)$, always ending in certain halting states, where n is its input size. The runtime of $O(n)$ is truly significant for languages in CFL. Likewise, we assume that, for any m-reduction npda, *all* computation paths should terminate (reaching halting inner states) within $O(n)$ time.

A language L over alphabet Σ is *many-one CFL-reducible* to another language A over alphabet Θ if there exists an m-reduction npda $M = (Q, \Sigma, \{\mathcal{c}, \$\}, \Theta, \Gamma, \delta, q_0, Z_0, Q_{acc}, Q_{rej})$ such that, for every input $x \in \Sigma^*$, (1) along each computation path $p \in ACC_M(x)$, M produces a valid query string $y_p \in \Theta^*$ on the query tape and (2) x is a member of L iff there is a computation path $p \in ACC_M(x)$ satisfying $y_p \in A$. For simplicity, we also say that M *reduces* (or *m-reduces*) L to A. With the use of this new reducibility, we make the notation CFL_m^A (or $CFL_m(A)$) express the family of all languages L that are many-one CFL-reducible to A, where the language A is customarily called an *oracle*. Given an oracle npda M and an oracle A, the notation $L(M, A)$ (or $L(M^A)$) denotes the set of strings accepted by M relative to A. For a class $\mathcal{C}$ of oracles, $CFL_m^{\mathcal{C}}$ (or $CFL_m(\mathcal{C})$) denotes the union $\bigcup_{A \in \mathcal{C}} CFL_m^A$.

Likewise, we define the relativized language family NFA_m^A (or $NFA_m(A)$) using "nfa's" as m-reduction machines instead of "npda's." To be more precise, an m-reduction nfa M for NFA_m^A is a tuple $(Q, \Sigma, \{\mathcal{c}, \$\}, \Theta, \delta, q_0, Q_{acc}, Q_{rej})$, where δ is a map from $(Q - Q_{halt}) \times (\check{\Sigma} \cup \{\lambda\})$ to $\mathcal{P}(Q \times (\Theta \cup \{\lambda\}))$. We also impose an $O(n)$ time-bound on all computation paths of M.

Making an analogy with "oracle Turing machine" that functions as a mechanism of reducing a language to another given target language A, we want to use the term "oracle npda" to mean an npda that is equipped with an extra write-only output tape (called a *query tape*) besides a read-only input tape. As noted before, we explicitly demand every oracle npda to terminate on all computation paths within $O(n)$ steps.

We will use an informal term of "guessing" when we refer to a nondeterministic choice (or a series of nondeterministic choices). For example, when we say that an npda M guesses a string z, we actually mean that M makes a series of nondeterministic choices that cause to produce z.

Example 1. As the first concrete example, setting $\Sigma = \{0, 1\}$, let us consider the language $Dup_2 = \{xx \mid x \in \Sigma^*\}$. This language is known to be non-context-free; however, it can be many-one CFL-reducible to CFL by the following M and A. An m-reduction (or oracle) npda M nondeterministically produces a query word $x^R \natural y$ (with a special symbol $\natural$) from each input of the form xy using a stack appropriately More formally, a transition function δ of this oracle npda M is given as follows: $\delta(q_0, \mathcal{c}, Z_0) = \{(q_0, Z_0, \lambda)\}$, $\delta(q_0, \$, Z_0) = \{(q_{acc}, Z_0, \natural)\}$, $\delta(q_0, \sigma, Z_0) = \{(q_1, \sigma Z_0, \lambda)\}$, $\delta(q_1, \sigma, \tau) = \{(q_1, \sigma\tau, \lambda), (q_2, \sigma\tau, \lambda)\}$, $\delta(q_2, \lambda, \tau) = \{(q_2, \lambda, \tau)\}$, $\delta(q_2, \lambda, Z_0) = \{(q_3, Z_0, \natural)\}$, $\delta(q_3, \lambda, Z_0) = \{(q_3, Z_0, \sigma)\}$, and $\delta(q_3, \$, Z_0) = \{(q_{acc}, Z_0, \lambda)\}$, where $\sigma, \tau \in \Sigma$. A CFL-oracle A is defined as $\{x^R \natural x \mid x \in \Sigma^*\}$; that is, the oracle A checks whether $x = y$ from the input $x^R \natural y$ using its own stack. In other words, Dup_2 belongs to CFL_m^A, which is included in CFL_m^{CFL}. Similarly, the non-context-free language $Dup_3 = \{xxx \mid x \in \Sigma^*\}$ also falls into CFL_m^{CFL}. For this case, we design an m-reduction npda to produce $x^R \natural y \natural y^R \natural z$ from each input xyz and make a CFL-oracle check whether $x = y = z$

by using its stack twice. Another language $Match = \{x\#w \mid \exists u, v\,[w = uxv]\}$, where $\#$ is a separator not in x and w, also belongs to $\mathrm{CFL}_m^{\mathrm{CFL}}$. These examples prove that $\mathrm{CFL}_m^{\mathrm{CFL}} \neq \mathrm{CFL}$.

Example 2. The language $Sq = \{0^n 1^{n^2} \mid n \geq 1\}$ belongs to $\mathrm{CFL}_m^{\mathrm{CFL}}$. To see this fact, let us consider the following oracle npda N and oracle A. Given any input w, N first checks if w is of the form $0^i 1^j$. Simultaneously, N nondeterministically selects $(j_1, j_2, \ldots, j_k)$ satisfying (i) $j = j_1 + j_2 + \cdots + j_k$ and (ii) $j_1 = j_2$, $j_3 = j_4$, ..., and N produces on its query tape a string w' of the form $0^i \natural 1^{j_1} \natural 1^{j_2} \natural \cdots \natural 1^{j_k}$. The desired oracle A receives w' and checks if the following two conditions are all met: (i') $j_2 = j_3$, $j_4 = j_5$, ... and (ii') $i = k$ by first pushing 0^i into a stack and then counting the number of $\natural$. Clearly, A belongs to CFL. Therefore, Sq is in CFL_m^A, which is included in $\mathrm{CFL}_m^{\mathrm{CFL}}$. A similar idea proves that the language $Comp = \{0^n \mid n \text{ is a composite number}\}$ belongs to $\mathrm{CFL}_m^{\mathrm{CFL}}$. In symmetry, $Prim = \{0^n \mid n \text{ is a prime number}\}$ is a member of co-$(\mathrm{CFL}_m^{\mathrm{CFL}})$, where co-$\mathcal{C}$ denotes the *complement* of language family $\mathcal{C}$, namely, co-$\mathcal{C} = \{\overline{A} \mid A \in \mathcal{C}\}$.

A *Dyck language* L over alphabet $\Sigma = \{\sigma_1, \sigma_2, \ldots, \sigma_d\} \cup \{\sigma_1', \sigma_2', \ldots, \sigma_d'\}$ is a language generated by a deterministic context-free grammar whose production set is $\{S \to \lambda | SS | \sigma_i S \sigma_i' : i \in [d]\}$, where S is a start symbol. For convenience, denote by $DYCK$ the family of all Dyck languages.

Lemma 1. $\mathrm{CFL}_m^{\mathrm{CFL}} = \mathrm{CFL}_m^{\mathrm{DCFL}} = \mathrm{CFL}_m^{DYCK}$.

Proof Sketch. We first claim that (1) $\mathrm{CFL} = \mathrm{NFA}_m^{DYCK}$ and (2) $\mathrm{CFL}_m^A = \mathrm{CFL}_m(\mathrm{NFA}_m^A)$ for any oracle A. The first claim (1) can be seen as a different form of Chomsky-Schützenberger theorem. To show (1), we employ a simple but useful technique of guessing a correct *stack history* (namely, a series of popped and pushed symbols along a halting computation path) and verifying its correctness. With an appropriate encoding method, we can claim that a stack history is correct iff its encoding belongs to a ceratin fixed Dyck language. Whenever an oracle npda tries to either push down symbols into its stack or pop up a symbol from the stack, instead of using an actual stack, we write down an encoded series of those symbols on a write-only query tape and then ask an oracle to verify that the series indeed encodes a correct stack history. We skip (2) due to the page limit. By combining the claims (1)–(2), it follows that $\mathrm{CFL}_m^{\mathrm{CFL}} = \mathrm{CFL}_m(\mathrm{NFA}_m^{DYCK}) \subseteq \mathrm{CFL}_m^{DYCK}$. $\qquad\square$

Given each number $k \in \mathbb{N}^+$, the *k-conjunctive closure of CFL*, denoted $\mathrm{CFL}(k)$ in [12], is defined recursively as follows: $\mathrm{CFL}(1) = \mathrm{CFL}$ and $\mathrm{CFL}(k+1) = \mathrm{CFL}(k) \wedge \mathrm{CFL}$. These language families truly form an infinite hierarchy [7]. For convenience, we set $\mathrm{CFL}(\omega) = \bigcup_{k \in \mathbb{N}^+} \mathrm{CFL}(k)$. Hereafter, we will explore basic properties of $\mathrm{CFL}_m^{\mathrm{CFL}(k)}$.

The lack of the transitivity property of the many-one CFL-reducibility necessitates an introduction of a helpful abbreviation of a *k-fold application of the reductions*. For any given oracle A, we recursively set $\mathrm{CFL}_{m[1]}^A = \mathrm{CFL}_m^A$ and $\mathrm{CFL}_{m[k+1]}^A = \mathrm{CFL}_m(\mathrm{CFL}_{m[k]}^A)$ for each index $k \in \mathbb{N}^+$. Given any language family $\mathcal{C}$, the notation $\mathrm{CFL}_{m[k]}^{\mathcal{C}}$ denotes the union $\bigcup_{A \in \mathcal{C}} \mathrm{CFL}_{m[k]}^A$.

Theorem 1. *For every index* $k \in \mathbb{N}^+$, $\mathrm{CFL}_m^{\mathrm{CFL}(k)} = \mathrm{CFL}_{m[k]}^{\mathrm{CFL}}$.

Proof Sketch. When $k = 1$, it holds that $\mathrm{CFL}_{m[1]}^{\mathrm{CFL}} = \mathrm{CFL}_m^{\mathrm{CFL}} = \mathrm{CFL}_m^{\mathrm{CFL}(1)}$. Next, we will show that, for every index $k \geq 2$, $\mathrm{CFL}_{m[k]}^{\mathrm{CFL}} \subseteq \mathrm{CFL}_m^{\mathrm{CFL}(k)}$ holds. We are focused on the most important case of $k = 2$. This case follows from the claim that $\mathrm{CFL}_{m[2]}^{\mathrm{CFL}(r)} \subseteq \mathrm{CFL}_m^{\mathrm{CFL}(r) \wedge \mathrm{CFL}}$ for every index $r \in \mathbb{N}^+$. Let $L \in \mathrm{CFL}_m^B$ and $B \in \mathrm{CFL}_m^A$ for a certain set $A \in \mathrm{CFL}(r)$. Let M_1 and M_2 be two oracle npda's witnessing $L \in \mathrm{CFL}_m^B$ and $B \in \mathrm{CFL}_m^A$, respectively. Consider the following oracle npda N. Given input x, N simulates M_1 on x in the following way. Whenever M_1 tries to write a symbol, say, b on a query tape, N simulates, using an actual stack, several steps (including all consecutive λ-moves) of M_2 that can be made during reading b. By simulating M_2, N aims at producing an encoded stack history y of M_2 (on the upper track of a tape) and a query word z (on the lower track). Since the tape heads of M_2 on both input and query tapes may move in different speeds, we need to adjust their speeds by inserting a series of fresh symbol, say, $\natural$ between symbols of the stack history and the query word. For this purpose, it is useful to introduce a terminology to describe strings obtained by inserting $\natural$. Assuming that $\natural \notin \Sigma$, a $\natural$-*extension* of a given string x over Σ is a string $\tilde{x}$ over $\Sigma \cup \{\natural\}$ satisfying that x is obtained directly from $\tilde{x}$ simply by removing all occurrences of $\natural$ in $\tilde{x}$. For instance, if $x = 01101$, then $\tilde{x}$ may be $01\natural1\natural01$ or $011\natural\natural01\natural$. N actually produces $\left[\begin{smallmatrix} y \\ z \end{smallmatrix}\right]$ on the query tape. An appropriate oracle in $\mathrm{CFL}(r) \wedge \mathrm{CFL}$ can check its correctness. Thus, $L \in \mathrm{CFL}_m^{\mathrm{CFL}(r) \wedge \mathrm{CFL}}$. $\square$

An immediate consequence is that $\mathrm{CFL}_m^{\mathrm{CFL}(\omega)} = \bigcup_{k \in \mathbb{N}^+} \mathrm{CFL}_{m[k]}^{\mathrm{CFL}}$.

3.2   Turing Reducibility by Npda's

We define a notion of *Turing CFL-reducibility* using a model of npda with a write-only query tape and three extra inner states q_{query}, q_{no}, and q_{yes} that represent a query signal and two possible oracle answers, respectively. More specifically, when an oracle npda enters q_{query}, it triggers a query, by which a query word is automatically transferred to an oracle, a query tape becomes blank, and its tape head instantly returns to the start cell. When the oracle returns its answer, either 0 (no) or 1 (yes), it automatically sets the oracle npda's inner state to q_{no} or q_{yes}, respectively. Such a machine is called a *T-reduction npda* (or just an *oracle npda* as before) and it is used to reduce a language to another language. To be more precise, an oracle npda is a tuple $(Q, \Sigma, \{\mathcal{c}, \$\}, \Theta, \Gamma, \delta, q_0, Z_0, Q_{oracle}, Q_{acc}, Q_{rej})$, where $Q_{oracle} = \{q_{query}, q_{yes}, q_{no}\}$, Θ is a query alphabet and δ has the form: $\delta : (Q - Q_{halt} \cup \{q_{query}\}) \times (\check{\Sigma} \cup \{\lambda\}) \times \Gamma \to \mathcal{P}((Q - \{q_{yes}, q_{no}\}) \times \Gamma^* \times (\Theta \cup \{\lambda\}))$.

Unlike many-one CFL-reductions, a T-reduction npda's computation depends on a series of oracle answers. Since such an oracle npda, in general, cannot implement an internal clock to control its running time, certain oracle answers may lead to an extremely long computation, and thus the machine may recognize even "infeasible" languages. To avoid such a pitfall, we need to demand that, *no matter what oracle is provided*, its underlying oracle npda M must halt on *all* computation paths within $O(n)$ time, where n refers to input size.

Similarly to CFL_m^A and CFL_m^C, we introduce two new notations CFL_T^A and CFL_T^C. An associated deterministic version is denoted DCFL_T^C. A simple relationship between the Turing and many-one CFL-reducibilities is exemplified in Proposition 1. To describe the proposition, we need a notion of the *Boolean hierarchy over CFL*, which was introduced in [13] by setting $\mathrm{CFL}_1 = \mathrm{CFL}$, $\mathrm{CFL}_{2k} = \mathrm{CFL}_{2k-1} \wedge \text{co-CFL}$, and $\mathrm{CFL}_{2k+1} = \mathrm{CFL}_{2k} \vee \mathrm{CFL}$. For simplicity, we denote by BHCFL the union $\bigcup_{k \in \mathbb{N}^+} \mathrm{CFL}_k$. Notice that $\mathrm{CFL} \neq \mathrm{CFL}_2$ holds because co-CFL $\subseteq \mathrm{CFL}_2$ and co-CFL $\not\subseteq \mathrm{CFL}$.

Proposition 1. $\mathrm{CFL}_T^{\mathrm{CFL}} = \mathrm{CFL}_m^{\mathrm{CFL}_2} = \mathrm{NFA}_m^{\mathrm{CFL}_2}$.

Proof Sketch. We wish to demonstrate that (1) $\mathrm{CFL}_T^{\mathrm{CFL}} \subseteq \mathrm{CFL}_m^{\mathrm{CFL}_2}$, (2) $\mathrm{CFL}_m^{\mathrm{CFL}_2} \subseteq \mathrm{NFA}_m^{\mathrm{CFL}_2}$, and (3) $\mathrm{NFA}_m^{\mathrm{CFL}_2} \subseteq \mathrm{CFL}_T^{\mathrm{CFL}}$. If all are proven, then the proposition immediately follows. We will show only (1) and (3).

(1) We start with an arbitrary language L in CFL_T^A relative to a certain language A in CFL. Take a T-reduction npda M reducing L to A, and let M_A be an npda recognizing A. Hereafter, we will build a new m-reduction npda N_1 to show that $L \in \mathrm{CFL}_m^{\mathrm{CFL}_2}$. On input x, the machine N_1 tries to simulate M on x by running the following procedure. Along each computation path, before M begins producing the ith query word on a query tape, N_1 guesses its oracle answer b_i (either 0 or 1) and writes it down onto its query tape. While M writes the ith query word y_i, N_1 appends $y_i \natural$ to b_i. When M halts, N_1 produces a query word w of the form $b_1 y_1 \natural b_2 y_2 \natural \cdots \natural b_k y_k \natural$, where $k \in \mathbb{N}$. Let L_2 be a collection of those w's such that, for every index $i \in [k]$, if $b_i = 1$ then $y_i \in A$. Similarly, let L_3 be a collection of those w's such that, for every index $i \in [k]$, if $b_i = 0$ then $y_i \in \overline{A}$. It is not difficult to verify that N_1 m-reduces L to $L_2 \cap L_3$.

Next, we want to claim that L_2 and $\overline{L_3}$ are in CFL. This claim leads to a conclusion that L is included in $\mathrm{CFL}_m^{L_2 \cap L_3} \subseteq \mathrm{CFL}_m(\mathrm{CFL} \wedge \text{co-CFL}) = \mathrm{CFL}_m^{\mathrm{CFL}_2}$. Obviously, L_2 is in CFL. To see that $\overline{L_3} \in \mathrm{CFL}$, let $w = b_1 y_1 \natural b_2 y_2 \natural \cdots \natural b_k y_k \natural$. If $w \in \overline{L_3}$, then there exists an index $i \in [k]$ such that $b_i = 0$ and $y_i \in A$. This last property can be checked by running M_A sequentially on each y_i and emptying its stack after each run of M_A. Thus, $\overline{L_3}$ is in CFL.

(3) Choose an oracle A in CFL_2 and consider an arbitrary language L in CFL_m^A. Furthermore, take two languages $A_1, A_2 \in \mathrm{CFL}$ for which $A = A_1 \cap \overline{A_2}$. Let M be an oracle nfa that recognizes L relative to A. Notice that M has no stack. We will define another oracle npda N as follows. On input x, N first marks 0 on its query tape and start simulating M on x. Whenever M tries to write a symbol σ on its query tape, N writes it down on a query tape and simultaneously copies it into a stack. After M halts with a query word, say, w, N makes the first query with the query word $0w$. If its oracle answer is 0, then N rejects the input. Subsequently, N writes 1 on the query tape (provided that the tape automatically becomes blank), pops the stored string w^R from the stack, and copies it to the query tape. After making the second query with $1w^R$, if its oracle answer equals 1, then N rejects the input. When N has not entered any rejecting state, then N finally accepts the input. The corresponding oracle B is defined as $\{0w \mid w \in A_1\} \cup \{1w^R \mid w \in A_2\}$. It is easy to see that $x \in L$ if

and only if N accepts x relative to B. Since CFL is known to be closed under reversal, $\{1w^R \mid w \in A_2\}$ is context-free, and thus B is a member of CFL. We then conclude that $L \in \text{CFL}_T^B \subseteq \text{CFL}_T^{\text{CFL}}$.   □

4   The CFL Hierarchy

4.1   Reducibility and a Hierarchy

Applying Turing CFL-reductions to CFL level by level, we can build a useful hierarchy, called the *CFL hierarchy*, whose kth level consists of three language families Δ_k^{CFL}, Σ_k^{CFL}, and Π_k^{CFL}. To be more precise, for each level $k \geq 1$, we set $\Delta_1^{\text{CFL}} = \text{DCFL}$, $\Sigma_1^{\text{CFL}} = \text{CFL}$, $\Delta_{k+1}^{\text{CFL}} = \text{DCFL}_T(\Sigma_k^{\text{CFL}})$, $\Pi_k^{\text{CFL}} = \text{co-}\Sigma_k^{\text{CFL}}$, and $\Sigma_{k+1}^{\text{CFL}} = \text{CFL}_T(\Sigma_k^{\text{CFL}})$. Additionally, we set $\text{CFLH} = \bigcup_{k \in \mathbb{N}^+} \Sigma_k^{\text{CFL}}$. The CFL hierarchy can be used to categorize the complexity of typical non-context-free languages discussed in most introductory textbooks. We will review a few typical examples that fall into the CFL hierarchy.

Example 3. In Example 1, we have seen the languages $Dup_2 = \{xx \mid x \in \{0,1\}^*\}$ and $Dup_3 = \{xxx \mid x \in \{0,1\}\}$, which are both in $\text{CFL}_m^{\text{CFL}}$. Note that, since $\text{CFL}_m^A \subseteq \text{CFL}_T^A$ for any oracle A, every language in $\text{CFL}_m^{\text{CFL}}$ belongs to $\text{CFL}_T^{\text{CFL}} = \Sigma_2^{\text{CFL}}$. Therefore, Dup_2 and Dup_3 are in Σ_2^{CFL}. In addition, as shown in Example 2, the language $Sq = \{0^n 1^{n^2} \mid n \geq 1\}$ is in $\text{CFL}_m^{\text{CFL}}$ while $Prim = \{0^n \mid n \text{ is a prime number }\}$ is in $\text{co-}(\text{CFL}_m^{\text{CFL}})$. Therefore, we conclude that Sq is in Σ_2^{CFL} and $Prim$ is in Π_2^{CFL}. A similar but more involved example is the language $MulPrim = \{0^{mn} \mid m \text{ and } n \text{ are prime numbers }\}$. It is possible to show that $MulPrim$ belongs to $\text{CFL}_m(\text{co-}(\text{CFL}_m^{\text{co-CFL}}))$, which equals Σ_3^{CFL}.

Lemma 2. *Let k be any integer satisfying $k \geq 1$.*

1. $\text{CFL}_T(\Sigma_k^{\text{CFL}}) = \text{CFL}_T(\Pi_k^{\text{CFL}})$ *and* $\text{DCFL}_T(\Sigma_k^{\text{CFL}}) = \text{DCFL}_T(\Pi_k^{\text{CFL}})$.
2. $\Sigma_k^{\text{CFL}} \cup \Pi_k^{\text{CFL}} \subseteq \Delta_{k+1}^{\text{CFL}} \subseteq \Sigma_{k+1}^{\text{CFL}} \cap \Pi_{k+1}^{\text{CFL}}$.
3. $\text{CFLH} \subseteq \text{DSPACE}(O(n))$.

Hereafter, we will explore fundamental properties of our new hierarchy. Our starting point is a closure property under length-nondecreasing substitution, where a substitution $s : \Sigma \to \mathcal{P}(\Theta^*)$ is called *length nondecreasing* if $s(\sigma) \neq \varnothing$ and $\lambda \notin s(\sigma)$ for every symbol $\sigma \in \Sigma$. We expand s as follows. Define $s(\sigma_1 \sigma_2 \cdots \sigma_n) = \{x_1 x_2 \cdots x_n \mid \forall i \in [n](x_i \in s(\sigma_i))\}$ for $\sigma_1, \sigma_2, \ldots, \sigma_n \in \Sigma$ and let $s(L) = \bigcup_{x \in L} s(x)$ for language $L \subseteq \Sigma^*$. A homomorphism $h : \Sigma \to \Theta^*$ is called *λ-free* if $h(\sigma) \neq \lambda$ for every $\sigma \in \Sigma$. Note that the condition of length nondecreasing is necessary because every recursively enumerable language can be a homomorphic image of a certain language in $\text{CFL}_2 (\subseteq \Sigma_2^{\text{CFL}})$ [3].

Lemma 3. *1. (substitution property) Let $k \in \mathbb{N}^+$ and let s be any length-nondecreasing substitution on alphabet Σ satisfying $s(\sigma) \in \Sigma_k^{\text{CFL}}$ for each symbol $\sigma \in \Sigma$. For any language A over Σ, if L is in Σ_k^{CFL}, then $s(L)$ is also in Σ_k^{CFL}.*

2. *For each index $k \in \mathbb{N}^+$, the family Σ_k^{CFL} is closed under the following operations: concatenation, union, reversal, Kleene closure, λ-free homomorphism, and inverse homomorphism.*

We will show that the second level of the CFL hierarchy contains BHCFL.

Proposition 2. BHCFL $\subseteq \Sigma_2^{\mathrm{CFL}} \cap \Pi_2^{\mathrm{CFL}}$.

Proof Sketch. We will show that BHCFL $\subseteq \Sigma_2^{\mathrm{CFL}}$. Obviously, $\mathrm{CFL}_1 \subseteq \Sigma_2^{\mathrm{CFL}}$ holds. It is therefore enough to show that $\mathrm{CFL}_k \subseteq \Sigma_2^{\mathrm{CFL}}$ for every index $k \geq 2$.

We first claim that, for every index $k \geq 1$, $\mathrm{CFL}_{2k} = \bigvee_{i \in [k]} \mathrm{CFL}_2$ $(= \mathrm{CFL}_2 \vee$ $\mathrm{CFL}_2 \vee \cdots \vee \mathrm{CFL}_2$ with k repetitions of CFL_2) and $\mathrm{CFL}_{2k+1} = (\bigvee_{i \in [k]} \mathrm{CFL}_2) \vee$ CFL. This can be shown using an idea of [13, Claim 4]. Next, we claim that $\mathrm{CFL}_{2k}, \mathrm{CFL}_{2k+1} \subseteq \Sigma_2^{\mathrm{CFL}}$ for all indices $k \geq 1$. The proof of this claim proceeds by induction on $k \geq 1$. Furthermore, we will prove that BHCFL $\subseteq \Pi_2^{\mathrm{CFL}}$. It is possible to prove by induction on $k \in \mathbb{N}^+$ that co-$\mathrm{CFL}_k \subseteq \mathrm{CFL}_{k+1}$. From this inclusion, we obtain co-BHCFL $\subseteq$ BHCFL. By symmetry, BHCFL $\subseteq$ co-BHCFL holds. Thus, we conclude that BHCFL = co-BHCFL. $\square$

Let us turn our attention to $\mathrm{CFL}(\omega)$. A direct analysis of each language family $\mathrm{CFL}(k)$ shows that $\mathrm{CFL}(\omega)$ is included in BHCFL.

Proposition 3. *1. $\mathrm{CFL}(\omega) \subseteq$ BHCFL (thus, $\mathrm{CFL}(\omega) \subseteq \Sigma_2^{\mathrm{CFL}} \cap \Pi_2^{\mathrm{CFL}}$).*

2. $\mathrm{CFL}_m^{\mathrm{CFL}(\omega)} \subseteq \Sigma_3^{\mathrm{CFL}}$.

Proof Sketch. A key to the proof of the first part of this proposition is the following claim: for every index $k \geq 1$, $\mathrm{CFL}(k) \subseteq \mathrm{CFL}_{2k+1}$ holds. The first part then implies that $\mathrm{CFL}_m^{\mathrm{CFL}(\omega)}$ is included in $\mathrm{CFL}_m^{\mathrm{BHCFL}}$. Since BHCFL $\subseteq \Sigma_2^{\mathrm{CFL}} \cap \Pi_2^{\mathrm{CFL}}$ by Proposition 2, it follows that $\mathrm{CFL}_m^{\mathrm{BHCFL}}$ is included in $\mathrm{CFL}_m(\Pi_2^{\mathrm{CFL}})$, which is obviously a subclass of $\mathrm{CFL}_T(\Pi_2^{\mathrm{CFL}}) = \Sigma_3^{\mathrm{CFL}}$. $\square$

4.2 Structural Properties

We will further explore structural properties that characterize the CFL hierarchy. Moreover, we will present three alternative characterizations (Theorem 2 and Proposition 4) of the hierarchy. Let us consider a situation in which Boolean operations are applied to languages in the CFL hierarchy. In the following lemma, the third statement needs an extra attention. As we have seen, it holds that CFL $\wedge$ CFL = $\mathrm{CFL}(2) \neq$ CFL. Therefore, the equality $\Sigma_k^{\mathrm{CFL}} \wedge \Sigma_k^{\mathrm{CFL}} = \Sigma_k^{\mathrm{CFL}}$ does not hold in the first level (i.e., $k = 1$). Surprisingly, it is possible to prove that this equality actually holds for any level *more than* 1.

Lemma 4. *Let $k \geq 1$.*

1. $\Sigma_k^{\mathrm{CFL}} \vee \Sigma_k^{\mathrm{CFL}} = \Sigma_k^{\mathrm{CFL}}$ and $\Pi_k^{\mathrm{CFL}} \wedge \Pi_k^{\mathrm{CFL}} = \Pi_k^{\mathrm{CFL}}$.

2. $\Sigma_k^{\mathrm{CFL}} \wedge \Pi_k^{\mathrm{CFL}} \subseteq \Sigma_{k+1}^{\mathrm{CFL}} \cap \Pi_{k+1}^{\mathrm{CFL}}$ and $\Sigma_k^{\mathrm{CFL}} \vee \Pi_k^{\mathrm{CFL}} \subseteq \Sigma_{k+1}^{\mathrm{CFL}} \cap \Pi_{k+1}^{\mathrm{CFL}}$.

3. $\Sigma_k^{\mathrm{CFL}} \wedge \Sigma_k^{\mathrm{CFL}} = \Sigma_k^{\mathrm{CFL}}$ and $\Pi_k^{\mathrm{CFL}} \vee \Pi_k^{\mathrm{CFL}} = \Pi_k^{\mathrm{CFL}}$ for all levels $k \geq 2$.

Lemma 4(3) is not quite trivial and its proof follows from Theorem 2, in which we give two new characterizations of Σ_k^{CFL} in terms of many-one reducibilities. For our purpose, we introduce two extra many-one hierarchies. The *many-one CFL hierarchy* consists of language families $\Sigma_{m,k}^{CFL}$ and $\Pi_{m,k}^{CFL}$ ($k \in \mathbb{N}^+$) defined as follows: $\Sigma_{m,1}^{CFL} = CFL$, $\Pi_{m,k}^{CFL} = \text{co-}\Sigma_{m,k}^{CFL}$, and $\Sigma_{m,k+1}^{CFL} = CFL_m(\Pi_{m,k}^{CFL})$ for any $k \geq 1$, where the subscript "m" stands for "many-one." A *relativized many-one NFA hierarchy*, which was essentially formulated in [8], is defined as follows relative to oracle A: $\Sigma_{m,1}^{NFA,A} = NFA_m^A$, $\Pi_{m,k}^{NFA,A} = \text{co-}\Sigma_{m,k}^{NFA,A}$, and $\Sigma_{m,k+1}^{NFA,A} = NFA_m(\Pi_{m,k}^{NFA,A})$ for every index $k \geq 1$. Given a language family $\mathcal{C}$, $\Sigma_{m,k}^{NFA,\mathcal{C}}$ (or $\Sigma_{m,k}^{NFA}(\mathcal{C})$) denotes the union $\bigcup_{A \in \mathcal{C}} \Sigma_{m,k}^{NFA,A}$.

Theorem 2. $\Sigma_k^{CFL} = \Sigma_{m,k}^{CFL} = \Sigma_{m,k}^{NFA}(DYCK)$ *for every index* $k \geq 1$.

Proof Sketch. The first step toward the proof is to prove two key claims. (1) For every index $k \geq 1$, it holds that $\Sigma_{k+1}^{CFL} \subseteq CFL_m(\Sigma_k^{CFL} \wedge \Pi_k^{CFL}) \subseteq NFA_m(\Sigma_k^{CFL} \wedge \Pi_k^{CFL})$. (2) For any two indices $k \geq 1$ and $e \geq k-1$, it holds that $NFA_m(\Sigma_{m,k}^{CFL} \wedge \Pi_{m,e}^{CFL}) \subseteq CFL_m(\Pi_{m,e}^{CFL})$.

In the second step, we use induction on $k \geq 1$ to prove the theorem. Since Lemma 1 handles the base case $k = 1$, it is sufficient to assume that $k \geq 2$. The second equality of the theorem is shown as follows. If $k = 1$, then the claim is exactly the same as $CFL = NFA_m^{DYCK}$. In the case of $k \geq 2$, assume that $L \in CFL_m^A$ for a certain language A in $\Pi_{m,k-1}^{CFL}$. A proof similar to that of $CFL = NFA_m^{DYCK}$ demonstrates the existence of a certain Dyck language D satisfying that $CFL_m^A = NFA_m^B$, where B is of the form $\{[\begin{smallmatrix} \tilde{y} \\ \tilde{z} \end{smallmatrix}] \mid y \in D, z \in A\}$ and $\tilde{y}$ and $\tilde{z}$ are $\natural$-extensions of y and z, respectively. The definition places B into the language family $DCFL \wedge \Pi_{m,k-1}^{CFL}$, which equals $\Pi_{m,k-1}^{CFL}$ because of $k \geq 2$. By our induction hypothesis, $\Pi_{m,k-1}^{CFL} = \Pi_{m,k-1}^{NFA}(DYCK)$ holds. It thus follows that $NFA_m^B \subseteq NFA_m(\Pi_{m,k-1}^{NFA}(DYCK)) = \Sigma_{m,k}^{NFA}(DYCK)$, and therefore we obtain $L \in CFL_m^A \subseteq NFA_m^A \subseteq \Sigma_{m,k}^{NFA}(DYCK)$.

Next, we will establish the first equality given in the theorem. Clearly, $\Sigma_{m,k}^{CFL} \subseteq \Sigma_k^{CFL}$ holds since $CFL_m^A \subseteq CFL_T^A$ for any oracle A. Now, we target the opposite containment. By (1), it follows that $\Sigma_k^{CFL} \subseteq NFA_m(\Sigma_{k-1}^{CFL} \wedge \Pi_{k-1}^{CFL})$. Since $\Sigma_{k-1}^{CFL} = \Sigma_{m,k-1}^{CFL}$, we obtain $\Sigma_k^{CFL} \subseteq NFA_m(\Sigma_{m,k-1}^{CFL} \wedge \Pi_{m,k-1}^{CFL})$. Note that (2) implies the inclusion $NFA_m(\Sigma_{m,k-1}^{CFL} \wedge \Pi_{m,k-1}^{CFL}) \subseteq CFL_m(\Pi_{m,k-1}^{CFL}) = \Sigma_{m,k}^{CFL}$. In conclusion, $\Sigma_k^{CFL} \subseteq \Sigma_{m,k}^{CFL}$ holds. □

An *upward collapse property* holds for the CFL hierarchy except for the first level. Similar to the notation CFL_e expressing the eth level of the Boolean hierarchy over CFL, a new notation $\Sigma_{k,e}^{CFL}$ is introduced to denote the eth level of the Boolean hierarchy over Σ_k^{CFL}. Additionally, we set $BH\Sigma_k^{CFL} = \bigcup_{e \in \mathbb{N}^+} \Sigma_{k,e}^{CFL}$.

Lemma 5. *(upward collapse properties) Let* k *be any integer at least 2.*

1. $\Sigma_k^{CFL} = \Sigma_{k+1}^{CFL}$ *iff* $CFLH = \Sigma_k^{CFL}$.
2. $\Sigma_k^{CFL} = \Pi_k^{CFL}$ *iff* $BH\Sigma_k^{CFL} = \Sigma_k^{CFL}$.

3. $\Sigma_k^{CFL} = \Pi_k^{CFL}$ implies $\Sigma_k^{CFL} = \Sigma_{k+1}^{CFL}$.

From Lemma 5, if the Boolean hierarchy over Σ_k^{CFL} collapses to Σ_k^{CFL}, then the entire CFL hierarchy collapses. It is not clear, however, that a much weaker assumption like $\Sigma_{k,e}^{CFL} = \Sigma_{k,e+1}^{CFL}$ suffices to draw the collapse of the CFL hierarchy (for instance, $\Sigma_{k+1}^{CFL} = \Sigma_{k+2}^{CFL}$).

Theorem 2 also gives a logical characterization of Σ_k^{CFL}. For convenience, we define a function Ext as $Ext(\tilde{x}) = x$ for any $\natural$-extension $\tilde{x}$ of string x.

Proposition 4. Let $k \geq 1$. For any language $L \in \Sigma_k^{CFL}$ over alphabet Σ, there exists another language $A \in DCFL$ and a linear polynomial p with $p(n) \geq n$ for all $n \in \mathbb{N}$ that satisfy the following equivalence relation: for any number $n \in \mathbb{N}$ and any string $x \in \Sigma^n$, $x \in L$ if and only if

$$\exists \tilde{x}(|\tilde{x}| \leq p(n)) \exists y_1(|y_1| \leq p(n)) \forall y_2(|y_2| \leq p(n))$$
$$\cdots Q_k y_k(|y_k| \leq p(n)) \, [\, x = Ext(\tilde{x}) \wedge [\tilde{x}, y_1, y_2, \ldots, y_k]^T \in A \,],$$

where Q_k is $\exists$ ($\forall$, resp.) if k is odd (even, resp.) and $\tilde{x}$ is a $\natural$-extension of x.

Recall that the first and second levels of the CFL hierarchy are different. It is possible to prove that the rest of the hierarchy is infinite unless the polynomial hierarchy over NP collapses.

References

1. Berstel, J.: Transductions and Context-Free Languages. B. G. Teubner, Stuttgart (1979)
2. Du, D., Ko., K.: Theory of Computational Complexity. John Willey & Sons (2000)
3. Ginsburg, S., Greibach, S.A., Harrison, M.A.: One-way stack languages. J. ACM 14, 389–418 (1967)
4. Greibach, S.A.: The hardest context-free language. SIAM J. Comput. 2, 304–310 (1973)
5. Hromkovič, J., Schnitger, G.: On probabilistic pushdown automata. Inf. Comput. 208, 982–995 (2010)
6. Ladner, R., Lynch, N., Selman, A.: A comparison of polynomial-time reducibilities. Theor. Comput. Sci. 1, 103–123 (1975)
7. Liu, L.Y., Weiner, P.: An infinite hierarchy of intersections of context-free languages. Math. Systems Theory 7, 185–192 (1973)
8. Reinhardt, K.: Hierarchies over the context-free languages. In: Dassow, J., Kelemen, J. (eds.) IMYCS 1990. LNCS, vol. 464, pp. 214–224. Springer, Heidelberg (1990)
9. Tadaki, K., Yamakami, T., Lin, J.C.H.: Theory of one-tape linear-time Turing machines. Theor. Comput. Sci. 411, 22–43 (2010)
10. Yamakami, T.: Swapping lemmas for regular and context-free languages. Available at arXiv:0808.4122 (2008)
11. Yamakami, T.: The roles of advice to one-tape linear-time Turing machines and finite automata. Int. J. Found. Comput. Sci. 21, 941–962 (2010)
12. Yamakami, T.: Immunity and pseudorandomness of context-free languages. Theor. Comput. Sci. 412, 6432–6450 (2011)
13. Yamakami, T., Kato, Y.: The dissecting power of regular languages. Inf. Pross. Lett. 113, 116–122 (2013)
14. Younger, D.H.: Recognition and parsing of context-free languages in time n^3. Inf. Control 10, 189–208 (1967)

Author Index